Lectures in Mathematics
ETH Zürich
Department of Mathematics
Research Institute of Mathematics

Managing Editor:
Michael Struwe

Goran Peskir
Albert Shiryaev

Optimal Stopping and Free-Boundary Problems

Birkhäuser Verlag
Basel · Boston · Berlin

Authors:

Goran Peskir
School of Mathematics
The University of Manchester
Sackville Street
Manchester
M60 1QD
United Kingdom
e-mail: goran@maths.man.ac.uk

Albert Shiryaev
Steklov Mathematical Institute
8 ul. Gubkina
119991 Moscow
Russia
e-mail: albertsh@mi.ras.ru

and

Moscow State University
GSP-2
Leninskie Gory
119992 Moscow
Russia

2000 Mathematical Subject Classification 60G40, 35R35, 45G10, 62L15, 91B28 (primary); 60J25, 60J60, 60J65, 60G42, 60J05, 60G44, 35K20, 45D05, 62L10, 62M02 (secondary)

A CIP catalogue record for this book is available from the Library of Congress, Washington D.C., USA

Bibliographic information published by Die Deutsche Bibliothek
Die Deutsche Bibliothek lists this publication in the Deutsche Nationalbibliografie; detailed bibliographic data is available in the Internet at <http://dnb.ddb.de>.

ISBN 3-7643-2419-8 Birkhäuser Verlag, Basel – Boston – Berlin

Part of Springer Science+Business Media
Printed on acid-free paper produced from chlorine-free pulp. TCF ∞
Printed in Germany
ISBN-10: 3-7643-2419-8
ISBN-13: 978-3-7643-2419-3
e-ISBN: 3-7643-7390-3

9 8 7 6 5 4 3 2 1

www.birkhauser.ch

Preface

The present monograph, based mainly on studies of the authors and their co-authors, and also on lectures given by the authors in the past few years, has the following particular aims:

> To present basic results (with proofs) of optimal stopping theory in both *discrete* and *continuous time* using both *martingale* and *Markovian approaches*;
>
> To select a series of *concrete* problems of *general interest* from the theory of probability, mathematical statistics, and mathematical finance that can be reformulated as *problems of optimal stopping* of stochastic processes and solved by reduction to *free-boundary problems* of real analysis (*Stefan problems*).

The table of contents found below gives a clearer idea of the material included in the monograph. Credits and historical comments are given at the end of each chapter or section. The bibliography contains a material for further reading.

Acknowledgements. The authors thank L. E. Dubins, S. E. Graversen, J. L. Pedersen and L. A. Shepp for useful discussions. The authors are grateful to T. B. Tolozova for the excellent editorial work on the monograph. Financial support and hospitality from ETH, Zürich (Switzerland), MaPhySto (Denmark), MIMS (Manchester) and Thiele Centre (Aarhus) are gratefully acknowledged. The authors are also grateful to INTAS and RFBR for the support provided under their grants. The grant NSh-1758.2003.1 is gratefully acknowledged. Large portions of the text were presented in the "School and Symposium on Optimal Stopping with Applications" that was held in Manchester, England from 17th to 27th January 2006. The authors are grateful to EPSRC and LMS for the sponsorship and financial support provided under their grants (EP/D035333/1).

7th March 2006 G.P. & A.Sh.

Contents

Introduction

1. The following scheme illustrates the kind of *concrete* problems of *general interest* that will be studied in the monograph:

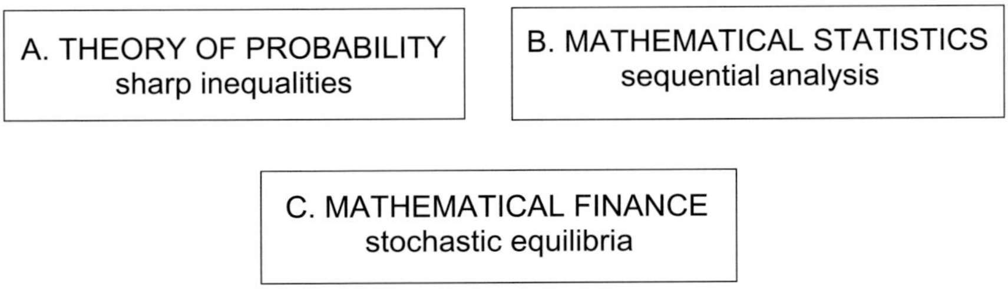

The solution method for problems A, B, C consists of *reformulation* to an optimal stopping problem and *reduction* to a free-boundary problem as follows:

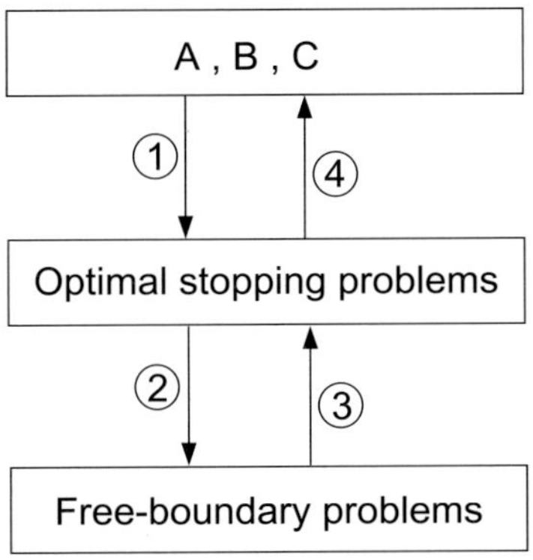

Steps 1 and 2 indicate the way of reformulation and reduction. Steps 3 and 4 indicate the way of finding a solution to the initial problem.

2. To get some idea of the character of problems A, B, C that will be studied, let us briefly consider the following simple examples.

(A) If $B = (B_t)_{t \geq 0}$ is a standard Brownian motion, then it is well known that the following *maximal equality* holds:

$$\mathsf{E}\Big(\max_{0 \leq t \leq T} |B_t|\Big) = \sqrt{\frac{\pi}{2}\, T} \tag{1}$$

for every deterministic time T. Suppose now that instead of the deterministic time T we are given some (random) stopping time τ of B. The question then arises naturally of how to determine $\mathsf{E}\,(\max_{0\le t\le\tau}|B_t|)$. On closer inspection, however, it becomes clear that it is virtually impossible to compute this expectation for every stopping time τ of B. Thus, as the second best thing, one can try to bound the expectation with a quantity which is easier to compute. A natural candidate for the latter is $\mathsf{E}\tau$ at least when finite. In this way a problem A has appeared. This problem then leads to the following *maximal inequality*:

$$\mathsf{E}\Big(\max_{0\le t\le\tau}|B_t|\Big)\le C\sqrt{\mathsf{E}\tau} \tag{2}$$

which is valid for all stopping times τ of B with the best constant C equal to $\sqrt{2}$.

We will see in Chapter V that the problem A just formulated can be solved in the form (2) by reformulation to the following optimal stopping problem:

$$V_* = \sup_\tau \mathsf{E}\Big(\max_{0\le t\le\tau}|B_t| - c\tau\Big) \tag{3}$$

where the supremum is taken over all stopping times τ of B satisfying $\mathsf{E}\tau<\infty$, and the constant $c>0$ is given and fixed. It constitutes Step 1 in the diagram above.

If $V_* = V_*(c)$ can be computed, then from (3) we get

$$\mathsf{E}\Big(\max_{0\le t\le\tau}|B_t|\Big)\le V_*(c) + c\,\mathsf{E}\tau \tag{4}$$

for all stopping times τ of B and all $c>0$. Hence we find

$$\mathsf{E}\Big(\max_{0\le t\le\tau}|B_t|\Big)\le \inf_{c>0}\big(V_*(c) + c\,\mathsf{E}\tau\big) \tag{5}$$

for all stopping times τ of B. The right-hand side in (5) defines a function of $\mathsf{E}\tau$ that, in view of (3), provides a sharp bound of the left-hand side.

We will see in Chapter IV that the optimal stopping problem (3) can be reduced to a free-boundary problem. This constitutes Step 2 in the diagram above. Solving the free-boundary problem one finds that $V_*(c) = 1/2c$. Inserting this into (5) yields

$$\inf_{c>0}\mathsf{E}\big(V_*(c) + c\,\mathsf{E}\tau\big) = \sqrt{2\,\mathsf{E}\tau} \tag{6}$$

so that the inequality (5) reads as follows:

$$\mathsf{E}\Big(\max_{0\le t\le\tau}|B_t|\Big)\le \sqrt{2\,\mathsf{E}\tau} \tag{7}$$

for all stopping times τ of B. This is exactly the inequality (2) above with $C=\sqrt{2}$. From the formulation of the optimal stopping problem (3) it is not surprising that equality in (7) is attained at a stopping time for which both sides in (7) are non-zero. This shows that the constant $\sqrt{2}$ is best possible in (7) as claimed in (2) above. The solution of (3) and its use in (2) just explained constitute Steps 3 and 4 in the diagram above and complete the solution to the initial problem.

Chapter V studies similar sharp inequalities for other stochastic processes using ramifications of the method just exposed. Apart from being able to derive sharp versions of known inequalities the method can also be used to derive new inequalities.

(B) The classic example of a problem in *sequential analysis* is the problem of sequential testing of two statistical hypotheses

$$H_0 : \mu = \mu_0 \quad \text{and} \quad H_1 : \mu = \mu_1 \tag{8}$$

about the drift parameter $\mu \in \mathbb{R}$ of the observed process

$$X_t = \mu t + B_t \tag{9}$$

for $t \geq 0$ where $B = (B_t)_{t \geq 0}$ is a standard Brownian motion.

Another classic example of a problem in *sequential analysis* is the problem of sequential testing of two statistical hypotheses

$$H_0 : \lambda = \lambda_0 \quad \text{and} \quad H_1 : \lambda = \lambda_1 \tag{10}$$

about the intensity parameter $\lambda > 0$ of the observed process

$$X_t = N_t^\lambda \tag{11}$$

for $t \geq 0$ where $N = (N_t)_{t \geq 0}$ is a standard Poisson process.

The basic problem in both cases seeks to find the optimal decision rule (τ_*, d_*) in the class $\Delta(\alpha, \beta)$ consisting of decision rules (d, τ), where τ is the time of stopping and accepting H_1 if $d = d_1$ or accepting H_0 if $d = d_0$, such that the probability errors of the first and second kind satisfy:

$$\mathsf{P}(\text{accept } H_1 \mid \text{true } H_0) \leq \alpha, \tag{12}$$

$$\mathsf{P}(\text{accept } H_0 \mid \text{true } H_1) \leq \beta \tag{13}$$

and the mean times of observation $\mathsf{E}_0 \tau$ and $\mathsf{E}_1 \tau$ are as small as possible. It is assumed above that $\alpha > 0$ and $\beta > 0$ with $\alpha + \beta < 1$.

It turns out that with this (*variational*) problem one may associate an optimal stopping (*Bayesian*) problem which in turn can be reduced to a free-boundary problem. This constitutes Steps 1 and 2 in the diagram above. Solving the free-boundary problem leads to an optimal decision rule (τ_*, d_*) in the class $\Delta(\alpha, \beta)$

satisfying (12) and (13) as well as the following two identities:

$$\mathsf{E}_0\mathcal{T} = \inf_{(\tau,d)} \mathsf{E}_0\tau, \tag{14}$$

$$\mathsf{E}_1\mathcal{T} = \inf_{(\tau,d)} \mathsf{E}_1\tau \tag{15}$$

where the infimum is taken over all decision rules (τ, d) in $\Delta(\alpha, \beta)$. This constitutes Steps 3 and 4 in the diagram above.

While the methodology just described is the same for both problems (8) and (10), it needs to be pointed out that the solution of the Bayesian problem in the Poisson case is more difficult than in the Brownian case. This is primarily due to the fact that, unlike in the Brownian case, the sample paths of the observed process are discontinuous in the Poisson case.

Chapter VI studies these as well as closely related problems of *quickest detection*. Two of the prime findings of this chapter, which also reflect the historical development of these ideas, are the *principles of smooth and continuous fit*, respectively.

(C) One of the best-known specific problems of *mathematical finance*, that has a direct connection with optimal stopping problems, is the problem of determining the arbitrage-free price of the *American put option*.

Consider the Black–Scholes model where the stock price $X = (X_t)_{t \geq 0}$ is assumed to follow a geometric Brownian motion

$$X_t = x \exp\Big(\sigma B_t + (r - \sigma^2/2)\, t\Big) \tag{16}$$

where $x > 0$, $\sigma > 0$, $r > 0$ and $B = (B_t)_{t \geq 0}$ is a standard Brownian motion. By Itô's formula one finds that the process X solves

$$dX_t = rX_t\, dt + \sigma X_t\, dB_t \tag{17}$$

with $X_0 = x$. General theory of financial mathematics makes it clear that the initial problem of determining the arbitrage-free price of the American put option can be reformulated as the following optimal stopping problem:

$$V_* = \sup_{\tau} \mathsf{E}\, e^{-r\tau}(K - X_\tau)^+ \tag{18}$$

where the supremum is taken over all stopping times τ of X. This constitutes Step 1 in the diagram above. The constant $K > 0$ is called the 'strike price'. It has a certain financial meaning which we set aside for now.

It turns out that the optimal stopping problem (18) can be reduced to a free-boundary problem which can be solved explicitly. It yields the existence of a constant b_* such that the stopping time

$$\tau_* = \inf\{\, t \geq 0 : X_t \leq b_* \,\} \tag{19}$$

is optimal in (18). This constitutes Steps 2 and 3 in the diagram above. Both the optimal stopping point b_* and the arbitrage-free price V_* can be expressed explicitly in terms of the other parameters in the problem. A financial interpretation of these expressions constitutes Step 4 in the diagram above.

In the formulation of the problem (18) above no restriction was imposed on the class of admissible stopping times, i.e. for certain reasons of simplicity it was assumed there that τ belongs to the class of stopping times

$$\mathfrak{M} = \{ \tau : 0 \le \tau < \infty \} \tag{20}$$

without any restriction on their size.

A more realistic requirement on a stopping time in search for the arbitrage-free price leads to the following optimal stopping problem:

$$V_*^T = \sup_{\tau \in \mathfrak{M}^T} \mathsf{E}\, e^{-r\tau}(K - X_\tau)^+ \tag{21}$$

where the supremum is taken over all τ belonging to the class of stopping times

$$\mathfrak{M}^T = \{ \tau : 0 \le \tau \le T \} \tag{22}$$

with the horizon T being finite.

The optimal stopping problem (21) can be reduced to a free-boundary problem that apparently cannot be solved explicitly. Its study yields that the stopping time

$$\tau_* = \inf \{ 0 \le t \le T : X_t \le b_*(t) \} \tag{23}$$

is optimal in (21), where $b_* : [0, T] \to \mathbb{R}$ is an increasing continuous function. A nonlinear Volterra integral equation can be derived which characterizes the optimal stopping boundary $t \mapsto b_*(t)$ and can be used to compute its values numerically as accurate as desired. The comments on Steps 1–4 in the diagram above made in the infinite horizon case carry over to the finite horizon case without any change.

Chapter VII studies these and other similar problems that arise from various financial interpretations of options. Chapter VIII studies *optimal prediction* problems. Fuller understanding of their scope is still incomplete at present.

3. So far we have only discussed problems A, B, C and their reformulations as optimal stopping problems. Now we want to address the methods of solution of optimal stopping problems and their reduction to free-boundary problems.

There are essentially two equivalent approaches to finding a solution of the optimal stopping problem. The first one deals with the problem

$$V_* = \sup_{\tau \in \mathfrak{M}} \mathsf{E}\, G_\tau \tag{24}$$

in the case of *infinite horizon*, or the problem

$$V_*^T = \sup_{\tau \in \mathfrak{M}^T} \mathsf{E} G_\tau \tag{25}$$

in the case of *finite horizon*, where $\mathfrak{M}$ and $\mathfrak{M}^T$ are the classes of stopping times defined in (20) and (22), respectively.

In this formulation it is important to realize that $G = (G_t)_{t \geq 0}$ is an arbitrary stochastic process defined on a filtered probability space $(\Omega, \mathcal{F}, (\mathcal{F}_t)_{t \geq 0}, \mathsf{P})$, where it is assumed that G is adapted to the filtration $(\mathcal{F}_t)_{t>0}$ which in turn makes each τ from $\mathfrak{M}$ or $\mathfrak{M}^T$ a stopping time. Since the method of solution to the problems (24) and (25) is based on results from the theory of martingales, the method itself is often referred to as the *martingale method*.

On the other hand, if we are to take a state space $(E, \mathcal{B})$ large enough, then one obtains the "Markov representation" $G_t = G(X_t)$ for some measurable function G, where $X = (X_t)_{t \geq 0}$ is a Markov process with values in E. Moreover, following the contemporary theory of Markov processes it is convenient to adopt the definition of a Markov process X as the *family* of Markov processes

$$((X_t)_{t \geq 0}, (\mathcal{F}_t)_{t \geq 0}, (\mathsf{P}_x)_{x \in E}) \tag{26}$$

where $\mathsf{P}_x(X_0 = x) = 1$, meaning that the process X starts at x under P_x. Such a point of view is convenient, for example, when dealing with the Kolmogorov forward or backward equations, which presuppose that the process can start at any point in the state space. Likewise, it is a profound attempt, developed in stages, to study optimal stopping problems through functions of initial points in the state space.

In this way we have arrived at the second approach which deals with the problem

$$V(x) = \sup_\tau \mathsf{E}_x G(X_\tau) \tag{27}$$

where the supremum is taken over $\mathfrak{M}$ or $\mathfrak{M}^T$ as above. Thus, if the Markov representation of the initial problem is valid, we will refer to the *Markovian method* of solution. Its elements will now be exposed in some detail.

Intuitively, it is clear in the Markovian setting that at time t the decision "to stop" or "to continue" with the observation should depend only on the present state X_t of the process and not on its past states X_s for $0 \leq s < t$. It is a great simplification and advantage of the Markovian setting to the general one. Indeed, in this case the problem of optimal stopping in essence becomes a problem of optimal stopping for a random path in the state space E, as opposed to the probability space Ω in the general case, while in many cases of interest we have $E = \mathbb{R}^n$ for some $n \geq 1$. This point of view makes it also clear that many Markovian problems of optimal stopping can be reformulated as problems for elliptic or parabolic equations in $\mathbb{R}^n$, because the transition density of a random path, for example in the case of a diffusion, satisfies the Kolmogorov equations. The connection to elliptic and parabolic equations just addressed will be further clarified by means of the *strong Markov property* in Chapter III.

4. To make the exposed facts more transparent, let us consider the optimal stopping problem (3) in more detail. Denote

$$X_t = |x+B_t| \tag{28}$$

for $x \geq 0$, and enable the maximum process to start at any point by setting

$$S_t = s \vee \Big(\max_{0 \leq r \leq t} X_r \Big) \tag{29}$$

for $s \geq x$. The process $S = (S_t)_{t \geq 0}$ is not Markov, but the pair $(X, S) = (X_t, S_t)_{t \geq 0}$ forms a Markov process with the state space $E = \{ (x,s) \in \mathbb{R}^2 : 0 \leq x \leq s \}$. The value V_* from (3) above coincides with the value function

$$V_*(x,s) = \sup_{\tau} \mathsf{E}_{x,s} \big(S_\tau - c\tau \big) \tag{30}$$

when $x = s = 0$. The problem thus needs to be solved in this more general form.

The general theory of optimal stopping for Markov processes makes it clear that the optimal stopping time in (30) can be written in the form

$$\tau_* = \inf \{ t \geq 0 : (X_t, S_t) \in D_* \} \tag{31}$$

where D_* is the *stopping set*, and $C_* = E \setminus D_*$ is the *continuation set*. In other words, if the observation of X was not stopped before time t since $X_s \in C_*$ for all $0 \leq s < t$, and we have that $X_t \in D_*$, then it is optimal to stop the observation at time t. On the other hand, if it happens that $X_t \in C_*$ as well, then the observation of X should be continued.

Heuristic considerations about the shape of the sets C_* and D_* makes it plausible to guess that there exist a point $s_* \geq 0$ and a continuous increasing function $s \mapsto g_*(s)$ with $g_*(s_*) = 0$ such that

$$D_* = \{ (x,s) \in \mathbb{R}^2 : 0 \leq x \leq g_*(s) \, , \; s \geq s_* \} \tag{32}$$

(see *Figure* 1). Note that such a guess about the shape of the set D_* can be made using the following intuitive arguments. If the process (X,S) starts from a point (x,s) with small x and large s, then it is reasonable to stop immediately, because to increase the value s one needs a large time τ which in the formula (30) appears with a minus sign. At the same time it is easy to see that if x is close or equal to s, then it is reasonable to continue the observation, at least for small time Δ, because s will increase for the value $\sqrt{\Delta}$ while the cost for using this time will be $c\Delta$, and thus $\sqrt{\Delta} - c\Delta > 0$ if Δ is small enough.

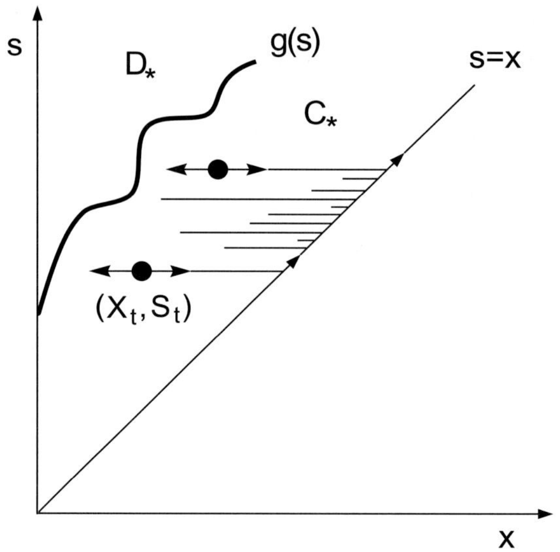

Figure 1: An illustration of the kinematics of the space-maximum process $(X_t, S_t)_{t \ge 0}$ in relation to the optimal stopping boundary g separating the continuation set C_* and the stopping set D_* .

Such an a priori analysis of the shape of the boundary between the stopping set C_* and the continuation set D_* is typical of the act of finding a solution to the optimal stopping problem. The art of guessing in this context very often plays a crucial role in solving the problem.

Having guessed that the stopping set D_* in the optimal stopping problem (30) takes the form (32), it follows that τ_* from (32) attains the supremum i.e.

$$V_*(x,s) = \mathsf{E}_{x,s}\big(S_{\tau_*} - c\tau_*\big) \tag{33}$$

for all $(x,s) \in E$. Denote by $\mathbb{L}_X = (1/2)\,\partial^2/\partial x^2$ the infinitesimal operator of the process X and consider $V_*(x,s)$ as defined by the right-hand side of (33) for (x,s) in the continuation set

$$C_* = C_*^1 \cup C_*^2 \tag{34}$$

where the two subsets are defined as follows:

$$C_*^1 = \{\, (x,s) \in \mathbb{R}^2 : 0 \le x \le s < s_* \,\}, \tag{35}$$

$$C_*^2 = \{\, (x,s) \in \mathbb{R}^2 : g_*(s) < x \le s\,,\ s \ge s_* \,\}. \tag{36}$$

By the strong Markov property one finds that V_* solves the following equation:

$$\mathbb{L}_X V_*(x,s) = c \tag{37}$$

for (x, s) in C_*. Note that if the process (X, S) starts at a point (x, s) with $x < s$, then during a positive time interval the second component S of the process does not change and remains equal to s. This explains why the infinitesimal operator of the process (X, S) reduces to the infinitesimal operator of the process X in the interior of C_*. On the other hand, from the structure of the process (X, S) it follows that at the diagonal in $\mathbb{R}^2_+$ the following condition of *normal reflection* holds:

$$\frac{\partial V_*}{\partial s}(x, s)\Big|_{x=s-} = 0. \tag{38}$$

Moreover, it is clear that for $(x, s) \in D_*$ the following condition of *instantaneous stopping* holds:

$$V_*(x, s) = s. \tag{39}$$

Finally, either by guessing or providing rigorous arguments, it is found that at the optimal boundary g_* the condition of *smooth fit* holds

$$\frac{\partial V_*}{\partial x}(x, s)\Big|_{x=g_*(s)+} = 0. \tag{40}$$

The condition of smooth fit embodies the key principle of optimal stopping that will be discussed extensively and used frequently in the sequel.

This analysis indicates that the value function V_* and the optimal stopping boundary g_* can be obtained by searching for the *pair of functions* (V, g) solving the following *free-boundary problem*:

$$\mathbb{L}_X V(x, s) = c \quad \text{for } (x, s) \text{ in } C_g, \tag{41}$$

$$\frac{\partial V}{\partial s}(x, s)\Big|_{x=s-} = 0 \quad \text{(normal reflection)}, \tag{42}$$

$$V(x, s) = s \quad \text{for } (x, s) \text{ in } D_g \text{ (instantaneous stopping)}, \tag{43}$$

$$\frac{\partial V}{\partial x}(x, s)\Big|_{x=g(s)+} = 0 \quad \text{(smooth fit)}, \tag{44}$$

where the two sets are defined as follows:

$$C_g = \{\, (x, s) \in \mathbb{R}^2 : 0 \le x \le s < s_0 \ \text{ or } \ g(s) < x \le s \ \text{ for } \ s \ge s_0 \,\}, \tag{45}$$

$$D_g = \{\, (x, s) \in \mathbb{R}^2 : 0 \le x \le g(s)\,,\ s \ge s_0 \,\} \tag{46}$$

with $g(s_0) = 0$. It turns out that this system does not have a unique solution so that an additional criterion is needed to make it unique in general (Chapter IV).

Let us briefly show how to solve the free-boundary problem (41)–(44) by picking the right solution. For more details see Chapters IV and V.

From (41) one finds that for (x, s) in C_g we have

$$V(x, s) = cx^2 + A(s)\, x + B(s) \tag{47}$$

where A and B are some functions of s. To determine A and B as well as g we can use the three conditions (42)–(44) which yield

$$g'(s) = \frac{1}{2(s - g(s))} \tag{48}$$

for $s \geq s_0$. It is easily verified that the linear function

$$g(s) = s - \frac{1}{2c} \tag{49}$$

solves (48). In this way a candidate for the optimal stopping boundary g_* is obtained.

For all $(x, s) \in E$ with $s \geq 1/2c$ one can determine $V(x, s)$ explicitly using (47) and (49). This in particular gives that $V(1/2c, 1/2c) = 3/4c$. For other points $(x, s) \in E$ when $s < 1/2c$ one can determine $V(x, s)$ using that the observation must be continued. In particular for $x = s = 0$ this yields that

$$V(0,0) = V(1/2c, 1/2c) - c\,\mathsf{E}_{0,0}(\sigma) \tag{50}$$

where σ is the first hitting time of the process (X, S) to the point $(1/2c, 1/2c)$. Because $\mathsf{E}_{0,0}(\sigma) = \mathsf{E}_{0,0}(X_\sigma^2) = (1/2c)^2$ and $V(1/2c, 1/2c) = 3/4c$, we find that

$$V(0,0) = \frac{1}{2c} \tag{51}$$

as already indicated prior to (6) above. In this way a candidate for the value function V_* is obtained.

The key role in the proof of the fact that $V = V_*$ and $g = g_*$ is played by Itô's formula (stochastic calculus) and the optional sampling theorem (martingale theory). This step forms a *verification theorem* that makes it clear that the solution of the free-boundary problem coincides with the solution of the optimal stopping problem.

5. The important point to be made in this context is that the verification theorem is usually not difficult to prove in the cases when a candidate solution to the free-boundary problem is obtained explicitly. This is quite typical for one-dimensional problems with infinite horizon, or some simpler two-dimensional problems, as the one just discussed. In the case of problems with finite horizon, however, or other multidimensional problems, the situation can be radically different. In these cases, in a manner quite opposite to the previous ones, the general results of optimal stopping can be used to prove the existence of a solution to the free-boundary problem, thus providing an alternative to analytic methods. Studies of this type will be presented in Chapters VII and VIII.

6. From the material exposed above it is clear that our basic interest concerns the case of continuous time. The theory of optimal stopping in the case

of continuous time is considerably more complicated than in the case of discrete time. However, since the former theory uses many basic ideas from the latter, we have chosen to present the case of discrete time first, both in the *martingale* and *Markovian* setting, which is then likewise followed by the case of continuous time. The two theories form Chapter I. As the methods employed throughout deal extensively with martingales and Markov processes, we have collected some of the basic facts from these theories in Chapter II. In Chapters III and IV we examine the relationship between optimal stopping problems and free-boundary problems. Finally, in Chapters V–VIII we study a number of *concrete* optimal stopping problems of *general interest* as discussed above.

Notes. To conclude the introduction we make a remark of general character about the two approaches used in optimal stopping problems (martingale and Markovian). Their similarities as well as distinction are mostly revealed by how they describe probabilistic evolution of stochastic processes which underly the optimal stopping problem.

To describe the probabilistic structure of a stochastic process in terms of general theory one commonly chooses between the following two methods which may naturally be thought of as *unconditional* and *conditional.*

In the first method one determines the probabilistic structure of a process $X = (X_t)_{t \ge 0}$ by its (unconditional) finite dimensional distributions which generate the corresponding probability distribution $P^X = \text{Law}(X)$ (on the space of trajectories of X). When speaking about optimal stopping of such processes we refer to the *martingale* approach. This terminology is justified by the fact that the appropriate techniques of solution are based on concepts and methods from the theory of martingales (the most important of which is the concept of 'Snell envelope' discussed in Chapter I below).

In the second method one does not begin with the finite-dimensional distributions (which are rather complicated formations) but with a consistent family of (conditional) *transition functions* taking into account the initial state x from where the trajectories of X start. Having its origin in the 1931 paper by A. N. Kolmogorov "Analytical methods in the theory of probability" (see [111]) and leading to the (Markovian) family of probability distributions $P_x^X = \text{Law}(X \mid X_0 = x)$, this approach proves to be very effective in optimal stopping problems due to powerful analytic tools provided by the theory of Markov processes (Kolmogorov forward and backward equations, theory of potential, stochastic differential equations, etc.). It is therefore natural to refer to this approach to optimal stopping problems as the *Markovian* approach.

When solving concrete problems of optimal stopping one may use either of the two approaches, and the choice certainly depends on special features of the problem. When dealing with this issue, however, it should be kept in mind that: on the one hand, any Markov process may be thought of as a special case of processes determined by unconditional probabilities; on the other hand, every process may be considered as Markov by introducing a complex state space whose elements are defined by the "past" of the underlying process.

In the present monograph we generally follow the Markovian approach since it allows us to use the well-developed analytical apparatus arising from theory and problems of differential (and integral) equations. Thus in the sequel we shall be mostly interested in free-boundary problems (Stefan problems) which arise from solving optimal stopping problems via the Markovian approach.

Chapter I.

Optimal stopping: General facts

The aim of the present chapter is to exhibit basic results of general theory of optimal stopping. Both *martingale* and *Markovian* approaches are studied first in discrete time and then in continuous time. The discrete time case, being direct and intuitively clear, provides a number of important insights into the continuous time case.

1. Discrete time

The aim of the present section is to exhibit basic results of optimal stopping in the case of discrete time. We first consider a martingale approach. This is then followed by a Markovian approach.

1.1. Martingale approach

1. Let $G = (G_n)_{n \geq 0}$ be a sequence of random variables defined on a filtered probability space $(\Omega, \mathcal{F}, (\mathcal{F}_n)_{n \geq 0}, \mathsf{P})$. We interpret G_n as the *gain* obtained if the observation of G is stopped at time n. It is assumed that G is adapted to the filtration $(\mathcal{F}_n)_{n \geq 0}$ in the sense that each G_n is $\mathcal{F}_n$-measurable. Recall that each $\mathcal{F}_n$ is a σ-algebra of subsets of Ω such that $\mathcal{F}_0 \subset \mathcal{F}_1 \subset \cdots \subset \mathcal{F}$. Typically $(\mathcal{F}_n)_{n \geq 0}$ coincides with the natural filtration $(\mathcal{F}_n^G)_{n \geq 0}$ but generally may also be larger. We interpret $\mathcal{F}_n$ as the *information* available up to time n. All our decisions in regard to optimal stopping at time n must be based on this information only (no anticipation is allowed).

The following definition formalizes the previous requirement and plays a key role in the study of optimal stopping.

Definition 1.1. A random variable $\tau : \Omega \to \{0, 1, \ldots, \infty\}$ is called a *Markov time* if $\{\, \tau \leq n \,\} \in \mathcal{F}_n$ for all $n \geq 0$. A Markov time is called a *stopping time* if $\tau < \infty$ P-a.s.

The family of all stopping times will be denoted by $\mathfrak{M}$, and the family of all Markov times will be denoted by $\bar{\mathfrak{M}}$. The following subfamilies of $\mathfrak{M}$ will be used in the present chapter:

$$\mathfrak{M}_n^N = \{\, \tau \in \mathfrak{M} : n \le \tau \le N \,\} \tag{1.1.1}$$

where $0 \le n \le N$. For simplicity we will set $\mathfrak{M}^N = \mathfrak{M}_0^N$ and $\mathfrak{M}_n = \mathfrak{M}_n^\infty$.

The *optimal stopping problem* to be studied seeks to solve

$$V_* = \sup_\tau \mathsf{E}\, G_\tau \tag{1.1.2}$$

where the supremum is taken over a family of stopping times. Note that (1.1.2) involves two tasks: (i) to compute the *value function* V_* as explicitly as possible; (ii) to exhibit an *optimal* stopping time τ_* at which the supremum is attained.

To ensure the existence of $\mathsf{E}\, G_\tau$ in (1.1.2) we need to impose additional conditions on G and τ. If the following condition is satisfied (with $G_N \equiv 0$ when $N = \infty$):

$$\mathsf{E}\Big(\sup_{n \le k \le N} |G_k| \Big) < \infty \tag{1.1.3}$$

then $\mathsf{E}\, G_\tau$ is well defined for all $\tau \in \mathfrak{M}_n^N$. Although for many results below it is possible to go beyond this condition and replace $|G_k|$ above by G_k^- or G_k^+ (or even consider only those τ for which $\mathsf{E}\, G_\tau$ is well defined) we will for simplicity assume throughout that (1.1.3) is satisfied. A more careful inspection of the proofs will easily reveal how the condition (1.1.3) can be relaxed.

With the subfamilies of stopping times $\mathfrak{M}_n^N$ introduced in (1.1.1) above we will associate the following value functions:

$$V_n^N = \sup_{\tau \in \mathfrak{M}_n^N} \mathsf{E}\, G_\tau \tag{1.1.4}$$

where $0 \le n \le N$. Again, for simplicity, we will set $V^N = V_0^N$ and $V_n = V_n^\infty$. Likewise, we will set $V = V_0^\infty$ when the supremum is taken over all τ in $\mathfrak{M}$. The main purpose of the present subsection is to study the optimal stopping problem (1.1.4) using a martingale approach.

Sometimes it is also of interest to admit that τ in (1.1.2) takes the value ∞ with positive probability, so that τ belongs to $\bar{\mathfrak{M}}$. In such a case we need to make an agreement about the value of G_τ on $\{\tau = \infty\}$. Clearly, if $\lim_{n\to\infty} G_n$ exists, then G_∞ is naturally set to take this value. Another possibility is to let G_∞ take an arbitrary but fixed value. Finally, for certain reasons of convenience, it is useful to set $G_\infty = \limsup_{n\to\infty} G_n$. In general, however, none of these choices is better than the others, and a preferred choice should always be governed by the meaning of a specific problem studied.

2. The method of backward induction. The first method for solving the problem (1.1.4) when $N < \infty$ uses backward induction first to construct a sequence of random variables $(S_n^N)_{0 \le n \le N}$ that solves the problem in a stochastic sense. Taking expectation then solves the problem in the original mean-valued sense.

Consider the optimal stopping problem (1.1.4) when $N < \infty$. Recall that (1.1.4) reads more explicitly as follows:

$$V_n^N = \sup_{n \le \tau \le N} \mathsf{E}\, G_\tau \tag{1.1.5}$$

where τ is a stopping time and $0 \le n \le N$. To solve the problem we can let the time go backward and proceed recursively as follows.

For $n = N$ we have to stop immediately and our gain S_N^N equals G_N. For $n = N-1$ we can either stop or continue. If we stop our gain S_{N-1}^N will be equal to G_{N-1}, and if we continue optimally our gain S_{N-1}^N will be equal to $\mathsf{E}\,(S_N^N \mid \mathcal{F}_{N-1})$. The latter conclusion reflects the fact that our decision about stopping or continuation at time $n = N-1$ must be based on the information contained in $\mathcal{F}_{N-1}$ only. It follows that if $G_{N-1} \ge \mathsf{E}\,(S_N^N \mid \mathcal{F}_{N-1})$ then we need to stop at time $n = N-1$, and if $G_{N-1} < \mathsf{E}\,(S_N^N \mid \mathcal{F}_{N-1})$ then we need to continue at time $n = N-1$. For $n = N-2, \ldots, 0$ the considerations are continued analogously.

The method of backward induction just explained leads to a sequence of random variables $(S_n^N)_{0 \le n \le N}$ defined recursively as follows:

$$S_n^N = G_N \quad \text{for } n = N, \tag{1.1.6}$$

$$S_n^N = \max\big(G_n, \mathsf{E}\,(S_{n+1}^N \mid \mathcal{F}_n)\big) \quad \text{for } n = N-1, \ldots, 0. \tag{1.1.7}$$

The method also suggests that we consider the following stopping time:

$$\tau_n^N = \inf\{\, n \le k \le N : S_k^N = G_k \,\} \tag{1.1.8}$$

for $0 \le n \le N$. Note that the infimum in (1.1.8) is always attained.

The first part of the following theorem shows that S_n^N and τ_n^N solve the problem in a stochastic sense. The second part of the theorem shows that this leads to a solution of the initial problem (1.1.5). The third part of the theorem provides a supermartingale characterization of the solution. The method of backward induction and the results presented in the theorem play a central role in the theory of optimal stopping.

Theorem 1.2. (Finite horizon) *Consider the optimal stopping problem* (1.1.5) *upon assuming that the condition* (1.1.3) *holds. Then for all* $0 \le n \le N$ *we have:*

$$S_n^N \ge \mathsf{E}\,(G_\tau \mid \mathcal{F}_n) \quad \textit{for each } \tau \in \mathfrak{M}_n^N, \tag{1.1.9}$$

$$S_n^N = \mathsf{E}\,(G_{\tau_n^N} \mid \mathcal{F}_n). \tag{1.1.10}$$

Moreover, if $0 \le n \le N$ *is given and fixed, then we have:*

The stopping time τ_n^N *is optimal in* (1.1.5). (1.1.11)

If τ_* *is an optimal stopping time in* (1.1.5) *then* $\tau_n^N \le \tau_*$ P-*a.s.* (1.1.12)

The sequence $(S_k^N)_{n \le k \le N}$ *is the smallest supermartingale which dominates* $(G_k)_{n \le k \le N}$. (1.1.13)

The stopped sequence $(S_{k \wedge \tau_n^N}^N)_{n \le k \le N}$ *is a martingale.* (1.1.14)

Proof. (1.1.9)–(1.1.10): The proof will be carried out by induction over $n = N$, $N-1, \ldots, 0$. Note that both relations are trivially satisfied when $n = N$ due to (1.1.6) above. Let us thus assume that (1.1.9) and (1.1.10) hold for $n = N, N-1, \ldots, k$ where $k \ge 1$, and let us show that (1.1.9) and (1.1.10) must then also hold for $n = k-1$.

(1.1.9): Take $\tau \in \mathfrak{M}_{k-1}^N$ and set $\bar{\tau} = \tau \vee k$. Then $\bar{\tau} \in \mathfrak{M}_k^N$ and since $\{\tau \ge k\} \in \mathcal{F}_{k-1}$ it follows that

$$\begin{aligned} \mathsf{E}(G_\tau \mid \mathcal{F}_{k-1}) &= \mathsf{E}\big(I(\tau = k-1)\, G_{k-1} \mid \mathcal{F}_{k-1}\big) + \mathsf{E}\big(I(\tau \ge k)\, G_{\bar{\tau}} \mid \mathcal{F}_{k-1}\big) \\ &= I(\tau = k-1)\, G_{k-1} + I(\tau \ge k)\, \mathsf{E}\big(\mathsf{E}(G_{\bar{\tau}} \mid \mathcal{F}_k) \mid \mathcal{F}_{k-1}\big). \end{aligned} \tag{1.1.15}$$

By the induction hypothesis the inequality (1.1.9) holds for $n = k$. Since $\bar{\tau} \in \mathfrak{M}_k^N$ this implies that $\mathsf{E}(G_{\bar{\tau}} \mid \mathcal{F}_k) \le S_k^N$. On the other hand, from (1.1.7) we see that $G_{k-1} \le S_{k-1}^N$ and $\mathsf{E}(S_k^N \mid \mathcal{F}_{k-1}) \le S_{k-1}^N$. Applying the preceding three inequalities to the right-hand side of (1.1.15) we get

$$\begin{aligned} \mathsf{E}(G_\tau \mid \mathcal{F}_{k-1}) &\le I(\tau = k-1)\, S_{k-1}^N + I(\tau \ge k)\, \mathsf{E}(S_k^N \mid \mathcal{F}_{k-1}) \\ &\le I(\tau = k-1)\, S_{k-1}^N + I(\tau \ge k)\, S_{k-1}^N = S_{k-1}^N. \end{aligned} \tag{1.1.16}$$

This shows that (1.1.9) holds for $n = k-1$ as claimed.

(1.1.10): To prove that (1.1.10) holds for $n = k-1$ it is enough to check that all inequalities in (1.1.15) and (1.1.16) remain equalities when $\tau = \tau_{k-1}^N$. For this, note from (1.1.8) that $\tau_{k-1}^N = \tau_k^N$ on $\{\tau_{k-1}^N \ge k\}$, so that from (1.1.15) with $\tau = \tau_{k-1}^N$ and the induction hypothesis (1.1.10) for $n = k$, we get

$$\begin{aligned} \mathsf{E}(G_{\tau_{k-1}^N} \mid \mathcal{F}_{k-1}) &= I(\tau_{k-1}^N = k-1)\, G_{k-1} \\ &\quad + I(\tau_{k-1}^N \ge k)\; \mathsf{E}\big(\mathsf{E}(G_{\tau_k^N} \mid \mathcal{F}_k) \mid \mathcal{F}_{k-1}\big) \\ &= I(\tau_{k-1}^N = k-1)\, G_{k-1} + I(\tau_{k-1}^N \ge k)\, \mathsf{E}(S_k^N \mid \mathcal{F}_{k-1}) \\ &= I(\tau_{k-1}^N = k-1)\, S_{k-1}^N + I(\tau_{k-1}^N \ge k)\, S_{k-1}^N = S_{k-1}^N \end{aligned} \tag{1.1.17}$$

where in the second last equality we use that $G_{k-1} = S_{k-1}^N$ on $\{\tau_{k-1}^N = k-1\}$ by (1.1.8) as well as that $\mathsf{E}(S_k^N \mid \mathcal{F}_{k-1}) = S_{k-1}^N$ on $\{\tau_{k-1}^N \ge k\}$ by (1.1.8) and (1.1.7). This shows that (1.1.10) holds for $n = k-1$ as claimed.

(1.1.11): Taking E in (1.1.9) we find that $\mathsf{E}\,S_n^N \ge \mathsf{E}\,G_\tau$ for all $\tau \in \mathfrak{M}_n^N$ and hence by taking the supremum over all $\tau \in \mathfrak{M}_n^N$ we see that $\mathsf{E}\,S_n^N \ge V_n^N$. On the other hand, taking the expectation in (1.1.10) we get $\mathsf{E}\,S_n^N = \mathsf{E}\,G_{\tau_n^N}$ which shows that $\mathsf{E}\,S_n^N \le V_n^N$. The two inequalities give the equality $V_n^N = \mathsf{E}\,S_n^N$, and since $\mathsf{E}\,S_n^N = \mathsf{E}\,G_{\tau_n^N}$, we see that $V_n^N = \mathsf{E}\,G_{\tau_n^N}$ implying the claim.

(1.1.12): We claim that the optimality of τ_* implies that $S_{\tau_*}^N = G_{\tau_*}$ P-a.s. Indeed, if this would not be the case, then using that $S_k^N \ge G_k$ for all $n \le k \le N$ by (1.1.6)–(1.1.7), we see that $S_{\tau_*}^N \ge G_{\tau_*}$ with $\mathsf{P}(S_{\tau_*}^N > G_{\tau_*}) > 0$. It thus follows that $\mathsf{E}\,G_{\tau_*} < \mathsf{E}\,S_{\tau_*}^N \le \mathsf{E}\,S_n^N = V_n^N$ where the second inequality follows by the optional sampling theorem (page 60) and the supermartingale property of $(S_k^N)_{n\le k\le N}$ established in (1.1.13) below, while the final equality follows from the proof of (1.1.11) above. The strict inequality, however, contradicts the fact that τ_* is optimal. Hence $S_{\tau_*}^N = G_{\tau_*}$ P-a.s. as claimed and the fact that $\tau_n^N \le \tau_*$ P-a.s. follows from the definition (1.1.8).

(1.1.13): From (1.1.7) it follows that

$$S_k^N \ge \mathsf{E}\,(S_{k+1}^N \mid \mathcal{F}_k) \tag{1.1.18}$$

for all $n \le k \le N-1$ showing that $(S_k^N)_{n\le k\le N}$ is a supermartingale. From (1.1.6) and (1.1.7) it follows that $S_k^N \ge G_k$ P-a.s. for all $n \le k \le N$ meaning that $(S_k^N)_{n\le k\le N}$ dominates $(G_k)_{n\le k\le N}$. Moreover, if $(\widetilde{S}_k)_{n\le k\le N}$ is another supermartingale which dominates $(G_k)_{n\le k\le N}$, then the claim that $\widetilde{S}_k \ge S_k^N$ P-a.s. can be verified by induction over $k = N, N-1, \ldots, l$. Indeed, if $k = N$ then the claim follows by (1.1.6). Assuming that $\widetilde{S}_k \ge S_k^N$ P-a.s. for $k = N, N-1, \ldots, l$ with $l \ge n+1$ it follows by (1.1.7) that $S_{l-1}^N = \max(G_{l-1}, \mathsf{E}\,(S_l^N \mid \mathcal{F}_{l-1})) \le \max(G_{l-1}, \mathsf{E}\,(\widetilde{S}_l \mid \mathcal{F}_{l-1})) \le \widetilde{S}_{l-1}$ P-a.s. using the supermartingale property of $(\widetilde{S}_k)_{n\le k\le N}$ and proving the claim.

(1.1.14): To verify the martingale property

$$\mathsf{E}\,\big(S_{(k+1)\wedge\tau_n^N}^N \mid \mathcal{F}_k\big) = S_{k\wedge\tau_n^N}^N \tag{1.1.19}$$

with $n \le k \le N-1$ given and fixed, note that

$$\begin{aligned} &\mathsf{E}\,\big(S_{(k+1)\wedge\tau_n^N}^N \mid \mathcal{F}_k\big) \\ &\quad= \mathsf{E}\,\big(I(\tau_n^N \le k)\, S_{k\wedge\tau_n^N}^N \mid \mathcal{F}_k\big) + \mathsf{E}\,\big(I(\tau_n^N \ge k{+}1)\, S_{k+1}^N \mid \mathcal{F}_k\big) \\ &\quad= I(\tau_n^N \le k)\, S_{k\wedge\tau_n^N}^N + I(\tau_n^N \ge k{+}1)\, \mathsf{E}\,(S_{k+1}^N \mid \mathcal{F}_k) \\ &\quad= I(\tau_n^N \le k)\, S_{k\wedge\tau_n^N}^N + I(\tau_n^N \ge k{+}1)\, S_k^N = S_{k\wedge\tau_n^N}^N \end{aligned} \tag{1.1.20}$$

where the second last equality follows from the fact that $S_k^N = \mathsf{E}\,(S_{k+1}^N \mid \mathcal{F}_k)$ on $\{\tau_n^N \ge k+1\}$, while $\{\tau_n^N \ge k+1\} \in \mathcal{F}_k$ since τ_n^N is a stopping time. This establishes (1.1.19) and the proof of the theorem is complete. □

Note that (1.1.9) can also be derived from the supermartingale property (1.1.13), and that (1.1.10) can also be derived from the martingale property (1.1.14), both by means of the optional sampling theorem (page 60).

It follows from Theorem 1.2 that the optimal stopping problem V_0^N is solved inductively by solving the problems V_n^N for $n = N, N-1, \ldots, 0$. Moreover, the optimal stopping rule τ_n^N for V_n^N satisfies $\tau_n^N = \tau_k^N$ on $\{\tau_n^N \geq k\}$ for $0 \leq n \leq k \leq N$ where τ_k^N is the optimal stopping rule for V_k^N. This, in other words, means that if it was not optimal to stop within the time set $\{n, n+1, \ldots, k-1\}$ then the same optimality rule applies in the time set $\{k, k+1, \ldots, N\}$. In particular, when specialized to the problem V_0^N, the following general principle is obtained: If the stopping rule τ_0^N is optimal for V_0^N and it was not optimal to stop within the time set $\{0, 1, \ldots, n-1\}$, then starting the observation at time n and being based on the information $\mathcal{F}_n$, the same stopping rule is still optimal for the problem V_n^N. This principle of solution for optimal stopping problems has led to the general principle of *dynamic programming* in the theory of optimal stochastic control (often referred to as *Bellman's principle*).

3. The method of essential supremum. The method of backward induction by its nature requires that the horizon N be finite so that the case of infinite horizon N remains uncovered. It turns out, however, that the random variables S_n^N defined by the recurrent relations (1.1.6)–(1.1.7) above admit a different characterization which can be directly extended to the case of infinite horizon N. This characterization forms the basis for the second method that will now be presented.

With this aim note that (1.1.9) and (1.1.10) in Theorem 1.2 above suggest that the following identity should hold:

$$S_n^N = \sup_{\tau \in \mathfrak{M}_n^N} \mathsf{E}\big(G_\tau \,|\, \mathcal{F}_n\big). \tag{1.1.21}$$

A difficulty arises, however, from the fact that both (1.1.9) and (1.1.10) hold only P-a.s. so that the exceptional P-null set may depend on the given $\tau \in \mathfrak{M}_n^N$. Thus, if the supremum in (1.1.21) is taken over uncountably many τ, then the right-hand side need not define a measurable function, and the identity (1.1.21) may fail as well. To overcome this difficulty it turns out that the concept of *essential supremum* proves useful.

Lemma 1.3. (Essential supremum) *Let $\{Z_\alpha : \alpha \in I\}$ be a family of random variables defined on $(\Omega, \mathcal{G}, \mathsf{P})$ where the index set I can be arbitrary. Then there exists a countable subset J of I such that the random variable $Z^* : \Omega \to \bar{\mathbb{R}}$ defined by*

$$Z^* = \sup_{\alpha \in J} Z_\alpha \tag{1.1.22}$$

satisfies the following two properties:

$$\mathsf{P}(Z_\alpha \le Z^*) = 1 \;\; \textit{for each } \alpha \in I. \tag{1.1.23}$$

If $\tilde{Z} : \Omega \to \bar{\mathbb{R}}$ *is another random variable satisfying* (1.1.23) *in place of* Z^*, *then* $\mathsf{P}(Z^* \le \tilde{Z}) = 1$. (1.1.24)

The random variable Z^* is called the *essential supremum* of $\{ Z_\alpha : \alpha \in I \}$ relative to P and is denoted by $Z^* = \operatorname{ess\,sup}_{\alpha\in I} Z_\alpha$. It is determined by the properties (1.1.23) and (1.1.24) uniquely up to a P-null set.

Moreover, if the family $\{Z_\alpha : \alpha \in I\}$ *is upwards directed in the sense that*

For any α *and* β *in* I *there exists* γ *in* I *such that* $Z_\alpha \vee Z_\beta \le Z_\gamma$ P*-a.s.* (1.1.25)

then the countable set $J = \{\alpha_n : n \ge 1\}$ *can be chosen so that*

$$Z^* = \lim_{n\to\infty} Z_{\alpha_n} \quad \mathsf{P}\text{-}a.s. \tag{1.1.26}$$

where $Z_{\alpha_1} \le Z_{\alpha_2} \le \cdots$ P*-a.s.*

Proof. Since $x \mapsto (2/\pi)\arctan(x)$ is a strictly increasing function from $\bar{\mathbb{R}}$ to $[-1,1]$, it is no restriction to assume that $|Z_\alpha| \le 1$ for all $\alpha \in I$. Otherwise, replace Z_α by $(2/\pi)\arctan(Z_\alpha)$ for $\alpha \in I$ and proceed as in the rest of the proof.

Let $\mathcal{C}$ denote the family of all countable subsets C of I. Choose an increasing sequence $\{ C_n : n \ge 1 \}$ in $\mathcal{C}$ such that

$$a = \sup_{C\in\mathcal{C}} \mathsf{E}\Big(\sup_{\alpha\in C} Z_\alpha \Big) = \sup_{n\ge1} \mathsf{E}\Big(\sup_{\alpha\in C_n} Z_\alpha \Big). \tag{1.1.27}$$

Then $J := \bigcup_{n=1}^\infty C_n$ is a countable subset of I and we claim that Z^* defined by (1.1.22) satisfies (1.1.23) and (1.1.24).

To verify these claims take $\alpha \in I$ arbitrarily and note the following. If $\alpha \in J$ then $Z_\alpha \le Z^*$ so that (1.1.23) holds. On the other hand, if $\alpha \notin J$ and we assume that $\mathsf{P}(Z_\alpha > Z^*) > 0$, then $a < \mathsf{E}(Z^* \vee Z_\alpha) \le a$ since $a = \mathsf{E}Z^* \in [-1,1]$ (by the monotone convergence theorem) and $J \cup \{\alpha\}$ belongs to $\mathcal{C}$. As the strict inequality is clearly impossible, we see that (1.1.23) holds for all $\alpha \in I$ as claimed. Moreover, it is obvious that (1.1.24) follows from (1.1.22) and (1.1.23) since J is countable.

Finally, if (1.1.25) is satisfied then the initial countable set $J = \{\alpha_1^0, \alpha_2^0, \dots\}$ can be replaced by a new countable set $J = \{ \alpha_1, \alpha_2, \dots \}$ if we initially set $\alpha_1 = \alpha_1^0$, and then inductively choose $\alpha_{n+1} \ge \alpha_n \vee \alpha_{n+1}^0$ for $n \ge 1$, where $\gamma \ge \alpha \vee \beta$ corresponds to Z_α, Z_β and Z_γ such that $Z_\gamma \ge Z_\alpha \vee Z_\beta$ P-a.s.

The concluding claim in (1.1.26) is then obvious, and the proof of the lemma is complete. □

With the concept of essential supremum we may now rewrite (1.1.9) and (1.1.10) in Theorem 1.2 above as follows:

$$S_n^N = \operatorname*{ess\,sup}_{n\le\tau\le N} \mathsf{E}\,(G_\tau \mid \mathcal{F}_n) \tag{1.1.28}$$

for all $0 \le n \le N$. This identity provides an additional characterization of the sequence of random variables $(S_n^N)_{0\le n\le N}$ introduced initially by means of the recurrent relations (1.1.6)–(1.1.7). Its advantage in comparison with the recurrent relations lies in the fact that the identity (1.1.28) can naturally be extended to the case of infinite horizon N. This programme will now be described.

Consider the optimal stopping problem (1.1.4) when $N=\infty$. Recall that (1.1.4) reads more explicitly as follows:

$$V_n = \sup_{\tau\ge n} \mathsf{E}\,G_\tau \tag{1.1.29}$$

where τ is a stopping time and $n \ge 0$. To solve the problem we will consider the sequence of random variables $(S_n)_{n\ge 0}$ defined as follows:

$$S_n = \operatorname*{ess\,sup}_{\tau\ge n} \mathsf{E}\,(G_\tau \mid \mathcal{F}_n) \tag{1.1.30}$$

as well as the following stopping time:

$$\tau_n = \inf\,\{\, k \ge n : S_k = G_k \} \tag{1.1.31}$$

for $n \ge 0$ where $\inf \varnothing = \infty$ by definition. The sequence $(S_n)_{n\ge 0}$ is often referred to as the *Snell envelope* of G.

The first part of the following theorem shows that $(S_n)_{n\ge 0}$ satisfies the same recurrent relations as $(S_n^N)_{0\le n\le N}$. The second part of the theorem shows that S_n and τ_n solve the problem in a stochastic sense. The third part of the theorem shows that this leads to a solution of the initial problem (1.1.29). The fourth part of the theorem provides a supermartingale characterization of the solution.

Theorem 1.4. (Infinite horizon) *Consider the optimal stopping problem* (1.1.29) *upon assuming that the condition* (1.1.3) *holds. Then the following recurrent relations hold:*

$$S_n = \max\big(G_n, \mathsf{E}\,(S_{n+1} \mid \mathcal{F}_n)\big) \tag{1.1.32}$$

for all $n \ge 0$. *Assume moreover when required below that*

$$\mathsf{P}(\tau_n < \infty) = 1 \tag{1.1.33}$$

where $n \geq 0$. *Then for all* $n \geq 0$ *we have:*

$$S_n \geq \mathsf{E}(G_\tau \mid \mathcal{F}_n) \;\; \textit{for each } \tau \in \mathfrak{M}_n, \tag{1.1.34}$$

$$S_n = \mathsf{E}(G_{\tau_n} \mid \mathcal{F}_n). \tag{1.1.35}$$

Moreover, if $n \geq 0$ *is given and fixed, then we have:*

The stopping time τ_n *is optimal in* (1.1.29). (1.1.36)

If τ_* *is an optimal stopping time in* (1.1.29) *then* $\tau_n \leq \tau_*$ P*-a.s.* (1.1.37)

The sequence $(S_k)_{k\geq n}$ *is the smallest supermartingale which dominates* $(G_k)_{k\geq n}$. (1.1.38)

The stopped sequence $(S_{k\wedge\tau_n})_{k\geq n}$ *is a martingale.* (1.1.39)

Finally, if the condition (1.1.33) *fails so that* $\mathsf{P}(\tau_n = \infty) > 0$, *then there is no optimal stopping time* (*with probability* 1) *in* (1.1.29).

Proof. (1.1.32): Let us first show that the left-hand side is smaller than the right-hand side when $n \geq 0$ is given and fixed.

For this, take $\tau \in \mathfrak{M}_n$ and set $\bar{\tau} = \tau \vee (n+1)$. Then $\bar{\tau} \in \mathfrak{M}_{n+1}$ and since $\{\tau \geq n+1\} \in \mathcal{F}_n$ we have

$$\begin{aligned} \mathsf{E}(G_\tau \mid \mathcal{F}_n) &= \mathsf{E}\big(I(\tau = n)\, G_n \mid \mathcal{F}_n\big) + \mathsf{E}\big(I(\tau \geq n+1)\, G_{\bar{\tau}} \mid \mathcal{F}_n\big) \\ &= I(\tau = n)\, G_n + I(\tau \geq n+1)\, \mathsf{E}(G_{\bar{\tau}} \mid \mathcal{F}_n) \\ &= I(\tau = n)\, G_n + I(\tau \geq n+1)\, \mathsf{E}\big(\mathsf{E}(G_{\bar{\tau}} \mid \mathcal{F}_{n+1}) \mid \mathcal{F}_n\big) \\ &\leq I(\tau = n)\, G_n + I(\tau \geq n+1)\, \mathsf{E}(S_{n+1} \mid \mathcal{F}_n) \\ &\leq \max\big(G_n, \mathsf{E}(S_{n+1} \mid \mathcal{F}_n)\big). \end{aligned} \tag{1.1.40}$$

From this inequality it follows that

$$\operatorname*{ess\,sup}_{\tau \geq n} \mathsf{E}(G_\tau \mid \mathcal{F}_n) \leq \max\big(G_n, \mathsf{E}(S_{n+1} \mid \mathcal{F}_n)\big) \tag{1.1.41}$$

which is the desired inequality.

To prove the reverse inequality, let us first note that $S_n \geq G_n$ P-a.s. by the definition of S_n so that it is enough to show that

$$S_n \geq \mathsf{E}(S_{n+1} \mid \mathcal{F}_n) \tag{1.1.42}$$

which is the supermartingale property of $(S_n)_{n\geq 0}$. To verify this inequality, let us first show that the family $\{\, \mathsf{E}(G_\tau \mid \mathcal{F}_{n+1}) : \tau \in \mathfrak{M}_{n+1} \,\}$ is upwards directed in the sense that (1.1.25) is satisfied. For this, note that if σ_1 and σ_2 are from $\mathfrak{M}_{n+1}$ and we set $\sigma_3 = \sigma_1 I_A + \sigma_2 I_{A^c}$ where $A = \{\, \mathsf{E}(G_{\sigma_1} \mid \mathcal{F}_{n+1}) \geq \mathsf{E}(G_{\sigma_2} \mid \mathcal{F}_{n+1}) \,\}$, then σ_3 belongs to $\mathfrak{M}_{n+1}$ and we have

$$\begin{aligned} \mathsf{E}(G_{\sigma_3} \mid \mathcal{F}_{n+1}) &= \mathsf{E}(G_{\sigma_1} I_A + G_{\sigma_2} I_{A^c} \mid \mathcal{F}_{n+1}) \\ &= I_A\, \mathsf{E}(G_{\sigma_1} \mid \mathcal{F}_{n+1}) + I_{A^c}\, \mathsf{E}(G_{\sigma_2} \mid \mathcal{F}_{n+1}) \\ &= \mathsf{E}(G_{\sigma_1} \mid \mathcal{F}_{n+1}) \vee \mathsf{E}(G_{\sigma_2} \mid \mathcal{F}_{n+1}) \end{aligned} \tag{1.1.43}$$

implying (1.1.25) as claimed. Hence by (1.1.26) there exists a sequence $\{\sigma_k : k \geq 1\}$ in $\mathfrak{M}_{n+1}$ such that

$$\operatorname*{ess\,sup}_{\tau \geq n+1} \mathsf{E}\,(G_\tau \,|\, \mathcal{F}_{n+1}) = \lim_{k\to\infty} \mathsf{E}\,(G_{\sigma_k} \,|\, \mathcal{F}_{n+1}) \tag{1.1.44}$$

where $\mathsf{E}\,(G_{\sigma_1} \,|\, \mathcal{F}_{n+1}) \leq \mathsf{E}\,(G_{\sigma_2} \,|\, \mathcal{F}_{n+1}) \leq \cdots$ P-a.s. Since the left-hand side in (1.1.44) equals S_{n+1}, by the conditional monotone convergence theorem we get

$$\begin{aligned}\mathsf{E}\,(S_{n+1} \,|\, \mathcal{F}_n) &= \mathsf{E}\Big(\lim_{k\to\infty} \mathsf{E}\,(G_{\sigma_k} \,|\, \mathcal{F}_{n+1}) \,|\, \mathcal{F}_n\Big) \\ &= \lim_{k\to\infty} \mathsf{E}\big(\mathsf{E}\,(G_{\sigma_k} \,|\, \mathcal{F}_{n+1}) \,|\, \mathcal{F}_n\big) \\ &= \lim_{k\to\infty} \mathsf{E}\,(G_{\sigma_k} \,|\, \mathcal{F}_n) \leq S_n\end{aligned} \tag{1.1.45}$$

where the final inequality follows from the definition of S_n. This establishes (1.1.42) and the proof of (1.1.32) is complete.

(1.1.34): This inequality follows directly from the definition (1.1.30).

(1.1.35): The proof of (1.1.39) below shows that the stopped sequence $(S_{k\wedge\tau_n})_{k\geq n}$ is a martingale. Moreover, setting $G_n^* = \sup_{k\geq n} |G_k|$ we have

$$|S_k| \leq \operatorname*{ess\,sup}_{\tau\geq k} \mathsf{E}\big(|G_\tau| \,|\, \mathcal{F}_k\big) \leq \mathsf{E}\,(G_n^* \,|\, \mathcal{F}_k) \tag{1.1.46}$$

for all $k \geq n$. Since G_n^* is integrable due to (1.1.3), it follows from (1.1.46) that $(S_k)_{k\geq n}$ is uniformly integrable. Thus the optional sampling theorem (page 60) can be applied to the martingale $(M_k)_{k\geq n} = (S_{k\wedge\tau_n})_{k\geq n}$ and the stopping time τ_n yielding

$$M_n = \mathsf{E}\,(M_{\tau_n} \,|\, \mathcal{F}_n). \tag{1.1.47}$$

Since $M_n = S_n$ and $M_{\tau_n} = S_{\tau_n}$ we see that (1.1.47) is the same as (1.1.35).

(1.1.36): This is proved using (1.1.34) and (1.1.35) in exactly the same way as (1.1.11) above using (1.1.9) and (1.1.10).

(1.1.37): This is proved in exactly the same way as (1.1.12) above.

(1.1.38): It was shown in (1.1.42) that $(S_k)_{k\geq n}$ is a supermartingale. Moreover, it follows from (1.1.30) that $S_k \geq G_k$ P-a.s. for all $k \geq n$ meaning that $(S_k)_{k\geq n}$ dominates $(G_k)_{k\geq n}$. Finally, if $(\tilde{S}_k)_{k\geq n}$ is another supermartingale which dominates $(G_k)_{k\geq n}$, then by (1.1.35) we find

$$S_k = \mathsf{E}\,(G_{\tau_k} \,|\, \mathcal{F}_k) \leq \mathsf{E}\,(\tilde{S}_{\tau_k} \,|\, \mathcal{F}_k) \leq \tilde{S}_k \tag{1.1.48}$$

for all $k \geq n$ where the final inequality follows by the optional sampling theorem (page 60) being applicable since $\tilde{S}_k^- \leq G_k^- \leq G_n^*$ for all $k \geq n$ with G_n^* integrable.

(1.1.39): This is proved in exactly the same way as (1.1.14) above.

Finally, note that the final claim follows directly from (1.1.37). This completes the proof of the theorem. □

4. In the last part of this subsection we will briefly explore a connection between the two methods above when the horizon N tends to infinity in the former.

For this, note from (1.1.28) that $N \mapsto S_n^N$ and $N \mapsto \tau_n^N$ are increasing, so that

$$S_n^\infty = \lim_{N\to\infty} S_n^N \quad \text{and} \quad \tau_n^\infty = \lim_{N\to\infty} \tau_n^N \tag{1.1.49}$$

exist P-a.s. for each $n \ge 0$. Note also from (1.1.5) that $N \mapsto V_n^N$ is increasing, so that

$$V_n^\infty = \lim_{N\to\infty} V_n^N \tag{1.1.50}$$

exists for each $n \ge 0$. From (1.1.28) and (1.1.30) we see that

$$S_n^\infty \le S_n \quad \text{and} \quad \tau_n^\infty \le \tau_n \tag{1.1.51}$$

P-a.s. for each $n \ge 0$. Similarly, from (1.1.10) and (1.1.35) we find that

$$V_n^\infty \le V_n \tag{1.1.52}$$

for each $n \ge 0$. The following simple example shows that in the absence of the condition (1.1.3) above the inequalities in (1.1.51) and (1.1.52) can be strict.

Example 1.5. Let $G_n = \sum_{k=0}^n \varepsilon_k$ for $n \ge 0$ where $(\varepsilon_k)_{k\ge 0}$ is a sequence of independent and identically distributed random variables with $\mathsf{P}(\varepsilon_k = -1) = \mathsf{P}(\varepsilon_k = 1) = 1/2$ for $k \ge 0$. Setting $\mathcal{F}_n = \sigma(\varepsilon_1, \ldots, \varepsilon_n)$ for $n \ge 0$ it follows that $(G_n)_{n\ge 0}$ is a martingale with respect to $(\mathcal{F}_n)_{n\ge 0}$. From (1.1.28) using the optional sampling theorem (page 60) one sees that $S_n^N = G_n$ and hence $\tau_n^N = n$ as well as $V_n^N = 0$ for all $0 \le n \le N$. On the other hand, if we make use of the stopping times $\sigma_m = \inf\{\, k \ge n : G_k = m \,\}$ upon recalling that $\mathsf{P}(\sigma_m < \infty) = 1$ whenever $m \ge 1$, it follows by (1.1.30) that $S_n \ge m$ P-a.s. for all $m \ge 1$. From this one sees that $S_n = \infty$ P-a.s. and hence $\tau_n = \infty$ P-a.s. as well as $V_n = \infty$ for all $n \ge 0$. Thus, in this case, all inequalities in (1.1.51) and (1.1.52) are strict.

Theorem 1.6. (From finite to infinite horizon) *Consider the optimal stopping problems* (1.1.5) *and* (1.1.29) *upon assuming that the condition* (1.1.3) *holds. Then equalities in* (1.1.51) *and* (1.1.52) *hold for all* $n \ge 0$.

Proof. Letting $N \to \infty$ in (1.1.7) and using the conditional monotone convergence theorem one finds that the following recurrent relations hold:

$$S_n^\infty = \max\big(G_n, \mathsf{E}\,(S_{n+1}^\infty \mid \mathcal{F}_n)\big) \tag{1.1.53}$$

for all $n \geq 0$. In particular, it follows that $(S_n^\infty)_{n\geq 0}$ is a supermartingale. Since $S_n^\infty \geq G_n$ P-a.s. we see that $(S_n^\infty)^- \leq G_n^- \leq \sup_{n\geq 0} G_n^-$ P-a.s. for all $n \geq 0$ from where by means of (1.1.3) we see that $((S_n^\infty)^-)_{n\geq 0}$ is uniformly integrable. Thus by the optional sampling theorem (page 60) we get

$$S_n^\infty \geq \mathsf{E}\,(S_\tau^\infty \mid \mathcal{F}_n) \tag{1.1.54}$$

for all $\tau \in \mathfrak{M}_n$. Moreover, since $S_k^\infty \geq G_k$ P-a.s. for all $k \geq n$, it follows that $S_\tau^\infty \geq G_\tau$ P-a.s. for all $\tau \in \mathfrak{M}_n$, and hence

$$\mathsf{E}\,(S_\tau^\infty \mid \mathcal{F}_n) \geq \mathsf{E}\,(G_\tau \mid \mathcal{F}_n) \tag{1.1.55}$$

for all $\tau \in \mathfrak{M}_n$. Combining (1.1.54) and (1.1.55) we see by (1.1.30) that $S_n^\infty \geq S_n$ P-a.s. for all $n \geq 0$. Since the reverse inequality holds in general as shown in (1.1.51) above, this establishes that $S_n^\infty = S_n$ P-a.s. for all $n \geq 0$. From this it also follows that $\tau_n^\infty = \tau_n$ P-a.s. for all $n \geq 0$. Finally, the third identity $V_n^\infty = V_n$ follows by the monotone convergence theorem. The proof of the theorem is complete. □

1.2. Markovian approach

In this subsection we will present basic results of optimal stopping when the time is discrete and the process is Markovian. (Basic definitions and properties of such processes are given in Subsections 4.1 and 4.2.)

1. Throughout we consider a time-homogeneous Markov chain $X = (X_n)_{n\geq 0}$ defined on a filtered probability space $(\Omega, \mathcal{F}, (\mathcal{F}_n)_{n\geq 0}, \mathsf{P}_x)$ and taking values in a measurable space $(E, \mathcal{B})$ where for simplicity we assume that $E = \mathbb{R}^d$ for some $d \geq 1$ and $\mathcal{B} = \mathcal{B}(\mathbb{R}^d)$ is the Borel σ-algebra on $\mathbb{R}^d$. It is assumed that the chain X starts at x under P_x for $x \in E$. It is also assumed that the mapping $x \mapsto \mathsf{P}_x(F)$ is measurable for each $F \in \mathcal{F}$. It follows that the mapping $x \mapsto \mathsf{E}_x(Z)$ is measurable for each random variable Z. Finally, without loss of generality we assume that $(\Omega, \mathcal{F})$ equals the canonical space $(E^{\mathbb{N}_0}, \mathcal{B}^{\mathbb{N}_0})$ so that the shift operator $\theta_n : \Omega \to \Omega$ is well defined by $\theta_n(\omega)(k) = \omega(n+k)$ for $\omega = (\omega(k))_{k\geq 0} \in \Omega$ and $n, k \geq 0$. (Recall that $\mathbb{N}_0$ stands for $\mathbb{N} \cup \{0\}$.)

2. Given a measurable function $G : E \to \mathbb{R}$ satisfying the following condition (with $G(X_N) = 0$ if $N = \infty$):

$$\mathsf{E}_x\Big(\sup_{0\leq n\leq N} |G(X_n)|\Big) < \infty \tag{1.2.1}$$

for all $x \in E$, we consider the optimal stopping problem

$$V^N(x) = \sup_{0\leq \tau\leq N} \mathsf{E}_x G(X_\tau) \tag{1.2.2}$$

where $x \in E$ and the supremum is taken over all stopping times τ of X. The latter means that τ is a stopping time with respect to the natural filtration of X given by $\mathcal{F}_n^X = \sigma(X_k : 0 \leq k \leq n)$ for $n \geq 0$. Since the same results remain valid if we take the supremum in (1.2.2) over stopping times τ with respect to $(\mathcal{F}_n)_{n\geq 0}$, and this assumption makes final conclusions more powerful (at least formally), we will assume in the sequel that the supremum in (1.2.2) is taken over this larger class of stopping times. Note also that in (1.2.2) we admit that N can be $+\infty$ as well. In this case, however, we still assume that the supremum is taken over stopping times τ, i.e. over Markov times τ satisfying $\tau < \infty$ P-a.s. In this way any specification of $G(X_\infty)$ becomes irrelevant for the problem (1.2.2).

3. To solve the problem (1.2.2) in the case when $N < \infty$ we may note that by setting

$$G_n = G(X_n) \tag{1.2.3}$$

for $n \geq 0$ the problem (1.2.2) reduces to the problem (1.1.5) where instead of P and E we have P_x and E_x for $x \in E$. Introducing the expectation in (1.2.2) with respect to P_x under which $X_0 = x$ and studying the resulting problem by means of the mapping $x \mapsto V^N(x)$ for $x \in E$ constitutes a profound step which most directly aims to exploit the Markovian structure of the problem. (The same remark applies in the theory of optimal stochastic control in contrast to classical methods developed in calculus of variations.)

Having identified the problem (1.2.2) as the problem (1.1.5) we can apply the method of backward induction (1.1.6)–(1.1.7) which leads to a sequence of random variables $(S_n^N)_{0\leq n\leq N}$ and a stopping time τ_n^N defined in (1.1.8). The key identity is

$$S_n^N = V^{N-n}(X_n) \tag{1.2.4}$$

for $0 \leq n \leq N$. This will be established in the proof of the next theorem. Once (1.2.4) is known to hold, the results of Theorem 1.2 translate immediately into the present setting and get a more transparent form as follows.

In the sequel we set

$$C_n = \{\, x \in E : V^{N-n}(x) > G(x) \,\}, \tag{1.2.5}$$

$$D_n = \{\, x \in E : V^{N-n}(x) = G(x) \,\} \tag{1.2.6}$$

for $0 \leq n \leq N$. We define

$$\tau_D = \inf\{\, 0 \leq n \leq N : X_n \in D_n \,\}. \tag{1.2.7}$$

Finally, the transition operator T of X is defined by

$$TF(x) = \mathsf{E}_x F(X_1) \tag{1.2.8}$$

for $x \in E$ whenever $F : E \to \mathbb{R}$ is a measurable function so that $F(X_1)$ is integrable with respect to P_x for all $x \in E$.

Theorem 1.7. (Finite horizon: The time-homogeneous case) *Consider the optimal stopping problem* (1.2.2) *upon assuming that the condition* (1.2.1) *holds. Then the value function* V^n *satisfies the* Wald–Bellman equations

$$V^n(x) = \max(G(x), TV^{n-1}(x)) \qquad (x \in E) \tag{1.2.9}$$

for $n = 1, \ldots, N$ *where* $V^0 = G$. *Moreover, we have:*

The stopping time τ_D *is optimal in* (1.2.2). (1.2.10)

If τ_* *is an optimal stopping time in* (1.2.2) *then* $\tau_D \leq \tau_*$ P_x*-a.s. for every* $x \in E$. (1.2.11)

The sequence $(V^{N-n}(X_n))_{0 \leq n \leq N}$ *is the smallest supermartingale which dominates* $(G(X_n))_{0 \leq n \leq N}$ *under* P_x *for* $x \in E$ *given and fixed.* (1.2.12)

The stopped sequence $(V^{N-n\wedge\tau_D}(X_{n\wedge\tau_D}))_{0 \leq n \leq N}$ *is a martingale under* P_x *for every* $x \in E$. (1.2.13)

Proof. To verify (1.2.4) recall from (1.1.10) that

$$S_n^N = \mathsf{E}_x\big(G(X_{\tau_n^N}) \,|\, \mathcal{F}_n\big) \tag{1.2.14}$$

for $0 \leq n \leq N$. Since $S_k^{N-n} \circ \theta_n = S_{n+k}^N$ we get that τ_n^N satisfies

$$\tau_n^N = \inf\{\, n \leq k \leq N : S_k^N = G(X_k)\,\} = n + \tau_0^{N-n} \circ \theta_n \tag{1.2.15}$$

for $0 \leq n \leq N$. Inserting (1.2.15) into (1.2.14) and using the Markov property we obtain

$$\begin{aligned} S_n^N &= \mathsf{E}_x\big(G(X_{n+\tau_0^{N-n}\circ\theta_n}) \,|\, \mathcal{F}_n\big) = \mathsf{E}_x\big(G(X_{\tau_0^{N-n}}) \circ \theta_n \,|\, \mathcal{F}_n\big) \\ &= \mathsf{E}_{X_n} G(X_{\tau_0^{N-n}}) = V^{N-n}(X_n) \end{aligned} \tag{1.2.16}$$

where the final equality follows by (1.1.9)–(1.1.10) which imply

$$\mathsf{E}_x S_0^{N-n} = \mathsf{E}_x G(X_{\tau_0^{N-n}}) = \sup_{0 \leq \tau \leq N-n} \mathsf{E}_x G(X_\tau) = V^{N-n}(x) \tag{1.2.17}$$

for $0 \leq n \leq N$ and $x \in E$. Thus (1.2.4) holds as claimed.

To verify (1.2.9) note that (1.1.7) using (1.2.4) and the Markov property reads as follows:

$$\begin{aligned} V^{N-n}(X_n) &= \max\big(G(X_n), \mathsf{E}_x\big(V^{N-n-1}(X_{n+1}) \,|\, \mathcal{F}_n\big)\big) \\ &= \max\big(G(X_n), \mathsf{E}_x\big(V^{N-n-1}(X_1) \circ \theta_n \,|\, \mathcal{F}_n\big)\big) \\ &= \max\big(G(X_n), \mathsf{E}_{X_n}\big(V^{N-n-1}(X_1)\big)\big) \\ &= \max\big(G(X_n), TV^{N-n-1}(X_n)\big) \end{aligned} \tag{1.2.18}$$

for all $0 \le n \le N$. Letting $n = 0$ and using that $X_0 = x$ under P_x we see that (1.2.18) yields (1.2.9).

The remaining statements of the theorem follow directly from Theorem 1.2 above. The proof is complete. $\square$

4. The Wald–Bellman equations (1.2.9) can be written in a more compact form as follows. Introduce the operator Q by setting

$$QF(x) = \max(G(x), TF(x)) \tag{1.2.19}$$

for $x \in E$ where $F : E \to \mathbb{R}$ is a measurable function for which $F(X_1) \in L^1(\mathsf{P}_x)$ for $x \in E$. Then (1.2.9) reads as follows:

$$V^n(x) = Q^n G(x) \tag{1.2.20}$$

for $1 \le n \le N$ where Q^n denotes the n-th power of Q. The recursive relations (1.2.20) form a constructive method for finding V^N when $\text{Law}(X_1 \mid \mathsf{P}_x)$ is known for $x \in E$.

5. Let us now discuss the case when X is a time-inhomogeneous Markov chain. Setting $Z_n = (n, X_n)$ for $n \ge 0$ one knows that $Z = (Z_n)_{n \ge 0}$ is a time-homogeneous Markov chain. Given a measurable function $G : \{0, 1, \dots, N\} \times E \to \mathbb{R}$ satisfying the following condition:

$$\mathsf{E}_{n,x}\Big(\sup_{0 \le k \le N-n} |G(n+k, X_{n+k})| \Big) < \infty \tag{1.2.1$'$}$$

for all $0 \le n \le N$ and $x \in E$, the optimal stopping problem (1.2.2) therefore naturally extends as follows:

$$V^N(n,x) = \sup_{0 \le \tau \le N-n} \mathsf{E}_{n,x} G(n+\tau, X_{n+\tau}) \tag{1.2.2$'$}$$

where the supremum is taken over stopping times τ of X and $X_n = x$ under $\mathsf{P}_{n,x}$ with $0 \le n \le N$ and $x \in E$ given and fixed.

As above one verifies that

$$S^N_{n+k} = V^N(n+k, X_{n+k}) \tag{1.2.21}$$

under $\mathsf{P}_{n,x}$ for $0 \le n \le N - n$. Moreover, inserting this into (1.1.7) and using the Markov property one finds

$$\begin{aligned} &V^N(n+k, X_{n+k}) && (1.2.22)\\ &\quad = \max\Big(G(n+k, X_{n+k}), \mathsf{E}_{n,x}\big(V^N(n+k+1, X_{n+k+1}) \mid \mathcal{F}_{n+k}\big)\Big)\\ &\quad = \max\Big(G(Z_{n+k}), \mathsf{E}_z\big(V^N(Z_{n+k+1}) \mid \mathcal{F}_{n+k}\big)\Big)\\ &\quad = \max\Big(G(Z_{n+k}), \mathsf{E}_z\big(V^N(Z_1) \circ \theta_{n+k} \mid \mathcal{F}_{n+k}\big)\Big)\\ &\quad = \max\Big(G(Z_{n+k}), \mathsf{E}_{Z_{n+k}}\big(V^N(Z_1)\big)\Big) \end{aligned}$$

for $0 \le k \le N-n-1$ where $z=(n,x)$ with $0 \le n \le N$ and $x \in E$. Letting $k=0$ and using that $Z_n = z = (n,x)$ under P_z, one gets

$$V^N(n,x) = \max\big(G(n,x), TV^N(n,x)\big) \tag{1.2.23}$$

for $n = N-1,\dots,1,0$ where $TV^N(N-1,x) = \mathsf{E}_{N-1,x}G(N,X_N)$ and T is the transition operator of Z given by

$$TF(n,x) = \mathsf{E}_{n,x}F(n+1,X_{n+1}) \tag{1.2.24}$$

for $0 \le n \le N$ and $x \in E$ whenever the right-hand side in (1.2.24) is well defined (finite).

The recursive relations (1.2.23) are the *Wald–Bellman equations* corresponding to the time-inhomogeneous problem (1.2.2′). Note that when X is time-homogeneous (and $G = G(x)$ only) we have $V^N(n,x) = V^{N-n}(x)$ and (1.2.23) reduces to (1.2.9). In order to present a reformulation of the property (1.2.12) in Theorem 1.7 above we will proceed as follows.

6. The following definition plays a fundamental role in finding a solution to the optimal stopping problem (1.2.2′).

Definition 1.8. A measurable function $F : \{0,1,\dots,N\} \times E \to \mathbb{R}$ is said to be *superharmonic* if

$$TF(n,x) \le F(n,x) \tag{1.2.25}$$

for all $(n,x) \in \{0,1,\dots,N\} \times E$.

It is assumed in (1.2.25) that $TF(n,x)$ is well defined i.e. that $F(n+1,X_{n+1}) \in L^1(\mathsf{P}_{n,x})$ for all (n,x) as above. Moreover, if $F(n+k,X_{n+k}) \in L^1(\mathsf{P}_{n,x})$ for all $0 \le k \le N-n$ and all (n,x) as above, then one verifies directly by the Markov property that the following stochastic characterization of superharmonic functions holds:

$$\begin{aligned}&F \text{ is superharmonic if and only if } (F(n+k,X_{n+k}))_{0\le k\le N-n} \text{ is}\\ &\text{a supermartingale under } \mathsf{P}_{n,x} \text{ for all } (n,x) \in \{0,1,\dots,N-1\} \times E.\end{aligned} \tag{1.2.26}$$

The proof of this fact is simple and will be given in a more general case following (1.2.40) below.

Introduce the continuation set

$$C = \big\{(n,x) \in \{0,1,\dots,N\} \times E : V(n,x) > G(n,x)\big\} \tag{1.2.27}$$

and the stopping set

$$D = \big\{(n,x) \in \{0,1,\dots,N\} \times E : V(n,x) = G(n,x)\big\}. \tag{1.2.28}$$

Introduce the first entry time τ_D into D by setting

$$\tau_D = \inf\{\, n \le k \le N : (n+k, X_{n+k}) \in D \,\} \tag{1.2.29}$$

under $\mathsf{P}_{n,x}$ where $(n,x) \in \{0,1,\dots,N\} \times E$.

The preceding considerations may now be summarized in the following extension of Theorem 1.7.

Theorem 1.9. (Finite horizon: The time-inhomogeneous case) *Consider the optimal stopping problem* (1.2.2′) *upon assuming that the condition* (1.2.1′) *holds. Then the value function* V^N *satisfies the* Wald–Bellman equations

$$V^N(n,x) = \max\big(G(n,x), TV^N(n,x)\big) \tag{1.2.30}$$

for $n = N-1, \ldots, 1, 0$ *where* $TV^N(N-1,x) = \mathsf{E}_{N-1,x} G(N, X_N)$ *and* $x \in E$. *Moreover, we have*:

The stopping time τ_D *is optimal in* (1.2.2′). (1.2.31)

If τ_* *is an optimal stopping time in* (1.2.2′) *then* $\tau_D \le \tau_*$ $\mathsf{P}_{n,x}$*-a.s. for every* $(n,x) \in \{0,1,\ldots,N\} \times E$. (1.2.32)

The value function V^N *is the smallest superharmonic function which dominates the gain function* G *on* $\{0,1,\ldots,N\} \times E$. (1.2.33)

The stopped sequence $\big(V^N((n+k) \wedge \tau_D, X_{(n+k)\wedge\tau_D})\big)_{0 \le k \le N-n}$ *is a martingale under* $\mathsf{P}_{n,x}$ *for every* $(n,x) \in \{0,1,\ldots,N\} \times E$. (1.2.34)

Proof. The proof is carried out in exactly the same way as the proof of Theorem 1.7 above. The key identity linking the problem (1.2.2′) to the problem (1.1.5) is (1.2.21). This yields (1.2.23) i.e. (1.2.30) as shown above. Note that (1.2.33) refines (1.2.12) and follows by (1.2.26). The proof is complete. □

7. Consider the optimal stopping problem (1.2.2) when $N = \infty$. Recall that (1.2.2) reads as follows:

$$V(x) = \sup_\tau \mathsf{E}_x G(X_\tau) \tag{1.2.35}$$

where τ is a stopping time of X and $\mathsf{P}_x(X_0 = x) = 1$ for $x \in E$.

Introduce the continuation set

$$C = \{\, x \in E : V(x) > G(x) \,\} \tag{1.2.36}$$

and the stopping set

$$D = \{\, x \in E : V(x) = G(x) \,\}. \tag{1.2.37}$$

Introduce the first entry time τ_D into D by setting

$$\tau_D = \inf\{\, t \ge 0 : X_t \in D \,\}. \tag{1.2.38}$$

8. The following definition plays a fundamental role in finding a solution to the optimal stopping problem (1.2.35). Note that Definition 1.8 above may be viewed as a particular case of this definition.

Definition 1.10. A measurable function $F : E \to \mathbb{R}$ is said to be *superharmonic* if

$$TF(x) \le F(x) \tag{1.2.39}$$

for all $x \in E$.

It is assumed in (1.2.39) that $TF(x)$ is well defined by (1.2.8) above i.e. that $F(X_1) \in L^1(\mathsf{P}_x)$ for all $x \in E$. Moreover, if $F(X_n) \in L^1(\mathsf{P}_x)$ for all $n \geq 0$ and all $x \in E$, then the following stochastic characterization of superharmonic functions holds (recall (1.2.26) above):

F is superharmonic if and only if $(F(X_n))_{n\geq 0}$ is a supermartingale under P_x for every $x \in E$. (1.2.40)

The proof of this equivalence relation is simple. Suppose first that F is superharmonic. Then (1.2.39) holds for all $x \in E$ and therefore by the Markov property we get

$$\begin{aligned} TF(X_n) &= \mathsf{E}_{X_n}(F(X_1)) = \mathsf{E}_x(F(X_1)\circ\theta_n \mid \mathcal{F}_n) \\ &= \mathsf{E}_x(F(X_{n+1}) \mid \mathcal{F}_n) \leq F(X_n) \end{aligned} \tag{1.2.41}$$

for all $n \geq 0$ proving the supermartingale property of $(F(X_n))_{n\geq 0}$ under P_x for every $x \in E$. Conversely, if $(F(X_n))_{n\geq 0}$ is a supermartingale under P_x for every $x \in E$, then the final inequality in (1.2.41) holds for all $n \geq 0$. Letting $n = 0$ and taking E_x on both sides gives (1.2.39). Thus F is superharmonic as claimed.

9. In the case of infinite horizon (i.e. when $N = \infty$ in (1.2.2) above) it is not necessary to treat the time-inhomogeneous case separately from the time-homogeneous case as we did it for clarity in the case of finite horizon (i.e. when $N < \infty$ in (1.2.2) above). This is due to the fact that the state space E may be general anyway (two-dimensional) and the passage from the time-inhomogeneous process $(X_n)_{n\geq 0}$ to the time-homogeneous process $(n, X_n)_{n\geq 0}$ does not affect the time set in which the stopping times of X take values (by altering the remaining time).

Theorem 1.11. (Infinite horizon) *Consider the optimal stopping problem* (1.2.35) *upon assuming that the condition* (1.2.1) *holds. Then the value function V satisfies the* Wald–Bellman equation

$$V(x) = \max(G(x), TV(x)) \tag{1.2.42}$$

for $x \in E$. Assume moreover when required below that

$$\mathsf{P}_x(\tau_D < \infty) = 1 \tag{1.2.43}$$

for all $x \in E$. Then we have:

The stopping time τ_D is optimal in (1.2.35). (1.2.44)

If τ_* is an optimal stopping time in (1.2.35) *then $\tau_D \leq \tau_*$ P_x-a.s. for every $x \in E$.* (1.2.45)

The value function V is the smallest superharmonic function which dominates the gain function G on E. (1.2.46)

The stopped sequence $(V(X_{n\wedge\tau_D}))_{n\geq 0}$ is a martingale under P_x for every $x \in E$. (1.2.47)

Finally, if the condition (1.2.43) *fails so that* $\mathsf{P}_x(\tau_D = \infty) > 0$ *for some* $x \in E$, *then there is no optimal stopping time* (*with probability* 1) *in* (1.2.35).

Proof. The key identity in reducing the problem (1.2.35) to the problem (1.1.29) is

$$S_n = V(X_n) \tag{1.2.48}$$

for $n \geq 0$. This can be proved by passing to the limit for $N \to \infty$ in (1.2.4) and using the result of Theorem 1.6 above. In exactly the same way one derives (1.2.42) from (1.2.9). The remaining statements follow from Theorem 1.4 above. Note also that (1.2.46) refines (1.1.38) and follows by (1.2.40). The proof is complete. □

Corollary 1.12. (Iterative method) *Under the initial hypothesis of Theorem* 1.11 *we have*

$$V(x) = \lim_{n \to \infty} Q^n G(x) \tag{1.2.49}$$

for all $x \in E$.

Proof. It follows from (1.2.9) and Theorem 1.6 above. □

The relation (1.2.49) offers a constructive method for finding the value function V. (Note that $n \mapsto Q^n G(x)$ is increasing on $\{0, 1, 2, \dots\}$ for every $x \in E$.)

10. We have seen in Theorem 1.7 and Theorem 1.9 that the Wald–Bellman equations (1.2.9) and (1.2.30) characterize the value function V^N when the horizon N is finite (i.e. these equations cannot have other solutions). This is due to the fact that V^N equals G in the "end of time" N. When the horizon N is infinite, however, this characterization is no longer true for the Wald–Bellman equation (1.2.42). For example, if G is identically equal to a constant c then any other constant C larger than c will define a function solving (1.2.42). On the other hand, it is evident from (1.2.42) that every solution of this equation is superharmonic and dominates G. By (1.2.46) we thus see that a minimal solution of (1.2.42) will coincide with the value function. This "minimality condition" (over all points) can be replaced by a single condition as the following theorem shows. From the standpoint of finite horizon such a "boundary condition at infinity" is natural.

Theorem 1.13. (Uniqueness in the Wald–Bellman equation)
Under the hypothesis of Theorem 1.11 *suppose that* $F : E \to \mathbb{R}$ *is a function solving the* Wald–Bellman equation

$$F(x) = \max(G(x), TF(x)) \tag{1.2.50}$$

for $x \in E$. (*It is assumed that* F *is measurable and* $F(X_1) \in L^1(\mathsf{P}_x)$ *for all* $x \in E$.) *Suppose moreover that* F *satisfies*

$$\mathsf{E}\Big(\sup_{n \geq 0} F(X_n)\Big) < \infty. \tag{1.2.51}$$

Then F *equals the value function* V *if and only if the following "boundary condition at infinity" holds:*

$$\limsup_{n\to\infty} F(X_n) = \limsup_{n\to\infty} G(X_n) \qquad \mathsf{P}_x\text{-}a.s. \tag{1.2.52}$$

for every $x \in E$. *(In this case the* $\limsup$ *on the left-hand side of* (1.2.52) *equals the* $\liminf$, *i.e. the sequence* $(F(X_n))_{n\ge 0}$ *is convergent* P_x *-a.s. for every* $x \in E$.)

Proof. If $F = V$ then by (1.2.46) we know that F is the smallest superharmonic function which dominates G on E . Let us show (the fact of independent interest) that any such function F must satisfy (1.2.52). Note that the condition (1.2.51) is not needed for this implication.

Since $F \ge G$ we see that the left-hand side in (1.2.52) is evidently larger than the right-hand side. To prove the reverse inequality, consider the function $H : E \to \mathbb{R}$ defined by

$$H(x) = \mathsf{E}_x\Big(\sup_{n\ge 0} G(X_n)\Big) \tag{1.2.53}$$

for $x \in E$. Then the key property of H stating that

$$H \text{ is superharmonic} \tag{1.2.54}$$

can be verified as follows. By the Markov property we have

$$\begin{aligned} TH(x) &= \mathsf{E}_x H(X_1) = \mathsf{E}_x\Big(\mathsf{E}_{X_1}\Big(\sup_{n\ge 0} G(X_n)\Big)\Big) \\ &= \mathsf{E}_x\Big(\mathsf{E}_x\Big(\sup_{n\ge 0} G(X_n)\circ\theta_1 \,\Big|\, \mathcal{F}_1\Big)\Big) = \mathsf{E}_x\Big(\sup_{n\ge 0} G(X_{n+1})\Big) \\ &\le H(x) \end{aligned} \tag{1.2.55}$$

for all $x \in E$ proving (1.2.54). Moreover, since $X_0 = x$ under P_x we see that $H(x) \ge G(x)$ for all $x \in E$. Hence $F(x) \le H(x)$ for all $x \in E$ by assumption. By the Markov property it thus follows that

$$\begin{aligned} F(X_n) \le H(X_n) &= \mathsf{E}_{X_n}\Big(\sup_{k\ge 0} G(X_k)\Big) = \mathsf{E}_x\Big(\sup_{k\ge 0} G(X_k)\circ\theta_n \,\Big|\, \mathcal{F}_n\Big) \\ &= \mathsf{E}_x\Big(\sup_{k\ge 0} G(X_{k+n}) \,\Big|\, \mathcal{F}_n\Big) \le \mathsf{E}_x\Big(\sup_{l\ge m} G(X_l) \,\Big|\, \mathcal{F}_n\Big) \end{aligned} \tag{1.2.56}$$

for any $m \le n$ given and fixed where $x \in E$. The final expression in (1.2.56) defines a (generalized) martingale for $n \ge 1$ under P_x which is known to converge P_x -a.s. as $n \to \infty$ for every $x \in E$ with the limit satisfying the following inequality:

$$\lim_{n\to\infty} \mathsf{E}_x\Big(\sup_{l\ge m} G(X_l) \,\Big|\, \mathcal{F}_n\Big) \le \mathsf{E}_x\Big(\sup_{l\ge m} G(X_l) \,\Big|\, \mathcal{F}_\infty\Big) = \sup_{l\ge m} G(X_l) \tag{1.2.57}$$

where the final identity follows from the fact that $\sup_{l \ge m} G(X_l)$ is $\mathcal{F}_\infty$-measurable. Letting $n \to \infty$ in (1.2.56) and using (1.2.57) we find

$$\limsup_{n\to\infty} F(X_n) \le \sup_{l\ge m} G(X_l) \qquad \mathsf{P}_x\text{-a.s.} \tag{1.2.58}$$

for all $m \ge 0$ and $x \in E$. Letting finally $m \to \infty$ in (1.2.58) we end up with (1.2.52). This ends the first part of the proof.

Conversely, suppose that F satisfies (1.2.50)–(1.2.52) and let us show that F must then be equal to V. For this, first note that (1.2.50) implies that F is superharmonic and that $F \ge G$. Hence by (1.2.46) we see that $V \le F$. To show that $V \ge F$ consider the stopping time

$$\tau_{D_\varepsilon} = \inf\{\, n \ge 0 : F(X_n) \le G(X_n) + \varepsilon \,\} \tag{1.2.59}$$

where $\varepsilon > 0$ is given and fixed. Then by (1.2.52) we see that $\tau_{D_\varepsilon} < \infty$ P_x-a.s. for $x \in E$. Moreover, we claim that $\big(F(X_{\tau_{D_\varepsilon}\wedge n})\big)_{n\ge 0}$ is a martingale under P_x for $x \in E$. For this, note that the Markov property and (1.2.50) imply

$$\begin{aligned}
&\mathsf{E}_x\big(F(X_{\tau_{D_\varepsilon}\wedge n}) \,|\, \mathcal{F}_{n-1}\big) \qquad\qquad (1.2.60)\\
&\quad = \mathsf{E}_x\big(F(X_n) I(\tau_{D_\varepsilon} \ge n) \,|\, \mathcal{F}_{n-1}\big) + \mathsf{E}_x\big(F(X_{\tau_{D_\varepsilon}}) I(\tau_{D_\varepsilon} < n) \,|\, \mathcal{F}_{n-1}\big)\\
&\quad = \mathsf{E}_x\big(F(X_n) \,|\, \mathcal{F}_{n-1}\big) I(\tau_{D_\varepsilon} \ge n) + \mathsf{E}_x\big(\textstyle\sum_{k=0}^{n-1} F(X_k) I(\tau_{D_\varepsilon} = k) \,|\, \mathcal{F}_{n-1}\big)\\
&\quad = \mathsf{E}_{X_{n-1}}\big(F(X_1)\big)\, I(\tau_{D_\varepsilon} \ge n) + \textstyle\sum_{k=0}^{n-1} F(X_k)\, I(\tau_{D_\varepsilon} = k)\\
&\quad = TF(X_{n-1})\, I(\tau_{D_\varepsilon} \ge n) + F(X_{\tau_{D_\varepsilon}})\, I(\tau_{D_\varepsilon} < n)\\
&\quad = F(X_{n-1})\, I(\tau_{D_\varepsilon} \ge n) + F(X_{\tau_{D_\varepsilon}})\, I(\tau_{D_\varepsilon} < n)\\
&\quad = F(X_{\tau_{D_\varepsilon}\wedge(n-1)})\, I(\tau_{D_\varepsilon} \ge n) + F(X_{\tau_{D_\varepsilon}\wedge(n-1)})\, I(\tau_{D_\varepsilon} < n)\\
&\quad = F(X_{\tau_{D_\varepsilon}\wedge(n-1)})
\end{aligned}$$

for all $n \ge 1$ and $x \in E$ proving the claim. Hence

$$\mathsf{E}_x\big(F(X_{\tau_{D_\varepsilon}\wedge n})\big) = F(x) \tag{1.2.61}$$

for all $n \ge 0$ and $x \in E$. Next note that

$$\mathsf{E}_x\big(F(X_{\tau_{D_\varepsilon}\wedge n})\big) = \mathsf{E}_x\big(F(X_{\tau_{D_\varepsilon}})\, I(\tau_{D_\varepsilon} \le n)\big) + \mathsf{E}_x\big(F(X_n)\, I(\tau_{D_\varepsilon} > n)\big) \tag{1.2.62}$$

for all $n \ge 0$. Letting $n \to \infty$, using (1.2.51) and (1.2.1) with $F \ge G$, we get

$$\mathsf{E}_x\big(F(X_{\tau_{D_\varepsilon}})\big) = F(x) \tag{1.2.63}$$

for all $x \in E$. This fact is of independent interest.

Finally, since V is superharmonic, we find using (1.2.63) that

$$V(x) \geq \mathsf{E}_x V(X_{\tau_{D_\varepsilon}}) \geq \mathsf{E}_x G(X_{\tau_{D_\varepsilon}}) > \mathsf{E}_x F(X_{\tau_{D_\varepsilon}}) - \varepsilon = F(x) - \varepsilon \tag{1.2.64}$$

for all $\varepsilon > 0$ and $x \in E$. Letting $\varepsilon \downarrow 0$ we get $V \geq F$ as needed and the proof is complete. □

11. Given $\alpha \in (0,1]$ and (bounded) measurable functions $g : E \to \mathbb{R}$ and $c : E \to \mathbb{R}_+$, consider the optimal stopping problem

$$V(x) = \sup_\tau \mathsf{E}_x \Big(\alpha^\tau g(X_\tau) - \sum_{k=1}^{\tau} \alpha^{k-1} c(X_{k-1}) \Big) \tag{1.2.65}$$

where τ is a stopping time of X and $\mathsf{P}_x(X_0 = x) = 1$.

The value $c(x)$ is interpreted as the cost of making the next observation of X when X equals x. The sum in (1.2.65) by definition equals 0 when τ equals 0.

The problem formulation (1.2.65) goes back to a problem formulation due to Bolza in classic calculus of variation (a more detailed discussion will be given in Chapter III below). Let us briefly indicate how the problem (1.2.65) can be reduced to the setting of Theorem 1.11 above.

For this, let $\widetilde{X} = (\widetilde{X}_n)_{n \geq 0}$ denote the Markov chain X killed at rate α. It means that the transition operator of $\widetilde{X}$ is given by

$$\widetilde{T} F(x) = \alpha \, T F(x) \tag{1.2.66}$$

for $x \in E$ whenever $F(X_1) \in L^1(\mathsf{P}_x)$. The problem (1.2.65) then reads

$$V(x) = \sup_\tau \mathsf{E}_x \Big(g(\widetilde{X}_\tau) - \sum_{k=1}^{\tau} c(\widetilde{X}_{k-1}) \Big) \tag{1.2.65$'$}$$

where τ is a stopping time of $\widetilde{X}$ and $\mathsf{P}_x(\widetilde{X}_0 = x) = 1$.

Introduce the sequence

$$\widetilde{I}_n = a + \sum_{k=1}^{n} c(\widetilde{X}_{k-1}) \tag{1.2.67}$$

for $n \geq 1$ with $\widetilde{I}_0 = a$ in $\mathbb{R}$. Then $\widetilde{Z}_n = (\widetilde{X}_n, \widetilde{I}_n)$ defines a Markov chain for $n \geq 0$ with $\widetilde{Z}_0 = (\widetilde{X}_0, \widetilde{I}_0) = (x, a)$ under P_x so that we may write $\mathsf{P}_{x,a}$ instead of P_x. (The latter can be justified rigorously by passage to the canonical probability space.) The transition operator of $\widetilde{Z} = (\widetilde{X}, \widetilde{I})$ equals

$$T_{\widetilde{Z}} F(x,a) = \mathsf{E}_{x,a} F(\widetilde{X}_1, \widetilde{I}_1) \tag{1.2.68}$$

for $(x,a) \in E \times \mathbb{R}$ whenever $F(\widetilde{X}_1, \widetilde{I}_1) \in L^1(\mathsf{P}_{x,a})$.

The problem (1.2.65′) may now be rewritten as follows:

$$W(x,a) = \sup_{\tau} \mathsf{E}_{x,a} G(Z_\tau) \tag{1.2.65″}$$

where we set

$$G(z) = g(x) - a \tag{1.2.69}$$

for $z = (x,a) \in E \times \mathbb{R}$. Obviously by subtracting a on both sides of (1.2.65′) we set that

$$W(x,a) = V(x) - a \tag{1.2.70}$$

for all $(x,a) \in E \times \mathbb{R}$.

The problem (1.2.65″) is of the same type as the problem (1.2.35) above and thus Theorem 1.11 is applicable. To write down (1.2.42) more explicitly note that

$$\begin{aligned} T_{\widetilde{Z}} W(x,a) &= \mathsf{E}_{x,a} W(\widetilde{X}_1, \widetilde{I}_1) = \mathsf{E}_{x,a}\big(V(\widetilde{X}_1) - \widetilde{I}_1\big) \\ &= \mathsf{E}_x V(\widetilde{X}_1) - a - c(x) = \alpha T V(x) - a - c(x) \end{aligned} \tag{1.2.71}$$

so that (1.2.42) reads

$$V(x) - a = \max\big(g(x) - a\,,\, \alpha T V(x) - a - c(x)\big) \tag{1.2.72}$$

where we used (1.2.70), (1.2.69) and (1.2.71). Clearly a can be removed from (1.2.72) showing finally that the Wald–Bellman equation (1.2.42) takes the following form:

$$V(x) = \max\big(g(x)\,,\, \alpha T V(x) - c(x)\big) \tag{1.2.73}$$

for $x \in E$. Note also that (1.2.39) takes the following form:

$$\alpha T F(x) - c(x) \le F(x) \tag{1.2.74}$$

for $x \in E$. Thus F satisfies (1.2.74) if and only if $(x,a) \mapsto F(x) - a$ is superharmonic relative to the Markov chain $\widetilde{Z} = (\widetilde{X}, \widetilde{I})$. Having (1.2.73) and (1.2.74) set out explicitly the remaining statements of Theorem 1.11 and Corollary 1.12 are directly applicable and we shall omit further details. It may be noted above that $L = T - I$ is the generator of the Markov chain X. More general problems of this type (involving also the maximum functional) will be discussed in Chapter III below. We will conclude this section by giving an illustrative example.

12. The following example illustrates general results of optimal stopping theory for Markov chains when applied to a nontrivial problem in order to determine the value function and an optimal Markov time (in the class $\bar{\mathfrak{M}}$).

Example 1.14. Let $\xi, \xi_1, \xi_2, \ldots$ be independent and identically distributed random variables, defined on a probability space $(\Omega, \mathcal{F}, \mathsf{P})$, with $\mathsf{E}\xi < 0$. Put $S_0 = 0$,

$S_n = \xi_1 + \cdots + \xi_n$ for $n \geq 1$; $X_0 = x$, $X_n = x + S_n$ for $n \geq 1$, and $M = \sup_{n \geq 0} S_n$. Let P_x be the probability distribution of the sequence $(X_n)_{n \geq 0}$ with $X_0 = x$ from $\mathbb{R}$. It is clear that the sequence $(X_n)_{n \geq 0}$ is a Markov chain started at x.

For any $n \geq 1$ define the gain function $G_n(x) = (x^+)^n$ where $x^+ = \max(x, 0)$ for $x \in \mathbb{R}$, and let

$$V_n(x) = \sup_{\tau \in \mathfrak{M}} \mathsf{E}_x G_n(X_\tau) \tag{1.2.75}$$

where the supremum is taken over the class $\mathfrak{M}$ of all Markov (stopping) times τ satisfying $\mathsf{P}_x(\tau < \infty) = 1$ for all $x \in \mathbb{R}$. Let us also denote

$$\bar{V}_n(x) = \sup_{\tau \in \bar{\mathfrak{M}}} \mathsf{E}_x G_n(X_\tau) I(\tau < \infty) \tag{1.2.76}$$

where the supremum is taken over the class $\bar{\mathfrak{M}}$ of all Markov times.

The problem of finding the value functions $V_n(x)$ and $\bar{V}_n(x)$ is of interest for the theory of American options because these functions represent arbitrage-free (fair, rational) prices of "Power options" under the assumption that any exercise time τ belongs to the class $\mathfrak{M}$ or $\bar{\mathfrak{M}}$ respectively. In the present case we have $V_n(x) = \bar{V}_n(x)$ for $n \geq 1$ and $x \in \mathbb{R}$, and it will be clear from what follows below that an optimal Markov time exists in the class $\bar{\mathfrak{M}}$ (but does not belong to the class $\mathfrak{M}$ of stopping times).

We follow [144] where the authors solved the formulated problems (see also [119]). First of all let us introduce the notion of the Appell polynomial which will be used in the formulation of the basic results.

Let $\eta = \eta(\omega)$ be a random variable with $\mathsf{E} e^{\lambda|\eta|} < \infty$ for some $\lambda > 0$. Consider the Esscher transform

$$x \rightsquigarrow \frac{e^{ux}}{\mathsf{E} e^{u\eta}} \qquad |u| \leq \lambda, \quad x \in \mathbb{R}, \tag{1.2.77}$$

and the decomposition

$$\frac{e^{ux}}{\mathsf{E} e^{u\eta}} = \sum_{k=0}^{\infty} \frac{u^k}{k!} Q_k^{(\eta)}(x). \tag{1.2.78}$$

Polynomials $Q_k^{(\eta)}(x)$ are called the *Appell polynomials* for the random variable η. (If $\mathsf{E}|\eta|^n < \infty$ for some $n \geq 1$ then the polynomials $Q_k^{(\eta)}(x)$ are uniquely defined for all $k \leq n$.)

The polynomials $Q_k^{(\eta)}(x)$ can be expressed through the semi-invariants $\varkappa_1$, $\varkappa_2, \ldots$ of the random variable η. For example,

$$\begin{aligned} &Q_0^{(\eta)}(x) = 1, && Q_2^{(\eta)}(x) = (x - \varkappa_1)^2 - \varkappa_2, \\ &Q_1^{(\eta)}(x) = x - \varkappa_1, && Q_3^{(\eta)}(x) = (x - \varkappa_1)^3 - 3\varkappa_2(x - \varkappa_1) - \varkappa_3, \end{aligned} \quad \ldots \tag{1.2.79}$$

where (as is well known) the semi-invariants $\varkappa_1, \varkappa_2, \ldots$ are expressed via the moments $\mu_1, \mu_2, \ldots$ of η:

$$\varkappa_1 = \mu_1, \quad \varkappa_2 = \mu_2 - \mu_1^2, \quad \varkappa_3 = 2\mu_1^3 - 3\mu_1\mu_2 + \mu_3, \quad \ldots . \tag{1.2.80}$$

Let us also mention the following property of the Appell polynomials: if $\mathsf{E}\,|\eta|^n < \infty$ then for $k \le n$ we have

$$\frac{d}{dx} Q_k^{(\eta)}(x) = k Q_{k-1}^{(\eta)}(x), \tag{1.2.81}$$

$$\mathsf{E}\, Q_k^{(\eta)}(x+\eta) = x^k. \tag{1.2.82}$$

For simplicity of notation we will use $\mathbb{Q}_k(s)$ to denote the polynomials $Q_k^{(M)}(x)$ for the random variable $M = \sup_{n \ge 0} S_n$. Every polynomial $\mathbb{Q}_k(x)$ has a unique positive root a_k^*. Moreover, $\mathbb{Q}_k(x) \le 0$ for $0 \le x < a_k^*$ and $\mathbb{Q}_k(x)$ increases for $x \ge a_k^*$.

In accordance with the characteristic property (1.2.46) recall that *the value function* $V_n(x)$ *is the smallest superharmonic (excessive) function which dominates the gain function* $G_n(x)$ *on* $\mathbb{R}$. Thus, one method to find $V_n(x)$ is to search for the smallest excessive majorant of the function $G_n(x)$. In [144] this method is realized as follows.

For every $a \ge 0$ introduce the Markov time

$$\tau_a = \inf\{n \ge 0 : X_n \ge a\} \tag{1.2.83}$$

and for each $n \ge 1$ consider the new optimal stopping problem:

$$\widehat{V}(x) = \sup_{a \ge 0} \mathsf{E}_x G_n(X_{\tau_a}) I(\tau_a < \infty). \tag{1.2.84}$$

It is clear that $G_n(X_{\tau_a}) = (X_{\tau_a}^+)^n = X_{\tau_a}^n$ (on the set $\{\tau_a < \infty\}$). Hence

$$\widehat{V}(x) = \sup_{a \ge 0} \mathsf{E}_x X_{\tau_a}^n I(\tau_a < \infty). \tag{1.2.85}$$

The identity (1.2.82) prompts that the following property should be valid: if $\mathsf{E}\,|M|^n < \infty$ then

$$\mathsf{E}\, \mathbb{Q}_n(x+M) I(x+M \ge a) = \mathsf{E}_x X_{\tau_a}^n I(\tau_a < \infty). \tag{1.2.86}$$

This formula and properties of the Appell polynomials imply that

$$\widehat{V}(x) = \sup_{a \ge 0} \mathsf{E}\, \mathbb{Q}_n(x+M) I(x+M \ge a) = \mathsf{E}\, \mathbb{Q}_n(x+M) I(x+M \ge a_n^*). \tag{1.2.87}$$

From this we see that $\tau_{a_n^*}$ is an optimal Markov time for the problem (1.2.84).

It is clear that $\bar{V}_n(x) \geq \widehat{V}_n(x)$. From (1.2.87) and properties of the Appell polynomials we obtain that $\widehat{V}_n(x)$ is an excessive majorant of the gain function ($\widehat{V}_n(x) \geq \mathsf{E}_x \widehat{V}_n(X_1)$ and $\widehat{V}_n(x) \geq G_n(x)$ for $x \in \mathbb{R}$). But $\bar{V}_n(x)$ is the *smallest* excessive majorant of $G_n(x)$. Thus $\bar{V}_n(x) \leq \widehat{V}_n(x)$.

On the whole we obtain the following result (for further details see [144]):

Suppose that $\mathsf{E}\,(\xi^+)^{n+1} < \infty$ and a_n^* is the largest root of the equation $\mathbb{Q}_n(x) = 0$ for $n \geq 1$ fixed. Denote $\tau_n^* = \inf\{\, k \geq 0 : X_k \geq a_n^* \,\}$. Then the Markov time τ_n^* is optimal:

$$V_n(x) = \sup_{\tau \in \bar{\mathfrak{M}}} \mathsf{E}_x (X_\tau^+)^n I(\tau < \infty) = \mathsf{E}_x \big(X_{\tau_n^*}^+\big)^n I(\tau < \infty). \tag{1.2.88}$$

Moreover,

$$V_n(x) = \mathsf{E}\,\mathbb{Q}_n(x+M) I(x+M \geq a_n^*). \tag{1.2.89}$$

Remark 1.15. In the cases $n = 1$ and $n = 2$ we have

$$a_1^* = \mathsf{E}\,M \quad \text{and} \quad a_2^* = \mathsf{E}\,M + \sqrt{\mathsf{D}M}. \tag{1.2.90}$$

Remark 1.16. If $\mathsf{P}(\xi = 1) = p$, $\mathsf{P}(\xi = -1) = q$ and $p < q$, then $M := \sup_{n \geq 0} S_n$ (with $S_0 = 0$ and $S_n = \xi_1 + \cdots + \xi_n$) has geometric distribution:

$$\mathsf{P}(M \geq k) = \Big(\frac{p}{q}\Big)^k \tag{1.2.91}$$

for $k \geq 0$. Hence

$$\mathsf{E}\,M = \frac{q}{q-p}. \tag{1.2.92}$$

2. Continuous time

The aim of the present section is to exhibit basic results of optimal stopping in the case of continuous time. We first consider a martingale approach (cf. Subsection 1.1 above). This is then followed by a Markovian approach (cf. Subsection 1.2 above).

2.1. Martingale approach

1. Let $G = (G_t)_{t \geq 0}$ be a stochastic process defined on a filtered probability space $(\Omega, \mathcal{F}, (\mathcal{F}_t)_{t \geq 0}, \mathsf{P})$. We interpret G_t as the *gain* if the observation of G is stopped at time t. It is assumed that G is adapted to the filtration $(\mathcal{F}_t)_{t \geq 0}$ in the sense that each G_t is $\mathcal{F}_t$-measurable. Recall that each $\mathcal{F}_t$ is a σ-algebra of subsets of Ω such that $\mathcal{F}_s \subseteq \mathcal{F}_t \subseteq \mathcal{F}$ for $s \leq t$. Typically $(\mathcal{F}_t)_{t \geq 0}$ coincides with the natural filtration $(\mathcal{F}_t^G)_{t \geq 0}$ but generally may also be larger. We interpret $\mathcal{F}_t$ as the *information* available up to time t. All our decisions in regard to optimal stopping at time t must be based on this information only (no anticipation is allowed).

The following definition formalizes the previous requirement and plays a key role in the study of optimal stopping (cf. Definition 1.1).

Definition 2.1. A random variable $\tau : \Omega \to [0, \infty]$ is called a *Markov time* if $\{\tau \le t\} \in \mathcal{F}_t$ for all $t \ge 0$. A Markov time is called a *stopping time* if $\tau < \infty$ P-a.s.

In the sequel we will only consider stopping times. We refer to Subsection 1.1 above for other similar comments which translate to the present setting of continuous time without major changes.

2. We will assume that the process G is right-continuous and left-continuous over stopping times (if τ_n and τ are stopping times such that $\tau_n \uparrow \tau$ as $n \to \infty$ then $G_{\tau_n} \to G_\tau$ P-a.s. as $n \to \infty$). We will also assume that the following condition is satisfied (with $G_T = 0$ when $T = \infty$):

$$\mathsf{E}\Big(\sup_{0 \le t \le T} |G_t| \Big) < \infty. \tag{2.1.1}$$

Just as in the case of discrete time (Subsection 1.1) here too it is possible to go beyond this condition in both theory and applications of optimal stopping, however, none of the conclusions will essentially be different and we thus work with (2.1.1) throughout.

In order to invoke a theorem on the existence of a right-continuous modification of a given supermartingale, we will assume in the sequel that the filtration $(\mathcal{F}_t)_{t \ge 0}$ is right-continuous and that each $\mathcal{F}_t$ contains all P-null sets from $\mathcal{F}$. This is a technical requirement and its enforcement has no significant impact on interpretations of the optimal stopping problem under consideration and its solution to be presented.

3. We consider the optimal stopping problem

$$V_t^T = \sup_{t \le \tau \le T} \mathsf{E}\, G_\tau \tag{2.1.2}$$

where τ is a stopping time and $0 \le t \le T$. In (2.1.2) we admit that T can be $+\infty$ as well. In this case, however, we assume that the supremum is still taken over stopping times τ, i.e. over Markov times τ satisfying $t \le \tau < \infty$. In this case we will set $V_t = V_t^\infty$ for $t \ge 0$. Moreover, for certain reasons of convenience we will also drop T from V_t^T in (2.1.1) even if the horizon T is finite.

4. By analogy with the results of Subsection 1.1 above (discrete time case) there are two possible ways to tackle the problem (2.1.2). The first method consists of replacing the time interval $[0, T]$ by sets $\mathcal{D}_n = \{t_0^n, t_1^n, \ldots, t_n^n\}$ where $\mathcal{D}_n \uparrow \mathcal{D}$

as $n \to \infty$ and $\mathcal{D}$ is a (countable) dense subset of $[0, T]$, applying the results of Subsection 1.1 (the method of backward induction) to each $G^n = (G_{t_i^n})_{0 \le i \le n}$, and then passing to the limit as $n \to \infty$. In this context it is useful to know that each stopping time τ can be obtained as a decreasing limit of the discrete stopping times $\tau_n = \sum_{i=1}^n t_i^n \, I(t_{i-1}^n \le \tau < t_i^n)$ as $n \to \infty$. The methodology described becomes useful for getting numerical approximations for the solution but we will omit further details in this direction. The second method aims directly to extend the method of essential supremum in Subsection 1.1 above from the discrete time case to the continuous time case. This programme will now be addressed.

5. Since there is no essential difference in the treatment of either finite or infinite horizon T, we will treat both cases at the same time by setting

$$V_t = V_t^T \tag{2.1.3}$$

for simplicity of notation.

To solve the problem (2.1.2) we will (by analogy with the results of Subsection 1.1) consider the process $S = (S_t)_{t \ge 0}$ defined as follows:

$$S_t = \operatorname*{ess\,sup}_{\tau \ge t} \mathsf{E}\,(G_\tau \,|\, \mathcal{F}_t) \tag{2.1.4}$$

where τ is a stopping time. In the case of a finite horizon T we also require in (2.1.4) that τ is smaller than or equal to T. We will see in the proof of Theorem 2.2 below that there is no restriction to assume that the process S is right-continuous. The process S is often referred to as the *Snell envelope* of G.

For the same reasons we will consider the following stopping time:

$$\tau_t = \inf\{\, s \ge t : S_s = G_s \,\} \tag{2.1.5}$$

for $t \ge 0$ where $\inf \varnothing = \infty$ by definition. In the case of a finite horizon T we also require in (2.1.5) that s is smaller than or equal to T.

Regarding the initial part of Theorems 1.2 and 1.4 (the Wald–Bellman equation) one should observe that Theorem 2.2 below implies that

$$S_t \ge \max\big(G_t, \mathsf{E}\,(S_s \,|\, \mathcal{F}_t)\big) \tag{2.1.6}$$

for $s \ge t$. The reverse inequality, however, is not true in general. The reason roughly speaking lies in the fact that, unlike in discrete time, in continuous time there is no smallest unit of time, so that no matter how close s to t is (when strictly larger) the values S_u can still wander far away from S_t when $u \in (t, s)$. Note however that Theorem 2.2 below implies that the following refinement of the Wald–Bellman equation still holds:

$$S_t = \max\big(G_t, \mathsf{E}\,(S_{\sigma \wedge \tau_t} \,|\, \mathcal{F}_t)\big) \tag{2.1.7}$$

for every stopping time σ larger than or equal to t (note that σ can also be identically equal to any $s \geq t$) where τ_t is given in (2.1.5) above.

The other three parts of Theorems 1.2 and 1.4 (pages 3 and 8) extend to the present case with no significant change. Thus the first part of the following theorem shows that $(S_s)_{s \geq t}$ and τ_t solve the problem in a stochastic sense. The second part of the theorem shows that this leads to a solution of the initial problem (2.1.2). The third part of the theorem provides a supermartingale characterization of the solution.

Theorem 2.2. *Consider the optimal stopping problem* (2.1.2) *upon assuming that the condition* (2.1.1) *holds. Assume moreover when required below that*

$$\mathsf{P}(\tau_t < \infty) = 1 \tag{2.1.8}$$

where $t \geq 0$. *(Note that this condition is automatically satisfied when the horizon* T *is finite.) Then for all* $t \geq 0$ *we have:*

$$S_t \geq \mathsf{E}\,(G_\tau \mid \mathcal{F}_t) \quad \textit{for each } \tau \in \mathfrak{M}_t, \tag{2.1.9}$$

$$S_t = \mathsf{E}\,(G_{\tau_t} \mid \mathcal{F}_t) \tag{2.1.10}$$

where $\mathfrak{M}_t$ *denotes the family of all stopping times* τ *satisfying* $\tau \geq t$ *(being also smaller than or equal to* T *when the latter is finite). Moreover, if* $t \geq 0$ *is given and fixed, then we have:*

The stopping time τ_t *is optimal in* (2.1.2). (2.1.11)

If τ_* *is an optimal stopping time in* (2.1.2) *then* $\tau_t \leq \tau_*$ P-*a.s.* (2.1.12)

The process $(S_s)_{s \geq t}$ *is the smallest right-continuous supermartingale which dominates* $(G_s)_{s \geq t}$. (2.1.13)

The stopped process $(S_{s \wedge \tau_t})_{s \geq t}$ *is a right-continuous martingale.* (2.1.14)

Finally, if the condition (2.1.8) *fails so that* $\mathsf{P}(\tau_t = \infty) > 0$, *then there is no optimal stopping time (with probability* 1) *in* (2.1.2).

Proof. 1°. Let us first show that $S = (S_t)_{t \geq 0}$ defined by (2.1.4) above is a supermartingale. For this, fix $t \geq 0$ and let us show that the family $\{\mathsf{E}\,(G_\tau \mid \mathcal{F}_t) : \tau \in \mathfrak{M}_t\}$ is upwards directed in the sense that (1.1.25) is satisfied. Indeed, note that if σ_1 and σ_2 are from $\mathfrak{M}_t$ and we set $\sigma_3 = \sigma_1 I_A + \sigma_2 I_{A^c}$ where $A = \{\mathsf{E}\,(G_{\sigma_1} \mid \mathcal{F}_t) \geq \mathsf{E}\,(G_{\sigma_2} \mid \mathcal{F}_t)\}$, then σ_3 belongs to $\mathfrak{M}_t$ and we have

$$\begin{aligned} \mathsf{E}\,(G_{\sigma_3} \mid \mathcal{F}_t) &= \mathsf{E}\,(G_{\sigma_1} I_A + G_{\sigma_2} I_{A^c} \mid \mathcal{F}_t) \\ &= I_A \mathsf{E}\,(G_{\sigma_1} \mid \mathcal{F}_t) + I_{A^c} \mathsf{E}\,(G_{\sigma_2} \mid \mathcal{F}_t) \\ &= \mathsf{E}\,(G_{\sigma_1} \mid \mathcal{F}_t) \vee \mathsf{E}\,(G_{\sigma_2} \mid \mathcal{F}_t) \end{aligned} \tag{2.1.15}$$

implying (1.1.25) as claimed. Hence by (1.1.26) there exists a sequence $\{\sigma_k : k \geq 1\}$ in $\mathfrak{M}_t$ such that

$$\operatorname*{ess\,sup}_{\tau \in \mathfrak{M}_t} \mathsf{E}\,(G_\tau \mid \mathcal{F}_t) = \lim_{k \to \infty} \mathsf{E}\,(G_{\sigma_k} \mid \mathcal{F}_t) \tag{2.1.16}$$

where $\mathsf{E}(G_{\sigma_1} \,|\, \mathcal{F}_t) \le \mathsf{E}(G_{\sigma_2} \,|\, \mathcal{F}_t) \le \cdots$ P-a.s. Since the left-hand side in (2.1.16) equals S_t, by the conditional monotone convergence theorem using (2.1.1) above, we find for any $s \in [0,t]$ that

$$\begin{aligned} \mathsf{E}(S_t \,|\, \mathcal{F}_s) &= \mathsf{E}\Big(\lim_{k\to\infty} \mathsf{E}(G_{\sigma_k} \,|\, \mathcal{F}_t) \,|\, \mathcal{F}_s\Big) \\ &= \lim_{k\to\infty} \mathsf{E}\big(\mathsf{E}(G_{\sigma_k} \,|\, \mathcal{F}_t) \,|\, \mathcal{F}_s\big) \\ &= \lim_{k\to\infty} \mathsf{E}(G_{\sigma_k} \,|\, \mathcal{F}_s) \le S_s \end{aligned} \tag{2.1.17}$$

where the final inequality follows by the definition of S_s given in (2.1.4) above. This shows that $(S_t)_{t\ge 0}$ is a supermartingale as claimed. Note also that (2.1.4) and (2.1.16) using the monotone convergence theorem and (2.1.1) imply that

$$\mathsf{E}\,S_t = \sup_{\tau \ge t} \mathsf{E}\,G_\tau \tag{2.1.18}$$

where τ is a stopping time and $t \ge 0$.

2°. Let us next show that the supermartingale S admits a right-continuous modification $\widetilde{S} = (\widetilde{S}_t)_{t\ge 0}$. A well-known result in martingale theory (see e.g. [134]) states that the latter is possible to achieve if and only if

$$t \mapsto \mathsf{E}\,S_t \quad \text{is right-continuous on } \mathbb{R}_+ . \tag{2.1.19}$$

To verify (2.1.19) note that by the supermartingale property of S we have $\mathsf{E}\,S_t \ge \cdots \ge \mathsf{E}\,S_{t_2} \ge \mathsf{E}\,S_{t_1}$ so that $L := \lim_{n\to\infty} \mathsf{E}\,S_{t_n}$ exists and $\mathsf{E}\,S_t \ge L$ whenever $t_n \downarrow t$ as $n \to \infty$ is given and fixed. To prove the reverse inequality, fix $\varepsilon > 0$ and by means of (2.1.18) choose $\sigma \in \mathfrak{M}_t$ such that

$$\mathsf{E}\,G_\sigma \ge \mathsf{E}\,S_t - \varepsilon. \tag{2.1.20}$$

Fix $\delta > 0$ and note that there is no restriction to assume that $t_n \in [t, t+\delta]$ for all $n \ge 1$. Define a stopping time σ_n by setting

$$\sigma_n = \begin{cases} \sigma & \text{if} \quad \sigma > t_n, \\ t+\delta & \text{if} \quad \sigma \le t_n \end{cases} \tag{2.1.21}$$

for $n \ge 1$. Then for all $n \ge 1$ we have

$$\mathsf{E}\,G_{\sigma_n} = \mathsf{E}\,G_\sigma I(\sigma > t_n) + \mathsf{E}\,G_{t+\delta} I(\sigma \le t_n) \le \mathsf{E}\,S_{t_n} \tag{2.1.22}$$

since $\sigma_n \in \mathfrak{M}_{t_n}$ and (2.1.18) holds. Letting $n \to \infty$ in (2.1.22) and using (2.1.1) we get

$$\mathsf{E}\,G_\sigma I(\sigma > t) + \mathsf{E}\,G_{t+\delta} I(\sigma = t) \le L \tag{2.1.23}$$

for all $\delta > 0$. Letting now $\delta \downarrow 0$ and using that G is right-continuous we finally obtain

$$\mathsf{E}\,G_\sigma I(\sigma > t) + \mathsf{E}\,G_t I(\sigma = t) = \mathsf{E}\,G_\sigma \le L. \tag{2.1.24}$$

From (2.1.20) and (2.1.24) we see that $L \geq \mathsf{E}\, S_t - \varepsilon$ for all $\varepsilon > 0$. Hence $L \geq \mathsf{E}\, S_t$ and thus $L = \mathsf{E}\, S_t$ showing that (2.1.19) holds. It follows that S admits a right-continuous modification $\widetilde{S} = (\widetilde{S}_t)_{t \geq 0}$ which we also denote by S throughout.

3°. Let us show that (2.1.13) holds. For this, let $\widehat{S} = (\widehat{S}_s)_{s \geq t}$ be another right-continuous supermartingale which dominates $G = (G_s)_{s \geq t}$. Then by the optional sampling theorem (page 60) using (2.1.1) above we have

$$\widehat{S}_s \geq \mathsf{E}\,(\widehat{S}_\tau \mid \mathcal{F}_s) \geq \mathsf{E}\,(G_\tau \mid \mathcal{F}_s) \tag{2.1.25}$$

for all $\tau \in \mathfrak{M}_s$ when $s \geq t$. Hence by the definition of S_s given in (2.1.4) above we find that $S_s \leq \widehat{S}_s$ P-a.s. for all $s \geq t$. By the right-continuity of S and $\widehat{S}$ this further implies that $\mathsf{P}(S_s \leq \widehat{S}_s \text{ for all } s \geq t) = 1$ as claimed.

4°. Noticing that (2.1.9) follows at once from (2.1.4) above, let us now show that (2.1.10) holds. For this, let us first consider the case when $G_t \geq 0$ for all $t \geq 0$.

For each $\lambda \in (0,1)$ introduce the stopping time

$$\tau_t^\lambda = \inf\{\, s \geq t : \lambda S_s \leq G_s \,\} \tag{2.1.26}$$

where $t \geq 0$ is given and fixed. For further reference note that by the right-continuity of S and G we have:

$$\lambda S_{\tau_t^\lambda} \leq G_{\tau_t^\lambda}, \tag{2.1.27}$$

$$\tau_{t+}^\lambda = \tau_t^\lambda \tag{2.1.28}$$

for all $\lambda \in (0,1)$. In exactly the same way we find:

$$S_{\tau_t} = G_{\tau_t}, \tag{2.1.29}$$

$$\tau_{t+} = \tau_t \tag{2.1.30}$$

for τ_t defined in (2.1.5) above.

Next note that the optional sampling theorem (page 60) using (2.1.1) above implies

$$S_t \geq \mathsf{E}\,(S_{\tau_t^\lambda} \mid \mathcal{F}_t) \tag{2.1.31}$$

since τ_t^λ is a stopping time greater than or equal to t. To prove the reverse inequality

$$S_t \leq \mathsf{E}\,(S_{\tau_t^\lambda} \mid \mathcal{F}_t) \tag{2.1.32}$$

consider the process

$$R_t = \mathsf{E}\,(S_{\tau_t^\lambda} \mid \mathcal{F}_t) \tag{2.1.33}$$

for $t \geq 0$. We claim that $R = (R_t)_{t \geq 0}$ is a supermartingale. Indeed, for $s < t$ we have

$$\mathsf{E}\,(R_t \mid \mathcal{F}_s) = \mathsf{E}\big(\mathsf{E}\,(S_{\tau_t^\lambda} \mid \mathcal{F}_t) \mid \mathcal{F}_s\big) = \mathsf{E}\,(S_{\tau_t^\lambda} \mid \mathcal{F}_s) \leq \mathsf{E}\,(S_{\tau_s^\lambda} \mid \mathcal{F}_s) = R_s \tag{2.1.34}$$

where the inequality follows by the optional sampling theorem (page 60) using (2.1.1) above since $\tau_t^\lambda \geq \tau_s^\lambda$ when $s < t$. This shows that R is a supermartingale as claimed. Hence $\mathsf{E}\, R_{t+h}$ increases when h decreases and $\lim_{h \downarrow 0} \mathsf{E}\, R_{t+h} \leq \mathsf{E}\, R_t$. On the other hand, note by Fatou's lemma using (2.1.1) above that

$$\lim_{h \downarrow 0} \mathsf{E}\, R_{t+h} = \lim_{h \downarrow 0} \mathsf{E}\, S_{\tau_{t+h}^\lambda} \geq \mathsf{E}\, S_{\tau_t^\lambda} = \mathsf{E}\, R_t \tag{2.1.35}$$

where we also use (2.1.28) above together with the facts that τ_{t+h}^λ decreases when h decreases and S is right-continuous. This shows that $t \mapsto \mathsf{E}\, R_t$ is right-continuous on $\mathbb{R}_+$ and hence R admits a right-continuous modification which we also denote by R in the sequel. It follows that there is no restriction to assume that the supermartingale R is right-continuous.

To prove (2.1.32) i.e. that $S_t \leq R_t$ P-a.s. consider the right-continuous supermartingale defined as follows:

$$L_t = \lambda S_t + (1-\lambda) R_t \tag{2.1.36}$$

for $t \geq 0$. We then claim that

$$L_t \geq G_t \quad \mathsf{P}\text{-a.s.} \tag{2.1.37}$$

for all $t \geq 0$. Indeed, we have

$$\begin{aligned} L_t &= \lambda S_t + (1-\lambda) R_t = \lambda S_t + (1-\lambda) R_t I(\tau_t^\lambda = t) \\ &\quad + (1-\lambda) R_t I(\tau_t^\lambda > t) \\ &= \lambda S_t + (1-\lambda) \mathsf{E}\big(S_t I(\tau_t^\lambda = t) \,|\, \mathcal{F}_t\big) + (1-\lambda) R_t I(\tau_t^\lambda > t) \\ &= \lambda S_t I(\tau_t^\lambda = t) + (1-\lambda) S_t I(\tau_t^\lambda = t) + \lambda S_t I(\tau_t^\lambda > t) \\ &\quad + (1-\lambda) R_t I(\tau_t^\lambda > t) \\ &\geq S_t I(\tau_t^\lambda = t) + \lambda S_t I(\tau_t^\lambda > t) \geq G_t I(\tau_t^\lambda = t) + G_t I(\tau_t^\lambda > t) = G_t \end{aligned} \tag{2.1.38}$$

where in the second last inequality we used that $R_t \geq 0$ and in the last inequality we used the definition of τ_t^λ given in (2.1.26) above. Thus (2.1.37) holds as claimed. Finally, since S is the smallest right-continuous supermartingale which dominates G, we see that (2.1.37) implies that

$$S_t \leq L_t \quad \mathsf{P}\text{-a.s.} \tag{2.1.39}$$

from where by (2.1.36) we conclude that $S_t \leq R_t$ P-a.s. Thus (2.1.32) holds as claimed. Combining (2.1.31) and (2.1.32) we get

$$S_t = \mathsf{E}\,(S_{\tau_t^\lambda} \,|\, \mathcal{F}_t) \tag{2.1.40}$$

for all $\lambda \in (0,1)$. From (2.1.40) and (2.1.27) we find

$$S_t \leq \frac{1}{\lambda} \mathsf{E}\,(G_{\tau_t^\lambda} \,|\, \mathcal{F}_t) \tag{2.1.41}$$

for all $\lambda \in (0,1)$. Letting $\lambda \uparrow 1$, using the conditional Fatou's lemma and (2.1.1) above together with the fact that G is left-continuous over stopping times, we obtain

$$S_t \le \mathsf{E}\,(G_{\tau_t^1} \,|\, \mathcal{F}_t) \tag{2.1.42}$$

where τ_t^1 is a stopping time given by

$$\tau_t^1 = \lim_{\lambda \uparrow 1} \tau_t^\lambda. \tag{2.1.43}$$

(Note that τ_t^λ increases when λ increases.) Since by (2.1.4) we know that the reverse inequality in (2.1.42) is always fulfilled, we may conclude that

$$S_t = \mathsf{E}\,(G_{\tau_t^1} \,|\, \mathcal{F}_t) \tag{2.1.44}$$

for all $t \ge 0$. Thus to complete the proof of (2.1.10) it is enough to verify that

$$\tau_t^1 = \tau_t \tag{2.1.45}$$

where τ_t is defined in (2.1.5) above. For this, note first that $\tau_t^\lambda \le \tau_t$ for all $\lambda \in (0,1)$ so that $\tau_t^1 \le \tau_t$. On the other hand, if $\tau_t(\omega) > t$ (the case $\tau_t(\omega) = t$ being obvious) then there exists $\varepsilon > 0$ such that $\tau_t(\omega) - \varepsilon > t$ and $S_{\tau_t(\omega)-\varepsilon} > G_{\tau_t(\omega)-\varepsilon} \ge 0$. Hence one can find $\lambda \in (0,1)$ (close enough to 1) such that $\lambda S_{\tau_t(\omega)-\varepsilon} > G_{\tau_t(\omega)-\varepsilon}$ showing that $\tau_t^\lambda(\omega) \ge \tau_t(\omega) - \varepsilon$. Letting first $\lambda \uparrow 1$ and then $\varepsilon \downarrow 0$ we conclude that $\tau_t^1 \ge \tau_t$. Hence (2.1.45) holds as claimed and the proof of (2.1.10) is complete in the case when $G_t \ge 0$ for all $t \ge 0$.

5°. In the case of general G satisfying (2.1.1) we can set

$$H = \inf_{t \ge 0} G_t \tag{2.1.46}$$

and introduce the right-continuous martingale

$$M_t = \mathsf{E}\,(H \,|\, \mathcal{F}_t) \tag{2.1.47}$$

for $t \ge 0$ so as to replace the initial gain process G by a new gain process $\widetilde{G} = (\widetilde{G}_t)_{t \ge 0}$ defined by

$$\widetilde{G}_t = G_t - M_t \tag{2.1.48}$$

for $t \ge 0$. Note that $\widetilde{G}$ need not satisfy (2.1.1) due to the existence of M, but M itself is uniformly integrable since $H \in L^1(\mathsf{P})$. Similarly, $\widetilde{G}$ is right-continuous and not necessarily left-continuous over stopping times due to the existence of M, but M itself is a (uniformly integrable) martingale so that the optional sampling theorem (page 60) is applicable. Finally, it is clear that $\widetilde{G}_t \ge 0$ and the optional sampling theorem implies that

$$\widetilde{S}_t = \operatorname*{ess\,sup}_{\tau \in \mathfrak{M}_t} \mathsf{E}\left(\widetilde{G}_\tau \,|\, \mathcal{F}_t\right) = \operatorname*{ess\,sup}_{\tau \in \mathfrak{M}_t} \mathsf{E}\,(G_\tau - M_\tau \,|\, \mathcal{F}_t) = S_t - M_t \tag{2.1.49}$$

for all $t \geq 0$. A closer inspection based on the new properties of $\widetilde{G}$ displayed above instead of the old ones imposed on G when $G_t \geq 0$ for all $t \geq 0$ shows that the proof above can be applied to $\widetilde{G}$ and $\widetilde{S}$ to yield the same conclusions implying (2.1.10) in the general case.

6°. Noticing that (2.1.11) follows by taking expectation in (2.1.10) and using (2.1.18), let us now show that (2.1.12) holds. We claim that the optimality of τ_* implies that $S_{\tau_*} = G_{\tau_*}$ P-a.s. Indeed, if this would not be the case then we would have $S_{\tau_*} \geq G_{\tau_*}$ P-a.s. with $\mathsf{P}(S_{\tau_*} > G_{\tau_*}) > 0$. It would then follow that $\mathsf{E}\,G_{\tau_*} < \mathsf{E}\,S_{\tau_*} \leq \mathsf{E}\,S_t = V_t$ where the second inequality follows by the optional sampling theorem (page 60) and the supermartingale property of $(S_s)_{s\geq t}$ using (2.1.1) above, while the final equality is stated in (2.1.18) above. The strict inequality, however, contradicts the fact that τ_* is optimal. Hence $S_{\tau_*} = G_{\tau_*}$ P-a.s. as claimed and the fact that $\tau_t \leq \tau_*$ P-a.s. follows from the definition (2.1.5) above.

7°. To verify the martingale property (2.1.14) it is enough to prove that

$$\mathsf{E}\,S_{\sigma\wedge\tau_t} = \mathsf{E}\,S_t \tag{2.1.50}$$

for all (bounded) stopping times σ greater than or equal to t. For this, note first that the optional sampling theorem (page 60) using (2.1.1) above implies

$$\mathsf{E}\,S_{\sigma\wedge\tau_t} \leq \mathsf{E}\,S_t. \tag{2.1.51}$$

On the other hand, from (2.1.10) and (2.1.29) we likewise see that

$$\mathsf{E}\,S_t = \mathsf{E}\,G_{\tau_t} = \mathsf{E}\,S_{\tau_t} \leq \mathsf{E}\,S_{\sigma\wedge\tau_t}. \tag{2.1.52}$$

Combining (2.1.51) and (2.1.52) we see that (2.1.50) holds and thus $(S_{s\wedge\tau_t})_{s\geq t}$ is a martingale (right-continuous by (2.1.13) above). This completes the proof of (2.1.14).

Finally, note that the final claim follows directly from (2.1.12). This completes the proof of the theorem. □

2.2. Markovian approach

In this subsection we will present basic results of optimal stopping when the time is continuous and the process is Markovian. (Basic definitions and properties of such processes are given in Subsection 4.3.)

1. Throughout we will consider a strong Markov process $X = (X_t)_{t\geq 0}$ defined on a filtered probability space $(\Omega, \mathcal{F}, (\mathcal{F}_t)_{t\geq 0}, \mathsf{P}_x)$ and taking values in a measurable space $(E, \mathcal{B})$ where for simplicity we will assume that $E = \mathbb{R}^d$ for some $d \geq 1$ and $\mathcal{B} = \mathcal{B}(\mathbb{R}^d)$ is the Borel σ-algebra on $\mathbb{R}^d$. It is assumed that the process X starts at x under P_x for $x \in E$ and that the sample paths of X are right-continuous and left-continuous over stopping times (if $\tau_n \uparrow \tau$ are stopping times, then $X_{\tau_n} \to X_\tau$ P_x-a.s. as $n \to \infty$). It is also assumed that the filtration $(\mathcal{F}_t)_{t\geq 0}$ is right-continuous (implying that the first entry times to open and closed sets are stopping times). In addition, it is assumed that the mapping $x \mapsto \mathsf{P}_x(F)$ is measurable for each $F \in \mathcal{F}$. It follows that the mapping $x \mapsto \mathsf{E}_x(Z)$ is measurable for each (bounded or non-negative) random variable Z. Finally, without loss of generality we will assume that $(\Omega, \mathcal{F})$ equals the canonical space $(E^{[0,\infty)}, \mathcal{B}^{[0,\infty)})$ so that the shift operator $\theta_t : \Omega \to \Omega$ is well defined by $\theta_t(\omega)(s) = \omega(t+s)$ for $\omega = (\omega(s))_{s\geq 0} \in \Omega$ and $t, s \geq 0$.

2. Given a measurable function $G : E \to \mathbb{R}$ satisfying the following condition (with $G(X_T) = 0$ if $T = \infty$):

$$\mathsf{E}_x\Big(\sup_{0\leq t\leq T} |G(X_t)|\Big) < \infty \tag{2.2.1}$$

for all $x \in E$, we consider the optimal stopping problem

$$V(x) = \sup_{0\leq \tau\leq T} \mathsf{E}_x G(X_\tau) \tag{2.2.2}$$

where $x \in E$ and the supremum is taken over stopping times τ of X. The latter means that τ is a stopping time with respect to the natural filtration of X given by $\mathcal{F}_t^X = \sigma(X_s : 0 \leq s \leq t)$ for $t \geq 0$. Since the same results remain valid if we take the supremum in (2.2.2) over stopping times τ with respect to $(\mathcal{F}_t)_{t\geq 0}$, and this assumption makes certain conclusions more elegant (the optimal stopping time will be attained), we will assume in the sequel that the supremum in (2.2.2) is taken over this larger class of stopping times. Note also that in (2.2.2) we admit that T can be ∞ as well (infinite horizon). In this case, however, we still assume that the supremum is taken over stopping times τ, i.e. over Markov times satisfying $0 \leq \tau < \infty$. In this way any specification of $G(X_\infty)$ becomes irrelevant for the problem (2.2.2).

3. Recall that V is called the *value function* and G is called the *gain function*. To solve the optimal stopping problem (2.2.2) means two things. Firstly, we need to exhibit an *optimal* stopping time, i.e. a stopping time τ_* at which the supremum is attained. Secondly, we need to compute the value $V(x)$ for $x \in E$ as explicitly as possible.

Let us briefly comment on what one expects to be a solution to the problem (2.2.2) (recall also Subsection 1.2 above). For this note that being Markovian means that the process X always starts afresh. Thus following the sample path $t \mapsto X_t(\omega)$ for $\omega \in \Omega$ given and fixed and evaluating $G(X_t(\omega))$ it is naturally expected that at each time t we shall be able optimally to decide either to continue with the observation or to stop it. In this way the state space E naturally splits into the *continuation set* C and the *stopping set* $D = E \setminus C$. It follows that as soon as the observed value $X_t(\omega)$ enters D, the observation should be stopped and an optimal stopping time is obtained. The central question thus arises as how to determine the sets C and D. (Note that the same arguments also hold in the discrete-time case of Subsection 1.2 above.)

In comparison with the general optimal stopping problem of Subsection 2.1 above, it may be noted that the description of the optimal stopping time just given does not involve any probabilistic construction (of a new stochastic process $S = (S_t)_{t \ge 0}$) but is purely deterministic (obtained by splitting E into two disjoint subsets defined by the deterministic functions G and V).

4. In the sequel we will treat the *finite horizon* formulation ($T < \infty$) and the *infinite horizon* formulation ($T = \infty$) of the optimal stopping problem (2.2.2) at the same time. It should be noted that in the former case ($T < \infty$) we need to replace the process X_t by the process $Z_t = (t, X_t)$ for $t \ge 0$ so that the problem reads

$$V(t,x) = \sup_{0 \le \tau \le T-t} \mathsf{E}_{t,x} G(t+\tau, X_{t+\tau}) \tag{2.2.2$'$}$$

where the "rest of time" $T - t$ changes when the initial state $(t,x) \in [0,T] \times E$ changes in its first argument. It turns out, however, that no argument below is more seriously affected by this change, and the results obtained for the problem (2.2.2) with $T = \infty$ will automatically hold for the problem (2.2.2$'$) if we simply think of X to be Z (with a new "two-dimensional" state space E equal to $\mathbb{R}_+ \times E$). Moreover, it may be noted in (2.2.2$'$) that at time T we have the "terminal" condition $V(T,x) = G(T,x)$ for all $x \in E$ so that the first entry time of Z to the stopping set D, denoted below by τ_D, will always be smaller than or equal to T and thus finite. This works to a technical advantage of the finite horizon formulation (2.2.2$'$) over the infinite horizon formulation (2.2.2) (where instead of the condition $V(T,x) = G(T,x)$ for all $x \in E$ another "boundary condition at infinity" such as (2.2.52) may hold).

5. Consider the optimal stopping problem (2.2.2) when $T = \infty$. Recall that (2.2.2) reads as follows:

$$V(x) = \sup_{\tau} \mathsf{E}_x G(X_\tau) \tag{2.2.3}$$

where τ is a stopping time (with respect to $(\mathcal{F}_t)_{t \ge 0}$) and $\mathsf{P}_x(X_0 = x) = 1$ for $x \in E$. Introduce the continuation set

$$C = \{x \in E : V(x) > G(x)\} \tag{2.2.4}$$

and the stopping set

$$D = \{x \in E : V(x) = G(x)\} \tag{2.2.5}$$

Note that if V is lsc (lower semicontinuous) and G usc (upper semicontinuous) then C is open and D is closed. Introduce the first entry time τ_D of X into D by setting

$$\tau_D = \inf\{\, t \ge 0 : X_t \in D \,\}. \tag{2.2.6}$$

Note that τ_D is a stopping (Markov) time with respect to $(\mathcal{F}_t)_{t\geq 0}$ when D is closed since both X and $(\mathcal{F}_t)_{t\geq 0}$ are right-continuous.

6. The following definition plays a fundamental role in solving the optimal stopping problem (2.2.3).

Definition 2.3. A measurable function $F : E \to \mathbb{R}$ is said to be *superharmonic* if

$$\mathsf{E}_x F(X_\sigma) \leq F(x) \tag{2.2.7}$$

for all stopping times σ and all $x \in E$.

It is assumed in (2.2.7) that the left-hand side is well defined (and finite) i.e. that $F(X_\sigma) \in L^1(\mathsf{P}_x)$ for all $x \in E$ whenever σ is a stopping time. Moreover, it will be verified in the proof of Theorem 2.4 below that the following stochastic characterization of superharmonic functions holds (recall also (1.2.40)):

F is superharmonic if and only if $(F(X_t))_{t\geq 0}$ is a right-continuous supermartingale under P_x for every $x \in E$ (2.2.8)

whenever F is lsc and $(F(X_t))_{t\geq 0}$ is uniformly integrable.

7. The following theorem presents *necessary conditions* for the existence of an optimal stopping time.

Theorem 2.4. *Let us assume that there exists an optimal stopping time* τ_* *in* (2.2.3), *i.e. let*

$$V(x) = \mathsf{E}_x G(X_{\tau_*}) \tag{2.2.9}$$

for all $x \in E$. *Then we have:*

The value function V *is the smallest superharmonic function which dominates the gain function* G *on* E. (2.2.10)

Let us in addition to (2.2.9) *assume that* V *is lsc and* G *is usc. Then we have:*

The stopping time τ_D *satisfies* $\tau_D \leq \tau_*$ P_x*-a.s. for all* $x \in E$ *and is optimal in* (2.2.3). (2.2.11)

The stopped process $(V(X_{t\wedge\tau_D}))_{t\geq 0}$ *is a right-continuous martingale under* P_x *for every* $x \in E$. (2.2.12)

Proof. (2.2.10): To show that V is superharmonic note that by the strong Markov property we have:

$$\begin{aligned} \mathsf{E}_x V(X_\sigma) &= \mathsf{E}_x \mathsf{E}_{X_\sigma} G(X_{\tau_*}) = \mathsf{E}_x \mathsf{E}_x\big(G(X_{\tau_*}) \circ \theta_\sigma \mid \mathcal{F}_\sigma\big) \\ &= \mathsf{E}_x G(X_{\sigma+\tau_*\circ\theta_\sigma}) \leq \sup_\tau \mathsf{E}_x G(X_\tau) = V(x) \end{aligned} \tag{2.2.13}$$

for each stopping time σ and all $x \in E$. This establishes (2.2.7) and proves the initial claim.

Let F be a superharmonic function which dominates G on E. Then we have

$$\mathsf{E}_x G(X_\tau) \le \mathsf{E}_x F(X_\tau) \le F(x) \tag{2.2.14}$$

for each stopping time τ and all $x \in E$. Taking the supremum over all τ in (2.2.14) we find that $V(x) \le F(x)$ for all $x \in E$. Since V is superharmonic itself, this proves the final claim.

(2.2.11): We claim that $V(X_{\tau_*}) = G(X_{\tau_*})$ P_x-a.s. for all $x \in E$. Indeed, if $\mathsf{P}_x\big(V(X_{\tau_*}) > G(X_{\tau_*})\big) > 0$ for some $x \in E$, then $\mathsf{E}_x G(X_{\tau_*}) < \mathsf{E}_x V(X_{\tau_*}) \le V(x)$ since V is superharmonic, leading to a contradiction with the fact that τ_* is optimal. From the identity just verified it follows that $\tau_D \le \tau_*$ P_x-a.s. for all $x \in E$ as claimed.

To make use of the previous inequality we may note that setting $\sigma \equiv s$ in (2.2.7) and using the Markov property we get

$$V(X_t) \ge \mathsf{E}_{X_t} V(X_s) = \mathsf{E}_x\big(V(X_{t+s}) \,|\, \mathcal{F}_t\big) \tag{2.2.15}$$

for all $t, s \ge 0$ and all $x \in E$. This shows:

$$\text{The process } (V(X_t))_{t\ge 0} \text{ is a supermartingale under } \mathsf{P}_x \text{ for each } x \in E. \tag{2.2.16}$$

Moreover, to indicate the argument as clearly as possible, let us for the moment assume that V is continuous. Then obviously it follows that $(V(X_t))_{t\ge 0}$ is right-continuous. Thus, by the optional sampling theorem (page 60) using (2.2.1) above, we see that (2.2.7) extends as follows:

$$\mathsf{E}_x V(X_\tau) \le \mathsf{E}_x V(X_\sigma) \tag{2.2.17}$$

for stopping times σ and τ such that $\sigma \le \tau$ P_x-a.s. with $x \in E$. In particular, since $\tau_D \le \tau_*$ P_x-a.s. by (2.2.17) we get

$$V(x) = \mathsf{E}_x G(X_{\tau_*}) = \mathsf{E}_x V(X_{\tau_*}) \le \mathsf{E}_x V(X_{\tau_D}) = \mathsf{E}_x G(X_{\tau_D}) \le V(x) \tag{2.2.18}$$

for $x \in E$ upon using that $V(X_{\tau_D}) = G(X_{\tau_D})$ since V is lsc and G is usc. This shows that τ_D is optimal if V is continuous. Finally, if V is only known to be lsc, then by Proposition 2.5 below we know that $(V(X_t))_{t\ge 0}$ is right-continuous P_x-a.s. for each $x \in E$, and the proof can be completed as above. This shows that τ_D is optimal if V is lsc as claimed.

(2.2.12): By the strong Markov property we have

$$\begin{aligned} \mathsf{E}_x\big(V(X_{t\wedge\tau_D})\,|\,\mathcal{F}_{s\wedge\tau_D}\big) &= \mathsf{E}_x\Big(\mathsf{E}_{X_{t\wedge\tau_D}}G(X_{\tau_D})\,|\,\mathcal{F}_{s\wedge\tau_D}\Big) \\ &= \mathsf{E}_x\Big(\mathsf{E}_x\big(G(X_{\tau_D})\circ\theta_{t\wedge\tau_D}\,|\,\mathcal{F}_{t\wedge\tau_D}\big)\,|\,\mathcal{F}_{s\wedge\tau_D}\Big) \\ &= \mathsf{E}_x\Big(\mathsf{E}_x\big(G(X_{\tau_D})\,|\,\mathcal{F}_{t\wedge\tau_D}\big)\,|\,\mathcal{F}_{s\wedge\tau_D}\Big) = \mathsf{E}_x\big(G(X_{\tau_D})\,|\,\mathcal{F}_{s\wedge\tau_D}\big) \\ &= \mathsf{E}_{X_{s\wedge\tau_D}}\big(G(X_{\tau_D})\big) = V(X_{s\wedge\tau_D}) \end{aligned} \tag{2.2.19}$$

for all $0 \le s \le t$ and all $x \in E$ proving the martingale property. The right-continuity of $\big(V(X_{t\wedge\tau_D})\big)_{t\ge0}$ follows from the right-continuity of $(V(X_t))_{t\ge0}$ and the proof is complete. □

The following fact was needed in the proof above to extend the result from continuous to lsc V.

Proposition 2.5. *If a superharmonic function $F : E \to \mathbb{R}$ is lsc (lower semicontinuous), then the supermartingale $(F(X_t))_{t\ge0}$ is right-continuous P_x-a.s. for every $x \in E$.*

Proof. Firstly, we will show that

$$F(X_\tau) = \lim_{h\downarrow0} F(X_{\tau+h}) \qquad \mathsf{P}_x\text{-a.s.} \tag{2.2.20}$$

for any given stopping time τ and $x \in E$. For this, note that the right-continuity of X and the ls-continuity of F, we get

$$F(X_\tau) \le \liminf_{h\downarrow0} F(X_{\tau+h}) \qquad \mathsf{P}_x\text{-a.s.} \tag{2.2.21}$$

To prove the reverse inequality we will first derive it for $\tau \equiv 0$, i.e. we have

$$\limsup_{h\downarrow0} F(X_h) \le F(x) \qquad \mathsf{P}_x\text{-a.s.} \tag{2.2.22}$$

For this, note by Blumenthal's 0-1 law (cf. page 97) that $\limsup_{h\downarrow0} F(X_h)$ is equal P_x-a.s. to a constant $c \in \mathbb{R}$. Let us assume that $c > F(x)$. Then there is $\varepsilon > 0$ such that $c > F(x)+\varepsilon$. Set $A_\varepsilon = \{\, y \in E : F(y) > F(x)+\varepsilon \,\}$ and consider the stopping time $\tau_\varepsilon = \inf\{\, h \ge 0 : X_h \in A_\varepsilon \,\}$. By definition of c and A_ε we see that $\tau_\varepsilon = 0$ P_x-a.s. Note however that A_ε is open (since F is lsc) and that we cannot claim a priori that X_{τ_ε}, i.e. x, belongs to A_ε as one would like to reach a contradiction. For this reason choose an increasing sequence of closed sets K_n for $n \ge 1$ such that $\bigcup_{n=1}^\infty K_n = A_\varepsilon$. Consider the stopping time $\tau_n = \inf\{\, h \ge 0 : X_h \in K_n \,\}$ for $n \ge 1$. Then $\tau_n \downarrow \tau_\varepsilon$ as $n \to \infty$ and since K_n is closed we see that $X_{\tau_n} \in K_n$ for all $n \ge 1$. Hence $X_{\tau_n} \in A_\varepsilon$ i.e. $F(X_{\tau_n}) > F(x) + \varepsilon$ for all $n \ge 1$. Using that F is superharmonic this implies

$$\begin{aligned} F(x) &\ge \mathsf{E}_x F(X_{\tau_n\wedge1}) = \mathsf{E}_x F(X_{\tau_n}) I(\tau_n \le 1) + \mathsf{E}_x F(X_1) I(\tau_n > 1) \\ &\ge (F(x)+\varepsilon)\mathsf{P}(\tau_n \le 1) + \mathsf{E}_x F(X_1) I(\tau_n > 1) \to F(x) + \varepsilon \end{aligned} \tag{2.2.23}$$

as $n \to \infty$ since $\tau_n \downarrow 0$ P_x-a.s. as $n \to \infty$ and $F(X_1) \in L^1(\mathsf{P}_x)$. As clearly (2.2.23) is impossible, we may conclude that (2.2.22) holds as claimed.

To treat the case of a general stopping time τ, take E_x on both sides of (2.2.22) and insert $x = X_\tau$. This by the strong Markov property gives

$$\begin{aligned} F(X_\tau) &\ge \mathsf{E}_{X_\tau}\Big(\limsup_{h\downarrow 0} F(X_h)\Big) = \mathsf{E}_x\Big(\limsup_{h\downarrow 0} F(X_h)\circ\theta_\tau \mid \mathcal{F}_\tau\Big) \\ &= \mathsf{E}_x\Big(\limsup_{h\downarrow 0} F(X_{\tau+h}) \mid \mathcal{F}_\tau\Big) = \limsup_{h\downarrow 0} F(X_{\tau+h}) \quad \mathsf{P}_x\text{-a.s.} \end{aligned} \tag{2.2.24}$$

since $\limsup_{h\downarrow 0} F(X_{\tau+h})$ is $\mathcal{F}_{\tau+}$-measurable and $\mathcal{F}_\tau = \mathcal{F}_{\tau+}$ by the right-continuity of $(\mathcal{F}_t)_{t\ge 0}$. Combining (2.2.21) and (2.2.24) we get (2.2.20). In particular, taking $\tau \equiv t$ we see that

$$\lim_{h\downarrow 0} F(X_{t+h}) = F(X_t) \qquad \mathsf{P}_x\text{-a.s.} \tag{2.2.25}$$

for all $t \ge 0$. Note that the P_x-null set in (2.2.25) does depend on the given t.

Secondly, by means of (2.2.20) we will now show that a single P_x-null set can be selected so that the convergence relation in (2.2.25) holds on its complement simultaneously for all $t \ge 0$. For this, set $\tau_0 = 0$ and define the stopping time

$$\tau_n = \inf\{\, t \ge \tau_{n-1} : |F(X_t) - F(X_{\tau_{n-1}})| > \varepsilon/2\,\} \tag{2.2.26}$$

for $n = 1, 2, \ldots$ where $\varepsilon > 0$ is given and fixed. By (2.2.20) we see that for each $n \ge 1$ there is a P_x-null set N_n such that $\tau_n > \tau_{n-1}$ on $\Omega \setminus N_n$. Continuing the procedure (2.2.26) by transfinite induction over countable ordinals (there can be at most countably many disjoint intervals in $\mathbb{R}_+$) and calling the union of the countably many P_x-null set by N_ε, it follows that for each $\omega \in \Omega \setminus N_\varepsilon$ and each $t \ge 0$ there is a countable ordinal α such that $\tau_\alpha(\omega) \le t < \tau_{\alpha+1}(\omega)$. Hence for every $s \in [\tau_\alpha(\omega), \tau_{\alpha+1}(\omega))$ we have $|F(X_t(\omega)) - F(X_s(\omega))| \le |F(X_t(\omega)) - F(X_{\tau_\alpha}(\omega))| + |F(X_s(\omega)) - F(X_\alpha(\omega))| \le \varepsilon/2 + \varepsilon/2 = \varepsilon$. This shows that $\limsup_{s\downarrow t} |F(X_t) - F(X_s)| \le \varepsilon$ on $\Omega \setminus N_\varepsilon$. Setting $N = \bigcup_{n=1}^\infty N_{1/n}$ we see that $\mathsf{P}_x(N) = 0$ and $\lim_{s\downarrow t} F(X_s) = F(X_t)$ on $\Omega \setminus N$ completing the proof. □

Remark 2.6. The result and proof of Theorem 2.4 above extend in exactly the same form (by slightly changing the notation only) to the finite horizon problem (2.2.2′). We will omit further details in this direction.

8. The following theorem provides *sufficient condition* for the existence of an optimal stopping time.

Theorem 2.7. *Consider the optimal stopping problem* (2.2.3) *upon assuming that the condition* (2.2.1) *is satisfied. Let us assume that there exists the smallest superharmonic function* $\widehat{V}$ *which dominates the gain function* G *on* E. *Let us in* *addition assume that* $\widehat{V}$ *is lsc and* G *is usc. Set* $D = \{x \in E : \widehat{V}(x) = G(x)\}$ *and let* τ_D *be defined by* (2.2.6) *above. We then have*:

If $\mathsf{P}_x(\tau_D < \infty) = 1$ *for all* $x \in E$, *then* $\widehat{V} = V$ *and* τ_D *is optimal in* (2.2.3). (2.2.27)

If $\mathsf{P}_x(\tau_D < \infty) < 1$ *for some* $x \in E$, *then there is no optimal stopping time* (*with probability* 1) *in* (2.2.3). (2.2.28)

Proof. Since $\widehat{V}$ is superharmonic, we have

$$\mathsf{E}_x G(X_\tau) \le \mathsf{E}_x \widehat{V}(X_\tau) \le \widehat{V}(x) \tag{2.2.29}$$

for all stopping times τ and all $x \in E$. Taking the supremum in (2.2.17) over all τ we find that

$$G(x) \le V(x) \le \widehat{V}(x) \tag{2.2.30}$$

for all $x \in E$. Assuming that $\mathsf{P}_x(\tau_D < \infty) = 1$ for all $x \in E$, we will now present two *different* proofs of the fact that $\widehat{V} = V$ implying also that τ_D is optimal in (2.2.3).

First proof. Let us assume that G is *bounded.* With $\varepsilon > 0$ given and fixed, consider the sets:

$$C_\varepsilon = \{\, x \in E : \widehat{V}(x) > G(x) + \varepsilon \,\}, \tag{2.2.31}$$

$$D_\varepsilon = \{\, x \in E : \widehat{V}(x) \le G(x) + \varepsilon \,\}. \tag{2.2.32}$$

Since $\widehat{V}$ is lsc and G is usc we see that C_ε is open and D_ε is closed. Moreover, it is clear that $C_\varepsilon \uparrow C$ and $D_\varepsilon \downarrow D$ as $\varepsilon \downarrow 0$ where C and D are defined by (2.2.4) and (2.2.5) above respectively.

Define the stopping time

$$\tau_{D_\varepsilon} = \inf \{\, t \ge 0 : X_t \in D_\varepsilon \}. \tag{2.2.33}$$

Since $D \subseteq D_\varepsilon$ and $\mathsf{P}_x(\tau_D < \infty) = 1$ for all $x \in E$, we see that $\mathsf{P}_x(\tau_{D_\varepsilon} < \infty) = 1$ for all $x \in E$. The latter fact can also be derived directly (without assuming the former fact) by showing that $\limsup_{t\to\infty} \widehat{V}(X_t) = \limsup_{t\to\infty} G(X_t)$ P_x-a.s. for all $x \in E$. This can be done in exactly the same way as in the first part of the proof of Theorem 1.13.

In order to show that

$$\mathsf{E}_x \widehat{V}(X_{\tau_{D_\varepsilon}}) = \widehat{V}(x) \tag{2.2.34}$$

for all $x \in E$, we will first show that

$$G(x) \le \mathsf{E}_x \widehat{V}(X_{\tau_{D_\varepsilon}}) \tag{2.2.35}$$

for all $x \in E$. For this, set

$$c = \sup_{x \in E} \big(G(x) - \mathsf{E}_x \widehat{V}(X_{\tau_{D_\varepsilon}}) \big) \tag{2.2.36}$$

and note that

$$G(x) \le c + \mathsf{E}_x \widehat{V}\big(X_{\tau_{D_\varepsilon}}\big) \tag{2.2.37}$$

for all $x \in E$. (Observe that c is finite since G is bounded implying also that $\widehat{V}$ is bounded.)

Next by the strong Markov property we find

$$\begin{aligned} \mathsf{E}_x \mathsf{E}_{X_\sigma} \widehat{V}\big(X_{\tau_{D_\varepsilon}}\big) &= \mathsf{E}_x \mathsf{E}_x \Big(\widehat{V}\big(X_{\tau_{D_\varepsilon}}\big) \circ \theta_\sigma \mid \mathcal{F}_\sigma \Big) \\ &= \mathsf{E}_x \mathsf{E}_x \Big(\widehat{V}\big(X_{\sigma + \tau_{D_\varepsilon} \circ \theta_\sigma}\big) \mid \mathcal{F}_\sigma \Big) \\ &= \mathsf{E}_x \widehat{V}\big(X_{\sigma + \tau_{D_\varepsilon} \circ \theta_\sigma}\big) \le \mathsf{E}_x \widehat{V}\big(X_{\tau_{D_\varepsilon}}\big) \end{aligned} \tag{2.2.38}$$

using that $\widehat{V}$ is superharmonic and lsc (recall Proposition 2.5 above) and $\sigma + \tau_{D_\varepsilon} \circ \theta_\sigma \ge \tau_{D_\varepsilon}$ since τ_{D_ε} is the first entry time to a set. This shows that the function

$$x \mapsto \mathsf{E}_x \widehat{V}\big(X_{\tau_{D_\varepsilon}}\big) \quad \text{is superharmonic} \tag{2.2.39}$$

from E to $\mathbb{R}$. Hence the function of the right-hand side of (2.2.37) is also superharmonic so that by the definition of $\widehat{V}$ we can conclude that

$$\widehat{V}(x) \le c + \mathsf{E}_x \widehat{V}\big(X_{\tau_{D_\varepsilon}}\big) \tag{2.2.40}$$

for all $x \in E$.

Given $0 < \delta \le \varepsilon$ choose $x_\delta \in E$ such that

$$G(x_\delta) - \mathsf{E}_{x_\delta} \widehat{V}\big(X_{\tau_{D_\varepsilon}}\big) \ge c - \delta. \tag{2.2.41}$$

Then by (2.2.40) and (2.2.41) we get

$$\widehat{V}(x_\delta) \le c + \mathsf{E}_{x_\delta} \widehat{V}\big(X_{\tau_{D_\varepsilon}}\big) \le G(x_\delta) + \delta \le G(x_\delta) + \varepsilon. \tag{2.2.42}$$

This shows that $x_\delta \in D_\varepsilon$ and thus $\tau_{D_\varepsilon} \equiv 0$ under P_{x_δ}. Inserting the latter conclusion into (2.2.41) we get

$$c - \delta \le G(x_\delta) - \widehat{V}(x_\delta) \le 0. \tag{2.2.43}$$

Letting $\delta \downarrow 0$ we see that $c \le 0$ thus proving (2.2.35) as claimed. Using the definition of $\widehat{V}$ and (2.2.39) we see that (2.2.34) follows directly from (2.2.35).

Finally, from (2.2.34) we get

$$\widehat{V}(x) = \mathsf{E}_x \widehat{V}\big(X_{\tau_{D_\varepsilon}}\big) \le \mathsf{E}_x G\big(X_{\tau_{D_\varepsilon}}\big) + \varepsilon \le V(x) + \varepsilon \tag{2.2.44}$$

for all $x \in E$ upon using that $\widehat{V}(X_{\tau_{D_\varepsilon}}) \le G(X_{\tau_{D_\varepsilon}}) + \varepsilon$ since $\widehat{V}$ is lsc and G is usc. Letting $\varepsilon \downarrow 0$ in (2.2.44) we see that $\widehat{V} \le V$ and thus by (2.2.30) we can conclude that $\widehat{V} = V$. From (2.2.44) we thus also see that

$$V(x) \le \mathsf{E}_x G(X_{\tau_{D_\varepsilon}}) + \varepsilon \tag{2.2.45}$$

for all $x \in E$.

Letting $\varepsilon \downarrow 0$ and using that $D_\varepsilon \downarrow D$ we see that $\tau_{D_\varepsilon} \uparrow \tau_0$ where τ_0 is a stopping time satisfying $\tau_0 \le \tau_D$. Since V is lsc and G is usc it is easily seen from the definition of τ_{D_ε} that $V(X_{\tau_{D_\varepsilon}}) \le G(X_{\tau_{D_\varepsilon}}) + \varepsilon$ for all $\varepsilon > 0$. Letting $\varepsilon \downarrow 0$ and using that X is left-continuous over stopping times it follows that $V(X_{\tau_0}) \le G(X_{\tau_0})$ since V is lsc and G is usc. This shows that $V(X_{\tau_0}) = G(X_{\tau_0})$ and therefore $\tau_D \le \tau_0$ showing that $\tau_0 = \tau_D$. Thus $\tau_{D_\varepsilon} \uparrow \tau_D$ as $\varepsilon \downarrow 0$.

Making use of the latter fact in (2.2.34) upon letting $\varepsilon \downarrow 0$ and applying Fatou's lemma, we get

$$\begin{aligned} V(x) &\le \limsup_{\varepsilon \downarrow 0} \mathsf{E}_x G(X_{\tau_{D_\varepsilon}}) \le \mathsf{E}_x \limsup_{\varepsilon \downarrow 0} G(X_{\tau_{D_\varepsilon}}) \\ &\le \mathsf{E}_x G\Big(\limsup_{\varepsilon \downarrow 0} X_{\tau_{D_\varepsilon}}\Big) = \mathsf{E}_x G(X_{\tau_D}) \end{aligned} \tag{2.2.46}$$

using that G is usc. This shows that τ_D is optimal in the case when G is bounded.

Second proof. We will divide the second proof in two parts depending on if G is bounded (from below) or not.

1°. Let us assume that G is *bounded from below*. It means that $c := \inf_{x \in E} G(x) > -\infty$. Replacing G by $G - c$ and $\widehat{V}$ by $\widehat{V} - c$ when $c < 0$ we see that there is no restriction to assume that $G(x) \ge 0$ for all $x \in E$.

By analogy with (2.2.31) and (2.2.32), with $0 < \lambda < 1$ given and fixed, consider the sets

$$C_\lambda = \{\, x \in E : \lambda \widehat{V}(x) > G(x) \,\}, \tag{2.2.47}$$

$$D_\lambda = \{\, x \in E : \lambda \widehat{V}(x) \le G(x) \,\}. \tag{2.2.48}$$

Since $\widehat{V}$ is lsc and G is usc we see that C_λ is open and D is closed. Moreover, it is clear that $C_\lambda \uparrow C$ and $D_\lambda \downarrow D$ as $\lambda \uparrow 1$ where C and D are defined by (2.2.4) and (2.2.5) above respectively.

Define the stopping time

$$\tau_{D_\lambda} = \inf\{\, t \ge 0 : X_t \in D_\lambda \,\}. \tag{2.2.49}$$

Since $D \subseteq D_\lambda$ and $\mathsf{P}_x(\tau_D < \infty) = 1$ for all $x \in E$, we see that $\mathsf{P}_x(\tau_{D_\lambda} < \infty) = 1$ for all $x \in E$. (The latter fact can also be derived directly as in the remark following (2.2.33) above.)

In order to show that

$$\mathsf{E}_x \widehat{V}\big(X_{\tau_{D_\lambda}}\big) = \widehat{V}(x) \tag{2.2.50}$$

for all $x \in E$, we will first note that

$$G(x) \le \lambda \widehat{V}(x) + (1-\lambda)\, \mathsf{E}_x \widehat{V}\big(X_{\tau_{D_\lambda}}\big) \tag{2.2.51}$$

for all $x \in E$. Indeed, if $x \in C_\lambda$ then $G(x) < \lambda \widehat{V}(x) \le \lambda \widehat{V}(x) + (1-\lambda) \mathsf{E}_x \widehat{V}\big(X_{\tau_{D_\lambda}}\big)$ since $\widehat{V} \ge G \ge 0$ on E. On the other hand, if $x \in D_\lambda$ then $\tau_{D_\lambda} \equiv 0$ and (2.2.51) follows since $G \le \widehat{V}$ on E.

Next in exactly the same way as in (2.2.38) above one verifies that the function

$$x \mapsto \mathsf{E}_x \widehat{V}\big(X_{\tau_{D_\lambda}}\big) \quad \text{is superharmonic} \tag{2.2.52}$$

from E to $\mathbb{R}$. Hence the function on the right-hand side of(2.2.51) is superharmonic so that by the definition of V we can conclude that

$$\widehat{V}(x) \le \lambda \widehat{V}(x) + (1-\lambda)\, \mathsf{E}_x \widehat{V}\big(X_{\tau_{D_\lambda}}\big) \tag{2.2.53}$$

for all $x \in E$. This proves (2.2.50) as claimed.

From (2.2.50) we get

$$\widehat{V}(x) = \mathsf{E}_x \widehat{V}\big(X_{\tau_{D_\lambda}}\big) \le \frac{1}{\lambda}\, \mathsf{E}_x G\big(X_{\tau_{D_\lambda}}\big) \le \frac{1}{\lambda}\, V(x) \tag{2.2.54}$$

for all $x \in E$ upon using that $\widehat{V}\big(X_{\tau_{D_\lambda}}\big) \le (1/\lambda)\, G\big(X_{\tau_{D_\lambda}}\big)$ since $\widehat{V}$ is lsc and G is usc. Letting $\lambda \uparrow 1$ in (2.2.54) we see that $\widehat{V} \le V$ and thus by (2.2.30) we can conclude that $\widehat{V} = V$. From (2.2.54) we thus see that

$$V(x) \le \frac{1}{\lambda}\, \mathsf{E}_x G\big(X_{\tau_{D_\lambda}}\big) \tag{2.2.55}$$

for all $x \in E$ and all $0 \le \lambda < 1$.

Letting $\lambda \uparrow 1$ and using that $D_\lambda \downarrow D$ we see that $\tau_{D_\lambda} \uparrow \tau_1$ where τ_1 is a stopping time satisfying $\tau_1 \le \tau_D$. Since V is lsc and G is usc it is easily seen from the definition of τ_{D_λ} that $V(\tau_{D_\lambda}) \le (1/\lambda)\, G(\tau_{D_\lambda})$ for all $0 < \lambda < 1$. Letting $\lambda \uparrow 1$ and using that X is left-continuous over stopping times it follows that $V(X_{\tau_1}) \le G(X_{\tau_1})$ since V is lsc and G is usc. This shows that $V(X_{\tau_1}) = G(X_{\tau_1})$ and therefore $\tau_D \le \tau_1$ showing that $\tau_1 = \tau_D$. Thus $\tau_{D_\lambda} \uparrow \tau_1$ as $\lambda \uparrow 1$.

Making use of the latter fact in (2.2.55) upon letting $\lambda \uparrow 1$ and applying Fatou's lemma, we get

$$\begin{aligned} V(x) &\le \limsup_{\lambda \uparrow 1} \mathsf{E}_x G\big(X_{\tau_{D_\lambda}}\big) \le \mathsf{E}_x \limsup_{\lambda \uparrow 1} G\big(X_{\tau_{D_\lambda}}\big) \\ &\le \mathsf{E}_x G\Big(\limsup_{\lambda \uparrow 1} X_{\tau_{D_\lambda}}\Big) = \mathsf{E}_x G(X_{\tau_D}) \end{aligned} \tag{2.2.56}$$

using that G is usc. This shows that τ_D is optimal in the case when G is bounded from below.

2°. Let us assume that G is a (general) measurable function satisfying (2.2.1) (i.e. not necessarily bounded or bounded from below). Then Part 1° of the proof can be extended by means of the function $h : E \to \mathbb{R}$ defined by

$$h(x) = \mathsf{E}_x\Big(\inf_{t\geq 0} G(X_t)\Big) \tag{2.2.57}$$

for $x \in E$. The key observation is that $-h$ is superharmonic which is seen as follows (recall (2.2.57)):

$$\begin{aligned}\mathsf{E}_x(-h(X_\sigma)) &= \mathsf{E}_x\mathsf{E}_{X_\sigma}\sup_{t\geq 0}(-G(X_t)) = \mathsf{E}_x\mathsf{E}_x\Big(\sup_{t\geq 0}(-G(X_t))\circ\theta_\sigma \mid \mathcal{F}_\sigma\Big) \\ &= \mathsf{E}_x\mathsf{E}_x\Big(\sup_{t\geq 0}(-G(X_{\sigma+t}))\Big) \leq -h(x)\end{aligned} \tag{2.2.58}$$

for all $x \in E$ proving the claim. Moreover, it is obvious that $\widehat{V} - h \geq G - h \geq 0$ on E. Knowing this we can define sets C_λ and D_λ by extending (2.2.47) and (2.2.48) as follows:

$$C_\lambda = \big\{\, x \in E : \lambda\big(\widehat{V}(x) - h(x)\big) > G(x) - h(x)\,\big\} \tag{2.2.59}$$

$$D_\lambda = \big\{\, x \in E : \lambda\big(\widehat{V}(x) - h(x)\big) \leq G(x) - h(x)\,\big\} \tag{2.2.60}$$

for $0 < \lambda < 1$.

We then claim that

$$G(x) - h(x) \leq \lambda\big(\widehat{V}(x) - h(x)\big) + (1-\lambda)\,\mathsf{E}_x\big(\widehat{V}(X_{\tau_{D_\lambda}}) - h(X_{\tau_{D_\lambda}})\big) \tag{2.2.61}$$

for all $x \in E$. Indeed, if $x \in C_\lambda$ then (2.2.61) follows by the fact that $\widehat{V} \geq h$ on E. On the other hand, if $x \in D_\lambda$ then $\tau_{D_\lambda} = 0$ and the inequality (2.2.61) reduces to the trivial inequality that $G \leq \widehat{V}$. Thus (2.2.61) holds as claimed.

Since $-h$ is superharmonic we have

$$-h(x) \geq -\lambda h(x) + (1-\lambda)\,\mathsf{E}_x\big(-h(X_{\tau_{D_\lambda}})\big) \tag{2.2.62}$$

for all $x \in E$. From (2.2.61) and (2.2.62) we see that

$$G(x) \leq \lambda\widehat{V}(x) + (1-\lambda)\,\mathsf{E}_x\widehat{V}(X_{\tau_{D_\lambda}}) \tag{2.2.63}$$

for all $x \in E$. Upon noting that $D_\lambda \downarrow D$ as $\lambda \uparrow 1$ the rest of the proof can be carried out in exactly the same way as in Part 1° above. (If h does not happen to be lsc, then C_λ and D_λ are still measurable sets and thus τ_{D_λ} is a stopping time (with respect to the completion of $(\mathcal{F}_t^X)_{t\geq 0}$ by the family of all P_x-null sets from $\mathcal{F}_\infty^X$ for $x \in E$). Moreover, it is easily verified using the strong Markov property of X and the conditional Fatou lemma that

$$h(X_{\tau_D}) \le \limsup_{\lambda \downarrow 0} h\big(X_{\tau_{D_\lambda}}\big) \quad \mathsf{P}_x\text{-a.s.} \tag{2.2.64}$$

for all $x \in E$, which is sufficient for the proof.)

The final claim of the theorem follows from (2.2.11) in Theorem 2.4 above. The proof is complete. □

Remark 2.8. The result and proof of Theorem 2.7 above extend in exactly the same form (by slightly changing the notation only) to the finite horizon problem (2.2.2′). Note moreover in this case that $\tau_D \le T < \infty$ (since $V(T,x) = G(T,x)$ and thus $(T,x) \in D$ for all $x \in E$) so that the condition $\mathsf{P}_x(\tau_D < \infty) = 1$ is automatically satisfied for all $x \in E$ and need not be assumed.

9. The following corollary is an elegant tool for tackling the optimal stopping problem in the case when one can prove *directly* from definition of V that V is lsc. Note that the result is particularly useful in the case of finite horizon since it provides the existence of an optimal stopping time τ_* by simply identifying it with τ_D from (2.2.6) above.

Corollary 2.9. (The existence of an optimal stopping time)

Infinite horizon. *Consider the optimal stopping problem* (2.2.3) *upon assuming that the condition* (2.2.1) *is satisfied. Suppose that* V *is lsc and* G *is usc. If* $\mathsf{P}_x(\tau_D < \infty) = 1$ *for all* $x \in E$ *, then* τ_D *is optimal in* (2.2.3). *If* $\mathsf{P}_x(\tau_D < \infty) < 1$ *for some* $x \in E$ *, then there is no optimal stopping time* (*with probability* 1) *in* (2.2.3).

Finite horizon. *Consider the optimal stopping problem* (2.2.2′) *upon assuming that the corresponding condition* (2.2.1) *is satisfied. Suppose that* V *is lsc and* G *is usc. Then* τ_D *is optimal in* (2.2.2′) .

Proof. The case of finite horizon can be proved in exactly the same way as the case of infinite horizon if the process (X_t) is replaced by the process (t, X_t) for $t \ge 0$. A proof in the case of infinite horizon can be given as follows.

The key is to show that V is superharmonic. For this, note that V is measurable (since it is lsc) and thus so is the mapping

$$V(X_\sigma) = \sup_\tau \mathsf{E}_{X_\sigma} G(X_\tau) \tag{2.2.65}$$

for any stopping time σ which is given and fixed. On the other hand, by the strong Markov property we have

$$\mathsf{E}_{X_\sigma} G(X_\tau) = \mathsf{E}_x\big(G(X_{\sigma+\tau\circ\theta_\sigma}) \,|\, \mathcal{F}_\sigma\big) \tag{2.2.66}$$

for every stopping time τ and $x \in E$. From (2.2.65) and (2.2.66) we see that

$$V(X_\sigma) = \operatorname*{ess\,sup}_{\tau} \mathsf{E}_x\big(G(X_{\sigma+\tau\circ\theta_\sigma}) \,|\, \mathcal{F}_\sigma\big) \tag{2.2.67}$$

under P_x where $x \in E$ is given and fixed.

Next we will show that the family

$$\big\{\, \mathsf{E}_x(X_{\sigma+\tau\circ\theta_\sigma} \,|\, \mathcal{F}_\sigma\big) : \tau \text{ is a stopping time} \big\} \tag{2.2.68}$$

is upwards directed in the sense of (1.1.25). Indeed, if τ_1 and τ_2 are stopping times given and fixed, set $\rho_1 = \sigma + \tau_1 \circ \theta_\sigma$ and $\rho_2 = \sigma + \tau_2 \circ \theta_\sigma$, and define

$$B = \big\{\, \mathsf{E}_x(X_{\rho_1} \,|\, \mathcal{F}_\sigma) \geq \mathsf{E}_x(X_{\rho_2} \,|\, \mathcal{F}_\sigma) \,\big\}. \tag{2.2.69}$$

Then $B \in \mathcal{F}_\sigma$ and the mapping

$$\rho = \rho_1 I_B + \rho_2 I_{B^c} \tag{2.2.70}$$

is a stopping time. To verify this let us note that $\{\rho \leq t\} = (\{\rho_1 \leq t\} \cap B) \cup (\{\rho_2 \leq t\} \cap B^c) = (\{\rho_1 \leq t\} \cap B \cap \{\sigma \leq t\}) \cup (\{\rho_2 \leq t\} \cap B^c \cap \{\sigma \leq t\}) \in \mathcal{F}_t$ since B and B^c belong to $\mathcal{F}_\sigma$ proving the claim. Moreover, the stopping time ρ can be written as

$$\rho = \sigma + \tau \circ \theta_\sigma \tag{2.2.71}$$

for some stopping time τ. Indeed, setting

$$A = \big\{\, \mathsf{E}_{X_0} G(X_{\tau_1}) \geq \mathsf{E}_{X_0} G(X_{\tau_2}) \,\big\} \tag{2.2.72}$$

we see that $A \in \mathcal{F}_0$ and $B = \theta_\sigma^{-1}(A)$ upon recalling (2.2.66). Hence from (2.2.70) we get

$$\begin{aligned} \rho &= (\sigma + \tau_1 \circ \theta_\sigma) I_B + (\sigma + \tau_2 \circ \theta_\sigma) I_{B^c} \\ &= \sigma + \big((\tau_1 \circ \theta_\sigma)(I_A \circ \theta_\sigma) + (\tau_2 \circ \theta_\sigma)(I_{A^c} \circ \theta_\sigma)\big) \\ &= \sigma + (\tau_1 I_A + \tau_2 I_{A^c}) \circ \theta_\sigma \end{aligned} \tag{2.2.73}$$

which implies that (2.2.71) holds with the stopping time $\tau = \tau_1 I_A + \tau_2 I_{A^c}$. (The latter is a stopping time since $\{\tau \leq t\} = (\{\tau_1 \leq t\} \cap A) \cup (\{\tau_2 \leq t\} \cap A^c) \in \mathcal{F}_t$ for all $t \geq 0$ due to the fact that $A \in \mathcal{F}_0 \subseteq \mathcal{F}_t$ for all $t \geq 0$.) Finally, we have

$$\begin{aligned} \mathsf{E}(X_\rho \,|\, \mathcal{F}_\sigma) &= \mathsf{E}(X_{\rho_1} \,|\, \mathcal{F}_\sigma) I_B + \mathsf{E}(X_{\rho_2} \,|\, \mathcal{F}_\sigma) I_{B^c} \\ &= \mathsf{E}(X_{\rho_1} \,|\, \mathcal{F}_\sigma) \vee \mathsf{E}(X_{\rho_2} \,|\, \mathcal{F}_\sigma) \end{aligned} \tag{2.2.74}$$

proving that the family (2.2.68) is upwards directed as claimed.

From the latter using (1.1.25) and (1.1.26) we can conclude that there exists a sequence of stopping times $\{\tau_n : n \geq 1\}$ such that

$$V(X_\sigma) = \lim_{n\to\infty} \mathsf{E}_x\big(G(X_{\sigma+\tau_n\circ\theta_\sigma}) \,|\, \mathcal{F}_\sigma\big) \tag{2.2.75}$$

where the sequence $\{ \mathsf{E}_x(G(X_{\sigma+\tau_n\circ\theta_\sigma}) \mid \mathcal{F}_\sigma) : n \geq 1 \}$ is increasing P_x-a.s. By the monotone convergence theorem using (2.2.1) above we can therefore conclude

$$\mathsf{E}_x V(X_\sigma) = \lim_{n\to\infty} \mathsf{E}_x G(X_{\sigma+\tau_n\circ\theta_\sigma}) \leq V(x) \tag{2.2.76}$$

for all stopping times σ and all $x \in E$. This proves that V is superharmonic. (Note that the only a priori assumption on V used so far is that V is measurable.) As evidently V is the smallest superharmonic function which dominates G on E (recall (2.2.14) above) we see that the remaining claims of the corollary follow directly from Theorem 2.7 above. This completes the proof. □

Remark 2.10. Note that the assumption of lsc on V and usc on G is natural, since the supremum of lsc functions defines an lsc function, and since every usc function attains its supremum on a compact set. To illustrate the former claim note that if the function

$$x \mapsto \mathsf{E}_x G(X_\tau) \tag{2.2.77}$$

is continuous (or lsc) for every stopping time τ, then $x \mapsto V(x)$ is lsc and the results of Corollary 2.9 are applicable. This yields a powerful existence result by simple means (both in finite and infinite horizon). We will exploit the latter in our study of finite horizon problems in Chapters VI–VIII below. On the other hand, if X is a one-dimensional diffusion, then V is continuous whenever G is measurable (see Subsection 9.3 below). Note finally that if X_t converges to X_∞ as $t \to \infty$ then there is no essential difference between infinite and finite horizon, and the second half of Corollary 2.9 above (Finite horizon) applies in this case as well, no matter if τ_D is finite or not. In the latter case one sees that τ_D is an optimal Markov time (recall Example 1.14 above).

Remark 2.11. Theorems 2.4 and 2.7 above have shown that the optimal stopping problem (2.2.2) is equivalent to the problem of finding the smallest superharmonic function $\widehat{V}$ which dominates G on E. Once $\widehat{V}$ is found it follows that $V = \widehat{V}$ and τ_D from (2.2.6) is optimal (if no obvious contradiction arises).

There are two traditional ways for finding $\widehat{V}$:

(i) *Iterative procedure* (constructive but non-explicit),

(ii) *Free-boundary problem* (explicit or non-explicit).

Note that Corollary 2.9 and Remark 2.10 present yet another way for finding $\widehat{V}$ simply by identifying it with V when the latter is known to be sufficiently regular (lsc).

The book [196, Ch. 3] provides numerous examples of (i) under various conditions on G and X. For example, it is known that if G is lsc and $\mathsf{E}_x \inf_{t\geq 0} G(X_t) > -\infty$ for all $x \in E$, then $\widehat{V}$ can be computed as follows:

$$Q_n G(x) := G(x) \vee \mathsf{E}_x G(X_{1/2^n}), \tag{2.2.78}$$

$$\widehat{V}(x) = \lim_{n\to\infty} \lim_{N\to\infty} Q_n^N G(x) \tag{2.2.79}$$

for $x \in E$ where Q_n^N is the N-th power of Q_n. The method of proof relies upon discretization of the time set $\mathbb{R}_+$ and making use of discrete-time results of optimal stopping reviewed in Subsection 1.2 above. It follows that

$$\text{If } G \text{ is continuous and } X \text{ is a Feller process, then } V \text{ is lsc.} \tag{2.2.80}$$

The present book studies various examples of (ii). The basic idea (following from the results of Theorems 2.4 and 2.7) is that $\widehat{V}$ and C (or D) should solve the free-boundary problem:

$$\mathbb{L}_X \widehat{V} \le 0 \quad (\widehat{V} \text{ minimal}), \tag{2.2.81}$$

$$\widehat{V} \ge G \quad (\widehat{V} > G \text{ on } C \ \& \ \widehat{V} = G \text{ on } D) \tag{2.2.82}$$

where $\mathbb{L}_X$ is the characteristic (infinitesimal) operator of X (cf. Chapter II below).

Assuming that G is smooth in a neighborhood of ∂C the following "rule of thumb" is valid. If X after starting at ∂C enters immediately into int(D) (e.g. when X is a diffusion process and ∂C is sufficiently nice) then the condition (2.2.81) (under (2.2.82) above) splits into the two conditions:

$$\mathbb{L}_X \widehat{V} = 0 \quad \text{in } C, \tag{2.2.83}$$

$$\frac{\partial \widehat{V}}{\partial x}\Big|_{\partial C} = \frac{\partial G}{\partial x}\Big|_{\partial C} \quad (\textit{smooth fit}). \tag{2.2.84}$$

On the other hand, if X after starting at ∂C does not enter immediately into int(D) (e.g. when X has jumps and no diffusion component while ∂C may still be sufficiently nice) then the condition (2.2.81) (under (2.2.82) above) splits into the two conditions:

$$\mathbb{L}_X \widehat{V} = 0 \quad \text{in } C, \tag{2.2.85}$$

$$\widehat{V}\big|_{\partial C} = G\big|_{\partial C} \quad (\textit{continuous fit}). \tag{2.2.86}$$

A more precise meaning of these conditions will be discussed in Chapter IV below (and through numerous examples throughout).

Remark 2.12. (Linear programming) A *linear programming problem* may be defined as the problem of maximizing or minimizing a linear function subject to linear constraints.

Optimal stopping problems may be viewed as linear programming problems (cf. [55, p. 107]). Indeed, we have seen in Theorems 2.4 and 2.7 that the optimal stopping problem (2.2.2) is equivalent to finding the smallest superharmonic function $\hat{V}$ which dominates G on E. Letting L denote the linear space of all superharmonic functions, letting the *constrained set* be defined by $L_G = \{V \in L : V \ge G\}$, and letting the *objective function* be defined by

$F(V) = V$ for $V \in L$, the optimal stopping problem (2.2.2) is equivalent to the linear programming problem

$$\hat{V} = \inf_{V \in L_G} F(V)\,. \tag{2.2.87}$$

Clearly, this formulation/interpretation extends to the martingale setting of Section 2.1 (where instead of superharmonic functions we need to deal with supermartingales) as well as to discrete time of both martingale and Markovian settings (Sections 1.1 and 1.2). Likewise, the free-boundary problem (2.2.81)–(2.2.82) may be viewed as a linear programming problem.

A *dual problem* to the *primal problem* (2.2.87) can be obtained using the fact that the first hitting time τ_* of $\hat{S}_t = \hat{V}(X_t)$ to $G_t = G(X_t)$ is optimal, so that

$$\sup_t \,(G_t - \hat{S}_t) = 0 \tag{2.2.88}$$

since $\hat{S}_t \ge G_t$ for all t. It follows that

$$\inf_S \mathsf{E} \sup_t \,(G_t - S_t) = 0 \tag{2.2.89}$$

where the infimum is taken over all supermartingales S satisfying $S_t \ge G_t$ for all t. (Note that (2.2.89) holds without the expectation sign as well.) Moreover, the infimum in (2.2.89) can equivalently be taken over all supermartingales S such that $\mathsf{E}S_0 = \mathsf{E}\hat{S}_0$ (where we recall that $\mathsf{E}\hat{S}_0 = \sup_\tau \mathsf{E}G_\tau$). Indeed, this follows since by the supermartingale property we have $\mathsf{E}S_{\tau_*} \le \mathsf{E}S_0$ so that

$$\mathsf{E} \sup_t \,(G_t - S_t) \ge \mathsf{E}(G_{\tau_*} - S_{\tau_*}) \ge \mathsf{E}G_{\tau_*} - \mathsf{E}S_0 = \mathsf{E}G_{\tau_*} - \mathsf{E}\hat{S}_0 = 0\,. \tag{2.2.90}$$

Finally, since $(\hat{S}_{t\wedge\tau_*})_{t\ge 0}$ is a martingale, we see that (2.2.89) can also be written as

$$\inf_M \mathsf{E} \sup_t \,(G_t - M_t) = 0 \tag{2.2.91}$$

where the infimum is taken over all martingales M satisfying $\mathsf{E}M_0 = \mathsf{E}\hat{S}_0$. In particular, the latter claim can be rewritten as

$$\sup_\tau \mathsf{E}G_\tau = \inf_M \mathsf{E} \sup_t \,(G_t - M_t) \tag{2.2.92}$$

where the infimum is taken over all martingales M satisfying $\mathsf{E}M_0 = 0$.

Notes. Optimal stopping problems originated in Wald's sequential analysis [216] representing a method of statistical inference (sequential probability ratio test) where the number of observations is not determined in advance of the experiment (see pp. 1–4 in the book for a historical account). Snell [206] formulated a general optimal stopping problem for discrete-time stochastic processes (sequences), and using the methods suggested in the papers of Wald & Wolfowitz [219] and Arrow, Blackwell & Girshick [5], he characterized the solution by means of the smallest supermartingale (called *Snell's envelope)* dominating the gain sequence. Studies in this direction (often referred to as *martingale methods*) are summarized in [31].

The key equation $V(x) = \max(G(x), \mathsf{E}_x V(X_1))$ was first stated explicitly in [5, p. 219] (see also the footnote on page 214 in [5] and the book [18, p. 253]) but was already characterized implicitly by Wald [216]. It is the simplest equation of "dynamic programming" developed by Bellman (cf. [15], [16]). This equation is often referred to as the *Wald–Bellman equation* (the term which we use too) and it was derived in the text above by a dynamic programming principle of "backward induction". For more details on optimal stopping problems in the discrete-time case see [196, pp. 111–112].

Following initial findings by Wald, Wolfowitz, Arrow, Blackwell and Girshick in discrete time, studies of sequential testing problems for *continuous-time* processes (including Wiener and Poisson processes) was initiated by Dvoretzky, Kiefer & Wolfowitz [51], however, with no advance to optimal stopping theory.

A transparent connection between optimal stopping and free-boundary problems first appeared in the papers by Mikhalevich [135] and [136] where he used the "principle of smooth fit" in an ad hoc manner. In the beginning of the 1960's several authors independently (from each other and from Mikhalevich) also considered free-boundary problems (with "smooth-fit" conditions) for solving various problems in sequential analysis, optimal stopping, and optimal stochastic control. Among them we mention Chernoff [29], Lindley [126], Shiryaev [187], [188], [190], Bather [10], Whittle [222], Breakwell & Chernoff [22] and McKean [133]. While in the papers from the 1940's and 50's the 'stopped' processes were either sums of independent random variables or processes with independent increments, the 'stopped' processes in these papers had a more general Markovian structure.

Dynkin [52] formulated a general optimal stopping problem for Markov processes and characterized the solution by means of the smallest superharmonic function dominating the gain function. Dynkin treated the case of discrete time in detail and indicated that the analogous results also hold in the case of continuous time. (For a connection of these results with Snell's results [206] see the corresponding remark in [52].)

The 1960's and 70's were years of an intensive development of the general theory of optimal stopping both in the "Markovian" and "martingale" setting as well as both in the discrete and continuous time. Among references dealing mainly with continuous time we mention [191], [88], [87], [193], [202], [210], [194], [184], [117], [62], [63], [211], [59], [60], [61], [141]. The book by Shiryaev [196] (see also [195]) provides a detailed presentation of the general theory of optimal stopping in the "Markovian" setting both for discrete and continuous time. The book by Chow, Robbins & Siegmund [31] gives a detailed treatment of optimal stopping problems for general stochastic processes in discrete time using the "martingale" approach. The present Chapter I is largely based on results exposed in these books and other papers quoted above. Further developments of optimal stopping following the 1970's and extending to more recent times will be addressed in the present monograph. Among those not mentioned explicitly below we refer to [105] and [153] for optimal stopping of diffusions, [171] and [139] for diffusions with jumps, [120] and [41] for passage from discrete to continuous time, and [147] for optimal stopping with delayed information. The facts of dual problem (2.2.88)–(2.2.92) were used by a number of authors in a more or less disguised form (see [36], [13], [14], [176], [91], [95]).

Remark on terminology. In general theory of *Markov processes* the term 'stopping time' is less common and one usually prefers the term '*Markov time*' (see e.g. [53]) originating from the fact that the strong Markov property remains preserved for such times. Nevertheless in general theory of stochastic processes, where the strong Markov property is not primary, one mostly uses the term '*stopping*' (or 'optional') time allowing it to take either finite or infinite values. In the present monograph we deal with both Markov processes and processes of general structure, and we are mainly interested in optimal stopping problems for which the *finite* stopping times are of central interest. This led us to use the "combined" terminology reserving the term 'Markov' for all and 'stopping' for finite times (the latter corresponding to "real stopping" before the "end of time").

Chapter II.

Stochastic processes: A brief review

From the table of contents of the monograph one sees that the basic processes we deal with are

MARTINGALES

(and related processes — supermartingales, submartingales, local martingales, semimartingales, etc.) and

MARKOV PROCESSES.

We will mainly be interested in the case of continuous time. The case of discrete time can be considered as its particular case (by embedding). However, we shall consider the case of discrete time separately because of its simplicity in comparison with the continuous-time case where there are many "technical" difficulties of the measure-theoretic character (for example, the existence of "good" modifications, and similar).

3. Martingales

3.1. Basic definitions and properties

1. At the basis of all our probability considerations a crucial role belongs to the notion of *stochastic basis*

$$(\Omega, \mathcal{F}, (\mathcal{F}_t)_{t\ge 0}, \mathsf{P}) \tag{3.1.1}$$

which is a probability space $(\Omega, \mathcal{F}, \mathsf{P})$ equipped with an additional structure $(\mathcal{F}_t)_{t\ge 0}$ called 'filtration'. A filtration is a *nondecreasing* (conforming to tradition we say "increasing") and *right-continuous* family of sub- σ -algebras of $\mathcal{F}$ (in other words $\mathcal{F}_s \subseteq \mathcal{F}_t$ for all $0 \le s \le t$ and $\mathcal{F}_t = \bigcap_{s>t} \mathcal{F}_s$ for all $t \ge 0$). We interpret $\mathcal{F}_t$ as the "information" (a family of events) obtained during the time interval $[0, t]$.

Without further mention we assume that the stochastic basis $(\Omega, \mathcal{F}, (\mathcal{F}_t)_{t\geq 0}, \mathsf{P})$ satisfies the *usual* conditions i.e. the σ-algebra $\mathcal{F}$ is P-complete and every $\mathcal{F}_t$ contains all P-null sets from $\mathcal{F}$.

Instead of the term 'stochastic basis' one often uses the term 'filtered probability space'.

If $X = (X_t)_{t\geq 0}$ is a family of random variables defined on $(\Omega, \mathcal{F})$ taking values in some measurable space $(E, \mathcal{E})$ (i.e. such that E-valued variables $X_t = X_t(\omega)$ are $\mathcal{F}/\mathcal{E}$-measurable for each $t \geq 0$) then one says that $X = (X_t)_{t\geq 0}$ is a *stochastic* (*random*) *process* with values in E.

Very often it is convenient to consider the process X as a *random element* with values in E^T where $T = [0, \infty)$. From such a point of view a trajectory $t \rightsquigarrow X_t(\omega)$ is an element i.e. "point" in E^T for each $\omega \in \Omega$.

All processes X considered in the monograph will be assumed to have their trajectories *continuous* ($X \in C$, the space of continuous functions) or right-continuous for $t \geq 0$ with left-hand limits for $t > 0$ ($X \in D$, the space of càdlàg functions; the French abbreviation càdlàg means *c*ontinu *à d*roite avec des *l*imites *à g*auche).

As usual we assume that for each $t \geq 0$ the random variable $X_t = X_t(\omega)$ is $\mathcal{F}_t$-measurable. To emphasize this property we often use the notation $X = (X_t, \mathcal{F}_t)_{t\geq 0}$ or $X = (X_t, \mathcal{F}_t)$ and say that the process X is *adapted* to the filtration $(\mathcal{F}_t)_{t\geq 0}$ (or simply adapted).

2. Throughout the monograph a key role belongs to the notion of a Markov time, i.e. a random variable, say $\tau = \tau(\omega)$, with values in $[0, \infty]$ such that

$$\{\omega : \tau(\omega) \leq t\} \in \mathcal{F}_t \tag{3.1.2}$$

for all $t \geq 0$.

If $\tau(\omega) < \infty$ for all $\omega \in \Omega$ or P-almost everywhere, then the Markov time τ is said to be finite. Usually such Markov times are called *stopping times*.

The property (3.1.2) has clear meaning: for each $t \geq 0$ a decision "to stop or not to stop" depends only on the "past and present information" $\mathcal{F}_t$ obtained on the interval $[0, t]$ and not depending on the "future".

With the given process $X = (X_t)_{t\geq 0}$ and a Markov time τ we associate a "stopped" process

$$X^\tau = (X_{t\wedge\tau})_{t\geq 0} \tag{3.1.3}$$

where $X_{t\wedge\tau} = X_{t\wedge\tau(\omega)}(\omega)$. It is clear that $X^\tau = X$ on the set $\{\omega : \tau(\omega) = \infty\}$.

If trajectories of the process $X = (X_t, \mathcal{F}_t)$ belong to the space D, then variables $X_\tau I(\tau < \infty)$ are $\mathcal{F}_\tau$-measurable where by definition the σ-algebra

$$\mathcal{F}_\tau = \big\{A \in \mathcal{F} : \{\tau \leq t\} \cap A \in \mathcal{F} \text{ for all } t \geq 0\big\}. \tag{3.1.4}$$

The notion of the stopped process plays a crucial role in defining the notions of "local classes" and "localization procedure".

Namely, let $\mathcal{X}$ be some class of processes. We say that a process X belongs to the "localized" class $\mathcal{X}_{\text{loc}}$ if there exists an increasing sequence $(\tau_n)_{n\geq 1}$ of Markov times (depending on X) such that $\lim_n \tau_n = \infty$ P-a.s. and each stopped process X^{τ_n} belongs to $\mathcal{X}$. The sequence $(\tau_n)_{n\geq 1}$ is called a *localizing sequence* for X (relative to $\mathcal{X}$).

3. The process $X = (X_t, \mathcal{F}_t)_{t\geq 0}$ is called a *martingale* [respectively *supermartingale* or *submartingale*] if $X \in D$ and

$\mathsf{E}|X_t| < \infty$ for $t \geq 0$; (3.1.5)

$\mathsf{E}(X_t \mid \mathcal{F}_s) = [\,\leq$ or $\geq\,]\, X_t$ for $s \leq t$ (a martingale [respectively supermartingale or submartingale] property). (3.1.6)

We denote the classes of martingales, supermartingales and submartingales by $\mathcal{M}$, $\sup\mathcal{M}$ and $\operatorname{sub}\mathcal{M}$, respectively. It is clear that if $X \in \sup\mathcal{M}$ then $-X \in \operatorname{sub}\mathcal{M}$. The processes $X = (X_t, \mathcal{F}_t)_{t\geq 0}$ which belong to the classes $\mathcal{M}_{\text{loc}}$, $(\sup\mathcal{M})_{\text{loc}}$ and $(\operatorname{sub}\mathcal{M})_{\text{loc}}$ are called local martingales, local supermartingales and local submartingales, respectively.

The notion of a local martingale is important for definition of a semimartingale.

We say that the process $X = (X_t, \mathcal{F}_t)_{t\geq 0}$ with càdlàg trajectories is a *semimartingale* ($X \in \operatorname{Semi}\mathcal{M}$) if this process admits a representation (generally speaking, not unique) of the form

$$X = X_0 + M + A \tag{3.1.7}$$

where X_0 is a finite-valued and $\mathcal{F}_0$-measurable random variable, $M = (M_t, \mathcal{F}_t)$ is a local martingale ($M \in \mathcal{M}_{\text{loc}}$) and $A = (A_t, \mathcal{F}_t)$ is a process of bounded variation ($A \in \mathcal{V}$), i.e. $\int_0^t |dA_s(\omega)| < \infty$ for $t > 0$ and $\omega \in \Omega$.

4. In the whole semimartingale theory the notion of *predictability* plays also (together with notions of Markov time, martingales, etc.) an essential role, being some kind of stochastic "determinancy".

Consider a space $\Omega \times \mathbb{R}_+ = \{(\omega, t) : \omega \in \Omega,\, t \geq 0\}$ and a process $Y = (Y_t(\omega), \mathcal{F}_t)_{t\geq 0}$ with left-continuous (càg = continuité à gauche) trajectories.

The *predictable* σ-algebra is the algebra $\mathcal{P}$ on $\Omega \times \mathbb{R}_+$ that is generated by all càg adapted processes Y considered as mappings $(\omega, t) \rightsquigarrow Y_t(\omega)$ on $\Omega \times \mathbb{R}_+$.

One may define $\mathcal{P}$ in the following equivalent way: the σ-algebra $\mathcal{P}$ is generated by

(i) the system of sets

$$\begin{aligned} &A \times \{0\} \ \text{ where } \ A \in \mathcal{F}_0 \ \text{ and} \\ &A \times (s,t] \ \text{ where } \ A \in \mathcal{F}_s \ \text{ for } \ 0 \le s \le t \end{aligned} \tag{3.1.8}$$

or

(ii) the system of

$$\begin{aligned} &\text{sets } \ A \times \{0\} \ \text{ where } \ A \in \mathcal{F}_0 \ \text{ and} \\ &\text{stochastic intervals } \ [\![0,\tau]\!] = \{(\omega,t) : 0 \le t \le \tau(\omega)\} \end{aligned} \tag{3.1.9}$$

where τ are Markov times.

Every adapted process $X = (X_t(\omega))_{t\ge 0}$, $\omega \in \Omega$, which is $\mathcal{P}$-measurable is called a *predictable process*.

5. The following theorem plays a fundamental role in stochastic calculus. In particular, it provides a famous example of a semimartingale. Recall that a process X belongs to the Dirichlet class (D) if the family of random variables $\{X_\tau : \tau \text{ is a finite stopping time}\}$ is uniformly integrable.

Theorem 3.1. (Doob–Meyer decomposition)
(a) *Every submartingale* $X = (X_t, \mathcal{F}_t)_{t\ge 0}$ *admits the decomposition*

$$X_t = X_0 + M_t + A_t, \qquad t \ge 0 \tag{3.1.10}$$

where $M \in \mathcal{M}_{\mathrm{loc}}$ *and* $A = (A_t, \mathcal{F}_t)_{t\ge 0}$ *is an increasing predictable locally integrable process* ($A \in \mathcal{P} \cap \mathcal{A}^+_{\mathrm{loc}}$).

(b) *Every submartingale* $X = (X_t, \mathcal{F}_t)_{t\ge 0}$ *of the Dirichlet class* (D) *admits the decomposition* $X_t = X_0 + M_t + A_t$, $t \ge 0$, *with a uniformly integrable martingale* M *and an increasing predictable integrable process* A ($\in \mathcal{P} \cap \mathcal{A}^+$).

Given decompositions are unique up to stochastic indistinguishability (i.e. if $X_t = X_0 + M'_t + A'_t$ *is another decomposition of the same type, then the sets* $\{\omega : \exists t \text{ with } M_t(\omega) \ne M'_t(\omega)\}$ *and* $\{\omega : \exists t \text{ with } A_t(\omega) \ne A'_t(\omega)\}$ *are* P*-null.*

Note two important corollaries of the Doob–Meyer decomposition.

(A) Every predictable local martingale $M = (M_t, \mathcal{F}_t)_{t\ge 0}$ with $M_0 = 0$ which at the same time belongs to the class $\mathcal{V}$ ($M \in \mathcal{M}_{\mathrm{loc}} \cap \mathcal{V}$) is equal to zero (up to stochastic indistinguishability).

(B) Suppose that a process $A \in \mathcal{A}^+_{\mathrm{loc}}$. Then there exists a process $\widetilde{A} \in \mathcal{P} \cap \mathcal{A}^+_{\mathrm{loc}}$ (unique up to stochastic indistinguishability) such that $A - \widetilde{A} \in \mathcal{M}_{\mathrm{loc}}$. The process $\widetilde{A}$ is called a *compensator* of the process A. This process may also be characterized as a predictable increasing process $\widetilde{A}$ such that for any (finite) stopping time τ one has

$$\mathsf{E} A_\tau = \mathsf{E} \widetilde{A}_\tau \tag{3.1.11}$$

or as a predictable increasing process $\widetilde{A}$ such that

$$\mathsf{E}\,(H \cdot A)_\infty = \mathsf{E}\,(H \cdot \widetilde{A})_\infty \tag{3.1.12}$$

for any non-negative increasing predictable process $H = (H_t, \mathcal{F}_t)_{t\ge 0}$. (Here $(H{\cdot}A)_t$ is the Lebesgue–Stieltjes integral $\int_0^t H_s(\omega)\, dA_s(\omega)$ for $\omega \in \Omega$.)

6. The notions introduced above for the case of continuous time admit the corresponding analogues also for the case of discrete time.

Here the stochastic basis is a filtered probability space $(\Omega, \mathcal{F}, (\mathcal{F}_n)_{n\ge 0}, \mathsf{P})$ with a family of σ-algebras $(\mathcal{F}_n)_{n\ge 0}$ such that $\mathcal{F}_0 \subseteq \mathcal{F}_1 \subseteq \cdots \subseteq \mathcal{F}$. (The notion of right-continuity loses its meaning in the case of discrete time.) The notions of martingales and related processes are introduced in a similar way.

With every "discrete" stochastic basis $(\Omega, \mathcal{F}, (\mathcal{F}_n)_{n\ge 0}, \mathsf{P})$ one may associate a "continuous" stochastic basis $(\Omega, \mathcal{F}, (\widehat{\mathcal{F}}_t)_{t\ge 0}, \mathsf{P})$ by setting $\widehat{\mathcal{F}}_t = \mathcal{F}_{[t]}$ for $t \ge 0$. In particular $\widehat{\mathcal{F}}_n = \mathcal{F}_n$, $\widehat{\mathcal{F}}_{n-} = \widehat{\mathcal{F}}_{n-1} = \mathcal{F}_{n-1}$ for $n \ge 1$.

In a similar way with any process $X = (X_n, \mathcal{F}_n)_{n\ge 0}$ we may associate the corresponding process $\widehat{X} = (\widehat{X}_t, \widehat{\mathcal{F}}_t)_{t\ge 0}$ in continuous time by setting $\widehat{X}_t = X_{[t]}$ for $t \ge 0$.

Observe also that the notion of predictability of a process $X = (X_n)_{n\ge 0}$ on the stochastic basis $(\Omega, \mathcal{F}, (\mathcal{F}_n)_{n\ge 0}, \mathsf{P})$ takes a very simple form: X_n is $\mathcal{F}_{n-1}$-measurable for each $n \ge 1$.

7. Along with the σ-algebra $\mathcal{P}$ of predictable sets on $\Omega \times \mathbb{R}_+$ an important role in the general theory of stochastic processes belongs to the notion of *optional* σ-algebra $\mathcal{O}$ that is a minimal σ-algebra generated by all adapted processes $Y = (Y_t(\omega))_{t\ge 0}$, $\omega \in \Omega$ (considered as a mapping $(\omega, t) \rightsquigarrow Y_t(\omega)$) with right-continuous (for $t \ge 0$) trajectories which have left-hand limits (for $t > 0$).

It is clear that $\mathcal{P} \subseteq \mathcal{O}$. If a process $X = (X_t, \mathcal{F}_t)$ is $\mathcal{O}$-measurable then we say that this process is *optional.* With an agreement that all our processes (martingales, supermartingales, etc.) have càdlàg trajectories, we see that they are optional. The important property of such processes is the following: for any open set B from $\mathcal{B}(\mathbb{R})$ the random variable

$$\tau = \inf\{\, t \ge 0 : X_t \in B \,\} \tag{3.1.13}$$

is a Markov time (we put $\inf \varnothing = \infty$ as usual). Note that this property also holds for adapted càd processes.

Another important property of optional processes X is the following: any "stopped" process $X^\tau = (X_{t\wedge\tau}, \mathcal{F}_t)$ where τ is a Markov time is optional again and the random variable $X_\tau I(\tau < \infty)$ is $\mathcal{F}_\tau$-measurable.

All adapted càd processes $X = (X_t, \mathcal{F}_t)$ are such that for each $t > 0$ the set

$$\{(\omega, s) : \omega \in \Omega, s \leq t \text{ and } X_s(\omega) \in B\} \in \mathcal{F}_t \otimes \mathcal{B}([0,t]) \tag{3.1.14}$$

where $B \in \mathcal{B}(\mathbb{R})$. This property is called the "property of progressive measurability". The importance of such processes may be demonstrated by the fact that then the process

$$I_t(\omega) = \int_0^t f(X_s(\omega))\, ds, \qquad t \geq 0 \tag{3.1.15}$$

for a measurable bounded function f will be adapted, i.e. $I_t(\omega)$ is $\mathcal{F}_t$-measurable for each $t \geq 0$.

8. Although in the monograph we deal mainly with continuous processes it will be useful to consider the structure of general local martingales from the standpoint of its decomposition into "continuous" and "discontinuous" components.

We say that two local martingales M and N are *orthogonal* if their product MN is a local martingale. A local martingale M is called *purely discontinuous* if $M_0 = 0$ and M is orthogonal to all continuous local martingales.

First decomposition of a local martingale $M = (M_t, \mathcal{F}_t)_{t \geq 0}$ states that there exists a unique (up to indistinguishability) decomposition

$$M = M_0 + M^c + M^d \tag{3.1.16}$$

where $M_0^c = M_0^d = 0$, M^c is a continuous local martingale, and M^d is a purely discontinuous one.

Second decomposition of a local martingale $M = (M_t, \mathcal{F}_t)_{t \geq 0}$ states that M admits a (non-unique) decomposition

$$M = M_0 + M' + M'' \tag{3.1.17}$$

where M' and M'' are local martingales with $M'_0 = M''_0 = 0$, M' has finite variation and $|\Delta M''| \leq a$ (i.e. $|\Delta M''_t| \leq a$ for all $t > 0$ where $\Delta M''_t = M''_t - M''_{t-}$).

From this decomposition we conclude that every local martingale can be written as a sum of a local martingale of bounded variation and a locally square-integrable martingale (because M'' is a process of such a type).

With each pair M and N of locally square-integrable martingales ($M, N \in \mathcal{M}^2_{\mathrm{loc}}$) one can associate (by the Doob–Meyer decomposition) a predictable process $\langle M, N \rangle$ of bounded variation such that

$$MN - \langle M, N \rangle \in \mathcal{M}_{\mathrm{loc}}. \tag{3.1.18}$$

If $N = M$ then we get $M^2 - \langle M \rangle \in \mathcal{M}_{\mathrm{loc}}$ where $\langle M \rangle$ stands for the increasing predictable process $\langle M, M \rangle$.

The process $\langle M,N\rangle$ is called the *predictable quadratic covariation* and the "angle bracket" process $\langle M\rangle$ is called the *predictable quadratic variation* or *quadratic characteristic.*

9. Let $X=(X_t,\mathcal{F}_t)$ be a semimartingale ($X\in \mathrm{Semi}\,\mathcal{M}$) with a decomposition

$$X_t = X_0 + M_t + A_t, \qquad t\ge 0 \tag{3.1.19}$$

where $M=(M_t,\mathcal{F}_t)\in\mathcal{M}_{\mathrm{loc}}$ and $A=(A_t,\mathcal{F}_t)\in\mathcal{V}$.

In the class $\mathrm{Semi}\,\mathcal{M}$ of semimartingales a special role is played by the class $\text{Sp-Semi}\,\mathcal{M}$ of *special semimartingales* i.e. semimartingales X for which one can find a decomposition $X=X_0+M+A$ with predictable process A of bounded variation. If we have also another decomposition $X=X_0+M'+A'$ with *predictable* A' then $A=A'$ and $M=M'$. So the "predictable" decomposition of a semimartingale is unique.

The typical example of a special semimartingale is a semimartingale with bounded jumps ($|\Delta X|\le a$). For such semimartingales one has $|\Delta A|\le a$ and $|\Delta M|\le 2a$. In particular, if X is a continuous semimartingale then A and M are also continuous.

Every local martingale M has the following property: for all $t>0$,

$$\sum_{s\le t}|\Delta M_s|^2<\infty \quad \mathsf{P}\text{-a.s.} \tag{3.1.20}$$

Because in a semimartingale decomposition $X=X_0+M+A$ the process A has bounded variation, we have for any $t>0$,

$$\sum_{s\le t}|\Delta X_s|^2<\infty \quad \mathsf{P}\text{-a.s.} \tag{3.1.21}$$

10. From the first decomposition of a local martingale $M=M^c+M^d$ with $M_0=0$ we conclude that if a semimartingale X has a decomposition $X=X_0+M+A$ then

$$X = X_0 + M^c + M^d + A. \tag{3.1.22}$$

If moreover $X=X_0+\bar{M}^c+\bar{M}^d+\bar{A}$ is another decomposition then

$$(M^c-\bar{M}^c)+(M^d-\bar{M}^d)=\bar{A}-A. \tag{3.1.23}$$

Since the process $\bar{A}-A\in\mathcal{V}$ it follows that the process $(M^c-\bar{M}^c)+(M^d-\bar{M}^d)\in\mathcal{V}$.

But every local martingale, which at the same time is a process from the class $\mathcal{V}$, is purely discontinuous. So the continuous local martingale $M^c-\bar{M}^c$ is at the same time purely discontinuous and therefore P-a.s. equal to zero, i.e. $M^c=\bar{M}^c$.

In other words the continuous martingale component of a semimartingale is defined uniquely. It explains why this component of X is usually denoted by X^c.

3.2. Fundamental theorems

There are three fundamental results in the martingale theory ("three pillars of martingale theory"):

A. The optional sampling theorem;
B. Martingale convergence theorem;
C. Maximal inequalities.

The basic statements here belong to J. Doob and there are many different modifications of his results.

Let us state basic results from A, B and C.

A. The optional sampling theorem

(A1) *Doob's stopping time theorem.* Suppose that $X = (X_t, \mathcal{F}_t)_{t\ge 0}$ is a submartingale (martingale) and τ is a Markov time. Then the "stopped" process $X^\tau = (X_{t\wedge\tau}, \mathcal{F}_t)_{t\ge 0}$ is also a submartingale (martingale).

(A2) *Hunt's stopping time theorem.* Let $X = (X_t, \mathcal{F}_t)_{t\ge 0}$ be a submartingale (martingale). Assume that $\sigma = \sigma(\omega)$ and $\tau = \tau(\omega)$ are bounded stopping times and $\sigma(\omega) \le \tau(\omega)$ for $\omega \in \Omega$. Then

$$X_\sigma \le (=)\, \mathsf{E}(X_\tau \mid \mathcal{F}_\sigma) \quad \mathsf{P}\text{-a.s.} \tag{3.2.1}$$

The statements of these theorems remain valid also for unbounded stopping times under the additional assumption that the family of random variables $\{X_t : t \ge 0\}$ is uniformly integrable.

The results (A1) and (A2) are particular cases of the following general proposition:

(A3) Let $X = (X_t, \mathcal{F}_t)_{t\ge 0}$ be a submartingale (martingale) and let σ and τ be two stopping times for which

$$\mathsf{E}|X_\sigma| < \infty, \quad \mathsf{E}|X_\tau| < \infty \quad \text{and} \quad \liminf_{t\to\infty} \mathsf{E}\, I(\tau > t)|X_t| = 0. \tag{3.2.2}$$

Then on the set $\{\tau \ge \sigma\}$

$$\mathsf{E}(X_\tau \mid \mathcal{F}_\sigma) \ge (=)\, X_\sigma \quad \mathsf{P}\text{-a.s.} \tag{3.2.3}$$

If in addition $\mathsf{P}(\tau \ge \sigma) = 1$ then

$$\mathsf{E}\, X_\tau \ge (=)\, \mathsf{E}\, X_\sigma. \tag{3.2.4}$$

If $B = (B_t)_{t \geq 0}$ is a standard Brownian motion and τ is a stopping time, then from (A3) one may obtain the *Wald identities*:

$$\mathsf{E} B_\tau = 0 \quad \text{if } \mathsf{E}\sqrt{\tau} < \infty, \tag{3.2.5}$$

$$\mathsf{E} B_\tau^2 = \mathsf{E}\tau \quad \text{if } \mathsf{E}\tau < \infty. \tag{3.2.6}$$

B. Martingale convergence theorem

(B1) *Doob's convergence theorem.* Let $X = (X_t, \mathcal{F}_t)_{t \geq 0}$ be a submartingale with

$$\sup_t \mathsf{E}|X_t| < \infty \quad (\text{equivalently: } \sup_t \mathsf{E} X_t^+ < \infty\,). \tag{3.2.7}$$

Then there exists an $\mathcal{F}_\infty$-measurable random variable X_∞ (where $\mathcal{F}_\infty \equiv \sigma(\bigcup_{t \geq 0} \mathcal{F}_t)$) with $\mathsf{E}|X_\infty| < \infty$ such that

$$X_t \to X_\infty \quad \mathsf{P}\text{-a.s.} \quad \text{as } t \to \infty\,. \tag{3.2.8}$$

If the condition (3.2.7) is strengthened to uniform integrability, then the P-a.s. convergence in (3.2.7) also takes place in L^1, i.e.:

(B2) If the family $\{X_t : t \geq 0\}$ is uniformly integrable, then

$$\mathsf{E}|X_t - X_\infty| \to 0 \quad \text{as } t \to \infty. \tag{3.2.9}$$

(B3) *Lévy's convergence theorem.* Let $(\Omega, \mathcal{F}, (\mathcal{F}_t)_{t \geq 0}, \mathsf{P})$ be a stochastic basis and ξ an integrable $\mathcal{F}$-measurable random variable. Put $\mathcal{F}_\infty = \sigma\big(\bigcup_{t \geq 0} \mathcal{F}_t\big)$. Then P-a.s. and in L^1,

$$\mathsf{E}(\xi \mid \mathcal{F}_t) \to \mathsf{E}(\xi \mid \mathcal{F}_\infty) \quad \text{as } t \to \infty. \tag{3.2.10}$$

C. Maximal inequalities

The following two classical inequalities of Kolmogorov and Khintchine gave rise to the field of the so-called 'martingale inequalities' (in probability and in mean) for random variables of type

$$\sup_{t \leq T} X_t, \quad \sup_{t \leq T} |X_t| \qquad \text{and} \qquad \sup_{t \geq 0} X_t, \quad \sup_{t \geq 0} |X_t|. \tag{3.2.11}$$

(C1) *Kolmogorov's inequalities.* Suppose that $S_n = \xi_1 + \dots + \xi_n$, $n \geq 1$, where $\xi_1, \xi_2, \dots$ are independent random variables with $\mathsf{E}\xi_k = 0$, $\mathsf{E}\xi_k^2 < \infty$, $k \geq 1$. Then for any $\varepsilon > 0$ and arbitrary $n \geq 1$,

$$\mathsf{P}\Big(\max_{1 \leq k \leq n} |S_k| \geq \varepsilon\Big) \leq \frac{\mathsf{E} S_n^2}{\varepsilon^2}. \tag{3.2.12}$$

If additionally $\mathsf{P}(|\xi_k| \leq c) = 1$, $k \geq 1$, then

$$\mathsf{P}\Big(\max_{1 \leq k \leq n} |S_k| \geq \varepsilon\Big) \geq 1 - \frac{(c+\varepsilon)^2}{\mathsf{E} S_n^2}. \tag{3.2.13}$$

(C2) *Khintchine's inequalities.* Suppose that $\xi_1, \xi_2, \dots$ are independent Bernoulli random variables with $\mathsf{P}(\xi_j = 1) = \mathsf{P}(\xi_j = -1) = 1/2$, $j \geq 1$, and (c_j) is a sequence of real numbers. Then for any $p > 0$ and any $n \geq 1$,

$$A_p\biggl(\sum_{j=1}^{n} c_j^2\biggr)^{p/2} \leq \mathsf{E}\biggl|\sum_{j=1}^{n} c_j\xi_j\biggr|^p \leq B_p\biggl(\sum_{j=1}^{n} c_j^2\biggr)^{p/2} \tag{3.2.14}$$

where A_p and B_p are some universal constants.

Note that in the inequalities of Kolmogorov and Khintchine the sequences $(S_n)_{n\geq 1}$ and $\bigl(\sum_{j=1}^{n} c_j\xi_j\bigr)_{n\geq 1}$ form martingales.

Generalizations of these inequalities are the following inequalities.

(C3) *Doob's inequalities (in probability).* Let $X = (X_t, \mathcal{F}_t)_{t\geq 0}$ be a submartingale. Then for any $\varepsilon > 0$ and each $T > 0$,

$$\mathsf{P}\Bigl(\sup_{t\leq T} X_t \geq \varepsilon\Bigr) \leq \frac{1}{\varepsilon}\,\mathsf{E}\Bigl[X_T^+\, I\Bigl(\sup_{t\leq T} X_t \geq \varepsilon\Bigr)\Bigr] \leq \frac{1}{\varepsilon}\,\mathsf{E}\,X_T^+ \tag{3.2.15}$$

and

$$\mathsf{P}\Bigl(\sup_{t\leq T} |X_t| \geq \varepsilon\Bigr) \leq \frac{1}{\varepsilon}\,\sup_{t\leq T} \mathsf{E}\,|X_t|. \tag{3.2.16}$$

If $X = (X_t, \mathcal{F}_t)_{t\geq 0}$ is a martingale then for all $p \geq 1$,

$$\mathsf{P}\Bigl(\sup_{t\leq T} |X_t| \geq \varepsilon\Bigr) \leq \frac{1}{\varepsilon^p}\,\mathsf{E}\,|X_T|^p \tag{3.2.17}$$

and, in particular, for $p = 2$,

$$\mathsf{P}\Bigl(\sup_{t\leq T} |X_t| \geq \varepsilon\Bigr) \leq \frac{1}{\varepsilon^2}\,\mathsf{E}\,|X_T|^2. \tag{3.2.18}$$

(C4) *Doob's inequalities (in mean).* Let $X = (X_t, \mathcal{F}_t)_{t\geq 0}$ be a non-negative submartingale. Then for $p > 1$ and any $T > 0$,

$$\mathsf{E}\,X_T^p \leq \mathsf{E}\Bigl(\sup_{t\leq T} X_t\Bigr)^p \leq \Bigl(\frac{p}{p-1}\Bigr)^p\,\mathsf{E}\,X_T^p \tag{3.2.19}$$

and for $p = 1$

$$\mathsf{E}\,X_T \leq \mathsf{E}\,\sup_{t\leq T} X_t \leq \frac{e}{e-1}\Bigl[1 + \mathsf{E}\,(X_T \log^+ X_T)\Bigr]. \tag{3.2.20}$$

In particular, if $X = (X_t, \mathcal{F}_t)_{t\geq 0}$ is a square-integrable martingale, then

$$\mathsf{E}\,\sup_{t\leq T} X_t^2 \leq 4\,\mathsf{E}\,X_T^2. \tag{3.2.21}$$

(C5) *Burkholder–Davis–Gundy's inequalities.* Suppose that $X = (X_t, \mathcal{F}_t)_{t\geq 0}$ is a martingale. Then for each $p \geq 1$ there exist universal constants A_p^* and B_p^* such that for any stopping time T,

$$A_p^* \,\mathsf{E}\,[X]_T^{p/2} \leq \mathsf{E} \sup_{t\leq T} |X_t|^p \leq B_p^* \,\mathsf{E}\,[X]_T^{p/2} \tag{3.2.22}$$

where $[X]_t = \langle X^c\rangle_t + \sum_{s\leq t}(\Delta X_s)^2$ is the quadratic variation of X.

In the case $p > 1$ these inequalities are equivalent (because of Doob's inequalities in mean for $p > 1$) to the following inequalities:

$$A_p \,\mathsf{E}\,[X]_T^{p/2} \leq \mathsf{E}\,|X_T|^p \leq B_p \,\mathsf{E}\,[X]_T^{p/2} \tag{3.2.23}$$

with some universal constants A_p and B_p (whenever $\mathsf{E}\,[X]_T^{p/2} < \infty$).

3.3. Stochastic integral and Itô's formula

1. The class of semimartingales is rather wide and rich because it is invariant with respect to many transformations — "stopping", "localization", "change of time", "absolute continuous change of measure", "change of filtration", etc. It is remarkable and useful that for a semimartingale X the notion of stochastic integral $H \cdot X$, that is a cornerstone of stochastic calculus, may be defined for a very large class of integrands H.

Let us briefly give the basic ideas and ways of construction of the stochastic integral.

Suppose that a function $H = (H_t)_{t\geq 0}$ is "very simple":

$$H = \begin{cases} YI_{[\![0]\!]} & \text{where } Y \text{ is } \mathcal{F}_0\text{-measurable,} \\ \text{or} & \\ YI_{]\!]r,s]\!]} & \text{where } Y \text{ is } \mathcal{F}_r\text{-measurable} \end{cases} \tag{3.3.1}$$

with $[\![0]\!] = \{(\omega, t) : \omega \in \Omega\,,\, t = 0\}$ and $]\!]r, s]\!] = \{(\omega, t) : \omega \in \Omega\,,\, r < t \leq s\}$ for $r < s$.

For such "very simple" functions a natural definition of the stochastic integral $H \cdot X = \{(H \cdot X)_t : t \geq 0\}$ should apparently be the following:

$$(H \cdot X)_t = \begin{cases} 0 & \text{if } H = YI_{[\![0]\!]}, \\ Y(X_{s\wedge t} - X_{r\wedge t}) & \text{if } H = YI_{]\!]r,s]\!]}. \end{cases} \tag{3.3.2}$$

By linearity one can extend this definition to the class of "simple" functions which are linear combination of "very simple" functions.

It is more interesting that the *stochastic integral* $H \cdot X$ defined in such a way can be extended to the class of

$$\textit{locally bounded predictable processes } H \tag{3.3.3}$$

so that the following properties remain valid:

(a) the process $H \cdot X$ is càdlàg;

(b) the mapping $H \rightsquigarrow H \cdot X$ is linear (i.e. $(aH' + H'') \cdot X = aH' \cdot X + H'' \cdot X$) up to stochastic indistinguishability;

(c) if a sequence (H^n) of predictable processes converges pointwise uniformly on $[0,t]$ for each $t > 0$ to a predictable process H and $|H^n| \le K$, where K is a locally bounded predictable process, then

$$\sup_{s \le t} |(H^n \cdot X)_s - (H \cdot X)_s| \xrightarrow{\mathsf{P}} 0 \quad \text{for each } t > 0\,. \tag{3.3.4}$$

The stochastic integral $H \cdot X$ constructed has many natural properties which are usually associated with the notion of 'integral':

(1) the mapping $H \rightsquigarrow H \cdot X$ is linear;

(2) the process $H \cdot X$ is a semimartingale;

(3) if $X \in \mathcal{M}_{\text{loc}}$ then $H \cdot X \in \mathcal{M}_{\text{loc}}$;

(4) if $X \in \mathcal{V}$ then $H \cdot X \in \mathcal{V}$;

(5) $(H \cdot X)_0 = 0$ and $H \cdot X = H \cdot (X - X_0)$ i.e. the stochastic integral is not "sensitive" to the initial value X_0;

(6) $\Delta(H \cdot X) = H\Delta X$;

(7) the stopped process $X^\tau = (X_t^\tau)_{t \ge 0}$ can be written in the form

$$X_t^\tau = X_0 + (I_{[\![0,\tau]\!]} \cdot X)_t. \tag{3.3.5}$$

Note that very often we use a more transparent notation $\int_0^t H_s\, dX_s$ for the stochastic integral $(H \cdot X)_t$.

One can also extend the stochastic integral to a class of predictable processes which are not locally bounded. (Note that some "natural" properties of the type $X \in \mathcal{M}_{\text{loc}} \implies H \cdot X \in \mathcal{M}_{\text{loc}}$ may fail to hold.)

To present this extension we need a series of new notions.

Let $X = (X_t, \mathcal{F}_t)$ be a semimartingale with a continuous martingale part X^c. Because $X^c \in \mathcal{M}^2_{\text{loc}}$ the predictable process $\langle X^c \rangle$ (called the *quadratic characteristic* of X) such that

$$(X^c)^2 - \langle X^c \rangle \in \mathcal{M}_{\text{loc}} \tag{3.3.6}$$

does exist. Define

$$[X] = \langle X^c \rangle + \sum_{s \leq \cdot} (\Delta X_s)^2. \tag{3.3.7}$$

The latter process (called the *quadratic variation* of the semimartingale X) can also be defined in terms of the stochastic integral $H \cdot X$ introduced above by taking $H_t = X_{t-}$ for $t > 0$. Moreover,

$$X^2 - X_0^2 - 2X_- \cdot X = [X]. \tag{3.3.8}$$

(Sometimes this identity is taken as a definition of $[X]$.)

Suppose that $X = X_0 + A + M$ is a decomposition of X with $A \in \mathcal{V}$, $M \in \mathcal{M}_{\text{loc}}$, and let H be a predictable process.

We say that

$$H \in L_{\text{var}}(A) \tag{3.3.9}$$

if $\int_0^t |H_s(\omega)| \, d(\text{Var}(A))_s < \infty$ for all $t > 0$ and $\omega \in \Omega$.

We also say that

$$H \in L'_{\text{loc}}(M) \tag{3.3.10}$$

if the process

$$\left(\left(\int_0^t H_s^2 \, d[M]_s \right)^{1/2} \right)_{t \geq 0} \in \mathcal{A}^+_{\text{loc}} \tag{3.3.11}$$

i.e. is locally integrable.

Finally, we say that

$$H \in L(X) \tag{3.3.12}$$

if *one can find* a decomposition $X = X_0 + A + M$ such that $H \in L_{\text{var}}(A) \cap L'_{\text{loc}}(M)$.

For functions $H \in L(X)$ by definition the stochastic integral is set to be

$$H \cdot X = H \cdot A + H \cdot M \tag{3.3.13}$$

where $H \cdot A$ is the Lebesgue–Stieltjes integral and $H \cdot M$ is an integral with respect to the local martingale M, which for functions $H \in L'_{\text{loc}}(M)$ is defined via a limiting procedure using the well-defined stochastic integrals $H^n \cdot X$ for bounded predictable functions $H^n = HI(|H| \leq n)$.

Let us emphasize that the definition of $H \cdot X$ given above assumes the existence of a decomposition $X = X_0 + A + M$ such that $H \in L_{\text{var}}(A) \cap L'_{\text{loc}}(M)$.

At a first glance this definition could appear a little strange. But its correctness is obtained from the following result: if there exists another decomposition $X = X_0 + A' + M'$ such that

$$H \in \big(L_{\mathrm{var}}(A) \cap L'_{\mathrm{loc}}(M)\big) \cap \big(L_{\mathrm{var}}(A') \cap L'_{\mathrm{loc}}(M')\big) \tag{3.3.14}$$

then

$$H \cdot A + H \cdot M = H \cdot A' + H \cdot M' \tag{3.3.15}$$

i.e. both definitions lead to the same (up to indistinguishability) integral.

2. Along with the quadratic variation $[X]$, an important role in stochastic calculus belongs to the notion of the *quadratic covariation* $[X, Y]$ of two semimartingales X and Y that is defined by

$$[X, Y] = \langle X^c, Y^c \rangle + \sum_{s \le \cdot} \Delta X_s \Delta Y_s . \tag{3.3.16}$$

Here $\langle X^c, Y^c \rangle$ is a predictable quadratic covariation of two continuous martingales X^c and Y^c i.e. $\langle X^c, Y^c \rangle = \frac{1}{4}(\langle X^c + Y^c \rangle - \langle X^c - Y^c \rangle)$.

The quadratic covariation has the following properties:

(a) If $X \in \mathcal{M}_{\mathrm{loc}}$ then $[X, X]^{1/2} \in \mathcal{A}_{\mathrm{loc}}$.

(b) If $X \in \mathcal{M}_{\mathrm{loc}}$ is continuous and $Y \in \mathcal{M}_{\mathrm{loc}}$ is purely discontinuous, then $[X, Y] = 0$.

(c) If $X, Y \in \mathcal{M}^c_{\mathrm{loc}}$ are orthogonal (i.e. $XY \in \mathcal{M}^c_{\mathrm{loc}}$) then $[X, Y] = \langle X, Y \rangle = 0$.

(d) If X and Y are semimartingales and H is a bounded predictable process, then $[H \cdot X, Y] = H \cdot [X, Y]$.

(e) If X is a continuous (purely discontinuous) local martingale, and H is a locally bounded predictable process, then $H \cdot X$ is a continuous (purely discontinuous) local martingale.

3. One of the central results of stochastic calculus is the celebrated *Itô formula.*

Let $X = (X^1, \dots, X^d)$ be a d-dimensional semimartingale (whose components are semimartingales). Suppose that a function $F \in C^2$ and denote by $D_i F$ and $D_{ij} F$ partial derivatives $\frac{\partial F}{\partial x_i}$ and $\frac{\partial^2 F}{\partial x_i \, \partial x_j}$. Then the process $Y = F(X)$ is a semimartingale again and the following *Itô's change-of-variable formula* holds:

$$\begin{aligned} F(X_t) = F(X_0) &+ \sum_{i\le d} D_i F(X_-)\cdot X^i \\ &+ \frac{1}{2}\sum_{i,j\le d} D_{i,j}F(X_-)\cdot\langle X^{i,c},X^{j,c}\rangle \\ &+ \sum_{s\le t}\Big(F(X_s)-F(X_{s-})-\sum_{i\le d} D_iF(X_{s-})\Delta X^i_s\Big). \end{aligned} \tag{3.3.17}$$

Let us note some particular cases of this formula and some useful corollaries.

A) If $X=(X^1,\dots,X^d)$ is a continuous semimartingale then

$$F(X_t) = F(X_0) + \sum_{i\le d} D_iF(X)\cdot X^i + \frac{1}{2}\sum_{i,j\le d} D_{i,j}F(X)\cdot\langle X^i,X^j\rangle. \tag{3.3.18}$$

B) If X and Y are two continuous semimartingales then the following *formula of integration by parts* holds:

$$X_tY_t = X_0Y_0 + \int_0^t X_s\,dY_s + \int_0^t Y_s\,dX_s + \langle X,Y\rangle_t. \tag{3.3.19}$$

In particular, we have

$$X_t^2 = X_0^2 + 2\int_0^t X_s\,dX_s + \langle X\rangle_t. \tag{3.3.20}$$

4. From Itô's formula for functions $F\in C^2$ one can get by limiting procedures its extensions for functions F satisfying less restrictive assumptions than $F\in C^2$.

(I) The first result in this direction was the *Tanaka* (or *Itô–Tanaka*) formula for a Brownian motion $X=B$ and function $F(x)=|x-a|$:

$$|B_t-a| = |B_0-a| + \int_0^t \operatorname{sgn}(B_s-a)\,dB_s + L_t^a \tag{3.3.21}$$

where L_t^a is the local time that the Brownian notion $B=(B_t)_{t\ge0}$ "spends" at the level a:

$$L_t^a = \lim_{\varepsilon\downarrow0}\frac{1}{2\varepsilon}\int_0^t I(|B_s-a|\le\varepsilon)\,ds. \tag{3.3.22}$$

(II) The second result was the *Itô–Tanaka–Meyer formula*: if the derivative $F'(x)$ is a function of bounded variation then

$$F(B_t) = F(B_0) + \int_0^t F'(B_s)\,dB_s + \frac{1}{2}\int_{\mathbb{R}} L_t^a\,F''(da) \tag{3.3.23}$$

where $F''(da)$ is defined as a sign measure on $\mathbb{R}$ in the sense of distributions.

Subsequently the following two formulae were derived:

(III) *The Bouleau–Yor formula* states that if the derivative $F'(x)$ is a locally bounded function then

$$F(B_t) = F(B_0) + \int_0^t F'(B_s)\,dB_s - \frac{1}{2}\int_{\mathbb{R}} F'(a)\,d_a L_t^a. \tag{3.3.24}$$

(IV) *The Föllmer–Protter–Shiryaev formula* states that if the derivative $F' \in L^2_{\mathrm{loc}}$, i.e. $\int_{|x|\le M}(F'(x))^2\,dx < \infty$ for all $M \ge 0$, then

$$F(B_t) = F(B_0) + \int_0^t F'(B_s)\,dB_s + \frac{1}{2}[F'(B),B]_t \tag{3.3.25}$$

where $[F'(B),B]$ is the quadratic covariation of $F'(B)$ and B:

$$\begin{aligned}&[F'(B),B]_t \\ &\quad = \mathsf{P}\text{-}\lim_{n\to\infty}\sum_k \Big(F'\big(B_{t_{k+1}^n\wedge t}\big) - F'\big(B_{t_k^n\wedge t}\big)\Big)\big(B_{t_{k+1}^n\wedge t} - B_{t_k^n\wedge t}\big)\end{aligned} \tag{3.3.26}$$

with $\sup_k(t_{k+1}^n \wedge t - t_k^n \wedge t) \to 0$.

Between the formulae (I)–(IV) there are the following relationships:

$$\text{(I)} \subseteq \text{(II)} \subseteq \text{(III)} \subseteq \text{(IV)} \tag{3.3.27}$$

in the sense of expanding to the classes of functions F to which the corresponding formulae are applicable.

There are some generalizations of the results of type (I)–(IV) for semimartingale case.

For example, if X is a *continuous* semimartingale then the Itô–Tanaka formula takes the following form similar to the case of a Brownian motion:

$$|X_t - a| = |X_0 - a| + \int_0^t \operatorname{sgn}(X_s - a)\,dX_s + L_t^a(X) \tag{3.3.28}$$

where

$$L_t^a(X) = \lim_{\varepsilon\downarrow 0}\frac{1}{2\varepsilon}\int_0^t I(|X_s| \le \varepsilon)\,d\langle X\rangle_s. \tag{3.3.29}$$

If a function $F = F(x)$ is concave (convex or the difference of the two) and X is a continuous semimartingale, then the Itô–Tanaka–Meyer formula takes the following form:

$$F(X_t) = F(X_0) + \int_0^t \tfrac{1}{2}\big(F'_+(X_s) + F'_-(X_s)\big)\,dX_s + \frac{1}{2}\int_{\mathbb{R}} L_t^a F''(da). \tag{3.3.30}$$

An important corollary of this formula is the following *occupation times formula*: If $\Phi = \Phi(x)$ is a positive Borel function then for every continuous semimartingale X

$$\int_0^t \Phi(X_s)\, d\langle X\rangle_s = \int_{\mathbb{R}} \Phi(a)\, d_a L_t^a. \tag{3.3.31}$$

For many problems of stochastic analysis (and, in particular, for optimal stopping problems) it is important to have analogues of Itô's formula for $F(t, X_t)$ where $F(t, x)$ is a continuous function whose derivatives in t and x are not continuous. A particular formula of this kind (derived by Peskir in [163]) will be given in Subsection 3.5 below (see (3.5.5) and (3.5.9)).

5. Stochastic canonical representation for semimartingales. (a) Probabilistic and analytic methods developed for semimartingales can be considered in some sense as a natural extension of the methods created in the theory of processes with independent increments. Therefore it is reasonable to recall some results of this theory.

A stochastic (random) process $X = (X_t, \mathcal{F}_t)$ is a *Process with Independent Increments* ($X \in$ PII) if $X_0 = 0$ and random variables $X_t - X_s$ for $t > s$ are independent from σ-algebras $\mathcal{F}_s$. Such process is called a *process with stationary independent increments* ($X \in$ PIIS) if distributions of $X_t - X_s$ depend only on difference $t - s$. Note that every deterministic process is a (degenerated) PII-process. In particular every deterministic function of unbounded variation is such a process and so it is not a semimartingale. We shall exclude this uninteresting case in order to stay within the semimartingale scheme. (The process X with independent increments is a semimartingale if and only if for each $\lambda \in \mathbb{R}$ the characteristic function $\varphi(t) = \mathsf{E}\, e^{i\lambda X_t}$, $t \ge 0$, is a function of bounded variation.)

It is well known that with every process $X \in$ PII one can associate a (deterministic) triplet (B, C, ν) where $B = (B_t)_{t\ge 0}$, $C = (C_t)_{t\ge 0}$, and $\nu = \nu((0,t]\times A)$ for $A \in \mathcal{B}(\mathbb{R}\setminus\{0\})$, $t \ge 0$, such that

$$\nu((0,t] \times A) = \mathsf{E}\, \mu(\omega; (0,t] \times A) \tag{3.3.32}$$

with the measure of jumps

$$\mu(\omega; (0,t] \times A) = \sum_{0<s\le t} I(\Delta X_s(\omega) \in A) \tag{3.3.33}$$

of the process X. One has $B_t \in \mathbb{R}$, $C_t \ge 0$ and $C_t \ge C_s$ for $t \ge s \ge 0$.

Measure ν has a special name — *Lévy measure* and satisfies the following property:

$$\int_{\mathbb{R}} \min(x^2 \wedge 1)\, \nu((0,t] \times dx) < \infty \tag{3.3.34}$$

for all $t > 0$.

This property enables us to introduce the *cumulant function*

$$K_t(\lambda) = i\lambda B_t - \frac{\lambda^2}{2}\, C_t + \int \bigl(e^{i\lambda x} - 1 - i\lambda h(x)\bigr)\, \nu((0,t] \times dx) \tag{3.3.35}$$

where $h = h(x)$ is a "truncation" function

$$h(x) = xI(|x| \le 1). \tag{3.3.36}$$

With this definition for each process $X \in \mathrm{PII}$ we have for the characteristic function

$$G_t(\lambda) = \mathsf{E}\, e^{i\lambda X_t}, \qquad \lambda \in \mathbb{R}, \quad t \ge 0 \tag{3.3.37}$$

the following equation:

$$dG_t(\lambda) = G_{t-}(\lambda)\, dK_t(\lambda), \qquad G_0(\lambda) = 1. \tag{3.3.38}$$

The solution of this equation is given by the formula

$$G_t(\lambda) = e^{K_t(\lambda)} \prod_{0<s\le t} (1 + \Delta K_s(\lambda)) e^{-\Delta K_s(\lambda)} \tag{3.3.39}$$

which gives a representation of the characteristic function $G_t(\lambda) = \mathsf{E}\, e^{i\lambda X_t}$ via cumulant function $K_t(\lambda)$.

The formula for $G_t(\lambda)$ takes the simple form:

$$G_t(\lambda) = e^{K_t(\lambda)} \tag{3.3.40}$$

in the case of continuous (in probability) processes X. In this case the functions B_t, C_t and $\nu = \nu((0,t] \times A)$ for $A \in \mathcal{B}(\mathbb{R} \setminus \{1\})$ are continuous in time, so that $\Delta K_t(\lambda) = 0$ which leads to the given formula (3.3.40).

For processes $X \in \mathrm{PII} \cap \mathrm{Semi}\,\mathcal{M}$ the following stochastic integral representation holds (sometimes also called the *Itô–Lévy representation*):

$$X_t = X_0 + B_t + X_t^c + \int_0^t \!\! \int h(x)\, d(\mu - \nu) + \int_0^t \!\! \int (x - h(x))\, d\mu \tag{3.3.41}$$

where $(X_t^c)_{t\ge 0}$ is a continuous Gaussian martingale with $\mathsf{E}\,(X_t^c)^2 = C_t$.

(b) It is interesting and quite remarkable that for semimartingales it is also possible to give analogues of the formulae given above.

Let us first introduce some notation.

Let $\mu = \mu(\omega; (0,t] \times A)$ be the measure of jumps of a semimartingale X, and let $\nu = \nu(\omega; (0,t] \times A)$ be a compensator of μ, which can be characterized as a predictable random measure such that for every non-negative predictable (in t for fixed $x \in \mathbb{R}$) function $W(\omega, t, x)$

$$\mathsf{E} \int_0^\infty \!\! \int_{\mathbb{R}} W(\omega, t, x)\, \mu(\omega; dt \times dx) = \mathsf{E} \int_0^\infty \!\! \int_{\mathbb{R}} W(\omega, t, x)\, \nu(\omega; dt \times dx). \tag{3.3.42}$$

Here the process

$$\left(\int_0^t \int_{\mathbb{R}} \min(x^2, 1)\, \nu(\omega; ds \times dx)\right) \in \mathcal{A}_{\mathrm{loc}}^+. \tag{3.3.43}$$

With every semimartingale X and the "truncation" function $h(x) = xI(|x| \le 1)$ one may associate a new semimartingale $\check{X}(h) = (X(h)_t)_{t\ge 0} = \sum_{s\le t}[\Delta X_s - h(\Delta X_s)]$ which is a sum of "big" jumps of X. Then process $X(h) = X - \check{X}(h)$ has bounded jumps and consequently is a special semimartingale which (as any process of such type) admits a decomposition

$$X(h) = X_0 + B(h) + M(h) \tag{3.3.44}$$

with a predictable process $B(h)$ and a local martingale $M(h)$.

Local martingale $M(h)$ can be written as the sum $M^c(h) + M^d(h)$ where $M^c(h)$ is a continuous local martingale and $M^d(h)$ is a purely discontinuous local martingale. Therefore

$$X = X_0 + B(h) + M^c(h) + M^d(h) + \check{X}(h). \tag{3.3.45}$$

For a given "truncation" function $h(x) = xI(|x| \le 1)$ (or any other bounded function $h = h(x)$ with a compact support and with linear behavior in the vicinity of $x = 0$) we obtain the following canonical representation of X:

$$X_t = X_0 + B_t + X_t^c + \int_0^t \int_{\mathbb{R}} h(x)\, d(\mu - \nu) + \int_0^t \int (x - h(x))\, d\mu \tag{3.3.46}$$

where we denoted $B = B(h)$ and $M^c(h) = X^c$.

Because $X^c \in \mathcal{M}_{\mathrm{loc}}^{c,2}$ we get by Doob–Meyer decomposition that there exists a predictable process $C = (C_t)_{t\ge 0}$ such that $(X^c)^2 - C \in \mathcal{M}_{\mathrm{loc}}^c$.

By analogy with the case of processes of the class PII we call the set (B, C, ν) a *triplet* of the predictable characteristics of the semimartingale X. Let us emphasize that for a process of the class PII a triplet (B, C, ν) consists of *deterministic* objects but for a general semimartingale the triplet (B, C, ν) consists of random objects of *predictable* character (that is some kind of "stochastic determinancy").

6. Now let X be a semimartingale with the triplet (B, C, ν). For this process, by analogy with (3.3.35), introduce a (predictable) cumulant process

$$K_t(\lambda) = i\lambda B_t - \frac{\lambda^2}{2} C_t + \int_{\mathbb{R}} \left(e^{i\lambda x} - 1 - i\lambda h(x)\right) \nu(\omega; (0, t] \times dx) \tag{3.3.47}$$

where $B_t = B_t(\omega)$, $C_t = C_t(\omega)$, $\nu(\omega; (0, t] \times dx)$ are predictable. For the process $(K_t(\lambda))_{t\ge 0}$ we consider the stochastic differential equation

$$d\mathcal{E}_t(\lambda) = \mathcal{E}_{t-}(\lambda)\, dK_t(\lambda), \qquad \mathcal{E}_0(\lambda) = 1. \tag{3.3.48}$$

The solution of this equation is given by the well-known stochastic exponential

$$\mathcal{E}_t(\lambda) = e^{K_t(\lambda)} \prod_{s\le t} (1 + \Delta K_s(\lambda))e^{-\Delta K_s(\lambda)}. \tag{3.3.49}$$

It is remarkable that in the case when $\Delta K_t(\lambda) \neq -1$, $t > 0$, the process

$$\left(\frac{e^{i\lambda X_t}}{\mathcal{E}_t(\lambda)}\right)_{t\ge 0} \in \mathcal{M}_{\text{loc}}. \tag{3.3.50}$$

This formula can be considered as a *semimartingale analogue* of the Kolmogorov–Lévy–Khintchine formula for processes with independent increments (see Subsection 4.6).

3.4. Stochastic differential equations

In the class of semimartingales $X = (X_t)_{t\ge 0}$ let us distinguish a certain class of processes which have (for the fixed truncation function h) the triplets of the following special form:

$$B_t(\omega) = \int_0^t b(s, X_s(\omega))\, ds, \tag{3.4.1}$$

$$C_t(\omega) = \int_0^t c(s, X_s(\omega))\, ds, \tag{3.4.2}$$

$$\nu(\omega; dt \times dy) = dt\, K_t(X_t(\omega); dy) \tag{3.4.3}$$

where b and c are Borel functions and $K_t(x; dy)$ is a *transition kernel* with $K_t(x; \{0\}) = 0$.

Such processes X are usually called *diffusion processes with jumps.* If $\nu = 0$ then the process X is called a *diffusion process.*

How can one construct such processes (starting with stochastic processes and measures of "simple" structure)?

By a process of "simple" structure we shall consider a standard Wiener process (also called a Brownian motion) and as a random measure we shall take a Poisson random measure $p(dt, dx)$ with the compensator (intensity) $q(dt, dx) = dt\, F(dx)$, $x \in \mathbb{R}$, where $F = F(dx)$ is a positive σ-finite measure. (It is assumed that these objects can be defined on the given stochastic basis.)

We shall consider stochastic differential equations of the following form:

$$\begin{aligned} dY_t = \beta(t, Y_t)\, dt + \gamma(t, Y_t)\, dW_t + h\big(\delta(t, Y_{t-}; z)\big)\big(p(dt, dz) - q(dt, dz)\big) \\ + h'\big(\delta(t, Y_{t-}; z)\big)\, p(dt, dz) \end{aligned} \tag{3.4.4}$$

where $h'(x) = x - h(x)$, $h(x) = xI(|x| \le 1)$ and $\beta(t, y)$, $\gamma(t, y)$, $\delta(t, y; z)$ are Borel functions. (Notice that if the measure p has a jump at a point (t, z) then

$\Delta Y_t = \delta(t, Y_{t-}; z)$.) Although the equation (3.4.4) is said to be "differential" one should, in fact, understand it in the sense that the corresponding integral equation holds:

$$\begin{aligned} Y_t = Y_0 &+ \int_0^t \beta(s, Y_s)\, ds + \int_0^t \gamma(s, Y_s)\, dW_s \\ &+ \int_0^t \int_{\mathbb{R}} h\big(\delta(s, Y_{s-}; z)\big)\big(p(ds, dz) - q(ds, dz)\big) \\ &+ \int_0^t \int_{\mathbb{R}} h'\big(\delta(s, Y_{s-}; z)\big)\, p(ds, dz). \end{aligned} \tag{3.4.5}$$

When considering the question about the existence and uniqueness of solutions to such equations it is common to distinguish two types of solutions:

(a) *solutions-processes* (*or strong solutions*) and
(b) *solutions-measures* (*or weak solutions*).

In the sequel in connection with the optimal stopping problems we shall consider only strong solutions. Therefore we cite below the results only for the case of strong solutions. (For more details about strong and weak solutions see [106] and [127]–[128].)

In the diffusion case when there is no jump component, the following classical result is well known.

Theorem 3.2. *Consider the stochastic differential equation*

$$dY_t = \beta(t, Y_t)\, dt + \gamma(t, Y_t)\, dW_t \tag{4.2$'$}$$

where $Y_0 = \text{const}$ *and the coefficients* $\beta(t, y)$ *and* $\gamma(t, y)$ *satisfy the* local Lipschitz condition *and the* condition of linear growth *respectively:*

(1) *for any* $n \geq 1$ *there exists a constant* θ_n *such that*

$$|\beta(s, y) - \beta(s, y')| \leq \theta_n |y - y'|, \quad |\gamma(s, y) - \gamma(s, y')| \leq \theta_n |y - y'| \tag{3.4.6}$$

for $s \leq n$ *and* $|y| \leq n$, $|y'| \leq n$; *and*

(2) *for any* $n \geq 1$

$$|\beta(s, y)| \leq \theta_n (1 + |y|), \quad |\gamma(s, y)| \leq \theta_n (1 + |y|) \tag{3.4.7}$$

for $s \leq n$ *and* $|y| \leq n$.

Then (*on any stochastic basis* $(\Omega, \mathcal{F}, (\mathcal{F}_t)_{t \geq 0}, \mathsf{P})$ *on which the Wiener process is defined*) *a strong solution* (*i.e. solution* $Y = (Y_t)_{t \geq 0}$ *such that* Y_t *is* $\mathcal{F}_t$ *-measurable for each* $t \geq 0$) *exists and is unique* (*up to stochastic indistinguishability*).

Theorem 3.3. *In the general case of equation* (3.4.4) (*in the presence of jumps*) *let us suppose, in addition to the assumptions of the previous theorem, that there exist functions* $\rho_n(x)$, $n \geq 1$, *with* $\int \rho_n^2(x)\, F(dx) < \infty$ *such that*

$$|h \circ \delta(s, y, x) - h \circ \delta(s, y', x)| \leq \rho_n(x)|y - y'|, \tag{3.4.8}$$
$$|h' \circ \delta(s, y, x) - h' \circ \delta(s, y', x)| \leq \rho_n^2(x)|y - y'|, \tag{3.4.9}$$
$$|h \circ \delta(s, y, x)| \leq \rho_n(x)(1 + |y|), \tag{3.4.10}$$
$$|h' \circ \delta(s, y, x)| \leq \big(\rho_n^2(x) \wedge \rho_n^4(x)\big)(1 + |y|) \tag{3.4.11}$$

for $s \leq n$ *and* $|y|, |y'| \leq n$.

Then (*on any stochastic basis* $(\Omega, \mathcal{F}, (\mathcal{F}_t)_{t\geq 0}, \mathsf{P})$ *on which the Wiener process and the Poisson random measure are defined*) *a strong solution exists and is unique* (*up to stochastic indistinguishability*).

3.5. A local time-space formula

Let $X = (X_t)_{t\geq 0}$ be a continuous semimartingale, let $c : \mathbb{R}_+ \to \mathbb{R}$ be a continuous function of bounded variation, and let $F : \mathbb{R}_+ \times \mathbb{R} \to \mathbb{R}$ be a continuous function satisfying:

$$F \quad \text{is} \quad C^{1,2} \quad \text{on} \quad \bar{C}_1, \tag{3.5.1}$$
$$F \quad \text{is} \quad C^{1,2} \quad \text{on} \quad \bar{C}_2 \tag{3.5.2}$$

where C_1 and C_2 are given as follows:

$$C_1 = \{\, (t, x) \in \mathbb{R}_+ \times \mathbb{R} \,:\, x > c(t) \,\}, \tag{3.5.3}$$
$$C_2 = \{\, (t, x) \in \mathbb{R}_+ \times \mathbb{R} \,:\, x < c(t) \,\}. \tag{3.5.4}$$

Then the following *change-of-variable formula* holds (for a proof see [163]):

$$\begin{aligned} F(t, X_t) = F(0, X_0) &+ \int_0^t \frac{1}{2}\Big(F_t(s, X_s+) + F_t(s, X_s-)\Big)\, ds \\ &+ \int_0^t \frac{1}{2}\Big(F_x(s, X_s+) + F_x(s, X_s-)\Big)\, dX_s \\ &+ \frac{1}{2}\int_0^t F_{xx}(s, X_s)\ I(X_s \neq c(s))\ d\langle X, X\rangle_s \\ &+ \frac{1}{2}\int_0^t \Big(F_x(s, X_s+) - F_x(s, X_s-)\Big)\, I(X_s = c(s))\ d\ell_s^c(X) \end{aligned} \tag{3.5.5}$$

where $\ell_s^c(X)$ is the local time of X at the curve c given by

$$\ell_s^c(X) = \mathsf{P}\text{-}\lim_{\varepsilon \downarrow 0} \frac{1}{2\varepsilon} \int_0^s I\big(c(r) - \varepsilon < X_r < c(r) + \varepsilon\big)\, d\langle X, X\rangle_r \tag{3.5.6}$$

and $d\ell_s^c(X)$ refers to integration with respect to the continuous increasing function $s \mapsto \ell_s^c(X)$.

Moreover, if X solves the stochastic differential equation

$$dX_t = b(X_t)\,dt + \sigma(X_t)\,dB_t \tag{3.5.7}$$

where b and σ are locally bounded and $\sigma \geq 0$, then the following condition:

$$\mathsf{P}\big(X_s = c(s)\big) = 0 \quad \text{for} \quad s \in (0,t] \tag{3.5.8}$$

implies that the first two integrals in (3.5.5) can be simplified to read

$$\begin{aligned} F(t,X_t) = F(0,X_0) &+ \int_0^t (F_t + \mathbb{L}_X F)(s,X_s)\, I(X_s \neq c(s))\, ds \\ &+ \int_0^t F_x(s,X_s)\, \sigma(X_s)\, I(X_s \neq c(s))\, dB_s \\ &+ \frac{1}{2}\int_0^t \Big(F_x(s,X_s+) - F_x(s,X_s-)\Big)\, I(X_s = c(s))\, d\ell_s^c(X) \end{aligned} \tag{3.5.9}$$

where $\mathbb{L}_X F = bF_x + (\sigma^2/2)F_{xx}$ is the action of the infinitesimal generator $\mathbb{L}_X$ on F.

Let us briefly discuss some extensions of the formulae (3.5.5) and (3.5.9) needed below. Assume that X solves (3.5.7) and satisfies (3.5.8), where $c : \mathbb{R}_+ \to \mathbb{R}$ is a continuous function of bounded variation, and let $F : \mathbb{R}_+ \times \mathbb{R} \to \mathbb{R}$ be a continuous function satisfying the following conditions instead of (3.5.1)–(3.5.2) above:

$$F \quad \text{is} \quad C^{1,2} \quad \text{on} \quad C_1 \cup C_2, \tag{3.5.10}$$

$$F_t + \mathbb{L}_X F \quad \text{is locally bounded,} \tag{3.5.11}$$

$$x \mapsto F(t,x) \quad \text{is convex,} \tag{3.5.12}$$

$$t \mapsto F_x(t,b(t)\pm) \quad \text{is continuous.} \tag{3.5.13}$$

Then it can be proved that the change-of-variable formula (3.5.9) still holds (cf. [163]). In this case, even if F_t is to diverge when the boundary c is approached within C_1, this deficiency is counterbalanced by a similar behaviour of F_{xx} through (3.5.11), and consequently the first integral in (3.5.9) is still well defined and finite. [When we say in (3.5.11) that $F_t + \mathbb{L}_X F$ is locally bounded, we mean that $F_t + \mathbb{L}_X F$ is bounded on $K \cap (C_1 \cup C_2)$ for each compact set in $\mathbb{R}_+ \times \mathbb{R}$.] The condition (3.5.12) can further be relaxed to the form where $F_{xx} = F_1 + F_2$ on $C_1 \cup C_2$ where F_1 is non-negative and F_2 is continuous on $\mathbb{R}_+ \times \mathbb{R}$. This will be referred to in Chapter VII as the relaxed form of (3.5.10)–(3.5.13). For more details on this and other extensions see [163]. For an extension of the change-of-variable formula (3.5.5) to *general semimartingales* (with jumps) and *local time on surfaces* see [166].

4. Markov processes

4.1. Markov sequences (chains)

1. A traditional approach to the notion of *Markov sequence* (*chain*) i.e. *discrete-time* Markov process—as well as a martingale approach—assumes that we are given a filtered probability space

$$(\Omega, \mathcal{F}, (\mathcal{F}_n)_{n\geq 0}, \mathsf{P}) \tag{4.1.1}$$

and a phase (state) space $(E, \mathcal{E})$, i.e. a measurable space E with a σ-algebra $\mathcal{E}$ of its subsets such that one-point sets $\{x\}$ belong to $\mathcal{E}$ for all $x \in E$.

A stochastic sequence $X = (X_n, \mathcal{F}_n)_{n\geq 0}$ is called a *Markov chain* (*in a wide sense*) if the random variables X_n are $\mathcal{F}_n/\mathcal{E}$-measurable and the following *Markov property* (*in a wide sense*) holds:

$$\mathsf{P}(X_{n+1} \in B \mid \mathcal{F}_n)(\omega) = \mathsf{P}(X_{n+1} \in B \mid X_n)(\omega) \quad \mathsf{P}\text{-a.s.} \tag{4.1.2}$$

for all $n \geq 0$ and $B \in \mathcal{E}$ (instead of $\mathsf{P}(X_{n+1} \in B \mid X_n)(\omega)$ one often writes $\mathsf{P}(X_{n+1} \in B \mid X_n(\omega))$).

When $\mathcal{F}_n = \mathcal{F}_n^X \equiv \sigma(X_0, X_1, \ldots, X_n)$, one calls the property (4.1.2) a *Markov property* (*in a strict sense*) and $X = (X_n)_{n\geq 0}$ a *Markov chain*.

From now on it is assumed that the phase space $(E, \mathcal{E})$ is Borel (see e.g. [106]). Under this assumption, it is well known (see [199, Chap. II, § 7]) that there exists a *regular* conditional distribution $P_n(x; B)$ such that (P-a.s.)

$$\mathsf{P}(X_n \in B \mid X_{n-1}(\omega)) = P_n(X_{n-1}(\omega); B), \qquad B \in \mathcal{E}, \quad n \geq 1. \tag{4.1.3}$$

In the Markov theory, functions $P_n(x; B)$ are called (Markov) *transition functions* (from E into $\mathcal{E}$) or *Markov kernels*. If $P_n(x; B)$, $n \geq 1$, do not depend on n ($= P(x; B)$), the Markov chain (in a wide or strict sense) is said to be *time-homogeneous*.

Besides a transition function, another important characteristic of a Markov chain is its *initial distribution* $\pi = \pi(B)$, $B \in \mathcal{E}$:

$$\pi(B) = \mathsf{P}(X_0 \in B). \tag{4.1.4}$$

It is clear that the collection $(\pi, P_1, P_2, \ldots)$ (or (π, P) in the time-homogeneous case) determines uniquely the probability distribution i.e. $\mathrm{Law}(X \mid \mathsf{P})$ of a Markov sequence $X = (X_0, X_1, \ldots)$.

2. In Chapter I (Section 1.2), when exposing results on optimal stopping, we have not taken a traditional but more "up-to-date" approach based on the idea that the object to start with is neither $(\Omega, \mathcal{F}, (\mathcal{F}_n)_{n\geq 0}, \mathsf{P})$ nor $X = (X_0, X_1, \ldots)$, but a collection of "transition functions" $(P_1, P_2, \ldots)$, $P_n = P_n(x; B)$, which map E into $\mathcal{E}$ where $(E, \mathcal{E})$ is a phase space. (In the "time-homogeneous" case one has to fix only one transition function $P = P(x; B)$.)

Starting from the collection $(P_1, P_2, \ldots)$ one can construct a family of probability measures $\{P_x : x \in E\}$ on the space $(\Omega, \mathcal{F}) = (E^\infty, \mathcal{E}^\infty)$ (e.g. by the Ionescu-Tulcea theorem) with respect to which the sequence $X = (X_0, X_1, \ldots)$, such that $X_n(\omega) = x_n$ if $\omega = (x_0, x_1, \ldots)$, is a Markov chain (in a strict sense) for each fixed $x \in E$, and for which $P_x(X_0 = x) = 1$ (the Markov chain starts at x). If $\pi = \pi(B)$ is a certain "initial" distribution we denote by P_π the new distribution given by $P_\pi(A) = \int_E P_x(A)\, \pi(dx)$ for $A \in \mathcal{E}^\infty$.

Relative to P_π it is natural to call the sequence X a Markov chain with the initial distribution π (i.e. $P_\pi(X_0 \in B) = \pi(B)$, $B \in \mathcal{E}$).

3. To expose the theory of Markov chains the following notions of shift operator θ and their iterations θ_n and θ_τ (τ is a Markov time) prove to be very useful.

An operator $\theta : \Omega \to \Omega$ is called a *shift operator* if for each $\omega = (x_0, x_1, \ldots)$

$$\theta(\omega) = (x_1, x_2, \ldots) \tag{4.1.5}$$

or in other words

$$(x_0, x_1, \ldots) \xrightarrow{\theta} (x_1, x_2, \ldots) \tag{4.1.6}$$

(i.e. θ shifts the trajectory $(x_0, x_1, \ldots)$ to the left for one position).

Let $\theta_0 = \mathbb{I}$ where $\mathbb{I}$ is the unit (identical) transformation (i.e. $\theta_0(\omega) = \omega$). We define the n-th ($n \geq 1$) iteration θ_n of an operator θ by the formula

$$\theta_n = \theta_{n-1} \circ \theta \quad (= \theta \circ \theta_{n-1}) \tag{4.1.7}$$

i.e. $\theta_n(\omega) = \theta_{n-1}(\theta(\omega))$.

If $\tau = \tau(\omega)$ is a Markov time ($\tau(\omega) \leq \infty$), one denotes by θ_τ the operator which acts only on the set $\Omega_\tau = \{\omega : \tau(\omega) < \infty\}$ so that $\theta_\tau = \theta_n$ if $\tau = n$, i.e. for all ω such that $\tau(\omega) = n$ one has

$$\theta_\tau(\omega) = \theta_n(\omega). \tag{4.1.8}$$

If $H = H(\omega)$ is an $\mathcal{F}$-measurable function (e.g. $\tau = \tau(\omega)$ or $X_m = X_m(\omega)$) one denotes by $H \circ \theta_n$ the function

$$(H \circ \theta_n)(\omega) = H(\theta_n(\omega)). \tag{4.1.9}$$

For an $\mathcal{F}$-measurable function $H = H(\omega)$ and a Markov time $\sigma = \sigma(\omega)$ one defines $H \circ \theta_\sigma$ only on the set $\Omega_\sigma = \{\omega : \sigma(\omega) < \infty\}$ and in such a way that if $\sigma(\omega) = n$ then

$$H \circ \theta_\sigma = H \circ \theta_n, \tag{4.1.10}$$

i.e. if $\omega \in \{\sigma(\omega) = n\}$ then

$$(H \circ \theta_\sigma)(\omega)(H \circ \theta_n)(\omega) = H(\theta_n(\omega)). \tag{4.1.11}$$

In particular

$$X_m \circ \theta_n = X_{m+n}, \qquad X_m \circ \theta_\sigma = X_{m+\sigma} \quad (\text{on } \Omega_\sigma\,) \tag{4.1.12}$$

and for finite Markov times τ and σ,

$$X_\tau \circ \theta_\sigma = X_{\tau \circ \theta_\sigma + \sigma}. \tag{4.1.13}$$

With operators $\theta_n : \Omega \to \Omega$ one can associate the inverse operators $\theta_n^{-1} : \mathcal{F} \to \mathcal{F}$ acting in such a way that if $A \in \mathcal{F}$ then

$$\theta_n^{-1}(A) = \{\omega : \theta_n(\omega) \in A\}. \tag{4.1.14}$$

In particular, if $A = \{\omega : X_m(\omega) \in B\}$ with $B \in \mathcal{E}$, then

$$\theta_n^{-1}(X_m^{-1}(B)) = X_{m+n}^{-1}(B). \tag{4.1.15}$$

4. The Markov property (4.1.2) in the case $\mathcal{F}_n = \mathcal{F}_n^X$ (i.e. the Markov property in a strict sense) and $\mathsf{P} = \mathsf{P}_x$, $x \in E$, can be written in a little bit more general form:

$$\mathsf{P}_x\big(X_{n+m} \in B \mid \mathcal{F}_n^X\big)(\omega) = \mathsf{P}_{X_n(\omega)}(X_m \in B) \quad \mathsf{P}_x\text{-a.s.} \tag{4.1.16}$$

If we use the notation of (4.1.3) above and put $H(\omega) = I_B(X_m(\omega))$ where $I_B(x)$ is the indicator of the set B, then, because of

$$(H \circ \theta_n)(\omega) = H\big(\theta_n(\omega)\big) = I_B\big(X_m(\theta_n(\omega))\big) = I_B\big(X_{n+m}(\omega)\big) \tag{4.1.17}$$

we get that

$$\mathsf{E}_x(H \circ \theta_n \mid \mathcal{F}_n^X)(\omega) = \mathsf{E}_{X_n(\omega)} H \quad \mathsf{P}_x\text{-a.s.} \tag{4.1.18}$$

From this, using standard "monotone class" arguments one obtains the following *generalized Markov property* which is very useful: if $H = H(\omega)$ is a bounded (or non-negative) $\mathcal{F}$-measurable function, then for *any* initial distribution π and for any $n \ge 0$ and $x \in E$,

$$\mathsf{E}_\pi(H \circ \theta_n \mid \mathcal{F}_n^X)(\omega) = \mathsf{E}_{X_n(\omega)} H \quad \mathsf{P}_x\text{-a.s.} \tag{4.1.19}$$

It is worth emphasizing that $\mathsf{E}_{X_n(\omega)} H$ should be understood as follows: first we take $\psi(x) = \mathsf{E}_x H$ and then, by definition, assume that $\mathsf{E}_{X_n(\omega)} H = \psi(X_n(\omega))$.

Integrating both sides of (4.1.16) we get the celebrated *Chapman–Kolmogorov* (or *Kolmogorov–Chapman*) *equations* (see the original papers [28] and [111]):

$$\mathsf{P}_x(X_{n+m} \in B) = \int_E \mathsf{P}_y(X_m \in B)\, \mathsf{P}_x(X_n \in dy) \tag{4.1.20}$$

for $x \in E$ and $B \in \mathcal{E}$.

5. The property (4.1.16) admits further a very useful generalization called the *strong Markov property* which is formulated as follows. Let $(H_n)_{n \geq 0}$ be a sequence of bounded (or non-negative) $\mathcal{F}$-measurable functions, and let τ be a finite Markov time, then for any initial distribution π,

$$\mathsf{E}_\pi(H_\tau \circ \theta_\tau \mid \mathcal{F}_\tau^X)(\omega) = \psi(\tau(\omega), X_{\tau(\omega)}(\omega)) \quad \mathsf{P}_\pi\text{-a.s.} \tag{4.1.21}$$

where $\psi(n,x) = \mathsf{E}_x H_n$. (Here $H_\tau \circ \theta_\tau$ means that if $\tau(\omega) = n$ then $(H_\tau \circ \theta_\tau)(\omega) = (H_n \circ \theta_n)(\omega)$.)

6. Note also two properties of stopping times which prove to be useful in different proofs related to the problems of optimal stopping.

Suppose that $B \in \mathcal{E}$ and

$$\tau_B = \inf\{\, n \geq 0 : X_n \in B \,\}, \qquad \sigma_B = \inf\{\, n > 0 : X_n \in B \,\} \tag{4.1.22}$$

are finite and γ is a stopping time. Then τ_B and σ_B are stopping times and so are

$$\gamma + \tau_B \circ \theta_\gamma = \inf\{\, n \geq \gamma : X_n \in B \,\}, \tag{4.1.23}$$

$$\gamma + \sigma_B \circ \theta_\gamma = \inf\{\, n > \gamma : X_n \in B \,\}. \tag{4.1.24}$$

In particular, if $\gamma \leq \tau_B$ then from (4.1.23) we get the following formula:

$$\gamma + \tau_B \circ \theta_\gamma = \tau_B. \tag{4.1.25}$$

(These properties are also valid when γ, τ_B and σ_B can take infinite values and the sets in (4.1.23) and (4.1.24) are empty.)

4.2. Elements of potential theory (discrete time)

1. It was mentioned above that, in the modern theory of time-homogeneous Markov chains $X = (X_n)_{n \geq 0}$ with values in a certain phase space $(E, \mathcal{E})$, the probability distribution of X is uniquely defined by its initial distribution $\pi = \pi(dx)$ and transition function $P = P(x; B)$, $x \in E$, $B \in \mathcal{E}$. Moreover, the probability distribution P_x on $(E^\infty, \mathcal{E}^\infty)$ is uniquely defined by the transition function $P = P(x; B)$ itself.

It is noteworthy that the notion of transition function (or Markov kernel) underlies the field of mathematical analysis (analytic non-probabilistic) which is called *potential theory*. Thus it is not surprising that there exists a close connection between this theory and the theory of time-homogeneous Markov chains, and that this connection proves to be mutually fruitful.

From the standpoint of optimal stopping problems that we are interested in, a discussion of some aspects of the potential theory can be very useful since both the material of Chapter I and our further exposition demonstrate that the Dirichlet problem in potential theory is related to the Stefan problem (with moving boundary) in optimal stopping theory. We can go even further and say that optimal stopping problems can be interpreted as optimization problems of potential theory (in particular for the Dirichlet problem)!

Let us associate with a transition function $P = P(x; B)$, $x \in E$, $B \in \mathcal{E}$, the linear (one-step) *transition operator* $\mathbb{P}g$ acting on functions $g = g(x)$ as follows:

$$(\mathbb{P}g)(x) = \int_E g(y)\, P(x; dy) \tag{4.2.1}$$

(one often writes $\mathbb{P}g(x)$). As the domain $\mathcal{D}_{\mathbb{P}}$ of the operator $\mathbb{P}$ we consider the set of those $\mathcal{E}$-measurable functions $g = g(x)$ for which the integral $\int_E g(y)\, P(x; dy)$ is well defined for all $x \in E$. For example, the set $\mathfrak{E}_+$ of non-negative $\mathcal{E}$-measurable functions is contained in $\mathcal{D}_{\mathbb{P}}$, and so is the set $\mathrm{b}\mathfrak{E}_+$ of bounded $\mathcal{E}$-measurable functions.

With the notation $\mathbb{I}$ for the unit (identical) operator ($\mathbb{I}g(x) = g(x)$) one can introduce the (n-step) transition operators $\mathbb{P}_n$, $n \geq 1$, by the formula $\mathbb{P}_n = \mathbb{P}(\mathbb{P}_{n-1})$ with $\mathbb{P}_0 = \mathbb{I}$. It is clear that for $g \in \mathcal{D}_{\mathbb{P}}$,

$$\mathbb{P}_n g(x) = \mathsf{E}_x g(X_n). \tag{4.2.2}$$

If τ is a Markov time (with respect to the filtration $(\mathcal{F}_n^X)_{n\geq 0}$), we shall denote by $\mathbb{P}_\tau$ the operator acting on functions $g \in \mathcal{D}_{\mathbb{P}}$ by the formula

$$\mathbb{P}_\tau g(x) = \mathsf{E}_x\big[I(\tau < \infty) g(X_\tau)\big]. \tag{4.2.3}$$

If $g(x) \equiv 1$ then $\mathbb{P}_\tau 1(x) = \mathsf{P}_x\{\tau < \infty\}$. From operators $\mathbb{P}_n$, $n \geq 0$, one can construct the important (in general unbounded) operator

$$\mathbb{U} = \sum_{n\geq 0} \mathbb{P}_n \tag{4.2.4}$$

called the *potential of the operator* $\mathbb{P}$ (or of the corresponding *Markov chain*).

If $g \in \mathcal{E}_+$ it is clear that

$$\mathbb{U}g = \sum_{n\geq 0} \mathbb{P}_n g = (\mathbb{I} + \mathbb{P}\mathbb{U})g \tag{4.2.5}$$

or in other words

$$\mathbb{U} = \mathbb{I} + \mathbb{P}\mathbb{U}. \tag{4.2.6}$$

The function $\mathbb{U}g$ is usually referred to as the *potential of the function* g.

Putting $g(x) = I_B(x)$ we find that

$$\mathbb{U}I_B(x) = \sum_{n\ge 0} \mathsf{E}_x I_B(X_n) = \mathsf{E}_x N_B \tag{4.2.7}$$

where N_B is the *number of visits* by X to the set $B \in \mathcal{E}$. When $x \in E$ is fixed, the function $U(x,B) = \mathbb{U}I_B(x)$ is a measure on $(E,\mathcal{E})$. It is called a *potential measure.* If $B = \{y\}$ i.e. B is a one-point set where $y \in E$, the function $U(x,\{y\})$ is usually denoted by $G(x,y)$ and called the *Green function.* The illustrative meaning of the Green function is clear:

$$G(x,y) = \mathsf{E}_x N_{\{y\}} \tag{4.2.8}$$

i.e. the *average number of visits* to a state y given that $X_0 = x$. It is clear that the Green function $G(x,y)$ admits the representation

$$G(x,y) = \sum_{n\ge 0} p(n;x,y) \tag{4.2.9}$$

where $p = (n;x,y) = \mathsf{P}_x(X_n = y)$ and consequently for $g(y) \ge 0$ the potential $\mathbb{U}g$ of a function g is defined by the formula

$$\mathbb{U}g = \sum_y g(y)G(x,y). \tag{4.2.10}$$

Remark 4.1. The explanation for the name "potential of the function g" given to $\mathbb{U}g(x)$ lies in analogy of $\mathbb{U}g(x)$ with the Newton potential $f(x)$ for the "mass" distribution with density $g(y)$, which, e.g. in the case $x \in \mathbb{R}^3$, has the form

$$f(x) = \frac{1}{2\pi}\int_{\mathbb{R}^3} \frac{g(y)\,dy}{\|x-y\|} \tag{4.2.11}$$

where $\|x-y\|$ is the distance between the points x and y. (According to the law of Newtonian attraction, the "mass" in $\mathbb{R}^3$ exerts influence upon a "unit mass" at point x which is proportional to the gradient of the function $f(x)$. Under nonrestrictive assumptions on the function $g(x)$ the potential $f(x)$ solves the *Poisson equation*

$$\frac{1}{2}\Delta f(x) = -g(x) \tag{4.2.12}$$

where Δ is the Laplace operator.)

In the case of simple symmetrical random walks on the lattice $\mathbb{Z}^d = \{0, \pm 1, \pm 2, \dots\}^d$ one has

$$G(x,y) \sim \frac{c_3}{\|x-y\|}, \qquad c_3 = \text{const.} \tag{4.2.13}$$

for large $\|x-y\|$, and consequently, according to the formula $\mathbb{U}g(x)=\sum_y g(y)$ $G(x,y)$ given above, we find that for $\|x\|\to\infty$, at least when the function $g(y)$ does not vanishes everywhere except a finite numbers of points, one finds that

$$\mathbb{U}g(x)\sim c_3\sum_y \frac{g(y)}{\|x-y\|}. \tag{4.2.14}$$

Thus the behavior of the potential $\mathbb{U}g(x)$ for large $\|x\|$ is analogous to that of the Newton potential $f(x)$.

More details regarding the preceding considerations may be found in [55, Chap. 1, § 5].

2. Let us relate to the operator $\mathbb{P}$ another important operator

$$\mathbb{L}=\mathbb{P}-\mathbb{I}. \tag{4.2.15}$$

In the theory of Markov processes this operator is called a *generating operator* (of a time-homogeneous Markov chain with the transition function $P=P(x;B)$). The domain $\mathcal{D}_{\mathbb{L}}$ of the operator $\mathbb{L}$ is the set of those $\mathcal{E}$-measurable functions $g=g(x)$ for which the expression $\mathbb{P}g-g$ is well defined.

Let a function h belong to $\bar{\mathfrak{E}}_+$ (i.e. h is $\mathcal{E}$-measurable and takes its values in $\bar{\mathbb{R}}_+$). Its potential $H=\mathbb{U}h$ satisfies the relation

$$H=h+\mathbb{P}H \tag{4.2.16}$$

(because $\mathbb{U}=\mathbb{I}+\mathbb{P}\mathbb{U}$). Thus if $H\in\mathcal{D}_{\mathbb{L}}$ then the potential H solves the *Poisson equation*

$$\mathbb{L}V=-h. \tag{4.2.17}$$

Suppose that $W\in\bar{\mathfrak{E}}_+$ is another solution to the equation $W=h+\mathbb{P}W$ (or to the equation $\mathbb{L}W=-h$ with $W\in\mathcal{D}_{\mathbb{L}}$). Because $W=h+\mathbb{P}W\ge h$, by induction we find that

$$W\ge\sum_{k=0}^{n}\mathbb{P}_k h\quad\text{for all}\quad n\ge1 \tag{4.2.18}$$

and so $W\ge H$. Thus the potential H is recognizable by its property to provide the *minimal* solution to the system $V=h+\mathbb{P}V$. Recall once again that

$$H=\mathbb{U}h=\mathsf{E}_x\sum_{k=0}^{\infty}h(X_k). \tag{4.2.19}$$

3. In the theory of optimal stopping a significant role is played by another important notion, namely the notion of excessive function.

A function $f = f(x)$, $x \in E$, belonging to the class $\bar{\mathfrak{E}}_+$ is said to be *excessive* or *superharmonic* for an operator $\mathbb{P}$ (or $\mathbb{P}$-excessive or $\mathbb{P}$-superharmonic) if

$$\mathbb{P}f \le f \qquad (\mathbb{L}f \le 0 \text{ if } \mathbb{L}f \text{ is well defined}) \tag{4.2.20}$$

i.e. $\mathsf{E}_x f(X_1) \le f(x)$, $x \in E$.

According to this definition the potential $H = \mathbb{U}h$ of a function $h \in \bar{\mathfrak{E}}_+$ is an excessive function (since by (4.2.16) we have $H = h + \mathbb{P}H \ge \mathbb{P}H$).

A function $f = f(x)$, $x \in E$, from the class $\bar{\mathfrak{E}}_+$ is called *harmonic* (or *invariant*) if

$$\mathbb{P}f = f \qquad (\mathbb{L}f = 0 \text{ if } \mathbb{L}f \text{ is well defined}) \tag{4.2.21}$$

i.e. $\mathsf{E}_x f(X_1) = f(x)$, $x \in E$.

It is important to be aware of the following connection between the notions of potential theory, theory of Markov chains, and martingale theory.

Let $X = (X_n)_{n \ge 0}$ be a one-dimensional time-homogeneous Markov chain with initial distribution π and transition function $P = P(x; B)$ generating the probability distribution P_π on $(E^\infty, \mathcal{E}^\infty)$, and let $f = f(x)$ be a $\mathbb{P}$-excessive function. Then *the sequence*

$$Y = (Y_n, \mathcal{F}_n^X, \mathsf{P}_\pi)_{n \ge 0} \tag{4.2.22}$$

with $Y_n = f(X_n)$ *is a non-negative (generalized) supermartingale*, i.e.

$$Y_n \text{ is } \mathcal{F}_n^X\text{-measurable and } Y_n \ge 0 \text{ so that } \mathsf{E}_\pi Y_n \text{ exists in } [0, \infty]; \tag{4.2.23}$$

$$\mathsf{E}_\pi(Y_{n+1} \mid \mathcal{F}_n^X) \le Y_n \quad \mathsf{P}_\pi\text{-a.s.} \tag{4.2.24}$$

for all $n \ge 0$. If $\mathsf{E}_\pi Y_n < \infty$ for all $n \ge 0$, then this sequence is an (ordinary) supermartingale.

4. The potential $H(x) = \mathbb{U}h(x)$ of a non-negative function $h = h(x)$ (from the class $\mathfrak{E}_+$ or $\bar{\mathfrak{E}}_+$) satisfies (4.2.16) and thus solves the *Wald–Bellman inequality*

$$H(x) \ge \max(h(x), \mathbb{P}H(x)), \tag{4.2.25}$$

i.e. the potential $H(x)$ of the function $h(x)$ is

(1) a majorant for the function $h(x)$, and

(2) an excessive function.

In other words, *the potential $H(x)$ of the function $h(x)$ is an excessive majorant of this function.*

In Chapter I we have already seen that *minimal* excessive majorants play an extremely important role in the theory of optimal stopping. We now show how the potential theory answers the question as how to find a minimal excessive majorant of the given non-negative $\mathcal{E}$-measurable function $g = g(x)$.

To this end introduce the operator $\mathbb{Q}$ acting on such functions by the formula

$$\mathbb{Q}g(x) = \max\big(g(x), \mathbb{P}g(x)\big). \tag{4.2.26}$$

Then the minimal excessive majorant $s(x)$ of the function $g(x)$ is given by

$$s(x) = \lim_n \mathbb{Q}^n g(x). \tag{4.2.27}$$

(See Corollary 1.12 where instead of $\mathbb{Q}$, g and s we used the notation Q, G and V.)

Note that $s = s(x)$ satisfies the *Wald–Bellman equation*

$$s(x) = \max\big(g(x), \mathbb{P}s(x)\big) \tag{4.2.28}$$

(cf. the Wald–Bellman inequality (4.2.25)). The equation (4.2.28) implies, in particular, that if the function $s \in \mathcal{D}_{\mathbb{L}}$ and

$$C_g = \{x : s(x) > g(x)\}, \tag{4.2.29}$$

$$D_g = \{x : s(x) \le g(x)\} \quad (= E \setminus C_g) \tag{4.2.30}$$

then

$$\begin{aligned} \mathbb{L}s(x) &= 0, && x \in C_g, \\ s(x) &= g(x), && x \in D_g. \end{aligned} \tag{4.2.31}$$

The system (4.2.31) is directly connected with the optimal stopping problem

$$s(x) = \sup \mathsf{E}_x g(X_\tau) \tag{4.2.32}$$

considered throughout the monograph as already illustrated in Chapter I (see e.g. Theorem 1.11).

5. In potential theory a lot of attention is paid to solving the *Dirichlet problem* (the "first boundary problem") for an operator $\mathbb{P}$: *Find a non-negative function* $V = V(x)$, $x \in E$ (from one or another class of functions, $\bar{\mathfrak{E}}_+$, $\mathfrak{E}_+$, $\mathrm{b}\mathfrak{E}_+$ etc.) *such that*

$$V(x) = \begin{cases} \mathbb{P}V(x) + h(x), & x \in C, \\ g(x), & x \in D. \end{cases} \tag{4.2.33}$$

Here C is a given subset of E ("domain"), $D = E \setminus C$, and h as well as g are non-negative $\mathcal{E}$-measurable functions.

If we consider only solutions V which belong to $\mathcal{D}_{\mathbb{L}}$, then the system (4.2.33) is equivalent to the following system:

$$\begin{aligned} \mathbb{L}V(x) &= -h(x), && x \in C, \\ V(x) &= g(x), && x \in D. \end{aligned} \tag{4.2.34}$$

The equation $\mathbb{L}V(x) = -h(x)$, $x \in C$, as it was mentioned above, bears the name of a *Poisson equation* in the set C and the problem (4.2.34) itself is called a *Dirichlet problem for the Poisson equation* (in the set C) with a *given* function g (in the set D).

It is remarkable that the solution to this (analytic i.e. non-probabilistic) problem can be obtained by a *probabilistic* method if we consider a Markov chain $X = (X_n)_{n \geq 0}$ constructed from the same transition function $P = P(x; B)$ that has been used to construct the operator $\mathbb{P}$.

Namely, let $X = (X_n)_{n \geq 0}$ be such a Markov chain, and let

$$\tau(D) = \inf\{n \geq 0 : X_n \in D\}. \tag{4.2.35}$$

(As usual, throughout we put $\tau(D) = \infty$ if the set in (4.2.35) is empty.)

From the theory of Markov processes it is known (see [53], [55]) that if the functions h and g belong to the class $\bar{\mathfrak{E}}_+$, then a solution to the Dirichlet problem (4.2.34) does exist and its *minimal* (non-negative) solution $V_D(x)$ is given by

$$V_D(x) = \mathsf{E}_x\big[I(\tau(D) < \infty) g(X_{\tau(D)})\big] + I_C(x)\, \mathsf{E}_x\Bigg[\sum_{k=0}^{\tau(D)-1} h(X_k)\Bigg]. \tag{4.2.36}$$

It is useful to mention some particular cases of the problem (4.2.34) when $g(x) \geq 0$.

(a) If $h = 0$, we seek a function $V = V(x)$ which is harmonic in C (i.e. $\mathbb{L}V(x) = 0$) and coincides with the function g in D. In this case the minimal non-negative solution $V_D(x)$ is given by

$$V_D(x) = \mathsf{E}_x\big[I(\tau(D) < \infty) g(X_{\tau(D)})\big]. \tag{4.2.37}$$

In particular, if $g(x) \equiv 1$, $x \in D$, then

$$V_D(x) = \mathsf{P}_x\{\tau(D) < \infty\}. \tag{4.2.38}$$

This result is interesting in the "reverse" sense: the probability $\mathsf{P}_x\{\tau(D) < \infty\}$ to reach the set D in finite time, under the assumption $X_0 = x \in C$, is *harmonic* (as a function of $x \in C$).

(b) If $g(x) = 0$, $x \in D$, and $h(x) = 1$, $x \in C$, i.e. we consider the system

$$V(x) = \begin{cases} \mathbb{P}V(x) + 1, & x \in C, \\ 0, & x \in D, \end{cases} \tag{4.2.39}$$

or, equivalently, the system

$$\begin{aligned} \mathbb{L}V(x) &= -1, \quad x \in C, \\ V(x) &= 0, \quad\;\; x \in D, \end{aligned} \tag{4.2.40}$$

with $V \in \mathcal{D}_{\mathbb{L}}$, then the minimal non-negative solution $V_D(x)$ is given by

$$V_D(x) = I_C(x)\,\mathsf{E}_x\Bigg[\sum_{k=0}^{\tau(D)-1} 1\Bigg] = \begin{cases} \mathsf{E}_x\tau(D), & x \in C, \\ 0, & x \in D. \end{cases} \tag{4.2.41}$$

Thus the expectation $\mathsf{E}_x\tau(D)$ of the time $\tau(D)$ of the first entry into the set D is the minimal non-negative solution of the system (4.2.40).

6. In the class of Markov chains that describe random walks in the phase space $(E, \mathcal{E})$, a special place (especially due to analogies with Brownian motion) is taken by *simple symmetrical random walks* in

$$E = \mathbb{Z}^d = \{0 \pm 1, \pm 2, \ldots\}^d \tag{4.2.42}$$

where $d = 1, 2, \ldots$. One can define such walks $X = (X_n)_{n \ge 0}$ constructively by specifying

$$X_n = x + \xi_1 + \cdots + \xi_n \tag{4.2.43}$$

where the random d-dimensional vectors $\xi_1, \xi_2, \ldots$ defined on a certain probability space are independent and identically distributed with

$$\mathsf{P}(\xi_1 = e) = (2d)^{-1} \tag{4.2.44}$$

($e = (e_1, \ldots, e_d)$ is a standard basis unit vector in $\mathbb{R}^d$ i.e. each e_i equals either ± 1 or 0 and $\|e\| \equiv |e_1| + \cdots + |e_d| = 1$).

The corresponding operator $\mathbb{P}$ has the very simple structure

$$\mathbb{P}f(x) = \mathsf{E}_x f(x + \xi_1) = \frac{1}{2d} \sum_{\|e\|=1} f(x + e), \tag{4.2.45}$$

and, consequently, the generating operator $\mathbb{L} = \mathbb{P} - \mathbb{I}$ (called a discrete *Laplacian* and denoted by Δ) has the following form:

$$\Delta f(x) = \frac{1}{2d} \sum_{|e|=1} (f(x + e) - f(x)). \tag{4.2.46}$$

Here it is natural to reformulate the Dirichlet problem stated above by taking into account that the exit from the set $C \subseteq \mathbb{Z}^d$ is only possible through the "boundary" set

$$\partial C = \{x \in \mathbb{Z}^d : x \notin C \text{ and } \|x - y\| = 1 \text{ for some } y \in C\}. \tag{4.2.47}$$

This fact leads to the following standard formulation of the *Dirichlet problem*: Given a set $C \subseteq \mathbb{Z}^d$ and functions $h = h(x)$, $x \in C$, $g = g(x)$, $x \in \partial C$, find a function $V = V(x)$ such that

$$\begin{aligned} \Delta V(x) &= -h(x), \qquad & x \in C, \\ V(x) &= g(x), \qquad & x \in \partial C. \end{aligned} \tag{4.2.48}$$

If the set C consists of a *finite* number of points then $\mathsf{P}_x(\tau(\partial C) < \infty) = 1$ for all $x \in C$ where $\tau(\partial C) = \inf\{\, n \geq 0 : X_n \in \partial C \,\}$. This allows one to prove that a unique solution to the problem (4.2.48) for $x \in C \cup \partial C$ is given by the following formula:

$$V_{\partial C}(x) = \mathsf{E}_x g(X_{\tau(\partial C)}) + I_C(x)\, \mathsf{E}_x \Bigg[\sum_{k=0}^{\tau(\partial C)-1} h(X_k) \Bigg]. \tag{4.2.49}$$

In particular, if $h = 0$ then a unique function which is harmonic in C and equal to $g(x)$ for $x \in \partial C$ is given by

$$V_{\partial C}(x) = \mathsf{E}_x g(X_{\tau(\partial C)}). \tag{4.2.50}$$

Let us also cite some results for the (classical) *Dirichlet problem*:

$$\begin{aligned} \Delta V(x) &= 0, && x \in C, \\ V(x) &= g(x), && x \in \partial C, \end{aligned} \tag{4.2.51}$$

when the set C is *unbounded.*

If $d \leq 2$ then $\mathsf{P}_x(\tau(\partial C) < \infty) = 1$ by the well-known Pólya ("recurrency/transiency") theorem, and for a *bounded* function $g = g(x)$, the solution in the class of *bounded* functions on $C \cup \partial C$ exists, is unique, and can be given by the same formula as in (4.2.50) above.

It should be noted that even in the case of bounded functions $g = g(x)$ the problem (4.2.51) can have (more than) one *unbounded* solution. The following example is classical.

Let $d = 1$, $C = \mathbb{Z} \setminus \{0\}$ and consequently $\partial C = \{0\}$. Taking $g(0) = 0$ we see that every *unbounded* function $V(x) = \alpha x$, $\alpha \in \mathbb{R}$, solves the Dirichlet problem $\Delta V(x) = 0$, $x \in \mathbb{Z} \setminus \{0\}$, and $V(0) = g(0)$.

When $d \geq 3$, the answer to the question on the existence and uniqueness of a solution to the Dirichlet problem ($\Delta V(x) = 0$, $x \in C$, and $V(x) = g(x)$, $x \in \partial C$), even in the case of bounded functions, depends essentially on whether the condition $\mathsf{P}_x\{\tau(\partial C) < \infty\} = 1$ is fulfilled for all $x \in C$. If this is the case, then in the class of bounded functions a solution exists, is unique, and can be given by the same formula as in (4.2.50) above.

However, if the condition $\mathsf{P}_x\{\tau(\partial C) < \infty\} = 1$, $x \in C$, does not hold then (in the case of bounded functions $g = g(x)$, $x \in \partial C$) all bounded solutions to the Dirichlet problem ($\Delta V(x) = 0$, $x \in C$, and $V(x) = g(x)$, $x \in \partial C$) are described by functions of the following form:

$$V^{(\alpha)}_{\partial C}(x) = \mathsf{E}_x\big[I(\tau(\partial C) < \infty) g(X_{\tau(\partial C)}) \big] + \alpha \mathsf{P}_x\{\tau(\partial C) = \infty\} \tag{4.2.52}$$

where $\alpha \in \mathbb{R}$. (For more details see e.g. [122].)

4.3. Markov processes (continuous time)

1. Foundations of the general theory of Markov processes were laid down in the well-known paper [111] of A. N. Kolmogorov entitled *"On analytical methods of probability theory"* (published in 1931).

This was the first work which clarified the deep connection between probability theory and mathematical analysis and initiated the construction and development of the theory of Markov processes (in continuous time).

In [111] Kolmogorov did not deal with trajectories of the *Markov* (as we say now) *process* under consideration directly. For him the main object were *transition probabilities*

$$P(s,x;t,A), \qquad 0 \le s \le t, \quad x \in E, \quad A \in \mathcal{E} \tag{4.3.1}$$

where $(E,\mathcal{E})$ is a phase (state) space ($E=\mathbb{R}^d$ as a rule).

The transition probability $P(s,x;t,A)$ is interpreted as "the probability for a 'system' to get at time t to the set A given that at time $s \le t$ the system was in a state x". The main requirement on the collection of *transition probabilities* $\{P(s,x;t,A)\}$—which determines the Markovian character of system's evolution—is the assumption that the *Chapman–Kolmogorov equation* holds:

$$P(s,x;t,A) = \int_E P(s,x;u,dy)P(u,y;t,A) \qquad (0 \le s < u < t). \tag{4.3.2}$$

2. We now assume that $E=\mathbb{R}$ and use the notation $F(s,x;t,y) = P(s,x;t,(-\infty,y])$. Suppose that the density (in y)

$$f(s,x;t,y) = \frac{\partial F(s,x;t,y)}{\partial y} \tag{4.3.3}$$

exists as well as the following limits:

$$\lim_{\Delta\downarrow 0} \frac{1}{\Delta} \int_{-\infty}^{\infty} (y-x) f(s,x;s+\Delta,y)\,dy \quad (=b(s,x)), \tag{4.3.4}$$

$$\lim_{\Delta\downarrow 0} \frac{1}{\Delta} \int_{-\infty}^{\infty} (y-x)^2 f(s,x;s+\Delta,y)\,dy \quad (=\sigma^2(s,x)), \tag{4.3.5}$$

$$\lim_{\Delta\downarrow 0} \frac{1}{\Delta} \int_{-\infty}^{\infty} |y-x|^{2+\delta} f(s,x;s+\Delta,y)\,dy \quad \text{for some } \delta>0. \tag{4.3.6}$$

The coefficients $b(s,x)$ and $\sigma^2(s,x)$ are called *differential characteristics* (or *drift coefficient* and *diffusion coefficient* respectively) of the corresponding Markov system whose evolution has the "diffusion" character. Under these assumptions Kolmogorov derived the *backward* parabolic differential equation (in (s,x)):

$$-\frac{\partial f}{\partial s} = b(s,x)\frac{\partial f}{\partial x} + \frac{1}{2}\sigma^2(s,x)\frac{\partial^2 f}{\partial x^2} \tag{4.3.7}$$

and the *forward* parabolic differential equation (in (t, y)):

$$\frac{\partial f}{\partial t} = -\frac{\partial}{\partial y}\big[b(t,y)f\big] + \frac{1}{2}\frac{\partial^2}{\partial y^2}\big[\sigma^2(t,y)f\big]. \tag{4.3.8}$$

(Special cases of the latter equation had been considered earlier by A. D. Fokker [68] and M. Planck [172].)

Kolmogorov also obtained the corresponding equations for Markov systems with finite or countable set of states (for details see [111]).

It is due to all these equations that the approach proposed by Kolmogorov was named *'analytical approach'* as reflected in the title of [111].

The 1940–60s saw a considerable progress in investigations of Markov systems. First of all one should cite the works by K. Itô [97]–[99], J. L. Doob [40], E. B. Dynkin [53] and W. Feller [64] in which, along with transition functions, the trajectories (of Markov processes) had begun to play an essential role.

Starting with the differential characteristics $b(s,x)$ and $\sigma^2(s,x)$ from *Analytical methods*, K. Itô [97]–[99] constructed processes $X = (X_t)_{t\geq 0}$ as solutions to stochastic differential equations

$$X_t = b(t, X_t)\, dt + \sigma(t, X_t)\, dB_t \tag{4.3.9}$$

where the "driving" process $B = (B_t)_{t\geq 0}$ is a standard Brownian motion (see Subsection 4.4 below) and $\sigma = +\sqrt{\sigma^2}$.

The main contribution of K. Itô consists in proving the following: If the differential characteristics b and σ satisfy the Lipschitz condition and increase linearly (in the space variable) then the equation (4.3.9) has a unique (strong) solution $X = (X_t)_{t\geq 0}$ which under certain conditions (e.g. if the differential characteristics are continuous in both variables) is a diffusion Markov process (in the Kolmogorov sense) such that the differential characteristics of the corresponding transition function

$$P(s,x;t,A) = \mathsf{P}(X_t \in A \mid X_s = x) \tag{4.3.10}$$

are just the same as b and σ that are involved in the equation (4.3.9).

Actually K. Itô considered d-dimensional processes $X = (X^1, \dots, X^d)$ such that the corresponding differential equations are of the form

$$dX_t^i = b_i(t, X_t)\, dt + \sum_{j=1}^{d} \sigma_{ij}(t, X_t)\, dB_t^j \tag{4.3.11}$$

for $1 \le i \le d$. If we use the notation

$$a_{ij} = \sum_{k=1}^{d} \sigma_{ik}\sigma_{kj} \tag{4.3.12}$$

$$L(s,x) = \sum_{i=1}^{d} b_i(s,x)\,\frac{\partial f}{\partial x_i} + \frac{1}{2}\sum_{i,j=1}^{d} a_{ij}(s,x)\,\frac{\partial^2 f}{\partial x_i\,\partial x_j} \tag{4.3.13}$$

$$L^*(t,y) = -\sum_{i=1}^{d} \frac{\partial}{\partial y_i}\big[b_i(t,y)f\big] + \frac{1}{2}\sum_{i,j=1}^{d} \frac{\partial^2}{\partial y_i\,\partial y_j}\big[a_{ij}(t,y)f\big] \tag{4.3.14}$$

then the *backward* and *forward* Kolmogorov equations take respectively the following form:

$$-\frac{\partial f}{\partial s} = L(s,x)f, \tag{4.3.15}$$

$$\frac{\partial f}{\partial t} = L^*(t,y)f. \tag{4.3.16}$$

It is important to notice that in the time-homogeneous case—when a_{ij} and b_i do not depend on the time parameter ($a_{ij} = a_{ij}(x)$, $b_i = b_i(x)$)—the following equality for the transition function holds:

$$f(s,x;t,y) = f(0,x;t-s,y), \qquad 0 \le s < t. \tag{4.3.17}$$

Putting

$$g(x;t,y) = f(0,x;t,y) \tag{4.3.18}$$

we find from the backward equation (4.3.15) that $g = g(x;t,y)$ as a function of (x,t) solves the following *parabolic equation*:

$$\frac{\partial g}{\partial t} = L(x)g \tag{4.3.19}$$

where

$$L(x)g = \sum_{i=1}^{d} b_i(x)\,\frac{\partial g}{\partial x_i} + \frac{1}{2}\sum_{i,j=1}^{d} a_{ij}(x)\frac{\partial^2 g}{\partial x_i\,\partial x_j}. \tag{4.3.20}$$

3. Let us now address a commonly used definition of Markov process [53]. When defining such notions as martingale, semimartingale, and similar, we start (see Subsection 4.1 above) from the fact that all considerations take place on a certain filtered probability space

$$(\Omega, \mathcal{F}, (\mathcal{F}_t)_{t\ge 0}, \mathsf{P}) \tag{4.3.21}$$

and the processes $X = (X_t)_{t \geq 0}$ considered are such that their trajectories are right-continuous (for $t \geq 0$), have limits from the left (for $t > 0$) and for every $t \geq 0$ the random variable X_t is $\mathcal{F}_t$-measurable (i.e. X is adapted).

When defining a (time-homogeneous) Markov process *in a wide sense* one also starts from a given filtered probability space $(\Omega, \mathcal{F}, (\mathcal{F}_t)_{t \geq 0}, \mathsf{P})$ and says that a stochastic process $X = (X_t)_{t \geq 0}$ taking values in a phase space is *Markov in a wide sense* if

$$\mathsf{P}(X_t \in B \mid \mathcal{F}_s)(\omega) = \mathsf{P}(X_t \in B \mid X_s)(\omega) \quad \mathsf{P}\text{-a.s.} \tag{4.3.22}$$

for all $s \leq t$.

If $\mathcal{F}_t = \mathcal{F}_t^X \equiv \sigma(X_s, s \leq t)$ then the stochastic process $X = (X_t)_{t \geq 0}$ is said to be *Markov in a strict sense.*

Just as in the discrete-time case (Subsection 4.1), the modern definition of a time-homogeneous Markov process places emphasis on both trajectories and transition functions as well as on their relation. To be more precise, we shall assume that the following objects are given:

(A) a phase space $(E, \mathcal{E})$;

(B) a family of probability spaces $(\Omega, \mathcal{F}, (\mathcal{F}_t)_{t \geq 0}; \mathsf{P}_x, x \in E)$ where each P_x is a probability measure on $(\Omega, \mathcal{F})$;

(C) a stochastic process $X = (X_t)_{t \geq 0}$ where each X_t is $\mathcal{F}_t/\mathcal{E}$-measurable.

Assume that the following conditions are fulfilled:

(a) the function $P(t, x; B) = \mathsf{P}_x(X_t \in B)$ is $\mathcal{E}$-measurable in x;

(b) $P(0, x; E \setminus \{x\}) = 0$, $x \in E$;

(c) for all $s, t \geq 0$ and $B \in \mathcal{E}$, the following (Markov) property holds:

$$\mathsf{P}_x(X_{t+s} \in B \mid \mathcal{F}_s) = P(s, X_t; B) \quad \mathsf{P}\text{-a.s.}; \tag{4.3.23}$$

(d) the space Ω is rich enough in the sense that for any $\omega \in \Omega$ and $h > 0$ there exists $\omega' \in \Omega$ such that $X_{t+h}(\omega) = X_t(\omega')$ for all $t \geq 0$.

Under these assumptions the process $X = (X_t)_{t \geq 0}$ is said to be a (time-homogeneous) *Markov process* defined on $(\Omega, \mathcal{F}, (\mathcal{F}_t)_{t \geq 0}; \mathsf{P}_x, x \in E)$ and the function $P(t, x; B)$ is called a *transition function* of this process.

The conditions (a) and (c) imply that P_x-a.s.

$$\mathsf{P}_x(X_{t+s} \in B \mid \mathcal{F}_t) = \mathsf{P}_{X_t}(X_s \in B), \quad x \in E, \quad B \in \mathcal{E}; \tag{4.3.24}$$

this property is called the *Markov property* of a process $X = (X_t)_{t \geq 0}$ satisfying the conditions (a)–(d).

In general theory of Markov processes an important role is played by those processes which, in addition to the Markov property, have the following *strong Markov property*: for any Markov time $\tau = \tau(\omega)$ (with respect to $(\mathcal{F}_t)_{t\ge 0}$)

$$\mathsf{P}_x(X_{\tau+s} \in B \mid \mathcal{F}_\tau) = P(s, X_\tau; B) \quad (\mathsf{P}_x\text{-a.s. on } \{\tau < \infty\}) \tag{4.3.25}$$

where $\mathcal{F}_\tau = \{A \in \mathcal{F} : A \cap \{\tau \le t\} \in \mathcal{F}_t \text{ for all } t \ge 0\}$ is a σ-algebra of events observed on the time interval $[0, \tau]$.

Remark 4.2. For $X_{\tau(\omega)}(\omega)$ to be $\mathcal{F}_\tau/\mathcal{E}$-measurable we have to impose an additional restriction—that of measurability—on the process X. For example, it suffices to assume that for every $t \ge 0$ the function $X_s(\omega)$, $s \le t$, defines a measurable mapping from $([0,t] \times \Omega, \mathcal{B}([0,t] \times \mathcal{F}_t)$ into the measurable space $(E, \mathcal{E})$.

4. In the case of discrete time and a canonical space Ω whose elements are sequences $\omega = (x_0, x_1, \ldots)$ with $x_i \in E$ we have introduced shift operators θ_n acting onto $\omega = (x_0, x_1, \ldots)$ by formulae $\theta_n(\omega) = \omega'$ where $\omega' = (x_n, x_{n+1}, \ldots)$, i.e. $\theta_n(x_0, x_1, \ldots) = (x_n, x_{n+1}, \ldots)$.

Likewise in a canonical space Ω which consists of functions $\omega = (x_s)_{s\ge 0}$ with $x_s \in E$ it is also useful to introduce shift operators θ_t, $t \ge 0$, acting onto $\omega = (x_s)_{s\ge 0}$ by formulae $\theta_t(\omega) = \omega'$ where $\omega' = (x_{s+t})_{s\ge 0}$ i.e. $\theta_t\big[(x_s)_{s\ge 0}\big] = (x_{s+t})_{s\ge 0}$.

In the subsequent considerations we shall assume that stochastic processes $X = (X_t(\omega))_{t\ge 0}$ are given on the canonical space Ω, which consists of functions $\omega = (x_s)_{s\ge 0}$, and that $X_s(\omega) = x_s$.

The notions introduced above imply that the "composition" $X_s \circ \theta_t(\omega) = X_s(\theta_t(\omega)) = X_{s+t}(\omega)$, and thus the Markov property (4.3.24) takes the form

$$\mathsf{P}_x(X_s \circ \theta_t \in B \mid \mathcal{F}_t) = \mathsf{P}_{X_t}(X_s \in B) \quad \mathsf{P}_x\text{-a.s.} \tag{4.3.26}$$

for every $x \in E$ and $B \in \mathcal{E}$ with $\mathcal{F}_t = \sigma(X_s, s \le t)$.

Similarly, the strong Markov property (4.3.25) assumes the form: for any Markov time τ,

$$\mathsf{P}_x(X_s \circ \theta_\tau \in B \mid \mathcal{F}_\tau) = P_{X_\tau}(X_s \in B) \quad (\mathsf{P}_x\text{-a.s. on } \{\tau < \infty\}) \tag{4.3.27}$$

for every $x \in E$ and $B \in \mathcal{E}$ where $\theta_\tau(\omega)$ by definition equals $\theta_{\tau(\omega)}(\omega)$ if $\tau(\omega) < \infty$.

The following useful property can be deduced from the strong Markov property (4.3.27):

$$\mathsf{E}_\pi(H \circ \theta_\tau \mid \mathcal{F}_\tau) = \mathsf{E}_{X_\tau} H \quad (\mathsf{P}_\pi\text{-a.s. on } \{\tau < \infty\}) \tag{4.3.28}$$

for any initial distribution π, any $\mathcal{F}$-measurable bounded (or non-negative) functional $H = H(\omega)$ and all Markov times τ. Similarly, from the strong Markov property one can deduce an analogue of the property (4.1.21).

We conclude this subsection with the remark that many properties presented in Subsection 4.1 in the case of Markov chains, for example properties (4.1.12), (4.1.13), (4.1.23)–(4.1.25), are also valid for Markov processes in the continuous-time case (with an evident change in notation).

Remark 4.3. Denoting $P(s,x;t,B) = \mathsf{P}(X_t \in B \mid X_s = x)$ recall that the conditional probabilities $\mathsf{P}(X_t \in B \mid X_s = x)$ are determined uniquely only up to a P_{X_s}-null set (where $\mathsf{P}_{X_s}(\cdot) = \mathsf{P}(X_s \in \cdot)$ is the law of X_s). This means that in principle there are different versions of transition functions $P(s,x;t,B)$ satisfying some or other "good" properties. Among such desired properties one is that the transition functions satisfy the Chapman–Kolmogorov equations (4.3.2). The Markov property (4.3.22) (for time-homogeneous or time-inhomogeneous processes) does not guarantee that (4.3.2) holds for all x but only for P_{X_s}-almost all x in E. In the case of discrete time and discrete state space, the Chapman–Kolmogorov equations ($p_{ij}^{(n+m)} = \sum_k p_{ik}^{(n)} p_{kj}^{(m)}$) are automatically satisfied when the Markov property holds (for the case of discrete time and *arbitrary* state space $(E,\mathcal{E})$ see [199, Vol. 2, Chap. VIII, § 1]). In the case of continuous time, however, the validity of the Chapman–Kolmogorov equations is far from being evident. It was shown in [118], nonetheless, that in the case of universally measurable (e.g. Borelian) space $(E,\mathcal{E})$ there always exist versions of transition probabilities such that the Chapman–Kolmogorov equations hold. Taking this into account, and without further mentioning it, in the sequel we shall consider only transition functions for which the Chapman–Kolmogorov equations are satisfied.

4.4. Brownian motion (Wiener process)

1. The process of Brownian motion, also called a Wiener process, is interesting from different points of view: this process is both martingale and Markov and has a magnitude of important applications. In this subsection we give only basic definitions and a number of fundamental properties which mainly relate these processes to various stopping times.

A one-dimensional (standard) *Wiener process* $W = (W_t)_{t\ge 0}$ is a process defined on a probability space $(\Omega,\mathcal{F},\mathsf{P})$ satisfying the following properties:

(a) $W_0 = 0$;

(b) the trajectories of $(W_t)_{t\ge 0}$ are continuous functions;

(c) the increments $W_{t_k} - W_{t_{k-1}}, W_{t_{k-1}} - W_{t_{k-2}}, \ldots, W_{t_1} - W_{t_0}$ are independent (for any $0 = t_0 < t_1 < \cdots < t_k$, $k \ge 1$);

(d) the random variables $W_t - W_s$, $s \le t$, have the normal distribution with

$$\mathsf{E}(W_t - W_s) = 0, \qquad \mathsf{D}(W_t - W_s) = t - s. \tag{4.4.1}$$

Thus a Wiener process $W = (W_t)_{t\geq 0}$ is, by definition, a Gaussian process with independent increments. It is clear that such a process is Markov (in a wide sense).

A *Brownian motion* is a process $B = (B_t)_{t\geq 0}$ defined on a filtered probability space $(\Omega, \mathcal{F}, (\mathcal{F}_t), \mathsf{P})$ such that:

(α) $B_0 = 0$;

(β) the trajectories of $B = (B_t)_{t\geq 0}$ are continuous functions;

(γ) the process $B = (B_t)_{t\geq 0}$ is a square-integrable martingale with respect to the filtration $(\mathcal{F}_t)_{t\geq 0}$ (i.e. each B_t is $\mathcal{F}_t$-measurable, $\mathsf{E}|B_t|^2 < \infty$ and $\mathsf{E}(B_t \mid \mathcal{F}_s) = B_s$ for $s \leq t$) such that P-a.s.

$$\mathsf{E}\big[(B_t - B_s)^2 \mid \mathcal{F}_s\big] = t - s, \qquad s \leq t. \tag{4.4.2}$$

The well-known "Lévy characterization theorem" (see e.g. [174, p. 150]) implies that such a process is Gaussian with independent increments as well as $\mathsf{E}(B_t - B_s) = 0$ and $\mathsf{E}(B_t - B_s)^2 = t - s$. Thus $B = (B_t)_{t\geq 0}$ is a Wiener process in the above sense.

The converse, in a certain sense, is also true: If $W = (W_t)_{t\geq 0}$ is a Wiener process then it can easily be checked that this process is a square-integrable Gaussian martingale with respect to the filtration $(\mathcal{F}_t^W)_{t\geq 0}$.

In the sequel we will not distinguish between these processes. Every time it will be clear from the context which of the two is meant (if at all relevant).

2. Let us list some basic properties of a Brownian motion $B = (B_t)_{t\geq 0}$ assumed to be defined on a filtered probability space $(\Omega, \mathcal{F}, (\mathcal{F}_t), \mathsf{P})$.

- The probability $\mathsf{P}(B_t \leq u)$ for $t > 0$ and $u \in \mathbb{R}$ is determined by

$$\mathsf{P}(B_t \leq u) = \int_{-\infty}^{u} \varphi_t(y)\, dy \tag{4.4.3}$$

where

$$\varphi_t(y) = \frac{1}{\sqrt{2\pi t}}\, e^{-y^2/(2t)} \tag{4.4.4}$$

is a fundamental solution to the Kolmogorov forward equation

$$\frac{\partial \varphi_t(y)}{\partial t} = \frac{1}{2}\frac{\partial^2 \varphi_t(y)}{\partial y^2}. \tag{4.4.5}$$

- The density

$$f(s, x; t, y) = \frac{\partial \mathsf{P}(B_t \leq y \mid B_s = x)}{\partial y}, \qquad 0 < s < t \tag{4.4.6}$$

is given by

$$f(s,x;t,y)=\frac{1}{\sqrt{2\pi(t-s)}}\exp\left\{-\frac{(y-x)^2}{2(t-s)}\right\} \tag{4.4.7}$$

and solves for $s<t$,

$$\frac{\partial f(s,x;t,y)}{\partial s}=-\frac{1}{2}\frac{\partial^2 f(s,x;t,y)}{\partial x^2} \quad \text{(backward equation)} \tag{4.4.8}$$

and for $t>s$,

$$\frac{\partial f(s,x;t,y)}{\partial t}=\frac{1}{2}\frac{\partial^2 f(s,x;t,y)}{\partial y^2} \quad \text{(forward equation).} \tag{4.4.9}$$

It is clear that $f(0,0;t,y)=\varphi_t(y)$.

- The joint density

$$\varphi_{t_1,\dots,t_n}(y_1,\dots,y_n)=\frac{\partial^n \mathsf{P}(B_{t_1}\le y_1,\dots,B_{t_n}\le y_n)}{\partial y_1\cdots\partial y_n} \tag{4.4.10}$$

for $0<t_1<t_2<\cdots<t_n$ is given by

$$\begin{aligned}&\varphi_{t_1,\dots,t_n}(y_1,\dots,y_n)\\&=\varphi_{t_1}(y_1)\varphi_{t_2-t_1}(y_2-y_1)\cdots\varphi_{t_n-t_{n-1}}(y_n-y_{n-1}).\end{aligned} \tag{4.4.11}$$

- $\mathsf{E}B_t=0$, $\mathsf{E}B_t^2=t$, $\operatorname{cov}(B_s,B_t)=\mathsf{E}B_sB_t=\min(s,t)$, $\mathsf{E}|B_t|=\sqrt{2t/\pi}$.
- The following property of *self-similarity* hold for any $a>0$:

$$\operatorname{Law}(B_{at};t\ge 0)=\operatorname{Law}(a^{1/2}B_t;t\ge 0). \tag{4.4.12}$$

- Together with a Brownian motion $B=(B_t)_{t\ge 0}$ the following processes:

$$B_t^{(1)}=-B_t \quad \text{for } t\ge 0, \tag{4.4.13}$$

$$B_t^{(2)}=tB_{1/t} \quad \text{for } t>0 \quad \text{with} \quad B_0^{(2)}=0, \tag{4.4.14}$$

$$B_t^{(3)}=B_{t+s}-B_s \quad \text{for } s\ge 0, \tag{4.4.15}$$

$$B_t^{(4)}=B_T-B_{T-t} \quad \text{for } 0\le t\le T \quad \text{with } T>0 \tag{4.4.16}$$

are also Brownian motions.

- For every fixed $t\ge 0$,

$$\operatorname{Law}\Big(\max_{s\le t}B_s\Big)=\operatorname{Law}(|B_t|) \tag{4.4.17}$$

and hence

$$\mathsf{E}\max_{s\le t} B_s = \mathsf{E}|B_t| = \sqrt{\frac{2t}{\pi}}. \tag{4.4.18}$$

The former assertion may be viewed as a variant of the *reflection principle for Brownian motion* (André [4], Bachelier [7, 1964 ed., p. 64], Lévy [125, p. 293] usually stated in the following form: for $t > 0$ and $x \ge 0$,

$$\mathsf{P}\Big(\max_{s\le t} B_s \ge x\Big) = 2\mathsf{P}(B_t \ge x) \quad (= \mathsf{P}(|B_t| \ge x)) \tag{4.4.19}$$

whence (4.4.17) can be written as

$$\mathsf{P}\Big(\max_{s\le t} B_s \le x\Big) = \Phi\Big(\frac{x}{\sqrt{t}}\Big) - \Phi\Big(\frac{-x}{\sqrt{t}}\Big). \tag{4.4.20}$$

The property (4.4.20) extends as follows (see [107, p. 368] or [197, pp. 759–760]): for $t > 0$, $x \ge 0$, $\mu \in \mathbb{R}$ and $\sigma > 0$,

$$\mathsf{P}\Big(\max_{s\le t}(\mu s + \sigma B_s) \le x\Big) = \Phi\Big(\frac{x-\mu t}{\sigma\sqrt{t}}\Big) - e^{2\mu x/\sigma^2}\Phi\Big(\frac{-x-\mu t}{\sigma\sqrt{t}}\Big). \tag{4.4.21}$$

This property implies the following useful facts:

$$\mathsf{P}\Big(\max_{t\ge 0}(\mu t + \sigma B_t) \le x\Big) = \exp\Big(\frac{2\mu x}{\sigma^2}\Big) \quad \text{if } \mu < 0, \tag{4.4.22}$$

$$\mathsf{P}\Big(\max_{t\ge 0}(\mu t + \sigma B_t) \le x\Big) = 0 \quad \text{if } \mu \ge 0. \tag{4.4.23}$$

• The following statement ("Lévy distributional theorem") is a natural extension of the property (4.4.17):

$$\mathrm{Law}\Big(\max_{s\le t} B_s - B_t, \max_{s\le t} B_s; t \ge 0\Big) = \mathrm{Law}\big(|B_t|, L_t; t \ge 0\big) \tag{4.4.24}$$

where L_t is the local time of a Brownian motion B on $[0,t]$:

$$L_t = \lim_{\varepsilon\downarrow 0}\frac{1}{2\varepsilon}\int_0^t I(|B_s| < \varepsilon)\,ds. \tag{4.4.25}$$

• Let $T_a = \inf\{t \ge 0 : B_t = a\}$ where $a > 0$. Then $\mathsf{P}(T_a < \infty) = 1$, $\mathsf{E}T_a = \infty$, and the density

$$\gamma_a(t) = \frac{d}{dt}\mathsf{P}(T_a \le t) \tag{4.4.26}$$

is given by

$$\gamma_a(t) = \frac{a}{\sqrt{2\pi t^3}}\, e^{-a^2/(2t)}. \tag{4.4.27}$$

It may be noted that

$$\gamma_a(t) = -\frac{\partial}{\partial a}\,\varphi_t(a). \tag{4.4.28}$$

• Let

$$T_{a,b} = \inf\{\, t \ge 0 : B_t = a + bt \,\} \tag{4.4.29}$$

where $a > 0$ and $b \in \mathbb{R}$. Then the density

$$\gamma_{a,b}(t) = \frac{d}{dt}\mathsf{P}(T_{a,b} \le t) \tag{4.4.30}$$

is given by

$$\gamma_{a,b}(t) = \frac{a}{\sqrt{2\pi t^3}}\,\exp\Big\{-\frac{(a+bt)^2}{2t}\Big\} \tag{4.4.31}$$

(see [40, p. 397], [130, p. 526] and also (4.6.17) below). If $b = 0$ then $\gamma_{a,0}(t) = \gamma_a(t)$ (see (4.4.27)).

When $b \ne 0$ one should separate the cases $b < 0$ and $b > 0$.

If $b < 0$ then $\mathsf{P}(T_{a,b} < \infty) = \int_0^\infty \gamma_{a,b}(t)\,dt = 1$.

If $b > 0$ then

$$\mathsf{P}(T_{a,b} < \infty) = \int_0^\infty \gamma_{a,b}(t)\,dt = e^{-2ab}. \tag{4.4.32}$$

• *Blumenthal's* 0-1 *law for Brownian motion.* Suppose that P_x, $x \in \mathbb{R}$, are measures on the measurable space of continuous functions $\omega = (\omega(t))_{t\ge 0}$ such that process $B_t(\omega) = \omega(t)$, $t \ge 0$, is a Brownian motion starting at x. Denote by $(\mathcal{F}_t^\circ)_{t\ge 0}$ the *natural* filtration of the process $B = (B_t(\omega))_{t\ge 0}$, i.e. $\mathcal{F}_t^\circ = \sigma\{B_s, s \le t\}$, $t \ge 0$, and let $(\mathcal{F}_t^+)_{t\ge 0}$ be the *right-continuous* filtration given by

$$\mathcal{F}_t^+ = \bigcap_{s>t} \mathcal{F}_s^\circ. \tag{4.4.33}$$

Then for every A from $\mathcal{F}_0^+$ and for all $x \in \mathbb{R}$ the probability $\mathsf{P}_x(A)$ takes only two values: either 0 or 1.

• *Laws of the iterated logarithm.* Let $B = (B_t)_{t\ge 0}$ be a standard Brownian motion (i.e. B starts from zero and the measure $\mathsf{P} = \mathsf{P}_0$ is such that $\mathsf{E}B_t = 0$ and $\mathsf{E}B_t^2 = t$ for $t \ge 0$).

The *law of the iterated logarithm at infinity* states that

$$\mathsf{P}\bigg(\limsup_{t\uparrow\infty} \frac{B_t}{\sqrt{2t\log\log t}} = 1,\ \liminf_{t\uparrow\infty} \frac{B_t}{\sqrt{2t\log\log t}} = -1\bigg) = 1. \tag{4.4.34}$$

The *law of the iterated logarithm at zero* states that

$$\mathsf{P}\bigg(\limsup_{t\downarrow 0} \frac{B_t}{\sqrt{2t\log\log(1/t)}} = 1,\ \liminf_{t\downarrow 0} \frac{B_t}{\sqrt{2t\log\log(1/t)}} = -1\bigg) = 1. \tag{4.4.35}$$

3. Consider a measurable space $(C, \mathcal{B}(C))$ consisting of continuous functions $\omega = (\omega_t)_{t\ge 0}$. If this space $(C, \mathcal{B}(C))$ is endowed with a Wiener measure P then the canonical process $B = (B_t)_{t\ge 0}$ with $B_t(\omega) = \omega_t$ becomes a Wiener process (Brownian motion) starting at time $t = 0$ from the point $\omega_0 = 0$.

Introduce, for all $x \in \mathbb{R}$, the processes $B^x = (B_t^x(\omega))_{t\ge 0}$ by setting

$$B_t^x(\omega) = x + B_t(\omega) \quad (= x + \omega_t). \tag{4.4.36}$$

Denote by P_x the measure on $(C, \mathcal{B}(C))$ induced by this process.

In order to keep ourselves within the framework of notation used in paragraph 3 of Subsection 4.3 while defining an "up-to-date" notion of a Markov process, from now on we denote $(\Omega, \mathcal{F}) = (C, \mathcal{B}(C))$ i.e. we assume that $(\Omega, \mathcal{F})$ is the measurable space of continuous functions $\omega = (\omega_t)_{t\ge 0}$ with the Borel σ-algebra $\mathcal{B}(C)$. Let $\mathcal{F}_t = \sigma\{\omega_s : s \le t\}$, $t \ge 0$, and let P_x be measure induced by a Wiener measure and mappings sending $(\omega_s)_{s\ge 0}$ to $(x + \omega_s)_{s\ge 0}$ as stated above.

On the constructed filtered spaces

$$(\Omega, \mathcal{F}, (\mathcal{F}_t); \mathsf{P}_x, x \in \mathbb{R}) \tag{4.4.37}$$

consider the canonical process $X = (X_t(\omega))_{t\ge 0}$ with $X_t(\omega) = \omega_t$ where $\omega = (\omega_t)_{t\ge 0}$ is a trajectory from Ω.

It follows immediately that the conditions (a)–(d), stated in paragraph 3 of Subsection 4.3 while defining the notion of a Markov process, are fulfilled. Indeed, the definition of measures P_x itself implies that the transition function

$$\mathsf{P}(t, x; B) = \mathsf{P}_x(X_t \in B) \tag{4.4.38}$$

coincides with the probability $\mathsf{P}(x + B_t \in B)$ which evidently is $\mathcal{B}$-measurable in x for any Borel set B and for any $t \ge 0$.

It is also clear that $P(0, x; \mathbb{R} \setminus \{x\}) = \mathsf{P}(x + B_0 \in \mathbb{R} \setminus \{x\}) = 0$ because $B_0 = 0$. To verify the property (c) it suffices to show that if f is a bounded function then for any $x \in \mathbb{R}$,

$$\mathsf{E}_x(f(X_{t+s}) \mid \mathcal{F}_t) = \mathsf{E}_{X_t} f(X_s) \quad \mathsf{P}_x\text{-a.s.} \tag{4.4.39}$$

Recall that $\mathsf{E}_{X_s} f(X_t)$ is the function $\psi(x) = \mathsf{E}_x f(X_t)$ with X_s inserted in place of x.

To this end it suffices in turn to prove a somewhat more general assertion: if $g(x, y)$ is a bounded function then

$$\mathsf{E}_x\big[g(X_t, X_{t+s} - X_t) \mid \mathcal{F}_t\big] = \psi_g(X_t) \tag{4.4.40}$$

where

$$\psi_g(x) = \int_{\mathbb{R}} g(x, y)\, \frac{1}{\sqrt{2\pi s}}\, e^{-y^2/(2s)}\, dy. \tag{4.4.41}$$

(The equality (4.4.39) results from (4.4.40) if we put $g(x,y) = f(x+y)$ and $x = X_t$.)

The standard technique for proving formulae of type (4.4.40) may be described as follows.

Assume first that $g(x,y)$ has the special form $g(x,y) = g_1(x)g_2(y)$. For such a function

$$\psi_g(x) = g_1(x) \int_{\mathbb{R}} g_2(y)\, \frac{1}{\sqrt{2\pi s}}\, e^{-y^2/(2s)}\, dy. \tag{4.4.42}$$

The left-hand side of (4.4.40) is equal to

$$g_1(X_t)\, \mathsf{E}_x\big[g_2(X_{t+s} - X_t) \mid \mathcal{F}_t\big]. \tag{4.4.43}$$

Below we shall prove that

$$\mathsf{E}_x\big[g_2(X_{t+s} - X_t) \mid \mathcal{F}_t\big] = \mathsf{E}_x g_2(X_{t+s} - X_t) \quad \mathsf{P}_x\text{-a.s.} \tag{4.4.44}$$

Then from (4.4.43) we find that for the function $g(x,y) = g_1(x)g_2(y)$,

$$\begin{aligned} \mathsf{E}_x\big[g(X_t, X_{t+s} - X_t) \mid \mathcal{F}_t\big] &= g_1(X_t)\, \mathsf{E}_x g_2(X_{t+s} - X_t) \\ &= g_1(X_t)\, \mathsf{E} g_2(B_{t+s} - B_t) = \psi_g(X_t) \quad \mathsf{P}_x\text{-a.s.} \end{aligned} \tag{4.4.45}$$

Using "monotone class" arguments we obtain that the property (4.4.45) remains valid for arbitrary measurable bounded functions $g(x,y)$. (For details see e.g. [50, Chap. 1, § 1] and [199, Chap. II, § 2].)

Thus it remains only to prove the property (4.4.44). Let $A \in \mathcal{F}_t$. According to the definition of conditional expectations we have to show that

$$\mathsf{E}_x\big[g_2(X_{t+s} - X_t); A\big] = \mathsf{P}_x(A)\mathsf{E}_x g_2(X_{t+s} - X_t). \tag{4.4.46}$$

For the sets A of the form

$$A = \{\omega : X_{t_1} \in C_1, \dots, X_{t_n} \in C_n\} \tag{4.4.47}$$

where $0 \le t_1 < \cdots < t_n \le t$ and C_i are Borel sets, (4.4.46) follows directly from the properties of Brownian motion and the fact that $\mathrm{Law}(X. \mid \mathsf{P}_x) = \mathrm{Law}(B. + x \mid \mathsf{P})$.

To pass from the special sets A just considered to arbitrary sets A from $\mathcal{F}_t$ one uses "monotone class" arguments (see again the above cited [50, Chap. 1, § 1] and [199, Chap. II, § 2]).

4. The above introduced process of Brownian motion X starting at an arbitrary point $x \in \mathbb{R}$ has, besides the established Markov property, the *strong Markov property* which can be formulated in the following (generalized) form (cf. (4.3.25) and (4.1.16)): If $H = H(\omega)$ is a bounded $\mathcal{F}$-measurable functional ($\mathcal{F} = \mathcal{B}(C)$), then for any $x \in \mathbb{R}$ and any Markov time τ,

$$\mathsf{E}_x(H \circ \theta_\tau \mid \mathcal{F}_\tau) = \mathsf{E}_{X_\tau} H \quad (\,\mathsf{P}_x\text{-a.s. on } \{\tau < \infty\}\,). \tag{4.4.48}$$

(For a proof see e.g. [50, Chap. 1, § 5].)

5. According to our definition, a time τ is a Markov time with respect to a filtration $(\mathcal{G}_t)_{t \geq 0}$ if for every $t \geq 0$,

$$\{\tau \leq t\} \in \mathcal{G}_t. \tag{4.4.49}$$

There are also other definitions. For example, sometimes a time τ is said to be Markov if for all $t > 0$,

$$\{\tau < t\} \in \mathcal{G}_t. \tag{4.4.50}$$

Since

$$\{\tau < t\} = \bigcup_n \{\tau \leq t - \tfrac{1}{n}\} \in \mathcal{G}_t \tag{4.4.51}$$

a Markov time in the sense of the first definition (i.e. one satisfying (4.4.49)) is necessarily Markov in the second sense. The inverse assertion in general is not true. Indeed,

$$\{\tau \leq t\} = \bigcap_n \{\tau < t + \tfrac{1}{n}\} \in \mathcal{G}_t^+ \tag{4.4.52}$$

where

$$\mathcal{G}_t^+ = \bigcap_{u>t} \mathcal{G}_u. \tag{4.4.53}$$

Therefore if $\mathcal{G}_t^+ \supset \mathcal{G}_t$ then the property to be a Markov time in the first sense (i.e. in the sense of (4.4.49)) in general does not imply this property in the second sense.

However from (4.4.53) it is clear that if the family $(\mathcal{G}_t)_{t \geq 0}$ is continuous from the right (i.e. $\mathcal{G}_t^+ = \mathcal{G}_t$ for all $t \geq 0$) then both definitions coincide.

Now let us consider a Brownian filtration $(\mathcal{F}_t)_{t \geq 0}$ where $\mathcal{F}_t = \sigma(X_s, s \leq t)$. Form a continuous-from-the-right filtration $(\mathcal{F}_t^+)_{t \geq 0}$ by setting $\mathcal{F}_t^+ = \bigcap_{u>t} \mathcal{F}_u$. It turns out that the Markov property (see paragraph 3 above) of a Brownian motion $X = (X_t)_{t \geq 0}$ with respect to the filtration $(\mathcal{F}_t)_{t \geq 0}$ remains valid for the larger filtration $(\mathcal{F}_t^+)_{t \geq 0}$. So when we deal with a Brownian motion there is no restriction to assume from the very beginning that the initial filtration is not $(\mathcal{F}_t)_{t \geq 0}$ but $(\mathcal{F}_t^+)_{t \geq 0}$. This assumption, as was explained above, simplifies a verification of whether one or another time τ is Markov. The strong Markov property also remains valid when we pass to the filtration $(\mathcal{F}_t^+)_{t \geq 0}$ (see e.g. [50, Chap. 1]).

4.5. Diffusion processes

1. *Diffusion* is widely interpreted as a (physical and mathematical) model describing the evolution of a "particle" moving continuously and chaotically. In this connection it is worthwhile to mention that the term *superdiffusion* is related to the random motion of a "cloud of particles" (see e.g. [54]). In the mathematical theory of stochastic processes it is the *Brownian motion* process (i.e. Wiener process) which is taken as a basic diffusion process.

In the modern theory of Markov processes, which was initiated, as we already mentioned above, by the Kolmogorov treatise *On analytical methods in probability theory* [111], the term 'diffusion' refers to a special class of continuous Markov processes specified as follows.

Let $X = (X_t)_{t \geq 0}$ be a time-homogeneous Markov process in a phase space $(E, \mathcal{E})$ defined on a filtered space $(\Omega, \mathcal{F}, (\mathcal{F}_t)_{t \geq 0}; \mathsf{P}_x, x \in E)$ with transition function $P(t, x; B)$. Let T_t be a shift operator acting on measurable functions $f = f(x)$, $x \in E$, by the formula:

$$T_t f(x) = \mathsf{E}_x f(X_t) \left(= \int_E f(y) \, P(t, x; dy) \right). \tag{4.5.1}$$

(The functions $f = f(x)$ are assumed to be such that $\mathsf{E}_x f(X_t)$ is well defined; very often the notation $\mathbb{P}_t f(x)$ is used instead of $T_t f(x)$.)

The Markov property implies that the operators T_t, $t \geq 0$, constitute a semi-group, i.e. $T_s T_t = T_{s+t}$ for $s, t \geq 0$.

The operator

$$\mathcal{A} f(x) = \lim_{t \downarrow 0} \frac{T_t f(x) - f(x)}{t} \tag{4.5.2}$$

is called the *infinitesimal operator* (of either the Markov process X or the semi-group $(T_t)_{t \geq 0}$ or the transition function $P(t, x; B)$).

Another important characteristic of the Markov process X is its *characteristic operator*

$$\mathfrak{A} f(x) = \lim_{U \downarrow x} \frac{T_{\tau(U)} f(x) - f(x)}{\mathsf{E}_x \tau(U)} \tag{4.5.3}$$

where

$$T_{\tau(U)} f(x) = \mathsf{E}_x f(X_{\tau(U)}), \tag{4.5.4}$$

$\tau(U)$ is the time of the first exit from the neighborhood U of a point x and "$U \downarrow x$" means that the limit is taken as the neighborhood U is diminishing into the point x. (More details about the operators introduced and their relation can be found in [53].)

2. Assume that $E = \mathbb{R}^d$. A continuous Markov process $X = (X_t)_{t \geq 0}$ satisfying the strong Markov property is said to be a *diffusion process* if its characteristic operator $\mathfrak{A}f(x)$ is well defined for any function $f \in C^2(x)$ i.e. for any function which is continuously differentiable in a neighborhood of the point x. It turns out (see [53, Chap. 5, § 5] that for diffusion processes and $f \in C^2(x)$ the operator $\mathfrak{A}f(x)$ is a second order operator

$$Lf(x) = \sum_{i,j=1}^{d} a_{ij}(x) \frac{\partial^2 f(x)}{\partial x_i \partial x_j} + \sum_{i=1}^{d} b_i(x) \frac{\partial f(x)}{\partial x_i} - c(x) f(x) \tag{4.5.5}$$

where $c(x) \geq 0$ and $\sum_{i,j=1}^{d} a_{ij}(x)\lambda_i\lambda_j \geq 0$ (the ellipticity condition) for any $\lambda_1, \dots, \lambda_d$. (Cf. Subsection 4.3.)

The functions $a_{ij}(x)$ and $b_i(x)$ are referred to as diffusion coefficients and drift coefficients respectively. The function $c(x)$ is called a *killing coefficient* (or *discounting rate*).

3. In Subsection 4.3 it was already noted that for given functions $a_{ij}(x)$ and $b_i(x)$ K. Itô provided a construction (based on a stochastic differential equation) of a time-homogeneous Markov process whose diffusion and drift coefficients coincide with the functions $a_{ij}(x)$ and $b_i(x)$. In Subsection 3.4 we also considered problems related to stochastic differential equations in the case of "diffusion with jumps".

4.6. Lévy processes

1. Brownian motion considered above provides an example of a process which, apart from having the Markov property, is remarkable for being a process with (stationary) independent increments. Lévy processes, which will be considered in this subsection, are also processes with (stationary) independent increments. Thus it is natural first to list basic definitions and properties of such processes taking also into account that these processes are semimartingales (discussed in Subsection 3.3).

Let $(\Omega, \mathcal{F}, (\mathcal{F}_t)_{t\geq 0}, \mathsf{P})$ be a filtered probability space. A stochastic process $X = (X_t)_{t\geq 0}$ is called a *process with independent increments* if

(α) $X_0 = 0$;

(β) the trajectories of $(X_t)_{t\geq 0}$ are right-continuous (for $t \geq 0$) and have limits from the left (for $t > 0$);

(γ) the variables X_t are $\mathcal{F}_t$-measurable, $t \geq 0$, and the increments $X_{t_1} - X_{t_0}, X_{t_2} - X_{t_1}, \dots, X_{t_n} - X_{t_{n-1}}$ are independent for all $0 \leq t_0 < t_1 < \dots < t_n$, $n \geq 1$.

Any such process X can be represented in the form $X = D + S$ where $D = (D_t)_{t\geq 0}$ is a deterministic function (maybe of unbounded variation) and $S = (S_t)_{t\geq 0}$ is a semimartingale (see [106, Chap. II, Theorem 5.1]).

Because D is a non-random process, and as such not interesting from the viewpoint of "stochastics", and the process S is both semimartingale and a process with independent increments, from now on we will assume that all processes with independent increments we consider are semimartingales.

Let (B, C, ν) be the triplet of predictable characteristics of a semimartingale X (for details see Subsection 3.3). A remarkable property of processes with independent increments (which are semimartingales as well) is that their triplets (B, C, ν) are *deterministic* (see [106, Chap. II, Theorem 4.15]).

As in Section 3.3 by means of the triplet (B, C, ν) define the *cumulant*

$$K_t(\lambda) = i\lambda B_t - \frac{\lambda^2}{2} C_t + \int_{\mathbb{R}} \big(e^{i\lambda x} - 1 - i\lambda h(x)\big)\, \nu\big((0,t] \times dx\big) \tag{4.6.1}$$

($\lambda \in \mathbb{R}$, $t \ge 0$) where $h = h(x)$ is a truncation function (the standard one is $h(x) = xI(|x| \le 1)$) and ν is the compensator of the measure of jumps of the process X. With the cumulant $K_t(\lambda)$, $t \ge 0$, we associate the *stochastic exponential* $\mathcal{E}(\lambda) = (\mathcal{E}_t(\lambda))_{t\ge 0}$ defined as a solution to the equation

$$d\mathcal{E}_t(\lambda) = \mathcal{E}_{t-}(\lambda)\, dK_t(\lambda), \qquad \mathcal{E}_0(\lambda) = 1. \tag{4.6.2}$$

A solution to this equation is the function

$$\mathcal{E}_t(\lambda) = e^{K_t(\lambda)} \prod_{s\le t} \big(1 + \Delta K_s(\lambda)\big) e^{-\Delta K_s(\lambda)}. \tag{4.6.3}$$

The Itô formula (page 67) immediately implies that for $\Delta K_t(\lambda) \ne -1$, $t \ge 0$, the process $M(\lambda) = (M_t(\lambda), \mathcal{F}_t)_{t\ge 0}$ given by

$$M_t(\lambda) = \frac{e^{i\lambda X_t}}{\mathcal{E}_t(\lambda)} \tag{4.6.4}$$

is a martingale for every $\lambda \in \mathbb{R}$, and so (since $\mathcal{E}_t(\lambda)$ is deterministic) the characteristic function of X_t equals $\mathcal{E}_t(\lambda)$, i.e.

$$\mathsf{E}\, e^{i\lambda X_t} = \mathcal{E}_t(\lambda). \tag{4.6.5}$$

In particular, if B_t, C_t and $\nu((0,t] \times A)$ are continuous functions in t then $\Delta K_t(\lambda) = 0$ and (4.6.5) becomes the well-known (generalized) *Kolmogorov–Lévy–Khintchine formula*:

$$\mathsf{E}\, e^{i\lambda X_t} = \exp\big(K_t(\lambda)\big). \tag{4.6.6}$$

2. Lévy processes are processes whose triplets have the very special structure:

$$B_t = bt, \quad C_t = ct \quad \text{and} \quad \nu(dt, dx) = dt\, F(dx) \tag{4.6.7}$$

where $F = F(dx)$ is a measure on $\mathbb{R}$ such that $F(\{0\}) = 0$ and

$$\int_{\mathbb{R}} \min(1, x^2)\, F(dx) < \infty. \tag{4.6.8}$$

A Lévy process has stationary independent increments, is continuous in probability, and has no fixed time of jump with probability 1.

Apart from Brownian motion, a classical example of a Lévy process is the Poisson process $N = (N_t)_{t \geq 0}$ (with parameter $a > 0$) which is characterized by its piecewise constant trajectories with unit jumps. If $T_1, T_2, \ldots$ are times of jumps then the random variables $T_1 - T_0$ (with $T_0 = 0$), $T_2 - T_1$, $T_3 - T_2$, ... are independent and identically distributed with $\mathsf{P}(T_1 > t) = e^{-at}$ (exponential distribution). It is clear that $N_t = \sum_{n \geq 0} I(T_n \leq t)$. For the Poisson process one has $b = a$, $c = 0$ $\nu(dt, dx) = a\, dt\, \delta_{\{1\}}(dx)$ where $\delta_{\{1\}}$ is a measure concentrated at the point $\{1\}$. Thus from (4.6.6) we find that

$$\mathsf{E}\, e^{i\lambda N_t} = \exp\big(at(e^{i\lambda} - 1)\big). \tag{4.6.9}$$

Another example of a jump-like Lévy process is given by the so-called *compound Poisson process*. By definition it is a process with the triplet $(0, 0, F)$ (with respect to the truncation function $h \equiv 0$) where the measure $F = F(dx)$ is such that $F(\mathbb{R}) < \infty$. Such a process X admits the following explicit construction:

$$X_t = \sum_{k=0}^{N_t} \xi_k \tag{4.6.10}$$

where $\xi_0 = 0$, $(\xi_k)_{k \geq 1}$ is a sequence of independent and identically distributed random variables with distribution $F(dx)/F(\mathbb{R})$ and N is a Poisson process with parameter $a = F(\mathbb{R})$.

3. An important subclass of the class of Lévy processes is formed by (strictly) α-stable processes $X = (X_t)_{t \geq 0}$ which are characterized by the following *self-similarity property*: for any $c > 0$,

$$\mathrm{Law}(X_{ct}; t \geq 0) = \mathrm{Law}(c^{1/\alpha} X_t; t \geq 0) \tag{4.6.11}$$

where $0 < \alpha \leq 2$.

For such processes the characteristic function is given by the following *Lévy–Khintchine representation*:

$$\mathsf{E}\, e^{i\lambda X_t} = \exp\big(t\psi(\lambda)\big) \tag{4.6.12}$$

where

$$\psi(\lambda) = \begin{cases} i\mu\lambda - \sigma^\alpha |\lambda|^\alpha \Big(1 - i\beta (\operatorname{sgn} \lambda) \tan \dfrac{\pi\alpha}{2}\Big), & \alpha \neq 1, \\ i\mu\lambda - \sigma|\lambda| \Big(1 + i\beta \dfrac{2}{\pi} (\operatorname{sgn} \lambda) \log |\lambda|\Big), & \alpha = 1 \end{cases} \tag{4.6.13}$$

with parameters

$$0 < \alpha \le 2, \quad |\beta| \le 1, \quad \sigma > 0, \quad \mu \in \mathbb{R}. \tag{4.6.14}$$

In the symmetrical case,

$$\psi(\lambda) = -\sigma^{\alpha}|\lambda|^{\alpha}. \tag{4.6.15}$$

Denote by $S_\alpha(\sigma, \beta, \mu)$ the stable laws with parameters α, σ, β, μ. Unfortunately the explicit form of this distribution is known only for a few special values of these parameters. These are as follows:

- $S_2(\sigma, 0, \mu) = \mathcal{N}(\mu, 2\sigma^2)$ — the normal distribution.
- $S_1(\sigma, 0, \mu)$ — Cauchy's distribution with the density

$$\frac{\sigma}{\pi((x-\mu)^2 + \sigma^2)}. \tag{4.6.16}$$

- $S_{1/2}(\sigma, 1, \mu)$ — the unilateral stable (Lévy–Smirnov) distribution with the density

$$\sqrt{\frac{\sigma}{2\pi}}\,\frac{1}{(x-\mu)^{3/2}}\exp\Bigl(-\frac{\sigma}{2(x-\mu)}\Bigr), \qquad x \in (\mu, \infty). \tag{4.6.17}$$

4. The exposed results on properties of the characteristic functions of processes with independent increments and Lévy processes are commonly referred to as results of *analytical* probability theory. There is another approach to studying such processes that is based on *stochastic analysis* of trajectories.

Let again $X = (X_t)_{t \ge 0}$ be a process with independent increments that is also a semimartingale. Then according to Subsection 3.3 the following *canonical representation* holds:

$$X_t = X_0 + B_t + X_t^c + \int_0^t \int_{\mathbb{R}} h(x)\, d(\mu - \nu) + \int_0^t \int (x - h(x))\, d\mu. \tag{4.6.18}$$

In this case:

B_t is a deterministic function;

X_t^c is a Gaussian process such that $\mathsf{D} X_t^c = C_t$;

$\nu = \nu((0,t] \times dx)$ is a deterministic function; and

$\mu = \mu(\omega; (0,t] \times dx)$ is a Poisson measure (see [106, Chap. II, § 1c]).

If the compensator ν has the form $\nu = \nu((0,t] \times dx) = t\,F(dx)$ (as in the case of Lévy processes), the measure μ is called a homogeneous Poisson measure. Then the random variable $\mu((0,t] \times A)$ has a Poisson distribution such that

$$\mathsf{E}\,\mu((0,t] \times A) = \nu((0,t] \times A). \tag{4.6.19}$$

The representation (4.6.18) is often called an *Itô–Lévy representation*. In the case of Lévy processes one has $B_t = bt$, $C_t = ct$ and, as already noted, $\nu((0,t]\times dx) = t\,F(dx)$.

5. Basic transformations

5.1. Change of time

1. When solving optimal stopping problems, if we want to obtain solutions in the closed form, we have to resort to various transformations both of the processes under consideration as well as of the equations determining one or another characteristic of these processes.

The best known and most important such transformations are:

(a) change of time (often applied simultaneously with (b) below);

(b) change of space;

(c) change of measure;

and others (killing, creating, ...).

2. In this subsection we concentrate on basic ideas related merely to *change of time*. In the next subsection we shall deal with problems of *change of space* together with methods of change of time because it is by combination of these methods that one succeeds to obtain results on transformation of "complicated" processes into "simple" ones.

The problem of change of time can be approached in the following way.

Imagine we have a process $X = (X_t)_{t\ge0}$ with rather complicated structure e.g. a process with differential $dX_t = \sigma(t, X_t)\,dW_t$ where $(W_t)_{t\ge0}$ is a Wiener process. Such a process, as we shall see, can be represented in the form

$$X = \widehat{X} \circ T \tag{5.1.1}$$

(i.e. in the form $X_t = \widehat{X}_{T(t)}$, $t \ge 0$) where $\widehat{X} = (\widehat{X}_\theta)_{\theta\ge0}$ is a certain "simple" process in the "new" time θ and $\theta = T(t)$ is a certain change of time exercising the transformation of the "old" time t into a "new" time θ.

Moreover the process $\widehat{X} = (\widehat{X}_\theta)_{\theta\ge0}$ can be chosen to be a very simple process—a Brownian motion.

This can be realized in the following way.

Let

$$T(t) = \int_0^t \sigma^2(u, X_u)\,du \tag{5.1.2}$$

and

$$\widehat{T}(\theta) = \inf\{t : T(t) > \theta\}. \tag{5.1.3}$$

Assume that $\sigma^2(s, X_s) > 0$, $\int_0^t \sigma^2(s, X_s)\,ds < \infty$, $t > 0$, and $\int_0^\infty \sigma^2(s, X_s)\,ds = \infty$ P-a.s. Then $T(t)$ is a continuous increasing process, $\widehat{T} = \inf\{t : T(t) = \theta\}$ and so

$$\int_0^{\widehat{T}(\theta)} \sigma^2(u, X_u)\,du = T\big(\widehat{T}(\theta)\big) = \theta. \tag{5.1.4}$$

Notice that the latter formula yields that

$$\frac{d\widehat{T}(\theta)}{d\theta} = \frac{1}{\sigma^2(\widehat{T}(\theta), X_{\widehat{T}}(\theta))}. \tag{5.1.5}$$

In the time-homogeneous case when σ depends only on x ($\sigma = \sigma(x)$) the measure $m = m(dx)$ defined by

$$m(dx) = \frac{1}{\sigma^2(x)}\,dx \tag{5.1.6}$$

is called the *speed measure* ("responsible" for the change of time).

In the "new" time θ let

$$\widehat{X}_\theta = X_{\widehat{T}(\theta)} = \int_0^{\widehat{T}(\theta)} \sigma(u, X_u)\,dW_u. \tag{5.1.7}$$

Immediately we see that

$$\mathsf{E}\,\widehat{X}_\theta = 0, \tag{5.1.8}$$

$$\mathsf{E}\,\widehat{X}_\theta^2 = \mathsf{E}\int_0^{\widehat{T}(\theta)} \sigma^2(u, X_u)\,du = \theta. \tag{5.1.9}$$

The process $\widehat{X} = (\widehat{X}_\theta)_{\theta\geq 0}$ is a martingale with respect to the filtration $(\mathcal{F}_{\widehat{T}(\theta)})_{\theta\geq 0}$ and thus—by the "Lévy characterization theorem" (page 94)—is a Brownian motion.

So the process

$$\widehat{X} = X \circ \widehat{T} \tag{5.1.10}$$

is a Brownian motion, and it is clear that the inverse passage—to the process X—is realized by the formula

$$X = \widehat{X} \circ T \tag{5.1.11}$$

that means, in more details, that

$$X_t = \widehat{X}_{T(t)} = \widehat{X}_{\int_0^t \sigma^2(s, X_s)ds}, \qquad t \geq 0. \tag{5.1.12}$$

In this way we have obtained the well-known result that a (local) martingale $X_t = \int_0^t \sigma(s, X_s)\,dW_s$ can be represented as a time change (by $\theta = T(t)$) of a new Brownian motion $\widehat{X}$, i.e.

$$X = \widehat{X} \circ T \tag{5.1.13}$$

(see Lemma 5.1 below).

Let us consider the same problem—on the representation of a "complicated" process $X_t = \int_0^t \sigma(s, X_s)\, dW_s$ by means of a "simple" process (in our case by means of a Brownian motion)—in terms of transition probabilities of a (Markov) process X.

Assume that the coefficient $\sigma = \sigma(s,x)$ is such that the equation $dX_t = \sigma(t, X_t)\, dW_t$ with $X_0 = 0$ has a unique strong solution which is a Markov process. (Sufficient conditions for this can be found in [94].) Denote by $f = f(s,x;t,y)$ its transition density:

$$f(s,x;t,y) = \frac{\partial \mathsf{P}(X_t \le y \mid X_s = x)}{\partial y}. \tag{5.1.14}$$

This density satisfies the forward equation (for $t > s$)

$$\frac{\partial f}{\partial t} = \frac{1}{2} \frac{\partial^2}{\partial y^2} \big(\sigma^2(t,y) f\big) \tag{5.1.15}$$

and the backward equation (for $s < t$)

$$\frac{\partial f}{\partial s} = -\frac{1}{2} \sigma^2(s,x) \frac{\partial^2 f}{\partial x^2}. \tag{5.1.16}$$

Leaving unchanged the space variable introduce the new time $\theta = T(t)$ ($= \int_0^t \sigma^2(u, X_u)\, du$) and denote $\theta' = T(s)$ for $s < t$. Under these assumptions let

$$f'(\theta', x; \theta, y) = f(s,x;t,y). \tag{5.1.17}$$

The function $f' = f'(\theta', x; \theta, y)$ is non-negative, satisfies the normalizing condition

$$\int f'(\theta', x; \theta, y)\, dy = 1 \qquad (\theta' < \theta) \tag{5.1.18}$$

and the Chapman–Kolmogorov equation (see (4.3.2)). It is not difficult to derive the corresponding forward and backward equations (in the "new θ-time"):

$$\frac{\partial f'}{\partial \theta} = \frac{1}{2} \frac{\partial^2 f'}{\partial y^2} \quad \text{and} \quad \frac{\partial f'}{\partial \theta'} = -\frac{1}{2} \frac{\partial^2 f'}{\partial y^2} \tag{5.1.19}$$

which follows from

$$\frac{\partial f'}{\partial \theta} = \frac{\partial f}{\partial t} \frac{1}{\frac{\partial \theta}{\partial t}} = \frac{\partial f}{\partial t} \frac{1}{\frac{\partial T(t)}{\partial t}} = \frac{1}{2} \sigma^2(t,y) \frac{\partial^2 f}{\partial y^2} \frac{1}{\sigma^2(t,y)} = \frac{1}{2} \frac{\partial^2 f}{\partial y^2} \tag{5.1.20}$$

(and in an analogous way for the backward equation).

It is clear that the transition function $f' = f'(\theta', x; \theta, y)$ is the transition function of a Brownian motion $\widehat{X} = (\widehat{X}_\theta)_{\theta \ge 0}$ given by the Laplace formula:

$$f'(\theta', x; \theta, y) = \frac{1}{\sqrt{2\pi(\theta - \theta')}} \exp\left(-\frac{(y-x)^2}{2(\theta - \theta')}\right). \tag{5.1.21}$$

3. Let us dwell on certain general principles of changing the "new" time θ into the "old" time t and vice versa. To this aim assume that we are given a filtered probability space $(\Omega, \mathcal{F}, (\mathcal{F}_t)_{t \ge 0}, \mathsf{P})$.

One says that a family of random variables $\widehat{T} = (\widehat{T}(\theta))_{\theta \ge 0}$ performs a change of time (to be more precise, a change of the "new" θ-time into the "old" t-time) if the following two conditions are fulfilled:

(a) $(\widehat{T}(\theta))_{\theta \ge 0}$ is a right-continuous non-decreasing family of random variables $\widehat{T}(\theta)$ which in general take their values in $[0, \infty]$;

(b) for every $\theta \ge 0$ the variable $\widehat{T}(\theta)$ is a Markov time, i.e. for all $t \ge 0$,

$$\{\widehat{T}(\theta) \le t\} \in \mathcal{F}_t. \tag{5.1.22}$$

Assume that the initial filtered probability space is the space $(\Omega, \mathcal{F}, (\widehat{\mathcal{F}}_\theta)_{\theta \ge 0}, \mathsf{P})$ and we are given a family of random variables $T = (T(t))_{t \ge 0}$ with the property analogous to (a) above and such that for all $t \ge 0$ the variables $T(t)$ are Markov times i.e. for all $\theta \ge 0$,

$$\{T(t) \le \theta\} \in \widehat{\mathcal{F}}_\theta. \tag{5.1.23}$$

Then we say that the family $T = (T(t))_{t \ge 0}$ changes the "old" t-time into the "new" θ-time. The primary method of construction of the system $(\widehat{T}(\theta))_{\theta \ge 0}$ (and, in an analogous way, of $T = (T(t))_{t \ge 0}$) consists in the following.

Let a process $A = (A_t)_{t \ge 0}$ defined on a filtered probability space $(\Omega, \mathcal{F}, (\mathcal{F}_t)_{t \ge 0}, \mathsf{P})$ be such that $A_0 = 0$, the variables A_t are $\mathcal{F}_t$-measurable and its trajectories are right-continuous (for $t \ge 0$) and have limits from the left (for $t > 0$).

Define

$$\widehat{T}(\theta) = \inf\{t \ge 0 : A_t > \theta\}, \qquad \theta \ge 0 \tag{5.1.24}$$

where as usual $\inf(\varnothing) = \infty$. It can be verified that the system $\widehat{T} = (\widehat{T}(\theta))_{\theta \ge 0}$ determines a change of time in the sense of the above definition.

In connection with transforming the process $X = (X_t)_{t \ge 0}$ into a new process $\widehat{X} = (\widehat{X}_\theta)_{\theta \ge 0}$ with $\widehat{X}_\theta = X_{\widehat{T}(\theta)}$ notice that to ensure $\mathcal{F}_{\widehat{T}(\theta)}$-measurability of the variables $X_{\widehat{T}(\theta)}$ one has to impose certain additional measurability conditions on the process X. For example it suffices to assume that the process $X = (X_t)_{t\ge 0}$ is progressively measurable, i.e. such that for any $t \ge 0$,

$$\{(\omega, s) \in \Omega \times [0,t] : X_s(\omega) \in B\} \in \mathcal{F}_t \times \mathcal{B}([0,t]). \tag{5.1.25}$$

(This property is guaranteed if X is right-continuous.)

Letting as above $T(t) = A_t$, $t \ge 0$, and $\widehat{T}(\theta) = \inf\{t : T(t) > \theta\}$ we easily find that the following useful properties hold:

(i) if the process $A = (A_t)_{t\ge 0}$ is increasing and right-continuous then

$$\widehat{T}(T(t)) = t, \quad T(\widehat{T}(\theta)) = \theta, \qquad \widehat{T}(\theta) = T^{-1}(\theta), \quad T(t) = \widehat{T}^{-1}(t); \tag{5.1.26}$$

(ii) for non-negative functions $f = f(t)$,

$$\int_0^{\widehat{T}(b)} f(t)\, d(T(t)) = \int_0^b f(\widehat{T}(\theta))\, d\theta, \tag{5.1.27}$$

$$\int_0^{T(a)} f(\theta)\, d(\widehat{T}(\theta)) = \int_0^a f(T(t))\, dt. \tag{5.1.28}$$

4. The following two lemmas are best known results on change of time in stochastic analysis.

Lemma 5.1. (Dambis–Dubins–Schwarz) *Let $X = (X_t)_{t\ge 0}$ be a continuous local martingale defined on a filtered probability space $(\Omega, \mathcal{F}, (\mathcal{F}_t)_{t\ge 0}, \mathsf{P})$ with $X_0 = 0$ and $\langle X\rangle_\infty = \infty$, and let*

$$T(t) = \langle X\rangle_t, \tag{5.1.29}$$

$$\widehat{T}(\theta) = \inf\{\, t \ge 0 : T(t) > \theta\,\}. \tag{5.1.30}$$

Then

(a) *the process $\widehat{X} = (\widehat{X}_\theta)_{\theta\ge 0}$ given by*

$$\widehat{X}_\theta = X_{\widehat{T}(\theta)} \tag{5.1.31}$$

is a Brownian motion;

(b) *the process X can be reconstructed from the Brownian motion $\widehat{X}$ by formulae:*

$$X_\theta = \widehat{X}_{T(t)}, \qquad t \ge 0. \tag{5.1.32}$$

Lemma 5.2. *Let $X = (X_t)_{t\ge 0}$ be a counting process, $X_0 = 0$, with compensator $A = (A_t)_{t\ge 0}$ and the Doob–Meyer decomposition $X = M + A$ where $M = (M_t)_{t\ge 0}$ is a local martingale. Let the compensator A be continuous (then $A = \langle M\rangle$).*

Define $T(t) = A_t \quad (= \langle M_t \rangle)$ *and*

$$\widehat{T}(\theta) = \inf\{\, t \geq 0 : T(t) > \theta \,\}. \tag{5.1.33}$$

Under the assumption that $T(t) \uparrow \infty$ *as* $t \to \infty$ *the process* $\widehat{X} = (\widehat{X}_\theta)_{\theta \geq 0}$ *given by*

$$\widehat{T}(\theta) = X_{\widehat{T}(\theta)} \tag{5.1.34}$$

is a standard Poisson process (with parameter 1*). The process* $X = (X_t)_{t \geq 0}$ *can be reconstructed from the process* $\widehat{X} = (\widehat{X}_\theta)_{\theta \geq 0}$ *by*

$$X_t = \widehat{X}_{T(t)}. \tag{5.1.35}$$

A proof can be found e.g. in [128].

5.2. Change of space

1. For the first time the question about transformations of (diffusion) processes into "simple" processes (like Brownian motion) was considered by A. N. Kolmogorov in "Analytical methods" [111]. He simultaneously considered both change of time $(\, t \to \theta\,,\ s \to \vartheta\,)$ and change of space $(\, x \to \xi\,,\ y \to \eta\,)$ with the purpose to transform the transition function $f = f(s, x; t, y)$ into a new transition functions $g = g(\vartheta, \xi; \theta, \eta)$ which satisfies simpler (forward and backward) differential equations than the equations for $f = f(s, z; t, y)$. To illustrate this consider an Ornstein–Uhlenbeck process $X = (X_t)_{t \geq 0}$ solving

$$dX_t = \big(\alpha(t) - \beta(t)X_t\big)\, dt + \sigma(t)\, dW_t, \qquad X_0 = x_0. \tag{5.2.1}$$

The coefficients of this equation are assumed to be such that

$$\int_0^t \left|\frac{\alpha(s)}{\gamma(s)}\right| ds < \infty, \quad \int_0^t \left|\frac{\sigma(s)}{\gamma(s)}\right|^2 ds < \infty \qquad \text{for } \ t > 0, \tag{5.2.2}$$

$$\int_0^t \left|\frac{\sigma(s)}{\gamma(s)}\right|^2 ds \uparrow \infty \quad \text{as } \ t \uparrow \infty \quad \text{where } \gamma(s) = \exp\Big(-\int_0^t \beta(u)\, du\Big). \tag{5.2.3}$$

It is not difficult to check, e.g. using Itô's formula, that

$$X_t = \gamma(t)\Big[x_0 + \int_0^t \frac{\alpha(s)}{\gamma(s)}\, ds + \int_0^t \frac{\sigma(s)}{\gamma(s)}\, dW_s\Big]. \tag{5.2.4}$$

Introducing the "new" time $\theta = T(t)$ with

$$T(t) = \int_0^t \left(\frac{\sigma(s)}{\gamma(s)}\right)^2 ds \tag{5.2.5}$$

and letting $\widehat{T}(\theta) = \inf\{t \geq 0 : T(t) > \theta\}$ as in Subsection 5.1 we find (from Lemma 5.1 above) that the process

$$\widehat{B}_\theta = \int_0^{\widehat{T}(\theta)} \frac{\sigma(s)}{\gamma(s)}\, dW_s \tag{5.2.6}$$

is a Brownian motion (Wiener process) and

$$X_t = \varphi(t) + \gamma(t)\widehat{B}_{T(t)} \tag{5.2.7}$$

where

$$\varphi(t) = \gamma(t)\left[x_0 + \int_0^t \frac{\alpha(s)}{\gamma(s)}\, ds\right]. \tag{5.2.8}$$

Following Kolmogorov [111] introduce the function

$$\Psi(t,y) = \frac{y}{\gamma(t)} - \int_0^t \frac{\alpha(u)}{\gamma(u)}\, du \tag{5.2.9}$$

and the function

$$g(\vartheta,\xi;\theta,\eta) = \frac{1}{\left(\frac{\partial \Psi(t,y)}{\partial y}\right)}\, f(s,x;t,y) \tag{5.2.10}$$

where $f(s,x;t,y)$ is the transition density of the process X (with $x_0 = 0$) and

$$\eta = \Psi(t,y), \quad \xi = \Psi(s,x), \tag{5.2.11}$$

$$\vartheta = \int_0^s \left(\frac{\sigma(u)}{\gamma(u)}\right)^2 du, \quad \theta = \int_0^t \left(\frac{\sigma(u)}{\gamma(u)}\right)^2 du.$$

Taking into account this notation it is not difficult to verify (see also the formulae (5.2.22) and (5.2.23) below) that

$$\frac{\partial g}{\partial \theta} = \frac{1}{2}\frac{\partial^2 g}{\partial \eta^2} \quad \text{and} \quad \frac{\partial g}{\partial \vartheta} = -\frac{1}{2}\frac{\partial^2 g}{\partial \xi^2}. \tag{5.2.12}$$

Thus the transformation of time

$$t \rightsquigarrow \theta = T(t) \qquad (s \rightsquigarrow \vartheta = T(s)) \tag{5.2.13}$$

and the transformation of space

$$y \rightsquigarrow \eta = \Psi(t,y) \qquad (x \rightsquigarrow \xi = \Psi(s,x)) \tag{5.2.14}$$

turn the function $f = f(s,x;t,y)$ into the function $g = g(\vartheta,\xi;\theta,\eta)$ which is the transition function of a Brownian motion. Since $\Psi(t,X_t) = \int_0^t \frac{\sigma(s)}{\gamma(s)}\, dW_s$ we have

$$\Psi(\widehat{T}(\theta), X_{\widehat{T}(\theta)}) = \int_0^{\widehat{T}(\theta)} \frac{\sigma(s)}{\gamma(s)}\, ds = \widehat{B}_\theta. \tag{5.2.15}$$

So the change of time $\widehat{T}(\theta)$ and the function $\Psi = \Psi(t, y)$ allow one to construct (in the "new" time θ) a Brownian motion $\widehat{B}$ (which in turn allows, by reverse change of time $T(t)$, to construct the process X according to (5.2.7)). All this makes the equations (5.2.12) transparent.

2. The preceding discussion about an Ornstein–Uhlenbeck process admits an extension to more general diffusion processes. Namely, following Kolmogorov [111], assume that $f = f(s, x; t, y)$ is a transition density satisfying the following forward and backward equation:

$$\frac{\partial f}{\partial t} = -\frac{\partial}{\partial y}\big[b(t,y)f\big] + \frac{1}{2}\frac{\partial^2}{\partial y^2}\big[a(t,y)f\big], \tag{5.2.16}$$

$$\frac{\partial f}{\partial s} = -b(s,x)\,\frac{\partial f}{\partial x} - \frac{1}{2}\,a(s,x)\,\frac{\partial^2 f}{\partial x^2} \tag{5.2.17}$$

respectively. In other words, we consider a diffusion Markov process $X = (X_t)_{t \ge 0}$ which is a strong solution to the stochastic differential equation

$$dX_t = b(t, X_t)\,dt + \sigma(t, X_t)\,dW_t \tag{5.2.18}$$

where $W = (W_t)_{t\ge 0}$ is a Wiener process and $\sigma^2(t,x) = a(t,x)$. (The coefficients $b(t,x)$ and $\sigma(t,x)$ are assumed to be such that the process X is Markov.)

As in the previous subsection introduce a change of time $\theta = T(t)$ ($\vartheta = T(s)$) and a change of the space variable $\eta = \Psi(t,y)$ ($\xi = \Psi(s,x)$).

We assume that $T(t)$ is an increasing function from the class C^1, the function $\Psi(t,y)$ belongs to the class $C^{1,2}$ and $\partial\Psi(t,y)/\partial y > 0$. Putting

$$g(\vartheta,\xi;\theta,\eta) = \frac{f(s,x;t,y)}{\frac{\partial\Psi(t,y)}{\partial y}}, \tag{5.2.19}$$

$$B(\theta,\eta) = \frac{\frac{1}{2}a(t,y)\frac{\partial^2\Psi(t,y)}{\partial y^2} + b(t,y)\frac{\partial\Psi(t,y)}{\partial y} + \frac{\partial\Psi(t,y)}{\partial t}}{\frac{\partial T(t)}{\partial t}}, \tag{5.2.20}$$

$$A(\theta,\eta) = \frac{a(t,y)\Big(\frac{\partial\Psi(t,y)}{\partial y}\Big)^2}{\frac{\partial T(t)}{\partial t}} \tag{5.2.21}$$

(and taking into account the connection between the variables introduced) we find that the non-negative function $g = g(\vartheta,\xi;\theta,\eta)$ satisfies the normalizing condition ($\int_{\mathbb{R}} g(\vartheta,\xi;\theta,\eta)\,d\eta = 1$) and solves the Chapman–Kolmogorov equation as well as the following forward and backward equations (in (θ,η) and (ϑ,ξ) respectively):

$$\frac{\partial g}{\partial \theta} = -\frac{\partial}{\partial \eta}\big[B(\theta,\eta)g\big] + \frac{1}{2}\frac{\partial^2}{\partial \eta^2}\big[A(\theta,\eta)g\big] \tag{5.2.22}$$

$$\frac{\partial g}{\partial \vartheta} = -B(\vartheta,\xi)\frac{\partial g}{\partial \xi} - \frac{1}{2}A(\vartheta,\eta)\frac{\partial^2 g}{\partial \xi^2}. \tag{5.2.23}$$

In other words, if the process X has characteristics (b, a) then the process $\widehat{X} = (\widehat{X}_\theta)_{\theta \ge 0}$ with

$$\widehat{X}_\theta = \Psi\big(\widehat{T}(\theta), X_{\widehat{T}(\theta)}\big) \tag{5.2.24}$$

is a (diffusion) process with characteristics (B, A) i.e. it satisfies the equation

$$d\widehat{X}_\theta = B(\theta, \widehat{X}_\theta)\, d\theta + \Sigma(\theta, \widehat{X}_\theta)\, d\widehat{W}_\theta \tag{5.2.25}$$

where $\Sigma^2 = A$ and $\widehat{W} = (\widehat{W}_\theta)_{\theta \ge 0}$ is a Wiener process.

3. Let us address the case when the initial process $X = (X_t)_{t \ge 0}$ is a *time-homogeneous* Markov diffusion process solving

$$dX_t = b(X_t)\, dt + \sigma(X_t)\, dW_t, \qquad \sigma(x) > 0. \tag{5.2.26}$$

In this case it is natural to choose Ψ to be homogeneous in the sense that $\Psi = \Psi(y)$; according to the tradition we shall denote this function by $S = S(y)$. From (5.2.20) we see that for the coefficient $B = B(\eta)$ to be equal to zero, the function $S(y)$ must satisfy the equation ($a = \sigma^2$),

$$\frac{1}{2} a(y) S''(y) + b(y) S'(y) = 0 \tag{5.2.27}$$

i.e.

$$\frac{S''(y)}{S'(y)} = -2\, \frac{b(y)}{a(y)}. \tag{5.2.28}$$

Solving this equation we find that

$$S(y) = c_1 + c_2 \int_{y_0}^{y} \exp\left\{ -\int_{y_0}^{z} \frac{2b(v)}{\sigma^2(v)}\, dv \right\} dz \tag{5.2.29}$$

where c_1 , c_2 and y_0 are constants.

This function (to be more precise any positive of these functions) is called a *scale function*. If we let $\theta = t$ (i.e. the “new” time coincides with the old one so that $T(t) = t$) then (5.2.21) yields

$$A(\eta) = \sigma^2(y)\big(S'(y)\big)^2. \tag{5.2.30}$$

Thus from (5.2.27) and (5.2.30) we conclude that for $Y_t = S(X_t)$

$$dY_t = \sigma(X_t) S'(X_t)\, d\widetilde{W}_t \tag{5.2.31}$$

where $\widetilde{W} = (\widetilde{W}_t)_{t \ge 0}$ is a Wiener process. The Dambis–Dubins–Schwarz lemma (page 110) implies that if

$$T_1(t) = \int_0^t \big(\sigma(X_u) S'(X_u)\big)^2\, du \quad \text{and} \quad \widehat{T}_1(\theta) = \inf\{\, t : T_1(t) > \theta \,\} \tag{5.2.32}$$

then in the "new" θ-time the process $\widehat{B} = (\widehat{B}_\theta)_{\theta \ge 0}$ given by

$$\widehat{B}_\theta = Y_{\widehat{T}_1(\theta)} = S\big(X_{\widehat{T}_1(\theta)}\big) = \int_0^{\widehat{T}_1(\theta)} \sigma(X_u) S'(X_u)\, d\widetilde{B}_u \tag{5.2.33}$$

is a Wiener process. Since $Y_t = S(X_t)$ we have $X_t = S^{-1}(Y_t)$. Here $Y_t = \widehat{B}_{T_1(t)}$. Therefore

$$X_t = S^{-1}\big(\widehat{B}_{T_1(t)}\big). \tag{5.2.34}$$

5.3. Change of measure

1. The essence of transformations which are related to a change of measure (and have also many applications in solving optimal stopping problems) may be described as follows.

The change of time aims to represent the "complicated" process $X = (X_t)_{t \ge 0}$ as a composition $X = \widehat{X} \circ T$ (i.e. $X_t = \widehat{X}_{T(t)}$, $t \ge 0$) where $\widehat{X} = (\widehat{X}_\theta)_{\theta \ge 0}$ is a "simple" process and $T = T(t)$ is a change of time.

Unlike the change of time (which, so to speak, makes the speed of movement along trajectories to vary) a change of measure does not deal with transformations of trajectories of X but transforms the initial probability measure P into another probability measure $\widetilde{\mathsf{P}}$ (which is equivalent to P) in such a way that

$$\mathrm{Law}(X \mid \widetilde{\mathsf{P}}) = \mathrm{Law}(\widetilde{X} \mid \mathsf{P}) \tag{5.3.1}$$

where $\widetilde{X} = (\widetilde{X}_t)_{t \ge 0}$ is a "simple" process (with respect to the measure P).

To illustrate this consider a filtered probability space $(\Omega, \mathcal{F}, (\mathcal{F}_t)_{t \ge 0}, \mathsf{P})$ with a Brownian motion $B = (B_t, \mathcal{F}_t)_{t \ge 0}$ and an adapted integrable process $b = (b_t, \mathcal{F}_t)_{t \ge 0}$ defined on this space.

Consider an Itô process $X = (X_t, \mathcal{F}_t)_{t \ge 0}$ solving

$$dX_t = b_t\, dt + dB_t. \tag{5.3.2}$$

Form a new measure $\widetilde{\mathsf{P}}$ such that its restriction $\widetilde{\mathsf{P}}_t = \widetilde{\mathsf{P}} | \mathcal{F}_t$ onto the σ-algebra $\mathcal{F}_t$ is given by

$$d\widetilde{\mathsf{P}}_t = Z_t\, d\mathsf{P} \tag{5.3.3}$$

where

$$Z_t = \exp\left\{ \int_0^t \big(\widetilde{b}_s(\omega) - b_s(\omega)\big)\, dB_s - \frac{1}{2} \int_0^t \big(\widetilde{b}_s(\omega) - b_s(\omega)\big)^2\, ds \right\} \tag{5.3.4}$$

and $\widetilde{b} = (\widetilde{b}_t, \mathcal{F}_t)_{t \ge 0}$ is another adapted integrable process. Assume that $\int_0^t \big(\widetilde{b}_s(\omega) - b_s(\omega)\big)^2\, ds < \infty$ P-a.s. and $\mathsf{E}\, Z_t = 1$ for every $t > 0$. Notice that the required measure $\widetilde{\mathsf{P}}$ satisfying the property $\widetilde{\mathsf{P}}_t = Z_t \, d\mathsf{P}$ for $t > 0$ can certainly be constructed on the σ-algebra $\bigvee_{t \geq 0} \mathcal{F}_t \; (= \sigma(\bigcup_{t \geq 0} \mathcal{F}_t))$ under the assumption that the set Ω is a space of functions $\omega = \omega(t)$, $t \geq 0$ which are right-continuous (for $t \geq 0$) and have limits from the left (for $t > 0$) and that all processes considered are canonical (see e.g. [209]).

According to the well-known *Girsanov theorem for Brownian motion* (see [77]) the process X with respect to the new measure $\widetilde{\mathsf{P}}$ becomes a Brownian motion; this can be expressed as

$$\operatorname{Law}(X \mid \widetilde{\mathsf{P}}) = \operatorname{Law}(B \mid \mathsf{P}). \tag{5.3.5}$$

In other words, the transition from the measure P to the measure $\widetilde{\mathsf{P}}$ "kills" the drift of the process X.

Now, keeping ourselves within the framework of Itô processes, consider the case when the change of measure $\mathsf{P} \rightsquigarrow \widetilde{\mathsf{P}}$ results in a change of drift $b \rightsquigarrow \widetilde{b}$ of the process X where $\widetilde{b} = (\widetilde{b}_t, \mathcal{F}_t)_{t \geq 0}$ is another adapted integrable process.

To this end introduce new measures $\widetilde{\mathsf{P}}_t$, $t \geq 0$ as above by letting $d\widetilde{\mathsf{P}}_t = Z_t \, d\mathsf{P}$ where Z_t is specified in (5.3.4) and the coefficients b and $\widetilde{b}$ are again such that $\mathsf{E} Z_t = 1$, $t > 0$. In this case the Girsanov result says that with respect to the new measure $\widetilde{\mathsf{P}}$ the process $\widetilde{B} = (\widetilde{B}_t, \mathcal{F}_t)_{t \geq 0}$ given by

$$\widetilde{B}_t = B_t - \int_0^t \left(\widetilde{b}_s - b_s \right) ds \tag{5.3.6}$$

is a Brownian motion and satisfies

$$dX_t = \widetilde{b}_t(\omega) \, dt + d\widetilde{B}_t. \tag{5.3.7}$$

This can also be expressed otherwise in the following way. Assume that, along with the process X, another process $\widetilde{X}$ solving

$$d\widetilde{X}_t = \widetilde{b}_t(\omega) \, dt + dB_t \tag{5.3.8}$$

is defined on the initial probability space. Then

$$\operatorname{Law}(X \mid \widetilde{\mathsf{P}}) = \operatorname{Law}(\widetilde{X} \mid \mathsf{P}). \tag{5.3.9}$$

In particular if $\widetilde{b} \equiv 0$ we obtain the result formulated earlier that the transition $\mathsf{P} \rightsquigarrow \widetilde{\mathsf{P}}$ "kills" the drift b of the process X (i.e. transforms b into $\widetilde{b} \equiv 0$). If $b \equiv 0$ (i.e. $X = B$) then the transition $\mathsf{P} \rightsquigarrow \widetilde{\mathsf{P}}$ by means of the process

$$Z_t = \exp\left\{ \int_0^t \widetilde{b}_s(\omega) \, dB_s - \frac{1}{2} \int_0^t \left(\widetilde{b}_s(\omega) \right)^2 ds \right\}, \qquad t \geq 0 \tag{5.3.10}$$

"creates" the drift of the Brownian motion, i.e. $B_{\cdot} \rightsquigarrow \int_0^{\cdot} \widetilde{b}_s(\omega) \, ds + B$.

The exposed results of I. V. Girsanov [77] gave rise to a wide cycle of results bearing the name of *Girsanov theorems* (for martingales, local martingales, semimartingales, etc.).

Let us cite several of them referring to the monograph [106, Chap. III, §§ 3b–3e] for details and proofs.

2. The case of local martingales. Assume that on the initial filtered probability space $(\Omega, \mathcal{F}, (\mathcal{F}_t)_{t\geq 0})$ we have two probability measures P and $\widetilde{\mathsf{P}}$ such that $\widetilde{\mathsf{P}} \overset{\text{loc}}{\ll} \mathsf{P}$ i.e. $\widetilde{\mathsf{P}}_t \ll \mathsf{P}_t$, $t \geq 0$ where $\mathsf{P}_t = \mathsf{P}|\mathcal{F}_t$ and $\widetilde{\mathsf{P}}_t = \widetilde{\mathsf{P}}|\mathcal{F}_t$, $t \geq 0$.

Let $Z_t = d\widetilde{\mathsf{P}}_t/d\mathsf{P}_t$. Assume that $M = (M_t, \mathcal{F}_t)_{t\geq 0}$ is a local martingale with $M_0 = 0$ such that the P-quadratic covariation $[M, Z]$ has P-locally integrable variation. In this case $[M, Z]$ has a P-compensator $\langle M, Z\rangle$ (see Subsection 3.3).

Lemma 5.3. (Girsanov's theorem for local martingales) *The process*

$$\widetilde{M} = M - \frac{1}{Z_-} \cdot \langle M, Z\rangle \tag{5.3.11}$$

i.e. the process $\widetilde{M} = (\widetilde{M}_t, \mathcal{F}_t)_{t\geq 0}$ *with* $\widetilde{M} = M_t - \int_0^t \frac{1}{Z_{s-}}\, d\langle M, Z\rangle_s$ *is a* $\widetilde{\mathsf{P}}$*-local martingale.*

This result can be interpreted in the following way: the process $\widetilde{M}$ obtained from a local P-martingale M by means of (5.3.11) is not in general a local martingale with respect to the measure P (this process is semimartingale) but with respect to the new measure $\widetilde{\mathsf{P}}$.

3. The case of semimartingales. Assume that the process X is a semimartingale (with respect to the measure P) with the triplet of predictable characteristics (B, C, ν). Consider a measure $\widetilde{\mathsf{P}}$ such that $\widetilde{\mathsf{P}} \overset{\text{loc}}{\ll} \mathsf{P}$.

Lemma 5.4. 1. *With respect to the measure* $\widetilde{\mathsf{P}}$ *the process* X *is again a semimartingale (with a triplet* $(\widetilde{B}, \widetilde{C}, \widetilde{\nu})$*).*

2. *The triplets* (B, C, ν) *and* $(\widetilde{B}, \widetilde{C}, \widetilde{\nu})$ *are related by the formulae*

$$\begin{aligned} \widetilde{B} &= B + \beta \cdot C + h(Y - 1) * \nu, \\ \widetilde{C} &= C, \\ \widetilde{\nu} &= Y \cdot \nu \end{aligned} \tag{5.3.12}$$

where

- $h = h(x)$ *is a truncation function (the standard one is* $h(x) = xI(|x| \leq 1)$*);*
- $\beta = (\beta_t(\omega))_{t\geq 0}$ *is an adapted process specified by*

$$\beta = \frac{d\langle Z^c, X^c\rangle}{d\langle X^c\rangle} \, \frac{I(Z_- > 0)}{Z_-} \tag{5.3.13}$$

where Z^c *is the continuous component of the process* $Z = (Z_t)_{t\geq 0}$ *with* $Z_t = d\widetilde{\mathsf{P}}_t/d\mathsf{P}_t$;

- $Y = Y(\omega; t, x)$ *is specified by*

$$Y = \mathsf{E}^{\mathsf{P}}_{\mu}\left(\frac{Z}{Z_-} I(Z_- > 0) \,\middle|\, \widetilde{\mathcal{P}}\right) \tag{5.3.14}$$

where $\mathsf{E}^{\mathsf{P}}_{\mu}$ *is averaging with respect to the measure* M^{P}_{μ} *on* $\big(\Omega\times\mathbb{R}_+\times\mathsf{P}, \mathcal{F}\otimes\mathcal{B}(\mathbb{R}_+)\otimes\mathcal{B}(\mathbb{R})\big)$, *specified by the formula* $X * M^{\mathsf{P}}_{\mu} = \mathsf{E}(W * \mu)$ *for all nonnegative measurable functions* $W = W(\omega, t, x)$, *and* $\widetilde{\mathcal{P}}$ *is the predictable* σ*-algebra,* $\widetilde{\mathcal{P}} = \mathcal{P}\otimes\mathcal{B}(\mathbb{R})$, *where* $\mathcal{P}$ *is the predictable* σ*-algebra on* $\Omega\times\mathbb{R}_+$.

(For more details see [106, Chap. III, § 3d].)

4. The above results answered one of the questions related to the change of measure, namely the question as what the initial process (or its characteristics) becomes if instead of the initial probability measure P we consider another measure $\widetilde{\mathsf{P}}$ such that $\widetilde{\mathsf{P}} \overset{\text{loc}}{\ll} \mathsf{P}$.

Another question naturally arises as how to describe all measures $\widetilde{\mathsf{P}}$ satisfying the property $\widetilde{\mathsf{P}} \overset{\text{loc}}{\ll} \mathsf{P}$.

We assume that the measures P and $\widetilde{\mathsf{P}}$ are probability distributions of a semimartingale $X = (X_t)_{t\geq 0}$ defined on the filtered space $(\Omega, \mathcal{F}, (\mathcal{F}_t)_{t\geq 0})$ where Ω consists of the functions $\omega = \omega(t)$, $t \geq 0$ which are right-continuous and have the limits from the left, and the process X is canonically defined (i.e. $X_t(\omega) = \omega(t)$ for $t \geq 0$).

General results from the theory of semimartingales (see [106, Chap. III, § 4d, Lemma 4.24] imply that the process $Z = (Z_t)_{t\geq 0}$ given by $Z_t = d\widetilde{\mathsf{P}}_t/d\mathsf{P}_t$ admits the representation

$$Z = H \cdot X^c + W * (\mu - \nu) + N \tag{5.3.15}$$

where the processes $H = (H_t(\omega))$ and $W = (W_t(\omega; t, x))$ with $t \geq 0$, $x \in \mathbb{R}$ satisfy certain integrability conditions, X^c is the continuous martingale component of X, μ is the measure of jumps of X and ν is the compensator of μ. In (5.3.15) it is the local martingale $N = (N_t)_{t\geq 0}$ which is troublesome because it lacks a simple description. In some simple cases $N \equiv 0$ and then the representation (5.3.15) takes the form

$$Z = H \cdot X^c + W * (\mu - \nu). \tag{5.3.16}$$

Starting from (5.3.16), under the assumption that $Z_t \geq 0$ and $\mathsf{E}\, Z_t = 1$ one can construct different measures $\widetilde{\mathsf{P}}$ by letting $d\widetilde{\mathsf{P}}_t = Z_t\, d\mathsf{P}_t$ for $t \geq 0$.

If the initial process X is a Brownian motion or, more generally, a process with independent increments (in particular a Lévy process) then the representation (5.3.16) is valid.

In the general case, when one cannot rely on getting the representation (5.3.16), one has to choose the way of construction of a concrete martingale Z that leads to the possibility of constructing a measure $\widetilde{\mathsf{P}}$ satisfying the property $\widetilde{\mathsf{P}} \overset{\text{loc}}{\ll} \mathsf{P}$.

Among these methods one far-famed is based on the Esscher transformation. The essence of this method may be described as follows.

Let $X = (X_t)_{t\ge 0}$ be a semimartingale on $(\Omega, \mathcal{F}, (\mathcal{F}_t)_{t\ge 0}, \mathsf{P})$ with the triplet (B, C, ν). Let $K(\lambda) = (K_t(\lambda))_{t\ge 0}$ be the cumulant process given by

$$K_t(\lambda) = i\lambda B_t - \frac{\lambda^2}{2} C_t + \int_{\mathbb{R}} \big(e^{i\lambda x} - 1 - i\lambda h(x)\big)\, \nu(\omega; (0,t] \times dx) \tag{5.3.17}$$

where $\lambda \in (-\infty, \infty)$, and form a new positive process $Z(\lambda) = (Z_t(\lambda))_{t\ge 0}$ by setting

$$Z_t(\lambda) = \exp\big\{\lambda X_t - \widehat{K}_t(\lambda)\big\} \tag{5.3.18}$$

where

$$\widehat{K}_t(\lambda) = \log \mathcal{E}_t(K(\lambda)) \tag{5.3.19}$$

and $\mathcal{E} = \mathcal{E}(K(\lambda))$ is the stochastic exponential ($d\mathcal{E}_t(K(\lambda)) = \mathcal{E}_{t-}(K(\lambda))\, dK_t(\lambda)$, $\mathcal{E}_0(K(\lambda)) = 1$). It turns out (see [106, second ed.]) that $Z_t(\lambda)$ admits the representation of the form

$$Z_t(\lambda) = \mathcal{E}\left(\lambda X^c + \frac{e^{\lambda x} - 1}{\widehat{W}(\lambda)} * (\mu^X - \nu)\right) \tag{5.3.20}$$

where $\widehat{W}_t(\lambda) = \int (e^{\lambda x} - 1)\, \nu(\{t\} \times dx)$ and μ^X is the measure of jumps of the process X.

From (5.3.20) it follows that the process $Z(\lambda) = (Z_t(\lambda))_{t\ge 0}$ is a positive local martingale (with respect to P). Thus if this process is a martingale then $\mathsf{E}\, Z_t(\lambda) = 1$ for each $t > 0$ and one may define a new probability measure $\widetilde{\mathsf{P}}(\lambda) \ll \mathsf{P}$ such that for each $t > 0$

$$\frac{d\widetilde{\mathsf{P}}_t(\lambda)}{d\mathsf{P}_t} = Z_t(\lambda). \tag{5.3.21}$$

The measure $\widetilde{\mathsf{P}}(\lambda)$ constructed in such a way is called the *Esscher measure.*

5.4. Killing (discounting)

The essence of transformations called “killing” (“discounting”) and “creation” is deeply rooted in the derivation of the diffusion equation (due to Fick [66] in 1855 upon mimicking Fourier’s 1822 derivation of the heat equation).

If $\gamma = \gamma(t,x)$ denotes the concentration of N Brownian particles suspended in a fluid (where N is large), and Fick's law of diffusion [66] applies, then the diffusion equation holds:

$$\begin{aligned}\gamma_t + K &= \operatorname{div}(D \operatorname{grad} \gamma) - \operatorname{div}(v\gamma) + C \\ &= \textstyle\sum_i (D\gamma_{x_i})_{x_i} - \sum_i (v\gamma)_{x_i} + C\end{aligned} \tag{5.4.1}$$

where $K = K(t,x)$ corresponds to the disappearance (killing) of Brownian particles, $D = D(t,x)$ is the diffusion coefficient, $v = v(t,x)$ is the velocity of the fluid, and $C = C(t,x)$ corresponds to the appearance (creation) of Brownian particles.

Since $p := \gamma/N$ for large N may be interpreted as the transition density of the position process $X = (X_t)_{t \ge 0}$, it follows that $p = p(t,x)$ solves the same diffusion equation:

$$p_t + K = \operatorname{div}(D \operatorname{grad} p) - \operatorname{div}(vp) + C. \tag{5.4.2}$$

To simplify the notation let us assume in the sequel that the setting is one-dimensional (i.e. $x \in \mathbb{R}$). Then (5.4.2) reads as follows:

$$p_t + K = (Dp_x)_x - (vp)_x + C. \tag{5.4.3}$$

Assuming for the moment that $K = C \equiv 0$ in (5.4.3) and that X is Markovian we know that $p = p(t,x)$ solves the Kolmogorov forward equation (cf. [111]):

$$p_t = -(\rho p)_x - \Big(\frac{\sigma^2}{2} p\Big)_{xx} \tag{5.4.4}$$

where $\rho = \rho(t,x)$ is the drift and $\sigma = \sigma(t,x) > 0$ is the (mathematical) diffusion coefficient. A direct comparison of (5.4.3) with $K = C \equiv 0$ and (5.4.4) shows that $\sigma = \sqrt{2D}$ and $\rho = v + D_x$. If the terms K and C are to be incorporated in (5.4.4) one may set $R = K - C$ and consider the following reformulation of the equation (5.4.3):

$$p_t + R = (Dp_x)_x - (vp)_x \tag{5.4.5}$$

where $R = R(t,x)$ may take both positive and negative values.

The following particular form of R is known to preserve the Markov property of a transformed (killed or created) process $\widetilde{X}$ to be defined:

$$R = \lambda p \tag{5.4.6}$$

where $\lambda = \lambda(x) > 0$ corresponds to killing and $\lambda = \lambda(x) < 0$ corresponds to creation. The equation (5.4.4) then reads as follows:

$$p_t + \lambda p = -(\rho p)_x + \Big(\frac{\sigma^2}{2} p\Big)_{xx}. \tag{5.4.7}$$

The process $\widetilde{X} = (\widetilde{X}_t)_{t \ge 0}$ is obtained by "killing" the sample paths of X at the rate $\lambda = \lambda(x)$ where $\lambda > 0$ and "creation" of new sample paths of X at the rate $\lambda = \lambda(x)$ where $\lambda < 0$. In probabilistic terms it means that the transition function of $\widetilde{X}$ is given by

$$\widetilde{P}_t(x, A) = \mathsf{P}_x(\widetilde{X}_t \in A) = \mathsf{E}_x\Big[e^{-\int_0^t \lambda(X_s)\, ds} I_A(X_t)\Big] \tag{5.4.8}$$

for $x \in \mathbb{R}$ and $A \in \mathcal{B}(\mathbb{R})$ with $t \ge 0$. The infinitesimal operator of $\widetilde{X}$ is given by

$$\mathbb{L}_{\widetilde{X}} = \mathbb{L}_X - \lambda I \tag{5.4.9}$$

where I is the identity operator.

To verify (5.4.9) note that

$$\begin{aligned} \frac{\mathsf{E}_x F(\widetilde{X}_t) - F(x)}{t} &= \frac{\mathsf{E}_x F(X_t) - F(x)}{t} + \frac{\mathsf{E}_x F(\widetilde{X}_t) - \mathsf{E}_x F(X_t)}{t} \\ &= \frac{\mathsf{E}_x F(X_t) - F(x)}{t} + \frac{\mathsf{E}_x\big[\exp\big(-\int_0^t \lambda(X_s)\, ds - 1\big) F(X_t)\big]}{t} \\ &\to \mathbb{L}_X F - \lambda F \end{aligned} \tag{5.4.10}$$

as $t \downarrow 0$ upon assuming that λ is continuous (at x) and that we can exchange the limit and the integral for the final convergence relation (sufficient conditions for the latter are well known).

Recalling that (5.4.3) can be written as

$$p_t = \mathbb{L}_X^* p \tag{5.4.11}$$

where $\mathbb{L}_X^*$ denotes the adjoint of $\mathbb{L}_X$, we see by (5.4.9) that (5.4.7) reads as follows:

$$p_t = \mathbb{L}_{\widetilde{X}}^* p \tag{5.4.12}$$

which is in agreement with preceding facts.

When $\lambda > 0$ is a constant there is a simple construction of the killed process $\widetilde{X}$. Let ζ be a random variable that is exponentially distributed with parameter λ (i.e. $\mathsf{P}_x(\zeta > t) = e^{-\lambda t}$ for $t > 0$) and independent of X under each P_x. The process $\widetilde{X}$ can then be defined as follows:

$$\widetilde{X}_t := \begin{cases} X_t & \text{if } t < \zeta, \\ \partial & \text{if } t \ge \zeta \end{cases} \tag{5.4.13}$$

where ∂ is a fictitous point ("cemetery") outside Ω. All functions defined on $\Omega \cup \{\partial\}$ are assumed to take value zero at ∂.

Notes. For further details on the material reviewed in Chapter II we refer to standard textbooks on stochastic processes found in the Bibliography.

Chapter III.

Optimal stopping and free-boundary problems

In the end of Chapter I we have seen that the optimal stopping problem for a Markov process X with the value function V is equivalent to the problem of finding the *smallest superharmonic function* $\widehat{V}$ which dominates the gain function G on the state space E. In this case, moreover, the first entry time of X into the stopping set $D = \{V = G\}$ is optimal. This yields the following representation:

$$V(x) = \mathsf{E}_x G(X_{\tau_D}) \tag{1}$$

for $x \in E$. Due to the Markovian structure of X, any function of the form (1) is intimately related to a deterministic equation which governs X in mean (parabolic/elliptic PDEs [partial differential equations] when X is continuous, or more general PIDEs [partial integro-differential equations] when X has jumps).

The main purpose of the present chapter is to unveil and describe the previous connection. This leads to differential or integro-differential equations which the function V solves. Since the optimal stopping set D is unknown and has to be determined among all possible candidate sets D in (1), it is clear that this connection and the equations obtained play a fundamental role in search for the solution to the problem. It is worthwhile to recall that when X is a Markov chain the analogous problems have been considered in Subsection 4.2 in the context of discrete-time "potential theory".

In order to focus on the equations only, and how these equations actually follow from the Markovian structure, we will adopt a formal point of view in the sequel where everything by definition is assumed to be sufficiently regular and valid as needed to make the given calculations possible. Most of the time such a set of sufficient conditions is easily specified. Sometimes, however, it may be more challenging to determine such sufficient conditions precisely. In any case, we will make use of the facts exposed in the present chapter mostly in a suggestive way throughout the monograph.

6. MLS formulation of optimal stopping problems

1. Throughout we will adopt the setting and notation of Subsection 2.2. Thus, we will consider a strong Markov process $X = (X_t)_{t\geq 0}$ defined on a filtered probability space $(\Omega, \mathcal{F}, (\mathcal{F}_t)_{t\geq 0}, \mathsf{P}_x)$ and taking values in a measurable space $(E, \mathcal{E})$ where for simplicity we will assume that $E = \mathbb{R}^d$ for some $d \geq 1$ and $\mathcal{E} = \mathcal{B}^d$ is the Borel σ-algebra on $\mathbb{R}^d$. It is assumed that the process X starts at x under P_x for $x \in E$ and that the sample paths of X are right-continuous and left-continuous over stopping times. It is also assumed that the filtration $(\mathcal{F}_t)_{t\geq 0}$ is right-continuous (implying that the first entry times to open and closed sets are stopping times). In addition, it is assumed that the mapping $x \mapsto \mathsf{P}_x(F)$ is measurable for each $F \in \mathcal{F}$. It follows that the mapping $x \mapsto \mathsf{E}_x(Z)$ is measurable for each random variable Z. Finally, without loss of generality we will assume that $(\Omega, \mathcal{F})$ equals the canonical space $(E^{[0,\infty)}, \mathcal{E}^{[0,\infty)})$ so that the shift operator $\theta_t : \Omega \to \Omega$ is well defined by $\theta_t(\omega)(s) = \omega(t+s)$ for $\omega \in \Omega$ with $t, s \geq 0$.

2. Given measurable (continuous) functions $M, L, K : E \to \mathbb{R}$ satisfying integrability conditions implying (2.2.1), consider the optimal stopping problem

$$V = \sup_{\tau} \mathsf{E}\Big(M(X_\tau) + \int_0^\tau L(X_t)\, dt + \sup_{0\leq t\leq \tau} K(X_t) \Big) \tag{6.0.1}$$

where the first supremum is taken over stopping times τ of X (or more generally with respect to $(\mathcal{F}_t)_{t\geq 0}$). In the case of a finite horizon $T \in [0, \infty)$ it is assumed that $0 \leq \tau \leq T$, and in the case of infinite horizon it is assumed that $0 \leq \tau < \infty$. In (6.0.1) we admit that any of the functions M, L or K may be identically equal to zero.

3. Clearly, the three terms following E in (6.0.1) provide three different performance measures. The first two are due to Mayer and Lagrange in the classical calculus of variations while the third one is more recent (see the notes in the end of the chapter for a historical account). Note that M in (6.0.1) stands for Mayer, L for Lagrange, and S for supremum (soon to be introduced below). This explains the term MLS in the title of the section.

4. For simplicity of exposition we will assume in the sequel that $K(x) = x$ for all $x \in E$ with $E = \mathbb{R}$. Note that when K is strictly monotone (and continuous) for example, then $K(X) = (K(X_t))_{t\geq 0}$ may define another Markov process, and by writing $M(X_t) = (M \circ K^{-1})(K(X_t))$ and $L(X_t) = (L \circ K^{-1})(K(X_t))$ we see no loss of generality in the previous assumption.

Introduce the following processes:

$$I_t = \int_0^t L(X_s)\, ds \qquad (\textit{integral process}), \tag{6.0.2}$$

$$S_t = \sup_{0\leq s\leq t} X_s \qquad (\textit{supremum process}) \tag{6.0.3}$$

for $t \geq 0$. Then the process $Z = (Z_t)_{t \geq 0}$ given by

$$Z_t = (I_t, X_t, S_t) \tag{6.0.4}$$

is Markovian (under general assumptions) and (6.0.1) reads as follows:

$$V = \sup_{\tau} \mathsf{E} G(Z_\tau) \tag{6.0.5}$$

where $G(z) = M(x) + a + s$ for $z = (a, x, s) \in \mathbb{R}^3$. We thus see that the optimal stopping problem (6.0.1) may be viewed as the optimal stopping problem (2.2.2) so that the general optimal stopping results of Subsection 2.2 are applicable. Note that the process Z is three-dimensional in general.

5. Despite the fact that the general results are applicable, it turns out that the specific form of stochastic processes $I = (I_t)_{t \geq 0}$ and $S = (S_t)_{t \geq 0}$ makes a direct approach to (6.0.1) or (6.0.5) more fruitful. The key feature of I is its linearity; to enable it to start at arbitrary points one sets

$$I_t^a = a + I_t \tag{6.0.6}$$

for $a \in \mathbb{R}$ and $t \geq 0$. The key feature of S is its constancy and a strict increase only at times when equal to X; to enable it to start at arbitrary points one sets

$$S_t^s = s \vee S_t \tag{6.0.7}$$

for $s \geq x$ in $\mathbb{R}$ and $t \geq 0$. With the choice of (6.0.6) and (6.0.7) the Markovian structure of Z remains preserved relative to P_x under which X starts at $x \in \mathbb{R}$. Moreover, if $X^x = (X_t^x)_{t \geq 0}$ starts at x under P and we set

$$Z_t^z = (I_t^a, X_t^x, S_t^s) \tag{6.0.8}$$

for $z = (a, x, s)$, then the family of probability measures $\mathsf{P}_z = \text{Law}(Z^z \mid \mathsf{P})$ defined on the canonical space is Markovian. This point of view proves useful in classifying ad-hoc solutions in terms of general theory (see e.g. Subsection 13.2). The problem (6.0.5) consequently reads

$$V(x) = \sup_{\tau} \mathsf{E}_z G(Z_\tau) \tag{6.0.9}$$

and the general optimal stopping theory of Chapter I is applicable.

6.1. Infinite and finite horizon problems

We have already pointed out in Chapter I that it is important to enable the process to start at arbitrary points in the state space (since the problem then can be studied by means of the value function). The fact whether the horizon T

in (6.0.1) is infinite or finite is closely related. Some of these issues will now be addressed.

1. Consider the case when the horizon T in (6.0.1) or (6.0.9) is *infinite.* Then, on the one hand, the problem is simpler than the finite horizon problem, however, only if we can come up with an explicit expression as the candidate for V (these expressions are obtained by solving the equations which govern Z in mean, typically ODEs [ordinary differential equations] in the infinite horizon case, and PDEs [partial differential equations] in the finite horizon case, at least when Z is continuous). On the other hand, if such explicit expressions are not available, then the infinite horizon problem may be more difficult (in terms of characterizing the solution via existence and uniqueness claims) than the corresponding finite horizon problem (see e.g. Section 27). This is due to the fact that the Wald–Bellman equations (cf. Chapter I) are not directly available in the infinite horizon case, while for example in the case of a diffusion process X we can characterize the optimal stopping boundaries in terms of nonlinear integral equations that can be quite similarly solved by backward induction (cf. Chapters VI and VII for details).

2. Consider the case when the horizon T in (6.0.1) or (6.0.9) is *finite.* Then the time variable becomes important (as the remaining time goes to zero) and (unless I_t itself is already of this type) needs to be added to (6.0.4) so that Z extended to $\widetilde{Z}$ reads as follows:

$$\widetilde{Z}_t = (t, I_t, X_t, S_t) \tag{6.1.1}$$

for $t \ge 0$. The process $\widetilde{Z} = (\widetilde{Z}_t)_{t\ge 0}$ is Markovian and the optimal stopping problem (6.0.5) i.e. (6.0.9) extends as follows:

$$V(t,z) = \sup_{0\le\tau\le T-t} \mathsf{E}_{t,z}\widetilde{G}(\widetilde{Z}_{t+\tau}) \tag{6.1.2}$$

where $Z_t = z$ under $\mathsf{P}_{t,z}$ and $\widetilde{G}(\tilde{z}) = \widetilde{G}(t,z)$ equals $G(z)$ for $\tilde{z} = (t,z)$, but note that $\widetilde{G}$ could also be a new function depending on both t and z. Note moreover that X itself could be of the form $X_t = (t, Y_t)$ for $t \ge 0$ where $Y = (Y_t)_{t\ge 0}$ is a Markov process, so that (6.1.2) even if $\widetilde{G} \equiv G$ reads

$$V(t,y) = \sup_{0\le\tau\le T-t} \mathsf{E}_{t,y} M(t+\tau, Y_{t+\tau}) \tag{6.1.3}$$

where $Y_t = y$ under $\mathsf{P}_{t,y}$ and M stands for G. Various particular cases of the problem (6.1.3) will be studied in Chapters VI–VIII below.

6.2. Dimension of the problem

Dimension of the problem refers to the minimal dimension of an underlying Markov process which leads to the solution. The latter Markov process does not need to be the same as the initial Markov process. For example, the problem (6.0.9) is three-dimensional generally since Z is a three-dimensional Markov process, but due to the linear structure of I (or Itô's formula to a similar end) it is also possible to view the problem as being two-dimensional, both being valid when the horizon is infinite. On the other hand, when the horizon is finite, then the same problem is four-dimensional generally, but due to the linear structure of I may also be viewed as three-dimensional.

To determine the dimension of a problem is not always a simple matter. We will see in Chapter IV how the initial dimension can be reduced by the method of time change (Section 10), the method of space change (Section 11), and the method of measure change (Section 12). These three methods are stochastic by their nature and each corresponds to a deterministic change of variables that reduces the initial more complicated equation (e.g. PDE) to a simpler equation (e.g. ODE). This equivalence is best tested and understood via specific examples (cf. Sections 10–12 and Sections 26–27).

6.3. Killed (discounted) problems

One is often more interested in the killed (discounted) version of the optimal stopping problem (6.0.1) that reads

$$V = \sup_{\tau} \mathsf{E}\left(e^{-\lambda_\tau} M(X_\tau) + \int_0^\tau e^{-\lambda_t} L(X_t)\,dt + e^{-\lambda_\tau} \sup_{0\le t\le\tau} K(X_t)\right) \tag{6.3.1}$$

where the killing (discounting) process $\lambda = (\lambda_t)_{t\ge 0}$ is given by

$$\lambda_t = \int_0^t \lambda(X_s)\,ds \tag{6.3.2}$$

for a measurable (continuous) function $\lambda : E \to \mathbb{R}_+$ called the killing (discounting) rate.

The problem (6.3.1) reduces to the initial problem (6.0.1) by replacing the underlying Markov process X with the new Markov process $\widetilde{X}$ which corresponds to the "killing" of the sample paths of X at the "rate" $\lambda(X)$ (cf. Subsection 5.4). The infinitesimal generator of $\widetilde{X}$ is given by

$$\mathbb{L}_{\widetilde{X}} = \mathbb{L}_X - \lambda I \tag{6.3.3}$$

where I is the identity operator (see (5.4.10)) and all what is said above or below for X extends to $\widetilde{X}$ by replacing $\mathbb{L}_X$ with $\mathbb{L}_{\widetilde{X}}$. Specific killed (discounted) problems are studied in Chapter VII below.

7. MLS functionals and PIDE problems

Throughout the section we will adopt the setting and notation of Subsection 2.2 (recalled in the beginning of Section 6 above).

Motivated by the representation (1) on page 123 let us assume that we are given a (bounded) open set $C \subseteq E$ and let us consider

$$\tau_D = \inf\{\, t \geq 0 : X_t \in D \,\} \tag{7.0.4}$$

where $D = C^c \; (= E \setminus C)$. Given a mapping $G : D \to \mathbb{R}$ the question then arises to determine a differential (or integro-differential) equation solved by

$$F(x) = \mathsf{E}_x G(X_{\tau_D}) \tag{7.0.5}$$

for $x \in E$. Moreover, when the representation (1) on page 123 gets the more specific form (6.0.1), we see that the question naturally splits into three new sub-questions (corresponding to M, L and K).

The purpose of the present section is to describe answers to these questions. These answers are based on a fundamental link between probability (Markov process) and analysis (differential/integral equation). When connected with the meaning of the representation (1) on page 123 this will lead to the formulation of a free-boundary problem in Section 8.

1. For further reference let us recall the following three facts playing the key role in the sequel.

• The strong Markov property of X can be expressed as

$$\mathsf{E}_x(H \circ \theta_\tau \mid \mathcal{F}_\tau) = \mathsf{E}_{X_\tau} H \tag{7.0.6}$$

where τ is a stopping time and H is a (bounded or non-negative) measurable functional (see (4.3.28)).

• If $\sigma \leq \tau$ where σ is a stopping time and τ is a hitting/entry time to a set, then

$$\tau = \sigma + \tau \circ \theta_\sigma . \tag{7.0.7}$$

• For all stopping times τ and σ we have

$$X_\tau \circ \theta_\sigma = X_{\sigma + \tau \circ \theta_\sigma} . \tag{7.0.8}$$

(For (7.0.6) recall (4.3.28), for (7.0.7) recall (4.1.25), and for (7.0.8) recall (4.1.13).)

2. The mean-value kinematics of the process X is described by the *characteristic operator* $\mathbb{L}_X$ defined on a function $F : E \to \mathbb{R}$ as follows:

$$\mathbb{L}_X F(x) = \lim_{U \downarrow x} \frac{\mathsf{E}_x F(X_{\tau_{U^c}}) - F(x)}{\mathsf{E}_x \tau_{U^c}} \tag{7.0.9}$$

where the limit is taken over a family of open sets U shrinking down to x in E.

Under rather general conditions it is possible to establish (see [53]) that the characteristic operator is an extension of the *infinitesimal operator* $\mathbb{L}_X$ defined on a function $F : E \to \mathbb{R}$ as follows:

$$\mathbb{L}_X F(x) = \lim_{t \downarrow 0} \frac{\mathsf{E}_x F(X_t) - F(x)}{t} \tag{7.0.10}$$

for $x \in E$.

We will not find it necessary to specify the domains of these two operators and for this reason the same symbol $\mathbb{L}_X$ will be used to denote both. Very often we will refer to $\mathbb{L}_X$ as the *infinitesimal generator* of the process X. Its infinitesimal role is uniquely determined through its action on sufficiently regular (smooth) functions F for which both limits (7.0.9) and (7.0.10) exist and coincide. This leads to the equations which will now be described.

3. From the general theory of Markov processes (see [53]) we know that (on a given relatively compact subset of E) the infinitesimal generator $\mathbb{L}_X$ takes the following integro-differential form:

$$\begin{aligned} \mathbb{L}_X F(x) = \lambda(x) F(x) + \sum_{i=1}^{d} \rho_i(x) \frac{\partial F}{\partial x_i}(x) + \sum_{i,j=1}^{d} \sigma_{ij}(x) \frac{\partial^2 F}{\partial x_i \, \partial x_j}(x) \\ + \int_{\mathbb{R}^d \setminus \{0\}} \Big(F(y) - F(x) - \sum_{i=1}^{d} (y_i - x_i) \frac{\partial F}{\partial x_i}(x) \Big) \, \nu(x, dy) \end{aligned} \tag{7.0.11}$$

where λ corresponds to killing (when positive) or creation (when negative), ρ is the drift coefficient, σ is the diffusion coefficient, and ν is the compensator of the measure μ of jumps of X.

If X is *continuous* then $\nu \equiv 0$ and $\mathbb{L}_X$ does not contain the final (integral) term in (7.0.11). If X cannot be killed or created then $\lambda \equiv 0$ and $\mathbb{L}_X$ does not contain the initial term in (7.0.11). Similar interpretations hold for the drift coefficient ρ and the diffusion coefficient σ. Each of the four terms λ, ρ, σ and ν in (7.0.11) has a transparent meaning (e.g. when X_t denotes the position of a particle in the fluid under external influence or the value of a stock price in a financial market).

4. *Regular boundary.* We will say that the boundary ∂C of C is *regular* (for D) if each point x from ∂C is *regular* (for D) in the sense that $\mathsf{P}_x(\sigma_D = 0) = 1$ where $\sigma_D = \inf \{ t > 0 : X_t \in D \}$. Thus, if X starts at a regular point for D, then X enters D immediately after taking off. It turns out that the notion of regularity of ∂C is intimately related to a regularity of the mapping (7.0.5) at ∂C (in the sense of continuity or smoothness).

To simplify the notation let us agree in the sequel that ∂C stands for D when X is discontinuous. It would be sufficient in fact to include only those points in D that can be reached by X when jumping from C (apart from the boundary points of C).

7.1. Mayer functional and Dirichlet problem

1. Given a continuous function $M : \partial C \to \mathbb{R}$ consider

$$F(x) = \mathsf{E}_x M(X_{\tau_D}) \tag{7.1.1}$$

for $x \in E$. *The function* F *solves the Dirichlet problem*:

$$\mathbb{L}_X F = 0 \quad \textit{in} \;\; C, \tag{7.1.2}$$

$$F\big|_{\partial C} = M \tag{7.1.3}$$

(cf. (4.2.50) and (4.2.51)).

Indeed, given $x \in C$ choose a (bounded) open set U such that $x \in U \subseteq C$. By the strong Markov property (7.0.6) with (7.0.7) and (7.0.8) we have

$$\begin{aligned} \mathsf{E}_x F(X_{\tau_{U^c}}) &= \mathsf{E}_x \mathsf{E}_{X_{\tau_{U^c}}} M(X_{\tau_D}) = \mathsf{E}_x \mathsf{E}_x \big(M(X_{\tau_D}) \circ \theta_{\tau_{U^c}} \, \big| \, \mathcal{F}_{\tau_{U^c}} \big) \\ &= \mathsf{E}_x M \big(X_{\tau_{U^c} + \tau_D \circ \theta_{\tau_{U^c}}} \big) = \mathsf{E}_x M(X_{\tau_D}) = F(x). \end{aligned} \tag{7.1.4}$$

Hence we see that

$$\lim_{U \downarrow x} \frac{\mathsf{E}_x F(X_{\tau_{U^c}}) - F(x)}{\mathsf{E}_x \tau_{U^c}} \equiv 0 \tag{7.1.5}$$

proving (7.1.2) as claimed. The condition (7.1.3) is evident. □

2. Continuity on $\bar{C}$. Let us assume that X is *continuous*, and let $x_n \in C$ converge to $x \in \partial C$ as $n \to \infty$. Then

$$F(x_n) = \mathsf{E}_{x_n} M(X_{\tau_D}) = \mathsf{E} M(X_{\tau_D}^{x_n}) \longrightarrow \mathsf{E} M(X_{\sigma_D}^{x}) = \mathsf{E}_x M(X_{\sigma_D}) \tag{7.1.6}$$

where $\sigma_D = \inf \{ t > 0 : X_t \in D \}$ and the convergence takes place since both X and M are continuous. Moreover, if x is regular for D , then $\sigma_D = 0$ under P_x and thus $\mathsf{E}_x M(X_{\sigma_D}) = M(x) = F(x)$. This shows:

If ∂C *is a regular boundary (for* D *), then* F *is continuous on* $\bar{C}$. (7.1.7)

In particular, if ∂C is a regular boundary (for D), then F is continuous at ∂C . Only special C however will have the power of making F smooth at ∂C . This is intimately related to the principle of smooth fit discussed in Subsection 9.1 below: Such a set C will be optimal.

If X is *discontinuous*, then F also may generally be discontinuous at ∂C . Only special C however will have the power of making F continuous at ∂C .

This is intimately related to the principle of continuous fit discussed in Subsection 9.2 below: Such a set will be optimal.

3. Smoothness in C. Let us assume that X is continuous and let us consider the Dirichlet problem (7.1.2)–(7.1.3) with $\mathbb{L}_X$ in (7.0.11) without the λ-term and the N-term. Standard PDE results then state that if ρ and σ are sufficiently smooth and ∂C is sufficiently regular, then there exists a solution F to (7.1.2)–(7.1.3) which is (equally) smooth in C and continuous on $\bar{C}$. (Frequently F will be smooth at least as ρ and σ.) To avoid any discussion of ∂C which generally is not easily accessible in free-boundary problems, we can apply the preceding PDE result locally around the given point $x \in C$ to $B = b(x,r) \subseteq C$ where $r > 0$ is sufficiently small. This gives the existence of a solution f to the *Dirichlet problem*:

$$\mathbb{L}_X f = 0 \quad \text{in} \ \ B, \tag{7.1.8}$$

$$f\big|_{\partial B} = F \tag{7.1.9}$$

such that f is smooth in B and continuous on $\bar{B}$. (Obviously ∂B is taken to be sufficiently regular.) Applying Itô's formula (page 67) to $f(X_t)$, setting $t = \tau_{B^c}$, taking E_x on both sides upon making use of (7.1.8) and the optional sampling theorem (page 60) using localization arguments if needed, we find by means of (7.1.9) that $f(x) = \mathsf{E}_x F(X_{\tau_{B^c}})$ which in turn equals $F(x)$ by (7.1.4) above. Thus F equals f on B and hence is smooth at x in C.

The preceding technique enables one to use the smoothness results from the PDE theory and carry them over to the function F defined in (7.1.1). Note that this requires only conditions on ρ and σ and that the smoothness of F holds only in the interior C of $\bar{C}$ and not at ∂C generally. Since in all examples to be studied below the remaining details above are easily verified, we will freely use the smoothness of F in C without further mention (cf. Chapters VI–VIII).

4. Killed version. Given a continuous function $M : \partial C \to \mathbb{R}$ consider

$$F(x) = \mathsf{E}_x\big(e^{-\lambda_{\tau_D}} M(X_{\tau_D})\big) \tag{7.1.10}$$

for $x \in E$ where $\lambda = (\lambda_t)_{t \ge 0}$ is given by

$$\lambda_t = \int_0^t \lambda(X_s)\, ds \tag{7.1.11}$$

for a measurable (continuous) function $\lambda : E \to \mathbb{R}_+$. *The function F solves the (killed) Dirichlet problem*:

$$\mathbb{L}_X F = \lambda F \quad \textit{in} \ \ C, \tag{7.1.12}$$

$$F\big|_{\partial C} = M. \tag{7.1.13}$$

Indeed, replacing X by the killed process $\widetilde{X}$ it follows that (7.1.10) reads as

$$F(x) = \mathsf{E}_x M(\widetilde{X}_{\tau_D}) \tag{7.1.14}$$

and the infinitesimal generator of $\widetilde{X}$ is given by

$$\mathbb{L}_{\widetilde{X}} = \mathbb{L}_X - \lambda I \tag{7.1.15}$$

(recall Subsection 6.3 above). Applying (7.1.2) and (7.1.3) to (7.1.14) and (7.1.15) we see that (7.1.12) and (7.1.13) hold as claimed. □

7.2. Lagrange functional and Dirichlet/Poisson problem

1. Given a continuous function $L : C \to \mathbb{R}$ consider

$$F(x) = \mathsf{E}_x \int_0^{\tau_D} L(X_t)\, dt \tag{7.2.1}$$

for $x \in E$. *The function* F *solves the Dirichlet/Poisson problem*:

$$\mathbb{L}_X F = -L \quad \textit{in} \;\; C, \tag{7.2.2}$$

$$F\big|_{\partial C} = 0 \tag{7.2.3}$$

(cf. (4.2.49) and (4.2.48)).

Indeed, given $x \in C$ choose a (bounded) open set U such that $x \in U \subseteq C$. By the strong Markov property (7.0.6) with (7.0.7) and (7.0.8) we have

$$\begin{aligned} \mathsf{E}_x F(X_{\tau_{U^c}}) &= \mathsf{E}_x\, \mathsf{E}_{X_{\tau_{U^c}}} \int_0^{\tau_D} L(X_t)\, dt \\ &= \mathsf{E}_x \mathsf{E}_x \Big(\int_0^{\tau_D} L(X_t)\, dt \circ \theta_{\tau_{U^c}} \,\Big|\, \mathcal{F}_{\tau_{U^c}} \Big) \\ &= \mathsf{E}_x \int_0^{\tau_D \circ \theta_{\tau_{U^c}}} L(X_t \circ \theta_{\tau_{U^c}})\, dt = \mathsf{E}_x \int_0^{\tau_D - \tau_{U^c}} L(X_{\tau_{U^c}+t})\, dt \\ &= \mathsf{E}_x \int_{\tau_{U^c}}^{\tau_D} L(X_s)\, ds = \mathsf{E}_x \int_0^{\tau_D} L(X_t)\, dt - \mathsf{E}_x \int_0^{\tau_{U^c}} L(X_t)\, dt \\ &= F(x) - \mathsf{E}_x \int_0^{\tau_{U^c}} L(X_t)\, dt \end{aligned} \tag{7.2.4}$$

upon substituting $\tau_{U^c} + t = s$. Hence we see that

$$\lim_{U \downarrow x} \frac{\mathsf{E}_x F(X_{\tau_{U^c}}) - F(x)}{\mathsf{E}_x \tau_{U^c}} = \lim_{U \downarrow x} \Big(-\frac{1}{\mathsf{E}_x \tau_{U^c}}\, \mathsf{E}_x \int_0^{\tau_{U^c}} L(X_t)\, dt \Big) = -L(x) \tag{7.2.5}$$

by the continuity of L. This proves (7.2.2) while (7.2.3) is evident. □

2. Continuity on $\bar{C}$. The same arguments as in Subsection 7.1 above show that if ∂C is a regular boundary (for D), then F *is continuous on* $\bar{C}$.

3. Smoothness in C. The same techniques as in Subsection 7.1 above enable one to use the smoothness results from the PDE theory and carry them over to the function F defined in (7.2.1).

4. Killed version. Given a continuous function $L : C \to \mathbb{R}$ consider

$$F(x) = \mathsf{E}_x \int_0^{\tau_D} e^{-\lambda_t} L(X_t)\, dt \tag{7.2.6}$$

for $x \in E$ where $\lambda = (\lambda_t)_{t \ge 0}$ is given by

$$\lambda_t = \int_0^t \lambda(X_s)\, ds \tag{7.2.7}$$

for a measurable (continuous) function $\lambda : E \to \mathbb{R}_+$. *The function F solves the (killed) Dirichlet/Poisson problem*:

$$\mathbb{L}_X F = \lambda F - L \quad \textit{in} \;\; C, \tag{7.2.8}$$

$$F\big|_{\partial C} = 0. \tag{7.2.9}$$

Indeed, replacing X by the killed process $\widetilde{X}$ it follows that (7.2.6) reads as

$$F(x) = \mathsf{E}_x \int_0^{\tau_D} L(\widetilde{X}_t)\, dt \tag{7.2.10}$$

and the infinitesimal generator of $\widetilde{X}$ is given by

$$\mathbb{L}_{\widetilde{X}} = \mathbb{L}_X - \lambda I \tag{7.2.11}$$

(recall Subsection 6.3 above). Applying (7.2.2) and (7.2.3) to (7.2.10) and (7.2.11) we see that (7.2.8) and (7.2.9) hold as claimed. □

7.3. Supremum functional and Neumann problem

In this subsection we will assume that X is *continuous* and takes values in $E = \mathbb{R}$.

Set

$$S_t = \max_{0 \le s \le t} X_s \tag{7.3.1}$$

for $t \ge 0$. Then $(X, S) = (X_t, S_t)_{t \ge 0}$ is a Markov process with the state space $\widetilde{E} = \{(x, s) \in E^2 : x \le s\}$ and S increases only when $X = S$ i.e. at the (main) diagonal of $\widetilde{E}$. We have $(X_0, S_0) = (x, s)$ under $\mathsf{P}_{x,s}$ for $(x, s) \in \widetilde{E}$.

Given a (bounded) open set $C \subseteq \widetilde{E}$ let us consider

$$\tau_D = \inf \{ t \ge 0 : (X_t, S_t) \in D \} \tag{7.3.2}$$

where $D = C^c$. Then (7.1.2)–(7.1.3) extends as follows.

1. Given a continuous function $M : \partial C \to \mathbb{R}$ consider

$$F(x,s) = \mathsf{E}_{x,s} M(X_{\tau_D}, S_{\tau_D}) \tag{7.3.3}$$

for $(x,s) \in \widetilde{E}$. *The function* F *solves the Neumann problem:*

$$\mathbb{L}_X F = 0 \quad \textit{for } x < s \quad \textit{with } s \textit{ fixed}, \tag{7.3.4}$$

$$\frac{\partial F}{\partial s}(x,s) = 0 \quad \textit{for } x = s, \tag{7.3.5}$$

$$F\big|_{\partial C} = M. \tag{7.3.6}$$

Indeed, since (X,S) can be identified with X when off the diagonal in $\widetilde{E}$, the identities (7.3.4) and (7.3.6) follow in the same way as (7.1.2) and (7.1.3).

To verify (7.3.5) we shall first note that the proof of (7.1.2) shows that

$$\lim_{U \downarrow (x,s)} \frac{\mathsf{E}_{x,s} F(X_{\tau_{U^c}}, S_{\tau_{U^c}}) - F(x,s)}{\mathsf{E}_{x,s} \tau_{U^c}} \equiv 0 \tag{7.3.7}$$

for $(x,s) \in \widetilde{E}$. In particular, this holds for all points (s,s) at the diagonal of $\widetilde{E}$.

Next, without loss of generality, assume that F is sufficiently smooth (e.g. $C^{2,1}$). Applying Itô's formula (page 67) to $F(X_t, S_t)$, taking $\mathsf{E}_{s,s}$ on both sides and applying the optional sampling theorem (page 60) to the continuous martingale (localized if needed) which appears in the identity obtained, we get

$$\begin{aligned} \frac{\mathsf{E}_{s,s} F(X_t, S_t) - F(s,s)}{t} &= \mathsf{E}_{s,s}\Big(\frac{1}{t}\int_0^t (\mathbb{L}_X F)(X_s, S_s)\, ds\Big) \\ &\quad + \mathsf{E}_{s,s}\Big(\frac{1}{t}\int_0^t \frac{\partial F}{\partial s}(X_s, S_s)\, ds\Big) \\ \longrightarrow \mathbb{L}_X F(s,s) + \frac{\partial F}{\partial s}(s,s) \lim_{t\downarrow 0} \frac{\mathsf{E}_{s,s}(S_t - s)}{t} \end{aligned} \tag{7.3.8}$$

as $t \downarrow 0$. Due to $\sigma > 0$ we have $t^{-1}\mathsf{E}_{s,s}(S_t - s) \to \infty$ as $t \downarrow 0$, and therefore the limit in (7.3.8) does not exist (and is finite) unless (7.3.5) holds. □

Facts on the continuity on $\bar{C}$, smoothness in C, and a killed version of (7.3.3) carry over from Subsections 7.1 and 7.2 above to the present case of function F without major changes. Further details in this direction will be omitted.

2. Given a measurable (continuous) function $L : E \to \mathbb{R}$ set

$$I_t = \int_0^t L(X_s)\, ds \tag{7.3.9}$$

for $t \geq 0$. Then $(I, X, S) = (I_t, X_t, S_t)_{t \geq 0}$ is a Markov process with the state space $\widehat{E} = \{(a,x,s) \in E^3 : x \leq s\}$. We have $(I_0, X_0, S_0) = (a,x,s)$ under $\mathsf{P}_{a,x,s}$.

Given a (bounded) open set $C \subseteq \widehat{E}$ let us consider

$$\tau_D = \inf\{\, t \ge 0 : (I_t, X_t, S_t) \in D \,\} \tag{7.3.10}$$

where $D = C^c$. A quick way to reduce this case to the preceding case above is to replace the Markov process X in the latter with the Markov process $\bar{X} = (I, X)$ coming out of the former. This leads to the following reformulation of (7.3.3)–(7.3.6) above.

Given a continuous function $M : \partial C \to \mathbb{R}$ consider

$$F(a, x, s) = \mathsf{E}_{a,x,s} M(I_{\tau_D}, X_{\tau_D}, S_{\tau_D}) \tag{7.3.11}$$

for $(a, x, s) \in \widehat{E}$. *The function F solves the Neumann problem*:

$$\mathbb{L}_X F = -L \quad \text{for } x < s \text{ with } s \text{ fixed}, \tag{7.3.12}$$

$$\frac{\partial F}{\partial s}(a, x, s) = 0 \quad \text{for } x = s, \tag{7.3.13}$$

$$F\big|_{\partial C} = M. \tag{7.3.14}$$

Indeed, replacing X by $\bar{X} = (I, X)$ we find that

$$\mathbb{L}_{\bar{X}} = \mathbb{L}_X + L. \tag{7.3.15}$$

Hence (7.3.12)–(7.3.14) reduce to (7.3.4)–(7.3.6). □

3. Facts on the continuity on $\bar{C}$, smoothness in C, and a killed version of (7.3.11) carry over from Subsections 7.1 and 7.2 above to the present case of function F without major changes. Further details in this direction will be omitted.

7.4. MLS functionals and Cauchy problem

1. Given a continuous function $M : E \to \mathbb{R}$ consider

$$F(t, x) = \mathsf{E}_x M(X_t) \tag{7.4.1}$$

for $(t, x) \in \mathbb{R}_+ \times E$. *The function F solves the Cauchy problem*:

$$\frac{\partial F}{\partial t} = \mathbb{L}_X F \quad \textit{in } \mathbb{R}_+ \times E, \tag{7.4.2}$$

$$F(0, x) = M(x) \quad \textit{for } x \in E. \tag{7.4.3}$$

Indeed, let us show how (7.4.1) reduces to (7.1.1) so that (7.1.2)–(7.1.3) become (7.4.2)–(7.4.3).

For this, define a new Markov process by setting

$$Y_s = (t-s, X_s) \tag{7.4.4}$$

for $s \geq 0$. Then the infinitesimal generator of $Y = (Y_s)_{s\geq 0}$ equals

$$\mathbb{L}_Y = -\frac{\partial}{\partial s} + \mathbb{L}_X. \tag{7.4.5}$$

Consider the exit time of Y from the open set $C = (0,\infty)\times E$ given by

$$\tau_D = \inf\{\, s \geq 0 : Y_s \in D \,\} \tag{7.4.6}$$

where $D = C^c$. Then obviously $\tau_D \equiv t$ and thus

$$F(t,x) = \mathsf{E}_{t,x}\widetilde{M}(Y_{\tau_D}) \tag{7.4.7}$$

where $\widetilde{M}(u,x) = M(x)$ for $u \geq 0$. In this way (7.4.1) has been reduced to (7.1.1) and thus by (7.1.2) we get

$$\mathbb{L}_Y F = 0. \tag{7.4.8}$$

From (7.4.5) we see that (7.4.8) is exactly (7.4.2) as claimed. The condition (7.4.3) is evident. □

2. Killed version (Mayer). Given a continuous function $M : E \to \mathbb{R}$ consider

$$F(t,x) = \mathsf{E}_x e^{-\lambda_t} M(X_t) \tag{7.4.9}$$

for $(t,x) \in \mathbb{R}_+ \times E$ where $\lambda = (\lambda_t)_{t\geq 0}$ is given by

$$\lambda_t = \int_0^t \lambda(X_s)\, ds \tag{7.4.10}$$

for a measurable (continuous) function $\lambda : E \to \mathbb{R}_+$. *The function F solves the (killed) Cauchy problem:*

$$\frac{\partial F}{\partial t} = \mathbb{L}_X F - \lambda F \quad \textit{in } \mathbb{R}_+ \times E, \tag{7.4.11}$$

$$F(0,x) = M(x) \quad \textit{for } x \in E. \tag{7.4.12}$$

Indeed, replacing X by the killed process $\widetilde{X}$ it follows that (7.4.9) reads as

$$F(t,x) = \mathsf{E}_x M(\widetilde{X}_t) \tag{7.4.13}$$

and the infinitesimal generator of $\widetilde{X}$ is given by

$$\mathbb{L}_{\widetilde{X}} = \mathbb{L}_X - \lambda I \tag{7.4.14}$$

(recall Subsection 6.3 above). Applying (7.4.2) and (7.4.3) to (7.4.13) and (7.4.14) we see that (7.4.11) and (7.4.12) hold as claimed. □

The expression (7.4.9) is often referred to as the *Feynman–Kac formula.*

3. Given a continuous function $L : E \to \mathbb{R}$ consider

$$F(t,x) = \mathsf{E}_x\left(\int_0^t L(X_s)\,ds\right) \tag{7.4.15}$$

for $(t,x) \in \mathbb{R}_+ \times E$. *The function F solves the Cauchy problem*:

$$\frac{\partial F}{\partial t} = \mathbb{L}_X F + L \quad \text{in } \mathbb{R}_+ \times E, \tag{7.4.16}$$

$$F(0,x) = 0 \quad \text{for } x \in E. \tag{7.4.17}$$

Indeed, this can be shown in exactly the same way as in the proof of (7.4.2)–(7.4.3) above by reducing (7.4.15) to (7.2.1) so that (7.2.2)–(7.2.3) become (7.4.16)–(7.4.17). □

4. Killed version (Lagrange). Given a continuous function $L : E \to \mathbb{R}$ consider

$$F(t,x) = \mathsf{E}_x \int_0^t e^{-\lambda_s} L(X_s)\,ds \tag{7.4.18}$$

for $(t,x) \in \mathbb{R}_+ \times E$ where $\lambda = (\lambda_t)_{t \ge 0}$ is given by

$$\lambda_t = \int_0^t \lambda(X_s)\,ds \tag{7.4.19}$$

for a measurable (continuous) function $\lambda : E \to \mathbb{R}_+$. *The function F solves the (killed) Cauchy problem*:

$$\frac{\partial F}{\partial t} = \mathbb{L}_X F - \lambda F - L, \tag{7.4.20}$$

$$F(0,x) = 0 \quad \text{for } x \in E. \tag{7.4.21}$$

Indeed, replacing X by the killed process $\widetilde{X}$ it follows that (7.4.18) reads as

$$F(t,x) = \mathsf{E}_x \int_0^t L(\widetilde{X}_s)\,ds \tag{7.4.22}$$

and the infinitesimal generator of $\widetilde{X}$ is given by

$$\mathbb{L}_{\widetilde{X}} = \mathbb{L}_X - \lambda I \tag{7.4.23}$$

(recall Subsection 6.3 above). Applying (7.4.16) and (7.4.17) to (7.4.22) and (7.4.23) we see that (7.4.20) and (7.4.21) hold as claimed. □

The expression (7.4.18) is often referred to as the *Feynman–Kac formula.*

5. Mixed case (Bolza). Given M, L and λ as above, consider

$$F(t,x) = \mathsf{E}_x\Big(e^{-\lambda_t}M(X_t) + \int_0^t e^{-\lambda_s}L(X_s)\,ds\Big) \tag{7.4.24}$$

for $(t,x)\in\mathbb{R}_+\times E$. *The function* F *solves the* (*killed*) *Cauchy problem*:

$$\frac{\partial F}{\partial t} = \mathbb{L}_X F - \lambda F + L \quad \text{in } \mathbb{R}_+\times E, \tag{7.4.25}$$

$$F(0,x) = M(x) \quad \text{for } x\in E. \tag{7.4.26}$$

Indeed, this follows from (7.4.11)–(7.4.12) and (7.4.20)–(7.4.21) using linearity. □

6. Mixed case (general). Given M, L and λ as above, consider

$$F(t,x,s) = \mathsf{E}_{x,s}\Big(e^{-\lambda_t}M(X_t) + \int_0^t e^{-\lambda_s}L(X_s)\,ds + e^{-\lambda_t}S_t\Big) \tag{7.4.27}$$

for $(t,x,s)\in\mathbb{R}_+\times\widetilde{E}$ where $(X_0,S_0)=(x,s)$ under $\mathsf{P}_{x,s}$. *The function* F *solves the* (*killed*) *Cauchy problem*:

$$\frac{\partial F}{\partial t} = \mathbb{L}_X F - \lambda F + L \quad \text{for } x<s \text{ with } s \text{ fixed}, \tag{7.4.28}$$

$$\frac{\partial F}{\partial s}(t,x,s) = 0 \quad \text{for } x=s \text{ with } t\ge 0, \tag{7.4.29}$$

$$F(0,x,s) = M(x)+s \quad \text{for } x\le s. \tag{7.4.30}$$

Indeed, replacing (X,S) by the killed process $(\widetilde{X},\widetilde{S})$ we see that (7.4.27) reads as follows:

$$F(t,x,s) = \mathsf{E}_{x,s}\Big(M(\widetilde{X}_t) + \int_0^t L(\widetilde{X}_s)\,ds + \widetilde{S}_t\Big) \tag{7.4.31}$$

and the infinitesimal operator of $(\widetilde{X},\widetilde{S})$ is given by:

$$\mathbb{L}_X - \lambda I \quad \text{for } x<s, \tag{7.4.32}$$

$$\frac{\partial}{\partial s} = 0 \quad \text{for } x=s. \tag{7.4.33}$$

Setting $\widetilde{M}(x,s) = M(x)+s$ we see that (7.4.31) reduces to (7.4.24) so that (7.4.28)–(7.4.30) follow by (7.4.25)–(7.4.26) using (7.4.32)–(7.4.33). □

7.5. Connection with the Kolmogorov backward equation

Recall that the *Kolmogorov backward equation* (in dimension one) reads (see (4.3.7))

$$\frac{\partial f}{\partial t} + \rho \frac{\partial f}{\partial x} + \frac{\sigma^2}{2} \frac{\partial^2 f}{\partial x^2} = 0 \tag{7.5.1}$$

where f is the transition density function of a diffusion process X given by

$$f(t,x;u,y) = \frac{d}{dy} \mathsf{P}(X_u \le y \mid X_t = x) \tag{7.5.2}$$

for $t < u$ in $\mathbb{R}_+$ and x, y in E.

1. Let us show that the equation (7.5.1) in relation to (7.5.2) has the same character as the equation (7.1.2) in relation to (7.1.1). For this, let us assume that we know (7.1.2) and let us show that this yields (7.5.1). If we define a new Markov process by setting

$$Z_s = (t+s, X_{t+s}) \tag{7.5.3}$$

with $Z_0 = (t,x)$ under $\mathsf{P}_{t,x}$, then the infinitesimal generator of $Z = (Z_s)_{s \ge 0}$ equals

$$\mathbb{L}_Z = \frac{\partial}{\partial s} + \mathbb{L}_X \tag{7.5.4}$$

where $\mathbb{L}_X = \rho\, \partial/\partial x + (\sigma^2/2)\, \partial^2/\partial x^2$.

Consider the exit time from the set $C = [0,u) \times E$ given by

$$\tau_D = \inf \{ s \ge 0 : Z_s \in D \} \tag{7.5.5}$$

where $D = C^c$. Then obviously $\tau_D \equiv u$ and thus

$$H(t,x) = \mathsf{E}_{t,x} \tilde{M}(Z_{\tau_D}) = \mathsf{E}_{t,x} M(Z_u) \tag{7.5.6}$$

satisfies the equation (7.1.2) where $\tilde{M}(t,x) = M(x)$ and M is a given function from E to $\mathbb{R}$. This yields

$$\mathbb{L}_Z H = 0 \tag{7.5.7}$$

which in view of (7.5.4) reduces to (7.5.1) by approximation.

In exactly the same way (choosing M in (7.5.6) to be an indicator function) one sees that

$$F(t,x;u,y) = \mathsf{P}(X_u \le y \mid X_t = x) \tag{7.5.8}$$

satisfies the equation (7.5.1) i.e.

$$\frac{\partial F}{\partial t} + \rho \frac{\partial F}{\partial x} + \frac{\sigma^2}{2} \frac{\partial^2 F}{\partial x^2} = 0 \tag{7.5.9}$$

and (7.5.1) follows by (formal) differentiation of (7.5.9) with respect to y (recall (7.5.2) above). Note also that M in (7.1.1) does not need to be continuous for (7.1.2) to be valid (inspect the proof).

On the other hand, note that (7.5.1) in terms of (7.5.3) reads as follows:

$$\mathbb{L}_Z f = 0 \tag{7.5.10}$$

and likewise (7.5.9) reads as follows:

$$\mathbb{L}_Z F = 0. \tag{7.5.11}$$

This shows that (7.5.1) and (7.5.9) are the same equations as (7.1.2) above.

2. The preceding considerations show that (7.5.1) in relation to (7.5.2) and (7.1.2) in relation to (7.1.1) are refinements of each other but the same equations. Given the central role that (7.1.1)–(7.1.2) play in the entire section above, where all equations under consideration can be seen as special cases of these relations, it is clear that the derivation of (7.1.2) via (7.1.4) above embodies the key Markovian principle which govern all these equations. This is a powerful unifying tool which everyone should be familiar with.

3. Time-homogeneous case (present-state and future-time mixed). When ρ and σ in (7.5.1) do not depend on time, i.e. when the process X is time-homogeneous, then

$$f(t,x;u,y) = f(0,x;u-t,y) \tag{7.5.12}$$

so that (7.5.1) becomes

$$-\frac{\partial f}{\partial s} + \rho \frac{\partial f}{\partial x} + \frac{\sigma^2}{2}\frac{\partial^2 f}{\partial x^2} = 0 \tag{7.5.13}$$

where $f = f(0,x;s,y)$. Note that this equation has the same form as the equation (7.4.2). Moreover, on closer inspection one also sees that the proof of (7.4.2) follows the same pattern as the proof of (7.5.1) via (7.1.2) above.

Finally, note also when the process X is time-homogeneous that the *semi-group* formulation of the Kolmogorov backward (and forward) equation (see (4.3.7) and (4.3.8)) has the following form:

$$\frac{d}{dt}\mathsf{P}_t f = \mathbb{L}_X \mathsf{P}_t f \ \ (= \mathsf{P}_t \mathbb{L}_X f) \tag{7.5.14}$$

where $\mathsf{P}_t f(x) = \mathsf{E}_x f(X_t)$ for a bounded (continuous) function $f : E \to \mathbb{R}$.

Notes. Stochastic control theory deals with three basic problem formulations which were inherited from classical calculus of variations (cf. [67, pp. 25–26]). Given the equation of motion

$$dX_t = \rho(X_t, u_t)\,dt + \sigma(X_t, u_t)\,dB_t \tag{7.5.15}$$

where $(B_t)_{t\ge 0}$ is standard Brownian motion, consider the optimal control problem

$$\inf_u \mathsf{E}_x\Big(\int_0^{\tau_D} L(X_t,u_t)\,dt + M(X_{\tau_D})\Big) \tag{7.5.16}$$

where the infimum is taken over all admissible controls $u=(u_t)_{t\ge 0}$ applied before the exit time $\tau_D = \inf\{t>0 : X_t \notin C\}$ for some open set $C=D^c$ and the process $(X_t)_{t\ge 0}$ starts at x under P_x. If $M\equiv 0$ and $L\neq 0$, the problem (7.5.16) is said to be *Lagrange* formulated. If $L\equiv 0$ and $M\neq 0$, the problem (7.5.16) is said to be *Mayer* formulated. If both $L\neq 0$ and $M\neq 0$, the problem (7.5.16) is said to be *Bolza* formulated.

The Lagrange formulation goes back to the 18th century, the Mayer formulation originated in the 19th century, and the Bolza formulation [20] was introduced in 1913. We refer to [19, pp. 187–189] with the references for a historical account of the Lagrange, Mayer and Bolza problems. Although the three problem formulations are formally known to be equivalent (see e.g. [19, pp. 189–193] or [67, pp. 25–26]), this fact is rarely proved to be essential when solving a concrete problem.

Setting $Z_t = L(X_t,u_t)$ or $Z_t = M(X_t)$, and focusing upon the sample path $t\mapsto Z_t$ for $t\in[0,\tau_D]$, we see that the three problem formulations measure the performance associated with a control u by means of the following two functionals:

$$\int_0^{\tau_D} Z_t\,dt \quad \& \quad Z_{\tau_D} \tag{7.5.17}$$

where the first one represents the surface area below (or above) the sample path, and the second one represents the sample-path's terminal value. In addition to these two functionals, it is suggested by elementary geometric considerations that the maximal value of the sample path

$$\max_{0\le t\le \tau_D} Z_t \tag{7.5.18}$$

provides yet another performance measure which, to a certain extent, is more sensitive than the previous two ones. Clearly, a sample path can have a small integral but still a large maximum, while a large maximum cannot be detected by the terminal value either.

A purpose of the present chapter was to point out that the problem formulations based on a maximum functional can be successfully added to optimal control theory (calculus of variations) and optimal stopping. This suggests a number of new avenues for further research upon extending the Bolza formulation (6.1.2) to optimize the following expression:

$$\mathsf{E}_x\Big(\int_0^{\tau_D} L(X_t,u_t)\,dt + M(X_{\tau_D}) + \max_{0\le t\le \tau_D} K(X_t,u_t)\Big) \tag{7.5.19}$$

where some of the maps K, L and M may also be identically zero.

Optimal stopping problems for the maximum process have been studied by a number of authors in the 1990's (see e.g. [103], [45], [185], [159], [85]) and the subject seems to be well understood now. The present monograph will expose some of these results in Chapters IV–VIII below.

Chapter IV.

Methods of solution

8. Reduction to free-boundary problem

Throughout we will adopt the setting and notation of Subsection 2.2. Thus $X = (X_t)_{t\ge 0}$ is a *strong Markov process* (right-continuous and left-continuous over stopping times) taking values in $E = \mathbb{R}^d$ for some $d \ge 1$.

1. Given a measurable function $G : E \to \mathbb{R}$ satisfying needed regularity conditions, consider the optimal stopping problem

$$V(x) = \sup_{\tau} \mathsf{E}_x G(X_\tau) \tag{8.0.1}$$

where the supremum is taken over all stopping times τ of X, and $X_0 = x$ under P_x with $x \in E$. In Chapter I we have seen that the problem (8.0.1) is equivalent to the problem of finding the smallest superharmonic function $\widehat{V} : E \to \mathbb{R}$ (to be equal to V) which dominate the gain function G on E. In this case, moreover, the first entry time τ_D of X into the stopping set $D = \{\widehat{V} = G\}$ is optimal, and $C = \{\widehat{V} > G\}$ is the continuation set. As already pointed out at the end of Chapter I, it follows that $\widehat{V}$ and C should solve the *free-boundary problem*

$$\mathbb{L}_X \widehat{V} \le 0 \quad (\widehat{V} \text{ minimal}), \tag{8.0.2}$$

$$\widehat{V} \ge G \quad (\widehat{V} > G \text{ on } C \;\;\&\;\; \widehat{V} = G \text{ on } D) \tag{8.0.3}$$

where $\mathbb{L}_X$ is the infinitesimal generator of X (cf. Chapter III above). It is important to realize that both $\widehat{V}$ and C are unknown in the system (8.0.2)–(8.0.3) (and both need to be determined).

2. Identifying $V = \widehat{V}$, upon invoking sufficient conditions at the end of Chapter I that make this identification possible, it follows that V admits the following representation:

$$V(x) = \mathsf{E}_x G(X_{\tau_D}) \tag{8.0.4}$$

for $x \in E$ where τ_D is the first entry time of X into D given by

$$\tau_D = \inf\{\, t \ge 0 : X_t \in D \,\}. \tag{8.0.5}$$

The results of Chapter III (for the Dirichlet problem) become then applicable and according to (7.1.2)–(7.1.3) it follows that

$$\mathbb{L}_X V = 0 \quad \text{in } \; C, \tag{8.0.6}$$

$$V\big|_D = G\big|_D. \tag{8.0.7}$$

Note that (8.0.6) stands in accordance with the general fact from Chapter I that $(V(X_{t\wedge\tau_D}))_{t\ge 0}$ is a martingale. Note also that X is multidimensional and thus can generally be equal to the time-integral-space-maximum process considered in Chapter III (including killed versions of these processes as well). In this way we see that the Dirichlet problem (8.0.6)–(8.0.7) embodies all other problems (Dirichlet/Poisson, Neumann, Cauchy) considered in Chapter III. Fuller details of these problem formulations are easily reconstructed in the present setting and for this reason will be omitted.

3. The condition (8.0.2) states that V is the smallest superharmonic function (which dominates G). The two properties "smallest" and "superharmonic" play a decisive role in the selection of the optimal boundary ∂C (i.e. sets C and D) in the sense that only special sets C (i.e. D) will qualify to meet these properties. Indeed, assuming that G is smooth (in a neighborhood of ∂C) the following general picture (stated more as a 'rule of thumb') is valid.

If X after starting at ∂C enters $\operatorname{int}(D)$ immediately (e.g. when X is a diffusion and ∂C is sufficiently regular e.g. Lipschitz) then the condition (8.0.2) leads to

$$\frac{\partial V}{\partial x}\Big|_{\partial C} = \frac{\partial G}{\partial x}\Big|_{\partial C} \quad (\textit{smooth fit}) \tag{8.0.8}$$

where $d = 1$ is assumed for simplicity (in the case $d > 1$ one should replace $\partial/\partial x$ in (8.0.8) by $\partial/\partial x_i$ for $1 \le i \le d$). However, if X after starting at ∂C does not enter $\operatorname{int}(D)$ immediately (e.g. when X has jumps and no diffusion component while ∂C may still be sufficiently regular e.g. Lipschitz) then the condition (8.0.2) leads to

$$V\big|_{\partial C} = G\big|_{\partial C} \quad (\textit{continuous fit}). \tag{8.0.9}$$

The more precise meaning of these conditions will be discussed in Section 9 below.

8.1. Infinite horizon

Infinite horizon problems in dimension one are generally easier than finite horizon problems since the equation (8.0.6) (or its killed version) can often be solved explicitly (in a closed form) yielding a candidate function to which a verification procedure (stochastic calculus) can be applied. Especially transparent in this context is the case of one-dimensional diffusions X where the existence of explicit solutions (scale function, speed measure) reduces the study of optimal stopping to the case of standard Brownian motion.

1. To illustrate the latter in more detail, let us assume that X is a one-dimensional diffusion solving the following SDE (stochastic differential equation):

$$dX_t = \rho(X_t)\,dt + \sigma(X_t)\,dB_t \tag{8.1.1}$$

and let us consider the optimal stopping problem

$$V(x) = \sup_{\tau} \mathsf{E}_x G(X_\tau) \tag{8.1.2}$$

where the supremum is taken over all stopping times τ of X, and $X_0 = x$ under P_x with $x \in \mathbb{R}$.

Denoting by S the scale function of X (see (5.2.29)) and writing

$$G(S_\tau) = G\big(S^{-1} \circ S(X_\tau)\big) = (G \circ S^{-1})(S(X_\tau)) = \widetilde{G}(M_\tau) = \widetilde{G}(\widetilde{B}_{\sigma_\tau}) \tag{8.1.3}$$

where $\widetilde{G} = G \circ S^{-1}$ is a new gain function, $M = S(X)$ is a continuous local martingale and $\widetilde{B}$ is a standard Brownian motion with $\sigma_t = \langle M, M \rangle_t$ (by Lemma 5.1), we see that (8.0.1) reads

$$V(x) = \sup_{\sigma} \mathsf{E}_{S(x)} \widetilde{G}(\widetilde{B}_\sigma) \tag{8.1.4}$$

where the supremum is taken over all stopping times σ of $\widetilde{B}$ (recall that τ is a stopping time of M if and only if σ_τ is a stopping time of $\widetilde{B}$). This shows that the optimal stopping problem (8.1.2) is equivalent to the optimal stopping problem (8.1.4). Moreover, it is easily verified by Itô's formula (page 67) that

$$\sigma_\tau = \langle M, M \rangle_\tau = \int_0^\tau S'(X_s)^2\, \sigma(X_s)^2\, ds \tag{8.1.5}$$

for every stopping time of X i.e. M. This identity establishes a transparent one-to-one correspondence between the optimal stopping time in the problem (8.1.2) and the optimal stopping time in the problem (8.1.4): having one of them given, we can reconstruct the other, and vice versa.

2. Recalling that the infinitesimal generator of the Brownian motion B equals $(1/2)\,\partial^2/\partial x^2$ we see that a smooth function $V : \mathbb{R} \to \mathbb{R}$ is superharmonic if and only if $V'' \le 0$ i.e. if and only if V is concave. This provides a transparent geometric interpretation of superharmonic functions for standard Brownian motion. Making use of the scale function S and exploiting the equivalence of (8.1.2)

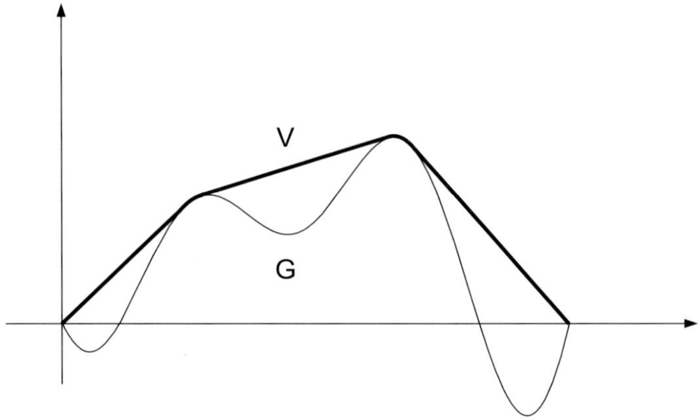

Figure IV.1: An obstacle G and the rope V depicting the superharmonic characterization.

and (8.1.4), this geometric interpretation (in somewhat less transparent form) extends from B to general one-dimensional diffusions X considered in (8.1.2). Since these details are evident but somewhat lengthy we shall omit further discussion (see Subsection 9.3 below).

The geometric interpretation of superharmonic functions for B leads to an appealing physical interpretation of the value function V associated with the gain function G: If G depicts an obstacle, and a rope is put above G with both ends pulled to the ground, the resulting shape of the rope coincides with V (see *Figure* IV.1). Clearly the fit of the rope and the obstacle should be smooth whenever the obstacle is smooth (smooth fit). A similar interpretation (as in *Figure* IV.1) extends to dimension two (membrane) and higher dimensions. This leads to a class of problems in mathematical physics called the "obstacle problems".

8.2. Finite horizon

Finite horizon problems (in dimension one or higher) are more difficult than infinite horizon problems since the equation (8.0.6) (or its killed version) contains the $\partial/\partial t$ term and most often cannot be solved explicitly (in a closed form). Thus, in this case it is not possible to produce a candidate function to which a verification procedure is to be applied. Instead one can try to characterize V and C (i.e. D) by means of the free-boundary problem derived above. A more refined method (just as in the case of two algebraic equations with two unknowns) aims at expressing V in terms of ∂C and then deriving a (nonlinear) equation for ∂C. This line of argument will be presented in more detail in Subsection 14.1 below, and examples of application will be given in Chapters VI–VIII below.

1. To illustrate the former method in more detail, let us consider the optimal stopping problem

$$V(t,x) = \sup_{0 \le \tau \le T-t} \mathsf{E}_{t,x} G(t+\tau, X_{t+\tau}) \tag{8.2.1}$$

where the supremum is taken over all stopping times τ of X, and $X_t = x$ under $\mathsf{P}_{t,x}$ for $(t,x) \in [0,T] \times E$. At this point it is useful to recall our discussion in Subsections 2.2 and 6.1 explaining why X needs to be replaced by the time-space process $Z_t = (t, X_t)$ in the finite-horizon formulation (8.0.1). It implies that the preceding discussion leading to the free-boundary problem (8.0.2)–(8.0.3) as well as (8.0.6)–(8.0.7) and (8.0.8) or (8.0.9) applies to the process Z instead of X. In the case when X is a diffusion, and when ∂C is sufficiently regular (e.g. Lipschitz), we see that (8.0.6)–(8.0.7) and (8.0.8) read:

$$V_t + \mathbb{L}_X V = 0 \quad \text{in } C, \tag{8.2.2}$$

$$V\big|_D = G\big|_D, \tag{8.2.3}$$

$$\frac{\partial V}{\partial x}\bigg|_{\partial C} = \frac{\partial G}{\partial x}\bigg|_{\partial C} \quad \text{(smooth fit)} \tag{8.2.4}$$

where $d = 1$ is assumed in (8.2.4) for simplicity (in the case $d > 1$ one should replace $\partial/\partial x$ in (8.2.4) by $\partial/\partial x_i$ for $1 \le i \le d$). It should be noted in (8.2.3) that all points (T,x) belong to D when $x \in E$. In the case when X has jumps and no diffusion component, and when ∂C may still be sufficiently nice (e.g. Lipschitz), the condition (8.2.4) needs to be replaced by

$$V\big|_{\partial C} = G\big|_{\partial C} \quad \text{(continuous fit)}. \tag{8.2.5}$$

The question of existence and uniqueness of the solution to the free-boundary problem (8.2.2)–(8.2.3) with (8.2.4) or (8.2.5) will be studied through specific examples in Chapters VI–VIII.

2. Another class of problems coming from mathematical physics fits into the free-boundary setting above. These are processes of melting and solidification leading to the *"Stefan free-boundary problem"*. Imagine a chunk of ice (at temperature G) immersed in water (at temperature V). Then the ice-water interface (as a function of time and space) will coincide with the optimal boundary (surface) ∂C. This illustrates a basic link between optimal stopping and the Stefan problem.

9. Superharmonic characterization

In this section we adopt the setting and notation from the previous section. Recall that the value function from (8.0.1) can be characterized as the smallest superharmonic function (relative to X) which dominates G (on E). As already pointed

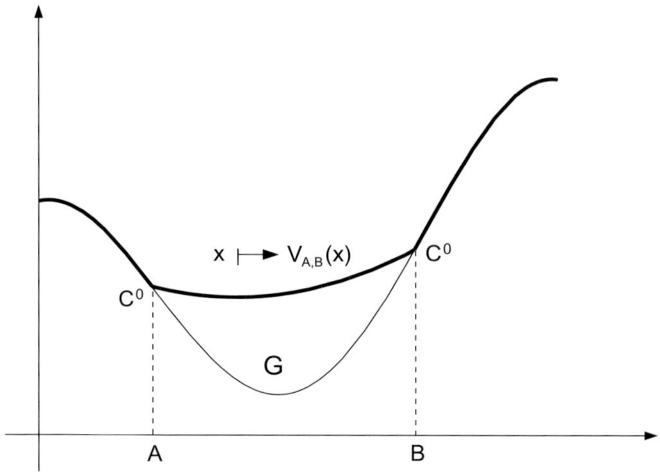

Figure IV.2: The function $x \mapsto V_{A,B}(x)$ from (9.0.7) above when X is a Markov process with diffusion component.

out in the previous section, the two properties "smallest" and "superharmonic" play a decisive role in the selection of the optimal boundary ∂C (i.e. sets C and D).

To illustrate the preceding fact in further detail, let us for simplicity assume that $E = \mathbb{R}$ and that C equals a bounded open interval in E (often this fact is evident from the form of X and G). By general theory (Chapter I) we then know that the exit time

$$\tau_{A,B} = \inf\{\, t \geq 0 : X_t \notin (A,B) \,\} \tag{9.0.6}$$

is optimal in (8.0.1) for some A and B to be found.

Given any two candidate points A and B and inserting $\tau_{A,B}$ into (8.0.1) as a candidate stopping time, we get the function

$$V_{A,B}(x) = \mathsf{E}_x G\big(X_{\tau_{A,B}}\big) \tag{9.0.7}$$

for $x \in E$. Clearly $V_{A,B}(x) = G(x)$ for $x \notin (A,B)$ and only those A and B are to be considered for which $V_{A,B}(x) \geq G(x)$ for all $x \in E$.

When X is a Markov process with diffusion component then $V_{A,B}$ will be smooth on (A,B) but only continuous at A and B as *Figure* IV.2 shows.

If we move A and B along the state space and examine what happens with the resulting function $V_{A,B}$ at A and B, typically we will see that only for a special (unique) pair of A and B, will the continuity of $V_{A,B}$ at A and B turn

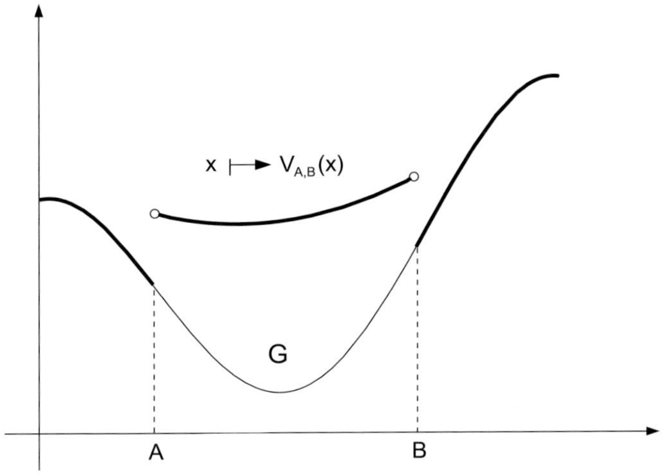

Figure IV.3: The function $x \mapsto V_{A,B}(x)$ from (9.0.7) above when X is a Markov process with jumps (without diffusion component).

into smoothness. This is a variational way to experience the *principle of smooth fit*.

When X has jumps and no diffusion component then $V_{A,B}$ will be continuous/smooth on (A, B) but only discountinuous at A and B as *Figure* IV.3 shows.

If we move A and B along the state space and examine what happens with the resulting function $V_{A,B}$ at A and B, typically we will see that only for a special (unique) pair of A and B, the discontinuity of $V_{A,B}$ at A and B will turn into continuity. (A mixed case of *Figure* IV.2 at A and *Figure* IV.3 at B, or vice versa, is possible as well.) This is a variational way to experience the *principle of continuous fit*.

Specific examples of *Figure* IV.2 and *Figure* IV.3 (including a mixed case) are studied in Chapter VI (see figures in Sections 23 and 24).

9.1. The principle of smooth fit

As already pointed out above, the principle of smooth fit (see (8.0.8)) states that the optimal stopping boundary (point) is selected so that the value function is smooth at that point. The aim of this subsection is to present two methods which (when properly modified if needed) can be used to *verify* the smooth fit principle.

For simplicity of exposition we will restrict our attention to the infinite horizon problems in dimension one. The second method extends to higher dimensions as well.

Given a *regular* diffusion process $X = (X_t)_{t \geq 0}$ with values in $E = \mathbb{R}$ and a measurable function $G : E \to \mathbb{R}$ satisfying the usual (or weakened) integrability condition, consider the optimal stopping problem

$$V(x) = \sup_{\tau} \mathsf{E}_x G(X_\tau) \tag{9.1.1}$$

where the supremum is taken over all stopping times τ of X, and $X_0 = x$ under P_x with $x \in E$. For simplicity let us assume that the continuation set C equals (b, ∞) and the stopping set D equals $(-\infty, b]$ where $b \in E$ is the *optimal stopping point.* We want to show (under natural conditions) that V is differentiable at b and that $V'(b) = G'(b)$ (smooth fit).

Method 1. First note that for $\varepsilon > 0$,

$$\frac{V(b+\varepsilon) - V(b)}{\varepsilon} \geq \frac{G(b+\varepsilon) - G(b)}{\varepsilon} \tag{9.1.2}$$

since $V \geq G$ and $V(b) = G(b)$. Next consider the exit time

$$\tau_\varepsilon = \inf \{ t \geq 0 : X_t \notin (b-\varepsilon, b+\varepsilon) \} \tag{9.1.3}$$

for $\varepsilon > 0$. Then

$$\begin{aligned} \mathsf{E}_b V(X_{\tau_\varepsilon}) &= V(b+\varepsilon)\, \mathsf{P}_b(X_{\tau_\varepsilon} = b+\varepsilon) + V(b-\varepsilon)\, \mathsf{P}_b(X_{\tau_\varepsilon} = b-\varepsilon) \\ &= V(b+\varepsilon)\, \mathsf{P}_b(X_{\tau_\varepsilon} = b+\varepsilon) + G(b-\varepsilon)\, \mathsf{P}_b(X_{\tau_\varepsilon} = b-\varepsilon). \end{aligned} \tag{9.1.4}$$

Moreover, since V is superharmonic (cf. Chapter I), we have

$$\mathsf{E}_b V(X_{\tau_\varepsilon}) \leq V(b) = V(b)\, \mathsf{P}_b(X_{\tau_\varepsilon} = b+\varepsilon) + G(b)\, \mathsf{P}_b(X_{\tau_\varepsilon} = b-\varepsilon). \tag{9.1.5}$$

Combining (9.1.4) and (9.1.5) we get

$$\begin{aligned} &\big(V(b+\varepsilon) - V(b)\big)\, \mathsf{P}_b(X_{\tau_\varepsilon} = b+\varepsilon) \\ &\leq \big(G(b) - G(b-\varepsilon)\big)\, \mathsf{P}_b(X_{\tau_\varepsilon} = b-\varepsilon). \end{aligned} \tag{9.1.6}$$

Recalling that $\mathsf{P}_b(X_{\tau_\varepsilon} = b+\varepsilon) = \big(S(b) - S(b-\varepsilon)\big) / \big(S(b+\varepsilon) - S(b-\varepsilon)\big)$ and $\mathsf{P}_b(X_{\tau_\varepsilon} = b-\varepsilon) = \big(S(b+\varepsilon) - S(b)\big) / \big(S(b+\varepsilon) - S(b-\varepsilon)\big)$, where S is the scale function of X, we see that (9.1.6) yields

$$\begin{aligned} \frac{V(b+\varepsilon) - V(b)}{\varepsilon} &\leq \frac{G(b) - G(b-\varepsilon)}{\varepsilon}\, \frac{S(b+\varepsilon) - S(b)}{S(b) - S(b-\varepsilon)} \\ &\longrightarrow G'(b)\, \frac{S'(b)}{S'(b)} = G'(b) \end{aligned} \tag{9.1.7}$$

as $\varepsilon \downarrow 0$ whenever G and S are differentiable at b (and $S'(b)$ is different from zero). Combining (9.1.7) and (9.1.2) and letting $\varepsilon \downarrow 0$ we see that V is differentiable at b and $V'(b) = G'(b)$. In this way we have verified that the following claim holds:

If G and S are differentiable at b, then V is differentiable at b and $V'(b) = G'(b)$ i.e. the smooth fit holds at b. (9.1.8)

The following example shows that differentiability of G at b cannot be omitted in (9.1.8). For a complementary discussion of Method 1 (filling the gaps of V being superharmonic and $S'(b)$ being different from zero) see Subsection 9.3 below.

Example 9.1. Let $X_t = x + B_t - t$ for $t \geq 0$ and $x \in \mathbb{R}$, and let $G(x) = 1$ for $x \geq 0$ and $G(x) = 0$ for $x < 0$. Consider the optimal stopping problem (9.1.1). Then clearly $V(x) = 1$ for $x \geq 0$, and the only candidate for an optimal stopping time when $x < 0$ is

$$\tau_0 = \inf\{\, t \geq 0 : X_t = 0 \,\} \tag{9.1.9}$$

where $\inf(\varnothing) = \infty$. Then (with $G(X_\infty) = G(-\infty) = 0$),

$$\begin{aligned} V(x) = \mathsf{E}_x G(X_{\tau_0}) &= \mathsf{E}_x\big(0 \cdot I(\tau_0 = \infty) + 1 \cdot I(\tau_0 < \infty)\big) \\ &= \mathsf{P}_x(\tau_0 < \infty). \end{aligned} \tag{9.1.10}$$

Using Doob's result

$$\mathsf{P}\Big(\sup_{t \geq 0}(B_t - \alpha t) \geq \beta\Big) = e^{-2\alpha\beta} \tag{9.1.11}$$

for $\alpha, \beta > 0$, it follows that

$$\begin{aligned} \mathsf{P}_x(\tau_0 < \infty) &= \mathsf{P}\Big(\sup_{t \geq 0}(x + B_t - t) \geq 0\Big) \\ &= \mathsf{P}\Big(\sup_{t \geq 0}(B_t - t) \geq -x\Big) = e^{2x} \end{aligned} \tag{9.1.12}$$

for $x < 0$. This shows that $V(x) = 1$ for $x \geq 0$ and $V(x) = e^{2x}$ for $x < 0$. Note that $V'(0+) = 0 \neq 2 = V'(0-)$. Thus the smooth fit does not hold at the optimal stopping point 0. Note that G is discontinuous at 0.

Moreover, if we take any continuous function $\widetilde{G} : \mathbb{R} \to \mathbb{R}$ satisfying $0 < \widetilde{G}(x) < V(x)$ for $x \in (-\infty, 0)$ and $\widetilde{G}(x) = 1$ for $x \in [0, \infty)$, and consider the optimal stopping problem (9.1.1) with $\widetilde{G}$ instead of G, then (due to $\mathsf{E}_x \widetilde{G}(X_{\tau_0}) \geq \mathsf{E}_x G(X_{\tau_0}) > \widetilde{G}(x)$ for $x \in (-\infty, 0)$) we see that it is never optimal to stop in $(-\infty, 0)$. Clearly it is optimal to stop in $[0, \infty)$, so that τ_0 is optimal again and we have

$$\widetilde{V}(x) = \mathsf{E}_x \widetilde{G}(X_{\tau_0}) = \mathsf{E}_x G(X_{\tau_0}) = V(x) \tag{9.1.13}$$

for all $x \in \mathbb{R}$. Thus, in this case too, we see that the smooth fit does not hold at the optimal stopping point 0. Note that $\widetilde{G}$ is not differentiable at $b = 0$.

Method 2. Assume that X equals the standard Brownian motion B [or any other (diffusion) process where the dependence on the initial point x is explicit and smooth]. Then as above we first see that (9.1.2) holds. Next let $\tau_*^\varepsilon = \tau_*(b+\varepsilon)$ denote the optimal stopping time for $V(b+\varepsilon)$, i.e. let

$$\tau_*^\varepsilon = \inf\{\, t \geq 0 : X_t \leq b \,\} \tag{9.1.14}$$

under $\mathsf{P}_{b+\varepsilon}$. Since $\text{Law}(X \mid \mathsf{P}_{b+\varepsilon}) = \text{Law}(X^{b+\varepsilon} \mid \mathsf{P})$ where $X_t^{b+\varepsilon} = b + \varepsilon + B_t$ under P, we see that τ_*^ε is equally distributed as $\sigma_*^\varepsilon = \inf\{\, t \geq 0 : b + \varepsilon + B_t \leq b \,\} = \inf\{\, t \geq 0 : B_t \leq -\varepsilon \,\}$ under P, so that $\tau_*^\varepsilon \downarrow 0$ and $\sigma_*^\varepsilon \downarrow 0$ as $\varepsilon \downarrow 0$. This reflects the fact that b is *regular* for D (relative to X) which is needed for the method to be applicable. We then have

$$\frac{V(b+\varepsilon) - V(b)}{\varepsilon} \leq \frac{\mathsf{E}G(b+\varepsilon+B_{\sigma_*^\varepsilon}) - \mathsf{E}G(b+B_{\sigma_*^\varepsilon})}{\varepsilon} \tag{9.1.15}$$

since $V(b+\varepsilon) = \mathsf{E}_{b+\varepsilon}G(X_{\tau_*^\varepsilon}) = \mathsf{E}G(b+\varepsilon+B_{\sigma_*^\varepsilon})$ and $V(b) \geq \mathsf{E}_b G(X_{\tau_*^\varepsilon}) = \mathsf{E}G(b+B_{\sigma_*^\varepsilon})$. By the mean value theorem we have

$$G(b+\varepsilon+B_{\sigma_*^\varepsilon}) - G(b+B_{\sigma_*^\varepsilon}) = G'\big(b+B_{\sigma_*^\varepsilon}+\theta\varepsilon\big)\varepsilon \tag{9.1.16}$$

for some $\theta \in (0,1)$. Inserting (9.1.16) into (9.1.15) and assuming that

$$|G'\big(b+B_{\sigma_*^\varepsilon}+\theta\varepsilon\big)| \leq Z \tag{9.1.17}$$

for all $\varepsilon > 0$ (small) with some $Z \in L^1(\mathsf{P})$, we see from (9.1.15) using (9.1.16) and the Lebesgue dominated convergence theorem that

$$\limsup_{\varepsilon \downarrow 0} \left(\frac{V(b+\varepsilon) - V(b)}{\varepsilon} \right) \leq G'(b). \tag{9.1.18}$$

Since $B_{\tau_*^\varepsilon} = -\varepsilon$ note that (9.1.17) is satisfied if G' is bounded (on a neighborhood containing b). From (9.1.2) it follows that

$$\liminf_{\varepsilon \downarrow 0} \left(\frac{V(b+\varepsilon) - V(b)}{\varepsilon} \right) \geq G'(b). \tag{9.1.19}$$

Combining (9.1.18) and (9.1.19) we see that V is differentiable at b and $V'(b) = G'(b)$. In this way we have verified that the following claim holds:

If G is C^1 (on a neighborhood containing b) then V is differentiable at b and $V'(b) = G'(b)$ i.e. the smooth fit holds at b. (9.1.20)

On closer inspection of the above proof, recalling that $B_{\sigma_*^\varepsilon} = -\varepsilon$, one sees that (9.1.16) actually reads

$$G(b+\varepsilon+B_{\sigma_*^\varepsilon}) - G(b+B_{\sigma_*^\varepsilon}) = G(b) - G(b-\varepsilon) \tag{9.1.21}$$

which is in line with (9.1.7) above since $S(x) = x$ for all x (at least when X is a standard Brownian motion). Thus, in this case (9.1.18) holds even if the additional hypotheses on Z and G' stated above are removed. The proof above, however, is purposely written in this more general form, since as such it also applies to more general (diffusion) processes X for which the (smooth) dependence on the initial point is expressed explicitly, as well as when C is not unbounded, i.e. when $C = (b, c)$ for some $c \in (b, \infty)$. In the latter case, for example, one can replace (9.1.14) by

$$\tau_*^\varepsilon = \inf\{\, t \geq 0 : \, X_t \leq b \text{ or } X_t \geq c\,\} \tag{9.1.22}$$

under $\mathsf{P}_{b+\varepsilon}$ and proceed as outlined above making only minor modifications.

The following example shows that regularity of the optimal point b for D (relative to X) cannot be omitted from the proof.

Example 9.2. Let $X_t = -t$ for $t \geq 0$, let $G(x) = x$ for $x \geq 0$, let $G(x) = H(x)$ for $x \in [-1, 0]$, and let $G(x) = 0$ for $x \leq -1$, where $H : [-1, 1] \to \mathbb{R}$ is a smooth function making G (continuous and) smooth (e.g. C^1) on $\mathbb{R}$. Assume moreover that $H(x) < 0$ for all $x \in (-1, 0)$ (with $H(-1) = H(0) = 0$). Then clearly $(-1, 0)$ is contained in C, and $(-\infty, -1] \cup [0, \infty)$ is contained in D. It follows that $V(x) = 0$ for $x \leq 0$ and $V(x) = x$ for $x > 0$. Hence V is not smooth at the optimal boundary point 0. Recall that G is smooth everywhere on $\mathbb{R}$ (at 0 as well) but 0 is not regular for D (relative to X).

Notes. The principle of smooth fit appears for the first time in the work of Mikhalevich [136]. Method 1 presented above was inspired by the method of Grigelionis and Shiryaev [88] (see also [196, pp. 159–161]) which uses a Taylor expansion of the value function at the optimal point (see Subsection 9.3 below for a deeper analysis). Method 2 presented above is due to Bather [11] (see also [215]). This method will be adapted and used in Chapters VI–VIII below. For comparison note that this method uses a Taylor expansion of the gain function which is given a priori. There are also other derivations of the smooth fit that rely upon the diffusion relation $X_t \sim \sqrt{t}$ for small t (see e.g. [30, p. 233] which also makes use of a Taylor expansion of the value function). Further references are given in the Notes to Subsection 9.3 and Section 25 below.

9.2. The principle of continuous fit

As already pointed out above, the principle of continuous fit states that the optimal stopping boundary (point) is selected so that the value function is continuous at that point. The aim of this subsection is to present a simple method which (when properly modified if needed) can be used to verify the continuous fit principle. For simplicity of exposition we will restrict our attention to the infinite horizon problems in dimension one.

Given a Markov process $X = (X_t)_{t \ge 0}$ (right-continuous and left-continuous over stopping times) taking value in $E = \mathbb{R}$ and a measurable function $G : E \to \mathbb{R}$ satisfying the usual (or weakened) integrability condition, consider the optimal stopping problem

$$V(x) = \sup_\tau \mathsf{E}_x G(X_\tau) \tag{9.2.1}$$

where the supremum is taken over all stopping times τ of X, and $X_0 = x$ under P_x with $x \in E$. For simplicity let us assume that the continuation set C equals (b, ∞) and the stopping set D equals $(-\infty, b]$ where $b \in E$ is the optimal stopping point. We want to show that (under natural conditions) V is continuous at b and that $V(b) = G(b)$ (continuous fit).

Method. First note that

$$V(b{+}\varepsilon) - V(b) \ge G(b{+}\varepsilon) - G(b) \tag{9.2.2}$$

for $\varepsilon > 0$. Next let τ_*^ε denote the optimal stopping time for $V(b+\varepsilon)$, i.e. let

$$\tau_*^\varepsilon = \inf\{\, t \ge 0 : X_t \le b\,\} \tag{9.2.3}$$

under $\mathsf{P}_{b+\varepsilon}$. Then we have

$$V(b{+}\varepsilon) - V(b) \le \mathsf{E}\, G\big(X_{\tau_*^\varepsilon}^{b+\varepsilon}\big) - \mathsf{E}\, G\big(X_{\tau_*^\varepsilon}^{b}\big) \tag{9.2.4}$$

since $V(b) \ge \mathsf{E}_b G\big(X_{\tau_*^\varepsilon}\big) = \mathsf{E}\, G\big(X_{\tau_*^\varepsilon}^b\big)$. Clearly $\tau_*^\varepsilon \downarrow \rho \ge 0$ as $\varepsilon \downarrow 0$. (In general, this ρ can be strictly positive which means that b is not regular and in turn can imply the breakdown of the smooth fit at b.) It follows that $X_{\tau_*^\varepsilon}^b \to X_\rho^b$ as $\varepsilon \downarrow 0$ by the right continuity of X. Moreover, if the following *time-space* (*Feller* motivated) *condition* on X holds:

$$X_{t+h}^{b+\varepsilon} \to X_t^b \quad \mathsf{P}\text{-a.s.} \tag{9.2.5}$$

as $\varepsilon \downarrow 0$ and $h \downarrow 0$, then we also have $X_{\tau_*^\varepsilon}^{b+\varepsilon} \to X_\rho^b$ P-a.s. as $\varepsilon \downarrow 0$. Combining these two convergence relations, upon assuming that G is continuous and

$$G\big(X_{\tau_*^\varepsilon}^{b+\varepsilon}\big) - G\big(X_{\tau_*^\varepsilon}^{b}\big) \le Z \tag{9.2.6}$$

for some $Z \in L^1(\mathsf{P})$, we see from (9.2.4) that

$$\limsup_{\varepsilon \downarrow 0} \big(V(b{+}\varepsilon) - V(b)\big) \le 0 \tag{9.2.7}$$

by Fatou's lemma. From (9.2.2) on the other hand it follows that

$$\liminf_{\varepsilon \downarrow 0} \big(V(b{+}\varepsilon) - V(b)\big) \ge 0. \tag{9.2.8}$$

Combining (9.2.7) and (9.2.8) we see that V is continuous at b (and $V(b) = G(b)$). In this way we have verified that the following claim holds:

If G is continuous at b and bounded (or (9.2.6) holds), and X satisfies (9.2.5), then V is continuous at b (and $V(b) = G(b)$), i.e. the continuous fit holds at b. (9.2.9)

Taking $G(x) = e^{-x}$ for $x \geq 0$ and $G(x) = -x$ for $x < 0$, and letting $X_t = -t$ for $t \geq 0$, we see that $V(x) = 1$ for $x \geq 0$ and $V(x) = -x$ for $x < 0$. Thus V is not continuous at the optimal stopping point 0. This shows that the continuity of G at b cannot be omitted in (9.2.9). Note that 0 is regular for D (relative to X).

Further (more illuminating) examples of the continuous fit principle (when X has jumps) will be given in Sections 23 and 24.

Notes. The principle of continuous fit was recognized as a key ingredient of the solution in [168] and [169]. The proof given above is new.

9.3. Diffusions with angles

The purpose of this subsection (following [167]) is to exhibit a complementary analysis of the smooth fit principle in the case of one-dimensional (regular) diffusions.

1. Recall that the *principle of smooth fit* states (see (8.0.8)) that the optimal stopping point b which separates the continuation set C from the stopping set D in the optimal stopping problem

$$V(x) = \sup_{\tau} \mathsf{E}_x G(X_\tau) \tag{9.3.1}$$

is characterized by the fact that $V'(b)$ exists and is equal to $G'(b)$. Typically, no other point $\tilde{b}$ separating the candidate sets $\tilde{C}$ and $\tilde{D}$ will satisfy this identity, and most often $V''(b)$ will either fail to exist or will not be equal to $G''(b)$. These unique features of the smooth fit principle make it a powerful tool in solving specific problems of optimal stopping. The same is true in higher dimensions but in the present subsection we focus on dimension one only.

Regular diffusion processes form a natural class of Markov processes X in (9.3.1) for which the smooth-fit principle is known to hold in great generality. On the other hand, it is easy to construct examples which show that the smooth fit $V'(b) = G'(b)$ can fail if the diffusion process X is not regular as well as that V need not be differentiable at b if G is not so (see Example 9.1 above). Thus *regularity* of the diffusion process X and *differentiability* of the gain function G are minimal conditions under which the smooth fit can hold in greater generality. In this subsection we address the question of their *sufficiency* (recall (9.1.8) above).

Our exposition can be summarized as follows. Firstly, we show that there exists a regular diffusion process X and a differentiable gain function G such that the smooth fit condition $V'(b) = G'(b)$ fails to hold at the optimal stopping point b (Example 3.5). Secondly, we show that the latter cannot happen if the scale function S is differentiable at b. In other words, if X is regular and both G and S are differentiable at b, then V is differentiable at b and $V'(b) = G'(b)$ (Theorem 3.3) [this was derived in (9.1.8) using similar means]. Thirdly, we give an example showing that the latter can happen even when $d^+G/dS < d^+V/dS < d^-V/dS < d^-V/dS$ at b (Example 3.2). The relevance of this fact will be reviewed shortly below.

A. N. Kolmogorov expressed the view that the principle of smooth fit holds because “diffusions do not like angles” (this is one of the famous tales of the second author). It hinges that there must be something special about the diffusion process X in the first example above since the gain function G is differentiable. We will briefly return to this point in the end of the present subsection.

2. Let $X = (X_t)_{t \geq 0}$ be a diffusion process with values in an interval J of $\mathbb{R}$. For simplicity we will assume that X can be killed only at the end-points of J which do not belong to J. Thus, if ζ denotes the death time of X, then X is a strong Markov process such that $t \mapsto X_t$ is continuous on $[0, \zeta)$, and the end-points of J at which X can be killed act as absorbing boundaries (once such a point is reached X stays there forever). We will denote by $I = (l, r)$ the interior of J.

Given $c \in J$ we will let

$$\tau_c = \inf \{ t > 0 : X_t = c \} \tag{9.3.2}$$

denote the hitting time of X to c. We will assume that X is *regular* in the sense that $\mathsf{P}_b(\tau_c < \infty) = 1$ for every $b \in I$ and all $c \in J$. It means that I cannot be decomposed into smaller intervals from which X could not exit. It also means that b is regular for both $D_1 = (l, b]$ and $D_2 = [b, r)$ in the sense that $\mathsf{P}_b(\tau_{D_i} = 0) = 1$ where $\tau_{D_i} = \inf \{ t > 0 : X_t \in D_i \}$ for $i = 1, 2$. In particular, each $b \in I$ is regular for itself in the sense that $\mathsf{P}_b(\tau_b = 0) = 1$.

Let S denote the scale function of X. Recall that $S : J \to \mathbb{R}$ is a strictly increasing continuous function such that

$$\mathsf{P}_x(\tau_a < \tau_b) = \frac{S(b) - S(x)}{S(b) - S(a)} \quad \& \quad \mathsf{P}_x(\tau_b < \tau_a) = \frac{S(x) - S(a)}{S(b) - S(a)} \tag{9.3.3}$$

for $a < x < b$ in J. Recall also that the scale function can be characterized (up to an affine transformation) as a continuous function $S : J \to \mathbb{R}$ such that $(S(X_{t \wedge \tau_l \wedge \tau_r}))_{t \geq 0}$ is a continuous local martingale.

3. Let $G : J \to \mathbb{R}$ be a measurable function satisfying

$$\mathsf{E} \sup_{0 \leq t < \zeta} |G(X_t)| < \infty. \tag{9.3.4}$$

Consider the optimal stopping problem

$$V(x) = \sup_{\tau} \mathsf{E}_x G(X_\tau) \tag{9.3.5}$$

for $x \in J$ where the supremum is taken over all stopping times τ of X (i.e. with respect to the natural filtration $\mathcal{F}_t^X = \sigma(X_s : 0 \le s \le t)$ generated by X for $t \ge 0$).

Following the argument of Dynkin and Yushkevich [55, p. 115], take $c < x < d$ in J and choose stopping times τ_1 and τ_2 such that $\mathsf{E}_c G(X_{\tau_1}) \ge V(c) - \varepsilon$ and $\mathsf{E}_d G(X_{\tau_2}) \ge V(d) - \varepsilon$ where ε is given and fixed. Consider the stopping time $\tau_\varepsilon = (\tau_c + \tau_1 \circ \theta_{\tau_c})\, I(\tau_c < \tau_d) + (\tau_d + \tau_2 \circ \theta_{\tau_d})\, I(\tau_d < \tau_c)$ obtained by applying τ_1 after hitting c (before d) and τ_2 after hitting d (before c). By the strong Markov property of X it follows that

$$\begin{aligned} V(x) &\ge \mathsf{E}_x G(X_{\tau_\varepsilon}) \\ &= \mathsf{E}_x G(X_{\tau_c + \tau_1 \circ \theta_{\tau_c}})\, I(\tau_c < \tau_d) + \mathsf{E}_x G(X_{\tau_d + \tau_2 \circ \theta_{\tau_d}})\, I(\tau_d < \tau_c) \\ &= \mathsf{E}_x G(X_{\tau_1}) \circ \theta_{\tau_c}\, I(\tau_c < \tau_d) + \mathsf{E}_x G(X_{\tau_2}) \circ \theta_{\tau_d}\, I(\tau_d < \tau_c) \\ &= \mathsf{E}_x\, \mathsf{E}_{X_{\tau_c}}(G(X_{\tau_1}))\, I(\tau_c < \tau_d) + \mathsf{E}_x\, \mathsf{E}_{X_{\tau_d}}(G(X_{\tau_2}))\, I(\tau_d < \tau_c) \\ &= \mathsf{E}_c(G(X_{\tau_1}))\, \mathsf{P}_x(\tau_c < \tau_d) + \mathsf{E}_d(G(X_{\tau_2}))\, \mathsf{P}_x(\tau_d < \tau_c) \\ &\ge (V(c) - \varepsilon)\, \frac{S(d) - S(x)}{S(d) - S(c)} + (V(d) - \varepsilon)\, \frac{S(x) - S(c)}{S(d) - S(c)} \\ &= V(c)\, \frac{S(d) - S(x)}{S(d) - S(c)} + V(d)\, \frac{S(x) - S(c)}{S(d) - S(c)} - \varepsilon \end{aligned} \tag{9.3.6}$$

where the first inequality follows by definition of V and the second inequality follows by the choice of τ_1 and τ_2. Letting $\varepsilon \downarrow 0$ in (9.3.6) one concludes that

$$V(x) \ge V(c)\, \frac{S(d) - S(x)}{S(d) - S(c)} + V(d)\, \frac{S(x) - S(c)}{S(d) - S(c)} \tag{9.3.7}$$

for $c < x < d$ in J. This means that V is S-concave (see e.g. [174, p. 546]). It is not difficult to verify that V *is superharmonic if and only if* V *is* S*-concave* (recall (2.2.8) above).

4. In exactly the same way as for concave functions (corresponding to $S(x) = x$ above) one then sees that (9.3.7) implies that

$$y \mapsto \frac{V(y) - V(x)}{S(y) - S(x)} \quad \text{is decreasing} \tag{9.3.8}$$

on J for every $x \in I$. It follows that

$$-\infty < \frac{d^+ V}{dS}(x) \le \frac{d^- V}{dS}(x) < +\infty \tag{9.3.9}$$

for every $x \in I$. It also follows that V is continuous on I since S is continuous. Note that *continuity* of V is obtained for *measurable* G generally. This links (9.3.6)–(9.3.7) with a powerful (geometric) argument applicable in one dimension only (compare it with the multidimensional results of Subsection 2.2).

5. Let us now assume that $b \in I$ is an optimal stopping point in the problem (9.3.5). Then $V(b) = G(b)$ and hence by (9.3.8) we get

$$\frac{G(b+\varepsilon)-G(b)}{S(b+\varepsilon)-S(b)} \le \frac{V(b+\varepsilon)-V(b)}{S(b+\varepsilon)-S(b)} \le \frac{V(b-\delta)-V(b)}{S(b-\delta)-S(b)} \le \frac{G(b-\delta)-G(b)}{S(b-\delta)-S(b)} \tag{9.3.10}$$

for $\varepsilon > 0$ and $\delta > 0$ where the first inequality follows since $G(b+\varepsilon) \le V(b+\varepsilon)$ and the third inequality follows since $-V(b-\delta) \le -G(b-\delta)$ (recalling also that S is strictly increasing). Passing to the limit for $\varepsilon \downarrow 0$ and $\delta \downarrow 0$ this immediately leads to

$$\frac{d^+G}{dS}(b) \le \frac{d^+V}{dS}(b) \le \frac{d^-V}{dS}(b) \le \frac{d^-G}{dS}(b) \tag{9.3.11}$$

whenever d^+G/dS and d^-G/dS exist at b. In this way we have reached the essential part of Salminen's result [180, p. 96]:

Theorem 9.3. (Smooth fit through scale) *If dG/dS exists at b, then dV/dS exists at b and*

$$\frac{dV}{dS}(b) = \frac{dG}{dS}(b) \tag{9.3.12}$$

whenever $V(b) = G(b)$ for $b \in I$.

In particular, if X is on natural scale (i.e. $S(x) = x$) then the smooth fit condition

$$\frac{dV}{dx}(b) = \frac{dG}{dx}(b) \tag{9.3.13}$$

holds at the optimal stopping point b as soon as G is differentiable at b.

The following example shows that equalities in (9.3.11) and (9.3.12) may fail to hold even though the smooth fit condition (9.3.13) holds.

Example 9.4. Let $X_t = F(B_t)$ where

$$F(x) = \begin{cases} x^{1/3} & \text{if } x \in [0,1], \\ -|x|^{1/3} & \text{if } x \in [-1,0) \end{cases} \tag{9.3.14}$$

and B is a standard Brownian motion in $(-1,1)$ absorbed (killed) at either -1 or 1. Since F is a strictly increasing and continuous function from $[-1,1]$ onto $[-1,1]$, it follows that X is a regular diffusion process in $(-1,1)$ absorbed (killed) at either -1 or 1.

Consider the optimal stopping problem (9.3.5) with

$$G(x) = 1 - x^2 \tag{9.3.15}$$

for $x \in (-1,1)$. Set $X_t^x = F(x+B_t)$ for $x \in (-1,1)$ and let B be defined on $(\Omega, \mathcal{F}, \mathsf{P})$ so that $B_0 = 0$ under P . Since F is increasing (and continuous) it can be verified that

$$\mathrm{Law}(X^x \,|\, \mathsf{P}) = \mathrm{Law}(C \,|\, \mathsf{P}_{F(x)}) \tag{9.3.16}$$

where $C_t(\omega) = \omega(t)$ is the coordinate process (on a canonical space) that is Markov under the family of probability measures P_c for $c \in (-1,1)$ with $\mathsf{P}_c(C_0 = c) = 1$ (note that each $c \in (-1,1)$ corresponds to $F(x)$ for some $x \in (-1,1)$ given and fixed).

In view of (9.3.16) let us consider the auxiliary optimal stopping problem

$$\tilde{V}(x) = \sup_\tau \mathsf{E}\tilde{G}(x+B_\tau) \tag{9.3.17}$$

where $\tilde{G} = G \circ F$ and the supremum is taken over all stopping times τ of B (up to the time of absorption at -1 or 1). Note that

$$\tilde{G}(x) = 1 - |x|^{2/3} \tag{9.3.18}$$

for $x \in (-1,1)$. Since $\tilde{V}$ is the smallest superharmonic (i.e. concave) function that dominates $\tilde{G}$ (cf. Chapter I), and clearly $\tilde{V}(-1) = \tilde{V}(1) = 0$, it follows that

$$\tilde{V}(x) = \begin{cases} 1-x & \text{if } \; x \in [0,1], \\ 1+x & \text{if } \; x \in [-1,0). \end{cases} \tag{9.3.19}$$

From (9.3.16) we see that $V(x) = \tilde{V}(F^{-1}(x))$ and since $F^{-1}(x) = x^3$, it follows that

$$V(x) = \begin{cases} 1-x^3 & \text{if } \; x \in [0,1], \\ 1+x^3 & \text{if } \; x \in [-1,0). \end{cases} \tag{9.3.20}$$

Comparing (9.3.20) with (9.3.15) we see that $b = 0$ is an optimal stopping point. Moreover, it is evident that the smooth fit (9.3.13) holds at $b = 0$, both derivatives being zero. However, noting that the scale function of X equals $S(x) = x^3$ for $x \in [-1,1]$ (since $S(X) = F^{-1}(F(B)) = B$ is a martingale), it is straightforwardly verified from (9.3.15) and (9.3.20) that

$$\frac{d^+G}{dS} = -\infty < \frac{d^+V}{dS} = -1 < \frac{d^-V}{dS} = 1 < \frac{d^-G}{dS} = +\infty \tag{9.3.21}$$

at the optimal stopping point $b = 0$.

6. Note that the scale function S in the preceding example is differentiable at the optimal stopping point b but that $S'(b) = 0$. This motivates the following extension of Theorem 9.3 above (recall (9.1.8) above).

Theorem 9.5. (Smooth fit) *If both dG/dx and dS/dx exist at b, then dV/dx exists at b and*

$$\frac{dV}{dx}(b) = \frac{dG}{dx}(b) \tag{9.3.22}$$

whenever $V(b) = G(b)$ for $b \in I$.

Proof. Assume first that $S'(b) \neq 0$. Multiplying by $(S(b+\varepsilon)-S(b))/\varepsilon$ in (9.3.10) we get

$$\begin{aligned}\frac{G(b+\varepsilon)-G(b)}{\varepsilon} &\leq \frac{V(b+\varepsilon)-V(b)}{\varepsilon} \\ &\leq \frac{S(b+\varepsilon)-S(b)}{\varepsilon}\,\frac{(G(b-\delta)-G(b))/(-\delta)}{(S(b-\delta)-S(b))/(-\delta)}\,.\end{aligned} \tag{9.3.23}$$

Passing to the limit for $\varepsilon \downarrow 0$ and $\delta \downarrow 0$, and using that $S'(b) \neq 0$, it follows that $d^+V/dx = dG/dx$ at b. (Note that one could take $\varepsilon = \delta$ in this argument.)

Similarly, multiplying by $(S(b-\delta)-S(b))/(-\delta)$ in (9.3.10) we get

$$\frac{(G(b+\varepsilon)-G(b))/\varepsilon}{(S(b+\varepsilon)-S(b))/\varepsilon}\,\frac{S(b-\delta)-S(b)}{-\delta} \leq \frac{V(b-\delta)-V(b)}{-\delta} \leq \frac{G(b-\delta)-G(b)}{-\delta}\,. \tag{9.3.24}$$

Passing to the limit for $\varepsilon \downarrow 0$ and $\delta \downarrow 0$, and using that $S'(b) \neq 0$, it follows that $d^-V/dx = dG/dx$ at b. (Note that one could take $\varepsilon = \delta$ in this argument.) Combining the two conclusions we see that dV/dx exists at b and (9.3.22) holds as claimed.

To treat the case $S'(b) = 0$ we need the following simple facts of real analysis.

Lemma 9.6. *Let $f : \mathbb{R}_+ \to \mathbb{R}$ and $g : \mathbb{R}_+ \to \mathbb{R}$ be two continuous functions satisfying:*

$$f(0) = 0 \;\; \textit{and} \;\; f(\varepsilon) > 0 \;\; \textit{for} \;\; \varepsilon > 0\,; \tag{9.3.25}$$

$$g(0) = 0 \;\; \textit{and} \;\; g(\delta) > 0 \;\; \textit{for} \;\; \delta > 0\,. \tag{9.3.26}$$

Then for every $\varepsilon_n \downarrow 0$ as $n \to \infty$ there are $\varepsilon_{n_k} \downarrow 0$ and $\delta_k \downarrow 0$ as $k \to \infty$ such that $f(\varepsilon_{n_k}) = g(\delta_k)$ for all $k \geq 1$. In particular, it follows that

$$\lim_{k\to\infty} \frac{f(\varepsilon_{n_k})}{g(\delta_k)} = 1\,. \tag{9.3.27}$$

Proof. Take any $\varepsilon_n \downarrow 0$ as $n \to \infty$. Since $f(\varepsilon_n) \to 0$ and $f(\varepsilon_n) > 0$ we can find a subsequence $\varepsilon_{n_k} \downarrow 0$ such that $x_{n_k} := f(\varepsilon_{n_k}) \downarrow 0$ as $k \to \infty$. Since $g(1) > 0$ there is no restriction to assume that $x_{n_1} < g(1)$. But then by continuity of g and the fact that $x_{n_1} \in (g(0), g(1))$ there must be $\delta_1 \in (0,1)$ such that $g(\delta_1) = x_{n_1}$. Since $x_{n_2} < x_{n_1}$ it follows that $x_{n_2} \in (g(0), g(\delta_1))$ and again by continuity of g there must be $\delta_2 \in (0, \delta_1)$ such that $g(\delta_2) = x_{n_2}$. Continuing likewise by induction we obtain a decreasing sequence $\delta_k \in (0,1)$ such that $g(\delta_k) = x_{n_k}$ for $k \geq 1$. Denoting $\delta = \lim_{k\to\infty} \delta_k$ we see that $g(\delta) = \lim_{k\to\infty} g(\delta_k) = \lim_{k\to\infty} x_{n_k} = 0$. Hence δ must be 0 by (9.3.26). This completes the proof of Lemma 9.6. □

Let us continue the proof of Theorem 9.5 in the case when $S'(b) = 0$. Take $\varepsilon_n \downarrow 0$ and by Lemma 9.6 choose $\delta_k \downarrow 0$ such that (9.3.27) holds with $f(\varepsilon) = (S(b+\varepsilon)-S(b))/\varepsilon$ and $g(\delta) = (S(b)-S(b-\delta))/\delta$. Then (9.3.23) reads

$$\frac{G(b+\varepsilon_{n_k})-G(b)}{\varepsilon_{n_k}} \leq \frac{V(b+\varepsilon_{n_k})-V(b)}{\varepsilon_{n_k}} \leq \frac{f(\varepsilon_{n_k})}{g(\delta_k)} \frac{G(b-\delta_k)-G(b)}{-\delta_k} \tag{9.3.28}$$

for all $k \geq 1$. Letting $k \to \infty$ and using (9.3.27) we see that $(V(b+\varepsilon_{n_k}) - V(b))/\varepsilon_{n_k} \to G'(b)$. Since this is true for any $\varepsilon_n \downarrow 0$ it follows that d^+V/dx exists and is equal to dG/dx at b.

Similarly, take $\varepsilon_n \downarrow 0$ and by Lemma 9.6 choose $\delta_k \downarrow 0$ such that (9.3.27) holds with $f(\varepsilon) = (S(b)-S(b-\varepsilon))/\varepsilon$ and $g(\delta) = (S(b+\delta)-S(b))/\delta$. Then (9.3.24) (with ε and δ traded) reads

$$\frac{G(b+\delta_k)-G(b)}{\delta_k} \frac{f(\varepsilon_{n_k})}{g(\delta_k)} \leq \frac{V(b-\varepsilon_{n_k})-V(b)}{-\varepsilon_{n_k}} \leq \frac{G(b-\varepsilon_{n_k})-G(b)}{-\varepsilon_{n_k}} \tag{9.3.29}$$

for all $k \geq 1$. Letting $k \to \infty$ and using (9.3.27) we see that $(V(b-\varepsilon_{n_k}) - V(b))/(-\varepsilon_{n_k}) \to G'(b)$. Since this is true for any $\varepsilon_n \downarrow 0$ it follows that d^-V/dx exists and is equal to dG/dx at b. Taken together with the previous conclusion on d^+V/dx this establishes (9.3.22) and the proof of Theorem 9.5 is complete. □

7. The question arising naturally from the previous considerations is whether differentiability of the gain function G and regularity of the diffusion process X imply the smooth fit $V'(b) = G'(b)$ at the optimal stopping point b.

The negative answer to this question is provided by the following example.

Example 9.7. Let $X_t = F(B_t)$ where

$$F(x) = \begin{cases} \sqrt{x} & \text{if } x \in [0,1], \\ -x^2 & \text{if } x \in [-1,0) \end{cases} \tag{9.3.30}$$

and B is a standard Brownian motion in $(-1,1)$ absorbed (killed) at either -1 or 1. Since F is a strictly increasing and continuous function from $[-1,1]$ onto $[-1,1]$, it follows that X is a regular diffusion process in $(-1,1)$ absorbed (killed) at either -1 or 1.

Consider the optimal stopping problem (9.3.5) with

$$G(x) = 1 - x \tag{9.3.31}$$

for $x \in (-1,1)$. Set $X_t^x = F(x+B_t)$ for $x \in (-1,1)$ and let B be defined on $(\Omega, \mathcal{F}, \mathsf{P})$ so that $B_0 = 0$ under P. Since F is increasing (and continuous) it follows that

$$\operatorname{Law}(X^x \,|\, \mathsf{P}) = \operatorname{Law}(C \,|\, \mathsf{P}_{F(x)}) \tag{9.3.32}$$

where $C_t(\omega) = \omega(t)$ is the coordinate process (on a canonical space) that is Markov under the family of probability measures P_c for $c \in (-1,1)$ with $\mathsf{P}_c(C_0 = c) = 1$ (note that each $c \in (-1,1)$ corresponds to $F(x)$ for some $x \in (-1,1)$ given and fixed).

In view of (9.3.32) let us consider the auxiliary optimal stopping problem

$$\tilde{V}(x) = \sup_{\tau} \mathsf{E}\, \tilde{G}(x+B_\tau) \tag{9.3.33}$$

where $\tilde{G} = G \circ F$ and the supremum is taken over all stopping times τ of B (up to the time of absorption at -1 or 1). Note that

$$\tilde{G}(x) = \begin{cases} 1 - \sqrt{x} & \text{if} \;\; x \in [0,1], \\ 1 + x^2 & \text{if} \;\; x \in [-1,0)\, . \end{cases} \tag{9.3.34}$$

Since $\tilde{V}$ is the smallest superharmonic (i.e. concave) function that dominates $\tilde{G}$ (cf. Chapter I), and clearly $\tilde{V}(-1) = 2$ and $\tilde{V}(1) = 0$, it follows that

$$\tilde{V}(x) = 1 - x \tag{9.3.35}$$

for $x \in [-1,1]$. From (9.3.32) we see that $V(x) = \tilde{V}(F^{-1}(x))$ and since

$$F^{-1}(x) = \begin{cases} x^2 & \text{if} \;\; x \in [0,1]\, , \\ -\sqrt{|x|} & \text{if} \;\; x \in [-1,0)\, , \end{cases} \tag{9.3.36}$$

it follows that

$$V(x) = \begin{cases} 1 - x^2 & \text{if} \;\; x \in [0,1]\, , \\ 1 + \sqrt{|x|} & \text{if} \;\; x \in [-1,0)\, . \end{cases} \tag{9.3.37}$$

Comparing (9.3.37) with (9.3.31) we see that $b = 0$ is an optimal stopping point. However, it is evident that the smooth fit $V'(b) = G'(b)$ fails at $b = 0$ (see *Figure* IV.4).

8. Note that the scale function S of X equals F^{-1} in (9.3.36) above (since $S(X) = F^{-1}(F(B)) = B$ is a martingale) so that $S'_+(0) = 0$ and $S'_-(0) = +\infty$. Note also from (9.3.30) above that X receives a "strong" push toward $(0,1]$ and a "mild" push toward $[-1,0)$ when at 0. The two extreme cases of $S'_+(0)$ and $S'_-(0)$ are not the only possible ones to ruin the smooth fit. Indeed, if we slightly modify F in (9.3.30) above by setting

$$F(x) = \begin{cases} \sqrt{x} & \text{if} \;\; x \in [0,1], \\ x & \text{if} \;\; x \in [-1,0), \end{cases} \tag{9.3.38}$$

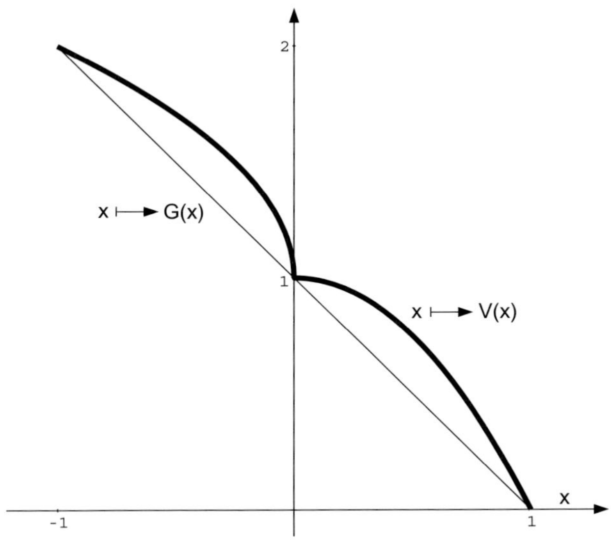

Figure IV.4: The gain function G and the value function V from Example 9.7. The smooth fit $V'(b) = G'(b)$ fails at the optimal stopping point $b = 0$.

then the same analysis as above shows that

$$V(x) = \begin{cases} 1 - x^2 & \text{if } x \in [0,1], \\ 1 - x & \text{if } x \in [-1,0), \end{cases} \tag{9.3.39}$$

so that the smooth fit $V'(b) = G'(b)$ still fails at the optimal stopping point $b = 0$. In this case the scale function S of X equals

$$F^{-1}(x) = \begin{cases} x^2 & \text{if } x \in [0,1], \\ x & \text{if } x \in [-1,0), \end{cases} \tag{9.3.40}$$

so that $S'_+(0) = 0$ and $S'_-(0) = 1$.

Moreover, any further speculation that the extreme condition $S'_+(0) = 0$ is needed to ruin the smooth fit is ruled out by the following modification of F in (9.3.30) above:

$$F(x) = \begin{cases} \frac{-1+\sqrt{1+8x}}{2} & \text{if } x \in [0,1], \\ x & \text{if } x \in [-1,0)\,. \end{cases} \tag{9.3.41}$$

Then the same analysis as above shows that

$$V(x) = \begin{cases} 1 - \frac{x^2+x}{2} & \text{if } x \in [0,1], \\ 1 - x & \text{if } x \in [-1,0), \end{cases} \tag{9.3.42}$$

so that the smooth fit $V'(b) = G'(b)$ still fails at the optimal stopping point $b = 0$. In this case the scale function S of X equals

$$F^{-1}(x) = \begin{cases} \frac{x^2+x}{2} & \text{if } x \in [0,1], \\ x & \text{if } x \in [-1,0), \end{cases} \tag{9.3.43}$$

so that $S'_+(0) = 1/2$ and $S'_-(0) = 1$.

9. In order to examine what is "angular" about the diffusion from the preceding example, let us recall that (9.3.3) implies that

$$\begin{aligned} \mathsf{P}_b(\tau_{b-\varepsilon} < \tau_{b+\varepsilon}) &= \frac{S(b+\varepsilon) - S(b)}{S(b+\varepsilon) - S(b-\varepsilon)} \\ &= \frac{(S(b+\varepsilon) - S(b))/\varepsilon}{(S(b+\varepsilon) - S(b))/\varepsilon + (S(b) - S(b-\varepsilon))/\varepsilon} \longrightarrow \frac{R}{R+L} \end{aligned} \tag{9.3.44}$$

as $\varepsilon \downarrow 0$ whenever $S'_+(b) =: R$ and $S'_-(b) =: L$ exist (and are assumed to be different from zero for simplicity). Likewise, one finds that

$$\begin{aligned} \mathsf{P}_b(\tau_{b+\varepsilon} < \tau_{b-\varepsilon}) &= \frac{S(b) - S(b-\varepsilon)}{S(b+\varepsilon) - S(b-\varepsilon)} \\ &= \frac{(S(b) - S(b-\varepsilon))/\varepsilon}{(S(b+\varepsilon) - S(b))/\varepsilon + (S(b) - S(b-\varepsilon))/\varepsilon} \longrightarrow \frac{L}{R+L} \end{aligned} \tag{9.3.45}$$

as $\varepsilon \downarrow 0$ whenever $S'_-(b) =: L$ and $S'_+(b) =: R$ exist (and are assumed to be different from zero for simplicity).

If S is differentiable at b then $R = L$ so that the limit probabilities in (9.3.44) and (9.3.45) are equal to $1/2$. Note that these probabilities correspond to X exiting b infinitesimally to either left or right respectively. On the other hand, if S is not differentiable at b, then the two limit probabilities $R/(R+L)$ and $L/(R+L)$ are different and this fact alone may ruin the smooth fit at b as Example 9.7 above shows. Thus, regularity of X itself is insufficient for the smooth fit to hold generally, and X requires this sort of "tuned regularity" instead (recall Theorem 9.5 above).

10. Another way of looking at such diffusions is obtained by means of stochastic calculus. The Itô–Tanaka–Meyer formula (page 67) implies that the process $X_t = F(B_t)$ solves the integral equation

$$\begin{aligned} X_t = X_0 &+ \int_0^t F' \circ F^{-1}(X_s)\, I(X_s \neq 0)\, dB_s \\ &+ \int_0^t \frac{1}{2} F'' \circ F^{-1}(X_s)\, I(X_s \neq 0)\, ds + \frac{1}{2}\big[F'_+(0) - F'_-(0)\big]\, \ell_t^0(B) \end{aligned} \tag{9.3.46}$$

where $\ell_t^0(B)$ is the local time of B at 0. Setting

$$A_t = \big[F'_+(0) - F'_-(0)\big]\, \ell_t^0(B) \tag{9.3.47}$$

we see that (9.3.46) reads

$$dX_t = \rho(X_t)\,dt + \sigma(X_t)\,dB_t + dA_t \tag{9.3.48}$$

where $(A_t)_{t\geq 0}$ is continuous, increasing (or decreasing), adapted to $(\mathcal{F}_t)_{t\geq 0}$ and satisfies

$$\int_0^t I(X_s \neq 0)\,dA_s = 0 \tag{9.3.49}$$

with $A_0 = 0$. These conditions usually bear the name of an *SDE with reflection* for (9.3.48). Note however that X is not necessarily non-negative as additionally required from solutions of SDEs with reflection.

Notes. A number of authors have contributed to understanding of the smooth-fit principle by various means. With no aim to review the full history of these developments, and in addition to the Notes to Subsection 9.1 above, we refer to [180], [145], [23], [146], [3], [37] and [2] (for Lévy processes). Further references are given in the Notes to Section 25 below.

10. The method of time change

The main goal of this section (following [155]) is to present a deterministic time-change method which enables one to solve some nonlinear optimal stopping problems explicitly. The basic idea is to transform the original (difficult) problem into a new (easier) problem. The method is firstly described (Subsection 10.1) and then illustrated through several examples (Subsection 10.2).

10.1. Description of the method

1. To explain the ideas in more detail, let $((X_t)_{t\geq 0}, \mathsf{P}_x)$ be a one-dimensional time-homogeneous diffusion associated with the infinitesimal generator

$$\mathbb{L}_X = b(x)\,\frac{\partial}{\partial x} + a^2(x)\,\frac{1}{2}\,\frac{\partial^2}{\partial x^2} \tag{10.1.1}$$

where $x \mapsto a(x) > 0$ and $x \mapsto b(x)$ are continuous. Assume moreover that there exists a standard Brownian motion $B = (B_t)_{t\geq 0}$ such that $X = (X_t)_{t\geq 0}$ solves the stochastic differential equation

$$dX_t = b(X_t)\,dt + a(X_t)\,dB_t \tag{10.1.2}$$

with $X_0 = x$ under P_x. The typical optimal stopping problem which appears under consideration below has the value function given by

$$V_*(t,x) = \sup_{\tau} \mathsf{E}_x\big(\alpha(t+\tau)\,X_\tau\big) \tag{10.1.3}$$

where the supremum is taken over a class of stopping times τ for X and α is a smooth but *nonlinear* function. This forces us to take $(t, X_t)_{t \geq 0}$ as the underlying diffusion in the problem, and thus by general optimal stopping theory (Chapter III) we know that the value function V_* should solve the following partial differential equation:

$$\frac{\partial V}{\partial t}(t,x) + \mathbb{L}_X V(t,x) = 0 \tag{10.1.4}$$

in the domain of continued observation. However, it is generally difficult to find a closed-form solution of the partial differential equation, and the basic idea of the time-change method is to transform the original problem into a new optimal stopping problem such that the new value function solves an ordinary differential equation.

2. To do so one is naturally led to find a deterministic time change $t \mapsto \sigma_t$ satisfying the following two conditions:

(i) $t \mapsto \sigma_t$ is continuous and strictly increasing;

(ii) there exists a one-dimensional time-homogeneous diffusion $Z = (Z_t)_{t \geq 0}$ with infinitesimal generator $\mathbb{L}_Z$ such that $\alpha(\sigma_t)\, X_{\sigma_t} = e^{-rt}\, Z_t$ for some $r \in \mathbb{R}$.

From general theory (Chapter III) we know that the new (time-changed) value function

$$W_*(z) = \sup_{\tau} \mathsf{E}_z\big(e^{-r\tau} Z_\tau\big), \tag{10.1.5}$$

where the supremum is taken over a class of stopping times τ for Z, should solve the ordinary differential equation

$$\mathbb{L}_Z W_*(z) = r\, W_*(z) \tag{10.1.6}$$

in the domain of continued observation. Note that under condition (i) there is a one-to-one correspondence between the original problem and the new problem, i.e. if τ is a stopping time for Z then σ_τ is a stopping time for X and vice versa.

3. Given the diffusion $X = (X_t)_{t \geq 0}$ the crucial point is to find the process $Z = (Z_t)_{t \geq 0}$ and the time change σ_t fulfilling conditions (i) and (ii) above. Itô's formula (page 67) offers an answer to these questions.

Setting $Y = (Y_t)_{t \geq 0} = (\beta(t) X_t)_{t \geq 0}$ where $\beta \neq 0$ is a smooth function, by Itô's formula we get

$$Y_t = Y_0 + \int_0^t \left(\frac{\beta'(u)}{\beta(u)}\, Y_u + \beta(u)\, b\Big(\frac{Y_u}{\beta(u)}\Big) \right) du + \int_0^t \beta(u)\, a\Big(\frac{Y_u}{\beta(u)}\Big)\, dB_u \tag{10.1.7}$$

and hence $Y = (Y_t)_{t \geq 0}$ has the infinitesimal generator

$$\mathbb{L}_Y = \left(\frac{\beta'(t)}{\beta(t)}\, y + \beta(t)\, b\Big(\frac{y}{\beta(t)}\Big) \right) \frac{\partial}{\partial y} + \beta^2(t)\, a^2\Big(\frac{y}{\beta(t)}\Big) \frac{1}{2} \frac{\partial^2}{\partial y^2}\,. \tag{10.1.8}$$

The time-changed process $Z = (Z_t)_{t\geq0} = (Y_{\sigma_t})_{t\geq0}$ has the infinitesimal generator (see [178, p. 175] and recall Subsection 5.1 above)

$$\mathbb{L}_Z = \frac{1}{\rho(t)}\,\mathbb{L}_Y \tag{10.1.9}$$

where σ_t is the time change given by

$$\sigma_t = \inf\left\{ r > 0 : \int_0^r \rho(u)\,du > t \right\} \tag{10.1.10}$$

for some $u \mapsto \rho(u) > 0$ (to be found) such that $\sigma_t \to \infty$ as $t \to \infty$.

The process $Z = (Z_t)_{t\geq0}$ and the time change σ_t will be fulfilling conditions (i) and (ii) above if the infinitesimal generator $\mathbb{L}_Z$ *does not depend on* t. In view of (10.1.8) this clearly imposes the following conditions on β (and α above) which make the method applicable:

$$b\left(\frac{y}{\beta(t)}\right) = \gamma(t)\,G_1(y), \tag{10.1.11}$$

$$a^2\left(\frac{y}{\beta(t)}\right) = \frac{\gamma(t)}{\beta(t)}\,G_2(y) \tag{10.1.12}$$

where $\gamma = \gamma(t)$, $G_1 = G_1(y)$ and $G_2 = G_2(y)$ are functions required to exist.

4. In our examples below the diffusion $X = (X_t)_{t\geq0}$ is given as Brownian motion $(B_t + x)_{t\geq0}$ started at x under P_x, and thus its infinitesimal generator is given by

$$\mathbb{L}_X = \frac{1}{2}\frac{\partial^2}{\partial x^2}\,. \tag{10.1.13}$$

By the foregoing observations we shall find a time change σ_t and a process $Z = (Z_t)_{t\geq0}$ satisfying conditions (i) and (ii) above. With the notation introduced above we see from (10.1.8) that the infinitesimal generator of $Y = (Y_t)_{t\geq0}$ in this case is given by

$$\mathbb{L}_Y = \frac{\beta'(t)}{\beta(t)}\,y\,\frac{\partial}{\partial y} + \beta^2(t)\,\frac{1}{2}\frac{\partial^2}{\partial y^2}\,. \tag{10.1.14}$$

Observe that conditions (10.1.11) and (10.1.12) are easily realized with $\gamma(t) = \beta(t)$, $G_1 = 0$ and $G_2 = 1$. Thus if β solves the differential equation $\beta'(t)/\beta(t) = -\beta^2(t)/2$, and we set $\rho = \beta^2/2$, then from (10.1.9) we see that $\mathbb{L}_Z$ does not depend on t. Noting that $\beta(t) = 1/\sqrt{1+t}$ solves this equation, and putting $\rho(t) = 1/2(1+t)$, we find that

$$\sigma_t = \inf\left\{ r > 0 : \int_0^r \rho(u)\,du > t \right\} = e^{2t} - 1\,. \tag{10.1.15}$$

Thus the time-changed process $Z = (Z_t)_{t\geq0}$ has the infinitesimal generator given by

$$\mathbb{L}_Z = -z\,\frac{\partial}{\partial z} + \frac{\partial^2}{\partial z^2} \tag{10.1.16}$$

and hence $Z = (Z_t)_{t \geq 0}$ is an Ornstein–Uhlenbeck process. While this fact is well known, the technique described may be applied in a similar context involving other diffusions (see Example 10.15 below).

5. In the next subsection we shall apply the time-change method described above and present solutions to several optimal stopping problems. Apart from the time-change arguments just described, the method of proof makes also use of Brownian scaling and the principle of smooth fit in a free-boundary problem. Once the guess is performed, Itô's calculus is used as a verification tool. The main emphasis of the section is on the method of proof and its unifying scope.

10.2. Problems and solutions

1. In this subsection we explicitly solve some nonlinear optimal stopping problems for a Brownian motion by applying the time-change method described in the previous subsection (recall also Subsection 5.1 above).

Throughout $B = (B_t)_{t \geq 0}$ denotes a standard Brownian motion started at zero under P, and the diffusion $X = (X_t)_{t \geq 0}$ is given as the Brownian motion $(B_t + x)_{t \geq 0}$ started at x under P_x.

Given the time change $\sigma_t = e^{2t} - 1$ from (10.1.15), we know that the time-changed process

$$Z_t = X_{\sigma_t} / \sqrt{1 + \sigma_t}, \qquad t \geq 0, \tag{10.2.1}$$

is an Ornstein–Uhlenbeck process satisfying

$$dZ_t = -Z_t\, dt + \sqrt{2}\, dB_t, \tag{10.2.2}$$

$$\mathbb{L}_Z = -z \frac{\partial}{\partial z} + \frac{\partial^2}{\partial z^2}. \tag{10.2.3}$$

With this notation we may now enter into the first example.

Example 10.1. Consider the optimal stopping problem with the value function

$$V_*(t, x) = \sup_{\tau} \mathsf{E}_x \Big(|X_\tau| - c \sqrt{t + \tau} \Big) \tag{10.2.4}$$

where the supremum is taken over all stopping times τ for X satisfying $\mathsf{E}_x \sqrt{\tau} < \infty$ and $c > 0$ is given and fixed. We shall solve this problem in five steps ($1°$–$5°$).

$1°$. In the first step we shall apply Brownian scaling and note that $\tilde{\tau} = \tau/t$ is a stopping time for the Brownian motion $s \mapsto t^{-1/2} B_{ts}$. If we now rewrite (10.2.4) as

$$\begin{aligned} V_*(t, x) &= \sup_{\tau} \mathsf{E} \Big(|B_\tau + x| - c \sqrt{t + \tau} \Big) \\ &= \sqrt{t} \sup_{\tau/t} \mathsf{E} \Big(\big| t^{-1/2} B_{t(\tau/t)} + x/\sqrt{t} \big| - c \sqrt{1 + \tau/t} \Big) \end{aligned} \tag{10.2.5}$$

we clearly see that

$$V_*(t,x) = \sqrt{t}\, V_*(1, x/\sqrt{t}) \tag{10.2.6}$$

and therefore we only need to look at $V_*(1,x)$ in the sequel. By using (10.2.6) we can also make the following observation on the optimal stopping boundary for the problem (10.2.4).

Remark 10.2. In the problem (10.2.4) the gain function equals $g(t,x) = |x| - c\sqrt{t}$ and the diffusion is identified with $\big(t+r, X_r\big)$. If a point (t_0, x_0) belongs to the boundary of the domain of continued observation, i.e. (t_0,x_0) is an instantaneously stopping point ($\tau \equiv 0$ is an optimal stopping time), then we get from (10.2.6) that $V_*(t_0,x_0) = |x_0| - c\sqrt{t_0} = \sqrt{t_0}\, V_*(1, x_0/\sqrt{t_0})$. Hence $V_*(1,x_0/\sqrt{t_0}) = |x_0|/\sqrt{t_0} - c$ and therefore the point $(1, x_0/\sqrt{t_0})$ is also instantaneously stopping. Set now $\gamma_0 = |x_0|/\sqrt{t_0}$ and note that if (t,x) is any point satisfying $|x|/\sqrt{t} = \gamma_0$, then this point is also instantaneously stopping. This offers a heuristic argument that the optimal stopping boundary should be $|x| = \gamma_0\sqrt{t}$ for some $\gamma_0 > 0$ to be found.

2°. In the second step we shall apply the time change $t \mapsto \sigma_t$ from (10.1.15) to the problem $V_*(1,x)$ and transform it into a new problem. From (10.2.1) we get

$$|X_{\sigma_\tau}| - c\sqrt{1+\sigma_\tau} = \sqrt{1+\sigma_\tau}\,\big(|Z_\tau| - c\big) = e^{\tau}\big(|Z_\tau| - c\big) \tag{10.2.7}$$

and the problem to determine $V_*(1,x)$ therefore reduces to computing

$$V_*(1,x) = W_*(x) \tag{10.2.8}$$

where W_* is the value function of the new (time-changed) optimal stopping problem

$$W_*(z) = \sup_{\tau} \mathsf{E}_z\big(e^{\tau}\big(|Z_\tau| - c\big)\big) \tag{10.2.9}$$

the supremum being taken over all stopping times τ for Z for which $\mathsf{E}_z e^{\tau} < \infty$. Observe that this problem is one-dimensional (see Subsection 6.2 above).

3°. In the third step we shall show how to solve the problem (10.2.9). From general optimal stopping theory (Chapter I) we know that the following stopping time should be optimal:

$$\tau_* = \inf\big\{\, t > 0 : |Z_t| \ge z_* \,\big\} \tag{10.2.10}$$

where $z_* \ge 0$ is the optimal stopping point to be found. Observe that this guess agrees with Remark 10.2. Note that the domain of continued observation $C = (-z_*, z_*)$ is assumed symmetric around zero since the Ornstein–Uhlenbeck process is symmetric, i.e. the process $-Z = (-Z_t)_{t\ge 0}$ is also an Ornstein–Uhlenbeck process started at $-z$. By using the same argument we may also argue that the value function W_* should be *even*.

To compute the value function W_* for $z \in (-z_*, z_*)$ and to determine the optimal stopping point z_*, in view of (10.2.9)–(10.2.10) it is natural (Chapter III) to formulate the following system:

$$\mathbb{L}_Z W(z) = -W(z) \quad \text{for } z \in (-z_*, z_*), \tag{10.2.11}$$

$$W(\pm z_*) = z_* - c \quad \text{(instantaneous stopping)}, \tag{10.2.12}$$

$$W'(\pm z_*) = \pm 1 \quad \text{(smooth fit)} \tag{10.2.13}$$

with $\mathbb{L}_Z$ in (10.2.3). The system (10.2.11)–(10.2.13) forms a *free-boundary problem*. The condition (10.2.13) is imposed since we expect that *the principle of smooth fit* should hold.

It is known (see pages 192–193 below) that the equation (10.2.11) admits the even solution (10.2.155) and the odd solution (10.2.156) as two linearly independent solutions. Since the value function should be even, we can forget the odd solution and from (10.2.155) we see that

$$W(z) = -A\, M(-\tfrac{1}{2}, \tfrac{1}{2}, \tfrac{z^2}{2}) \tag{10.2.14}$$

for some $A > 0$ to be found.

From *Figure* IV.5 we clearly see that only for $c \geq z_1^*$ can the two boundary conditions (10.2.12)–(10.2.13) be fulfilled, where z_1^* is the unique positive root of $M(-1/2, 1/2, z^2/2) = 0$. Thus by (10.2.12)–(10.2.13) and (10.2.157) when $c \geq z_1^*$ we find that $A = z_*^{-1}/M(1/2, 3/2, z_*^2/2)$ and that $z_* \leq z_1^*$ is the unique positive root of the equation

$$z^{-1} M(-\tfrac{1}{2}, \tfrac{1}{2}, \tfrac{z^2}{2}) = (c - z)\, M(\tfrac{1}{2}, \tfrac{3}{2}, \tfrac{z^2}{2})\,. \tag{10.2.15}$$

Note that for $c < z_1^*$ the equation (10.2.15) has no solution.

In this way we have obtained the following candidate for the value function W_* in the problem (10.2.9) when $c \geq z_1^*$:

$$W(z) = \begin{cases} -z_*^{-1} M(-\tfrac{1}{2}, \tfrac{1}{2}, \tfrac{z^2}{2}) / M(\tfrac{1}{2}, \tfrac{3}{2}, \tfrac{z_*^2}{2}) & \text{if } |z| < z_*, \\ |z| - c & \text{if } |z| \geq z_* \end{cases} \tag{10.2.16}$$

and the following candidate for the optimal stopping time τ_* when $c > z_1^*$:

$$\tau_{z_*} = \inf\{\, t > 0 : |Z_t| \geq z_* \,\}\,. \tag{10.2.17}$$

In the proof below we shall see that $\mathsf{E}_z(e^{\tau_{z_*}}) < \infty$ when $c > z_1^*$ (and thus $z_* < z_1^*$). For $c = z_1^*$ (and thus $z_* = z_1^*$) the stopping time τ_{z_*} fails to satisfy $\mathsf{E}_z(e^{\tau_{z_*}}) < \infty$, but clearly τ_{z_*} are approximately optimal if we let $c \downarrow z_1^*$ (and hence $z_* \uparrow z_1^*$). For $c < z_1^*$ we have $W(z) = \infty$ and it is never optimal to stop.

4°. To verify that these formulae are correct (with $c > z_1^*$ given and fixed) we shall apply Itô's formula (page 67) to the process $(e^t\, W(Z_t))_{t \geq 0}$. For this, note

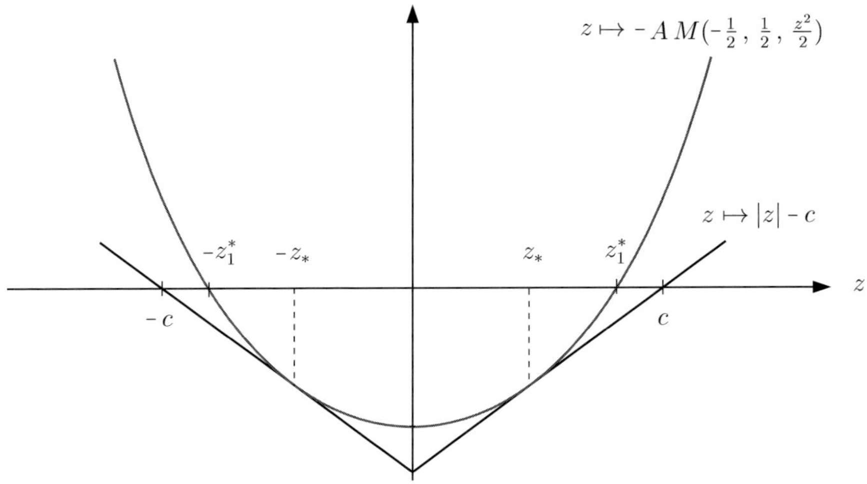

Figure IV.5: A computer drawing of the solution of the free-boundary problem (10.2.11)–(10.2.13). The solution equals $z \mapsto -A\,M(-1/2, 1/2, z^2/2)$ for $|z| < z_*$ and $z \mapsto |z| - c$ for $|z| \geq z_*$. The constant A is chosen (and z_* is obtained) such that the smooth fit holds at $\pm z_*$ (the first derivative of the solution is continuous at $\pm z_*$).

that $z \mapsto W(z)$ is C^2 everywhere but at $\pm z_*$. However, since Lebesgue measure of those u for which $Z_u = \pm z_*$ is zero, the values $W''(\pm z_*)$ can be chosen in the sequel arbitrarily. In this way by (10.2.2) we obtain

$$e^t\,W(Z_t) = W(z) + \int_0^t e^u \big(\mathbb{L}_Z W(Z_u) + W(Z_u)\big)\,du + M_t \tag{10.2.18}$$

where $M = (M_t)_{t \geq 0}$ is a continuous local martingale given by

$$M_t = \sqrt{2} \int_0^t e^u\,W'(Z_u)\,dB_u\,. \tag{10.2.19}$$

Using that $\mathbb{L}_Z W(z) + W(z) \leq 0$ for $z \neq \pm z_*$, hence we get

$$e^{-t}\,W(Z_t) \leq W(z) + M_t \tag{10.2.20}$$

for all t. Let τ be any stopping time for Z satisfying $\mathsf{E}_z e^{\tau} < \infty$. Choose a localization sequence (σ_n) of bounded stopping times for M. Clearly $W(z) \geq |z| - c$ for all z, and hence from (10.2.20) we find

$$\begin{aligned} \mathsf{E}_z\big(e^{\tau\wedge\sigma_n}(|Z_{\tau\wedge\sigma_n}| - c)\big) &\leq \mathsf{E}_z\big(e^{\tau\wedge\sigma_n}\,W(Z_{\tau\wedge\sigma_n})\big) \\ &\leq W(z) + \mathsf{E}_z M_{\tau\wedge\sigma_n} = W(z) \end{aligned} \tag{10.2.21}$$

for all $n \geq 1$. Letting $n \to \infty$ and using Fatou's lemma, and then taking supremum over all stopping times τ satisfying $\mathsf{E}_z e^{\tau} < \infty$, we obtain

$$W_*(z) \leq W(z)\,. \tag{10.2.22}$$

Finally, to prove that equality in (10.2.22) is attained, and that the stopping time (10.2.17) is optimal, it is enough to verify that

$$W(z) = \mathsf{E}_z\big(e^{\tau_{z_*}}\big(|Z_{\tau_{z_*}}| - c\big)\big) = (z_* - c)\,\mathsf{E}_z e^{\tau_{z_*}}\,. \tag{10.2.23}$$

However, from general Markov process theory (see Chapter III) we know that $w(z) = \mathsf{E}_z e^{\tau_{z_*}}$ solves (10.2.11), and clearly it satisfies $w(\pm z_*) = 1$. Thus (10.2.23) follows immediately from (10.2.16) and definition of z_* (see also Remark 10.7 below).

5°. In this way we have established that the formulae (10.2.16) and (10.2.17) are correct. Recalling by (10.2.6) and (10.2.8) that

$$V_*(t,x) = \sqrt{t}\,W_*(x/\sqrt{t}) \tag{10.2.24}$$

we have therefore proved the following result.

Theorem 10.3. *Let z_1^* denote the unique positive root of $M(-1/2\,,1/2\,,z^2/2) = 0$. The value function of the optimal stopping problem* (10.2.4) *for $c \geq z_1^*$ is given by*

$$V_*(t,x) = \begin{cases} -\sqrt{t}\,z_*^{-1}\,M(-\frac{1}{2}\,,\frac{1}{2}\,,\frac{x^2}{2t})/M(\frac{1}{2}\,,\frac{3}{2}\,,\frac{z_*^2}{2}) & \text{if } |x|/\sqrt{t} < z_*, \\ |x| - c\sqrt{t} & \text{if } |x|/\sqrt{t} \geq z_* \end{cases} \tag{10.2.25}$$

where z_* is the unique positive root of the equation

$$z^{-1}M(-\tfrac{1}{2}\,,\tfrac{1}{2}\,,\tfrac{z^2}{2}) = (c - z)\,M(\tfrac{1}{2}\,,\tfrac{3}{2}\,,\tfrac{z^2}{2}) \tag{10.2.26}$$

satisfying $z_* \leq z_1^*$. The optimal stopping time in (10.2.4) *for $c > z_1^*$ is given by* (see *Figure* IV.6)

$$\tau_* = \inf\{\, r > 0 : |X_r| \geq z_*\sqrt{t+r}\,\}\,. \tag{10.2.27}$$

For $c = z_1^*$ the stopping times τ_* are approximately optimal if we let $c \downarrow z_1^*$. For $c < z_1^*$ we have $V_*(t,x) = \infty$ and it is never optimal to stop.

Using $\sqrt{t+\tau} \leq \sqrt{t} + \sqrt{\tau}$ in (10.2.4) it is easily verified that $V_*(t,0) \to V_*(0,0)$ as $t \downarrow 0$. Hence we see that $V_*(0,0) = 0$ with $\tau_* \equiv 0$. Note also that $V_*(0,x) = |x|$ with $\tau_* \equiv 0$.

2. Let τ be any stopping time for B satisfying $\mathsf{E}\sqrt{\tau} < \infty$. Then from Theorem 10.3 we see that $\mathsf{E}|X_\tau| \leq c\,\mathsf{E}\sqrt{t+\tau} + V_*(t,0)$ for all $c > z_1^*$. Letting first $t \downarrow 0$, and then $c \downarrow z_1^*$, we obtain the following sharp inequality which was first derived by Davis [34].

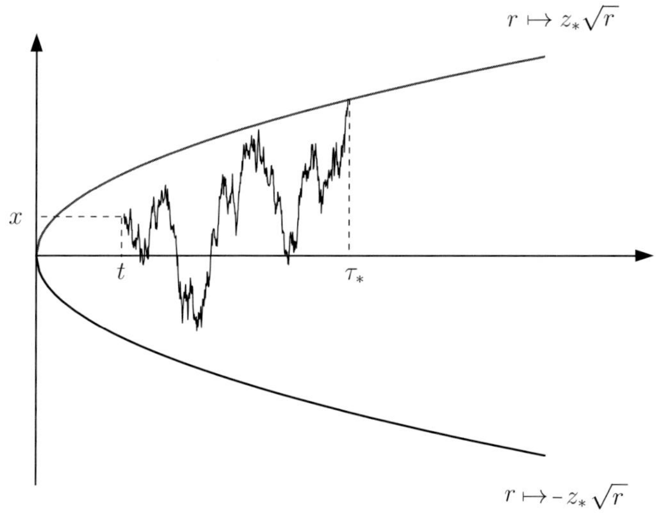

Figure IV.6: A computer simulation of the optimal stopping time τ_* in the problem (10.2.4) for $c > z_1^*$ as defined in (10.2.27). The process above is a standard Brownian motion which at time t starts at x. The optimal time τ_* is obtained by stopping the process as soon as it hits the area above or below the parabolic boundary $r \mapsto \pm z_* \sqrt{r}$.

Corollary 10.4. *Let $B = (B_t)_{t \geq 0}$ be a standard Brownian motion started at 0, and let τ be any stopping time for B. Then the following inequality is satisfied:*

$$\mathsf{E}\,|B_\tau| \leq z_1^* \,\mathsf{E}\sqrt{\tau} \tag{10.2.28}$$

with z_1^* being the unique positive root of $M(-1/2, 1/2, z^2/2) = 0$. The constant z_1^* is best possible. The equality is attained through the stopping times

$$\tau_* = \inf\left\{ r > 0 : |B_r| \geq z_* \sqrt{t+r} \right\} \tag{10.2.29}$$

when $t \downarrow 0$ and $c \downarrow z_1^*$, where z_* is the unique positive root of the equation

$$z^{-1} M(-\tfrac{1}{2}, \tfrac{1}{2}, \tfrac{z^2}{2}) = (c - z)\, M(\tfrac{1}{2}, \tfrac{3}{2}, \tfrac{z^2}{2}) \tag{10.2.30}$$

satisfying $z_* < z_1^*$. (Numerical calculations show that $z_1^* = 1.30693\ldots$)

3. The optimal stopping problem (10.2.4) can naturally be extended from the power 1 to all other $p > 0$. For this consider the optimal stopping problem with the value function

$$V_*(t,x) = \sup_\tau \mathsf{E}_x\Big(|X_\tau|^p - c\,\big(t+\tau\big)^{p/2} \Big) \tag{10.2.31}$$

where the supremum is taken over all stopping times τ for X satisfying $\mathsf{E}_x \tau^{p/2} < \infty$ and $c > 0$ is given and fixed.

Note that the case $p = 2$ is easily solved directly, since we have

$$V_*(t,x) = \sup_\tau \big((1-c)\,\mathsf{E}\,\tau + x^2 - c\,t \big) \tag{10.2.32}$$

due to $\mathsf{E}\,|B_\tau|^2 = \mathsf{E}\,\tau$ whenever $\mathsf{E}\,\tau < \infty$. Hence we see that $V_*(t,x) = +\infty$ if $c < 1$ (and it is never optimal to stop), and $V_*(t,x) = x^2 - ct$ if $c \geq 1$ (and it is optimal to stop instantly). Thus below we concentrate most on the cases when $p \neq 2$ (although the results formally extend to the case $p = 2$ by passing to the limit).

The following extension of Theorem 10.3 and Corollary 10.4 is valid. (Note that in the second part of the results we make use of parabolic cylinder functions $z \mapsto D_p(z)$ defined in (10.2.158) below.)

Theorem 10.5. (I): *For $0 < p < 2$ given and fixed, let z_p^* denote the unique positive root of $M(-p/2\,,1/2\,,z^2/2) = 0$. The value function of the optimal stopping problem* (10.2.31) *for $c \geq (z_p^*)^p$ is given by*

$$V_*(t,x) \tag{10.2.33}$$
$$= \begin{cases} -t^{p/2}\, z_*^{p-2}\, M(-\frac{p}{2}\,,\frac{1}{2}\,,\frac{x^2}{2t}) / M(1-\frac{p}{2}\,,\frac{3}{2}\,,\frac{z_*^2}{2}) & \textit{if } \ |x|/\sqrt{t} < z_*, \\ |x|^p - c\,t^{p/2} & \textit{if } \ |x|/\sqrt{t} \geq z_* \end{cases}$$

where z_* is the unique positive root of the equation

$$z^{p-2}\, M(-\tfrac{p}{2}\,,\tfrac{1}{2}\,,\tfrac{z^2}{2}) = (c - z^p)\, M(1-\tfrac{p}{2}\,,\tfrac{3}{2}\,,\tfrac{z^2}{2}) \tag{10.2.34}$$

satisfying $z_* \leq z_p^*$. The optimal stopping time in (10.2.31) *for $c > (z_p^*)^p$ is given by*

$$\tau_* = \inf\,\{\, r > 0 : |X_r| \geq z_* \sqrt{t+r}\,\}\,. \tag{10.2.35}$$

For $c = (z_p^*)^p$ the stopping times τ_* are approximately optimal if we let $c \downarrow (z_p^*)^p$. For $c < (z_p^*)^p$ we have $V_*(t,x) = \infty$ and it is never optimal to stop.

(II): *For $2 < p < \infty$ given and fixed, let z_p denote the largest positive root of $D_p(z) = 0$. The value function of the optimal stopping problem* (10.2.31) *for $c \geq (z_p)^p$ is given by*

$$V_*(t,x) = \begin{cases} t^{p/2}\, z_*^{p-1} e^{(x^2/4t)-(z_*^2/4)}\, \dfrac{D_p(|x|/\sqrt{t})}{D_{p-1}(z_*)} & \textit{if } \ |x|/\sqrt{t} > z_*, \\ |x|^p - c\,t^{p/2} & \textit{if } \ |x|/\sqrt{t} \leq z_* \end{cases} \tag{10.2.36}$$

where z_* is the unique root of the equation

$$z^{p-1}\, D_p(z) = (z^p - c)\, D_{p-1}(z) \tag{10.2.37}$$

satisfying $z_* \geq z_p$. *The optimal stopping time in* (10.2.31) *for* $c > (z_p)^p$ *is given by*

$$\tau_* = \inf\{ r > 0 : |X_r| \leq z_* \sqrt{t+r} \} . \tag{10.2.38}$$

For $c = (z_p)^p$ *the stopping times* τ_* *are approximately optimal if we let* $c \downarrow (z_p)^p$. *For* $c < (z_p)^p$ *we have* $V_*(t,x) = \infty$ *and it is never optimal to stop.*

Proof. The proof is an easy extension of the proof of Theorem 10.3, and we only present a few steps with differences for convenience.

By Brownian scaling we have

$$\begin{aligned} V_*(t,x) &= \sup_{\tau} \mathsf{E}_x \Big(|B_\tau + x|^p - c\,(t+\tau)^{p/2} \Big) \qquad (10.2.39)\\ &= t^{p/2} \sup_{\tau/t} \mathsf{E}_x \Big(\big|t^{-1/2} B_{t(\tau/t)} + x/\sqrt{t}\big|^p - c\,(1+\tau/t)^{p/2} \Big) \end{aligned}$$

and hence we see that

$$V_*(t,x) = t^{p/2}\, V_*(1, x/\sqrt{t}) . \tag{10.2.40}$$

By the time change $t \mapsto \sigma_t$ from (10.1.15) we find

$$|X_{\sigma_\tau}|^p - c\,(1+\sigma_\tau)^{p/2} = (1+\sigma_\tau)^{p/2} \big(|Z_\tau|^p - c\big) = e^{p\tau} \big(|Z_\tau|^p - c\big) \tag{10.2.41}$$

and the problem to determine $V_*(1,x)$ therefore reduces to computing

$$V_*(1,x) = W_*(x) \tag{10.2.42}$$

where W_* is the value function of the new (time-changed) optimal stopping problem

$$W_*(z) = \sup_{\tau} \mathsf{E}_z \Big(e^{p\tau} \big(|Z_\tau|^p - c\big) \Big) \tag{10.2.43}$$

the supremum being taken over all stopping times τ for Z for which $\mathsf{E}_z e^{p\tau} < \infty$.

To compute W_* we are naturally led to formulate the following free-boundary problem:

$$\begin{aligned} &\mathbb{L}_Z W(z) = -p\, W(z) && \text{for } z \in C, && (10.2.44)\\ &W(z) = |z|^p - c && \text{for } z \in \partial C \quad \text{(instantaneous stopping)}, && (10.2.45)\\ &W'(z) = \text{sign}(z)\, p |z|^{p-1} && \text{for } z \in \partial C \quad \text{(smooth fit)} && (10.2.46) \end{aligned}$$

where C is the domain of continued observation. Observe again that W_* should be even.

In the case $0 < p < 2$ we have $C = (-z_*, z_*)$ and the stopping time

$$\tau_* = \inf \big\{ t > 0 : |Z_t| \geq z_* \big\} \tag{10.2.47}$$

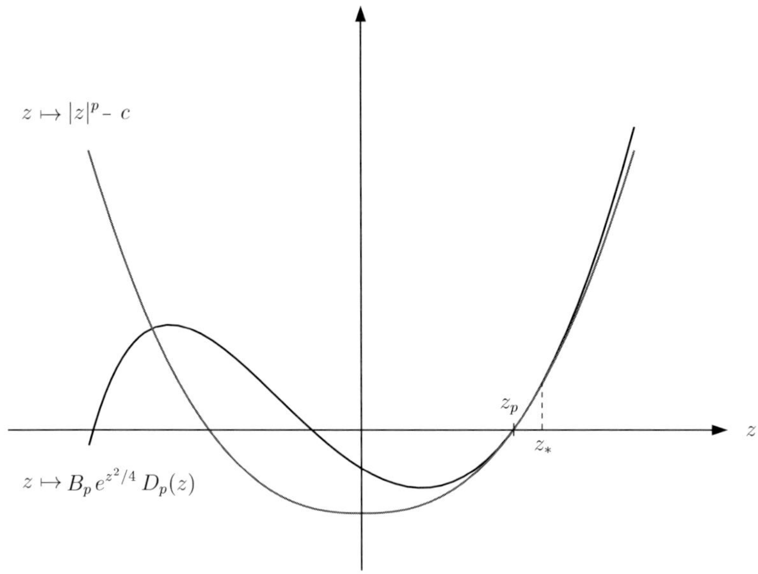

Figure IV.7: A computer drawing of the solution of the free-boundary problem (10.2.44)–(10.2.46) for positive z when $p = 2.5$. The solution equals $z \mapsto B_p \exp(z^2/4)\, D_p(z)$ for $z > z_*$ and $z \mapsto z^p - c$ for $0 \le z \le z_*$. The solution extends to negative z by mirroring to an even function. The constant B_p is chosen (and z_* is obtained) such that the smooth fit holds at z_* (the first derivative of the solution is continuous at z_*). A similar picture holds for all other $p > 2$ which are not even integers.

is optimal. The proof in this case can be carried out along exactly the same lines as above when $p = 1$. However, in the case $2 < p < \infty$ we have $C = (-\infty, -z_*) \cup (z_*, \infty)$ and thus the following stopping time:

$$\tau_* = \inf \left\{ t > 0 : |Z_t| \le z_* \right\} \tag{10.2.48}$$

is optimal. The proof in this case requires a small modification of the previous argument. The main difference is that the solution of (10.2.44) used above does not have the power of smooth fit (10.2.45)–(10.2.46) any longer. It turns out, however, that the solution $z \mapsto e^{z^2/4} D_p(z)$ has this power (see *Figure* IV.7 and *Figure* IV.8), and once this is understood, the proof is again easily completed along the same lines as above (see also Remark 10.7 below). □

Corollary 10.6. *Let $B = (B_t)_{t \ge 0}$ be a standard Brownian motion started at zero, and let τ be any stopping time for B.*

(I): *For $0 < p \le 2$ the following inequality is satisfied:*

$$\mathsf{E}\, |B_\tau|^p \le (z_p^*)^p \, \mathsf{E}\, \tau^{p/2} \tag{10.2.49}$$

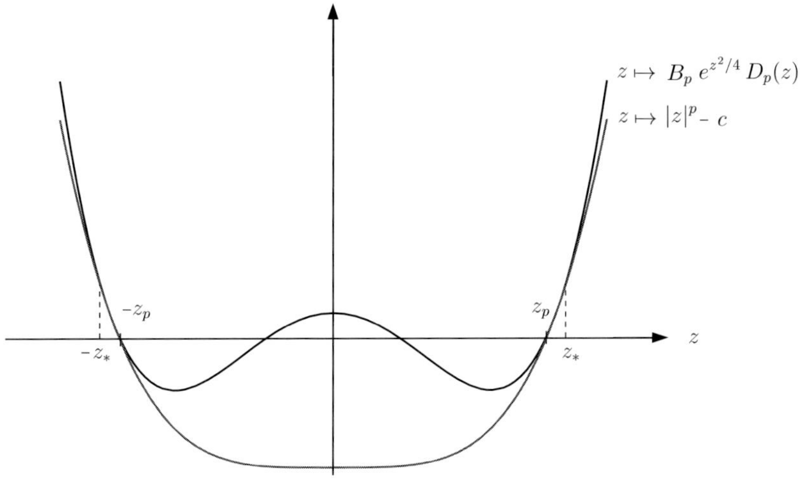

Figure IV.8: A computer drawing of the solution of the free-boundary problem (10.2.44)–(10.2.46) when $p = 4$. The solution equals $z \mapsto B_p \exp(z^2/4)\, D_p(z)$ for $|z| > z_*$ and $z \mapsto |z|^p - c$ for $|z| \le z_*$. The constant B_p is chosen (and z_* is obtained) such that the smooth fit holds at $\pm z_*$ (the first derivative of the solution is continuous at $\pm z_*$). A similar picture holds for all other $p > 2$ which are even integers.

with z_p^* being the unique positive root of $M(-p/2\,, 1/2\,, z^2/2) = 0$. The constant $(z_p^*)^p$ is best possible. The equality is attained through the stopping times

$$\tau_* = \inf\left\{ r > 0 : |B_r| \ge z_* \sqrt{t+r} \right\} \tag{10.2.50}$$

when $t \downarrow 0$ and $c \downarrow (z_p^*)^p$, where z_* is the unique positive root of the equation

$$z^{p-2} M(-\tfrac{p}{2}\,, \tfrac{1}{2}\,, \tfrac{z^2}{2}) = (c - z^p)\, M(1 - \tfrac{p}{2}\,, \tfrac{3}{2}\,, \tfrac{z^2}{2}) \tag{10.2.51}$$

satisfying $z_* < z_p^*$.

(II): *For $2 \le p < \infty$ the following inequality is satisfied:*

$$\mathsf{E}\,|B_\tau|^p \le (z_p)^p\, \mathsf{E}\,\tau^{p/2} \tag{10.2.52}$$

with z_p being the largest positive root of $D_p(z) = 0$. The constant $(z_p)^p$ is best possible. The equality is attained through the stopping times

$$\sigma_* = \inf\{ r > 0 : |B_r + x| \le z_* \sqrt{r} \} \tag{10.2.53}$$

when $x \downarrow 0$ and $c \downarrow (z_p)^p$, where z_* is the unique root of the equation

$$z^{p-1} D_p(z) = (z^p - c)\, D_{p-1}(z) \tag{10.2.54}$$

satisfying $z_* > z_p$.

Remark 10.7. The argument used above to verify (10.2.23) extends to the general setting of Theorem 10.5 and leads to the following explicit formulae for $0 < p < \infty$. (Note that these formulae are also valid for $-\infty < p < 0$ upon setting $z_p^* = +\infty$ and $z_p = -\infty$.)

1°. For $a > 0$ define the following stopping times:

$$\tau_a = \inf\big\{\, r > 0 : |Z_r| \geq a \,\big\}, \tag{10.2.55}$$

$$\gamma_a = \inf\big\{\, r > 0 : |X_r| \geq a\sqrt{t+r} \,\big\}. \tag{10.2.56}$$

By Brownian scaling and the time change (10.1.15) it is easily verified that

$$\mathsf{E}_x\big(\gamma_a + t\big)^{p/2} = t^{p/2}\, \mathsf{E}_{x/\sqrt{t}} e^{p\tau_a}. \tag{10.2.57}$$

The argument quoted above for $|z| < a$ then gives

$$\mathsf{E}_z e^{p\tau_a} = \begin{cases} M(-\frac{p}{2}, \frac{1}{2}, \frac{z^2}{2}) / M(-\frac{p}{2}, \frac{1}{2}, \frac{a^2}{2}) & \text{if } 0 < a < z_p^*, \\ \infty & \text{if } a \geq z_p^*. \end{cases} \tag{10.2.58}$$

Thus by (10.2.57) for $|x| < a\sqrt{t}$ we obtain

$$\mathsf{E}_x\big(\gamma_a + t\big)^{p/2} = \begin{cases} t^{p/2}\, M(-\frac{p}{2}, \frac{1}{2}, \frac{x^2}{2t}) / M(-\frac{p}{2}, \frac{1}{2}, \frac{a^2}{2}) & \text{if } 0 < a < z_p^*, \\ \infty & \text{if } a \geq z_p^*. \end{cases} \tag{10.2.59}$$

This formula is also derived in [183].

2°. For $a > 0$ define the following stopping times:

$$\widetilde{\tau}_a = \inf\big\{\, r > 0 : Z_r \leq a \,\big\}, \tag{10.2.60}$$

$$\widetilde{\gamma}_a = \inf\big\{\, r > 0 : X_r \leq a\sqrt{t+r} \,\big\}. \tag{10.2.61}$$

By precisely the same arguments for $z > a$ we get

$$\mathsf{E}_z e^{p\tilde{\tau}_a} = \begin{cases} e^{(z^2/4)-(a^2/4)}\, D_p(z)/D_p(a) & \text{if } a > z_p, \\ \infty & \text{if } a \leq z_p, \end{cases} \tag{10.2.62}$$

and for $x > a\sqrt{t}$ we thus obtain

$$\mathsf{E}_x\big(\widetilde{\gamma}_a + t\big)^{p/2} = \begin{cases} t^{p/2}\, e^{(x^2/4t)-(a^2/4)}\, D_p(x/\sqrt{t})/D_p(a) & \text{if } a > z_p, \\ \infty & \text{if } a \leq z_p. \end{cases} \tag{10.2.63}$$

This formula is also derived in [143].

Example 10.8. Consider the optimal stopping problem with the value function

$$V_*(t,x) = \sup_\tau \mathsf{E}_x\Big(\frac{X_\tau}{t+\tau}\Big) \tag{10.2.64}$$

where the supremum is taken over all stopping times τ for X. This problem was first solved by Shepp [184] and Taylor [210], and it was later extended by Walker [220] and Van Moerbeke [214]. To compute (10.2.64) we shall use the same arguments as in the proof of Theorem 10.3 above.

1°. In the first step we rewrite (10.2.64) as

$$V_*(t,x) = \sup_\tau \mathsf{E}\left(\frac{B_\tau + x}{t+\tau}\right) = \frac{1}{\sqrt{t}} \sup_{\tau/t} \mathsf{E}\left(\frac{t^{-1/2}B_{t(\tau/t)} + x/\sqrt{t}}{1+\tau/t}\right) \tag{10.2.65}$$

and note by Brownian scaling that

$$V_*(t,x) = \tfrac{1}{\sqrt{t}}\, V_*(1, x/\sqrt{t}) \tag{10.2.66}$$

so that we only need to look at $V_*(1,x)$ in the sequel. In exactly the same way as in Remark 10.2 above, from (10.2.66) we can heuristically conclude that the optimal stopping boundary should be $x = \gamma_0\sqrt{t}$ for some $\gamma_0 > 0$ to be found.

2°. In the second step we apply the time change $t \mapsto \sigma_t$ from (10.1.15) to the problem $V_*(1,x)$ and transform it into a new problem. From (10.2.1) we get

$$X_{\sigma_\tau}/(1+\sigma_\tau) = Z_\tau/\sqrt{1+\sigma_\tau} = e^{-\tau} Z_\tau \tag{10.2.67}$$

and the problem to determine $V_*(1,x)$ therefore reduces to computing

$$V_*(1,x) = W_*(x) \tag{10.2.68}$$

where W_* is the value function of the new (time-changed) optimal stopping problem

$$W_*(z) = \sup_\tau \mathsf{E}_z\big(e^{-\tau} Z_\tau\big) \tag{10.2.69}$$

the supremum being taken over all stopping times τ for Z.

3°. In the third step we solve the problem (10.2.69). From general optimal stopping theory (Chapter I) we know that the following stopping time should be optimal:

$$\tau_* = \inf\big\{\, t > 0 : Z_t \ge z_* \,\big\} \tag{10.2.70}$$

where z_* is the optimal stopping point to be found.

To compute the value function W_* for $z < z_*$ and to determine the optimal stopping point z_*, it is natural (Chapter III) to formulate the following free-boundary problem:

$$\mathbb{L}_Z W(z) = W(z) \quad \text{for} \quad z < z_*, \tag{10.2.71}$$

$$W(z_*) = z_* \quad \text{(instantaneous stopping)}, \tag{10.2.72}$$

$$W'(z_*) = 1 \quad \text{(smooth fit)} \tag{10.2.73}$$

with $\mathbb{L}_Z$ in (10.2.3).

The equation (10.2.71) is of the same type as the equation from Example 10.1. Since the present problem is not symmetrical, we choose its general solution in accordance with (10.2.160)–(10.2.161), i.e.

$$W(z) = A\, e^{z^2/4} D_{-1}(z) + B\, e^{z^2/4} D_{-1}(-z) \tag{10.2.74}$$

where A and B are unknown constants.

To determine A and B the following observation is crucial. Letting $z \to -\infty$ above, we see by (10.2.163) that $e^{z^2/4} D_{-1}(z) \to \infty$ and $e^{z^2/4} D_{-1}(-z) \to 0$. Hence we find that $A > 0$ would contradict the clear fact that $z \mapsto W_*(z)$ is increasing, while $A < 0$ would contradict the fact that $W_*(z) \geq z$ (by observing that $e^{z^2/4} D_{-1}(z)$ converges to ∞ faster than a polynomial). Therefore we must have $A = 0$. Moreover, from (10.2.163) we easily find that

$$e^{z^2/4}\, D_{-1}(-z) = e^{z^2/2} \int_{-\infty}^{z} e^{-u^2/2}\, du \tag{10.2.75}$$

and hence $W'(z) = z\, W(z) + B$. The boundary condition (10.2.73) implies that $1 = W'(z_*) = z_*\, W(z_*) + B = z_*^2 + B$, and hence we obtain $B = 1 - z_*^2$ (see *Figure* IV.9). Setting this into (10.2.72), we find that z_* is the root of the equation

$$z = (1 - z^2)\, e^{z^2/2} \int_{-\infty}^{z} e^{-u^2/2}\, du\, . \tag{10.2.76}$$

In this way we have obtained the following candidate for the value function W_*:

$$W(z) = \begin{cases} (1 - z_*^2)\, e^{z^2/2} \displaystyle\int_{-\infty}^{z} e^{-u^2/2}\, du & \text{if } z < z_*, \\ z & \text{if } z \geq z_*, \end{cases} \tag{10.2.77}$$

and the following candidate for the optimal stopping time:

$$\widehat{\tau}_{z_*} = \inf \left\{ t > 0 : Z_t \geq z_* \right\}. \tag{10.2.78}$$

4°. To verify that these formulae are correct, we can apply Itô's formula (page 67) to $(e^{-t}\, W(Z_t))_{t \geq 0}$, and in exactly the same way as in the proof of Theorem 10.3 above we can conclude

$$W_*(z) \leq W(z)\, . \tag{10.2.79}$$

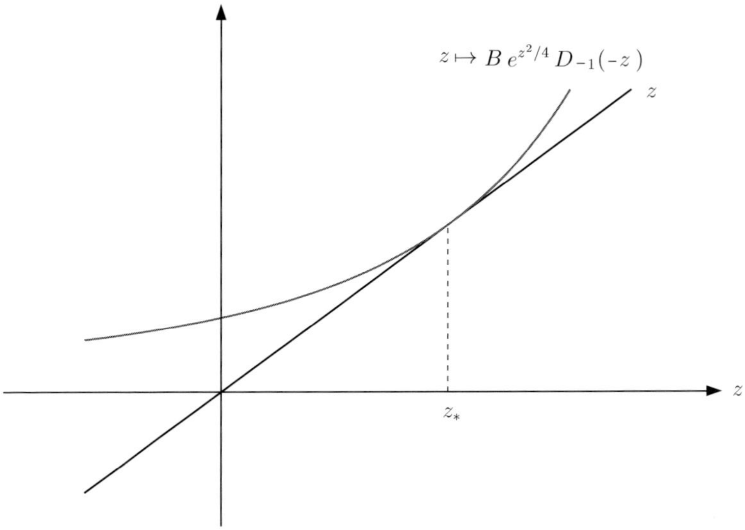

Figure IV.9: A computer drawing of the solution of the free-boundary problem (10.2.71)–(10.2.73). The solution equals $z \mapsto B\exp(z^2/4)\, D_{-1}(-z)$ for $z < z_*$ and $z \mapsto z$ for $z \geq z_*$. The constant B is chosen (and z_* is obtained) such that the smooth fit holds at z_* (the first derivative of the solution is continuous at z_*).

To prove that equality is attained at $\widehat{\tau}_{z_*}$ from (10.2.78), it is enough to show that

$$W(z) = \mathsf{E}_z\big(e^{-\hat{\tau}_{z_*}} Z_{\hat{\tau}_{z_*}}\big) = z_*\, \mathsf{E}_z\big(e^{-\hat{\tau}_{z_*}}\big)\, . \tag{10.2.80}$$

However, from general Markov process theory (Chapter III) we know that $w(z) = \mathsf{E}_z e^{-\hat{\tau}_{z_*}}$ solves (10.2.71), and clearly it satisfies $w(z_*) = 1$ and $w(-\infty) = 0$. Thus (10.2.80) follows from (10.2.77).

5°. In this way we have established that formulae (10.2.77) and (10.2.78) are correct. Recalling by (10.2.66) and (10.2.68) that

$$V_*(t,x) = \tfrac{1}{\sqrt{t}}\, W_*(x/\sqrt{t}) \tag{10.2.81}$$

we have therefore proved the following result.

Theorem 10.9. *The value function of the optimal stopping problem* (10.2.64) *is given by*

$$V_*(t,x) = \begin{cases} \dfrac{1}{\sqrt{t}}\,(1-z_*^2)\, e^{x^2/2t} \displaystyle\int_{-\infty}^{x/\sqrt{t}} e^{-u^2/2}\, du & \textit{if } \; x/\sqrt{t} < z_*, \\ x/t & \textit{if } \; x/\sqrt{t} \geq z_* \end{cases} \tag{10.2.82}$$

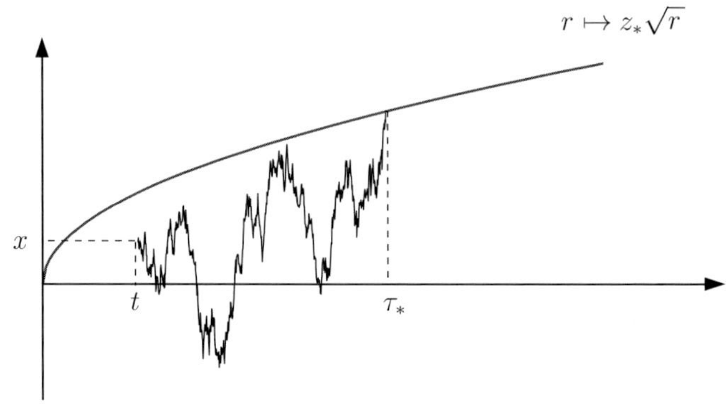

Figure IV.10: A computer simulation of the optimal stopping time τ_* in the problem (10.2.64) as defined in (10.2.84). The process above is a standard Brownian motion which at time t starts at x. The optimal time τ_* is obtained by stopping this process as soon as it hits the area above the parabolic boundary $r \mapsto z_* \sqrt{r}$.

where z_* *is the unique root of the equation*

$$z = (1 - z^2)\, e^{z^2/2} \int_{-\infty}^{z} e^{-u^2/2}\, du\,. \tag{10.2.83}$$

The optimal stopping time in 10.2.64 *is given by* (see *Figure* IV.10)

$$\tau_* = \inf\, \big\{\, r > 0 : X_r \geq z_* \sqrt{t+r}\, \big\}\,. \tag{10.2.84}$$

(*Numerical calculations show that* $z_* = 0.83992\dots$.)

4. Since the state space of the process $X = (X_t)_{t \geq 0}$ is $\mathbb{R}$ the most natural way to extend the problem (10.2.64) is to take $X = (X_t)_{t \geq 0}$ to the power of an odd integer (such that the state space again is $\mathbb{R}$). Consider the optimal stopping problem with the value function

$$V_*(t,x) = \sup_{\tau} \mathsf{E}_x \left(\frac{X_\tau^{2n-1}}{(t+\tau)^q} \right) \tag{10.2.85}$$

where the supremum is taken over all stopping times τ for X, and $n \geq 1$ and $q > 0$ are given and fixed. This problem was solved by Walker [220] in the case $n = 1$ and $q > 1/2$. We may now further extend Theorem 10.9 as follows.

Theorem 10.10. *Let* $n \geq 1$ *and* $q > 0$ *be taken to satisfy* $q > n - \frac{1}{2}$. *Then the value function of the optimal stopping problem* (10.2.85) *is given by*

$$V_*(t,x) \tag{10.2.86}$$
$$= \begin{cases} z_*^{2n-1}\, t^{n-q-1/2}\, e^{(x^2/4t)-(z_*^2/4)} \\ \qquad \times D_{2(n-q)-1}(-x/\sqrt{t}) / D_{2(n-q)-1}(-z_*) & \text{if } x/\sqrt{t} < z_*, \\ x^{2n-1}/t^q & \text{if } x/\sqrt{t} \geq z_* \end{cases}$$

where z_* *is the unique root of the equation*

$$(2n-1)\, D_{2(n-q)-1}(-z) = z\, \big(2(q-n)+1\big)\, D_{2(n-q-1)}(-z)\,. \tag{10.2.87}$$

The optimal stopping time in (10.2.85) *is given by*

$$\tau_* = \inf\big\{\, r > 0 : X_r \geq z_* \sqrt{t+r}\,\big\}\,. \tag{10.2.88}$$

(*Note that in the case* $q \leq n - 1/2$ *we have* $V_*(t,x) = \infty$ *and it is never optimal to stop.*)

Proof. The proof will only be sketched, since the arguments are the same as for the proof of Theorem 10.9. By Brownian scaling and the time change we find

$$V_*(t,x) = t^{n-q-1/2}\, W_*(x/\sqrt{t}) \tag{10.2.89}$$

where W_* is the value function of the new (time-changed) optimal stopping problem

$$W_*(z) = \sup_\tau \mathsf{E}_z \Big(e^{(2(n-q)-1)\tau}\, Z_\tau^{2n-1} \Big) \tag{10.2.90}$$

the supremum being taken over all stopping times τ for Z.

Again the optimal stopping time should be of the form

$$\tau_* = \inf\{\, t > 0 : Z_t \geq z_* \,\} \tag{10.2.91}$$

and therefore the value function W_* and the optimal stopping point z_* should solve the following free-boundary problem:

$$\mathbb{L}_Z W(z) = \big(1-2(n-q)\big)\, W(z) \quad \text{for } 4z < z_*, \tag{10.2.92}$$
$$W(z_*) = z_*^{2n-1} \quad \text{(instantaneous stopping)}, \tag{10.2.93}$$
$$W'(z_*) = (2n-1)\, z_*^{2(n-1)} \quad \text{(smooth fit)}. \tag{10.2.94}$$

Arguing as in the proof of Theorem 10.9 we find that the following solution of (10.2.92) should be taken into consideration:

$$W(z) = A\, e^{z^2/4}\, D_{2(n-q)-1}(-z) \tag{10.2.95}$$

where A is an unknown constant. The two boundary conditions (10.2.93)–(10.2.94) with (10.2.162) imply that $A = z_*^{2n-1} e^{-z_*^2/4} / D_{2(n-q)-1}(-z_*)$ where z_* is the root of the equation

$$(2n-1)\, D_{2(n-q)-1}(-z) = z\,(2(q-n)+1)\, D_{2(n-q-1)}(-z)\,. \tag{10.2.96}$$

Thus the candidate guessed for W_* is

$$W(z) = \begin{cases} z_*^{2n-1}\, e^{(z^2/4)-(z_*^2/4)}\, \dfrac{D_{2(n-q)-1}(-z)}{D_{2(n-q)-1}(-z_*)} & \text{if } z < z_*, \\ z^{2n-1} & \text{if } z \ge z_* \end{cases} \tag{10.2.97}$$

and the optimal stopping time is given by (10.2.91). By applying Itô's formula (page 67) as in the proof of Theorem 10.9 one can verify that these formulae are correct. Finally, inserting this back into (10.2.89) one obtains the result. □

Remark 10.11. By exactly the same arguments as in Remark 10.7 above, we can extend the verification of (10.2.80) to the general setting of Theorem 10.10, and this leads to the following explicit formulae for $0 < p < \infty$.

For $a > 0$ define the following stopping times:

$$\widehat{\tau}_a = \inf\{\, r > 0 : Z_r \ge a\,\}, \tag{10.2.98}$$

$$\widehat{\gamma}_a = \inf\{\, r > 0 : X_r \ge a\sqrt{t+r}\,\}\,. \tag{10.2.99}$$

Then for $z < a$ we get

$$\mathsf{E}_z e^{-p\hat{\tau}_a} = e^{(z^2/4)-(a^2/4)}\, \frac{D_{-p}(-z)}{D_{-p}(-a)} \tag{10.2.100}$$

and for $x < a\sqrt{t}$ we thus obtain

$$\mathsf{E}_x(\widehat{\gamma}_a + t)^{-p/2} = t^{-p/2}\, e^{(x^2/4t)-(a^2/4)}\, \frac{D_{-p}(-x/\sqrt{t})}{D_{-p}(-a)}\,. \tag{10.2.101}$$

Example 10.12. Consider the optimal stopping problem with the value function

$$V_*(t,x) = \sup_\tau \mathsf{E}_x\left(\frac{|X_\tau|}{t+\tau}\right) \tag{10.2.102}$$

where the supremum is taken over all stopping times τ for X. This problem (for the *reflected Brownian motion* $|X| = (|X_t|)_{t\ge 0}$) is a natural extension of the problem (10.2.64) and can be solved likewise.

By Brownian scaling and a time change we find

$$V_*(t,x) = \tfrac{1}{\sqrt{t}}\, W_*(x/\sqrt{t}) \tag{10.2.103}$$

where W_* is the value function of the new (time-changed) optimal stopping problem

$$W_*(z) = \sup_{\tau} \mathsf{E}_z \big(e^{-\tau} \, |Z_\tau| \big) \tag{10.2.104}$$

the supremum being taken over all stopping times for Z.

The problem (10.2.104) is symmetrical (recall the discussion about (10.2.10) above), and therefore the following stopping time should be optimal:

$$\tau_* = \inf \{ \, t > 0 : |Z_t| \geq z_* \, \} . \tag{10.2.105}$$

Thus it is natural (Chapter III) to formulate the following free-boundary problem:

$$\mathbb{L}_Z W(z) = W(z) \quad \text{for } \; z \in (-z_*, z_*), \tag{10.2.106}$$

$$W(\pm z_*) = |z_*| \qquad \text{(instantaneous stopping)}, \tag{10.2.107}$$

$$W'(\pm z_*) = \pm 1 \qquad \text{(smooth fit)}. \tag{10.2.108}$$

From the proof of Theorem 10.3 we know that the equation (10.2.106) admits an even and an odd solution which are linearly independent. Since the value function should be even, we can forget the odd solution, and therefore we must have

$$W(z) = A \, M(\tfrac{1}{2} \, , \tfrac{1}{2} \, , \tfrac{z^2}{2}) \tag{10.2.109}$$

for some $A > 0$ to be found. Note that $M(1/2 \, , 1/2 \, , z^2/2) = \exp(z^2/2)$ (see pages 192–193 below). The two boundary conditions (10.2.107) and (10.2.108) imply that $A = 1/\sqrt{e}$ and $z_* = 1$, and in this way we obtain the following candidate for the value function:

$$W(z) = e^{(z^2/2)-(1/2)} \tag{10.2.110}$$

for $z \in (-1, 1)$, and the following candidate for the optimal stopping time:

$$\tau = \inf \{ \, t > 0 : |Z_t| \geq 1 \, \} . \tag{10.2.111}$$

By applying Itô's formula (as in Example 10.8) one can prove that these formulae are correct. Inserting this back into (10.2.103) we obtain the following result.

Theorem 10.13. *The value function of the optimal stopping problem* (10.2.102) *is given by*

$$V_*(t,x) = \begin{cases} \frac{1}{\sqrt{t}} \, e^{(x^2/2t)-(1/2)} & \text{if } \; |x| < \sqrt{t}, \\ |x|/t & \text{if } \; |x| \geq \sqrt{t} \, . \end{cases} \tag{10.2.112}$$

The optimal stopping time in (10.2.102) *is given by*

$$\tau_* = \inf \{ \, r > 0 : |X_r| \geq \sqrt{t+r} \, \} . \tag{10.2.113}$$

As in Example 10.8 above, we can further extend (10.2.102) by considering the optimal stopping problem with the value function

$$V_*(t,x) = \sup_\tau \mathsf{E}_x \left(\frac{|X_\tau|^p}{(t+\tau)^q} \right) \tag{10.2.114}$$

where the supremum is taken over all finite stopping times τ for X, and p, $q>0$ are given and fixed. The arguments used to solve the problem (10.2.102) can be repeated, and in this way we obtain the following result (see [138]).

Theorem 10.14. *Let p, $q>0$ be taken to satisfy $q>p/2$. Then the value function of the optimal stopping problem* (10.2.114) *is given by*

$$\begin{aligned} &V_*(t,x) \\ &= \begin{cases} z_*^p \, t^{p/2-q} \, M(q-\frac{p}{2}, \frac{1}{2}, \frac{x^2}{2t}) / M(q-\frac{p}{2}, \frac{1}{2}, \frac{z_*^2}{2}) & \text{if } |x|/\sqrt{t} < z_*, \\ |x|^p/t^q & \text{if } |x|/\sqrt{t} \geq z_* \end{cases} \end{aligned} \tag{10.2.115}$$

where z_* is the unique root of the equation

$$p\, M(q-\tfrac{p}{2}, \tfrac{1}{2}, \tfrac{z^2}{2}) = z^2\,(2q-p)\, M(q+1-\tfrac{p}{2}, \tfrac{3}{2}, \tfrac{z^2}{2})\,. \tag{10.2.116}$$

The optimal stopping time in (10.2.114) *is given by*

$$\tau_* = \inf\{\, r>0 : X_r \geq z_* \sqrt{t+r}\,\}\,. \tag{10.2.117}$$

(Note that in the case $q \leq p/2$ we have $V_*(t,x)=\infty$ and it is never optimal to stop.)

Example 10.15. In this example we indicate how the problem and the results in Example 10.1 and Example 10.12 above can be extended from reflected Brownian motion to *Bessel processes* of arbitrary dimension $\alpha \geq 0$. To avoid the computational complexity which arises, we shall only indicate the essential steps towards solution.

1°. *The case $\alpha>1$.* The Bessel process of dimension $\alpha>1$ is a unique (non-negative) strong solution of the stochastic differential equation

$$dX_t = \frac{\alpha-1}{2X_t}\, dt + dB_t \tag{10.2.118}$$

satisfying $X_0=x$ for some $x \geq 0$. The boundary point 0 is *instantaneously reflecting* if $\alpha<2$, and is an *entrance* boundary point if $\alpha \geq 2$. (When $\alpha \in \mathbb{N} = \{1,2\dots\}$ the process $X=(X_t)_{t\geq 0}$ may be realized as the *radial part* of the α-dimensional Brownian motion.)

In the notation of Subsection 10.1 let us consider the process $Y=(Y_t)_{t\geq 0} = (\beta(t)X_t)_{t\geq 0}$ and note that $b(x)=(\alpha-1)/2x$ and $a(x)=1$. Thus conditions (10.1.11) and (10.1.12) may be realized with $\gamma(t)=\beta(t)$, $G_1(y)=(\alpha-1)/2y$

and $G_2(y) = 1$. Noting that $\beta(t) = 1/\sqrt{1+t}$ solves $\beta'(t)/\beta(t) = -\beta^2(t)/2$ and setting $\rho = \beta^2/2$, we see from (10.1.9) that

$$\mathbb{L}_Z = \left(-z + \frac{\alpha - 1}{z}\right)\frac{\partial}{\partial z} + \frac{\partial^2}{\partial z^2} \tag{10.2.119}$$

where $Z = (Z_t)_{t\geq 0} = (Y_{\sigma_t})_{t\geq 0}$ with $\sigma_t = e^{2t} - 1$. Thus $Z = (Z_t)_{t\geq 0}$ solves the equation

$$dZ_t = \left(-Z_t + \frac{\alpha - 1}{Z_t}\right) dt + \sqrt{2}\, dB_t. \tag{10.2.120}$$

Observe that the diffusion $Z = (Z_t)_{t\geq 0}$ may be seen as *the Euclidean velocity* of the α-dimensional Brownian motion whenever $\alpha \in \mathbb{N}$, and thus may be interpreted as the Euclidean velocity of the Bessel process $X = (X_t)_{t\geq 0}$ of any dimension $\alpha > 1$.

The Bessel process $X = (X_t)_{t\geq 0}$ of any dimension $\alpha \geq 0$ satisfies the *Brownian scaling* property $\mathrm{Law}\big((c^{-1}X_{c^2t})_{t\geq 0} \,|\, \mathsf{P}_{x/c}\big) = \mathrm{Law}\big((X_t)_{t\geq 0} \,|\, \mathsf{P}_x\big)$ for all $c > 0$ and all x. Thus the initial arguments used in Example 10.1 and Example 10.12 can be repeated, and the crucial point in the formulation of the corresponding free-boundary problem is the following analogue of the equations (10.2.11) and (10.2.106):

$$\mathbb{L}_Z W(z) = \rho\, W(z) \tag{10.2.121}$$

where $\rho \in \mathbb{R}$. In comparison with the equation (10.2.151) this reads as follows:

$$y''(x) - \left(x - \tfrac{\alpha-1}{x}\right) y'(x) - \rho\, y(x) = 0 \tag{10.2.122}$$

where $\rho \in \mathbb{R}$. By substituting $y(x) = x^{-(\alpha-1)/2} \exp(x^2/4)\, u(x)$ the equation (10.2.122) reduces to the following equation:

$$u''(x) - \left(\tfrac{x^2}{4} + \left(\rho - \tfrac{\alpha}{2}\right) + \tfrac{\alpha-1}{2}\left(\tfrac{\alpha-1}{2} - 1\right)\tfrac{1}{x^2}\right) u(x) = 0\,. \tag{10.2.123}$$

The unpleasant term in this equation is $1/x^2$, and the general solution is not immediately found in the list of special functions in [1]. Motivated by our considerations below when $0 \leq \alpha \leq 1$, we may substitute $\bar{y}(x^2) = y(x)$ and observe that the equation (10.2.122) is equivalent to

$$4z\, \bar{y}''(z) + 2(\alpha - z)\, \bar{y}'(z) - \rho\, \bar{y}(z) = 0 \tag{10.2.124}$$

where $z = x^2$. This equation now can be reduced to *Whittaker's equation* (see [1]) as described in (10.2.131) and (10.2.132) below. The general solution of Whittaker's equation is given by Whittaker's functions which are expressed in terms of Kummer's functions. This establishes a basic fact about the extension of the free-boundary problem from the reflected Brownian motion to the Bessel process of the dimension $\alpha > 1$. The problem then can be solved in exactly the same manner as before. It is interesting to observe that if the dimension α of the Bessel process $X = (X_t)_{t\ge 0}$ equals 3, then the equation (10.2.123) is of the form (10.2.152), and thus the optimal stopping problem is solved immediately by using the corresponding closed form solution given in Example 10.1 and Example 10.12 above.

2°. *The case* $0 \le \alpha \le 1$. The Bessel process of dimension $0 \le \alpha \le 1$ does not solve a stochastic differential equation in the sense of (10.2.118), and therefore it is convenient to look at the *squared Bessel process* $\bar{X} = (\bar{X}_t)_{t\ge 0}$ which is a unique (non-negative) strong solution of the stochastic differential equation

$$d\bar{X}_t = \alpha\, dt + 2\sqrt{\bar{X}_t}\, dB_t \tag{10.2.125}$$

satisfying $\bar{X}_0 = \bar{x}$ for some $\bar{x} \ge 0$. (This is true for all $\alpha \ge 0$.) The Bessel process $X = (X_t)_{t\ge 0}$ is then defined as the square root of $\bar{X} = (\bar{X}_t)_{t\ge 0}$. Thus

$$X_t = \sqrt{\bar{X}_t}. \tag{10.2.126}$$

The boundary point 0 is an *instantaneously reflecting boundary point* if $0 < \alpha \le 1$, and is a *trap* if $\alpha = 0$. (The Bessel process $X = (X_t)_{t\ge 0}$ may be realized as a reflected Brownian motion when $\alpha = 1$.)

In the notation of Subsection 10.1 let us consider the process $\bar{Y} = (\bar{Y}_t)_{t\ge 0} = (\beta(t)\bar{X}_t)_{t\ge 0}$ and note that $b(x) = \alpha$ and $a(x) = 2\sqrt{x}$. Thus conditions (10.1.11) and (10.1.12) may be realized with $\gamma(t) = 1$, $G_1(y) = \alpha$ and $G_2(y) = 4y$. Noting that $\beta(t) = 1/(1{+}t)$ solves $\beta'(t)/\beta(t) = -\beta(t)$ and setting $\rho = \beta/2$, we see from (10.1.9) that

$$\mathbb{L}_{\bar{Z}} = 2\big(-z + \alpha\big)\,\frac{\partial}{\partial z} + 4z\,\frac{\partial^2}{\partial z^2} \tag{10.2.127}$$

where $\bar{Z} = (\bar{Z}_t)_{t\ge 0} = (\bar{Y}_{\sigma_t})_{t\ge 0}$ with $\sigma_t = e^{2t} - 1$. Thus $\bar{Z} = (\bar{Z}_t)_{t\ge 0}$ solves the equation

$$d\bar{Z}_t = 2\big(-\bar{Z}_t + \alpha\big)\, dt + 2\sqrt{2\bar{Z}_t}\, dB_t\,. \tag{10.2.128}$$

It is interesting to observe that

$$\bar{Z}_t = \bar{Y}_{\sigma_t} = \frac{\bar{X}_{\sigma_t}}{1{+}\sigma_t} = \left(\frac{X_{\sigma_t}}{\sqrt{1{+}\sigma_t}}\right)^2 \tag{10.2.129}$$

and thus the process $\big(\sqrt{\bar{Z}_t}\,\big)_{t\ge 0}$ may be seen as *the Euclidean velocity* of the α-dimensional Brownian motion for $\alpha \in [0,1]$.

This enables us to reformulate the initial problem about $X = (X_t)_{t\ge 0}$ in terms of $\bar{X} = (\bar{X}_t)_{t\ge 0}$ and then after Brownian scaling and time change $t \mapsto \sigma_t$ in terms of the diffusion $\bar{Z} = (\bar{Z}_t)_{t\ge 0}$. The pleasant fact is hidden in the formulation of the corresponding free-boundary problem for $\bar{Z} = (\bar{Z}_t)_{t\ge 0}$:

$$\mathbb{L}_{\bar{Z}} W = \rho\, W \tag{10.2.130}$$

which in comparison with the equation (10.2.151) reads as follows:

$$4x\,y''(x) + 2(\alpha - x)\,y'(x) - \rho\,y(x) = 0\,. \tag{10.2.131}$$

Observe that this equation is of the same type as equation (10.2.124). By substituting $y(x) = x^{-\alpha/4}\exp(x/4)\,u(x)$ equation (10.2.131) reduces to

$$u''(x) + \left(-\frac{1}{16} + \frac{1}{4}\Big(\rho + \frac{\alpha}{2}\Big)\frac{1}{x} + \frac{\alpha}{4}\Big(1 - \frac{\alpha}{4}\Big)\frac{1}{x^2}\right)u(x) = 0 \tag{10.2.132}$$

which may be recognized as a *Whittaker's equation* (see [1]). The general solution of Whittaker's equation is given by Whittaker's functions which are expressed in terms of Kummer's functions. This again establishes a basic fact about extension of the free-boundary problem from the reflected Brownian motion to the Bessel process of dimension $0 \le \alpha < 1$. The problem then can be solved in exactly the same manner as before. Note also that the arguments about the passage to the squared Bessel process just presented are valid for all $\alpha \ge 0$. When $\alpha > 1$ it is a matter of taste which method to choose.

Example 10.16. In this example we show how to solve some *path-dependent* optimal stopping problems (i.e. problems with the gain function depending on the entire path of the underlying process up to the time of observation). For comparison with general theory recall Section 6 above.

Given an Ornstein–Uhlenbeck process $Z = (Z_t)_{t\ge 0}$ satisfying (10.2.2), started at z under P_z, consider the optimal stopping problem with the value function

$$\widetilde{W}_*(z) = \sup_{\tau} \mathsf{E}_z\left(\int_0^\tau e^{-u} Z_u\,du\right) \tag{10.2.133}$$

where the supremum is taken over all stopping time τ for Z. This problem is motivated by the fact that the integral appearing above may be viewed as a measure of the accumulated gain (up to the time of observation) which is assumed proportional to the velocity of the Brownian particle being discounted. We will first verify by Itô's formula (page 67) that this problem is in fact equivalent to the one-dimensional problem (10.2.69). Then by using the time change σ_t we shall show that these problems are also equivalent to yet another path-dependent optimal stopping problem which is given in (10.2.140) below.

1°. Applying Itô's formula (page 67) to the process $(e^{-t}Z_t)_{t\ge 0}$, we find by using (10.2.2) that

$$e^{-t}Z_t = z + M_t - 2\int_0^t e^{-u} Z_u\,du \tag{10.2.134}$$

where $M = (M_t)_{t\ge 0}$ is a continuous local martingale given by

$$M_t = \sqrt{2}\int_0^t e^{-u}\,dB_u\,. \tag{10.2.135}$$

If τ is a bounded stopping time for Z, then by the optional sampling theorem (page 60) we get

$$\mathsf{E}_z\Big(\int_0^\tau e^{-u} Z_u\, du\Big) = \frac{1}{2}\Big(z + \mathsf{E}_z\big(e^{-\tau}(-Z_\tau)\big)\Big)\,. \tag{10.2.136}$$

Taking the supremum over all bounded stopping times τ for Z , and using that $-Z = (-Z_t)_{t\geq 0}$ is an Ornstein–Uhlenbeck process starting from $-z$ under P_z , we obtain

$$\widetilde{W}_*(z) = \frac{1}{2}\big(z + W_*(-z)\big) \tag{10.2.137}$$

where W_* is the value function from (10.2.69). The explicit expression for W_* is given in (10.2.77), and inserting it in (10.2.137), we immediately obtain the following result.

Corollary 10.17. *The value function of the optimal stopping problem* (10.2.133) *is given by*

$$\widetilde{W}_*(z) = \begin{cases} \frac{1}{2}\Big(z+(1-z_*^2)\, e^{z^2/2} \displaystyle\int_z^\infty e^{-u^2/2}\, du\Big) & \text{if } z > -z_*, \\ 0 & \text{if } z \leq -z_* \end{cases} \tag{10.2.138}$$

where $z_* > 0$ *is the unique root of* (10.2.83). *The optimal stopping time in* (10.2.133) *is given by*

$$\tau_* = \inf\{\, t > 0 : Z_t \leq -z_* \,\}\,. \tag{10.2.139}$$

2°. Given the Brownian motion $X_t = B_t + x$ started at x under P_x , consider the optimal stopping problem with the value function

$$\widetilde{V}_*(t,x) = \sup_\tau \mathsf{E}_x\bigg(\int_0^\tau \frac{X_u}{(t+u)^2}\, du\bigg) \tag{10.2.140}$$

where the supremum is taken over all stopping times τ for X . It is easily verified by Brownian scaling that we have

$$\widetilde{V}_*(t,x) = \frac{1}{\sqrt{t}}\, \widetilde{V}_*(1, x/\sqrt{t})\,. \tag{10.2.141}$$

Moreover, by time change (10.1.15) we get

$$\begin{aligned} \int_0^{\sigma_\tau} \frac{X_u}{(1+u)^2}\, du &= \int_0^\tau \frac{X_{\sigma_u}}{(1+\sigma_u)^2}\, d\sigma_u \\ &= 2\int_0^\tau e^{2u}(1+\sigma_u)^{-3/2} Z_u\, du = 2\int_0^\tau e^{-u} Z_u\, du \end{aligned} \tag{10.2.142}$$

and the problem to determine $\widetilde{V}_*(1,x)$ therefore reduces to computing

$$\widetilde{V}_*(1,x) = \widetilde{W}_*(x) \tag{10.2.143}$$

where $\widetilde{W}_*$ is given by (10.2.133). From (10.2.141) and (10.2.143) we thus obtain the following result as an immediate consequence of Corollary 10.17.

Corollary 10.18. *The value function of the optimal stopping problem* (10.2.140) *is given by*

$$\widetilde{V}_*(t,x) = \begin{cases} \frac{x}{t} + \frac{1}{\sqrt{t}}(1 - z_*^2)\, e^{x^2/2t} \displaystyle\int_{x/\sqrt{t}}^{\infty} e^{-u^2/2}\, du & \text{if } \; x/\sqrt{t} > -z_*, \\ 0 & \text{if } \; x/\sqrt{t} \le -z_* \end{cases} \tag{10.2.144}$$

where $z_* > 0$ *is the unique root of* (10.2.83). *The optimal stopping time in* (10.2.140) *is given by*

$$\tau_* = \inf\{\, r > 0 : X_r \le -z_* \sqrt{t+r}\,\}\,. \tag{10.2.145}$$

3°. The optimal stopping problem (10.2.133) can be naturally extended by considering the optimal stopping problem with the value function

$$\widetilde{W}_*(z) = \sup_{\tau} \mathsf{E}_z\Big(\int_0^{\tau} e^{-pu} \mathrm{He}_n(Z_u)\, du \Big) \tag{10.2.146}$$

where the supremum is taken over all stopping times τ for Z and $x \mapsto \mathrm{He}_n(x)$ is the Hermite polynomial given by (10.2.166), with $p > 0$ given and fixed. The crucial fact is that $x \mapsto \mathrm{He}_n(x)$ solves the differential equation (10.2.151), and by Itô's formula (page 67) and (10.2.2) this implies

$$\begin{aligned} e^{-pt}\mathrm{He}_n(Z_t) &= \mathrm{He}_n(z) \\ &\quad + M_t + \int_0^t e^{-pu}\Big(\mathbb{L}_Z(\mathrm{He}_n)(Z_u) - p\mathrm{He}_n(Z_u) \Big)\, du \\ &= \mathrm{He}_n(z) + M_t - (n{+}p) \int_0^t e^{-pu}\mathrm{He}_n(Z_u)\, du \end{aligned} \tag{10.2.147}$$

where $M = (M_t)_{t \ge 0}$ is a continuous local martingale given by

$$M_t = \sqrt{2} \int_0^t e^{-pu} (\mathrm{He}_n)'(Z_u)\, du\,. \tag{10.2.148}$$

Again as above we find that

$$\widetilde{W}_*(z) = \tfrac{1}{n+p} \big(\mathrm{He}_n(z) + W_*(z) \big) \tag{10.2.149}$$

with W_* being the value function of the optimal stopping problem

$$W_*(z) = \sup_{\tau} \mathsf{E}_z\Big(e^{-p\tau}\big(-\mathrm{He}_n(Z_u) \big) \Big) \tag{10.2.150}$$

where the supremum is taken over all stopping times τ for Z. This problem is one-dimensional and can be solved by the method used in Example 10.1.

4°. Observe that the problem (10.2.146) with the arguments just presented can be extended from the Hermite polynomial to any solution of the differential equation (10.2.151).

5. Auxiliary results. In the examples above we need the general solution of the second-order differential equation

$$y''(x) - x\,y'(x) - \rho\,y(x) = 0 \tag{10.2.151}$$

where $\rho \in \mathbb{R}$. By substituting $y(x) = \exp(x^2/4)\,u(x)$ the equation (10.2.151) reduces to

$$u''(x) - \left(\frac{x^2}{4} + \left(\rho - \frac{1}{2}\right)\right) u(x) = 0\,. \tag{10.2.152}$$

The general solution of (10.2.152) is well known, and in the text above we make use of the following two pairs of linearly independent solutions (see [1]).

1°. The *Kummer confluent hypergeometric* function is defined by

$$M(a,b,x) = 1 + \frac{a}{b}\,x + \frac{a(a+1)}{b(b+1)}\,\frac{x^2}{2!} + \cdots\,. \tag{10.2.153}$$

Two linearly independent solutions of (10.2.152) can be expressed as

$$u_1(x) = e^{-x^2/4}\,M(\tfrac{\rho}{2}\,,\tfrac{1}{2}\,,\tfrac{x^2}{2}) \quad \& \quad u_2(x) = x\,e^{-x^2/4}\,M(\tfrac{\rho}{2}+\tfrac{1}{2}\,,\tfrac{3}{2}\,,\tfrac{x^2}{2}) \tag{10.2.154}$$

and therefore two linearly independent solutions of (10.2.151) are given by

$$y_1(x) = M(\tfrac{\rho}{2}\,,\tfrac{1}{2}\,,\tfrac{x^2}{2}), \tag{10.2.155}$$

$$y_2(x) = x\,M(\tfrac{\rho}{2}+\tfrac{1}{2}\,,\tfrac{3}{2}\,,\tfrac{x^2}{2})\,. \tag{10.2.156}$$

Observe that y_1 is *even* and y_2 is *odd.* Note also that

$$M'(a,b,x) = \tfrac{a}{b}\,M(a+1,b+1,x)\,. \tag{10.2.157}$$

2°. The *parabolic cylinder* function is defined by

$$D_\nu(x) = A_1\,e^{-x^2/4}\,M(-\tfrac{\nu}{2}\,,\tfrac{1}{2}\,,\tfrac{x^2}{2}) + A_2\,x\,e^{-x^2/4}\,M(-\tfrac{\nu}{2}+\tfrac{1}{2}\,,\tfrac{3}{2}\,,\tfrac{x^2}{2}) \tag{10.2.158}$$

where $A_1 = 2^{\nu/2}\pi^{-1/2}\cos(\nu\pi/2)\Gamma((1+\nu)/2)$ and $A_2 = 2^{(1+\nu)/2}\pi^{-1/2}\sin(\nu\pi/2)$ $\Gamma(1+\nu/2)$. Two linearly independent solutions of (10.2.152) can be expressed as

$$\widetilde{u}_1(x) = D_{-\rho}(x) \quad \& \quad \widetilde{u}_2(x) = D_{-\rho}(-x) \tag{10.2.159}$$

and therefore two linearly independent solutions of (10.2.151) are given by

$$\widetilde{y}_1(x) = e^{x^2/4}\, D_{-\rho}(x), \tag{10.2.160}$$

$$\widetilde{y}_2(x) = e^{x^2/4}\, D_{-\rho}(-x) \tag{10.2.161}$$

whenever $-\rho \notin \mathbb{N} \cup \{0\}$. Note that $\widetilde{y}_1$ and $\widetilde{y}_2$ are not symmetric around zero unless $-\rho \in \mathbb{N} \cup \{0\}$. Note also that

$$\frac{d}{dx}\big(e^{x^2/4} D_\nu(x)\big) = \nu\, e^{x^2/4} D_{\nu-1}(x)\,. \tag{10.2.162}$$

Moreover, the following integral representation is valid:

$$D_\nu(x) = \frac{e^{-x^2/4}}{\Gamma(-\nu)} \int_0^\infty u^{-\nu-1}\, e^{-xu-u^2/2}\, du \tag{10.2.163}$$

whenever $\nu < 0$.

3°. To identify zero points of the solutions above, it is useful to note that

$$M(-n\,, \tfrac{1}{2}\,, \tfrac{x^2}{2}) = \mathrm{He}_{2n}(x)/\mathrm{He}_{2n}(0), \tag{10.2.164}$$

$$e^{x^2/4} D_n(x) = \mathrm{He}_n(x) \tag{10.2.165}$$

where $x \mapsto \mathrm{He}_n(x)$ is the *Hermite polynomial*

$$\mathrm{He}_n(x) = (-1)^n e^{x^2/2} \frac{d^n}{dx^n}\big(e^{-x^2/2}\big) \tag{10.2.166}$$

for $n \ge 0$. For more information on the facts presented in this part we refer to [1].

11. The method of space change

In this section we adopt the setting and notation from Section 8 above. Given two state spaces E_1 and E_2, any measurable function $C : E_1 \to E_2$ is called a *space change*. It turns out that such space changes sometimes prove useful in solving optimal stopping problems. In this section we will discuss two simple examples of this type. It is important to notice (and keep in mind) that any change of space can be performed either on the process (probabilistic transformation) or on the equation of the infinitesimal generator (analytic transformation). The two transformations stand in one-to-one correspondence to each other and yield equivalent conclusions.

11.1. Description of the method

To illustrate two examples of space change, let us assume that $X = (X_t)_{t \ge 0}$ is a one-dimensional diffusion process solving

$$dX_t = \rho(X_t)\, dt + \sigma(X_t)\, dB_t \tag{11.1.1}$$

and let us consider the optimal stopping problem

$$V(x) = \sup_\tau \mathsf{E}_x G(X_\tau) \tag{11.1.2}$$

where the supremum is taken over all stopping times τ of X and $X_0 = x$ under P_x with $x \in \mathbb{R}$.

1. Change of scale. Given a strictly increasing smooth function $C : \mathbb{R} \to \mathbb{R}$, set

$$Z_t = C(X_t) \tag{11.1.3}$$

and note that we can write

$$G(X_t) = G\big(C^{-1} \circ C(X_t)\big) = (G \circ C^{-1})(Z_t) = \widetilde{G}(Z_t) \tag{11.1.4}$$

for $t \geq 0$ where we denote

$$\widetilde{G}(z) = G(C^{-1}(z)) \tag{11.1.5}$$

for $z \in \mathbb{R}$ (in the image of C). By Itô's formula (page 67) we get

$$C(X_t) = C(X_0) + \int_0^t (\mathbb{L}_X C)(X_s)\, ds + \int_0^t C'(X_s)\sigma(X_s)\, dB_s \tag{11.1.6}$$

or equivalently

$$Z_t = Z_0 + \int_0^t (\mathbb{L}_X C)(C^{-1}(Z_s))\, ds + \int_0^t C'(C^{-1}(Z_s))\sigma(C^{-1}(Z_s))\, dB_s \tag{11.1.7}$$

for $t \geq 0$ upon recalling that $\mathbb{L}_X C = \rho\, C' + (\sigma^2/2)\, C''$. From (11.1.4) and (11.1.7) we see that the problem (11.1.2) is equivalent to the following problem:

$$\widetilde{V}(z) = \sup_\tau \mathsf{E}_z \widetilde{G}(Z_\tau) \tag{11.1.8}$$

where $Z = (Z_t)_{t \geq 0}$ is a new one-dimensional diffusion process solving

$$dZ_t = \widetilde{\rho}(Z_t)\, dt + \widetilde{\sigma}(Z_t)\, dB_t \tag{11.1.9}$$

with $Z_0 = z$ under P_z and:

$$\widetilde{\rho} = (\mathbb{L}_X C) \circ C^{-1}, \tag{11.1.10}$$

$$\widetilde{\sigma} = (C'\sigma) \circ C^{-1}. \tag{11.1.11}$$

For some C the process Z may be simpler than the initial process X and this in turn may lead to a solution of the problem (11.1.8). This solution is then readily transformed back to a solution of the initial problem (11.1.2) using (11.1.3).

The best known example of such a function C is the scale function $S : \mathbb{R} \to \mathbb{R}$ of X solving

$$\mathbb{L}_X S = 0. \tag{11.1.12}$$

This yields the following explicit expression:

$$S(x) = \int_{\cdot}^{x} \exp\Big(- \int_{\cdot}^{y} \frac{2\rho(z)}{\sigma^2(z)} \, dz \Big) \, dy \tag{11.1.13}$$

which is determined uniquely up to an affine transformation ($\widetilde{S} = aS + b$ is also a solution to (11.1.12) when $a, b \in \mathbb{R}$). From (11.1.7) one sees that $Z = S(X)$ is a continuous (local) martingale (since the first integral term vanishes). The martingale property may then prove helpful in the search for a solution to (11.1.8). Moreover, yet another step may be to time change Z and reduce the setting to a standard Brownian motion as indicated in (8.1.3)–(8.1.5).

2. Change of variables. Assuming that G is smooth (e.g. C^2) we find by Itô's formula (page 67) that

$$G(X_t) = G(X_0) + \int_0^t (\mathbb{L}_X G)(X_s) \, ds + \int_0^t G'(X_s)\sigma(X_s) \, dB_s \tag{11.1.14}$$

where $M_t := \int_0^t G'(X_s)\sigma(X_s) \, dB_s$ is a continuous (local) martingale for $t \ge 0$. By the optional sampling theorem (page 60) upon localization if needed, we may conclude that $\mathsf{E}_x M_\tau = 0$ for all stopping times τ satisfying $\mathsf{E}_x \sqrt{\tau} < \infty$, given that G' and σ satisfy certain integrability conditions (e.g. both being bounded). In this case it follows from (11.1.14) that

$$\mathsf{E}_x G(X_\tau) = G(x) + \mathsf{E}_x \Big(\int_0^\tau (\mathbb{L}_X G)(X_s) \, ds \Big) \tag{11.1.15}$$

for all stopping times τ of X satisfying $\mathsf{E}_x \sqrt{\tau} < \infty$. Setting

$$L = \mathbb{L}_X G \tag{11.1.16}$$

we see that the Mayer formulated problem (11.1.2) is equivalent to the Lagrange formulated problem

$$\bar{V}(x) = \sup_\tau \mathsf{E}_x \Big(\int_0^\tau L(X_t) \, dt \Big) \tag{11.1.17}$$

where the supremum is taken over all stopping times τ of X. Moreover, if we are given $G + M$ instead of G in (11.1.2), where $M : \mathbb{R} \to \mathbb{R}$ is a measurable function satisfying the usual integrability condition, then (11.1.2) is equivalent to the Bolza formulated problem

$$\bar{V}(x) = \sup_\tau \mathsf{E}_x \Big(M(X_\tau) + \int_0^\tau L(X_t) \, dt \Big) \tag{11.1.18}$$

where the supremum is taken as in (11.1.17) above.

It should be noted that the underlying Markov process X in (11.1.2) transforms into the underlying Markov process (X, I) in (11.1.18) where $I_t = \int_0^t L(X_s)\,ds$ for $t \geq 0$. The latter process is more complicated and frequently, when (11.1.18) is to be solved, one applies the preceding transformation to reduce this problem to problem (11.1.2). One simple example of this type will be given in the next subsection. It should also be noted however that when we are given $N(I_\tau)$ instead of $I_\tau = \int_0^\tau L(X_t)\,dt$ in (11.1.18), where N is a nonlinear function (e.g. $N(x) = \sqrt{x}$), then the preceding transformation is generally not applicable (cf. Section 20 below). While in the former case ($N(x) = x$) we speak of *linear problems*, in the latter case we often speak of *nonlinear problems*.

11.2. Problems and solutions

Let us illustrate the preceding transformation (change of variables) with one simple example.

Example 11.1. Consider the optimal stopping problem

$$V = \sup_\tau \mathsf{E}\left(|B_\tau| - \tau\right) \tag{11.2.1}$$

where the supremum is taken over all stopping times τ of the standard Brownian motion B satisfying $\mathsf{E}\tau < \infty$. By Itô's formula (page 67) applied to $F(B_t) = B_t^2$, and the optional sampling theorem (page 60), we know that

$$\mathsf{E}\tau = \mathsf{E}B_\tau^2 \tag{11.2.2}$$

whenever $\mathsf{E}\tau < \infty$. It follows that problem (11.2.1) is equivalent to the problem

$$\bar{V} = \sup_\tau \mathsf{E}\left(|B_\tau| - |B_\tau|^2\right) \tag{11.2.3}$$

where the supremum is taken as in (11.2.1). Setting $Z = |B_\tau|$ with τ as above, we clearly have

$$\mathsf{E}\left(|B_\tau| - |B_\tau|^2\right) = \int_0^\infty (z - z^2)\,d\mathsf{P}_Z(z). \tag{11.2.4}$$

Observing that the function $z \mapsto z - z^2$ has a unique maximum on $[0, \infty)$ attained at $z_* = \frac{1}{2}$, we see that the supremum over all Z in (11.2.4) is attained at $Z \equiv \frac{1}{2}$. Recalling that $Z = |B_\tau|$ we see that

$$\tau_* = \inf\{\, t \geq 0 : |B_t| = 1/2 \,\} \tag{11.2.5}$$

is an optimal stopping time in (11.2.3). It follows that τ_* is an optimal stopping time in the initial problem (11.2.1), and we have $V = \frac{1}{2} - \frac{1}{4} = \frac{1}{4}$ as is seen from (11.2.4).

Further examples of this kind will be studied in Section 16 below.

12. The method of measure change

In this section we will adopt the setting and notation from Section 8 above. The basic idea of the method of measure change is to reduce the dimension of the problem by replacing the initial probability measure with a new probability measure which preserves the Markovian setting. Such a replacement is most often not possible but in some cases it works.

12.1. Description of the method

To illustrate the method in a general setting, let us assume that X is a one-dimensional diffusion process solving

$$dX_t = \rho(X_t)\,dt + \sigma(X_t)\,dB_t \tag{12.1.1}$$

and let us consider the optimal stopping problem

$$V = \sup_{0\le\tau\le T} \mathsf{E}\, G(Z_\tau) \tag{12.1.2}$$

where the supremum is taken over all stopping times τ of Z and G is a measurable function satisfying needed regularity conditions. Recall that $Z = (I, X, S)$ is a three-dimensional (strong) Markov process where I is the integral process of X and S is the maximum process of X (see (6.0.2) and (6.0.3)).

Introduce the exponential martingale

$$\mathcal{E}_t = \exp\Big(\int_0^t H_s\,dB_s - \frac{1}{2}\int_0^t H_s^2\,ds\Big) \tag{12.1.3}$$

for $t \ge 0$ where H is a suitable process making (12.1.3) well defined (and satisfying e.g. the Novikov condition $\mathsf{E}\exp\big(\frac{1}{2}\int_0^T H_s^2\,ds\big) < \infty$ which implies the martingale property). Rewrite the expectation in (12.1.2) as follows (when possible):

$$\mathsf{E}\, G(Z_\tau) = \mathsf{E}\Big(\mathcal{E}_\tau\,\frac{G(Z_\tau)}{\mathcal{E}_\tau}\Big) = \widetilde{\mathsf{E}}\Big(\frac{G(Z_\tau)}{\mathcal{E}_\tau}\Big) = \widetilde{\mathsf{E}}\,\widetilde{G}(Y_\tau) \tag{12.1.4}$$

where the symbol $\widetilde{\mathsf{E}}$ denotes the expectation under a new probabilities measure $\widetilde{\mathsf{P}}$ given by

$$d\widetilde{\mathsf{P}} = \mathcal{E}_T\,d\mathsf{P}; \tag{12.1.5}$$

the symbol $\widetilde{G}$ denotes a new gain function, and Y is a (strong) Markov process. Clearly, finding H which makes the latter possible is the key issue which makes the method applicable or not.

When this is possible (see Example 12.1 below to see how $\tilde{G}$ and Y can be chosen as suggested) it follows that problem (12.1.2) is equivalent to the problem

$$\widetilde{V} = \sup_{0\le\tau\le T} \widetilde{\mathsf{E}}\,\widetilde{G}(Y_\tau) \tag{12.1.6}$$

in the sense that having a solution to (12.1.6) we can reconstruct the corresponding solution to (12.1.2) using (12.1.5), and vice versa. The advantage of problem (12.1.6) over problem (12.1.2) is that the former is often only one-dimensional while the latter (i.e. the initial problem) may be two- or three-dimensional (recall our discussion in Subsection 6.2).

12.2. Problems and solutions

Let us illustrate the preceding discussion by one example.

Example 12.1. Consider the optimal stopping problem

$$V = \sup_{\tau} \mathsf{E}\left(e^{-\lambda\tau}(I_\tau - X_\tau)\right) \tag{12.2.1}$$

where X is a geometric Brownian motion solving

$$dX_t = \rho X_t\,dt + \sigma X_t\,dB_t \tag{12.2.2}$$

with $X_0 = 1$ and I is the integral process of X given by

$$I_t = \int_0^t X_s\,ds \tag{12.2.3}$$

for $t \ge 0$. In (12.2.1) and (12.2.2) we assume that $\lambda > 0$, $\rho \in \mathbb{R}$, $\sigma > 0$ and B is a standard Brownian motion.

Recall that the unique (strong) solution to (12.2.2) is given by

$$X_t = \exp\big(\sigma B_t + (\rho - \sigma^2/2)t\big) = e^{\rho t}\mathcal{E}_t \tag{12.2.4}$$

where $\mathcal{E}_t = \exp\big(\sigma B_t - (\sigma^2/2)t\big)$ is an exponential martingale for $t \ge 0$. It follows that (12.2.1) can be rewritten as follows:

$$\begin{aligned} V &= \sup_{\tau} \mathsf{E}\left(e^{-\lambda\tau}(I_\tau - X_\tau)\right) = \sup_{\tau} \mathsf{E}\left(X_\tau e^{-\lambda\tau}(I_\tau/X_\tau - 1)\right) \\ &= \sup_{\tau} \mathsf{E}\left(e^{-(\lambda-\rho)\tau}\mathcal{E}_\tau(I_\tau/X_\tau - 1)\right) = \sup_{\tau} \widetilde{\mathsf{E}}\left(e^{-r\tau}(Y_\tau - 1)\right) \end{aligned} \tag{12.2.5}$$

where $d\widetilde{\mathsf{P}} = \mathcal{E}_T\,d\mathsf{P}$, we set $r = \lambda - \rho$, and $Y_t = I_t/X_t$ for $t \ge 0$.

It turns out that Y is a (strong) Markov process. To verify the Markov property note that for

$$Y_t^y = \frac{y + I_t}{X_t} = \frac{1}{X_t}\left(y + \int_0^t X_s\,ds\right) \tag{12.2.6}$$

with $y \in \mathbb{R}$ we have

$$Y^y_{t+h} = \frac{y + \int_0^t X_s\,ds + \int_t^{t+h} X_s\,ds}{\exp\big(\sigma(B_{t+h} - B_t) + \sigma B_t + \tilde{\rho}(t+h-t) + \tilde{\rho}t\big)} \tag{12.2.7}$$
$$= \frac{1}{\exp(\sigma\widetilde{B}_h + \tilde{\rho}h)} \left[\frac{1}{X_t}\Big(y + \int_0^t X_s\,ds\Big) + \int_t^{t+h} \exp\big(\sigma(B_s - B_t) + \tilde{\rho}(s-t)\big)\,ds\right]$$

where $\tilde{\rho} = \rho - \sigma^2/2$ and $\widetilde{B}_h = B_{t+h} - B_t$ is a standard Brownian motion independent from $\mathcal{F}^X_t$ for $h \ge 0$ and $t \ge 0$. (The latter conclusion makes use of stationary independent increments of B.) From the final expression in (12.2.7) upon recalling (12.2.6) it is evident that Y is a (strong) Markov process. Moreover, using Itô's formula (page 67) it is easily checked that Y solves

$$dY_t = \big(1 + (\sigma^2 - \rho)\big)Y_t\,dt + \sigma Y_t\,d\widehat{B}_t \tag{12.2.8}$$

where $\widehat{B} = -B$ is also a standard Brownian motion. It follows that the infinitesimal generator of Y is given by

$$\mathbb{L}_Y = \big(1 + (\sigma^2 - \rho)\big)\frac{\partial}{\partial y} + \frac{\sigma^2 y^2}{2}\frac{\partial^2}{\partial y^2} \tag{12.2.9}$$

and the problem (12.2.5) can be treated by standard one-dimensional techniques (at least when the horizon is infinite).

Further examples of this kind (involving the maximum process too) will be studied in Sections 26 and 27.

13. Optimal stopping of the maximum process

13.1. Formulation of the problem

Let $X = (X_t)_{t\ge0}$ be a one-dimensional time-homogeneous *diffusion* process associated with the infinitesimal generator

$$\mathbb{L}_X = \rho(x)\frac{\partial}{\partial x} + \frac{\sigma^2(x)}{2}\frac{\partial^2}{\partial x^2} \tag{13.1.1}$$

where the *drift* coefficient $x \mapsto \rho(x)$ and the *diffusion* coefficient $x \mapsto \sigma(x) > 0$ are continuous. Assume moreover that there exists a standard *Brownian motion* $B = (B_t)_{t\ge0}$ defined on $(\Omega, \mathcal{F}, \mathsf{P})$ such that X solves the stochastic differential equation

$$dX_t = \rho(X_t)\,dt + \sigma(X_t)\,dB_t \tag{13.1.2}$$

with $X_0 = x$ under $\mathsf{P}_x := \text{Law}(X \mid \mathsf{P}, X_0 = x)$ for $x \in \mathbb{R}$. The state space of X is assumed to be $\mathbb{R}$.

With X we associate the maximum process

$$S_t = \Big(\max_{0 \le r \le t} X_r \Big) \vee s \tag{13.1.3}$$

started at $s \ge x$ under $\mathsf{P}_{x,s} := \text{Law}(X, S \mid \mathsf{P}, X_0 = x, S_0 = s)$. The main objective of this section is to present the solution to the optimal stopping problem with the value function

$$V_*(x,s) = \sup_\tau \mathsf{E}_{x,s}\Big(S_\tau - \int_0^\tau c(X_t)\, dt \Big) \tag{13.1.4}$$

where the supremum is taken over stopping times τ of X satisfying

$$\mathsf{E}_{x,s}\Big(\int_0^\tau c(X_t)\, dt \Big) < \infty, \tag{13.1.5}$$

and the *cost* function $x \mapsto c(x) > 0$ is continuous.

1. To state and prove the initial observation about (13.1.4), and for further reference, we need to recall a few general facts about one-dimensional diffusions (recall Subsection 4.5 and see e.g. [178, p. 270–303] for further details).

The *scale function* of X is given by

$$L(x) = \int^x \exp\Big(-\int^y \frac{2\rho(z)}{\sigma^2(z)}\, dz \Big)\, dy \tag{13.1.6}$$

for $x \in \mathbb{R}$. Throughout we denote

$$\tau_x = \inf\{\, t > 0 : X_t = x \,\} \tag{13.1.7}$$

and set $\tau_{x,y} = \tau_x \wedge \tau_y$. Then we have

$$\mathsf{P}_x\big(X_{\tau_{a,b}} = a\big) = \frac{L(b) - L(x)}{L(b) - L(a)}, \tag{13.1.8}$$

$$\mathsf{P}_x\big(X_{\tau_{a,b}} = b\big) = \frac{L(x) - L(a)}{L(b) - L(a)} \tag{13.1.9}$$

whenever $a \le x \le b$.

The *speed measure* of X is given by

$$m(dx) = \frac{2\, dx}{L'(x)\, \sigma^2(x)}. \tag{13.1.10}$$

The *Green function* of X on $[a,b]$ is defined by

$$G_{a,b}(x,y) = \begin{cases} \dfrac{(L(b) - L(x))(L(y) - L(a))}{(L(b) - L(a))} & \text{if } a \le y \le x, \\[2ex] \dfrac{(L(b) - L(y))(L(x) - L(a))}{(L(b) - L(a))} & \text{if } x \le y \le b. \end{cases} \tag{13.1.11}$$

If $f : \mathbb{R} \mapsto \mathbb{R}$ is a measurable function, then

$$\mathsf{E}_x\Big(\int_0^{\tau_{a,b}} f(X_t)\,dt\Big) = \int_a^b f(y)G_{a,b}(x,y)\,m(dy). \tag{13.1.12}$$

2. Due to the specific form of the optimal stopping problem (13.1.4), the following observation is nearly evident (see [45, p. 237–238]).

Proposition 13.1. *The process* $\bar{X}_t = (X_t, S_t)$ *cannot be optimally stopped on the diagonal of* $\mathbb{R}^2$.

Proof. Fix $x \in \mathbb{R}$, and set $l_n = x - 1/n$ and $r_n = x + 1/n$. Denoting $\tau_n = \tau_{l_n,r_n}$ it will be enough to show that

$$\mathsf{E}_{x,x}\Big(S_{\tau_n} - \int_0^{\tau_n} c(X_t)\,dt\Big) > x \tag{13.1.13}$$

for $n \geq 1$ large enough.

For this, note first by the strong Markov property and (13.1.8)–(13.1.9) that

$$\begin{aligned}\mathsf{E}_{x,x}(S_{\tau_n}) &\geq x\mathsf{P}_x(X_{\tau_n} = l_n) + r_n\mathsf{P}_x(X_{\tau_n} = r_n) \\ &= x\,\frac{L(r_n) - L(x)}{L(r_n) - L(l_n)} + r_n\,\frac{L(x) - L(l_n)}{L(r_n) - L(l_n)} \\ &= x + (r_n - x)\,\frac{L(x) - L(l_n)}{L(r_n) - L(l_n)} \\ &= x + (r_n - x)\,\frac{L'(\xi_n)(x - l_n)}{L'(\eta_n)(r_n - l_n)} \geq x + \frac{K}{n}\end{aligned} \tag{13.1.14}$$

since $L \in C^1$. On the other hand $K_1 := \sup_{l_n \leq z \leq r_n} c(z) < \infty$. Thus by (13.1.10)–(13.1.12) we get

$$\begin{aligned}\mathsf{E}_{x,x}\Big(\int_0^{\tau_n} c(X_t)\,dt\Big) &\leq K_1\mathsf{E}_x\tau_n = 2K_1\int_{l_n}^{r_n} G_{a,b}(x,y)\,\frac{dy}{\sigma^2(y)L'(y)} \\ &\leq K_2\Big(\int_{l_n}^{x} (L(y) - L(l_n))\,dy + \int_x^{r_n} (L(r_n) - L(y))\,dy\Big) \\ &\leq K_3\Big((x - l_n)^2 + (r_n - x)^2\Big) = \frac{2K_3}{n^2}\end{aligned} \tag{13.1.15}$$

since σ is continuous and $L \in C^1$. Combining (13.1.14) and (13.1.15) we clearly obtain (13.1.13) for $n \geq 1$ large enough. The proof is complete. □

13.2. Solution to the problem

In the setting of (13.1.1)–(13.1.3) consider the optimal stopping problem (13.1.4) where the supremum is taken over all stopping times τ of X satisfying (13.1.5).

Our main aim in this subsection is to present the solution to this problem (Theorem 13.2). We begin our exposition with a few observations on the underlying structure of (13.1.4) with a view to the Markovian theory of optimal stopping (Chapter I).

1. Note that $\bar{X}_t = (X_t, S_t)$ is a two-dimensional Markov process with the state space $D = \{ (x,s) \in \mathbb{R}^2 : x \le s \}$, which can change (increase) in the second coordinate only after hitting the diagonal $x = s$ in $\mathbb{R}^2$. Off the diagonal, the process $\bar{X} = (\bar{X}_t)_{t \ge 0}$ changes only in the first coordinate and may be identified with X. Due to its form and behaviour at the diagonal, we claim that the infinitesimal generator of $\bar{X}$ may thus be formally described as follows:

$$\mathbb{L}_{\bar{X}} = \mathbb{L}_X \quad \text{for } x < s, \tag{13.2.1}$$

$$\frac{\partial}{\partial s} = 0 \quad \text{at } x = s \tag{13.2.2}$$

with $\mathbb{L}_X$ as in (13.1.1). This means that the infinitesimal generator of $\bar{X}$ is acting on a space of C^2-functions f on D satisfying $(\partial f/\partial s)(s,s) = 0$. Observe that we do not tend to specify the domain of $\mathbb{L}_{\bar{X}}$ precisely, but will only verify that if $f : D \to \mathbb{R}$ is a C^2-function which belongs to the domain, then $(\partial f/\partial s)(s,s)$ must be zero.

To see this, we shall apply Itô's formula (page 67) to the process $f(X_t, S_t)$ and take the expectation under $\mathsf{P}_{s,s}$. By the optional sampling theorem (page 60) being applied to the continuous local martingale which appears in this process (localized if needed), we obtain

$$\begin{aligned} \frac{\mathsf{E}_{s,s} f(X_t, S_t) - f(s,s)}{t} &= \mathsf{E}_{s,s}\left(\frac{1}{t} \int_0^t (\mathbb{L}_X f)(X_r, S_r)\, dr \right) \\ &\quad + \mathsf{E}_{s,s}\left(\frac{1}{t} \int_0^t \frac{\partial f}{\partial s}(X_r, S_r)\, dS_r \right) \\ &\longrightarrow \mathbb{L}_X f(s,s) + \frac{\partial f}{\partial s}(s,s) \left(\lim_{t \downarrow 0} \frac{\mathsf{E}_{s,s}(S_t - s)}{t} \right) \end{aligned} \tag{13.2.3}$$

as $t \downarrow 0$. Due to $\sigma > 0$, we have $t^{-1} \mathsf{E}_{s,s}(S_t - s) \to \infty$ as $t \downarrow 0$, and therefore the limit above is infinite, unless $(\partial f/\partial s)(s,s) = 0$. This completes the claim (see also [45, p. 238–239]).

2. The problem (13.1.4) can be considered as a standard (i.e. of type (11.1.2)) optimal stopping problem for a d-dimensional Markov process by introducing the functional

$$A_t = a + \int_0^t c(X_r)\, dr \tag{13.2.4}$$

with $a \ge 0$ given and fixed, and noting that $Z_t = (A_t, X_t, S_t)$ is a Markov process which starts at (a, x, s) under P. Its infinitesimal generator is obtained by adding $c(x)\,(\partial/\partial a)$ to the infinitesimal generator of $\bar{X}$, which combined with (13.2.1) leads to the formal description

$$\mathbb{L}_Z = c(x)\,\frac{\partial}{\partial a} + \mathbb{L}_X \quad \text{in} \;\; x<s, \tag{13.2.5}$$
$$\frac{\partial}{\partial s} = 0 \quad \text{at} \;\; x=s$$

with $\mathbb{L}_X$ as in (13.1.1). Given $Z=(Z_t)_{t\ge 0}$, introduce the *gain function* $G(a,x,s) = s-a$, note that the value function (13.1.4) viewed in terms of the general theory ought to be defined as

$$\widetilde{V}_*(a,x,s) = \sup_{\tau} \mathsf{E}\, G(Z_\tau) \tag{13.2.6}$$

where the supremum is taken over all stopping times τ of Z satisfying $\mathsf{E}\, A_\tau < \infty$, and observe that

$$\widetilde{V}_*(a,x,s) = V_*(x,s) - a \tag{13.2.7}$$

where $V_*(x,s)$ is defined in (13.1.4). This identity is the main reason that we abandon the general formulation (13.2.6) and simplify it to the form (13.1.4), and that we speak of optimal stopping for the process $\bar{X}_t = (X_t, S_t)$ rather than the process $Z_t = (A_t, X_t, S_t)$.

Let us point out that the contents of this paragraph are used in the sequel merely to clarify the result and method in terms of the general theory (recall Section 6 above).

3. From now on our main aim will be to show that the problem (13.1.4) reduces to the problem of solving a first-order nonlinear differential equation (for the optimal stopping boundary). To derive this equation we shall first try to get a feeling for the points in the state space $\{\,(x,s)\in\mathbb{R}^2 : x\le s\,\}$ at which the process $\bar{X}_t = (X_t, S_t)$ can be optimally stopped (recall *Figure* 1 on page xviii above).

When on the horizontal level s, the process $\bar{X}_t = (X_t, S_t)$ stays at the same level until it hits the diagonal $x=s$ in $\mathbb{R}^2$. During that time $\bar{X}$ does not change (increase) in the second coordinate. Due to the strictly positive cost in (13.1.4), it is clear that we should not let the process $\bar{X}$ run too much to the left, since it could be "too expensive" to get back to the diagonal in order to offset the "cost" spent to travel all that way. More specifically, given s there should exist a point $g_*(s)\le s$ such that if the process (X,S) reaches the point $(g_*(s),s)$ we should stop it instantly. In other words, the stopping time

$$\tau_* = \inf\{\, t>0 : X_t \le g_*(S_t)\,\} \tag{13.2.8}$$

should be optimal for the problem (13.1.4). For this reason we call $s\mapsto g_*(s)$ an *optimal stopping boundary*, and our aim will be to prove its existence and to characterize it. Observe by Proposition 13.1 that we must have $g_*(s)<s$ for all s, and that $V_*(x,s)=s$ for all $x\le g_*(s)$.

4. To compute the value function $V_*(x,s)$ for $g_*(s) < x \le s$, and to find the optimal stopping boundary $s \mapsto g_*(s)$, we are led (recall Section 6 above) to formulate the following system:

$$(\mathbb{L}_X V)(x,s) = c(x) \qquad \text{for } g(s) < x < s \text{ with } s \text{ fixed}, \tag{13.2.9}$$

$$\frac{\partial V}{\partial s}(x,s)\Big|_{x=s-} = 0 \qquad (\textit{normal reflection}), \tag{13.2.10}$$

$$V(x,s)\big|_{x=g(s)+} = s \qquad (\textit{instantaneous stopping}), \tag{13.2.11}$$

$$\frac{\partial V}{\partial x}(x,s)\Big|_{x=g(s)+} = 0 \qquad (\textit{smooth fit}) \tag{13.2.12}$$

with $\mathbb{L}_X$ as in (13.1.1). Note that (13.2.9)–(13.2.10) are in accordance with the general theory (Section 6) upon using (13.2.5) and (13.2.7) above: the infinitesimal generator of the process being applied to the value function must be zero in the continuation set. The condition (13.2.11) is evident. The condition (13.2.12) is not part of the general theory; it is imposed since we believe that in the "smooth" setting of the problem (13.1.4) the principle of smooth fit should hold (recall Section 6 above). This belief will be vindicated after the fact, when we show in Theorem 13.2.1, that the solution of the system (13.2.9)–(13.2.12) leads to the value function of (13.1.4). The system (13.2.9)–(13.2.12) constitutes a free-boundary problem (see Chapter III above). It was derived for the first time by Dubins, Shepp and Shiryaev [45] in the case of Bessel processes.

5. To solve the system (13.2.9)–(13.2.12) we shall consider a stopping time of the form

$$\tau_g = \inf\{\, t>0 : X_t \le g(S_t)\,\} \tag{13.2.13}$$

and the map

$$V_g(x,s) = \mathsf{E}_{x,s}\Big(S_{\tau_g} - \int_0^{\tau_g} c(X_t)\,dt \Big) \tag{13.2.14}$$

associated with it, where $s \mapsto g(s)$ is a given function such that both $\mathsf{E}_{x,s} S_{\tau_g}$ and $\mathsf{E}_{x,s}(\int_0^{\tau_g} c(X_t)\,dt)$ are finite. Set $V_g(s) := V_g(s,s)$ for all s. Considering $\tau_{g(s),s} = \inf\{\, t>0 : X_t \notin (g(s),s)\}$ and using the strong Markov property of X at $\tau_{g(s),s}$, by (13.1.8)–(13.1.12) we find

$$\begin{aligned} V_g(x,s) = {} & s\,\frac{L(s)-L(x)}{L(s)-L(g(s))} + V_g(s)\,\frac{L(x)-L(g(s))}{L(s)-L(g(s))} \\ & - \int_{g(s)}^{s} G_{g(s),s}(x,y)\,c(y)\,m(dy) \end{aligned} \tag{13.2.15}$$

for all $g(s) < x < s$.

In order to determine $V_g(s)$, we shall rewrite (13.2.15) as follows:

$$V_g(s) - s \tag{13.2.16}$$
$$= \frac{L(s) - L(g(s))}{L(x) - L(g(s))} \left((V_g(x,s) - s) + \int_{g(s)}^{s} G_{g(s),s}(x,y)\, c(y)\, m(dy) \right)$$

and then divide and multiply through by $x - g(s)$ to obtain

$$\lim_{x \downarrow g(s)} \frac{V_g(x,s) - s}{L(x) - L(g(s))} = \frac{1}{L'(g(s))} \frac{\partial V_g}{\partial x}(x,s)\Big|_{x=g(s)+}. \tag{13.2.17}$$

It is easily seen by (13.2.12) that

$$\lim_{x \downarrow g(s)} \frac{L(s) - L(g(s))}{L(x) - L(g(s))} \int_{g(s)}^{s} G_{g(s),s}(x,y)\, c(y)\, m(dy) \tag{13.2.18}$$
$$= \int_{g(s)}^{s} \big(L(s) - L(y)\big)\, c(y)\, m(dy).$$

Thus, if *the condition of smooth fit*

$$\frac{\partial V_g}{\partial x}(x,s)\Big|_{x=g(s)+} = 0 \tag{13.2.19}$$

is satisfied, we see from (13.2.16)–(13.2.18) that the following identity holds:

$$V_g(s) = s + \int_{g(s)}^{s} \big(L(s) - L(y)\big)\, c(y)\, m(dy). \tag{13.2.20}$$

Inserting this into (13.2.15), and using (13.1.11)–(13.1.12), we get

$$V_g(x,s) = s + \int_{g(s)}^{x} \big(L(x) - L(y)\big)\, c(y)\, m(dy) \tag{13.2.21}$$

for all $g(s) \le x \le s$.

If we now forget the origin of $V_g(x,s)$ in (13.2.14), and consider it purely as defined by (13.2.21), then it is straightforward to verify that $(x,s) \mapsto V_g(x,s)$ solves the system (13.2.9)–(13.2.12) in the region $g(s) < x < s$ if and only if the C^1 -function $s \mapsto g(s)$ solves the following first-order nonlinear differential equation:

$$g'(s) = \frac{\sigma^2(g(s))\, L'(g(s))}{2\, c(g(s))\, [L(s) - L(g(s))]}\, . \tag{13.2.22}$$

Thus, to each solution $s \mapsto g(s)$ of the equation (13.2.22) corresponds a function $(x,s) \mapsto V_g(x,s)$ defined by (13.2.21) which solves the system (13.2.9)–(13.2.12) in the region $g(s) < x < s$, and coincides with the expectation in (13.2.14) whenever $\mathsf{E}_{x,s} S_{\tau_g}$ and $\mathsf{E}_{x,s} \int_0^{\tau_g} c(X_t)\, dt$ are finite (the latter is easily verified by Itô's formula). We shall use this fact in the proof of Theorem 13.2 below upon approximating the selected solution of (13.2.22) by solutions which hit the diagonal in $\mathbb{R}^2$.

6. Observe that among all possible functions $s \mapsto g(s)$, only those which satisfy (13.2.22) lead to the smooth-fit property (13.2.19) for $V_g(x,s)$ of (13.2.14), and vice versa. Thus the *differential equation* (13.2.22) *is obtained by the principle of smooth fit* in the problem (13.1.4). The fundamental question to be answered is how to choose the optimal stopping boundary $s \mapsto g_*(s)$ among all admissible candidates which solve (13.2.22).

Before passing to answer this question let us also observe from (13.2.21) that

$$\frac{\partial V_g}{\partial x}(x,s) = L'(x) \int_{g(s)}^{x} c(y)\, m(dy), \tag{13.2.23}$$

$$V_g'(s) = L'(s) \int_{g(s)}^{s} c(y)\, m(dy). \tag{13.2.24}$$

These equations show that, in addition to the continuity of the derivative of $V_g(x,s)$ along the vertical line across $g(s)$ in (13.2.19), we have obtained the continuity of $V_g(x,s)$ along the vertical line and the diagonal in $\mathbb{R}^2$ across the point where they meet. In fact, we see that the latter condition is equivalent to the former, and thus may be used as an alternative way of looking at the principle of smooth fit in this problem.

7. In view of the analysis of (13.2.8), we assign a constant value to $V_g(x,s)$ at all $x < g(s)$. The following properties of the solution $V_g(x,s)$ obtained are then straightforward:

$$V_g(x,s) = s \quad \text{for } x \le g(s), \tag{13.2.25}$$

$$x \mapsto V_g(x,s) \quad \text{is (strictly) increasing on } [\,g(s), s], \tag{13.2.26}$$

$$(x,s) \mapsto V_g(x,s) \quad \text{is } C^2 \text{ outside } \{(g(s),s) : s \in \mathbb{R}\}, \tag{13.2.27}$$

$$x \mapsto V_g(x,s) \quad \text{is } C^1 \text{ at } g(s). \tag{13.2.28}$$

Let us also make the following observations:

$g \mapsto V_g(x,s)$ is (strictly) *decreasing*. (13.2.29)

The function $(a,x,s) \mapsto V_g(x,s) - a$ is *superharmonic* for the Markov process $Z_t = (A_t, X_t, S_t)$ (with respect to stopping times τ satisfying (13.1.5)). (13.2.30)

The property (13.2.29) is evident from (13.2.21), whereas (13.2.30) is derived in the proof of Theorem 13.2 (see (13.2.38) below).

8. Combining (13.2.7) and (13.2.29)–(13.2.30) with the superharmonic characterization of the value function from the Markovian theory (see Theorem 2.4

and Theorem 2.7), and recalling the result of Proposition 13.1, we are led to the following *maximality principle* for determining the optimal stopping boundary (we say that $s \mapsto g_*(s)$ is an *optimal stopping boundary* for the problem (13.1.4), if the stopping time τ_* defined in (13.2.8) is optimal for this problem).

The Maximality Principle. *The optimal stopping boundary $s \mapsto g_*(s)$ for the problem* (13.1.4) *is the maximal solution of the differential equation* (13.2.22) *satisfying $g_*(s) < s$ for all s.*

This principle is equivalent to the superharmonic characterization of the value function (for the process $Z_t = (A_t, X_t, S_t)$), and may be viewed as its alternative (analytic) description. The proof of its validity is given in the next theorem, the main result of the subsection. (For simplicity of terminology we shall say that a function $g = g(s)$ is an *admissible function* if $g(s) < s$ for all s.)

Theorem 13.2. (Optimal stopping of the maximum process) *In the setting of* (13.1.1)–(13.1.3) *consider the optimal stopping problem* (13.1.4) *where the supremum is taken over all stopping times τ of X satisfying* (13.1.5).

(I): *Let $s \mapsto g_*(s)$ denote the maximal admissible solution of* (13.2.22) *whenever such a solution exists* (*see Figure* IV.11). Then we have:

1. *The value function is finite and is given by*

$$V_*(x,s) = s + \int_{g_*(s)}^{x} \big(L(x) - L(y)\big)\, c(y)\, m(dy) \tag{13.2.31}$$

for $g_*(s) \le x \le s$ and $V_*(x,s) = s$ for $x \le g_*(s)$.

2. *The stopping time*

$$\tau_* = \inf\{\, t > 0 : X_t \le g_*(S_t)\} \tag{13.2.32}$$

is optimal for the problem (13.1.4) *whenever it satisfies* (13.1.5); *otherwise it is "approximately" optimal in the sense described in the proof below.*

3. *If there exists an optimal stopping time σ in* (13.1.4) *satisfying* (13.1.5), *then $\mathsf{P}_{x,s}(\tau_* \le \sigma) = 1$ for all (x,s), and τ_* is an optimal stopping time for* (13.1.4) *as well.*

(II): *If there is no* (*maximal*) *admissible solution of* (13.2.22), *then $V_*(x,s) = +\infty$ for all (x,s), and there is no optimal stopping time.*

Proof. (I): Let $s \mapsto g(s)$ be any solution of (13.2.22) satisfying $g(s) < s$ for all s. Then, as indicated above, the function $V_g(x,s)$ defined by (13.2.21) solves the system (13.2.9)–(13.2.12) in the region $g(s) < x < s$. Due to (13.2.27) and (13.2.28), Itô's formula (page 67) can be applied to the process $V_g(X_t, S_t)$, and

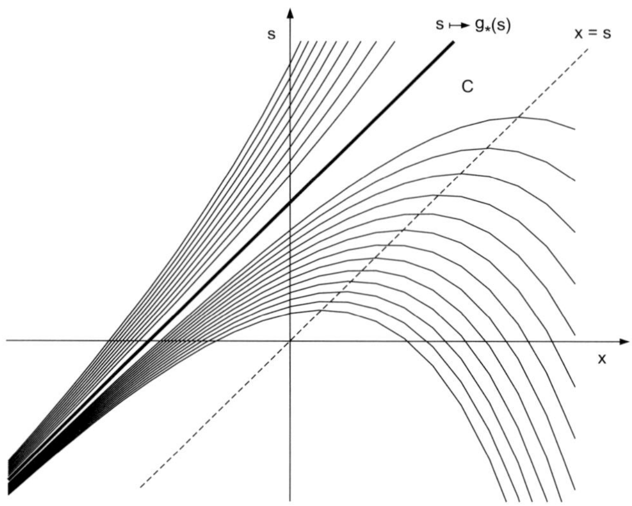

Figure IV.11: A computer drawing of solutions of the differential equation (13.2.22) in the case when $\rho \equiv 0$, $\sigma \equiv 1$ (thus $L(x) = x$) and $c \equiv 1/2$. The bold line $s \mapsto g_*(s)$ is the *maximal admissible* solution. (In this particular case $s \mapsto g_*(s)$ is a linear function.) By the maximality principle proved below, this solution is the optimal stopping boundary (the stopping time τ_* from (13.2.8) is optimal for the problem (13.1.4)).

in this way by (13.1.1)–(13.1.2) we get

$$\begin{aligned} V_g(X_t, S_t) &= V_g(x,s) + \int_0^t \frac{\partial V_g}{\partial x}(X_r, S_r)\, dX_r \\ &\quad + \int_0^t \frac{\partial V_g}{\partial s}(X_r, S_r)\, dS_r + \frac{1}{2}\int_0^t \frac{\partial^2 V_g}{\partial x^2}(X_r, S_r)\, d\langle X, X\rangle_r \\ &= V_g(x,s) + \int_0^t \sigma(X_r)\frac{\partial V_g}{\partial x}(X_r, S_r)\, dB_r + \int_0^t (\mathbb{L}_X V_g)(X_r, S_r)\, dr \end{aligned} \tag{13.2.33}$$

where the integral with respect to dS_r is zero, since the increment ΔS_r outside the diagonal in $\mathbb{R}^2$ equals zero, while at the diagonal we have (13.2.10).

The process $M = (M_t)_{t \geq 0}$ defined by

$$M_t = \int_0^t \sigma(X_r)\, \frac{\partial V_g}{\partial x}(X_r, S_r)\, dB_r \tag{13.2.34}$$

is a continuous local martingale. Introducing the increasing process

$$P_t = \int_0^t c(X_r)\, 1_{(X_r \leq g(S_r))}\, dr \tag{13.2.35}$$

and using the fact that the set of all t for which X_t is either $g(S_t)$ or S_t is of Lebesgue measure zero, the identity (13.2.33) can be rewritten as

$$V_g(X_t, S_t) - \int_0^t c(X_r)\,dr = V_g(x,s) + M_t - P_t \tag{13.2.36}$$

by means of (13.2.9) with (13.2.25). From this representation we see that the process $V_g(X_t, S_t) - \int_0^t c(X_r)\,dr$ is a local supermartingale.

Let τ be any stopping time of X satisfying (13.1.5). Choose a localization sequence $(\sigma_n)_{n\geq 1}$ of bounded stopping times for M. By means of (13.2.25) and (13.2.26) we see that $V_g(x,s) \geq s$ for all (x,s), so that from (13.2.36) it follows that

$$\begin{aligned} &\mathsf{E}_{x,s}\Big(S_{\tau\wedge\sigma_n} - \int_0^{\tau\wedge\sigma_n} c(X_t)\,dt\Big) \\ &\quad\leq \mathsf{E}_{x,s}\Big(V_g(X_{\tau\wedge\sigma_n}, S_{\tau\wedge\sigma_n}) - \int_0^{\tau\wedge\sigma_n} c(X_t)\,dt\Big) \\ &\quad\leq V_g(x,s) + \mathsf{E}_{x,s} M_{\tau\wedge\sigma_n} = V_g(x,s). \end{aligned} \tag{13.2.37}$$

Letting $n \to \infty$, and using Fatou's lemma with (13.1.5), we get

$$\mathsf{E}_{x,s}\Big(S_\tau - \int_0^\tau c(X_t)\,dt\Big) \leq V_g(x,s). \tag{13.2.38}$$

This proves (13.2.30). Taking the supremum over all such τ, and then the infimum over all such g, by means of (13.2.29) we may conclude

$$V_*(x,s) \leq \inf_g V_g(x,s) = V_{g_*}(x,s) \tag{13.2.39}$$

for all (x,s). From these considerations it clearly follows that the only possible candidate for the optimal stopping boundary is the maximal solution $s \mapsto g_*(s)$ of (13.2.22).

To prove that we have the equality in (13.2.39), and that the value function $V_*(x,s)$ is given by (13.2.31), assume first that the stopping time τ_* defined by (13.2.32) satisfies (13.1.5). Then, as pointed out when deriving (13.2.21), we have

$$V_{g_*}(x,s) = \mathsf{E}_{x,s}\Big(S_{\tau_{g_*}} - \int_0^{\tau_{g_*}} c(X_t)\,dt\Big) \tag{13.2.40}$$

so that $V_{g_*}(x,s) = V_*(x,s)$ in (13.2.39) and τ_* is an optimal stopping time. The explicit expression given in (13.2.31) is obtained by (13.2.21).

Assume now that τ_* fails to satisfy (13.1.5). Let $(g_n)_{n\geq 1}$ be a decreasing sequence of solutions of (13.2.22) satisfying $g_n(s) \downarrow g_*(s)$ as $n \to \infty$ for all s. Note that each such solution must hit the diagonal in $\mathbb{R}^2$, so the stopping times

τ_{g_n} defined as in (13.2.13) must satisfy (13.1.5). Moreover, since $S_{\tau_{g_n}}$ is bounded by a constant, we see that $V_{g_n}(x,s)$ defined as in (13.2.14) is given by (13.2.21) with $g = g_n$ for $n \ge 1$. By letting $n \to \infty$ we get

$$V_{g_*}(x,s) = \lim_{n\to\infty} V_{g_n}(x,s) = \lim_{n\to\infty} \mathsf{E}_{x,s}\Big(S_{\tau_{g_n}} - \int_0^{\tau_{g_n}} c(X_t)\,dt \Big). \tag{13.2.41}$$

This shows that the equality in (13.2.39) is attained through the sequence of stopping times $(\tau_{g_n})_{n\ge1}$, and the explicit expression in (13.2.31) is easily obtained as already indicated above.

To prove the final (uniqueness) statement, assume that σ is an optimal stopping time in (13.1.4) satisfying (13.1.5). Suppose that $\mathsf{P}_{x,s}(\sigma < \tau_*) > 0$. Note that τ_* can be written in the form

$$\tau_* = \inf\{\, t>0 : V_*(X_t, S_t) = S_t \,\} \tag{13.2.42}$$

so that $S_\sigma < V_*(X_\sigma, S_\sigma)$ on $\{\sigma < \tau_*\}$, and thus

$$\begin{aligned} \mathsf{E}_{x,s}\Big(S_\sigma - \int_0^\sigma c(X_t)\,dt \Big) &< \mathsf{E}_{x,s}\Big(V_*(X_\sigma, S_\sigma) - \int_0^\sigma c(X_t)\,dt \Big) \\ &\le V_*(x,s) \end{aligned} \tag{13.2.43}$$

where the latter inequality is derived as in (13.2.38), since the process $V_*(X_t,S_t) - \int_0^t c(X_r)\,dr$ is a local supermartingale. The strict inequality in (13.2.43) shows that $\mathsf{P}_{x,s}(\sigma < \tau_*) > 0$ fails, so we must have $\mathsf{P}_{x,s}(\tau_* \le \sigma) = 1$ for all (x,s).

To prove the optimality of τ_* in such a case, it is enough to note that if σ satisfies (13.1.5) then τ_* must satisfy it as well. Therefore (13.2.40) is satisfied, and thus τ_* is optimal. A straightforward argument can also be given by using the local supermartingale property of the process $V_*(X_t,S_t) - \int_0^t c(X_r)\,dr$. Indeed, since $\mathsf{P}_{x,s}(\tau_* \le \sigma) = 1$, we get

$$\begin{aligned} V_*(x,s) &= \mathsf{E}_{x,s}\Big(S_\sigma - \int_0^\sigma c(X_t)\,dt \Big) \\ &\le \mathsf{E}_{x,s}\Big(V_*(X_\sigma, S_\sigma) - \int_0^\sigma c(X_t)\,dt \Big) \\ &\le \mathsf{E}_{x,s}\Big(V_*(X_{\tau_*}, S_{\tau_*}) - \int_0^{\tau_*} c(X_t)\,dt \Big) = \mathsf{E}_{x,s}\Big(S_{\tau_*} - \int_0^{\tau_*} c(X_t)\,dt \Big) \end{aligned} \tag{13.2.44}$$

so τ_* is optimal for (13.1.4). The proof of the first part of the theorem is complete.

(II): Let $(g_n)_{n\ge1}$ be a decreasing sequence of solutions of (13.2.22) which satisfy $g_n(0) = -n$ for $n \ge 1$. Then each g_n must hit the diagonal in $\mathbb{R}^2$ at some $s_n > 0$ for which we have $s_n \uparrow \infty$ when $n \to \infty$. Since there is no solution of (13.2.22) which is less than s for all s, we must have $g_n(s) \downarrow -\infty$ as $n \to \infty$

for all s. Let τ_{g_n} denote the stopping time defined by (13.2.13) with $g = g_n$. Then τ_{g_n} satisfies (13.1.5), and since $S_{\tau_{g_n}} \le s \vee s_n$, we see that $V_{g_n}(x,s)$, defined by (13.2.14) with $g = g_n$, is given as in (13.2.21):

$$V_{g_n}(x,s) = s + \int_{g_n(s)}^{x} \big(L(x) - L(y)\big) c(y)\, m(dy) \tag{13.2.45}$$

for all $g_n(s) \le x \le s$. Letting $n \to \infty$ in (13.2.45), we see that the integral

$$I := \int_{-\infty}^{x} \big(L(x) - L(y)\big)\, c(y)\, m(dy) \tag{13.2.46}$$

plays a crucial role in the proof (independently of the given x and s).

Assume first that $I = +\infty$ (this is the case whenever $c(y) \ge \varepsilon > 0$ for all y, and $-\infty$ is a natural boundary point for X, see paragraph 11 below). Then from (13.2.45) we clearly get

$$V_*(x,s) \ge \lim_{n\to\infty} V_{g_n}(x,s) = +\infty \tag{13.2.47}$$

so the value function must be infinite.

On the other hand, if $I < \infty$, then (13.1.11)–(13.1.12) imply

$$\mathsf{E}_{x,s}\Big(\int_0^{\tau_{\hat{s}}} c(X_t)\, dt \Big) \le \int_{-\infty}^{\hat{s}} \big(L(\hat{s}) - L(y)\big)\, c(y)\, m(dy) < \infty \tag{13.2.48}$$

where $\tau_{\hat{s}} = \inf\{t > 0 : X_t = \hat{s}\}$ for $\hat{s} \ge s$. Thus, if we let the process (X_t, S_t) first hit $(\hat{s}, \hat{s})$, and then the boundary $\{(g_n(s), s) : s \in \mathbb{R}\}$ with $n \to \infty$, then by (13.2.45) (with $x = s = \hat{s}$) we see that the value function equals at least $\hat{s}$. More precisely, if the process (X_t, S_t) starts at (x,s), consider the stopping times $\tau_n = \tau_{\hat{s}} + \tau_{g_n} \circ \theta_{\tau_{\hat{s}}}$ for $n \ge 1$. Then by (13.2.48) we see that each τ_n satisfies (13.1.5), and by the strong Markov property of X we easily get

$$V_*(x,s) \ge \limsup_{n\to\infty} \mathsf{E}_{x,s}\Big(S_{\tau_n} - \int_0^{\tau_n} c(X_t)\, dt \Big) \ge \hat{s}. \tag{13.2.49}$$

By letting $\hat{s} \uparrow \infty$, we again find $V_*(x,s) = +\infty$. The proof of the theorem is complete. □

9. On the equation (13.2.22). Theorem 13.2 shows that the optimal stopping problem (13.1.4) reduces to the problem of solving the first-order nonlinear differential equation (13.2.22). If this equation has a maximal admissible solution, then this solution is an optimal stopping boundary. We may note that this equation is of the following *normal* form:

$$y' = \frac{F(y)}{G(x) - G(y)} \tag{13.2.50}$$

for $x > y$, where $y \mapsto F(y)$ is strictly positive, and $x \mapsto G(x)$ is strictly increasing. To the best of our knowledge the equation (13.2.50) has not been studied before in full generality, and in view of the result proved above we want to point out the need for its investigation. It turns out that its treatment depends heavily on the behaviour of the map G.

(i): If the process X is in natural scale, that is $L(x) = x$ for all x, we can completely characterize and describe the maximal admissible solution of (13.2.22). This can be done in terms of equation (13.2.50) with $G(x) = x$ and $F(y) = \sigma^2(y)/2c(y)$ as follows. Note that by passing to the inverse $z \mapsto y^{-1}(z)$, equation (13.2.50) in this case can be rewritten as

$$\big(y^{-1}\big)'(z) - \frac{1}{F(z)}\, y^{-1}(z) = -\frac{z}{F(z)}. \tag{13.2.51}$$

This is a first-order *linear* equation and its general solution is given by

$$y_\alpha^{-1}(z) = \exp\Big(\int_0^z \frac{dy}{F(y)}\Big)\Big(\alpha - \int_0^z \frac{y}{F(y)} \exp\Big(-\int_0^y \frac{du}{F(u)}\Big)dy\Big), \tag{13.2.52}$$

where α is a constant. Hence we see that, with $G(x) = x$, *the necessary and sufficient condition* for equation (13.2.50) to have a maximal admissible solution, is that

$$\alpha_* := \sup_{z\in\mathbb{R}} \Big(z \exp\Big(-\int_0^z \frac{dy}{F(y)}\Big) + \int_0^z \frac{y}{F(y)} \exp\Big(-\int_0^y \frac{du}{F(u)}\Big)dy\Big) < \infty, \tag{13.2.53}$$

and that this supremum is not attained at any $z \in \mathbb{R}$. In this case the maximal admissible solution $x \mapsto y_*(x)$ of (13.2.50) can be expressed explicitly through its inverse $z \mapsto y_{\alpha_*}^{-1}(z)$ given by (13.2.52).

Note also when $L(x) = G(x) = x^2 \operatorname{sgn}(x)$ that the same argument transforms (13.2.50) into a *Riccati equation*, which then can be further transformed into a linear homogeneous equation of second order by means of standard techniques. The trick of passing to the inverse in (13.2.22) is further used in [160] where a natural connection between the result of the present subsection and the Azéma–Yor solution of the Skorokhod-embedding problem [6] is described.

(ii): If the process X is not in natural scale, then the treatment of (13.2.50) is much harder, due to the lack of closed form solutions. In such cases it is possible to prove (or disprove) the existence of the maximal admissible solution by using Picard's method of successive approximations. The idea is to use Picard's theorem locally, step by step, and in this way show the existence of some global solution which is admissible. Then, by passing to the equivalent integral equation and using a monotone convergence theorem, one can argue that this implies the existence of the maximal admissible solution. This technique is described in detail in Section 3 of [81] in the case of $G(x) = x^p$ and $F(y) = y^{p+1}$ when $p > 1$. It is also seen there that during the construction one obtains tight bounds on the maximal solution which makes it possible to compute it numerically as accurate as desired (see [81] for details). In this process it is desirable to have a local existence and uniqueness of the solution, and these are provided by the following general facts.

From the general theory (Picard's method) we know that if the direction field $(x,y) \mapsto f(x,y) := F(y)/(G(x) - G(y))$ is (locally) continuous and (locally) Lipschitz in the second variable, then the equation (13.2.50) admits (locally) a unique solution. For instance, this will be so if along a (local) continuity of $(x,y) \mapsto f(x,y)$, we have a (local) continuity of $(x,y) \mapsto (\partial f/\partial y)(x,y)$. In particular, upon differentiating over y in $f(x,y)$ we see that (13.2.22) admits (locally) a unique solution whenever the map $y \mapsto \sigma^2(y)L'(y)/c(y)$ is (locally) C^1. It is also possible to prove that the equation (13.2.50) admits (locally) a solution, if only the (local) continuity of the direction field $(x,y) \mapsto F(y)/(G(x) - G(y))$ is verified. However, such a solution may fail to be (locally) unique.

Instead of entering further into such abstract considerations here, we shall rather confine ourselves to some concrete examples with applications in Chapter V below.

10. We have proved in Theorem 13.2 that τ_* is optimal for (13.1.4) whenever it satisfies (13.1.5). In Example 18.7 we will exhibit a stopping time τ_* which fails to satisfy (13.1.5), but nevertheless its value function is given by (13.2.31) as proved above. In this case τ_* is "approximately" optimal in the sense that (13.2.41) holds with $\tau_{g_n} \uparrow \tau_*$ as $n \to \infty$.

11. Other state spaces. The result of Theorem 13.2 extends to diffusions with other state spaces in $\mathbb{R}$. In view of many applications, we will indicate such an extension for non-negative diffusions.

In the setting of (13.1.1)–(13.1.3) assume that the diffusion X is non-negative, consider the optimal stopping problem (13.1.4) where the supremum is taken over all stopping times τ of X satisfying (13.1.5), and note that the result of Proposition 13.1 extends to this case provided that the diagonal is taken in $(0,\infty)^2$. In this context it is natural to assume that $\sigma(x) > 0$ for $x > 0$, and $\sigma(0)$ may be equal 0. Similarly, we shall see that the case of strictly positive cost function c differs from the case when c is strictly positive only on $(0,\infty)$. In any case, both $x \mapsto \sigma(x)$ and $x \mapsto c(x)$ are assumed continuous on $[0,\infty)$.

In addition to the infinitesimal characteristics from (13.1.1) which govern X in $(0,\infty)$, we must specify the boundary behaviour of X at 0. For this we shall consider the cases when 0 is a natural, exit, regular (instantaneously reflecting), and entrance boundary point (see [109, p. 226–250]).

The relevant fact in the case when 0 is either a *natural* or *exit* boundary point is that

$$\int_0^s \big(L(s) - L(y)\big)\, c(y)\, m(dy) = +\infty \tag{13.2.54}$$

for all $s > 0$ whenever $c(0) > 0$. In view of (13.2.31) this shows that for the maximal solution of (13.2.22) we must have $0 < g_*(s) < s$ for all $s > 0$ unless $V_*(s,s) = +\infty$. If $c(0) = 0$, then the integral in (13.2.54) can be finite, and we cannot state a similar claim; but from our method used below it will be clear how to handle such a case too, and therefore the details in this direction will be omitted for simplicity.

The relevant fact in the case when 0 is either a *regular* (*instantaneously reflecting*) or *entrance* boundary point is that

$$\mathsf{E}_{0,s}\Big(\int_0^{\tau_{s_*}} c(X_t)\, dt\Big) = \int_0^{s_*} \big(L(s_*) - L(y)\big)\, c(y)\, m(dy) \tag{13.2.55}$$

for all $s_* \geq s > 0$ where $\tau_{s_*} = \inf\{\, t > 0 : X_t = s_* \,\}$. In view of (13.2.31) this shows that it is never optimal to stop at $(0, s)$. Therefore, if the maximal solution of (13.2.22) satisfies $g_*(s_*) = 0$ for some $s_* > 0$ with $g_*(s) > 0$ for all $s > s_*$, then $\tau_* = \inf\{\, t > 0 : X_t \leq g_*(S_t) \,\}$ is to be the optimal stopping time, since X does not take negative values. If moreover $c(0) = 0$, then the value of $m(\{0\})$ does not play any role, and all regular behaviour (from absorption $m(\{0\}) = +\infty$, over sticky barrier phenomenon $0 < m(\{0\}) < +\infty$, to instantaneous reflection $m(\{0\}) = 0$) can be treated in the same way.

For simplicity in the next result we will assume that $c(0) > 0$ if 0 is either a natural (attracting or unattainable) or an exit boundary point, and will only consider the instantaneously-reflecting regular case. The remaining cases can be treated similarly.

Corollary 13.3. (Optimal stopping for non-negative diffusions) *In the setting of* (13.1.1)–(13.1.3) *assume that the diffusion X is non-negative, and that 0 is a natural, exit, instantaneously-reflecting regular, or entrance boundary point. Consider the optimal stopping problem* (13.1.4) *where the supremum is taken over all stopping times τ of X satisfying* (13.1.5).

(I): *Let $s \mapsto g_*(s)$ denote the maximal admissible solution of* (13.2.22) *in the following sense (whenever such a solution exists — see Figure* IV.12): *There exists a point $s_* \geq 0$ (with $s_* = 0$ if 0 is either a natural or an exit boundary point) such that $g_*(s_*) = 0$ and $g_*(s) > 0$ for all $s > s_*$; the map $s \mapsto g_*(s)$ solves* (13.2.22) *for $s > s_*$ and is admissible (i.e. $g_*(s) < s$ for all $s > s^*$); the map $s \mapsto g_*(s)$ is the maximal solution satisfying these two properties (the comparison of two maps is taken pointwise wherever they are both strictly positive). Then we have:*

1°. *The value function is finite and for* $s \geq s_*$ *is given by*

$$V_*(x,s) = s + \int_{g_*(s)}^{x} \big(L(x) - L(y)\big)\, c(y)\, m(dy) \tag{13.2.56}$$

for $g_*(s) \leq x \leq s$ *with* $V_*(x,s) = s$ *for* $0 \leq x \leq g_*(s)$ *, and for* $s \leq s_*$ *(when* 0 *is either an instantaneously-reflecting regular or an entrance boundary point) is given by*

$$V_*(x,s) = s_* + \int_{0}^{x} \big(L(x) - L(y)\big)\, c(y)\, m(dy) \tag{13.2.57}$$

for $0 \leq x \leq s$ *.*

2°. *The stopping time*

$$\tau_* = \inf\,\{t > 0 : S_t \geq s_*, X_t \leq g_*(S_t)\,\} \tag{13.2.58}$$

is optimal for the problem (13.1.4) *whenever it satisfies* (13.1.5); *otherwise, it is "approximately" optimal.*

3°. *If there exists an optimal stopping time* σ *in* (13.1.4) *satisfying* (13.1.5), *then* $\mathsf{P}_{x,s}(\tau_* \leq \sigma) = 1$ *for all* (x,s) *, and* τ_* *is an optimal stopping time for* (13.1.4) *as well.*

(II): *If there is no (maximal) solution of* (13.2.22) *in the sense of* (I) *above, then* $V_*(x,s) = +\infty$ *for all* (x,s) *, and there is no optimal stopping time.*

Proof. With only minor changes the proof can be carried out in exactly the same way as the proof of Theorem 13.2 upon using the additional facts about (13.2.54) and (13.2.55) stated above, and the details will be omitted. Note, however, that in the case when 0 is either an instantaneously-reflecting regular or an entrance boundary point, the strong Markov property of X at $\tau_{s_*} = \inf\,\{t > 0 : X_t = s_*\,\}$ gives

$$V_*(x,s) = s_* + \int_0^{s_*} \big(L(s_*) - L(y)\big)\, c(y)\, m(dy) - \mathsf{E}_{x,s}\bigg(\int_0^{\tau_{s_*}} c(X_t)\, dt\bigg) \tag{13.2.59}$$

for all $0 \leq x \leq s \leq s_*$. Hence formula (13.2.57) follows by applying (13.1.11)+(13.1.12) to the last term in (13.2.59). (In the instantaneous reflecting case one can make use of τ_{s_*,s_*} after extending L to $\mathbb{R}_-$ by setting $L(x) := -L(-x)$ for $x < 0$). The proof is complete. □

12. The "discounted" problem. One is often more interested in the discounted version of the optimal stopping problem (13.1.4). Such a problem can be reduced to the initial problem (13.1.4) by changing the underlying diffusion process.

Given a continuous function $x \mapsto \lambda(x) \geq 0$ called the *discounting rate*, in the setting of (13.1.1)–(13.1.3) introduce the functional

$$\Lambda(t) = \int_0^t \lambda(X_r)\, dr, \tag{13.2.60}$$

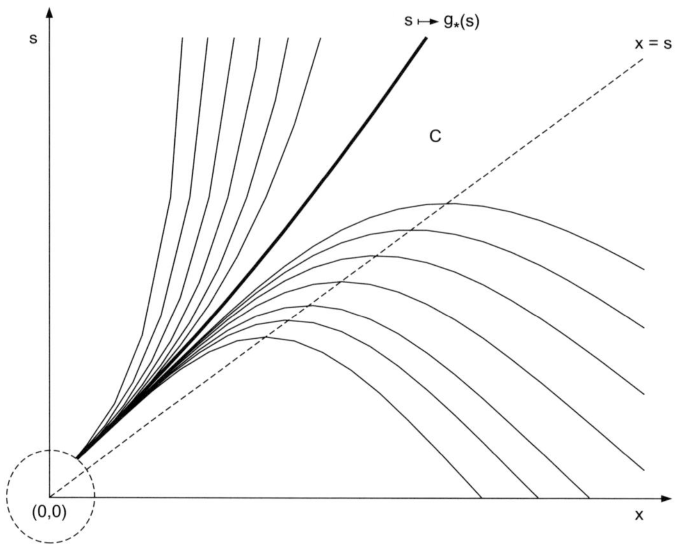

Figure IV.12: A computer drawing of solutions of the differential equation (13.2.22) in the case when X is a geometric Brownian motion from Example 18.9 with $\rho = -1$, $\sigma^2 = 2$ (thus $\Delta = 2$) and $c = 50$. The bold line $s \mapsto g_*(s)$ is the *maximal* admissible solution. (In this particular case there is no closed formula for $s \mapsto g_*(s)$, but it is proved that $s \mapsto g_*(s)$ satisfies (18.4.15).)

and consider the optimal stopping problem with the value function

$$V_*(x,s) = \sup_\tau \mathsf{E}_{x,s}\Big(e^{-\Lambda(\tau)}S_\tau - \int_0^\tau e^{-\Lambda(t)}c(X_t)\,dt\Big), \tag{13.2.61}$$

where the supremum is taken over all stopping times τ of X for which the integral has finite expectation, and the *cost* function $x \mapsto c(x) > 0$ is continuous.

The standard argument (Subsection 5.4) shows that the problem (13.2.61) is equivalent to the problem

$$V_*(x,s) = \sup_\tau \mathsf{E}_{x,s}\Big(\widetilde{S}_\tau - \int_0^\tau c(\widetilde{X}_t)\,dt\Big) \tag{13.2.62}$$

where $\widetilde{X} = (\widetilde{X}_t)_{t\ge 0}$ is a diffusion process which corresponds to the "killing" of the sample paths of X at the "rate" $\lambda(X)$. The infinitesimal generator of $\widetilde{X}$ is given by

$$\mathbb{L}_{\widetilde{X}} = -\lambda(x) + \rho(x)\frac{\partial}{\partial x} + \frac{\sigma^2(x)}{2}\frac{\partial^2}{\partial x^2}. \tag{13.2.63}$$

We conjecture that the maximality principle proved above also holds for this problem (see [185] and [151]). The main technical difficulty in a general treatment of this problem is the fact that the infinitesimal generator $\mathbb{L}_{\widetilde{X}}$ has the term $-\lambda(x)$, so that $\mathbb{L}_{\widetilde{X}} = 0$ may have no simple solution. Nonetheless, it is clear that the corresponding system (13.2.9)–(13.2.12) must be valid, and this system defines the (maximal) boundary $s \mapsto g_*(s)$ implicitly.

13. The "Markovian" cost problem. Yet another class of optimal stopping problems (Mayer instead of Lagrange formulated) reduces to the problem (13.1.4). Suppose that in the setting of (13.1.1)–(13.1.3) we are given a smooth function $x \mapsto D(x)$, and consider the optimal stopping problem with the value function

$$V_*(x,s) = \sup_{\tau} \mathsf{E}_{x,s}\big(S_\tau - D(X_\tau)\big) \tag{13.2.64}$$

where the supremum is taken over a class of stopping times τ of X. Then a variant of Itô's formula (page 67) applied to $D(X_t)$, the optional sampling theorem (page 60) applied to the continuous local martingale $M_t = \int_0^t D'(X_s)\sigma(X_s)\,dB_s$ localized if necessary, and uniform integrability conditions enable one to conclude

$$\mathsf{E}_{x,s} D(X_\tau) = D(x) + \mathsf{E}_{x,s}\bigg(\int_0^\tau (\mathbb{L}_X D)(X_s)\,ds\bigg). \tag{13.2.65}$$

Hence we see that the problem (13.2.64) reduces to the problem (13.1.4) with $x \mapsto c(x)$ replaced by $x \mapsto (\mathbb{L}_X D)(x)$ whenever non-negative. The conditions assumed above to make such a transfer possible are not restrictive in general (see Section 19 below).

Notes. Our main aim in this section (following [159]) is to present the solution to a problem of optimal stopping for the maximum process associated with a one-dimensional time-homogeneous diffusion. The solution found has a large number of applications, and may be viewed as the cornerstone in a general treatment of the maximum process.

In the setting of (13.1.1)–(13.1.3) we consider the optimal stopping problem (13.1.4), where the supremum is taken over all stopping times τ satisfying (13.1.5), and the cost function c is positive and continuous. The main result of the section is presented in Theorem 13.2, where it is proved that this problem has a solution (the value function is finite and there is an optimal stopping strategy) if and only if *the maximality principle* holds, i.e. the first-order nonlinear differential equation (13.2.22) has a maximal admissible solution (see *Figures* IV.11 and IV.12). The maximal admissible solution is proved to be an optimal stopping boundary, i.e. the stopping time (13.2.32) is optimal, and the value function is given explicitly by (13.2.31). Moreover, this stopping time is shown to be pointwise the smallest possible optimal stopping time. If there is no such maximal admissible solution of (13.2.22), the value function is proved to be infinite and there is no optimal stopping time. The examples given in Chapter V below are aimed to illustrate some applications of the result proved.

The optimal stopping problem (13.1.4) has been considered in some special cases earlier. Jacka [103] treats the case of reflected Brownian motion, while Dubins, Shepp and Shiryaev [45] treat the case of Bessel processes. In these papers the problem was solved effectively by guessing the nature of the optimal stopping boundary and making use of the principle of smooth fit. The same is true for the "discounted" problem (13.2.61) with $c \equiv 0$ in the case of geometric Brownian motion which in the framework of option pricing theory (Russian option) was solved by Shepp and Shiryaev in [185] (see also [186] and [79]). For the first time a strong need for additional arguments was felt in [81], where the problem (13.1.4) for geometric Brownian motion was considered with the cost function $c(x) \equiv c > 0$. There, by use of Picard's method of successive approximations, it was proved that the maximal admissible solution of (13.2.22) is an optimal stopping boundary, and since this solution could not be expressed in closed form, it really showed the full power of the method. Such nontrivial solutions were also obtained in [45] by a method which relies on estimates of the value function obtained a priori. Motivated by similar ideas, sufficient conditions for the maximality principle to hold for general diffusions are given in [82]. The method of proof used there relies on a transfinite induction argument. In order to solve the problem in general, the fundamental question was how to relate the maximality principle to the superharmonic characterization of the value function, which is the key result in the general theory (recall Theorems 2.4 and 2.7 above).

The most interesting point in our solution of the optimal stopping problem (13.1.4) relies on the fact that we have described this connection, and actually proved that the maximality principle is equivalent to the superharmonic characterization of the value function (for a three-dimensional process). The crucial observations in this direction are (13.2.29) and (13.2.30), which show that the only possible optimal stopping boundary is the maximal admissible solution (see (13.2.39) in the proof of Theorem 13.2). In the next step of proving that the maximal solution is indeed an optimal stopping boundary, it was crucial to make use of so-called "bad-good" solutions of (13.2.22), "bad" in the sense that they hit the diagonal in $\mathbb{R}^2$, and "good" in the sense that they are not too large (see *Figures* IV.11 and IV.12). These "bad-good" solutions are used to approximate the maximal solution in a desired manner, see the proof of Theorem 13.2 (starting from (13.2.41) onwards), and this turns out to be the key argument in completing the proof.

Our methodology adopts and extends earlier results of Dubins, Shepp and Shiryaev [45], and is, in fact, quite standard in the business of solving particular optimal stopping problems: (i) one tries to guess the nature of the optimal stopping boundary as a member of a "reasonable" family; (ii) computes the expected reward; (iii) maximizes this over the family; (iv) and then tries to argue that the resulting stopping time is optimal in general. This process is often facilitated by "ad hoc" principles, as the "principle of smooth fit" for instance. This procedure is used effectively in this section too, as opposed to results from the general theory of optimal stopping (Chapter I). It should be clear, however, that the maximality principle of the present section should rather be seen as a convenient reformulation of the basic principle on a superharmonic characterization from the general theory, than a new principle on its own (see also [154] for a related result).

For results on discounted problems see [151] and for similar optimal stopping problems of Poisson processes see [119].

14. Nonlinear integral equations

This section is devoted to nonlinear integral equations which play a prominent role in problems of optimal stopping (Subsection 14.1) and the first passage problem (Subsection 14.2). The two avenues are by no means independent and the purpose of this section is to highlight this fact without drawing parallels explicitly.

14.1. The free-boundary equation

In this subsection we will briefly indicate how the local time-space calculus (cf. Subsection 3.5) naturally leads to nonlinear integral equations which characterize the optimal stopping boundary within an admissible class of functions.

For simplicity of exposition, let us assume that X is a one-dimensional diffusion process solving

$$dX_t = \rho(X_t)\,dt + \sigma(X_t)\,dB_t \tag{14.1.1}$$

and let us consider the optimal stopping problem

$$V(t,x) = \sup_{0\le\tau\le T-t} \mathsf{E}_{t,x} G(t+\tau, X_{t+\tau}) \tag{14.1.2}$$

where $X_t = x$ under $\mathsf{P}_{t,x}$ and τ is a stopping time of X. Assuming further that G is smooth we know (cf. (8.2.2)–(8.2.4)) that (14.1.2) leads to the following free-boundary problem:

$$V_t + \mathbb{L}_X V = 0 \quad \text{in } C, \tag{14.1.3}$$

$$V = G \quad \text{in } D, \tag{14.1.4}$$

$$V_x = G_x \quad \text{at } \partial C \tag{14.1.5}$$

where $C = \{V > G\}$ is the continuation set, $D = \{V = G\}$ is the stopping set, and the stopping time $\tau_D = \inf\{\, s \in [0, T-t] : (t+s, X_{t+s}) \in D \,\}$ is optimal in (14.1.2) under $\mathsf{P}_{t,x}$.

For simplicity of exposition, let us further assume that

$$C = \big\{\, (t,x) \in [0,T] \times \mathbb{R} : x > b(t) \,\big\}, \tag{14.1.6}$$

$$D = \big\{\, (t,x) \in [0,T] \times \mathbb{R} : x \le b(t) \,\big\} \tag{14.1.7}$$

where $b : [0,T] \rightarrow \mathbb{R}$ is a continuous function of bounded variation. Then b is the optimal stopping boundary and the problem reduces to determining V and b. Thus (14.1.3)–(14.1.5) may be viewed as a system of equations for the two unknowns V and b.

Generally, if one is given to solve a system of two equations with two unknowns, a natural approach is to use the first equation in order to express the first unknown in terms of the second unknown, insert the resulting expression in the second equation, and consider the resulting equation in order to determine the second unknown. Quite similarly, this methodology extends to the system (14.1.3)–(14.1.5) as follows.

Assuming that sufficient conditions stated in Subsection 3.5 are satisfied, let us apply the change-of-variable formula (3.5.9) to $V(t+s, X_{t+s})$ under $\mathsf{P}_{t,x}$. This yields:

$$\begin{aligned} V(t+s, X_{t+s}) &= V(t,x) \\ &+ \int_0^s (V_t + \mathbb{L}_X V)(t+u, X_{t+u})\, I\big(X_{t+u} \neq b(t+u)\big)\, du \\ &+ \int_0^s V_x(t+u, X_{t+u})\, \sigma(X_{t+u})\, I\big(X_{t+u} \neq b(t+u)\big)\, dB_{t+u} \\ &+ \int_0^s \big(V_x(s, X_s+) - V_x(s, X_s-)\big)\, I\big(X_{t+u} = b(t+u)\big)\, d\ell^b_{t+u}(X) \end{aligned} \tag{14.1.8}$$

for $s \in [0, T-t]$ where $\mathbb{L}_X V = \rho V_x + (\sigma^2/2) V_{xx}$. Due to the smooth-fit condition (14.1.5) we see that the final integral in (14.1.8) must be zero. Moreover, since $V_t + \mathbb{L}_X V = 0$ in C by (14.1.3), and $V_t + \mathbb{L}_X V = G_t + \mathbb{L}_X G$ in D by (14.1.4), we see that (14.1.8) reads as follows:

$$\begin{aligned} V(t+s, X_{t+s}) &= V(t,x) \\ &+ \int_0^s (G_t + \mathbb{L}_X G)(t+u, X_{t+u})\, I\big(X_{t+u} < b(t+u)\big)\, du + M_s \end{aligned} \tag{14.1.9}$$

where $M_s := \int_0^s V_x(t+u, X_{t+u})\, \sigma(X_{t+u})\, I\big((X_{t+u} \neq b(t+u)\big)\, dB_{t+u}$ is a continuous (local) martingale for $s \in [0, T-t]$. The identity (14.1.9) may be viewed as an explicit semimartingale decomposition of the value function composed with the process (i.e. $V(t+s, X_{t+s})$ for $s \in [0, T-t]$) under $\mathsf{P}_{t,x}$.

Setting $s = T-t$ in (14.1.9), taking $\mathsf{E}_{t,x}$ on both sides, and using that $\mathsf{E}_{t,x}(M_{T-t}) = 0$ (whenever fulfilled), we get

$$\begin{aligned} \mathsf{E}_{t,x} G(T, X_T) &= V(t,x) \\ &+ \int_0^{T-t} \mathsf{E}_{t,x}\Big[(G_t + \mathbb{L}_X G)(t+u, X_{t+u})\, I\big(X_{t+u} < b(t+u)\big)\Big]\, du \end{aligned} \tag{14.1.10}$$

for all $t \in [0,T]$ and all $x \in \mathbb{R}$. When $x > b(t)$ then (14.1.10) is an equation containing both unknowns b and V. On the other hand, when $x \le b(t)$ then

$V(t,x) = G(t,x)$ is a known value so that (14.1.10) is an equation for b only. In particular, if we insert $x = b(t)$ in (14.1.10) and use (14.1.4), we see that (14.1.10) becomes

$$\mathsf{E}_{t,b(t)} G(T, X_T) = G(t, b(t)) \tag{14.1.11}$$
$$+ \int_0^{T-t} \mathsf{E}_{t,x} \Big[(G_t + \mathbb{L}_X G)(t+u, X_{t+u})\, I\big(X_{t+u} < b(t+u)\big) \Big]\, du$$

for $t \in [0,T]$. This is a nonlinear integral equation for b that we call the *free-boundary equation.*

We will study specific examples of the free-boundary equation (14.1.11) in Chapters VI–VIII below. It will be shown there that this equation characterizes the optimal stopping boundary within an admissible class of functions. This fact is far from being obvious at first glance and its establishment has led to the development of the local time-space calculus reviewed briefly in Subsection 3.5. On closer inspection it is instructive to note that the structure of the free-boundary equation (14.1.11) is rather similar to the structure of the *first-passage equation* treated in the following section.

14.2. The first-passage equation

1. Let $B = (B_t)_{t \ge 0}$ be a standard Brownian motion started at zero, let $g : (0,\infty) \to \mathbb{R}$ be a continuous function satisfying $g(0+) \ge 0$, let

$$\tau = \inf\{\, t > 0 : B_t \ge g(t)\,\} \tag{14.2.1}$$

be the first-passage time of B over g, and let F denote the distribution function of τ.

The *first-passage problem* seeks to determine F when g is given. The *inverse first-passage problem* seeks to determine g when F is given. Both the process B and the boundary g in these formulations may be more general, and our choice of Brownian motion is primarily motivated by the tractability of the exposition. The facts to be presented below can be extended to more general *Markov* processes and boundaries (such as two-sided ones) and the time may also be discrete.

2. Chapman–Kolmogorov equations of Volterra type. It will be convenient to divide our discussion into two parts depending on if the time set T of the Markov process $X = (X_t)_{t \in T}$ is either discrete (finite or countable) or continuous (uncountable). The state space E of the process may be assumed to be a subset of $\mathbb{R}$.

1°. *Discrete time and space.* Recall that $(X_n)_{n \ge 0}$ is a (time-homogeneous) Markov process if the following condition is satisfied:

$$\mathsf{E}_x(H \circ \theta_k \mid \mathcal{F}_k) = \mathsf{E}_{X_k}(H) \tag{14.2.2}$$

for all (bounded) measurable H and all k and x. (Recall that $X_0 = x$ under P_x, and that $X_n \circ \theta_k = X_{n+k}$.)

Then the *Chapman–Kolmogorov equation* (see (4.1.20)) holds:

$$\mathsf{P}_x(X_n = z) = \sum_{y \in E} \mathsf{P}_y(X_{n-k} = z)\, \mathsf{P}_x(X_k = y) \tag{14.2.3}$$

for x, z in E and $1 < k < n$ given and fixed, which is seen as follows:

$$\begin{aligned} \mathsf{P}_x(X_n = z) &= \sum_{y \in E} \mathsf{P}_x(X_n = z, X_k = y) \\ &= \sum_{y \in E} \mathsf{E}_x\Big(I(X_k = y)\mathsf{E}_x\big(I(X_{n-k} = z) \circ \theta_k \mid \mathcal{F}_k\big)\Big) \\ &= \sum_{y \in E} \mathsf{E}_x\big(I(X_k = y)\, \mathsf{E}_{X_k} I(X_{n-k} = z)\big) \\ &= \sum_{y \in E} \mathsf{P}_x(X_k = y)\, \mathsf{P}_y(X_{n-k} = z) \end{aligned} \tag{14.2.4}$$

upon using (14.2.2) with $Y = I(X_{n-k} = z)$.

A geometric interpretation of the Chapman–Kolmogorov equation (14.2.3) is illustrated in *Figure* IV.13 (note that the vertical line passing through k is given and fixed). Although for (14.2.3) we only considered the time-homogeneous Markov property (14.2.2) for simplicity, it should be noted that a more general Markov process creates essentially the same picture.

Imagine now on *Figure* IV.13 that the vertical line passing through k begins to move continuously and eventually transforms into a new curve still separating x from z as shown in *Figure* IV.14. The question then arises naturally how the Chapman–Kolmogorov equation (14.2.3) extends to this case.

An evident answer to this question is stated in the following Theorem 14.1. This fact is then extended to the case of continuous time and space in Theorem 14.2 below.

Theorem 14.1. *Let $X = (X_n)_{n \ge 0}$ be a Markov process (taking values in a countable set E), let x and z be given and fixed in E, let $g : \mathbb{N} \to E$ be a function separating x and z relative to X (i.e. if $X_0 = x$ and $X_n = z$ for some $n \ge 1$, then there exists $1 \le k \le n$ such that $X_k = g(k)$), and let*

$$\tau = \inf\{\, k \ge 1 : X_k = g(k)\,\} \tag{14.2.5}$$

be the first-passage time of X over g. Then the following sum equation holds:

$$\mathsf{P}_x(X_n = z) = \sum_{k=1}^{n} \mathsf{P}\big(X_n = z \mid X_k = g(k)\big)\, \mathsf{P}_x(\tau = k). \tag{14.2.6}$$

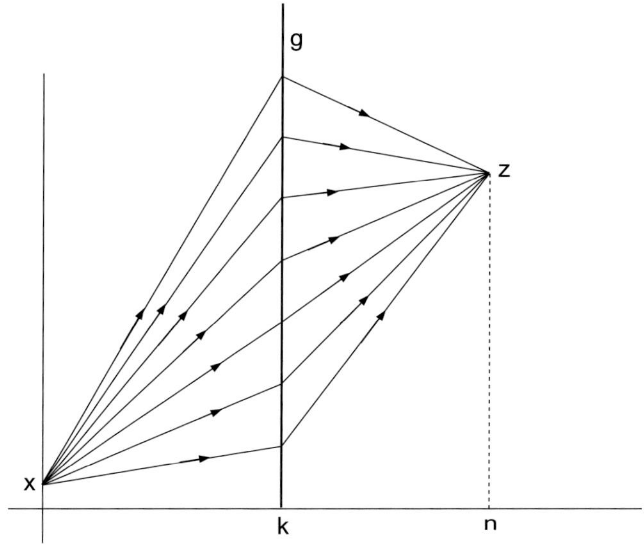

Figure IV.13: A symbolic drawing of the Chapman–Kolmogorov equation (14.2.3). The arrows indicate a time evolution of the sample paths of the process. The vertical line at k represents the state space of the process. The equations (14.2.12) have a similar interpretation.

Moreover, if the Markov process X *is time-homogeneous, then* (14.2.6) *reads as follows*:

$$\mathsf{P}_x(X_n = z) = \sum_{k=1}^{n} \mathsf{P}_{g(k)}(X_{n-k} = z)\, \mathsf{P}_x(\tau = k). \tag{14.2.7}$$

Proof. Since g separates x and z relative to X, we have

$$\mathsf{P}_x(X_n = z) = \sum_{k=1}^{n} \mathsf{P}_x(X_n = z, \tau = k). \tag{14.2.8}$$

On the other hand, by the Markov property:

$$\mathsf{P}_x(X_n = z \mid \mathcal{F}_k) = \mathsf{P}_{X_k}(X_n = z) \tag{14.2.9}$$

and the fact that $\{\tau = k\} \in \mathcal{F}_k$, we easily find

$$\mathsf{P}_x(X_n = z, \tau = k) = \mathsf{P}\big(X_n = z \mid X_k = g(k)\big)\, \mathsf{P}_x(\tau = k). \tag{14.2.10}$$

Inserting this into (14.2.8) we obtain (14.2.6). The time-homogeneous simplification (14.2.7) follows then immediately, and the proof is complete. □

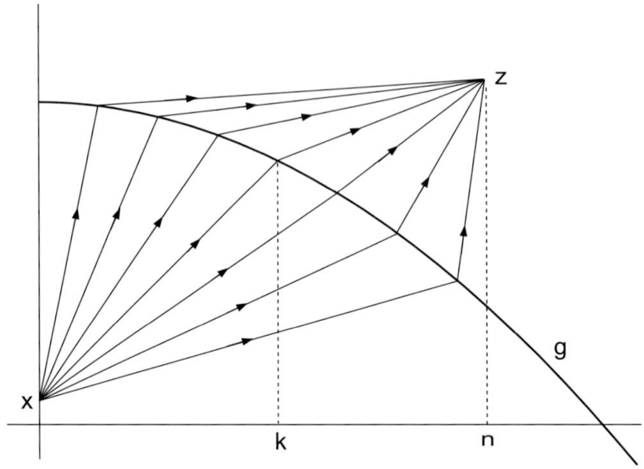

Figure IV.14: A symbolic drawing of the integral equation (14.2.6)–(14.2.7). The arrows indicate a time evolution of the sample paths of the process. The vertical line at k has been transformed into a time-dependent boundary g. The equations (14.2.17)–(14.2.18) have a similar interpretation.

The equations (14.2.6) and (14.2.7) extend to the case when the state space S is uncountable. In this case the relation "$= z$" in (14.2.6) and (14.2.7) can be replaced by "$\in G$" where G is any measurable set that is "separated" from the initial point x relative to X in the sense described above. The extensions of (14.2.6) and (14.2.7) obtained in this way will be omitted.

2. Continuous time and space. A passage from the discrete to the continuous case introduces some technical complications (e.g. regular conditional probabilities are needed) which we set aside in the sequel (see Subsection 4.3).

A process $(X_t)_{t\geq 0}$ is called a Markov process (in a wide sense) if the following condition is satisfied:

$$\mathsf{P}(X_t \in G \mid \mathcal{F}_s) = \mathsf{P}(X_t \in G \mid X_s) \tag{14.2.11}$$

for all measurable G and all $s < t$ (recall (4.1.2)). Then the *Chapman–Kolmogorov equation* (see (4.3.2)) holds:

$$P(s,x;t,A) = \int_E P(s,x;u,dy)\, P(u,y;t,A) \qquad (0 \leq s < u < t) \tag{14.2.12}$$

where $P(s,x;t,A) = \mathsf{P}(X_t \in A \mid X_s = x)$ and $s < u < t$ are given and fixed.

Kolmogorov [111] called (14.2.12) 'the fundamental equation', noted that (under a desired Markovian interpretation) it is satisfied if the state space E

is finite or countable (the 'total probability law'), and in the case when E is uncountable took it as a "new axiom".

If X_t under $X_s = x$ has a density function f satisfying

$$P(s,x;t,A) = \int_A f(s,x;t,y)\,dy \tag{14.2.13}$$

for all measurable sets A, then the equations (14.2.12) reduce to

$$f(s,x;t,z) = \int_E f(s,x;u,dy)\,f(u,y;t,z)\,dy \tag{14.2.14}$$

for x and z in E and $s < u < t$ given and fixed.

In [111] Kolmogorov proved that under some additional conditions f satisfies certain differential equations of parabolic type (*the forward and the backward equation* — see (4.3.7) and (4.3.8)). Note that in [112] Kolmogorov mentioned that this integral equation was studied by Smoluchowski [204], and in a footnote he acknowledged that these differential equations for certain particular cases were introduced by Fokker [68] and Planck [172] independently of the Smoluchowski integral equation. (The Smoluchowski integral equation [204] is a time-homogeneous version of (14.2.14). The Bachelier–Einstein equation (cf. [7], [56])

$$f(t+s,z) = \int_E f(t,z-x)\,f(s,x)\,dx \tag{14.2.15}$$

is a space-time homogeneous version of the Smoluchowski equation.)

Without going into further details on these facts, we will only note that the interpretation of the Chapman–Kolmogorov equation (14.2.3) described above by means of *Figure* IV.13 carries over to the general case of the equation (14.2.12), and the same is true for the question raised above by means of *Figure* IV.14. The following theorem extends the result of Theorem 14.1 on this matter.

Theorem 14.2. (cf. Schrödinger [182] and Fortet [69, p. 217]) *Let* $X = (X_t)_{t\geq 0}$ *be a strong Markov process with continuous sample paths started at* x , *let* $g : (0,\infty) \to \mathbb{R}$ *be a continuous function satisfying* $g(0+) \geq x$, *let*

$$\tau = \inf\{\, t > 0 : X_t \geq g(t)\} \tag{14.2.16}$$

be the first-passage time of X *over* g , *and let* F *denote the distribution function of* τ .

Then the following integral equation holds:

$$\mathsf{P}_x(X_t \in G) = \int_0^t \mathsf{P}\big(X_t \in G \mid X_s = g(s)\big)\,F(ds) \tag{14.2.17}$$

for each measurable set G *contained in* $[\,g(t),\infty)$.

Moreover, if the Markov process X *is time-homogeneous, then* (14.2.17) *reads as follows*:

$$\mathsf{P}_x(X_t \in G) = \int_0^t \mathsf{P}_{g(s)}\big(X_{t-s} \in G\big)\, F(ds). \tag{14.2.18}$$

for each measurable set G *contained in* $[\,g(t), \infty)$.

Proof. The key argument in the proof is to apply a strong Markov property at time τ (see (4.3.27) and (4.3.28)). This can be done informally (with $G \subseteq [\,g(t), \infty)$ given and fixed) as follows:

$$\begin{aligned}\mathsf{P}_x(X_t \in G) &= \mathsf{P}_x(X_t \in G, \tau \le t) = \mathsf{E}_x\Big(I\,(\tau \le t)\, \mathsf{E}_x\big(I\,(X_t \in G) \,|\, \tau\big)\Big) \\ &= \int_0^t \mathsf{E}_x\big(I(X_t \in G) \,|\, \tau = s\big)\, F(ds) \\ &= \int_0^t \mathsf{P}\big(X_t \in G \,|\, X_s = g(s)\big)\, F(ds)\end{aligned} \tag{14.2.19}$$

which is (14.2.17). In the last identity above we used that for $s \le t$ we have

$$\mathsf{E}_x\big(I(X_t \in G) \,|\, \tau = s\big) = \mathsf{P}\big(X_t \in G \,|\, X_s = g(s)\big) \tag{14.2.20}$$

which formally requires a precise argument. This is what we do in the rest of the proof.

For this, recall that if $Z = (Z_t)_{t \ge 0}$ is a strong Markov process then

$$\mathsf{E}_z(H \circ \theta_\sigma \,|\, \mathcal{F}_\sigma) = \mathsf{E}_{Z_\sigma}(H) \tag{14.2.21}$$

for all (bounded) measurable H and all stopping times σ .

For our proof we choose $Z_t = (t, X_t)$ and define

$$\sigma = \inf\{\, t > 0 : Z_t \notin C\,\}, \tag{14.2.22}$$

$$\beta = \inf\{\, t > 0 : Z_t \notin C \cup D\,\} \tag{14.2.23}$$

where $C = \{(s,y) : 0 < s < t,\, y < g(s)\}$ and $D = \{(s,y) : 0 < s < t,\, y \ge g(s)\}$, so that $C \cup D = \{(s,y) : 0 < s < t\}$. Thus $\beta = t$ under $\mathsf{P}_{(0,x)}$ i.e. P_x , and moreover $\beta = \sigma + \beta \circ \theta_\sigma$ since both σ and β are hitting times of the process Z to closed (open) sets, the second set being contained in the first one, so that $\sigma \le \beta$. (See (7.0.7) and (4.1.25) above.)

Setting $F(s,y) = 1_G(y)$ and $H = F(Z_\beta)$, we thus see that $H \circ \theta_\sigma = F(Z_\beta) \circ \theta_\sigma = F(Z_{\sigma + \beta \circ \sigma}) = F(Z_\beta) = H$, which by means of (14.2.21) implies that

$$\mathsf{E}_z(F(Z_\beta) \,|\, \mathcal{F}_\sigma) = \mathsf{E}_{Z_\sigma} F(Z_\beta). \tag{14.2.24}$$

In the special case $z = (0, x)$ this reads

$$\mathsf{E}_{(0,x)}\big(I\,(X_t \in G)\,|\,\mathcal{F}_\sigma\big) = \mathsf{E}_{(\sigma,g(\sigma))} I\,(X_t \in G) \tag{14.2.25}$$

where $\mathcal{F}_\sigma$ on the left-hand side can be replaced by σ since the right-hand side defines a measurable function of σ. It follows then immediately from such modified (14.2.25) that

$$\mathsf{E}_{(0,x)}\big(I(X_t \in G)\,|\,\sigma = s\big) = \mathsf{E}_{(s,g(s))}\big(I(X_t \in G)\big) \tag{14.2.26}$$

and since $\sigma = \tau \wedge t$ we see that (14.2.26) implies (14.2.20) for $s \le t$. Thus the final step in (14.2.19) is justified and therefore (14.2.17) is proved as well. The time-homogeneous simplification (14.2.18) is a direct consequence of (14.2.17), and the proof of the theorem is complete. □

The proof of Theorem 14.2 just presented is not the only possible one. The proof of Theorem 14.3 given below can easily be transformed into a proof of Theorem 14.2. Yet another quick proof can be given by applying the strong Markov property of the process (t, X_t) to establish (14.2.25) (multiplied by $I(\tau \le t)$ on both sides) with $\sigma = \tau \wedge t$ on the left-hand side and $\sigma = \tau$ on the right-hand side. The right-hand side then easily transforms to the right-hand side of (14.2.17) thus proving the latter.

In order to examine the scope of the equations (14.2.17) in a clearer manner, we will leave the realm of a general Markov process in the sequel, and consider the case of a standard Brownian motion instead. The facts and methodology presented below extend to the case of more general Markov processes (or boundaries) although some of the formulae may be less explicit.

3. The master equation. The following notation will be used throughout:

$$\varphi(x) = \frac{1}{\sqrt{2\pi}}\, e^{-x^2/2}, \quad \Phi(x) = \int_{-\infty}^{x} \varphi(z)\, dz, \quad \Psi(x) = 1 - \Phi(x) \tag{14.2.27}$$

for $x \in \mathbb{R}$. We begin this paragraph by recalling the result of Theorem 14.2. Thus, let $g : (0, \infty) \to \mathbb{R}$ be a continuous function satisfying $g(0+) \ge 0$, and let F denote the distribution function of τ from (14.2.16).

If specialized to the case of standard Brownian motion $(B_t)_{t \ge 0}$ started at zero, the equation (14.2.18) with $G = [\,g(t), \infty)$ reads as follows:

$$\Psi\left(\frac{g(t)}{\sqrt{t}}\right) = \int_0^t \Psi\left(\frac{g(t) - g(s)}{\sqrt{t-s}}\right) F(ds) \tag{14.2.28}$$

where the scaling property $B_t \sim \sqrt{t}\, B_1$ of B is used, as well as that $(z + B_t)_{t \ge 0}$ defines a standard Brownian motion started at z whenever $z \in \mathbb{R}$.

1. Derivation. It turns out that the equation (14.2.28) is just one in the sequence of equations that can be derived from a single master equation from

Theorem 14.3 below. This master equation can be obtained by taking $G = [z, \infty)$ in (14.2.18) with $z \ge g(t)$. We now present yet another proof of this derivation.

Theorem 14.3. (The Master Equation) *Let $B = (B_t)_{t \ge 0}$ be a standard Brownian motion started at zero, let $g : \langle 0, \infty \rangle \to \mathbb{R}$ be a continuous function satisfying $g(0+) \ge 0$, let*

$$\tau = \inf\{\, t > 0 : B_t \ge g(t) \} \tag{14.2.29}$$

be the first-passage time of B over g, and let F denote the distribution function of τ.

Then the following integral equation (called the Master Equation) holds:

$$\Psi\Big(\frac{z}{\sqrt{t}}\Big) = \int_0^t \Psi\Big(\frac{z - g(s)}{\sqrt{t-s}}\Big)\, F(ds) \tag{14.2.30}$$

for all $z \ge g(t)$ where $t > 0$.

Proof. We will make use of the strong Markov property of the process $Z_t = (t, B_t)$ at time τ. This makes the present argument close to the argument used in the proof of Theorem 14.2.

For each $t > 0$ let $z(t)$ from $[\, g(t), \infty)$ be given and fixed. Setting $f(t,x) = I(x \ge z(t))$ and $H = \int_0^\infty e^{-\lambda s} f(Z_s)\, ds$ by the strong Markov property (of the process Z) given in (14.2.21) with $\sigma = \tau$, and the scaling property of B, we find:

$$\begin{aligned} \int_0^\infty e^{-\lambda t} \mathsf{P}_0\big(B_t \ge z(t)\big)\, dt &= \mathsf{E}_0\Big(\int_0^\infty e^{-\lambda t} f(Z_t)\, dt\Big) \qquad (14.2.31)\\ &= \mathsf{E}_0\Big(\mathsf{E}_0\Big(\int_\tau^\infty e^{-\lambda t} f(Z_t)\, dt \,\Big|\, \mathcal{F}_\tau\Big)\Big)\\ &= \mathsf{E}_0\Big(\mathsf{E}_0\Big(\int_0^\infty e^{-\lambda(\tau+s)} f(Z_{\tau+s})\, ds \,\Big|\, \mathcal{F}_\tau\Big)\Big)\\ &= \mathsf{E}_0\big(e^{-\lambda\tau}\mathsf{E}_0(H\circ\theta_\tau \mid \mathcal{F}_\tau)\big) = \mathsf{E}_0(e^{-\lambda\tau}\mathsf{E}_{Z_\tau} H)\\ &= \int_0^\infty e^{-\lambda t}\mathsf{E}_{(t,g(t))}\Big(\int_0^\infty e^{-\lambda s} f(Z_s)\, ds\Big)\, F(dt)\\ &= \int_0^\infty e^{-\lambda t}\int_0^\infty e^{-\lambda s}\, \mathsf{P}_0\big(g(t)+B_s \ge z(t+s)\big)\, ds\, F(dt)\\ &= \int_0^\infty e^{-\lambda t}\int_0^\infty e^{-\lambda s}\, \Psi\Big(\frac{z(t+s)-g(t)}{\sqrt{s}}\Big)\, ds\, F(dt)\\ &= \int_0^\infty e^{-\lambda t}\int_t^\infty e^{-\lambda(r-t)}\, \Psi\Big(\frac{z(r)-g(t)}{\sqrt{r-t}}\Big)\, dr\, F(dt)\\ &= \int_0^\infty e^{-\lambda r}\int_0^r \Psi\Big(\frac{z(r)-g(t)}{\sqrt{r-t}}\Big)\, F(dt)\, dr \end{aligned}$$

for all $\lambda > 0$. By the uniqueness theorem for Laplace transform it follows that

$$\mathsf{P}_0\big(B_t \ge z(t)\big) = \int_0^t \Psi\left(\frac{z(t)-g(s)}{\sqrt{t-s}}\right) F(ds) \tag{14.2.32}$$

which is seen equivalent to (14.2.30) by the scaling property of B. The proof is complete. □

2. Constant and linear boundaries. It will be shown in Theorem 14.4 below that when g is C^1 on $(0,\infty)$ then there exists a continuous density $f = F'$ of τ. The equation (14.2.28) then becomes

$$\Psi\left(\frac{g(t)}{\sqrt{t}}\right) = \int_0^t \Psi\left(\frac{g(t)-g(s)}{\sqrt{t-s}}\right) f(s)\, ds \tag{14.2.33}$$

for $t > 0$. This is a linear *Volterra integral equation of the first kind* in f if g is known (it is a *nonlinear* equation in g if f is known). Its kernel

$$K(t,s) = \Psi\left(\frac{g(t)-g(s)}{\sqrt{t-s}}\right) \tag{14.2.34}$$

is *nonsingular* in the sense that the mapping $(s,t) \mapsto K(t,s)$ for $0 \le s < t$ is bounded.

If $g(t) \equiv c$ with $c \in \mathbb{R}$, then (14.2.28) or (14.2.33) reads as follows:

$$\mathsf{P}(\tau \le t) = 2\,\mathsf{P}(B_t \ge c) \tag{14.2.35}$$

and this is the *reflection principle* (see (4.4.19)).

If $g(t) = a + bt$ with $b \in \mathbb{R}$ and $a > 0$, then (14.2.33) reads as follows:

$$\Psi\left(\frac{g(t)}{\sqrt{t}}\right) = \int_0^t \Psi\big(b\sqrt{t-s}\big)\, f(s)\, ds \tag{14.2.36}$$

where we see that the kernel $K(t,s)$ is a function of the difference $t-s$ and thus of a convolution type. Standard Laplace transform techniques therefore can be applied to solve the equation (14.2.36) yielding the following explicit formula:

$$f(t) = \frac{a}{t^{3/2}}\,\varphi\left(\frac{a+bt}{\sqrt{t}}\right) \tag{14.2.37}$$

(see (4.4.31)).

The case of more general boundaries g will be treated using classic theory of integral equations in Theorem 14.7 below.

3. Numerical calculation. The fact that the kernel (14.2.34) of the equation (14.2.33) is nonsingular in the sense explained above makes this equation especially attractive to numerical calculations of f if g is given. This can be done using the simple idea of Volterra (dating back to 1896).

Setting $t_j = jh$ for $j = 0, 1, \ldots, n$ where $h = t/n$ and $n \geq 1$ is given and fixed, we see that the following approximation of the equation (14.2.33) is valid (when g is C^1 for instance):

$$\sum_{j=1}^{n} K(t, t_j)\, f(t_j)\, h = b(t) \tag{14.2.38}$$

where we set $b(t) = \Psi(g(t)/\sqrt{t})$. In particular, applying this to each $t = t_i$ yields

$$\sum_{j=1}^{i} K(t_i, t_j)\, f(t_j)\, h = b(t_i) \tag{14.2.39}$$

for $i = 1, \ldots, n$. Setting

$$a_{ij} = 2K(t_i, t_j), \qquad x_j = f(t_j), \qquad b_i = 2b(t_i)/h \tag{14.2.40}$$

we see that the system (14.2.39) reads as follows:

$$\sum_{j=1}^{i} a_{ij} x_j = b_i \qquad (i = 1, \ldots, n) \tag{14.2.41}$$

the simplicity of which is obvious (cf. [149]).

4. Remarks. It follows from (14.2.37) that for τ in (14.2.29) with $g(t) = a+bt$ we have

$$\mathsf{P}(\tau < \infty) = e^{-2\alpha\beta} \tag{14.2.42}$$

whenever $b \geq 0$ and $a > 0$. This shows that F in (14.2.30) does not have to be a proper distribution function but generally satisfies $F(+\infty) \in (0, 1]$.

On the other hand, recall that *Blumenthal's 0–1 law* implies that $\mathsf{P}(\tau = 0)$ is either 0 or 1 for τ in (14.2.29) and a continuous function $g : (0, \infty) \to \mathbb{R}$. If $\mathsf{P}(\tau = 0) = 0$ then g is said to be an *upper function* for B, and if $\mathsf{P}(\tau = 0) = 1$ then g is said to be a *lower function* for B. *Kolmogorov's test* (see e.g. [100, pp. 33–35]) gives sufficient conditions on g to be an upper or lower function. It follows by Kolmogorov's test that $\sqrt{2\,t \log\log 1/t}$ is a lower function for B, and $\sqrt{(2+\varepsilon)\,t \log\log 1/t}$ is an upper function for B for every $\varepsilon > 0$.

4. The existence of a continuous first-passage density. The equation (14.2.33) is a Volterra integral equation of the *first kind.* These equations are generally known to be difficult to deal with directly, and there are two standard ways of reducing them to Volterra integral equations of the *second kind.* The first method consists of differentiating both sides in (14.2.33) with respect to t, and the second method (Theorem 14.7) makes use of an integration by parts in (14.2.33) (see e.g. [92, pp. 40–41]). Our focus in this paragraph is on the first method.

Being led by this objective we now present a simple proof of the fact that F is C^1 when g is C^1 (compare the arguments given below with those given in [207, p. 323] or [65, p. 322]).

Theorem 14.4. *Let $B=(B_t)_{t\ge 0}$ be a standard Brownian motion started at zero, let $g:(0,\infty)\to\mathbb{R}$ be an upper function for B, and let τ in (14.2.29) be the first-passage time of B over g.*

If g is continuously differentiable on $(0,\infty)$ then τ has a continuous density f. Moreover, the following identity is satisfied:

$$\frac{\partial}{\partial t}\Psi\Big(\frac{g(t)}{\sqrt{t}}\Big)=\frac{1}{2}\,f(t)+\int_0^t \frac{\partial}{\partial t}\Psi\Big(\frac{g(t)-g(s)}{\sqrt{t-s}}\Big)\,f(s)\,ds \tag{14.2.43}$$

for all $t>0$.

Proof. 1°. Setting $G(t)=\Psi(g(t)/\sqrt{t})$ and $K(t,s)=\Psi((g(t)-g(s))/\sqrt{t-s})$ for $0\le s<t$ we see that (14.2.28) (i.e. (14.2.30) with $z=g(t)$) reads as follows:

$$G(t)=\int_0^t K(t,s)\,F(ds) \tag{14.2.44}$$

for all $t>0$. Note that $K(t,t-)=\psi(0)=1/2$ for every $t>0$ since $(g(t)-g(s))/\sqrt{t-s}\to 0$ as $s\uparrow t$ for g that is C^1 on $(0,\infty)$. Note also that

$$\frac{\partial}{\partial t}K(t,s)=\frac{1}{\sqrt{t-s}}\Big(\frac{1}{2}\,\frac{g(t)-g(s)}{t-s}-g'(t)\Big)\,\varphi\Big(\frac{g(t)-g(s)}{\sqrt{t-s}}\Big) \tag{14.2.45}$$

for $0<s<t$. Hence we see that $(\partial K/\partial t)(t,t-)$ is not finite (whenever $g'(t)\neq 0$), and we thus proceed as follows.

2°. Using (14.2.44) we find by Fubini's theorem that

$$\begin{aligned}
&\lim_{\varepsilon\downarrow 0}\int_{t_1}^{t_2}\Big(\int_0^{t-\varepsilon}\frac{\partial}{\partial t}K(t,s)\,F(ds)\Big)\,dt \qquad\qquad (14.2.46)\\
&=\lim_{\varepsilon\downarrow 0}\Big(\int_0^{t_2-\varepsilon}K(t_2,s)\,F(ds)\\
&\qquad -\int_0^{t_1-\varepsilon}K(t_1,s)\,F(ds)-\int_{t_1-\varepsilon}^{t_2-\varepsilon}K(s+\varepsilon,s)\,F(ds)\Big)\\
&=G(t_2)-G(t_1)-\frac{1}{2}\big(F(t_2)-F(t_1)\big)
\end{aligned}$$

for $0<t_1\le t\le t_2<\infty$. On the other hand, we see from (14.2.45) that

$$\Big|\int_0^{t-\varepsilon}\frac{\partial}{\partial t}K(t,s)\,F(ds)\Big|\le C\int_0^t\frac{F(ds)}{\sqrt{t-s}} \tag{14.2.47}$$

for all $t \in [t_1, t_2]$ and $\varepsilon > 0$, while again by Fubini's theorem it is easily verified that

$$\int_{t_1}^{t_2} \left(\int_0^t \frac{F(ds)}{\sqrt{t-s}} \right) dt < \infty. \tag{14.2.48}$$

We may thus by the dominated convergence theorem (applied twice) interchange the first limit and the first integral in (14.2.46) yielding

$$\int_{t_1}^{t_2} \left(\int_0^t \frac{\partial}{\partial t} K(t,s)\, F(ds) \right) dt = G(t_2) - G(t_1) - \frac{1}{2}\big(F(t_2) - F(t_1)\big) \tag{14.2.49}$$

at least for those $t \in [t_1, t_2]$ for which

$$\int_0^t \frac{F(ds)}{\sqrt{t-s}} < \infty. \tag{14.2.50}$$

It follows from (14.2.48) that the set of all $t > 0$ for which (14.2.50) fails is of Lebesgue measure zero.

3°. To verify (14.2.50) for all $t > 0$ we may note that a standard rule on the differentiation under an integral sign can be applied in (14.2.30), and this yields the following equation:

$$\frac{1}{\sqrt{t}}\, \varphi\Big(\frac{z}{\sqrt{t}}\Big) = \int_0^t \frac{1}{\sqrt{t-s}}\, \varphi\Big(\frac{z-g(s)}{\sqrt{t-s}}\Big)\, F(ds) \tag{14.2.51}$$

for all $z > g(t)$ with $t > 0$ upon differentiating in (14.2.30) with respect to z. By Fatou's lemma hence we get

$$\begin{aligned} &\int_0^t \frac{1}{\sqrt{t-s}}\, \varphi\Big(\frac{g(t)-g(s)}{\sqrt{t-s}}\Big)\, F(ds) \\ &\quad = \int_0^t \liminf_{z \downarrow g(t)} \frac{1}{\sqrt{t-s}}\, \varphi\Big(\frac{z-g(s)}{\sqrt{t-s}}\Big)\, F(ds) \\ &\quad \le \liminf_{z \downarrow g(t)} \int_0^t \frac{1}{\sqrt{t-s}}\, \varphi\Big(\frac{z-g(s)}{\sqrt{t-s}}\Big)\, F(ds) = \frac{1}{\sqrt{t}}\, \varphi\Big(\frac{g(t)}{\sqrt{t}}\Big) < \infty \end{aligned} \tag{14.2.52}$$

for all $t > 0$. Now for $s < t$ close to t we know that $\varphi((g(t)-g(s))/\sqrt{t-s})$ in (14.2.52) is close to $1/\sqrt{2\pi} > 0$, and this easily establishes (14.2.50) for all $t > 0$.

4°. Returning to (14.2.49) it is easily seen using (14.2.45) that $t \mapsto \int_0^t (\partial K/\partial t)\ (t,s)\, F(ds)$ is right-continuous at $t \in (t_1, t_2)$ if we have

$$\int_t^{t_n} \frac{F(ds)}{\sqrt{t_n - s}} \to 0 \tag{14.2.53}$$

for $t_n \downarrow t$ as $n \to \infty$. To check (14.2.53) we first note that by passing to the limit for $z \downarrow g(t)$ in (14.2.51), and using (14.2.50) with the dominated convergence theorem, we obtain (14.2.56) below for all $t > 0$. Noting that $(s,t) \mapsto \varphi((g(t)-g(s))/\sqrt{t-s})$ attains its strictly positive minimum $c > 0$ over $0 < t_1 \le t \le t_2$ and $0 \le s < t$, we may write

$$\int_t^{t_n} \frac{F(ds)}{\sqrt{t_n - s}} \le \frac{1}{c} \int_t^{t_n} \frac{1}{\sqrt{t_n - s}}\, \varphi\Big(\frac{g(t_n)-g(s)}{\sqrt{t_n - s}}\Big)\, F(ds) \tag{14.2.54}$$
$$= \frac{1}{c}\Big(\frac{1}{\sqrt{t_n}}\,\varphi\Big(\frac{g(t_n)}{\sqrt{t_n}}\Big) - \int_0^t \frac{1}{\sqrt{t_n - s}}\,\varphi\Big(\frac{g(t_n)-g(s)}{\sqrt{t_n - s}}\Big)\, F(ds)\Big)$$

where the final expression tends to zero as $n \to \infty$ by means of (14.2.56) below and using (14.2.50) with the dominated convergence theorem. Thus (14.2.53) holds and therefore $t \mapsto \int_0^t (\partial K/\partial t)(t,s)\, F(ds)$ is right-continuous. It can be similarly verified that this mapping is left-continuous at each $t \in (t_1, t_2)$ and thus continuous on $(0,\infty)$.

5°. Dividing finally by $t_2 - t_1$ in (14.2.49) and then letting $t_2 - t_1 \to 0$, we obtain

$$F'(t) = 2\Big(G'(t) - \int_0^t \frac{\partial}{\partial t} K(t,s)\, F(ds)\Big) \tag{14.2.55}$$

for all $t > 0$. Since the right-hand side of (14.2.55) defines a continuous function of $t > 0$, it follows that $f = F'$ is continuous on $(0,\infty)$, and the proof is complete. □

5. Derivation of known equations. In the previous proof we saw that the master equation (14.2.30) can be once differentiated with respect to z implying the equation (14.2.51), and that in (14.2.51) one can pass to the limit for $z \downarrow g(t)$ obtaining the following equation:

$$\frac{1}{\sqrt{t}}\,\varphi\Big(\frac{g(t)}{\sqrt{t}}\Big) = \int_0^t \frac{1}{\sqrt{t-s}}\,\varphi\Big(\frac{g(t)-g(s)}{\sqrt{t-s}}\Big)\, F(ds) \tag{14.2.56}$$

for all $t > 0$.

The purpose of this paragraph is to show how the equations (14.2.43) and (14.2.56) yield some known equations studied previously by a number of authors.

1°. We assume throughout that the hypotheses of Theorem 14.4 are fulfilled (and that $t > 0$ is given and fixed). Rewriting (14.2.43) more explicitly by computing derivatives on both sides gives

$$\Big(\frac{1}{2}\frac{g(t)}{t^{3/2}} - \frac{g'(t)}{\sqrt{t}}\Big)\varphi\Big(\frac{g(t)}{\sqrt{t}}\Big) = \frac{1}{2}\, f(t) \tag{14.2.57}$$
$$+ \int_0^t \Big(\frac{1}{2}\,\frac{g(t)-g(s)}{(t-s)^{3/2}} - \frac{g'(t)}{\sqrt{t-s}}\Big)\varphi\Big(\frac{g(t)-g(s)}{\sqrt{t-s}}\Big) f(s)\, ds.$$

Recognizing now the identity (14.2.56) multiplied by $g'(t)$ within (14.2.57), and multiplying the remaining part of the identity (14.2.57) by 2, we get

$$\frac{g(t)}{t^{3/2}}\,\varphi\Big(\frac{g(t)}{\sqrt{t}}\Big) = f(t) + \int_0^t \frac{g(t)-g(s)}{(t-s)^{3/2}}\,\varphi\Big(\frac{g(t)-g(s)}{\sqrt{t-s}}\Big) f(s)\,ds. \tag{14.2.58}$$

This equation has been derived and studied by Ricciardi et al. [175] using other means. Moreover, the same argument shows that the factor $1/2$ can be removed from (14.2.57) yielding

$$\Big(\frac{g(t)}{t^{3/2}} - \frac{g'(t)}{\sqrt{t}}\Big)\varphi\Big(\frac{g(t)}{\sqrt{t}}\Big) = f(t) \tag{14.2.59}$$
$$+ \int_0^t \Big(\frac{g(t)-g(s)}{(t-s)^{3/2}} - \frac{g'(t)}{\sqrt{t-s}}\Big)\varphi\Big(\frac{g(t)-g(s)}{\sqrt{t-s}}\Big) f(s)\,ds.$$

This equation has been derived independently by Ferebee [65] and Durbin [48]. Ferebee's derivation is, set aside technical points, the same as the one presented here. Williams [49] presents yet another derivation of this equation (assuming that f exists). [Multiplying both sides of (14.2.33) by $2r(t)$ and both sides of (14.2.56) by $2(k(t){+}g'(t))$, and adding the resulting two equations to the equation (14.2.58), we obtain the equation (14.2.10)+(14.2.30) in Buonocore et al. [24] derived by other means.]

2°. With a view to the inverse problem (of finding g if f is given) it is of interest to produce as many nonequivalent equations linking g to f as possible. (Recall that (14.2.33) is a *nonlinear* equation in g if f is known, and nonlinear equations are marked by a nonuniqueness of solutions.) For this reason it is tempting to derive additional equations to the one given in (14.2.56) starting with the master equation (14.2.30) and proceeding similarly to (14.2.51) above.

A standard rule on the differentiation under an integral sign can be inductively applied to (14.2.30), and this gives the following equations:

$$\frac{1}{t^{n/2}}\,\varphi^{(n-1)}\Big(\frac{z}{\sqrt{t}}\Big) = \int_0^t \frac{1}{(t-s)^{n/2}}\,\varphi^{(n-1)}\Big(\frac{z-g(s)}{\sqrt{t-s}}\Big)\,F(ds) \tag{14.2.60}$$

for all $z > g(t)$ and all $n \ge 1$ where $t > 0$. Recall that

$$\varphi^{(n)}(x) = (-1)^n h_n(x)\varphi(x) \tag{14.2.61}$$

for $x \in \mathbb{R}$ and $n \ge 1$ where h_n is a Hermite polynomial of degree n for $n \ge 1$.

Noting that $\varphi'(x) = -x\varphi(x)$ and recalling (14.2.58) we see that a passage to the limit for $z \downarrow g(t)$ in (14.2.60) is not straightforward when $n \ge 2$ but complicated. For this reason we will not pursue it in further detail here.

3°. The Chapman–Kolmogorov equation (14.2.12) is known to admit a reduction to the forward and backward equation (see [111] and Subsection 4.3) which are partial differential equations of parabolic type. No such derivation or reduction is generally possible in the entire-past dependent case of the equation (14.2.17) or (14.2.18), and the same is true for the master equation (14.2.30) in particular. We showed above how the differentiation with respect to z in the master equation (14.2.30) leads to the density equation (14.2.56), which together with the distribution equation (14.2.28) yields known equations (14.2.58) and (14.2.59). It was also indicated above that no further derivative with respect to z can be taken in the master equation (14.2.30) so that the passage to the limit for $z \downarrow g(t)$ in the resulting equation becomes straightforward.

6. Derivation of new equations. Expanding on the previous facts a bit further we now note that it is possible to proceed in a reverse order and integrate the master equation (14.2.30) with respect to z as many times as we please. This yields a whole spectrum of new nonequivalent equations, which taken together with (14.2.28) and (14.2.56), may play a fundamental role in the inverse problem (see page 240).

Theorem 14.5. *Let $B = (B_t)_{t \ge 0}$ be a standard Brownian motion started at zero, let $g : (0, \infty) \to \mathbb{R}$ be a continuous function satisfying $g(0+) \ge 0$, let τ in* (14.2.29) *be the first-passage time of B over g, and let F denote the distribution function of τ.*

Then the following system of integral equations is satisfied:

$$t^{n/2} H_n\left(\frac{g(t)}{\sqrt{t}}\right) = \int_0^t (t-s)^{n/2} H_n\left(\frac{g(t)-g(s)}{\sqrt{t-s}}\right) F(ds) \tag{14.2.62}$$

for $t > 0$ and $n = -1, 0, 1, \ldots,$ where we set

$$H_n(x) = \int_x^\infty H_{n-1}(z)\, dz \tag{14.2.63}$$

with $H_{-1} = \varphi$ being the standard normal density from (14.2.27).

Remark 14.6. For $n = -1$ the equation (14.2.62) is the density equation (14.2.56). For $n = 0$ the equation (14.2.62) is the distribution equation (14.2.28). All equations in (14.2.62) for $n \ne -1$ are *nonsingular* (in the sense that their kernels are bounded over the set of all (s,t) satisfying $0 \le s < t \le T$).

Proof. Let $t > 0$ be given and fixed. Integrating (14.2.30) we get

$$\int_z^\infty \Psi\left(\frac{z'}{\sqrt{t}}\right) dz' = \int_0^t \int_z^\infty \Psi\left(\frac{z'-g(s)}{\sqrt{t-s}}\right) dz'\, F(ds) \tag{14.2.64}$$

for all $z \ge g(t)$ by means of Fubini's theorem. Substituting $u = z'/\sqrt{t}$ and $v = (z'-g(s))/\sqrt{t-s}$ we can rewrite (14.2.64) as follows:

$$\sqrt{t} \int_{z/\sqrt{t}}^\infty \Psi(u)\, du = \int_0^t \sqrt{t-s} \int_{(z-g(s))/\sqrt{t-s}}^\infty \Psi(v)\, dv\, F(ds) \tag{14.2.65}$$

which is the same as the following identity:

$$\sqrt{t}\, H_1\Big(\frac{z}{\sqrt{t}}\Big) = \int_0^t \sqrt{t-s}\; H_1\Big(\frac{z-g(s)}{\sqrt{t-s}}\Big)\, F(ds) \tag{14.2.66}$$

for all $z \geq g(t)$ upon using that H_1 is defined by (14.2.63) above with $n=1$.

Integrating (14.2.66) as (14.2.30) prior to (14.2.64) above, and proceeding similarly by induction, we get

$$t^{n/2} H_n\Big(\frac{z}{\sqrt{t}}\Big) = \int_0^t (t-s)^{n/2} H_n\Big(\frac{z-g(s)}{\sqrt{t-s}}\Big)\, F(ds) \tag{14.2.67}$$

for all $z \geq g(t)$ and all $n \geq 1$. (This equation was also established earlier for $n=0$ in (14.2.30) and for $n=-1$ in (14.2.51).) Setting $z=g(t)$ in (14.2.67) above we obtain (14.2.62) for all $n \geq 1$. (Using that $\Psi(x) \leq \sqrt{2/\pi}\,\varphi(x)$ for all $x>0$ it is easily verified by induction that all integrals appearing in (14.2.62)–(14.2.67) are finite.) As the equation (14.2.62) was also proved earlier for $n=0$ in (14.2.28) and for $n=-1$ in (14.2.56) above, we see that the system (14.2.62) holds for all $n \geq -1$, and the proof of the theorem is complete. □

In view of our considerations in paragraph 5 above it is interesting to establish the analogues of the equations (14.2.58) and (14.2.59) in the case of other equations in (14.2.62).

For this, fix $n \geq 1$ and $t>0$ in the sequel, and note that taking a derivative with respect to t in (14.2.62) gives

$$\begin{aligned} &\frac{n}{2}\, t^{n/2-1} H_n\Big(\frac{g(t)}{\sqrt{t}}\Big) + t^{n/2} H_n'\Big(\frac{g(t)}{\sqrt{t}}\Big)\Big(\frac{g'(t)}{\sqrt{t}} - \frac{g(t)}{2t^{3/2}}\Big) \\ &\quad = \int_0^t \bigg(\frac{n}{2}\,(t-s)^{n/2-1} H_n\Big(\frac{g(t)-g(s)}{\sqrt{t-s}}\Big) \\ &\qquad\qquad + (t-s)^{n/2} H_n'\Big(\frac{g(t)-g(s)}{\sqrt{t-s}}\Big)\Big(\frac{g'(t)}{\sqrt{t-s}} - \frac{g(t)-g(s)}{2(t-s)^{3/2}}\Big)\bigg)\, F(ds). \end{aligned} \tag{14.2.68}$$

Recognizing now the identity (14.2.62) (with $n-1$ instead of n using that $H_n' = H_{n-1}$) multiplied by $g'(t)$ within (14.2.68), and multiplying the remaining part of the identity (14.2.68) by 2, we get

$$\begin{aligned} &t^{n/2-1}\bigg(nH_n\Big(\frac{g(t)}{\sqrt{t}}\Big) - \frac{g(t)}{\sqrt{t}}\, H_{n-1}\Big(\frac{g(t)}{\sqrt{t}}\Big)\bigg) \\ &\quad = \int_0^t (t-s)^{n/2-1}\bigg(nH_n\Big(\frac{g(t)-g(s)}{\sqrt{t-s}}\Big) \\ &\qquad\qquad - \frac{g(t)-g(s)}{\sqrt{t-s}}\, H_{n-1}\Big(\frac{g(t)-g(s)}{\sqrt{t-s}}\Big)\bigg)\, F(ds). \end{aligned} \tag{14.2.69}$$

Moreover, the same argument shows that the factor $1/2$ can be removed from (14.2.68) yielding

$$\begin{aligned} & t^{n/2}\left(\frac{n}{t}\,H_n\left(\frac{g(t)}{\sqrt{t}}\right) - \left(\frac{g(t)}{t^{3/2}} - \frac{g'(t)}{\sqrt{t}}\right)H_{n-1}\left(\frac{g(t)}{\sqrt{t}}\right)\right) \\ &= \int_0^t (t-s)^{n/2}\left(\frac{n}{(t-s)}\,H_n\left(\frac{g(t)-g(s)}{\sqrt{t-s}}\right)\right. \\ &\qquad \left. - \left(\frac{g(t)-g(s)}{(t-s)^{3/2}} - \frac{g'(t)}{\sqrt{t-s}}\right)H_{n-1}\left(\frac{g(t)-g(s)}{\sqrt{t-s}}\right)\right) F(ds). \end{aligned} \tag{14.2.70}$$

Each of the equations (14.2.69) and (14.2.70) is contained in the system (14.2.62). No equation of the system (14.2.62) is equivalent to another equation from the same system but itself.

7. A closed expression for the first-passage distribution. In this paragraph we briefly tackle the problem of finding F when g is given using classic theory of linear integral equations (see e.g. [92]). The key tool in this approach is *the fixed-point theorem* for *contractive* mappings, which states that a mapping $T : X \to X$, where (X, d) is a complete metric space, satisfying

$$d(T(x), T(y)) \le \beta\, d(x, y) \tag{14.2.71}$$

for all $x, y \in X$ with some $\beta \in (0, 1)$ has a unique fixed point in X, i.e. there exists a unique point $x_0 \in X$ such that $T(x_0) = x_0$.

Using this principle and some of its ramifications developed within the theory of integral equations, the papers [148] and [175] present explicit expressions for F in terms of g in the case when X is taken to be a Hilbert space L^2. These results will here be complemented by describing a narrow class of boundaries g that allow X to be the Banach space $B(\mathbb{R}_+)$ of all bounded functions $h : \mathbb{R}_+ \to \mathbb{R}$ equipped with the sup-norm

$$\|h\|_\infty = \sup_{t \ge 0} |h(t)|. \tag{14.2.72}$$

While examples from this class range from a constant to a square-root boundary, the approach itself is marked by simplicity of the argument.

Theorem 14.7. *Let $B = (B_t)_{t \ge 0}$ be a standard Brownian motion started at zero, let $g : \mathbb{R}_+ \to \mathbb{R}$ be a continuous function satisfying $g(0) > 0$, let τ in (14.2.29) be the first-passage time of B over g, and let F denote the distribution function of τ.*

Assume, moreover, that g is C^1 on $(0, \infty)$, increasing, concave, and that it satisfies

$$g(t) \le g(0) + c\sqrt{t} \tag{14.2.73}$$

for all $t \geq 0$ *with some* $c > 0$. *Then we have*

$$F(t) = h(t) + \sum_{n=1}^{\infty} \Big(\int_0^t K_n(t,s)\, h(s)\, ds \Big) \tag{14.2.74}$$

where the series converges uniformly over all $t \geq 0$, *and we set*

$$h(t) = 2\Psi\Big(\frac{g(t)}{\sqrt{t}}\Big), \tag{14.2.75}$$

$$K_1(t,s) = \frac{1}{\sqrt{t-s}} \Big(2\, g'(s) - \frac{g(t)-g(s)}{t-s} \Big)\, \varphi\Big(\frac{g(t)-g(s)}{\sqrt{t-s}} \Big), \tag{14.2.76}$$

$$K_{n+1}(t,s) = \int_s^t K_1(t,r) K_n(r,s)\, dr \tag{14.2.77}$$

for $0 \leq s < t$ *and* $n \geq 1$.

Moreover, introducing the function

$$R(t,s) = \sum_{n=1}^{\infty} K_n(t,s) \tag{14.2.78}$$

for $0 \leq s < t$, *the following representation is valid*:

$$F(t) = h(t) + \int_0^t R(t,s)\, h(s)\, ds \tag{14.2.79}$$

for all $t > 0$.

Proof. Setting $u = \Psi\big((g(t)-g(s))/\sqrt{t-s}\,\big)$ and $v = F(s)$ in the integral equation (14.2.28) and using the integration by parts formula, we obtain

$$\Psi\Big(\frac{g(t)}{\sqrt{t}}\Big) = \frac{1}{2}\, F(t) - \int_0^t \frac{\partial}{\partial s} \Psi\Big(\frac{g(t)-g(s)}{\sqrt{t-s}} \Big) F(s)\, ds \tag{14.2.80}$$

for each $t > 0$ that is given and fixed in the sequel. Using the notation of (14.2.75) and (14.2.76) above we can rewrite (14.2.80) as follows:

$$F(t) - \int_0^t K_1(t,s) F(s)\, ds = h(t). \tag{14.2.81}$$

Introduce a mapping T on $B(\mathbb{R}_+)$ by setting

$$(T(G))(t) = h(t) + \int_0^t K_1(t,s)\, G(s)\, ds \tag{14.2.82}$$

for $G \in B(\mathbb{R}_+)$. Then (14.2.81) reads as follows:

$$T(F) = F \tag{14.2.83}$$

and the problem reduces to solving (14.2.83) for F in $B(\mathbb{R}_+)$.

In view of the fixed-point theorem quoted above, we need to verify that T is a contraction from $B(\mathbb{R}_+)$ into itself with respect to the sup-norm (14.2.72). For this, note that

$$\begin{aligned} \|T(G_1)-T(G_2)\|_\infty &= \sup_{t\ge 0} |(T(G_1-G_2))(t)| \\ &= \sup_{t\ge 0} \Big| \int_0^t K_1(t,s)\,\big(G_1(s)-G_2(s)\big)\,ds \Big| \\ &\le \Big(\sup_{t\ge 0} \int_0^t |K_1(t,s)|\,ds \Big) \|G_1-G_2\|_\infty . \end{aligned} \tag{14.2.84}$$

Since $s \to g(s)$ is concave and increasing, it is easily verified that $s \mapsto (g(t)-g(s))/\sqrt{t-s}$ is decreasing and thus $s \mapsto \Psi\big((g(t)-g(s))/\sqrt{t-s}\big)$ is increasing on $(0,t)$. It implies that

$$\begin{aligned} \beta := \sup_{t\ge 0} \int_0^t \big|K_1(t,s)\big|\,ds &= \sup_{t\ge 0} \int_0^t \Big| 2\,\frac{\partial}{\partial s}\Psi\Big(\frac{g(t)-g(s)}{\sqrt{t-s}}\Big)\Big|\,ds \\ = \sup_{t\ge 0} \int_0^t 2\,\frac{\partial}{\partial s}\Psi\Big(\frac{g(t)-g(s)}{\sqrt{t-s}}\Big)\,ds & \\ = \sup_{t\ge 0} 2\Big(\frac{1}{2} - \Psi\Big(\frac{g(t)-g(0)}{\sqrt{t}}\Big)\Big) &\le 1-2\Psi(c) < 1 \end{aligned} \tag{14.2.85}$$

using the hypothesis (14.2.73). This shows that T is a contraction from the Banach space $B(\mathbb{R}_+)$ into itself, and thus by the fixed-point theorem there exists a unique F_0 in $B(\mathbb{R}_+)$ satisfying (14.2.83). Since the distribution function F of τ belongs to $B(\mathbb{R}_+)$ and satisfies (14.2.83), it follows that F_0 must be equal to F.

Moreover, the representation (14.2.74) follows from (14.2.81) and the well-known formula for the resolvent of the integral operator $K = T-h$ associated with the kernel K_1:

$$\big(I-K\big)^{-1} = \sum_{n=0}^{\infty} K^n \tag{14.2.86}$$

upon using Fubini's theorem to justify that K_{n+1} in (14.2.77) is the kernel of the integral operator K^{n+1} for $n \ge 1$. Likewise, the final claim about (14.2.78) and (14.2.79) follows by the Fubini–Tonelli theorem since all kernels in (14.2.76) and (14.2.77) are non-negative, and so are all maps $s \mapsto K_n(t,s)\,h(s)$ in (14.2.74) as well. This completes the proof. □

Leaving aside the question on usefulness of the multiple-integral series representation (14.2.74), it is an interesting mathematical question to find a similar expression for F in terms of g that would not require additional hypotheses on g such as (14.2.73) for instance. In this regard especially those g satisfying $g(0+) = 0$ seem problematic as they lead to singular (or weakly singular) kernels generating the integral operators that turn out to be noncontractive.

8. The inverse problem. In this paragraph we will reformulate the inverse problem of finding g when F is given using the result of Theorem 14.5. Recall from there that g and F solve

$$t^{n/2} H_n\left(\frac{g(t)}{\sqrt{t}}\right) = \int_0^t (t-s)^{n/2} H_n\left(\frac{g(t)-g(s)}{\sqrt{t-s}}\right) F(ds) \tag{14.2.87}$$

for $t > 0$ and $n \geq -1$ where $H_n(x) = \int_x^\infty H_{n-1}(z)\, dz$ with $H_{-1} = \varphi$. Then the inverse problem reduces to answer the following three questions:

Question 8.1. Does there exist a (continuous) solution $t \mapsto g(t)$ of the system (14.2.87)?

Question 8.2. Is this solution unique?

Question 8.3. Does the (unique) solution $t \mapsto g(t)$ solve the inverse first-passage problem i.e. is the distribution function of τ from (14.2.29) equal to F?

It may be noted that each equation in g of the system (14.2.87) is a *nonlinear Volterra integral equation of the second kind.* Nonlinear equations are known to lead to nonunique solutions, so it is hoped that the totality of countably many equations could counterbalance this deficiency.

Perhaps the main example one should have in mind is when F has a continuous density f. Note that in this case $f(0+)$ can be strictly positive (and finite). Some information on possible behaviour of g at zero for such f can be found in [162] (see also [207] for closely related results).

Notes. The first-passage problem has a long history and a large number of applications. Yet explicit solutions to the first-passage problem (for Brownian motion) are known only in a limited number of special cases including linear or quadratic g. The law of τ is also known for a square-root boundary g but only in the form of a Laplace transform (which appears intractable to inversion). The inverse problem seems even harder. For example, it is not known if there exists a boundary g for which τ is exponentially distributed (cf. [162]).

One way to tackle the problem is to derive an equation which links g and F. Motivated by this fact many authors have studied integral equations in connection with the first-passage problem (see e.g. [182], [205], [69], [201], [149], [65], [175], [48], [124]) under various hypotheses and levels of rigor. The main aim of this section (following [161]) is to present a unifying approach to the integral equations arising in the first-passage problem that is done in a rigorous fashion and with minimal tools.

The approach naturally leads to a system of integral equations for g and F (paragraph 6) in which the first two equations contain the previously known ones (paragraph 5). These equations are derived from a single *master equation* (Theorem 14.3) that can be viewed as a *Chapman–Kolmogorov equation of Volterra type* (see Theorem 14.2). The initial idea in the derivation of the master equation goes back to Schrödinger [182]. The master equation cannot be reduced to a partial differential equation of forward or backward type (cf. [111]). A key technical detail needed to connect the second equation of the system to known methods leads to a simple proof of the fact that F has a continuous density when g is continuously differentiable (Theorem 14.4). The problem of finding F when g is given is tackled using classic theory of linear integral equations (Theorem 14.7). The inverse problem is reduced to solving a system of *nonlinear* Volterra integral equations of the second kind (see (14.2.87)). General theory of such systems seems far from being complete at present.

Chapter V.

Optimal stopping in stochastic analysis

The aim of this chapter is to study a number of optimal stopping problems which are closely related to sharp inequalities arising in stochastic analysis. We will begin by giving a general overview of the methodology which will be applied throughout.

15. Review of problems

In stochastic analysis one often deals with a "complicated" random variable X^c and tries to say something about its properties in terms of a "simple" random variable X^s. For example, if $B = (B_t)_{t \geq 0}$ is a standard Brownian motion and we consider $X^c = \sup_{0 \leq t \leq \tau} |B_t|^2$ for a stopping time τ of B, then X^c may be a complicated random variable and, for example, it may be nontrivial to compute $\mathsf{E}X^c$ or even say something about its exact size. On the other hand, recalling that $\mathsf{E}X^c = \mathsf{E} \max_{0 \leq t \leq \tau} |B_t|^2 \leq 4\,\mathsf{E}\tau$ by Doob's inequality (and the optional sampling theorem) and setting $X^s = \tau$, we see that although X^s may not be that simple at all, it may be possible to say something about its expectation $\mathsf{E}X^s = \mathsf{E}\tau$ (e.g. show that it is finite) and in this way get some information about the "complicated" quantity $\mathsf{E}X^c = \mathsf{E} \max_{0 \leq t \leq \tau} |B_t|^2$ (i.e. conclude that it is finite). Even a more appealing choice in terms of simplicity for X^s is $|B_\tau|^2$ when the inequality $\mathsf{E}X^c = \mathsf{E} \max_{0 \leq t \leq \tau} |B_t|^2 \leq 4\mathsf{E}|B_\tau|^2 = 4\mathsf{E}X^s$ provides a rather strong conclusion on the size of the maximum of $|B_t|^2$ over all $t \in [0, \tau]$ in terms of the terminal value $|B_\tau|^2$.

This sort of reasoning is the main motivation for the present chapter. It turns out that optimal stopping techniques prove very helpful in deriving sharp inequalities of the preceding type.

To describe this in more detail, still keeping it very general, let us assume that $X = (X_t)_{t \geq 0}$ is a Markov process, and for given functions L and K let

$$I_t = \int_0^t L(X_s)\, ds, \tag{15.0.1}$$

$$S_t = \max_{0 \leq s \leq t} K(X_s) \tag{15.0.2}$$

be the integral process associated with $L(X)$ and the maximum process associated with $K(X)$ for $t \geq 0$. Given functions F and G we may and will consider the following optimal stopping problem:

$$V(c) = \sup_{\tau} \mathsf{E}\Big[F(I_\tau, X_\tau, S_\tau) - c\, G(I_\tau, X_\tau, S_\tau)\Big] \tag{15.0.3}$$

where the supremum is taken over a class of stopping/Markov times τ, and $c > 0$ is a given and fixed constant.

It follows from (15.0.3) that

$$\mathsf{E} F(I_\tau, X_\tau, S_\tau) \leq V(c) + c\, \mathsf{E} G(I_\tau, X_\tau, S_\tau) \tag{15.0.4}$$

for all stopping times τ and all $c > 0$. Hence

$$\begin{aligned} \mathsf{E} F(I_\tau, X_\tau, S_\tau) &\leq \inf_{c>0} \Big(V(c) + c\, \mathsf{E} G(I_\tau, X_\tau, S_\tau)\Big) \\ &:= H\big(\mathsf{E} G(I_\tau, X_\tau, S_\tau)\big) \end{aligned} \tag{15.0.5}$$

for all stopping times τ. In this way we have produced a function H which has the power of providing a sharp estimate of $\mathsf{E} F(I_\tau, X_\tau, S_\tau)$ in terms of $\mathsf{E} G(I_\tau, X_\tau, S_\tau)$. Note that when supremum in (15.0.3) is attained, then equality in (15.0.4) is attained, so that whenever this is true for all $c > 0$, it is in particular true for $c_* > 0$ at which the infimum in (15.0.5) is attained (or approximately attained), demonstrating that (15.0.5) is indeed a sharp inequality as claimed.

In what follows we will study a number of specific examples of the optimal stopping problem (15.0.3). Normally the functions F and G (as well as L and K) take a simple form. For example, in the case of Doob's inequality above we have $X = B$, $K(x) = x^2$, $L(x) \equiv 1$, $F(a, x, s) = s$ and $G(a, x, s) = a$. When G (or F to the same effect) is a nonlinear function of a (e.g. $G(a, x, s) = \sqrt{a}$) we speak of *nonlinear* problems. Note that such problems are also studied in Section 10 above (see also Section 20 below).

16. Wald inequalities

The aim of this section (following [78]) is to present the solution to a class of Wald type optimal stopping problems for Brownian motion, and from this deduce some sharp inequalities, which give bounds on the expectation of functionals of randomly stopped Brownian motion in terms of the expectation of the stopping time.

16.1. Formulation of the problem

Let $B = (B_t)_{t \ge 0}$ be a standard Brownian motion defined on a probability space $(\Omega, \mathcal{F}, \mathsf{P})$. In this section we solve all optimal stopping problems of the following form: *Maximize the expectation*

$$\mathsf{E}\big(G(|B_\tau|) - c\tau\big) \tag{16.1.1}$$

over all stopping times τ *of* B *with* $\mathsf{E}\,\tau < \infty$, *where the measurable function* $G : \mathbb{R}_+ \to \mathbb{R}$ *satisfies* $G(|x|) \le c|x|^2 + d$ *for all* $x \in \mathbb{R}$ *with some* $d \in \mathbb{R}$, *and* $c > 0$ *is given and fixed.*

It will be shown below (Theorem 16.1) that the (approximately) optimal stopping time is the first hitting time of the reflecting Brownian motion $|B| = (|B_t|)_{t \ge 0}$ to the set of all (approximate) maximum points of the function $x \mapsto G(|x|) - cx^2$ on $\mathbb{R}$. This leads to some sharp inequalities which will be discussed below.

16.2. Solution to the problem

In this subsection we present the solution to the optimal stopping problem (16.1.1). For simplicity, we will only consider the case where $G(|x|) = |x|^p$ for $0 < p \le 2$, and it will be clear from the proof below that the case of a general function G (satisfying the boundedness condition) could be treated analogously.

1. Thus, if $B = (B_t)_{t \ge 0}$ is a standard Brownian motion, then the problem under consideration is the following: *Maximize the expectation*

$$\mathsf{E}\big(|B_\tau|^p - c\tau\big) \tag{16.2.1}$$

over all stopping times τ *of* B *with* $\mathsf{E}\,\tau < \infty$, *where* $0 < p \le 2$ *and* $c > 0$ *are given and fixed.*

Firstly, it should be noted that in the case $p = 2$, we find by the classical Wald identity (see (3.2.6)) for Brownian motion ($\mathsf{E}\,|B_\tau|^2 = \mathsf{E}\,\tau$) that the expression in (16.2.1) equals $(1-c)\mathsf{E}\,\tau$. Thus, taking $\tau \equiv n$ or 0 for $n \ge 1$, depending on whether $0 < c < 1$ or $1 < c < \infty$, we see that the supremum equals $+\infty$ or 0 respectively. If $c = 1$, then the supremum equals 0, and any stopping time τ of B with $\mathsf{E}\,\tau < \infty$ is optimal. These facts solve the problem (16.2.1) in the case $p = 2$. The solution in the general case $0 < p < 2$ is formulated in the following theorem.

Theorem 16.1. (Wald's optimal stopping of Brownian motion) *Let* $B = (B_t)_{t \ge 0}$ *be standard Brownian motion and let* $0 < p < 2$ *and* $c > 0$ *be given and fixed. Consider the optimal stopping problem*

$$\sup_\tau \mathsf{E}\big(|B_\tau|^p - c\tau\big) \tag{16.2.2}$$

where the supremum is taken over all stopping times τ *of* B *with* $\mathsf{E}\tau<\infty$. *Then the optimal stopping time in* (16.2.2) (*the one at which the supremum is attained*) *is given by*

$$\tau_{p,c}^{*} = \inf\left\{ t>0 : |B_t| = \left(\frac{p}{2c}\right)^{1/(2-p)} \right\}. \tag{16.2.3}$$

Moreover, for all stopping times τ *of* B *with* $\mathsf{E}\tau<\infty$ *we have*

$$\mathsf{E}\left(|B_\tau|^p - c\tau\right) \le \left(\frac{2-p}{2}\right)\left(\frac{p}{2c}\right)^{p/(2-p)}. \tag{16.2.4}$$

The upper bound in (16.2.4) *is best possible.*

Proof. Given $0<p<2$ and $c>0$, denote

$$V_\tau(p,c) = \mathsf{E}\left(|B_\tau|^p - c\tau\right) \tag{16.2.5}$$

whenever τ is a stopping time of B with $\mathsf{E}\tau<\infty$. Then by Wald's identity for Brownian motion it follows that the expression in (16.2.5) may be equivalently written in the following form:

$$V_\tau(p,c) = \int_{-\infty}^{\infty} \left(|x|^p - cx^2\right) dP_{B_\tau}(x) \tag{16.2.6}$$

whenever τ is a stopping time of B with $\mathsf{E}\tau<\infty$. Our next step is to maximize the function $x \mapsto D(x) = |x|^p - cx^2$ over $\mathbb{R}$. For this, note that $D(-x)=D(x)$ for all $x\in\mathbb{R}$, and therefore it is enough to consider $D(x)$ for $x>0$. We have $D'(x) = px^{p-1} - 2cx$ for $x>0$, and hence we see that D attains its maximal value at the point $\pm(p/2c)^{1/(2-p)}$. Thus it is clear from (16.2.6) that the optimal stopping time in (16.2.2) is to be defined by (16.2.3). This completes the first part of the proof.

Finally, inserting $\tau^* = \tau_{p,c}^*$ from (16.2.3) into (16.2.6), we easily find that

$$V_{\tau^*}(p,c) = D\left(\left(\frac{p}{2c}\right)^{1/(2-p)}\right) = \left(\frac{2-p}{2}\right)\left(\frac{p}{2c}\right)^{p/(2-p)}. \tag{16.2.7}$$

This establishes (16.2.4) with the last statement of the theorem, and the proof is complete. □

Remark 16.2. The preceding proof shows that the solution to the problem (16.1.1) in the case of a general function G (satisfying the boundedness condition) could be found by using exactly the same method: *The (approximately) optimal stopping time is the first hitting time of the reflecting Brownian motion* $|B| = (|B_t|)_{t\ge0}$ *to the set of all (approximate) maximum points of the function* $x\mapsto D(x) = G(|x|) - cx^2$ *on* $\mathbb{R}$. Here "approximate" stands to cover the case (in an obvious manner) when D does not attain its least upper bound on the real line.

2. In the remainder of this subsection we will explore some consequences of the inequality (16.2.4) in more detail. For this, let a stopping time τ of B with $\mathsf{E}\,\tau < \infty$ and $0 < p < 2$ be given and fixed. Then from (16.2.4) we get

$$\mathsf{E}\,|B_\tau|^p \le \inf_{c>0} \left(c\mathsf{E}\,\tau + \Big(\frac{2-p}{2}\Big)\Big(\frac{p}{2c}\Big)^{p/(2-p)} \right). \tag{16.2.8}$$

It is elementary to compute that this infimum equals $(\mathsf{E}\,\tau)^{p/2}$. In this way we obtain

$$\mathsf{E}\,|B_\tau|^p \le (\mathsf{E}\,\tau)^{p/2} \qquad (0 < p \le 2) \tag{16.2.9}$$

with the constant 1 being best possible in all the inequalities. (Observe that this also follows by Wald's identity and Jensen's inequality in a straightforward way.)

Next let us consider the case $2 < p < \infty$. Thus we shall look at $-V_\tau(p,c)$ instead of $V_\tau(p,c)$ in (16.2.5) and (16.2.6). By the same argument as for (16.2.6) we obtain

$$-V_\tau(p,c) = \mathsf{E}\left(c\tau - |B_\tau|^p\right) = \int_{-\infty}^{\infty} \left(cx^2 - |x|^p\right) dP_{B_\tau}(x) \tag{16.2.10}$$

where $2 < p < \infty$. The same calculation as in the proof of Theorem 16.1 shows that the function $x \mapsto -D(x) = cx^2 - |x|^p$ attains its maximal value over $\mathbb{R}$ at the point $\pm(p/2c)^{1/(2-p)}$. Thus as in the proof of Theorem 16.1 we find

$$\mathsf{E}\left(c\tau - |B_\tau|^p\right) \le \Big(\frac{p-2}{2}\Big)\Big(\frac{p}{2c}\Big)^{p/(2-p)}. \tag{16.2.11}$$

From this inequality we get

$$\sup_{c>0} \left(c\mathsf{E}\,\tau + \Big(\frac{2-p}{2}\Big)\Big(\frac{p}{2c}\Big)^{p/(2-p)} \right) \le \mathsf{E}\,|B_\tau|^p. \tag{16.2.12}$$

The same calculation as for the proof of (16.2.9) shows that this supremum equals $(\mathsf{E}\,\tau)^{p/2}$. Thus as above for (16.2.9) we obtain

$$(\mathsf{E}\,\tau)^{p/2} \le \mathsf{E}\,|B_\tau|^p \qquad (2 \le p < \infty) \tag{16.2.13}$$

with the constant 1 being best possible in all the inequalities. (Observe again that this also follows by Wald's identity and Jensen's inequality in a straightforward way.)

3. The previous calculations together with the conclusions (16.2.9) and (16.2.13) indicate that the inequality (16.2.4)+(16.2.8) provide sharp estimates which are otherwise obtainable by a different method that relies upon convexity and Jensen's inequality (see Remark 16.4 below). This leads precisely to the main observation: *The previous procedure can be repeated for any measurable map G*

satisfying the boundedness condition. In this way we obtain a sharp estimate of the form

$$\mathsf{E}\,G\big(|B_\tau|\big) \le \gamma_{_G}(\mathsf{E}\,\tau) \tag{16.2.14}$$

where $\gamma_{_G}$ is a function to be found (by maximizing and minimizing certain real valued functions of real variables). We formulate this more precisely in the next corollary.

Corollary 16.3. *Let* $B=(B_t)_{t\ge 0}$ *be standard Brownian motion, and let* $G:\mathbb{R}\to\mathbb{R}$ *be a measurable map. Then for any stopping time* τ *of* B *the following inequality holds*:

$$\mathsf{E}\,G\big(|B_\tau|\big) \le \inf_{c>0}\Big(c\,\mathsf{E}\,\tau + \sup_{x\in\mathbb{R}}\big(G(|x|)-cx^2\big)\Big) \tag{16.2.15}$$

and is sharp whenever the right-hand side is finite. Similarly, if $H:\mathbb{R}\to\mathbb{R}$ *is a measurable map, then for any stopping time* τ *of* B *with* $\mathsf{E}\,\tau<\infty$ *the following inequality holds*:

$$\sup_{c>0}\Big(c\,\mathsf{E}\,\tau + \inf_{x\in\mathbb{R}}\big(H(|x|)-cx^2\big)\Big) \le \mathsf{E}\,H(|B_\tau|) \tag{16.2.16}$$

and is sharp whenever the left-hand side is finite.

Proof. It follows from the proof of Theorem 16.1 as indicated in Remark 16.2 and the lines above following it (or just straightforwardly by using Wald's identity). It should be noted that the boundedness condition on the maps G and H is contained in the nontriviality of the conclusions. □

Remark 16.4. If we set $H(x)=G(\sqrt{x})$ for $x\ge 0$, then

$$\sup_{x\in\mathbb{R}}\big(G(|x|)-cx^2\big) = -\inf_{x\ge 0}\big(cx-H(x)\big) =: -\widetilde{H}(c) \tag{16.2.17}$$

where $\widetilde{H}$ denotes the *concave conjugate* of H. Similarly, we have

$$\inf_{c>0}\Big(c\,\mathsf{E}\,\tau + \sup_{x\in\mathbb{R}}\big(G(|x|)-cx^2\big)\Big) = \inf\big(c\,\mathsf{E}\,\tau - \widetilde{H}(c)\big) = \widetilde{\widetilde{H}}(\mathsf{E}\,\tau). \tag{16.2.18}$$

Thus (16.2.15) reads as

$$\mathsf{E}\,H(|B_\tau|^2) \le \widetilde{\widetilde{H}}(\mathsf{E}\,\tau). \tag{16.2.19}$$

Moreover, since $\widetilde{\widetilde{H}}$ is the (smallest) concave function which dominates H, it is clear from a simple comparison that (16.2.19) also follows by Jensen's inequality. This provides an alternative way of looking at (16.2.15) and clarifies (16.2.8)–(16.2.9). (A similar remark may be directed to (16.2.16) with (16.2.12)–(16.2.13).) Note that (16.2.19) gets the form

$$\mathsf{E}\,G\big(|B_\tau|\big) \le G\big(\sqrt{\mathsf{E}\,\tau}\,\big) \tag{16.2.20}$$

whenever $x\mapsto G(\sqrt{x})$ is *concave* on $\mathbb{R}_+$.

Remark 16.5. By using the standard time-change method (see Subsection 5.1), one can generalize and extend the inequalities (16.2.15) and (16.2.16) to cover the case of all continuous local martingales. *Let* $M = (M_t)_{t\ge 0}$ *be a continuous local martingale with the quadratic variation process* $\langle M\rangle = (\langle M\rangle_t)_{t\ge 0}$ *(see* (3.3.6)) *such that* $M_0 = 0$ *, and let* $G, H : \mathbb{R}_+ \to \mathbb{R}$ *be measurable functions. Then for any* $t > 0$ *for which* $\mathsf{E}\,\langle M\rangle_t < \infty$ *the following inequalities hold:*

$$\mathsf{E}\,G(|M_t|) \le \inf_{c>0}\Big(c\,\mathsf{E}\,\langle M\rangle_t + \sup_{x\in\mathbb{R}}\big(G(|x|) - cx^2\big)\Big), \tag{16.2.21}$$

$$\sup_{c>0}\Big(c\,\mathsf{E}\,\langle M\rangle_t + \inf_{x\in\mathbb{R}}\big(H(|x|) - cx^2\big)\Big) \le \mathsf{E}\,H(|M_t|) \tag{16.2.22}$$

and are sharp whenever the right-hand side in (16.2.21) *and the left-hand side in* (16.2.22) *are finite.*

To prove the sharpness of (16.2.21) and (16.2.22) for any given and fixed $t > 0$, consider $M_t = B_{\alpha t+\tau_\beta}$ with $\alpha > 0$ and τ_β being the first hitting time of the reflecting Brownian motion $|B| = (|B_t|)_{t\ge 0}$ to some $\beta > 0$. Letting $\alpha \to \infty$ and using (integrability) properties of τ_β (in the context of Corollary 16.3), by the Burkholder-Davis-Gundy inequalities (see (C5) on page 63) and uniform integrability arguments one ends up with the inequalities (16.2.15) and (16.2.16) for optimal $\tau = \tau_\beta$, at least in the case when G allows that the limiting procedures required can be performed (the case of general G can then follow by approximation). Thus the sharpness of (16.2.21)–(16.2.22) follows from the sharpness of (16.2.15)–(16.2.16).

16.3. Applications

As an application of the methodology exposed above, we will present a simple proof of the Dubins–Jacka–Schwarz–Shepp–Shiryaev (square-root-of-two) maximal inequality for randomly stopped Brownian motion, which was first derived in [44] and independently in [103], and then proved by an entirely different method in [45]. We will begin by stating two inequalities to be proved (the second one being the "square-root-of-two" inequality).

Let $B = (B_t)_{t\ge 0}$ be a standard Brownian motion, and let τ be a stopping time of B with $\mathsf{E}\,\tau<\infty$. Then the following inequalities are sharp:

$$\mathsf{E}\Big(\max_{0\le t\le\tau} B_t\Big) \le \sqrt{\mathsf{E}\,\tau}, \tag{16.3.1}$$

$$\mathsf{E}\Big(\max_{0\le t\le\tau} |B_t|\Big) \le \sqrt{2}\,\sqrt{\mathsf{E}\,\tau}. \tag{16.3.2}$$

1°. We shall first deduce these inequalities by our method, and then show their sharpness by exhibiting the optimal stopping times (at which the equalities are attained). Our approach to the problem of establishing (16.3.1) is motivated by the fact that the process $(\max_{0\le s\le t} B_s - B_t)_{t\ge 0}$ is equally distributed as the reflecting Brownian motion process $(|B_t|)_{t\ge 0}$ for which we have found optimal bound (16.2.4) (from where by (16.2.8) we get (16.2.9) with $p=1$), while $\mathsf{E}\,B_\tau = 0$ whenever $\mathsf{E}\sqrt{\tau} < \infty$. These observations clearly lead us to (16.3.1), at least for some stopping times. To extend this to all stopping times, we shall use a simple martingale argument.

Proof of (16.3.1): Set $S_t = \max_{0\le s\le t} B_s$ for $t \ge 0$. Since $(B_t^2 - t)_{t\ge 0}$ is a martingale, and $(S_t - B_t)_{t\ge 0}$ is equally distributed as $(|B_t|)_{t\ge 0}$, we see that

$$Z_t = c\big((S_t - B_t)^2 - t\big) + 1/4c \tag{16.3.3}$$

is a martingale (with respect to the natural filtration which is known to be the same as the natural filtration of B). Using $\mathsf{E}\,B_\tau = 0$, by the optional sampling theorem (page 60) and the elementary inequality $x - ct \le c(x^2 - t) + 1/4c$, we find

$$\mathsf{E}\,(S_\tau - c\tau) = \mathsf{E}\,(S_\tau - B_\tau - c\tau) \le \mathsf{E}\,Z_\tau = \mathsf{E}\,Z_0 = 1/4c \tag{16.3.4}$$

for any bounded stopping time τ. Hence we get

$$\mathsf{E}\,S_\tau \le \inf_{c>0}\Big(c\,\mathsf{E}\,\tau + \frac{1}{4c}\Big) = \sqrt{\mathsf{E}\,\tau} \tag{16.3.5}$$

for any bounded stopping time τ. Passing to the limit, we obtain (16.3.1) for all stopping times with finite expectation. This completes the proof of (16.3.1).

2°. Next we extend (16.3.1) to any continuous local martingale $M = (M_t)_{t\ge 0}$ with $M_0 = 0$. For this, note that by the time change and (16.3.1) we obtain

$$\mathsf{E}\Big(\max_{0\le s\le t} M_s\Big) = \mathsf{E}\Big(\max_{0\le s\le t} B_{\langle M\rangle_s}\Big) = \mathsf{E}\Big(\max_{0\le s\le \langle M\rangle_t} B_s\Big) \le \sqrt{\mathsf{E}\,\langle M\rangle_t} \tag{16.3.6}$$

for all $t > 0$.

3°. In the next step we will apply (16.3.6) to the continuous martingale M defined by

$$M_t = \mathsf{E}\Big(|B_\tau| - \mathsf{E}\,|B_\tau| \,\big|\, \mathcal{F}_{t\wedge\tau}\Big) \tag{16.3.7}$$

for $t \ge 0$. In this way we get

$$\mathsf{E}\Big(\max_{0\le t<\infty} \mathsf{E}\,\big(|B_\tau| - \mathsf{E}\,|B_\tau| \,\big|\, \mathcal{F}_{t\wedge\tau}\big)\Big) \le \sqrt{\mathsf{E}\,\big(|B_\tau| - \mathsf{E}\,|B_\tau|\big)^2}. \tag{16.3.8}$$

We now pass to the proof of the “square-root-of-two” inequality.

Proof of (16.3.2): Since $\sqrt{A - x^2} + x \le \sqrt{2A}$ for $0 < x < \sqrt{A}$, by (16.3.8) we find

$$\mathsf{E}\Big(\max_{0\le t\le\tau}|B_t|\Big) = \mathsf{E}\Big(\max_{0\le t<\infty}|B_{t\wedge\tau}|\Big) \le \mathsf{E}\Big(\max_{0\le t<\infty}\mathsf{E}\big(|B_\tau|\,\big|\,\mathcal{F}_{t\wedge\tau}\big)\Big) \tag{16.3.9}$$

$$= \mathsf{E}\Big(\max_{0\le t<\infty}\mathsf{E}\big(|B_\tau|-\mathsf{E}|B_\tau|\,\big|\,\mathcal{F}_{t\wedge\tau}\big)\Big) + \mathsf{E}|B_\tau|$$

$$\le \sqrt{\mathsf{E}\left(|B_\tau|-\mathsf{E}|B_\tau|\right)^2} + \mathsf{E}|B_\tau|$$

$$= \sqrt{\mathsf{E}\tau-\left(\mathsf{E}|B_\tau|\right)^2} + \mathsf{E}|B_\tau| \le \sqrt{2\mathsf{E}\tau}.$$

This establishes (16.3.2) and completes the first part of the proof.

4°. To prove the sharpness of (16.3.1) one may take the stopping time

$$\tau_1^* = \inf\{\,t>0 : |B_t| = a\,\} \tag{16.3.10}$$

for any $a > 0$. Then the equality in (16.3.1) is attained. It follows by Wald's identity. Note that for any $a > 0$ the stopping time τ_1^* could be equivalently (in distribution) defined by

$$\tau_1^* = \inf\Big\{t>0 : \max_{0\le s\le t} B_s - B_t \ge a\Big\}. \tag{16.3.11}$$

5°. To prove the sharpness of (16.3.2) one may take the stopping time

$$\tau_2^* = \inf\Big\{t>0 : \max_{0\le s\le t}|B_s| - |B_t| \ge a\Big\} \tag{16.3.12}$$

for any $a > 0$. Then it is easily verified that $\mathsf{E}\big(\max_{0\le t\le\tau_2^*}|B_t|\big) = 2a$ and $\mathsf{E}\tau_2^* = 2a^2$ (see [45]). Thus the equality in (16.3.2) is attained, and the proof of the sharpness is complete.

17. Bessel inequalities

The aim of this section (following [45]) is to formulate and solve an optimal stopping problem for Bessel processes, and from this deduce a sharp maximal inequality which gives bounds in terms of the expectation of the stopping time.

17.1. Formulation of the problem

Recall that a Bessel process of dimension $\alpha \in \mathbb{R}$ is a Markov process $X = (X_t)_{t\ge0}$ with the state space $E = \mathbb{R}_+$ and continuous sample paths being associated with the infinitesimal generator

$$\mathbb{L}_X = \frac{\alpha-1}{2x}\frac{d}{dx} + \frac{1}{2}\frac{d^2}{dx^2} \tag{17.1.1}$$

and the boundary point 0 is a trap if $\alpha \leq 0$, instantaneously reflecting if $0 < \alpha < 2$, and entrance if $\alpha \geq 2$. (For a detailed study of Bessel processes see [132], [100], [53], [94], [174].)

In the case $\alpha = n \in \mathbb{N}$ the Bessel process X can be realised as the radial part of an n-dimensional Brownian motion $(B^1, \ldots, B^n)$, i.e.

$$X_t^x = \Big(\sum_{i=1}^{n} \big| a_i + B_t^i \big|^2 \Big)^{1/2} \tag{17.1.2}$$

for $t \geq 0$ where $X_0^x = x = \big(\sum_{i=1}^{n} |a_i|^2 \big)^{1/2}$.

Given a Bessel process $X = (X_t)_{t\geq 0}$ of dimension $\alpha \in \mathbb{R}$, let $S = (S_t)_{t \geq 0}$ be the maximum process associated with X, and let $\mathsf{P}_{x,s}$ be a probability measure under which $X_0 = x$ and $S_0 = s$ where $s \geq x \geq 0$. Recall that this is possible to achieve by setting

$$S_t = s \vee \max_{0 \leq u \leq t} X_u \tag{17.1.3}$$

where $X_0 = x$ under P_x.

The main purpose of the present section is to consider the following optimal stopping problem:

$$V(x,s) = \sup_{\tau} \mathsf{E}_{x,s}(S_\tau - c\tau) \tag{17.1.4}$$

where $0 \leq x \leq s$ and $c > 0$ are given and fixed, and the supremum is taken over all stopping times τ of X.

The solution to this problem is presented in the next subsection (Theorem 17.1). If we set $V(0,0) = V(c)$ to indicate the dependence on c in (17.1.4), then it follows that

$$\mathsf{E}\, S_\tau \leq \inf_{c>0} \big(V(c) + c\, \mathsf{E}\, \tau \big) \tag{17.1.5}$$

where $X_0 = S_0 = 0$ under P. This yields a sharp maximal inequality (Corollary 17.2) where the right-hand side defines a function of $\mathsf{E}\,\tau$. In the case $\alpha = 1$ this inequality reduces to the inequality (16.3.2).

17.2. Solution to the problem

Recalling our discussion on the kinematics of the process (X, S) given in Section 13 and applying similar arguments in the present setting one obtains the following result.

Theorem 17.1. *Consider the optimal stopping problem* (17.1.4) *where S is the maximum process associated with the Bessel process X of dimension $\alpha \in \mathbb{R}$ and $c > 0$ is given and fixed.*

The following stopping time is optimal in (17.1.4):

$$\tau_* = \inf \big\{ t \geq 0 : S_t \geq s_* \;\&\; X_t \leq g_*(S_t) \big\} \tag{17.2.1}$$

where $s \mapsto g_*(s)$ *is the maximal solution of the nonlinear differential equation:*

$$\frac{2c}{\alpha-2}\, g'(s)g(s)\left(1-\left(\frac{g(s)}{s}\right)^{\alpha-2}\right)=1 \tag{17.2.2}$$

satisfying $g_*(s) < s$ *for* $s > s_* \geq 0$ *where* $g_*(s_*) = 0$. *If* $\alpha = 2$ *then* (17.2.2) *reads:*

$$2c\, g'(s)\, g(s) \log\left(\frac{s}{g(s)}\right) = 1 \tag{17.2.3}$$

which is obtained from (17.2.2) *by passing to the limit as* $\alpha \to 2$. *The solution* g_* *to* (17.2.2) *or* (17.2.3) *may also be characterized by the boundary condition at infinity:*

$$\lim_{s\to\infty} \frac{g_*(s)}{s} = 1. \tag{17.2.4}$$

The value function V *in* (17.1.4) *is explicitly given as follows. Setting*

$$C_*^1 = \big\{(x,s) \in \mathbb{R}_+ \times \mathbb{R}_+ : s > s_* \ \&\ g_*(s) \leq x \leq s\big\}, \tag{17.2.5}$$

$$C_*^2 = \big\{(x,s) \in \mathbb{R}_+ \times \mathbb{R}_+ : 0 \leq x \leq s \leq s_*\big\} \quad (s_* \neq 0), \tag{17.2.6}$$

we denote by $C_* = C_*^1 \cup C_*^2$ *the continuation set and by* $D_* = \big\{(x,s) \in \mathbb{R}_+ \times \mathbb{R}_+ : 0 \leq x \leq s\big\} \setminus C_*$ *the stopping set.*

If $\alpha > 0$ *then*

$$V(x,s) = \begin{cases} s \quad \text{if } (x,s) \in D_*, \\ s + \dfrac{c}{\alpha}\big(x^2 - g_*^2(s)\big) + \dfrac{2cg_*^2(s)}{\alpha(\alpha-2)}\left[\left(\dfrac{g_*(s)}{x}\right)^{\alpha-2} - 1\right] \\ \qquad\qquad \text{if } (x,s) \in C_*^1 \text{ and } \alpha \neq 2, \\ s + \dfrac{c}{2}\big(x^2 - g_*^2(0)\big) + cg_*^2(s) \log\left(\dfrac{g_*(s)}{x}\right) \\ \qquad\qquad \text{if } (x,s) \in C_*^1 \text{ and } \alpha = 2, \\ \dfrac{c}{\alpha}\, x^2 + s_* \quad \text{if } (x,s) \in C_*^2. \end{cases} \tag{17.2.7}$$

If $\alpha = 0$ *then*

$$V(x,s) = \begin{cases} s \quad \text{if } (x,s) \in D_*, \\ s + \dfrac{c}{2}\big(g_*^2(s) - x^2\big) + cx^2 \log\left(\dfrac{x}{g_*(s)}\right) \\ \text{if } (x,s) \in C_*. \end{cases} \tag{17.2.8}$$

If $\alpha < 0$ *then*

$$V(x,s) = \begin{cases} s \quad \text{if } (x,s) \in D_*, \\ s + \dfrac{c}{\alpha}\big(x^2 - g_*^2(s)\big) + \dfrac{2cg_*^2(s)}{\alpha(\alpha-2)}\left[\left(\dfrac{g_*(s)}{x}\right)^{\alpha-2} - 1\right] \\ \quad \text{if } (x,s) \in C_*. \end{cases} \tag{17.2.9}$$

(*If* $\alpha \le 0$ *then* $s_* = 0$ *and* $(0,0) \in D_*$ *so that* $C_*^2 = \varnothing$.)

Proof. This can be derived using Corollary 13.3 (for remaining details see the original article [45]). □

Corollary 17.2. *Let* $X = (X_t)_{t\ge 0}$ *be a Bessel process of dimension* $\alpha > 0$, *and let* $S = (S_t)_{t\ge 0}$ *be the maximum process associated with* X *such that* $X_0 = S_0 = 0$ *under* P .

Then we have:

$$\mathsf{E}\Big(\max_{0\le t\le\tau} X_t\Big) \le \sqrt{4s_*(1,\alpha)}\,\sqrt{\mathsf{E}\,\tau} \tag{17.2.10}$$

for all stopping times τ *of* X , *where* $s_*(1,\alpha)$ *is the root of the equation* $g_*(s) = 0$ *and* $s \mapsto g_*(s)$ *is the maximal solution of* (17.2.2) *with* $c = 1$ *satisfying* $g_*(s) < s$ *for all* $s > s_*(1,\alpha)$. (*This solution may also be characterized by the boundary condition at infinity as in* (17.2.4) *above.*)

Proof. From (17.2.7) we see that

$$\sup_\tau \mathsf{E}\Big(\max_{0\le t\le\tau} X_t - c\tau\Big) = s_*(c,\alpha). \tag{17.2.11}$$

By self-similarity of X in the sense that

$$\left(\frac{X_{ct}^{\sqrt{c}x}}{\sqrt{c}}\right)_{t\ge 0} \overset{\text{law}}{=} \big(X_t^x\big)_{t\ge 0} \tag{17.2.12}$$

it is not difficult to verify that

$$s_*(c,\alpha) = \frac{1}{c}\, s_*(1,\alpha). \tag{17.2.13}$$

From (17.2.12) and (17.2.13) we get

$$\mathsf{E}\Big(\max_{0\le t\le\tau} X_t\Big) \le \inf_{c>0}\Big(\frac{s_*(1,\alpha)}{c} + c\,\mathsf{E}\,\tau\Big) = \sqrt{4s_*(1,\alpha)}\,\sqrt{\mathsf{E}\,\tau} \tag{17.2.14}$$

as claimed. □

Note that if $\alpha = 1$ then $g_*(s) = s - 1/2c$ so that $s_*(1,1) = 1/2$ and (17.2.10) becomes (16.3.2).

18. Doob inequalities

The main purpose of the section (following [80]) is to derive and examine a sharp maximal inequality of Doob type for one-dimensional Brownian motion which may start at any point.

18.1. Formulation of the problem

Let us assume that we are given a standard Brownian motion $B=(B_t)_{t\geq 0}$ which is defined on a probability space $(\Omega, \mathcal{F}, \mathsf{P})$ and which starts at 0 under P. Then the well-known Doob maximal inequality states:

$$\mathsf{E}\Big(\max_{0\leq t\leq \tau}|B_t|^2\Big)\leq 4\,\mathsf{E}\,|B_\tau|^2 \tag{18.1.1}$$

where τ may be any stopping time for B with $\mathsf{E}\tau<\infty$ (see [40, p. 353] and (C4) on page 62). The constant 4 is known to be best possible in (18.1.1). For this one can consider the stopping times

$$\sigma_{\lambda,\varepsilon}=\inf\Big\{\,t>0:\max_{0\leq s\leq t}|B_s|-\lambda|B_t|\geq\varepsilon\,\Big\} \tag{18.1.2}$$

where $\lambda,\,\varepsilon>0$. It is well known that $\mathsf{E}\,(\sigma_{\lambda,\varepsilon})^{p/2}<\infty$ if and only if $\lambda<p/(p-1)$ whenever $\varepsilon>0$ (see e.g. [221]). Applying Doob's maximal inequality with a general constant $K>0$ to the stopping time in (18.1.2) with some $\varepsilon>0$ when $0<\lambda<2$, we get

$$\mathsf{E}\Big(\max_{0\leq t\leq\sigma_{\lambda,\varepsilon}}|B_t|^2\Big)=\lambda^2\mathsf{E}\,|B_{\sigma_{\lambda,\varepsilon}}|^2+2\lambda\varepsilon\,\mathsf{E}\,|B_{\sigma_{\lambda,\varepsilon}}|+\varepsilon^2\leq K\mathsf{E}\,|B_{\sigma_{\lambda,\varepsilon}}|^2. \tag{18.1.3}$$

Dividing through in (18.3.1) by $\mathsf{E}\,|B_{\sigma_{\lambda,\varepsilon}}|^2$ and using that $\mathsf{E}\,|B_{\sigma_{\lambda,\varepsilon}}|^2=\mathsf{E}\,(\sigma_{\lambda,\varepsilon})\to\infty$ together with $\mathsf{E}\,|B_{\sigma_{\lambda,\varepsilon}}|/\mathsf{E}\,|B_{\sigma_{\lambda,\varepsilon}}|^2\leq 1/\sqrt{\mathsf{E}\,\sigma_{\lambda,\varepsilon}}\to 0$ as $\lambda\uparrow 4$, we see that $K\geq 4$.

Motivated by these facts our main aim in this section is to find an analogue of the inequality (18.1.1) when the Brownian motion B does not necessarily start from 0, but may start at any given point $x\geq 0$ under P_x. Thus $\mathsf{P}_x(B_0=x)=1$ for all $x\geq 0$, and we identify P_0 with P. Our main result (Theorem 18.1) is the inequality

$$\mathsf{E}_x\Big(\max_{0\leq t\leq \tau}|B_t|^2\Big)\leq 4\,\mathsf{E}_x|B_\tau|^2-2x^2 \tag{18.1.4}$$

which is valid for any stopping time τ for B with $\mathsf{E}_x\tau<\infty$, and which is shown to be sharp as such. This is obtained as a consequence of the following inequality:

$$\mathsf{E}_x\Big(\max_{0\leq t\leq \tau}|B_t|^2\Big)\leq c\,\mathsf{E}_x\tau+\frac{c}{2}\bigg(1-\sqrt{1-\frac{4}{c}}\,\bigg)\,x^2 \tag{18.1.5}$$

which is valid for all $c \geq 4$. If $c > 4$ then

$$\tau_c = \inf \left\{ t > 0 : \max_{0 \leq s \leq t} |B_s| - \frac{2}{1 + \sqrt{1 - 4/c}} |B_t| \geq 0 \right\} \tag{18.1.6}$$

is a stopping time at which equality in (18.1.5) is attained, and moreover we have

$$\mathsf{E}_x \tau_c = \frac{\left(1 - \sqrt{1 - 4/c}\right)^2}{4\sqrt{1 - 4/c}} \, x^2 \tag{18.1.7}$$

for all $x \geq 0$ and all $c > 4$.

In particular, if we consider the stopping time

$$\tau_{\lambda,\varepsilon} = \inf \left\{ t > 0 : \max_{0 \leq s \leq t} B_s - \lambda B_t \geq \varepsilon \right\} \tag{18.1.8}$$

then (18.1.7) can be rewritten to read as follows:

$$\mathsf{E}_0 \tau_{\lambda,\varepsilon} = \frac{\varepsilon^2}{\lambda(2 - \lambda)} \tag{18.1.9}$$

for all $\varepsilon > 0$ and all $0 < \lambda < 2$. Quite independently from this formula and its proof, below we present a simple argument for $\mathsf{E}\,\tau_{2,\varepsilon} = \infty$ which is based upon Tanaka's formula (page 67). Finally, since $\sigma_{\lambda,\varepsilon}$ defined by (18.1.2) is shown to be a convolution of $\tau_{\lambda,\lambda\varepsilon}$ and H_ε, where $H_\varepsilon = \inf\{ t > 0 : |B_t| = \varepsilon \}$, from (18.1.9) we obtain the formula

$$\mathsf{E}_0 \sigma_{\lambda,\varepsilon} = \frac{2\varepsilon^2}{2 - \lambda} \tag{18.1.10}$$

for all $\varepsilon > 0$ and all $0 < \lambda < 2$ (see Corollary 18.5 below).

18.2. Solution to the problem

In this subsection we will solve the problem formulated in the previous subsection. The main result is contained in the following theorem (see also Corollaries 18.2 and 18.3 below).

Theorem 18.1. *Let $B = (B_t)_{t \geq 0}$ be a standard Brownian motion started at x under P_x for $x \geq 0$, and let τ be any stopping time for B such that $\mathsf{E}_x \tau < \infty$. Then the following inequality is valid:*

$$\mathsf{E}_x \Big(\max_{0 \leq t \leq \tau} |B_t|^2 \Big) \leq 4\, \mathsf{E}_x |B_\tau|^2 - 2x^2. \tag{18.2.1}$$

The constants 4 and 2 are best possible.

Proof. We shall begin by considering the following optimal stopping problem:

$$V(x,s) = \sup_{\tau} \mathsf{E}_{x,s}(S_\tau - c\tau) \tag{18.2.2}$$

where the supremum is taken over all stopping times τ for B satisfying $\mathsf{E}_{x,s}\tau < \infty$, while the maximum process $S = (S_t)_{t\geq 0}$ is defined by

$$S_t = \Big(\max_{0\leq r\leq t} |B_r|^2 \Big) \vee s \tag{18.2.3}$$

where $s \geq x \geq 0$ are given and fixed. The expectation in (18.2.2) is taken with respect to the probability measure $\mathsf{P}_{x,s}$ under which S starts at s, and the process $X = (X_t)_{t\geq 0}$ defined by

$$X_t = |B_t|^2 \tag{18.2.4}$$

starts at x. The Brownian motion B from (18.2.3) and (18.2.4) may be realized as

$$B_t = \widetilde{B}_t + \sqrt{x} \tag{18.2.5}$$

where $\widetilde{B} = (\widetilde{B}_t)_{t\geq 0}$ is a standard Brownian motion started at 0 under P. Thus the (strong) Markov process (X,S) starts at (x,s) under P, and $\mathsf{P}_{x,s}$ may be identified with P.

By Itô's formula (page 67) we find

$$dX_t = dt + 2\sqrt{X_t}\, dB_t. \tag{18.2.6}$$

Hence we see that the infinitesimal operator of the (strong) Markov process X in $(0,\infty)$ acts like

$$\mathbb{L}_X = \frac{\partial}{\partial x} + 2x\,\frac{\partial^2}{\partial x^2} \tag{18.2.7}$$

while the boundary point 0 is a point of the instantaneous reflection.

If we assume that the supremum in (18.2.2) is attained at the exit time from an open set by the (strong) Markov process (X,S) which is degenerated in the second component, then by the general Markov processes theory (cf. Chapter III) it is plausible to assume that the value function $x \mapsto V(x,s)$ satisfies the following equation:

$$\mathbb{L}_X V(x,s) = c \tag{18.2.8}$$

for $x \in (g_*(s), s)$ with $s > 0$ given and fixed, where $s \mapsto g_*(s)$ is an optimal stopping boundary to be found (cf. Section 13). The boundary conditions which may be fulfilled are the following:

$$V(x,s)\Big|_{x=g_*(s)+} = s \quad (\textit{instantaneous stopping}), \tag{18.2.9}$$

$$\frac{\partial V}{\partial x}(x,s)\Big|_{x=g_*(s)+} = 0 \quad (\textit{smooth fit}), \tag{18.2.10}$$

$$\frac{\partial V}{\partial s}(x,s)\Big|_{x=s-} = 0 \quad (\textit{normal reflection}). \tag{18.2.11}$$

The general solution to the equation (18.2.8) for fixed s is given by

$$V(x,s) = A(s)\sqrt{x} + B(s) + cx \tag{18.2.12}$$

where $A(s)$ and $B(s)$ are unspecified constants. From (18.2.9)–(18.2.10) we find that

$$A(s) = -2\,c\,\sqrt{g_*(s)}, \tag{18.2.13}$$

$$B(s) = s + c\,g_*(s). \tag{18.2.14}$$

Inserting this into (18.2.12) gives

$$V(x,s) = -2\,c\,\sqrt{g_*(s)}\,\sqrt{x} + s + cg_*(s) + cx. \tag{18.2.15}$$

By (18.2.11) we find that $s \mapsto g_*(s)$ is to satisfy the (nonlinear) differential equation

$$cg'(s)\left(1 - \sqrt{\frac{s}{g(s)}}\right) + 1 = 0. \tag{18.2.16}$$

The general solution of the equation (18.2.16) can be expressed in closed form. Instead of going into this direction we shall rather note that this equation admits a linear solution of the form

$$g_*(s) = \alpha s \tag{18.2.17}$$

where the given $\alpha > 0$ is to satisfy

$$\alpha - \sqrt{\alpha} + 1/c = 0. \tag{18.2.18}$$

Motivated by the maximality principle (see Section 13) we shall choose the greater α satisfying (18.2.18) as our candidate:

$$\alpha = \left(\frac{1 + \sqrt{1 - 4/c}}{2}\right)^2. \tag{18.2.19}$$

Inserting this into (18.2.15) gives

$$V_*(x,s) = \begin{cases} -2c\sqrt{\alpha x s} + (1 + c\alpha)s + cx & \text{if } \alpha s \le x \le s, \\ s & \text{if } 0 \le x \le \alpha s \end{cases} \tag{18.2.20}$$

as a candidate for the value function $V(x,s)$ defined in (18.2.2). The optimal stopping time is then to be

$$\tau_* = \inf\left\{\, t > 0 : X_t \le g_*(S_t) \,\right\} \tag{18.2.21}$$

where $s \mapsto g_*(s)$ is defined by (18.2.17)+(18.2.19).

To verify that the formulae (18.2.20) and (18.2.21) are indeed correct, we shall use the Itô–Tanaka–Meyer formula (page 68) being applied two-dimensionally (see [81] for a formal justification of its use in this context — note that $(x,s) \mapsto V_*(x,s)$ is C^2 outside $\{\, (g_*(s),s) : s>0 \,\}$ while $x \mapsto V_*(x,s)$ is convex and C^2 on $(0,s)$ but at $g_*(s)$ where it is only C^1 whenever $s>0$ is given and fixed — for the standard one-dimensional case see (3.3.23)). In this way we obtain

$$V_*(X_t,S_t) = V_*(X_0,S_0) + \int_0^t \frac{\partial V_*}{\partial x}(X_r,S_r)\,dX_r \tag{18.2.22}$$
$$+ \int_0^t \frac{\partial V_*}{\partial s}(X_r,S_r)\,dS_r + \frac{1}{2}\int_0^t \frac{\partial^2 V_*}{\partial x^2}(X_r,S_r)\,d\langle X,X\rangle_r$$

where we set $(\partial^2 V_*/\partial x^2)(g_*(s),s) = 0$. Since the increment dS_r equals zero outside the diagonal $x=s$, and $V_*(x,s)$ at the diagonal satisfies (18.2.11), we see that the second integral in (18.2.22) is identically zero. Thus by (18.2.6)–(18.2.7) and the fact that $d\langle X,X\rangle_t = 4X_t\,dt$, we see that (18.2.22) can be equivalently written as follows:

$$V_*(X_t,S_t) = V_*(x,s) + \int_0^t \mathbb{L}_X V_*(X_r,S_r)\,dr \tag{18.2.23}$$
$$+ 2\int_0^t \sqrt{X_r}\,\frac{\partial V_*}{\partial x}(X_r,S_r)\,dB_r.$$

Next note that $\mathbb{L}_X V_*(y,s) = c$ for $g_*(s) < y < s$, and $\mathbb{L}_X V_*(y,s) = 0$ for $0 \le y \le g_*(s)$. Moreover, due to the normal reflection of X, the set of those $r>0$ for which $X_r = S_r$ is of Lebesgue measure zero. This by (18.2.23) shows that

$$V_*(X_\tau,S_\tau) \le V_*(x,s) + c\tau + M_\tau \tag{18.2.24}$$

for any stopping time τ for B, where $M=(M_t)_{t\ge 0}$ is a continuous local martingale defined by

$$M_t = 2\int_0^t \sqrt{X_r}\,\frac{\partial V_*}{\partial x}(X_r,S_r)\,dB_r. \tag{18.2.25}$$

Moreover, this also shows that

$$V_*(X_\tau,S_\tau) = V_*(x,s) + c\,\tau + M_\tau \tag{18.2.26}$$

for any stopping time τ for B satisfying $\tau \le \tau_*$.

Next we show that

$$\mathsf{E}_{x,s} M_\tau = 0 \tag{18.2.27}$$

whenever τ is a stopping time for B with $\mathsf{E}_{x,s}\tau < \infty$. For (18.2.27), by the Burkholder–Davis–Gundy inequality for continuous local martingales (see (C5) on page 63), it is sufficient to show that

$$\mathsf{E}_{x,s}\left(\int_0^\tau \left(\sqrt{X_r}\,\frac{\partial V_*}{\partial x}(X_r,S_r) \right)^2 1_{\{X_r \ge g_*(S_r)\}}\,dr \right)^{1/2} := I < \infty. \tag{18.2.28}$$

From (18.2.20) we compute:

$$\frac{\partial V_*}{\partial x}(y,s) = -\frac{c\sqrt{\alpha s}}{\sqrt{y}} + c \tag{18.2.29}$$

for $\alpha s \le y \le s$. Inserting this into (18.2.28) we get:

$$\begin{aligned} I &= c\,\mathsf{E}_{x,s}\bigg(\int_0^\tau \Big(\sqrt{X_r} - \sqrt{\alpha S_r}\Big)^2\, 1_{\{X_r \ge \alpha S_r\}}\, dr\bigg)^{1/2} \\ &\le c\,(1-\sqrt{\alpha})\,\mathsf{E}_{x,s}\bigg(\int_0^\tau S_r\, dr\bigg)^{1/2} \le c\,(1-\sqrt{\alpha})\,\mathsf{E}_{x,s}\Big(\sqrt{S_\tau}\sqrt{\tau}\Big) \\ &\le c\,(1-\sqrt{\alpha})\,\sqrt{\mathsf{E}_{x,s}S_\tau}\,\sqrt{\mathsf{E}_{x,s}\tau} \\ &= c\,(1-\sqrt{\alpha})\bigg(\mathsf{E}_{x,s}\Big(\Big(\max_{0\le t\le\tau}|\widetilde{B}_t+\sqrt{x}|^2\Big)\vee s\Big)\bigg)^{1/2}\sqrt{\mathsf{E}_{x,s}\tau} \\ &\le c\,(1-\sqrt{\alpha})\bigg(2\,\mathsf{E}_{x,s}\Big(\max_{0\le t\le\tau}|\widetilde{B}_t|^2\Big) + 2\,x + s\bigg)^{1/2}\sqrt{\mathsf{E}_{x,s}\tau} \\ &\le c\,(1-\sqrt{\alpha})\big(8\,\mathsf{E}_{x,s}\tau + 2x + s\big)^{1/2}\sqrt{\mathsf{E}_{x,s}\tau} < \infty \end{aligned} \tag{18.2.30}$$

where we used Hölder's inequality, Doob's inequality (18.1.1), and the fact that $\mathsf{E}_{x,s}|\widetilde{B}_\tau|^2 = \mathsf{E}_{x,s}\tau$ whenever $\mathsf{E}_{x,s}\tau < \infty$.

Since $V_*(x,s) \ge s$, from (18.2.24)+(18.2.27) we find

$$\begin{aligned} V(x,s) = \sup_\tau \mathsf{E}_{x,s}\big(S_\tau - c\tau\big) &\le \sup_\tau \mathsf{E}_{x,s}\big(S_\tau - V_*(X_\tau, S_\tau)\big) \\ &\quad + \sup_\tau \mathsf{E}_{x,s}\big(V_*(X_\tau,S_\tau) - c\tau\big) \le V_*(x,s). \end{aligned} \tag{18.2.31}$$

Moreover, from (18.2.26)–(18.2.27) with $\tau = \tau_*$ we see that

$$\mathsf{E}_{x,s}\big(S_{\tau_*} - c\,\tau_*\big) = \mathsf{E}_{x,s}\big(V_*(X_{\tau_*}, S_{\tau_*}) - c\,\tau_*\big) = V_*(x,s) \tag{18.2.32}$$

provided that $\mathsf{E}_{x,s}\tau_* < \infty$, which is known to be true if and only if $c > 4$ (see [221]). (Below we present a different proof of this fact and moreover compute the value $\mathsf{E}_{x,s}\tau_*$ exactly.) Matching (18.2.31) and (18.2.32) we see that the value function (18.2.2) is indeed given by the formula (18.2.20), and an optimal stopping time for (18.2.2) (at which the supremum is attained) is given by (18.2.21) with $s \mapsto g_*(s)$ from (18.2.17) and $\alpha \in (0,1)$ from (18.2.19).

In particular, note that from (18.2.20) with α from (18.2.19) we get

$$V_*(x,x) = \frac{c}{2}\bigg(1\sqrt{1-\frac{4}{c}}\bigg)x. \tag{18.2.33}$$

Applying the very definition of $V(x,x) = V_*(x,x)$ and letting $c \downarrow 4$, this yields

$$\mathsf{E}_x\Big(\max_{0\le t\le\tau}|B_t|^2\Big) \le 4\mathsf{E}_x\tau + 2x. \tag{18.2.34}$$

Finally, standard arguments show that

$$\mathsf{E}_x|B_\tau|^2 = \mathsf{E}_x|\widetilde{B}_\tau + \sqrt{x}|^2 = \mathsf{E}_x|\widetilde{B}_\tau|^2 + 2\sqrt{x}\,\mathsf{E}_x(\widetilde{B}_\tau) + x = \mathsf{E}_x\tau + x. \tag{18.2.35}$$

Inserting this into (18.2.34) we obtain (18.2.1). The sharpness clearly follows from the definition of the value function in (18.2.2) completing the proof of the theorem. □

The previous result and method easily extend to the case $p>1$. For reader's convenience we state this extension and sketch the proof.

Corollary 18.2. *Let $B = (B_t)_{t\ge 0}$ be a standard Brownian motion started at x under P_x for $x \ge 0$, let $p > 1$ be given and fixed, and let τ be any stopping time for B such that $\mathsf{E}_x\tau^{p/2} < \infty$. Then the following inequality is sharp:*

$$\mathsf{E}_x\Big(\max_{0\le t\le\tau}|B_t|^p\Big) \le \Big(\frac{p}{p-1}\Big)^p\,\mathsf{E}_x|B_\tau|^p - \Big(\frac{p}{p-1}\Big)x^p. \tag{18.2.36}$$

The constants $(p/(p-1))^p$ and $p/(p-1)$ are best possible.

Proof. In parallel to (18.2.2) let us consider the following optimal stopping problem:

$$V(x,s) = \sup_\tau \mathsf{E}_{x,s}\big(S_\tau - cI_\tau\big) \tag{18.2.37}$$

where the supremum is taken over all stopping times τ for B satisfying $\mathsf{E}_{x,s}\tau^{p/2} < \infty$, and the underlying processes are given as follows:

$$S_t = \Big(\max_{0\le r\le t} X_r\Big) \vee s, \tag{18.2.38}$$

$$I_t = \int_0^t (X_r)^{(p-2)/p}\,dr, \tag{18.2.39}$$

$$X_t = |B_t|^p, \tag{18.2.40}$$

$$B_t = \widetilde{B}_t + x^{1/p}, \tag{18.2.41}$$

where $\widetilde{B} = (\widetilde{B}_t)_{t\ge 0}$ is a standard Brownian motion started at 0 under $\mathsf{P} = \mathsf{P}_{x,s}$. This problem can be solved in exactly the same way as the problem (18.2.2) along the following lines.

The infinitesimal operator of X equals

$$\mathbb{L}_X = \frac{p(p-1)}{2}\,x^{1-2/p}\,\frac{\partial}{\partial x} + \frac{p^2}{2}\,x^{2-2/p}\,\frac{\partial^2}{\partial x^2}. \tag{18.2.42}$$

The analogue of the equation (18.2.8) is

$$\mathbb{L}_X V(x,s) = c\,x^{(p-2)/p}. \tag{18.2.43}$$

The conditions (18.2.9)–(18.2.11) are to be satisfied again. The analogue of the solution (18.2.15) is

$$V(x,s) = -\frac{2c}{p-1}\, g_*^{1-1/p}(s)\, x^{1/p} + s + \frac{2c}{p} g_*(s) + \frac{2c}{p(p-1)}\, x \tag{18.2.44}$$

where $s \mapsto g_*(s)$ is to satisfy the equation

$$\frac{2c}{p}\, g'(s)\Big(1 - \Big(\frac{s}{g(s)}\Big)^{1/p}\Big) + 1 = 0. \tag{18.2.45}$$

Again, as in (18.2.16), this equation admits a linear solution of the form

$$g_*(s) = \alpha s \tag{18.2.46}$$

where $0 < \alpha < 1$ is the maximal root (out of two possible ones) of the equation

$$\alpha - \alpha^{1-1/p} + p/2c = 0. \tag{18.2.47}$$

By standard arguments one can verify that (18.2.47) admits such a root if and only if $c \geq p^{p+1}/2(p-1)^{(p-1)}$. The optimal stopping time is then to be

$$\tau_* = \inf\{\, t>0 : X_t \leq g_*(S_t)\,\} \tag{18.2.48}$$

where $s \mapsto g_*(s)$ is from (18.2.46).

To verify that the guessed formulae (18.2.44) and (18.2.48) are indeed correct we can use exactly the same procedure as in the proof of Theorem 18.1. For this, it should be recalled that $\mathsf{E}_{x,s}\tau_*^{p/2} < \infty$ if and only if $c > p^{p+1}/2(p-1)^{(p-1)}$ (see [221]). Note also by Itô's formula (page 67) and the optional sampling theorem (page 60) that the analogue of (18.2.35) is given by

$$\mathsf{E}_{x,s}(X_\tau) = x + \frac{p(p-1)}{2}\, \mathsf{E}_{x,s}(I_\tau) \tag{18.2.49}$$

whenever $\mathsf{E}_{x,s}(\tau^{p/2}) < \infty$, which was the motivation for considering the problem (18.2.37) with (18.2.39). The remaining details are easily completed and will be left to the reader. □

Due to the universal role of Brownian motion in this context, the inequality (18.2.36) extends to all non-negative submartingales. This can be obtained by using the maximal embedding result of Jacka [101].

Corollary 18.3. *Let $X = (X_t)_{t\geq 0}$ be a non-negative càdlàg (right continuous with left limits) uniformly integrable submartingale started at $x \geq 0$ under P. Let X_∞ denote the P-a.s. limit of X_t for $t \to \infty$ (which exists by* (B1) *on page* 61*). Then the following inequality is satisfied and sharp:*

$$\mathsf{E}\Big(\sup_{t>0} X_t^p\Big) \leq \Big(\frac{p}{p-1}\Big)^p \mathsf{E}\, X_\infty^p - \frac{p}{p-1}\, x^p \tag{18.2.50}$$

for all $p > 1$.

Proof. Given such a submartingale $X = (X_t)_{t \ge 0}$ satisfying $\mathsf{E}\, X_\infty < \infty$, and a Brownian motion $B = (B_t)_{t \ge 0}$ started at $X_0 = x$ under P_x, by the result of Jacka [101] we know that there exists a stopping time τ for B, such that $|B_\tau| \sim X_\infty$ and $\mathsf{P}\{ \sup_{t \ge 0} X_t \ge \lambda \} \le \mathsf{P}_x\{ \max_{0 \le t \le \tau} |B_\tau| \ge \lambda \}$ for all $\lambda > 0$, with $(B_{t \wedge \tau})_{t \ge 0}$ being uniformly integrable. The result then easily follows from Corollary 18.2 by using the integration by parts formula. Note that by the submartingale property of $(|B_{t \wedge \tau}|)_{t \ge 0}$ we get $\sup_{t \ge 0} \mathsf{E}_x |B_{t \wedge \tau}|^p = \mathsf{E}_x |B_\tau|^p$ for all $p > 1$, so that $\mathsf{E}_x \tau^{p/2}$ is finite if and only if $\mathsf{E}_x |B_\tau|^p$ is so. □

Notes. There are other ways to derive the inequalites (18.2.36). Burkholder obtained these inequalities as a by-product from his new proof of Doob's inequality for discrete non-negative submartingales (see [25, p. 14]). While the proof given there in essence relies on a submartingale property, the proof given above is based on the (strong) Markov property. An advantage of the latter approach lies in its applicability to all diffusions (see [81]). Another advantage is that during the proof one explicitly writes down the optimal stopping times (those through which equality is attained). Cox [32] also derived the analogue of these inequalities for discrete martingales by a method which is based on results from the theory of moments. In his paper Cox notes that "the method does have the drawback of computational complexity, which sometimes makes it difficult or impossible to push the calculations through". Cox [32] also observed that equality in Doob's maximal inequality (18.2.36) cannot be attained by a non-zero (sub)martingale. It may be noted that this fact follows from the method and results above (equality in (18.2.36) is attained only in the limit). For an extension of the results in this subsection to Bessel processes see [150].

18.3. The expected waiting time

In this subsection we will derive an explicit formula for the expectation of the optimal stopping time τ_* constructed in the proof of Theorem 18.1 (or Corollary 18.2).

Throughout we will work within the setting and notation of Theorem 18.1 and its proof. By (18.2.21) with (18.2.17) we have

$$\tau_* = \inf \{ t > 0 : X_t \le \alpha S_t \} \tag{18.3.1}$$

where $\alpha = \alpha(c)$ is the constant given in (18.2.19) for $c > 4$. Note that $1/4 < \alpha(c) \uparrow 1$ as $c \uparrow \infty$. Our main task in this subsection is to compute explicitly the function

$$m(x, s) = \mathsf{E}_{x,s} \tau_* \tag{18.3.2}$$

for $0 \le x \le s$, where $\mathsf{E}_{x,s}$ denotes the expectation with respect to $\mathsf{P}_{x,s}$ under which X starts at x and S starts at s. Since clearly $m(x, s) = 0$ for $0 \le x \le \alpha s$, we shall assume throughout that $\alpha s < x \le s$ are given and fixed.

Because τ_* may be viewed as the exit time from an open set by the (strong) Markov process (X, S) which is degenerated in the second component, by the general Markov processes theory (see Subsection 7.2) it is plausible to assume that $x \mapsto m(x,s)$ satisfies the equation

$$\mathbb{L}_X m(x,s) = -1 \tag{18.3.3}$$

for $\alpha s < x < s$ with $\mathbb{L}_X$ given by (18.2.7). The following two boundary conditions are apparent:

$$m(x,s)\Big|_{x=\alpha s+} = 0 \quad (\textit{instantaneous stopping}), \tag{18.3.4}$$

$$\frac{\partial m}{\partial s}(x,s)\Big|_{x=s-} = 0 \quad (\textit{normal reflection}). \tag{18.3.5}$$

The general solution to (18.3.3) is given by

$$m(x,s) = A(s)\sqrt{x} + B(s) - x \tag{18.3.6}$$

where $A(s)$ and $B(s)$ are unspecified constants. By (18.3.4) and (18.3.5) we find

$$A(s) = Cs^{\Delta} + \frac{2\alpha}{2\sqrt{\alpha}-1}\sqrt{s}, \tag{18.3.7}$$

$$B(s) = -C\sqrt{\alpha}\, s^{\Delta+1/2} - \frac{\alpha}{2\sqrt{\alpha}-1}\, s \tag{18.3.8}$$

where $C = C(\alpha)$ is a constant to be determined, and where

$$\Delta = \frac{\sqrt{\alpha}}{2(1-\sqrt{\alpha})}. \tag{18.3.9}$$

In order to determine the constant C, we shall note by (18.3.6)–(18.3.9) that

$$m(x,x) = C(1-\sqrt{\alpha})\, x^{1/2(1-\sqrt{\alpha})} + \frac{(\sqrt{\alpha}-1)^2}{2\sqrt{\alpha}-1}\, x. \tag{18.3.10}$$

Observe now that the power $1/2(1-\sqrt{\alpha}) > 1$, due to the fact that $\alpha = \alpha(c) > 1/4$ when $c > 4$. However, the value function in (18.2.33) is linear and given by

$$V_*(x,x) := V_*(x;c) = K(c) \cdot x \tag{18.3.11}$$

where $K(c) = (c/2)(1 - \sqrt{1-4/c})$. This indicates that the constant C must be identically zero. Formally, this is verified as follows.

Since $c > 4$ there is $\lambda \in (0,1)$ such that $\lambda c > 4$. By definition of the value function we have

$$\begin{aligned} 0 < V_*(x;c) &= \mathsf{E}_{x,x}\big(S_{\tau_*(c)} - c\,\tau_*(c)\big) \\ &= \mathsf{E}_{x,x}\big(S_{\tau_*(c)} - \lambda c\,\tau_*(c)\big) - (1-\lambda)c\,\mathsf{E}_{x,x}\tau_*(c) \\ &\le V_*(x;\lambda c) - (1-\lambda)c\,\mathsf{E}_{x,x}\tau_*(c) \\ &\le K(\lambda c)\cdot x - (1-\lambda)c\,\mathsf{E}_{x,x}\tau_*(c). \end{aligned} \tag{18.3.12}$$

This shows that $x \mapsto m(x,x)$ is at most linear:

$$m(x,x) = \mathsf{E}_{x,x}\tau_*(c) \le \frac{K(\lambda c)}{(1-\lambda)c}\, x. \tag{18.3.13}$$

Looking back at (18.3.10) we may conclude that $C \equiv 0$.

Thus by (18.3.6)–(18.3.8) with $C \equiv 0$ we end up with the following candidate:

$$m(x,s) = \frac{2\alpha}{2\sqrt{\alpha}-1}\sqrt{xs} - \frac{\alpha}{2\sqrt{\alpha}-1}\, s - x \tag{18.3.14}$$

for $\mathsf{E}_{x,s}\tau_*$ when $\alpha s < x \le s$. In order to verify that this formula is indeed correct we shall use the Itô–Tanaka–Meyer formula (page 68) in the proof below.

Theorem 18.4. *Let $B = (B_t)_{t\ge0}$ be a standard Brownian motion, and let $X = (X_t)_{t\ge0}$ and $S = (S_t)_{t\ge0}$ be associated with B by formulae* (18.2.3)–(18.2.4). *Then for the stopping time τ_* defined in* (18.3.1) *we have:*

$$\mathsf{E}_{x,s}\tau_* = \begin{cases} \dfrac{2\alpha}{2\sqrt{\alpha}-1}\sqrt{xs} - \dfrac{\alpha}{2\sqrt{\alpha}-1}\, s - x & \text{if } \alpha s \le x \le s, \\ 0 & \text{if } 0 \le x \le \alpha s \end{cases} \tag{18.3.15}$$

where $\alpha > 1/4$.

Proof. Denote the function on the right-hand side of (18.3.15) by $m(x,s)$. Note that $x \mapsto m(x,s)$ is concave and non-negative on $[\alpha s, s]$ for each fixed $s > 0$.

By the Itô–Tanaka–Meyer formula (see [81] for a justification of its use) we get:

$$\begin{aligned} m(X_t,S_t) = m(X_0,S_0) &+ \int_0^t \mathbb{L}_X m(X_r,S_r)\,dr \\ &+ 2\int_0^t \sqrt{X_r}\,\frac{\partial m}{\partial x}(X_r,S_r)\,dB_r + \int_0^t \frac{\partial m}{\partial s}(X_r,S_r)\,dS_r. \end{aligned} \tag{18.3.16}$$

Due to (18.3.5) the final integral in (18.3.16) is identically zero.

In addition, let us consider the region $G = \{\, (x,s) : \alpha s < x < s+1 \,\}$. Given $(x,s) \in G$ choose bounded open sets $G_1 \subset G_2 \subset \cdots$ such that $\bigcup_{n=1}^{\infty} G_n = G$ and $(x,s) \in G_1$. Denote the exit time of (X,S) from G_n by τ_n. Then clearly $\tau_n \uparrow \tau_*$ as $n \to \infty$. Denote further the second integral in (18.3.16) by M_t. Then $M = (M_t)_{t\ge0}$ is a continuous local martingale, and we have

$$\mathsf{E}_{x,s} M_{\tau_n} = 0 \tag{18.3.17}$$

for all $n \ge 1$. For this (see page 60) note that

$$\mathsf{E}_{x,s}\Big(\int_0^{\tau_n} \Big(\sqrt{X_r}\,\frac{\partial m}{\partial x}(X_r,S_r)\Big)^2\,dr \Big) \le K\,\mathsf{E}_{x,s}\tau_n < \infty \tag{18.3.18}$$

with some $K > 0$, since $(x, s) \mapsto \sqrt{x}\,(\partial m/\partial x)(x, s)$ is bounded on the closure of G_n.

By (18.3.3) from (18.3.16)–(18.3.17) we find

$$\mathsf{E}_{x,s} m(X_{\tau_n}, S_{\tau_n}) = m(x, s) - \mathsf{E}_{x,s}\tau_n. \tag{18.3.19}$$

Since $(x, s) \mapsto m(x, s)$ is non-negative, hence first of all we may deduce

$$\mathsf{E}_{x,s}\tau_* = \lim_{n\to\infty} \mathsf{E}_{x,s}\tau_n \le m(x, s) < \infty. \tag{18.3.20}$$

This proves the finiteness of the expectation of τ_* (see [221] for another proof based on random walk).

Moreover, motivated by a uniform integrability argument we may note that

$$m(X_{\tau_n}, S_{\tau_n}) \le \frac{2\alpha}{2\sqrt{\alpha} - 1} \sqrt{X_{\tau_n} S_{\tau_n}} \le \frac{2\alpha}{2\sqrt{\alpha} - 1} S_{\tau_*} \tag{18.3.21}$$

uniformly over all $n \ge 1$. By Doob's inequality (18.1.1) and (18.3.20) we find

$$\mathsf{E}_{x,s} S_{\tau_*} \le 2\big(4\mathsf{E}_{x,s}\tau_* + x\big) + s < \infty. \tag{18.3.22}$$

Thus the sequence $(m(X_{\tau_n}, S_{\tau_n}))_{n\ge1}$ is uniformly integrable, while it clearly converges pointwise to zero. Hence we may conclude

$$\lim_{n\to\infty} \mathsf{E}_{x,s} m(X_{\tau_n}, S_{\tau_n}) = 0. \tag{18.3.23}$$

This shows that we have an equality in (18.3.20), and the proof is complete. □

Corollary 18.5. *Let $B = (B_t)_{t\ge0}$ be a standard Brownian motion started at 0 under P. Consider the stopping times*

$$\tau_{\lambda,\varepsilon} = \inf\Big\{\, t > 0 : \max_{0\le s\le t} B_s - \lambda B_t \ge \varepsilon \Big\}, \tag{18.3.24}$$

$$\sigma_{\lambda,\varepsilon} = \inf\Big\{\, t > 0 : \max_{0\le s\le t} |B_s| - \lambda |B_t| \ge \varepsilon \Big\} \tag{18.3.25}$$

for $\varepsilon > 0$ and $0 < \lambda < 2$. Then $\sigma_{\lambda,\varepsilon}$ is a convolution of $\tau_{\lambda,\lambda\varepsilon}$ and H_ε, where $H_\varepsilon = \inf\{\, t > 0 : |B_t| = \varepsilon \,\}$, and the following formulae are valid:

$$\mathsf{E}\,\tau_{\lambda,\varepsilon} = \frac{\varepsilon^2}{\lambda(2-\lambda)}, \tag{18.3.26}$$

$$\mathsf{E}\,\sigma_{\lambda,\varepsilon} = \frac{2\varepsilon^2}{2-\lambda} \tag{18.3.27}$$

for all $\varepsilon > 0$ and all $0 < \lambda < 2$.

Proof. Consider the definition rule for $\sigma_{\lambda,\varepsilon}$ in (18.3.25). Clearly $\sigma_{\lambda,\varepsilon} > H_\varepsilon$ and after hitting ε, the reflected Brownian motion $|B| = (|B_t|)_{t\ge 0}$ does not hit zero before $\sigma_{\lambda,\varepsilon}$. Thus its absolute value sign may be dropped out during the time interval between H_ε and $\sigma_{\lambda,\varepsilon}$, and the claim about the convolution identity follows by the reflection property and the strong Markov property of Brownian motion.

(18.3.26): Consider the stopping time τ_* defined in (18.3.1) for $s = x$. By the very definition it can be rewritten to read as follows:

$$\begin{aligned}\tau_* &= \inf\Big\{\, t>0 : |B_t|^2 \le \alpha \max_{0\le s\le t} |B_s|^2 \,\Big\} \\ &= \inf\Big\{\, t>0 : \max_{0\le s\le t} |B_s| - \frac{1}{\sqrt{\alpha}}\,|B_t| \ge 0 \,\Big\} \\ &= \inf\Big\{\, t>0 : \max_{0\le s\le t} |\widetilde{B}_s + \sqrt{x}| - \frac{1}{\sqrt{\alpha}}\,|\widetilde{B}_t + \sqrt{x}| \ge 0 \,\Big\} \\ &= \inf\Big\{\, t>0 : \max_{0\le s\le t} (\widetilde{B}_s + \sqrt{x}) - \frac{1}{\sqrt{\alpha}}\,(\widetilde{B}_t + \sqrt{x}) \ge 0 \,\Big\} \\ &= \inf\Big\{\, t>0 : \max_{0\le s\le t} \widetilde{B}_s - \frac{1}{\sqrt{\alpha}}\,\widetilde{B}_t \ge \Big(\frac{1}{\sqrt{\alpha}} - 1\Big)\sqrt{x} \,\Big\}.\end{aligned} \tag{18.3.28}$$

Setting $\lambda = 1/\sqrt{\alpha}$ and $\varepsilon = (1/\sqrt{\alpha} - 1)\sqrt{x}$, by (18.3.15) hence we find

$$\mathsf{E}\,\tau_{\lambda,\varepsilon} = \mathsf{E}_{x,x}\tau_* = \frac{(\sqrt{\alpha}-1)^2}{2\sqrt{\alpha}-1}\,x = \frac{\varepsilon^2}{\lambda(2-\lambda)}. \tag{18.3.29}$$

(18.3.27): Since $\mathsf{E}\,H_\varepsilon = \varepsilon^2$, by (18.3.26) we get

$$\mathsf{E}\,\sigma_{\lambda,\varepsilon} = \mathsf{E}\,\tau_{\lambda,\lambda\varepsilon} + \mathsf{E}\,H_\varepsilon = \frac{2\varepsilon^2}{2-\lambda}. \tag{18.3.30}$$

The proof is complete. □

Remark 18.6. Let $B = (B_t)_{t\ge 0}$ be a standard Brownian motion started at 0 under P. Consider the stopping time

$$\tau_{2,\varepsilon} = \inf\Big\{\, t>0 : \max_{0\le s\le t} B_s - 2B_t \ge \varepsilon \,\Big\} \tag{18.3.31}$$

for $\varepsilon \ge 0$. It follows from (18.3.26) in Corollary 18.5 that

$$\mathsf{E}\,\tau_{2,\varepsilon} = +\infty \tag{18.3.32}$$

if $\varepsilon > 0$. Here we present another argument based upon Tanaka's formula (page 67) which implies (18.3.32).

For this consider the process

$$\beta_t = \int_0^t \operatorname{sign}(B_s)\, dB_s \tag{18.3.33}$$

where $\operatorname{sign}(x) = -1$ for $x \le 0$ and $\operatorname{sign}(x) = 1$ for $x > 0$. Then $\beta = (\beta_t)_{t\ge 0}$ is a standard Brownian motion, and Tanaka's formula (page 67) states:

$$|B_t| = \beta_t + L_t \tag{18.3.34}$$

where $L = (L_t)_{t\ge 0}$ is the local time process of B at 0 given by

$$L_t = \max_{0\le s\le t}(-\beta_s). \tag{18.3.35}$$

Thus $\tau_{2,\varepsilon}$ is equally distributed as

$$\begin{aligned} \sigma &= \inf\Big\{ t > 0 : \max_{0\le s\le t}(-\beta_s) - 2(-\beta_t) \ge \varepsilon \Big\} \\ &= \inf\Big\{ t > 0 : |B_t| \ge \varepsilon - \beta_t \Big\}. \end{aligned} \tag{18.3.36}$$

Note that σ is an $(\mathcal{F}_t^\beta)$-stopping time, and since $\mathcal{F}_t^\beta = \mathcal{F}_t^{|B|} \subset \mathcal{F}_t^B$, we see that σ is an $(\mathcal{F}_t^B)$-stopping time too. Assuming now that $\mathsf{E}\,\tau_{2,\varepsilon}$ which equals $\mathsf{E}\,\sigma$ is finite, by the standard Wald identity for Brownian motion (see (3.2.6)) we obtain

$$\mathsf{E}\,\sigma = \mathsf{E}\,|B_\sigma|^2 = \mathsf{E}\,(\varepsilon - \beta_\sigma)^2 = \varepsilon^2 - 2\varepsilon\,\mathsf{E}\,\beta_\sigma + \mathsf{E}\,|\beta_\sigma|^2 = \varepsilon^2 + \mathsf{E}\,\sigma. \tag{18.3.37}$$

Hence we see that ε must be zero. This completes the proof of (18.3.32).

Notes. Theorem 18.4 extends a result of Wang [221] who showed that the expectation of τ_* is finite. Stopping times of the form τ_* have been studied by a number of people. Instead of going into a historical exposition on this subject we will refer the interested reader to the paper by Azéma and Yor [6] where further details in this direction can be found. One may note however that as long as one is concerned with the expectation of such a stopping time only, the Laplace transform method (developed in some of these works) may have the drawback of computational complexity in comparison with the method used above (see also [154] for a related result).

18.4. Further examples

The result of Theorem 18.1 and Corollary 18.2 can also be obtained directly from the maximality principle (see Section 13). We will present this line of argument through several examples.

Example 18.7. (The Doob inequality) Consider the optimal stopping problem (18.2.37) being the same as the optimal stopping problem (13.1.4) with $X_t = |B_t+x|^p$ and $c(x) = cx^{(p-2)/p}$ for $p > 1$. Then X is a non-negative diffusion having 0 as an instantaneously-reflecting regular boundary point, and the infinitesimal generator of X in $(0,\infty)$ is given by the expression

$$\mathbb{L}_X = \frac{p(p-1)}{2}\, x^{1-2/p} \frac{\partial}{\partial x} + \frac{p^2}{2}\, x^{2-2/p} \frac{\partial^2}{\partial x^2}. \tag{18.4.1}$$

The equation (13.2.22) takes the form

$$g'(s) = \frac{p g^{1/p}(s)}{2c\big(s^{1/p} - g^{1/p}(s)\big)}, \tag{18.4.2}$$

and its *maximal* admissible solution of (18.4.2) is given by

$$g_*(s) = \alpha s \tag{18.4.3}$$

where $0 < \alpha < 1$ is the maximal root (out of two possible ones) of the equation

$$\alpha - \alpha^{1-1/p} + \frac{p}{2c} = 0. \tag{18.4.4}$$

It can be verified that equation (18.4.4) admits such a root if and only if $c \geq p^{p+1}/2(p-1)^{(p-1)}$. Then by the result of Corollary 13.3, upon using (13.2.65) and letting $c \downarrow p^{p+1}/2(p-1)^{(p-1)}$, we get

$$\mathsf{E}\Big(\max_{0\le t\le\tau} |B_t + x|^p\Big) \le \Big(\frac{p}{p-1}\Big)^p \mathsf{E}\,|B_\tau + x|^p - \frac{p}{p-1}\, x^p \tag{18.4.5}$$

for all stopping times τ of B such that $\mathsf{E}\,\tau^{p/2} < \infty$. The constants $(p/(p-1))^p$ and $p/(p-1)$ are best possible, and equality in (18.4.5) is attained in the limit through the stopping times $\tau_* = \inf\{t>0 : X_t \le \alpha S_t\}$ when $c \downarrow p^{p+1}/2(p-1)^{(p-1)}$. These stopping times are pointwise the smallest possible with this property, and they satisfy $\mathsf{E}\,\tau_*^{p/2} < \infty$ if and only if $c > p^{p+1}/2(p-1)^{(p-1)}$. For more information and remaining details we refer to [80].

Example 18.8. (Further Doob type bounds) The inequality (18.4.5) can be further extended using the same method as follows (for simplicity we state this extension only for $x = 0$):

$$\mathsf{E}\Big(\max_{0\le t\le\tau} |B_t|^p\Big) \le \gamma^*_{p,q} \Big(\mathsf{E} \int_0^\tau |B_t|^{q-1}\, dt\Big)^{p/(q+1)} \tag{18.4.6}$$

for all stopping times τ of B, all $0 < p < 1+q$, and all $q > 0$, with the best possible value for the constant $\gamma^*_{p,q}$ being equal

$$\gamma^*_{p,q} = (1+\kappa)\Big(\frac{s_*}{\kappa^\kappa}\Big)^{1/(1+\kappa)} \tag{18.4.7}$$

where we set $\kappa = p/(q-p+1)$, and s_* is the zero point of the *maximal* admissible solution $s \mapsto g_*(s)$ of

$$g'(s) = \frac{p\, g^{(1-q/p)}(s)}{2(s^{1/p} - g^{1/p}(s))} \tag{18.4.8}$$

satisfying $0 < g_*(s) < s$ for all $s > s_*$. (This solution is also characterized by $g_*(s)/s \to 1$ for $s \to \infty$.) The equality in (18.4.6) is attained at the stopping time $\tau_* = \inf\{t > 0 : X_t = g_*(S_t)\}$ which is pointwise the smallest possible with this property. In the case $p = 1$ the closed form for $s \mapsto g_*(s)$ is given by

$$s \exp\Big(-\frac{2}{pq} g_*^q(s)\Big) + \frac{2}{p} \int_0^{g_*(s)} t^q \exp\Big(-\frac{2}{pq} t^q\Big)\, dt = \Big(\frac{pq}{2}\Big)^{1/q} \Gamma\Big(\frac{q+1}{q}\Big) \tag{18.4.9}$$

for $s \geq s_*$. This, in particular, yields

$$\gamma_{1,q}^* = \Big(\frac{q(1+q)}{2}\Big)^{1/(1+q)} \Big(\Gamma\Big(2 + \frac{1}{q}\Big)\Big)^{q/(1+q)} \tag{18.4.10}$$

for all $q > 0$. In the case $p \neq 1$ no closed form for $s \mapsto g_*(s)$ seems to exist. For more information and remaining details in this direction, as well as for the extension of inequality (18.4.6) to $x \neq 0$, we refer to [158] (see also [156]). To give a more familiar form to the inequality (18.4.6), note by Itô's formula (page 67) and the optional sampling theorem (page 60) that

$$\mathsf{E}\Big(\int_0^\tau |B_t|^{q-1}\, dt\Big) = \frac{2}{q(q+1)} \mathsf{E}\, |B_\tau|^{q+1} \tag{18.4.11}$$

whenever τ is a stopping time of B satisfying $\mathsf{E}\,(\tau^{(q+1)/2}) < \infty$ for $q > 0$. Hence we see that the right-hand side in (18.4.6) is the well-known Doob bound (see (C4) on page 62). The advantage of formulation (18.4.6) lies in its validity for all stopping times.

Notes. While the inequality (18.4.6) (with some constant $\gamma_{p,q} > 0$) can be derived quite easily, the question of its sharpness has gained interest. The case $p = 1$ was treated independently by Jacka [103] (probabilistic methods) and Gilat [75] (analytic methods) who both found the best possible value $\gamma_{1,q}^*$ for $q > 0$. This in particular yields $\gamma_{1,1}^* = \sqrt{2}$ which was independently obtained by Dubins and Schwarz [44], and later again by Dubins, Shepp and Shiryaev [45] who studied the more general case of Bessel processes. (A simple probabilistic proof of $\gamma_{1,1}^* = \sqrt{2}$ is given in [78] — see Subsection 16.3 above). The Bessel processes results are further extended in [150]. In the case $p = 1+q$ with $q > 0$, the inequality (18.4.6) reduces to the Doob maximal inequality (18.4.5). The best values $\gamma_{p,q}^*$ in (18.4.6) and the corresponding optimal stopping times τ^* for all $0 < p \leq 1+q$ and all $q > 0$ are given in [158]. A novel fact about (18.4.5) and (18.4.6) disclosed is that the optimal τ_* from (13.2.58) is pointwise the *smallest* possible stopping time at which the equalities in (18.4.5) (in the limit) and in (18.4.6) can be attained. The results about (18.4.5) and (18.4.6) extend to all non-negative submartingales. This can be obtained by using the maximal embedding result of Jacka [101] (for details see [80] and [158]).

Example 18.9. (A maximal inequality for geometric Brownian motion) Consider the optimal stopping problem (13.1.4) where X is geometric Brownian motion and $c(x) \equiv c$. Recall that X is a non-negative diffusion having 0 as an entrance boundary point, and the infinitesimal generator of X in $(0, \infty)$ is given by the expression

$$\mathbb{L}_X = \rho x \frac{\partial}{\partial x} + \frac{\sigma^2}{2} x^2 \frac{\partial^2}{\partial x^2} \tag{18.4.12}$$

where $\rho \in \mathbb{R}$ and $\sigma > 0$. The process X may be realized as

$$X_t = x \exp\left(\sigma B_t + \left(\rho - \frac{\sigma^2}{2}\right) t\right) \tag{18.4.13}$$

with $x \geq 0$. The equation (13.2.22) takes the form

$$g'(s) = \frac{\Delta \, \sigma^2 \, g^{\Delta+1}(s)}{2\, c\, (s^{\Delta} - g^{\Delta}(s))} \tag{18.4.14}$$

where $\Delta = 1 - 2\rho/\sigma^2$. By using Picard's method of successive approximations it is possible to prove that for $\Delta > 1$ the equation (18.4.14) admits the *maximal admissible* solution $s \mapsto g_*(s)$ satisfying

$$g_*(s) \sim s^{1-1/\Delta} \tag{18.4.15}$$

for $s \to \infty$ (see *Figure* IV.12 and [81] for further details). There seems to be no closed form for this solution. In the case $\Delta = 1$ it is possible to find the general solution of (18.4.14) in a closed form, and this shows that the only non-negative solution is the zero function (see [81]). By the result of Corollary 13.3 we may conclude that the value function (13.1.4) is finite if and only if $\Delta > 1$ (note that another argument was used in [81] to obtain this equivalence), and in this case it is given by

$$\begin{aligned} &V_*(x,s) \qquad\qquad (18.4.16)\\ &= \begin{cases} \dfrac{2c}{\Delta^2 \sigma^2} \left(\left(\dfrac{x}{g_*(s)} \right)^{\Delta} - \log \left(\dfrac{x}{g_*(s)} \right)^{\Delta} - 1 \right) + s & \text{if } \; g_*(s) < x \leq s, \\ s & \text{if } \; 0 < x \leq g_*(s). \end{cases} \end{aligned}$$

The optimal stopping time is given by (13.2.58) with $s_* = 0$. By using explicit estimates on $s \mapsto g_*(s)$ from (18.4.15) in (18.4.16), and then minimizing over all

$c > 0$, we obtain

$$\mathsf{E}\Big(\max_{0\le t\le\tau}\exp\Big(\sigma B_t+\Big(\rho-\frac{\sigma^2}{2}\Big)\,t\Big)\Big) \le 1-\frac{\sigma^2}{2\rho}+\frac{\sigma^2}{2\rho}\exp\Big(-\frac{(\sigma^2-2\rho)^2}{2\sigma^2}\mathsf{E}\,\tau-1\Big) \tag{18.4.17}$$

for all stopping times τ of B. This inequality extends the well-known estimates of Doob in a sharp manner from deterministic times to stopping times. For more information and remaining details we refer to [81]. Observe that the cost function $c(x) = cx$ in the optimal stopping problem (13.1.4) would imply that the maximal admissible solution of (13.2.22) is linear. This shows that such a cost function better suits the maximum process and therefore is more natural. Explicit formulae for the value function, and the maximal inequality obtained by minimizing over $c>0$, are also easily derived in this case from the result of Corollary 13.3.

19. Hardy–Littlewood inequalities

The main purpose of this section (following [83]) is to derive and examine sharp versions of the $L\log L$-inequality of Hardy and Littlewood for one-dimensional Brownian motion which may start at any point.

19.1. Formulation of the problem

Let $B = (B_t)_{t\ge0}$ be a standard Brownian motion defined on a probability space $(\Omega, \mathcal{F}, \mathsf{P})$ such that $B_0 = 0$ under P. The $L\log L$-inequality of Hardy and Littlewood [90] formulated in the optimal stopping setting of B states:

$$\mathsf{E}\Big(\max_{0\le t\le\tau}|B_t|\Big) \le C_1\Big(1+\mathsf{E}\left(|B_\tau|\log^+|B_\tau|\right)\Big) \tag{19.1.1}$$

for all stopping times τ of B with $\mathsf{E}\,\tau^r < \infty$ for some $r > 1/2$, where C_1 is a universal numerical constant (see [40]). The analogue of the problem considered by Gilat [74] may be stated as follows: *Determine the best value for the constant* C_1 *in* (19.1.1), *and find the corresponding optimal stopping time* (*the one at which equality in* (19.1.1) *is attained*).

It is well known that the inequality (19.1.1) remains valid if the plus sign is removed from the logarithm sign, so that we have

$$\mathsf{E}\Big(\max_{0\le t\le\tau}|B_t|\Big) \le C_2\Big(1+\mathsf{E}\left(|B_\tau|\log|B_\tau|\right)\Big) \tag{19.1.2}$$

for all stopping times τ of B with $\mathsf{E}\,\tau^r < \infty$ for some $r > 1/2$, where C_2 is a universal numerical constant. The problem about (19.1.1) stated above extends in exactly the same form to (19.1.2). It turns out that this problem is somewhat easier, however, both problems have some new features which make them interesting from the standpoint of optimal stopping theory.

To describe this in more detail, note that in both cases of (19.1.1) and (19.1.2) we are given an optimal stopping problem with the value function

$$V = \sup_{\tau} \mathsf{E}\left(S_\tau - cF(X_\tau)\right) \tag{19.1.3}$$

where $c > 0$, and in the first case $F(x) = x \log^+ x$, while in the second case $F(x) = x \log x$, with $X_t = |B_t|$ and $S_t = \max_{0 \le r \le t} |B_r|$. The interesting feature of the first problem is that the cost $x \mapsto cF(x)$ is somewhat artificially set to zero for $x \le 1$, while in the second problem the cost is not monotone all over as a function of time. Moreover, in the latter case the Itô formula (page 67) is not directly applicable to $F(X_t)$, due to the fact that $F''(x) = 1/x$ so that $\int_0^\tau F''(X_t)\, dt = \infty$ for all stopping times τ for which $X_\tau \neq 0$ P-a.s. This makes it difficult to find a "useful" increasing functional $t \mapsto I_t$ with the same expectation as the cost (the fact which enables one to write down a differential equation for the value function).

Despite these difficulties one can solve both optimal stopping problems and in turn get solutions to (19.1.1) and (19.1.2) as consequences. The first problem is solved by *guessing* and then verifying that the guess is correct (cf. Theorem 19.1 and Corollary 19.2). The second problem is solved by a *truncation* method (cf. Theorem 19.3 and Corollary 19.4). The obtained results extend to all non-negative submartingales (Corollary 19.6).

19.2. Solution to the problem

In this subsection we present the main results and proofs. Since the problem (19.1.2) is somewhat easier, we begin by stating the main results in this direction (Theorem 19.1). The facts obtained in the proof will be used later (Theorem 19.3) in the solution for the problem (19.1.1). It is instructive to compare these two proofs and notice the essential argument needed to conclude in the latter (note that $d\widetilde{F}/dx$ from the proof of Theorem 19.1 is continuous at $1/e$, while dF_+/dx from the proof of Theorem 19.3 is discontinuous at 1, thus bringing the local time of X at 1 into the game — this is the crucial difference between these two problems). The Gilat paper [74] finishes with a concluding remark where a gap between the $L \log L$ and $L \log^+ L$ case is mentioned. The discovery of the exact size of this gap is stressed to be the main point of his paper. The essential argument mentioned above offers a probabilistic explanation for this gap and gives its exact size in terms of optimal stopping strategies (compare (19.2.2) and (19.2.30) and notice the middle term in (19.2.45) in comparison with (19.2.28)).

Theorem 19.1. *Let $B = (B_t)_{t\ge 0}$ be standard Brownian motion started at zero under P. Then the following inequality is satisfied:*

$$\mathsf{E}\Big(\max_{0\le t\le\tau}|B_t|\Big) \le \frac{c^2}{e(c-1)} + c\,\mathsf{E}\big(|B_\tau|\log|B_\tau|\big) \tag{19.2.1}$$

for all $c>1$ and all stopping times τ of B satisfying $\mathsf{E}\tau^r<\infty$ for some $r>1/2$. This inequality is sharp: equality is attained at the stopping time

$$\sigma_* = \inf\big\{\, t>0 : S_t \ge v_*,\, X_t = \alpha S_t \,\big\} \tag{19.2.2}$$

where $v_* = c/e(c-1)$ and $\alpha = (c-1)/c$ for $c>1$ with $X_t = |B_t|$ and $S_t = \max_{0\le r\le t}|B_r|$.

Proof. Given $c>1$ consider the optimal stopping problem

$$V(x,s) = \sup_\tau \mathsf{E}_{x,s}\big(S_\tau - c\,F(X_\tau)\big) \tag{19.2.3}$$

where $F(x) = x\log x$ for $x\in\mathbb{R}$, $X_t = |B_t+x|$ and $S_t = s\vee\max_{0\le r\le t}|B_r+x|$ for $0\le x\le s$. Note that the (strong) Markov process (X,S) starts at (x,s) under $\mathsf{P} := \mathsf{P}_{x,s}$.

The main difficulty in this problem is that we cannot apply Itô's formula (page 67) to $F(X_t)$. We thus truncate $F(x)$ by setting $\widetilde{F}(x) = F(x)$ for $x\ge 1/e$ and $\widetilde{F}(x) = -1/e$ for $0\le x\le 1/e$. Then $\widetilde{F}\in C^1$ and $\widetilde{F}''$ exists everywhere but $1/e$. Since the time spent by X at $1/e$ is of Lebesgue measure zero, setting $\widetilde{F}''(1/e) := e$, by the Itô–Tanaka–Meyer formula (page 68) we get

$$\begin{aligned}\widetilde{F}(X_t) &= \widetilde{F}(x) + \int_0^t \widetilde{F}'(X_r)\,dX_r + \frac{1}{2}\int_0^t \widetilde{F}''(X_r)\,d\langle X,X\rangle_r \\ &= \widetilde{F}(x) + \int_0^t \widetilde{F}'(X_r)\,d(\beta_r+\ell_r) + \frac{1}{2}\int_0^t \widetilde{F}''(X_r)\,dr \\ &= \widetilde{F}(x) + M_t + \frac{1}{2}\int_0^t \widetilde{F}''(X_r)\,dr\end{aligned} \tag{19.2.4}$$

where $\beta = (\beta_t)_{t\ge 0}$ is a standard Brownian motion, $\ell = (\ell_t)_{t\ge 0}$ is the local time of X at zero, and $M_t = \int_0^t \widetilde{F}'(X_r)\,d\beta_r$ is a continuous local martingale, due to $\widetilde{F}'(0) = 0$ and the fact that $d\ell_r$ is concentrated at $\{\,t : X_t = 0\,\}$. By the optional sampling theorem (page 60) and the Burkholder–Davis–Gundy inequality for continuous local martingales (see (C5) on page 63), we easily find

$$\mathsf{E}_{x,s}\widetilde{F}(X_\tau) = \widetilde{F}(x) + \frac{1}{2}\,\mathsf{E}_{x,s}\Big(\int_0^\tau \widetilde{F}''(X_t)\,dt\Big) \tag{19.2.5}$$

for all stopping times τ of B satisfying $\mathsf{E}_{x,s}\tau^r<\infty$ for some $r>1/2$. By (19.2.5) we see that the value function $V(x,s)$ from (19.2.3) should be identical to $\widetilde{V}(x,s) := W(x,s) - c\,\widetilde{F}(x)$ where

$$W(x,s) = \sup_{\tau} \mathsf{E}_{x,s}\Big(S_\tau - \frac{c}{2} \int_0^\tau \widetilde{F}''(X_t)\,dt \Big) \tag{19.2.6}$$

for $0 \le x \le s$. For this, note that clearly $V(x,s) \le \widetilde{V}(x,s)$, so if we prove that the optimal stopping time $\tilde{\sigma}_*$ in (19.2.6) satisfies $X_{\tilde{\sigma}_*} \ge 1/e$, then due to $\mathsf{E}_{x,s}(S_{\tilde{\sigma}_*} - c\widetilde{F}(X_{\tilde{\sigma}_*})) = \mathsf{E}_{x,s}(S_{\tilde{\sigma}_*} - cF(X_{\tilde{\sigma}_*}))$ this will show that $V(x,s) = \widetilde{V}(x,s)$ with $\tilde{\sigma}_*$ being optimal in (19.2.3) too. In the rest of the proof we solve the optimal stopping problem (19.2.6) and show that the truncation procedure indicated above works as desired.

Supposing that the supremum in (19.2.6) is attained at the exit time of diffusion (X,S) from an open set, we see (cf. Section 7) that the value function $W(x,s)$ should satisfy

$$\mathbb{L}_X W(x,s) = \frac{c}{2}\,\widetilde{F}''(x) \qquad (g_*(s) < x < s) \tag{19.2.7}$$

where $\mathbb{L}_X = \partial^2/2\partial x^2$ is the infinitesimal operator of X in $(0,\infty)$ and $s \mapsto g_*(s)$ is the optimal stopping boundary to be found. To solve (19.2.7) in an explicit form, we shall make use of the following boundary conditions:

$$W(x,s)\Big|_{x=g_*(s)+} = s \quad (\textit{instantaneous stopping}), \tag{19.2.8}$$

$$\frac{\partial W}{\partial x}(x,s)\Big|_{x=g_*(s)+} = 0 \quad (\textit{smooth fit}), \tag{19.2.9}$$

$$\frac{\partial W}{\partial s}(x,s)\Big|_{x=s-} = 0 \quad (\textit{normal reflection}). \tag{19.2.10}$$

Note that (19.2.7)–(19.2.10) forms a problem with free boundary $s \mapsto g_*(s)$. The general solution of (19.2.7) is given by

$$W(x,s) = C(s)x + D(s) + c\,\widetilde{F}(x) \tag{19.2.11}$$

where $s \mapsto C(s)$ and $s \mapsto D(s)$ are unknown functions. By (19.2.8) and (19.2.9) we find

$$C(s) = -c\,\widetilde{F}'(g_*(s)), \tag{19.2.12}$$

$$D(s) = s + c\,g_*(s)\widetilde{F}'(g_*(s)) - c\,\widetilde{F}(g_*(s)). \tag{19.2.13}$$

Inserting (19.2.12) and (19.2.13) into (19.2.11) we obtain

$$W(x,s) = s - c\,\big(x - g_*(s)\big)\widetilde{F}'(g_*(s)) - c\,\widetilde{F}(g_*(s)) + c\widetilde{F}(x) \tag{19.2.14}$$

for $g_*(s) \le x \le s$. Clearly $W(x,s) = s$ for $0 \le x \le g_*(s)$. Finally, by (19.2.10) we find that $s \mapsto g_*(s)$ should satisfy

$$g_*'(s)\,\widetilde{F}''(g_*(s))\,\big(s - g_*(s)\big) = \frac{1}{c} \tag{19.2.15}$$

for $s > 0$. Note that this equation makes sense only for $F''(g_*(s)) > 0$ or equivalently $g_*(s) \ge 1/e$ when it reads as follows:

$$g_*'(s)\left(\frac{s}{g_*(s)} - 1\right) = \frac{1}{c} \tag{19.2.16}$$

for $s \ge v_*$ where $g_*(v_*) = 1/e$. Next observe that (19.2.16) admits a linear solution of the form

$$g_*(s) = \alpha s \tag{19.2.17}$$

for $s \ge v_*$ where $\alpha = (c-1)/c$. (Note that this solution is the maximal admissible solution to either (19.2.15) or (19.2.16). This is in accordance with the maximality principle (see Section 13) and is the main motivation for the candidate (19.2.17).) This in addition indicates that the formula (19.2.14) will be valid only if $s \ge v_*$, where v_* is determined from $g_*(v_*) = 1/e$, so that

$$v_* = c/e(c-1). \tag{19.2.18}$$

The corresponding candidate for the optimal stopping time is

$$\tilde{\sigma}_* = \inf\{\, t > 0 : X_t \le g_*(S_t)\,\} \tag{19.2.19}$$

where $s \mapsto g_*(s)$ is given by (19.2.17) for $s \ge v_*$. The candidate for the value function (19.2.6) given by the formula (19.2.14) for $g_*(s) \le x \le s$ with $s \ge v_*$ will be denoted by $W_*(x,s)$ in the sequel. Clearly $W_*(x,s) = s$ for $0 \le x \le g_*(s)$ with $s \ge v_*$. In the next step we verify that this candidate equals the value function (19.2.6), and that $\tilde{\sigma}_*$ from (19.2.19) is the optimal stopping time.

To verify this, we shall apply the (natural extension of the) Itô–Tanaka–Meyer formula (page 68) to $W_*(X_t,S_t)$. Since $\widetilde{F}'' \ge 0$, this gives

$$\begin{aligned} W_*(X_t,S_t) &= W_*(x,s) + \int_0^t \frac{\partial W_*}{\partial x}(X_r,S_r)\,dX_r \\ &\quad + \int_0^t \frac{\partial W_*}{\partial s}(X_r,S_r)\,dS_r + \frac{1}{2}\int_0^t \frac{\partial^2 W_*}{\partial x^2}(X_r,S_r)\,d\langle X,X\rangle_r \\ &\le W_*(x,s) + \int_0^t \frac{\partial W_*}{\partial x}(X_r,S_r)\,d(\beta_r + \ell_r) + \frac{c}{2}\int_0^t \widetilde{F}''(X_r)\,dr \\ &= W_*(x,s) + M_t + \frac{c}{2}\int_0^t \widetilde{F}''(X_r)\,dr \end{aligned} \tag{19.2.20}$$

with $M_t = \int_0^t (\partial W_*/\partial x)(X_r,S_r)\,d\beta_r$ being a continuous local martingale for $t \ge 0$, where we used that dS_r equals zero for $X_r < S_r$, so that by (19.2.10) the integral over dS_r is equal to zero, while due to $(\partial W_*/\partial x)(0,s) = 0$ the integral over $d\ell_r$ is equal to zero too. Now since $W_*(x,s) \ge s$ for all $x \ge g_*(s)$ (with equality if $x = g_*(s)$) it follows that

$$S_\tau - \frac{c}{2}\int_0^\tau \widetilde{F}''(X_t)\,dt \le W_*(x,s) + M_\tau \tag{19.2.21}$$

for all (bounded) stopping times τ of B with equality (in (19.2.20) as well) if $\tau = \tilde{\sigma}_*$. Taking the expectation on both sides we get

$$\mathsf{E}_{x,s}\Big(S_\tau - \frac{c}{2}\int_0^\tau \widetilde{F}''(X_t)\,dt\Big) \le W_*(x,s) \tag{19.2.22}$$

for all (bounded) stopping times τ of B satisfying

$$\mathsf{E}_{x,s}M_\tau = 0 \tag{19.2.23}$$

with equality in (19.2.22) under the validity of (19.2.23) if $\tau = \tilde{\sigma}_*$. We first show that (19.2.23) holds for all stopping times τ of B satisfying $\mathsf{E}_{x,s}\sqrt{\tau} < \infty$. For this, we compute

$$\begin{aligned} &\mathsf{E}_{x,s}\bigg(\int_0^\tau \Big(\frac{\partial W_*}{\partial x}(X_r,S_r)\Big)^2 dr\bigg)^{1/2} \\ &\quad = c\,\mathsf{E}_{x,s}\bigg(\int_0^\tau \log^2\Big(\frac{X_r}{g_*(S_r)}\Big)\,1_{\{g_*(S_r)\le X_r\}}\,dr\bigg)^{1/2} \\ &\quad = c\log\Big(\frac{1}{\alpha}\Big)\,\mathsf{E}_{x,s}\big(\sqrt{\tau}\big) \end{aligned} \tag{19.2.24}$$

so that (19.2.23) follows by the optional sampling theorem (page 60) and the Burkholder–Davis–Gundy inequality for continuous local martingales (see (C5) on page 63) whenever $\mathsf{E}_{x,s}\sqrt{\tau} < \infty$. Moreover, it is well known (see [221]) that $\mathsf{E}_{x,s}\tilde{\sigma}_*^r < \infty$ for all $r < c/2$. In particular $\mathsf{E}_{x,s}\sqrt{\tilde{\sigma}_*} < \infty$, so that (19.2.23) holds for $\tau = \tilde{\sigma}_*$, and thus we have equality in (19.2.22) for $\tau = \tilde{\sigma}_*$. This completes the proof that the value function (19.2.6) equals $W_*(x,s)$ for $0 \le x \le s$ with $s \ge v_*$, and that σ_* is the optimal stopping time.

Note that $X_{\tilde{\sigma}_*} \ge 1/e$ so that by (19.2.14) and the remark following (19.2.6) we get

$$\begin{aligned} V(x,s) = \widetilde{V}(x,s) &= W(x,s) - c\,\widetilde{F}(x) \\ &= s - cx - cx\log g_*(s) + c\,g_*(s) \end{aligned} \tag{19.2.25}$$

for all $g_*(s) \le x \le s$ with $s \ge v_*$, where $g_*(s) = \alpha s$ with $\alpha = (c-1)/c$ and $v_* = c/e(c-1)$. To complete the proof it remains to compute the value function $V(x,s)$ for $0 \le x \le s$ with $0 \le s < v_*$. A simple observation which motivates our formal move in this direction is as follows.

The best point to stop in the region $0 \le x \le s < v_*$ would be $(1/e\,,s)$ with s as close as possible to v_* , since the cost function $x \mapsto cx\log x$ attains its minimal value at $1/e$. The value function V equals (tends) to $v_*{+}c/e$ if the process (X,S) is started and stopped at $(1/e\,,s)$ with s being equal (tending) to v_* . However, it is easily seen that the value function $V(x,s)$ computed above for $s \ge v_*$ satisfies $V(v_*,v_*) = v_*{+}c/e = c^2/e(c-1)$. This indicates that in the region $0 \le x \le s < v_*$ there should be no point of stopping. This can be formally verified as follows. Given a (bounded) stopping time τ of B, define τ' to be τ on $\{\tau \ge \tau_{v_*}\}$ and $\tilde{\sigma}_*$ on $\{\tau < \tau_{v_*}\}$. Then τ' is a stopping time of B, and clearly $(X_{\tau'}, S_{\tau'})$ does not belong to the region $0 \le x \le s < v_*$. Moreover, by the strong Markov property,

$$\begin{aligned} &\mathsf{E}_{x,s}\big(S_{\tau'} - cF(X_{\tau'})\big) \\ &\quad = \mathsf{E}_{x,s}\big((S_\tau - cF(X_\tau))\, 1_{\{\tau \ge \tau_{v_*}\}}\big) \\ &\qquad\qquad + \mathsf{E}_{x,s}\big((S_{\tilde{\sigma}_*} - cF(X_{\tilde{\sigma}_*}))\, 1_{\{\tau < \tau_{v_*}\}}\big) \\ &\quad = \mathsf{E}_{x,s}\big((S_\tau - cF(X_\tau))\, 1_{\{\tau \ge \tau_{v_*}\}}\big) \\ &\qquad\qquad + \mathsf{E}_{x,s}\big(\mathsf{E}_{v_*,v_*}(S_{\tilde{\sigma}_*} - cF(X_{\tilde{\sigma}_*}))\, 1_{\{\tau < \tau_{v_*}\}}\big) \\ &\quad = \mathsf{E}_{x,s}\big((S_\tau - cF(X_\tau))\, 1_{\{\tau \ge \tau_{v_*}\}}\big) + V(v_*, v_*)\mathsf{P}_{x,s}\big(\tau < \tau_{v_*}\big) \\ &\quad \ge \mathsf{E}_{x,s}\big(S_\tau - cF(X_\tau)\big) \end{aligned} \tag{19.2.26}$$

for all $0 \le x \le s$ with $0 \le s < v_*$, where $\tau_{v_*} = \inf\{\, t > 0 : X_t = v_* \,\}$. Thus $V(x,s) = V(v_*, v_*) = c^2/e(c-1)$ for all $0 \le x \le v_*$, and noting that σ'_* in this case equals σ_* from (19.2.2), the proof is complete. □

The result of Theorem 19.1 extends to the case when Brownian motion B starts at points different from zero.

Corollary 19.2. *Let $B = (B_t)_{t \ge 0}$ be standard Brownian motion started at zero under P. Then the following inequality is satisfied:*

$$\mathsf{E}\Big(\max_{0 \le t \le \tau} |B_t + x|\Big) \le V(x; c) + c\,\mathsf{E}\big(|B_\tau + x| \log |B_\tau + x|\big) \tag{19.2.27}$$

for all $c > 1$ and all stopping times τ of B satisfying $\mathsf{E}\tau^r < \infty$ for some $r > 1/2$, where

$$V(x; c) = \begin{cases} \dfrac{c^2}{e(c-1)} & \text{if } 0 \le x \le v_*, \\ cx \log \dfrac{c}{x(c-1)} & \text{if } x \ge v_* \end{cases} \tag{19.2.28}$$

with $v_* = c/e(c-1)$. This inequality is sharp: for each $c > 1$ and $x \ge 0$ given and fixed, equality in (19.2.27) is attained at the stopping time σ_* defined in (19.2.2) with $X_t = |B_t + x|$ and $S_t = \max_{0 \le r \le t} |B_r + x|$.

Proof. It follows from the proof of Theorem 19.1. Note that $V(x;c)$ equals $V(x,x)$ in the notation of this proof, so that the explicit expression for $V(x;c)$ is given in (19.2.25). □

In the next theorem we present the solution in the $L \log^+ L$-case. The first part of the proof (i.e. the proof of (19.2.29)) is identical to the first part of the proof of Theorem 19.1, and therefore it is omitted.

Theorem 19.3. *Let* $B = (B_t)_{t \ge 0}$ *be standard Brownian motion started at zero under* P. *Then the following inequality is satisfied:*

$$\mathsf{E}\Big(\max_{0 \le t \le \tau} |B_t|\Big) \le 1 + \frac{1}{e^c(c-1)} + c\,\mathsf{E}\big(|B_\tau| \log^+ |B_\tau|\big) \tag{19.2.29}$$

for all $c > 1$ *and all stopping times* τ *of* B *satisfying* $\mathsf{E}\tau^r < \infty$ *for some* $r > 1/2$. *This inequality is sharp: equality is attained at the stopping time*

$$\tau_* = \inf\big\{\, t>0 : S_t \ge u_*,\ X_t = 1 \vee \alpha S_t \big\} \tag{19.2.30}$$

where $u_* = 1 + 1/e^c(c-1)$ *and* $\alpha = (c-1)/c$ *for* $c > 1$ *with* $X_t = |B_t|$ *and* $S_t = \max_{0 \le r \le t} |B_r|$.

Proof. Given $c > 1$ consider the optimal stopping problem

$$V_+(x,s) = \sup_\tau \mathsf{E}_{x,s}\big(S_\tau - cF_+(X_\tau)\big) \tag{19.2.31}$$

where $F_+(x) = x \log^+ x$, $X_t = |B_t + x|$ and $S_t = s \vee \max_{0 \le r \le t} |B_r + x|$ for $0 \le x \le s$. Since $F_+(x) = F(x)$ for all $x \ge 1$, it is clear that $V_+(x,s)$ coincides with the value function $V(x,s)$ from (19.2.3) (with the same optimal stopping time given by either (19.2.2) or (19.2.30)) for $0 \le x \le s$ with $s \ge s_*$, where s_* is determined from $g_*(s_*) = 1$ with $g_*(s) = \alpha s$ and $\alpha = (c-1)/c$, so that $s_* = 1/\alpha = c/(c-1)$. It is also clear that the process (X,S) cannot be optimally stopped at some τ with $X_\tau < 1$ since $F_+(x) = 0$ for $x < 1$. This shows that $V_+(0,0) = V_+(x,s) = V_+(1,1)$ for all $0 \le x \le s \le 1$. So it remains to compute the value function $V_+(x,s)$ for $0 \le x \le s$ with $1 \le s < s_*$. This evaluation is the main content of the proof. We begin by giving some intuitive arguments which are followed by a rigorous justification.

The best place to stop in the region $0 \le x \le s$ with $1 \le s \le s_*$ is clearly at $(1,s)$, so that there should exist a point $1 \le u_* \le s_*$ such that the process (X,S) should be stopped at the vertical line $\{\,(1,s) : u_* \le s \le s_*\,\}$, as well as to the left from it (if started there). We also expect that $V_+(u_*,u_*) = u_*$ (since we do not stop at $(1, u_* - \varepsilon)$ where the value function V_+ would be equal $u_* - \varepsilon$ for $\varepsilon > 0$ as small as desired). Clearly, the value function V_+ should be constant in the region $0 \le x \le s \le u_*$ (note that there is no running cost), and then (when restricted to the diagonal $x = s$ for $u_* \le s \le s_*$) it should decrease. Note from (19.2.25) that $V_+(s_*,s_*) = V(s_*,s_*) = 0$. So let us try to determine such a point u_*.

Thus we shall try to compute

$$\sup_{1 \le s \le s_*} \mathsf{E}_{s,s}\big(S_{\tau'_*} - cF_+(X_{\tau'_*})\big) \tag{19.2.32}$$

where $\tau'_* = \tau'_*(s) = \inf\{\, t > 0 : X_t = 1 \vee \alpha S_t\,\}$. For this, note by the strong Markov property (and $V_+(s_*,s_*) = 0$) that if τ'_* is to be an optimal stopping time (for some $s = u_*$), we should have

$$\begin{aligned} V_{+,*}(x,s) &:= \mathsf{E}_{x,s}\big(S_{\tau'_*} - cF_+(X_{\tau'_*})\big) \qquad (19.2.33)\\ &= \mathsf{E}_{x,s}\big(S_{\tau'_*}\, 1_{\{\tau'_*<\tau_{s_*}\}}\big) + \mathsf{E}_{x,s}\big((S_{\tau'_*} - c\,F_+(X_{\tau'_*}))\, 1_{\{\tau'_*\ge\tau_{s_*}\}}\big)\\ &= \mathsf{E}_{x,s}\big(S_{\tau'_*}\, 1_{\{\tau'_*<\tau_{s_*}\}}\big) + \mathsf{E}_{x,s}\big(\mathsf{E}_{s_*,s_*}(S_{\tau'_*} - c\,F_+(X_{\tau'_*}))\, 1_{\{\tau'_*\ge\tau_{s_*}\}}\big)\\ &= \mathsf{E}_{x,s}\big(S_{\tau'_*}\, 1_{\{\tau'_*<\tau_{s_*}\}}\big) \end{aligned}$$

for all $1 \le x \le s \le s_*$ where $\tau_{s_*} = \inf\{\, t>0 : X_t = s_* \,\}$. Note further that τ'_* (on $\{\,\tau'_* < \tau_{s_*}\}$) may be viewed as the exit time of (X,S) from an open set, so that $V_{+,*}(x,s)$ should satisfy

$$\mathbb{L}_X V_{+,*}(x,s) = 0 \quad (1 < x < s), \qquad (19.2.34)$$

$$V_{+,*}(x,s)\Big|_{x=1} = s \quad \text{(instantaneous stopping)}, \qquad (19.2.35)$$

$$\frac{\partial V_{+,*}}{\partial s}(x,s)\Big|_{x=s-} = 0 \quad \text{(normal reflection)}, \qquad (19.2.36)$$

$$V_{+,*}(x,s)\Big|_{x=s=s_*} = 0 \quad \text{(strong Markov property)} \qquad (19.2.37)$$

for $1 \le x \le s \le s_*$. System (19.2.34)–(19.2.37) has a unique solution given by

$$V_{+,*}(x,s) = s + (1-x)\log(s-1) + K(x-1) \qquad (19.2.38)$$

for $1 \le x \le s \le s_*$ where

$$K = -c - \log(c-1). \qquad (19.2.39)$$

Solving $(\partial V_{+,*}/\partial s)(s,s) = 0$ we find the point at which the supremum in (19.2.32) is to be attained:

$$u_* = 1 + e^K = 1 + 1/e^c(c-1). \qquad (19.2.40)$$

Thus the candidate $V_{+,*}(x,s)$ for the value function (19.2.31) is given by (19.2.38) for all $1 \le x \le s$ with $u_* \le s \le s_*$. Clearly we have to put $V_{+,*}(x,s) = s$ for $0 \le x \le 1$ with $u_* \le s \le s_*$. Note moreover that $V_{+,*}(x,s) = V_{+,*}(u_*,u_*) = u_* = 1+1/e^c(c-1)$ for all $0 \le x \le s \le u_*$ as suggested above. So if we show in the sequel that this candidate is indeed the value function with the optimal stopping time $\tau_* = \tau'_*(u_*)$, the proof of the theorem will be complete.

That there should be no point of stopping in the region $0 \le x \le s \le u_*$ is verified in exactly the same way as in (19.2.26). So let us concentrate on the case when $u_* \le s \le s_*$. To apply Itô's formula (page 67) we shall redefine $V_{+,*}(x,s)$ for $x < 1$ by (19.2.38). This extension will be denoted by $\widetilde{V}_{+,*}(x,s)$. Obviously $\widetilde{V}_{+,*} \in C^2$ and $\widetilde{V}_{+,*}(x,s) = V_{+,*}(x,s)$ for $1 \le x \le s$ with $u_* \le s \le s_*$. Applying Itô's formula (page 67) to $\widetilde{V}_{+,*}(X_t,S_t)$ and noting that $(\partial\widetilde{V}_{+,*}/\partial x)(0,s) \le 0$ for

$u_* \le s \le s_*$ (any such C^2-extension would do) we get

$$\begin{aligned}\widetilde{V}_{+,*}(X_t,S_t) &= \widetilde{V}_{+,*}(x,s) + \int_0^t \frac{\partial \widetilde{V}_{+,*}}{\partial x}(X_r,S_r)\,dX_r \\ &\quad + \int_0^t \frac{\partial \widetilde{V}_{+,*}}{\partial s}(X_r,S_r)\,dS_r + \frac{1}{2}\int_0^t \frac{\partial^2 \widetilde{V}_{+,*}}{\partial x^2}(X_r,S_r)\,d\langle X,X\rangle_r \\ &= \widetilde{V}_{+,*}(x,s) + \int_0^t \frac{\partial \widetilde{V}_{+,*}}{\partial x}(X_r,S_r)\,d(\beta_r+\ell_r) \\ &= \widetilde{V}_{+,*}(x,s) + M_t + \int_0^t \frac{\partial \widetilde{V}_{+,*}}{\partial x}(0,S_r)\,d\ell_r \le V_{+,*}(x,s) + M_t\end{aligned} \tag{19.2.41}$$

for all $0 \le x \le s$ with $u_* \le s \le s_*$ where $M_t = \int_0^t (\partial \widetilde{V}_{+,*}/\partial x)(X_r,S_r)\,d\beta_r$ is a continuous local martingale. Moreover, hence we find

$$\widetilde{V}_{+,*}(X_\tau,S_\tau) \le \widetilde{V}_{+,*}(x,s) + M_\tau \tag{19.2.42}$$

for all stopping times τ of B with equality if $\tau \le \tau_*$. Due to $ey \le e^y$ it is easily verified that $\widetilde{V}_{+,*}(x,s) = V_{+,*}(x,s) \ge s - cF_+(x)$ for $1 \le x \le s$ with $u_* \le s \le s_*$ (with equality if $x = 1$). Now given a stopping time τ of B, define τ' to be τ on $\{X_\tau \ge 1\}$ and τ_1 on $\{X_\tau < 1\}$, where $\tau_1 = \inf\{t > 0 : X_t = 1\}$. Then τ' is a new stopping time of B, and by (19.2.42) and the remark following it, we clearly have

$$\begin{aligned}\mathsf{E}_{x,s}\big(S_\tau - c\,F_+(X_\tau)\big) &= \mathsf{E}_{x,s}\big(S_{\tau'} - c\,F_+(X_{\tau'})\big) \\ &\le \mathsf{E}_{x,s}\big(V_{+,*}(X_{\tau'},S_{\tau'})\big) \le V_{+,*}(x,s) + \mathsf{E}_{x,s}M_{\tau'} = V_{+,*}(x,s)\end{aligned} \tag{19.2.43}$$

for all $1 \le x \le s$ with $u_* \le s \le s_*$ whenever $\mathsf{E}_{x,s}(\tau')^r < \infty$ for some $r > 1/2$ with equalities if $\tau = \tau_*$ (recall that $\mathsf{E}_{x,s}\tau_*^r < \infty$ for all $r < c/2$). The proof of optimality of the stopping time τ_* defined in (19.2.30) above is complete. □

The result of Theorem 19.3 also extends to the case when Brownian motion B starts at points different from zero.

Corollary 19.4. *Let $B = (B_t)_{t\ge 0}$ be standard Brownian motion started at zero under P. Then the following inequality is satisfied:*

$$\mathsf{E}\Big(\max_{0\le t\le \tau} |B_t+x|\Big) \le V_+(x;c) + c\mathsf{E}\big(|B_\tau+x|\,\log^+|B_\tau+x|\big) \tag{19.2.44}$$

for all $c > 1$ and all stopping times τ of B satisfying $\mathsf{E}\tau^r < \infty$ for some $r > 1/2$, where

$$V_+(x;c) = \begin{cases} 1 + 1/e^c(c-1) & \text{if } 0 \le x \le u_*, \\ x + (1-x)\log(x-1) \\ \qquad -(c+\log(c-1))(x-1) & \text{if } u_* \le x \le s_*, \\ cx\log\big(c/x(c-1)\big) & \text{if } x \ge s_* \end{cases} \tag{19.2.45}$$

with $u_* = 1 + 1/e^c(c-1)$ *and* $s_* = c/(c-1)$. *This inequality is sharp: for each* $c > 1$ *and* $x \ge 0$ *given and fixed, equality in* (19.2.44) *is attained at the stopping time* τ_* *defined in* (19.2.30) *with* $X_t = |B_t + x|$ *and* $S_t = \max_{0 \le r \le t} |B_r + x|$.

Proof. It follows from the proof of Theorem 19.3. Note that $V_+(x;c)$ equals $V_+(x,x)$ in the notation of this proof, so that the explicit expression for $V_+(x;c)$ is given in (19.2.38). □

Remark 19.5. The distribution law of X_{τ_*} and S_{τ_*} from Theorem 19.1 (Corollary 19.2) and Theorem 19.3 (Corollary 19.4) can be computed explicitly (see [6]). For this one can use the fact that $H(S_t) - (S_t - X_t)H'(S_t)$ is a (local) martingale before X hits zero for sufficiently many functions H. We will omit further details.

Due to the universal role of Brownian motion in this context, the inequalities (19.2.27) and (19.2.44) extend to all non-negative submartingales. This can be obtained by using the maximal embedding result of Jacka [101].

Corollary 19.6. *Let* $X = (X_t)_{t \ge 0}$ *be a non-negative càdlàg (right continuous with left limits) uniformly integrable submartingale started at* $x \ge 0$ *under* P. *Let* X_∞ *denote the* P*-a.s. limit of* X *for* $t \to \infty$ *(which exists by* (B1) *on page* 61*). Then the following inequality is satisfied:*

$$\mathsf{E} \sup_{t>0} X_t \le W_G(x;c) + c\,\mathsf{E}\,G(X_\infty) \tag{19.2.46}$$

for all $c > 1$, *where* $G(y)$ *is either* $y \log y$ *and in this case* $W_G(x;c)$ *is given by* (19.2.28), *or* $G(y)$ *is* $y \log^+ y$ *and in this case* $W_G(x;c)$ *is given by* (19.2.45). *This inequality is sharp.*

Proof. Given such a submartingale $X = (X_t)_{t \ge 0}$ satisfying $\mathsf{E}\,G(X_\infty) < \infty$, and a Brownian motion $B = (B_t)_{t \ge 0}$ started at $X_0 = x$ under P_x, by the result of Jacka [101] we know that there exists a stopping time τ of B, such that $|B_\tau| \sim X_\infty$ and $\mathsf{P}\{\sup_{t \ge 0} X_t \ge \lambda\} \le \mathsf{P}_x\{\max_{0 \le t \le \tau} |B_t| \ge \lambda\}$ for all $\lambda > 0$, with $(B_{t \wedge \tau})_{t \ge 0}$ being uniformly integrable. The inequality (19.2.46) then easily follows from Corollary 19.2 and Corollary 19.4 by using the integration by parts formula. Note that by the submartingale property of $(|B_{t \wedge \tau}|)_{t \ge 0}$ we have $\sup_{t \ge 0} \mathsf{E}_x G(|B_{t \wedge \tau}|) = \mathsf{E}_x G(|B_\tau|)$. This completes the proof. □

Notes. This section is motivated by the paper of Gilat [74] where he settles a question raised by Dubins and Gilat [43], and later by Cox [32], on the $L \log L$-inequality of Hardy and Littlewood. Instead of recalling his results in the analytic framework of the Hardy–Littlewood theory, we shall rather refer the reader to Gilat's paper [74] where a splendid historical exposition on the link between the Hardy–Littlewood theory and probability (martingale theory) can be found too. Despite the fact that Gilat's paper finishes with a comment on the use of his result in the martingale theory, his proof is entirely analytic. The main aim of this section is to present a new probabilistic solution to this problem. While Gilat's result gives the best value for C_1, it does not detect the optimal stopping strategy τ_* in (19.1.1), but rather gives the distribution law of B_{τ_*} and S_{τ_*} (see Remark 19.5). In contrast to this, the proof above does both, and together with the extension to the case when B starts at any point, this detection (of the optimal stopping strategy) forms the principal result of the section.

19.3. Further examples

The result of Theorem 19.1 and Theorem 19.3 can also be obtained directly from the maximality principle (see Section 13). We will illustrate this line of argument by one more example.

Example 19.7. (A sharp integral inequality of the L log L-type) Consider the optimal stopping problem (13.1.4) with $X_t = |B_t+x|$ and $c(x) = c/(1+x)$ for $x \geq 0$ and $c > 0$. Then X is a non-negative diffusion having 0 as an instantaneously-reflecting regular boundary point, and the infinitesimal generator of X in $(0, \infty)$ is given by (18.4.1) with $p = 1$. The equation (13.2.22) takes the form

$$g'(s) = \frac{1+g(s)}{2c(s-g(s))}, \tag{19.3.1}$$

and its *maximal* admissible solution is given by

$$g_*(s) = \alpha s - \beta \tag{19.3.2}$$

where $\alpha = (2c-1)/2c$ and $\beta = 1/2c$. By applying the result of Corollary 13.3 we get

$$\mathsf{E}\Big(\max_{0\leq t\leq\tau} |B_t+x|\Big) \leq W(x;c) + c\,\mathsf{E}\bigg(\int_0^\tau \frac{dt}{1+|B_t+x|}\bigg) \tag{19.3.3}$$

for all stopping times τ of B, all $c > 1/2$ and all $x \geq 0$, where

$$W(x;c) = \begin{cases} \dfrac{1}{2c-1} + 2c\Big((1+x)\log(1+x) - x\Big) & \text{if } x \leq \dfrac{1}{(2c-1)}, \\ 2c(1+x)\log\Big(1+\dfrac{1}{2c-1}\Big) - 1 & \text{if } x > \dfrac{1}{(2c-1)}. \end{cases} \tag{19.3.4}$$

This inequality is sharp, and for each $c > 1/2$ and $x \geq 0$ given and fixed, equality in (19.3.4) is attained at the stopping time

$$\tau_* = \inf\big\{\, t > 0 : S_t - \alpha X_t \geq \beta \,\big\} \tag{19.3.5}$$

which is pointwise the smallest possible with this property. By minimizing over all $c > 1/2$ on the right-hand side in (19.3.3) we get a sharp inequality (equality is attained at each stopping time τ_* from (19.3.5) whenever $c > 1/2$ and $x \geq 0$). In particular, this for $x = 0$ yields

$$\mathsf{E}\Big(\max_{0\leq t\leq\tau}|B_t|\Big) \leq \frac{1}{2}\,\mathsf{E}\bigg(\int_0^\tau \frac{dt}{1+|B_t|}\bigg) + \sqrt{2}\,\bigg(\mathsf{E}\int_0^\tau \frac{dt}{1+|B_t|}\bigg)^{1/2} \tag{19.3.6}$$

for all stopping times τ of B. This inequality is sharp, and equality in (19.3.6) is attained at each stopping time τ_* from (19.3.5). Note by Itô's formula (page 67) and the optional sampling theorem (page 60) that

$$\mathsf{E}\bigg(\int_0^\tau \frac{dt}{1+|B_t|}\bigg) = 2\,\mathsf{E}\Big(\big(1+|B_\tau|\big)\log\big(1+|B_\tau|\big) - |B_\tau|\Big) \tag{19.3.7}$$

for all stopping times τ of B satisfying $\mathsf{E}\,\tau^r < \infty$ for some $r > 1/2$. This shows that the inequality (19.3.6) in essence is of the $L\log L$-type. The advantage of (19.3.6) upon the classical Hardy–Littlewood $L\log L$-inequality is its sharpness for small stopping times as well (note that equality in (19.3.6) is attained for $\tau \equiv 0$). For more information on this inequality and remaining details we refer to [157].

20. Burkholder–Davis–Gundy inequalities

All optimal stopping problems considered so far in this chapter were linear in the sense that the gain function is a linear function of time (recall our discussion in the end of Section 15 above). In this section we will briefly consider a nonlinear problem in order to address difficulties which such problems carry along.

Let $B = (B_t)_{t\geq 0}$ be a standard Brownian motion defined on a probability space $(\Omega, \mathcal{F}, \mathsf{P})$, and let $p > 0$ be given and fixed. Then the Burkholder–Davis–Gundy inequalities (see (C5) on page 63) state that

$$c_p\,\mathsf{E}\,\tau^{p/2} \leq \mathsf{E}\max_{0\leq t\leq\tau}|B_t|^p \leq C_p\,\mathsf{E}\,\tau^{p/2} \tag{20.0.8}$$

for all stopping times τ of B where $c_p > 0$ and $C_p > 0$ are universal constants.

The question of finding the best possible values for c_p and C_p in (20.0.8) appears to be of interest. Its emphasis is not so much on having these values but more on finding a method of proof which can deliver them. Clearly, the case $p = 2$ reduces trivially to Doob's maximal inequality treated in Section 18 above and $C_2 = 4$ is the best possible constant in (20.0.8) when $p = 2$. Likewise, it is easily seen that $c_2 = 1$ is the best constant in (20.0.8) when $p = 2$ (consider $\tau = \inf\{t \geq 0 : |B_t| = 1\}$ for instance). In the case of other p however the situation is much more complicated. For example, if $p = 1$ then (20.0.8) reads as follows:

$$c_1\,\mathsf{E}\sqrt{\tau} \leq \mathsf{E}\max_{0\leq t\leq\tau}|B_t| \leq C_2\,\mathsf{E}\sqrt{\tau} \tag{20.0.9}$$

and the best constants c_1 and C_2 in (20.0.9), being valid for all stopping times τ of B, seem to be unknown to date.

To illustrate difficulties which are inherently present in tackling these questions let us concentrate on the problem of finding the best value for C_2. For this, consider the optimal stopping problem

$$V = \sup_{\tau} \mathsf{E}\Big(\max_{0 \le t \le \tau} |B_t| - c\sqrt{\tau}\Big) \tag{20.0.10}$$

where the supremum is taken over all stopping times τ of B satisfying $\mathsf{E}\sqrt{\tau} < \infty$, and $c > 0$ is a given and fixed constant.

In order to solve this problem we need to determine its dimension (see Subsection 6.2) and (if possible) try to reduce it by using some of the available transformations (see Sections 10–12). Leaving the latter aside for the moment note that the underlying Markov process is $Z_t = (t, X_t, S_t)$ where $X_t = |B_t|$ and $S_t = \max_{0 \le s \le t} |B_s|$ for $t \ge 0$. Due to the existence of the square root in (20.0.10) it is not possible to remove the time component t from Z_t and therefore the *nonlinear* problem (20.0.10) is inherently *three-dimensional.*

Recalling our discussion in Subsection 13.2 it is plausible to assume that the following optimal stopping time should be optimal in (20.0.10):

$$\tau_* = \inf\big\{\, t \ge 0 : X_t \le g_*(t, S_t)\,\big\} \tag{20.0.11}$$

for $c > C_1$, where $(t, s) \mapsto g_*(t, s)$ is the optimal stopping time which now depends on both time t and maximum s so that its explicit determination becomes much more delicate. (To be more precise one should consider (20.0.11) under $\mathsf{P}_{u,x,s}$ where $\mathsf{P}_{u,x,s}(X_u = x, S_u = s) = 1$ for $u \ge 0$ and $s > x > 0$.)

Note that when $\max_{0 \le t \le \tau} |B_t|$ is replaced by $|B_\tau|$ in (20.0.11) then the problem can be successfully tackled by the method of time change (see Section 10). We will omit further details in this direction.

Chapter VI.

Optimal stopping in mathematical statistics

21. Sequential testing of a Wiener process

In the Bayesian formulation of the problem it is assumed that we observe a trajectory of the Wiener process (Brownian motion) $X = (X_t)_{t \geq 0}$ with drift $\theta\mu$ where the random variable θ may be 1 or 0 with probability π or $1-\pi$, respectively.

1. For a precise probabilistic formulation of the Bayesian problem it is convenient to assume that all our considerations take place on a *probability-statistical* space $(\Omega; \mathcal{F}; \mathsf{P}_\pi, \pi \in [0,1])$ where the probability measure P_π has the following structure:

$$\mathsf{P}_\pi = \pi \mathsf{P}_1 + (1-\pi)\mathsf{P}_0 \tag{21.0.1}$$

for $\pi \in [0,1]$. (Sometimes $(\Omega; \mathcal{F}; \mathsf{P}_\pi, \pi \in [0,1])$ is called a *statistical experiment.*)

Let θ be a random variable taking two values 1 and 0 with probabilities $\mathsf{P}_\pi(\theta = 1) = \pi$ and $\mathsf{P}_\pi(\theta = 0) = 1 - \pi$, and let $W = (W_t)_{t \geq 0}$ be a standard Wiener process started at zero under P_π. It is assumed that θ and W are independent.

It is further assumed that we observe a process $X = (X_t)_{t \geq 0}$ of the form

$$X_t = \theta\mu t + \sigma W_t \tag{21.0.2}$$

where $\mu \neq 0$ and $\sigma^2 > 0$ are given and fixed. Thus $\mathsf{P}_\pi(X \in \cdot \mid \theta = i) = \mathsf{P}_i(X \in \cdot)$ is the distribution law of a Wiener process with drift $i\mu$ and diffusion coefficient $\sigma^2 > 0$ for $i = 0, 1$, so that π and $1-\pi$ play the role of a priori probabilities of the statistical hypotheses

$$H_1\colon\ \theta = 1 \quad \text{and} \quad H_0\colon\ \theta = 0 \tag{21.0.3}$$

respectively.

Being based upon the continuous observation of X our task is to test sequentially the hypotheses H_1 and H_0 with a minimal loss. For this, we consider a sequential decision rule (τ, d), where τ is a stopping time of the observed process X (i.e. a stopping time with respect to the natural filtration $\mathcal{F}_t^X = \sigma(X_s : 0 \le s \le t)$ generated by X for $t \ge 0$), and d is an $\mathcal{F}_\tau^X$-measurable random variable taking values 0 and 1. After stopping the observation at time τ, the terminal decision function d indicates which hypothesis should be accepted according to the following rule: if $d = 1$ we accept H_1, and if $d = 0$ we accept H_0. The problem then consists of computing the *risk* function

$$V(\pi) = \inf_{(\tau,d)} \mathsf{E}_\pi\big(\tau + aI(d=0,\theta=1) + bI(d=1,\theta=0)\big) \tag{21.0.4}$$

and finding the optimal decision rule (τ_*, d_*) at which the infimum in (21.0.4) is attained. Here $\mathsf{E}_\pi \tau$ is the average loss due to a cost of the observations, and $a\mathsf{P}_\pi(d=0,\theta=1) + b\mathsf{P}_\pi(d=1,\theta=0)$ is the average loss due to a wrong terminal decision, where $a > 0$ and $b > 0$ are given constants.

2. By means of standard arguments (see [196, pp. 166–167]) one can reduce the Bayesian problem (21.0.4) to the optimal stopping problem

$$V(\pi) = \inf_{\tau} \mathsf{E}_\pi\big(\tau + a\pi_\tau \wedge b(1-\pi_\tau)\big) \tag{21.0.5}$$

for the *a posteriori probability process* $\pi_t = \mathsf{P}_\pi(\theta = 1 \,|\, \mathcal{F}_t^X)$ with $t \ge 0$ and $\mathsf{P}_\pi(\pi_0 = \pi) = 1$ (where $x \wedge y = \min\{x, y\}$). Setting $c = b/(a+b)$ the optimal decision function is then given by $d_* = 1$ if $\pi_{\tau_*} \ge c$ and $d_* = 0$ if $\pi_{\tau_*} < c$.

3. It can be shown (see [196, pp. 180–181]) that the *likelihood ratio process* $(\varphi_t)_{t\ge 0}$, defined as the Radon–Nikodým derivative

$$\varphi_t = \frac{d(\mathsf{P}_1 | \mathcal{F}_t^X)}{d(\mathsf{P}_0 | \mathcal{F}_t^X)}, \tag{21.0.6}$$

admits the following representation:

$$\varphi_t = \exp\Big(\frac{\mu}{\sigma^2}\Big(X_t - \frac{\mu}{2}t\Big)\Big) \tag{21.0.7}$$

while the *a posteriori probability process* $(\pi_t)_{t\ge 0}$ can be expressed as

$$\pi_t = \Big(\frac{\pi}{1-\pi}\,\varphi_t\Big) \Big/ \Big(1 + \frac{\pi}{1-\pi}\,\varphi_t\Big) \tag{21.0.8}$$

and hence solves the stochastic differential equation

$$d\pi_t = \frac{\mu}{\sigma}\pi_t(1-\pi_t)\, d\bar{W}_t \qquad (\pi_0 = \pi) \tag{21.0.9}$$

where the *innovation* process $(\bar{W}_t)_{t\ge 0}$ defined by

$$\bar{W}_t = \frac{1}{\sigma}\Big(X_t - \mu \int_0^t \pi_s\, ds\Big) \tag{21.0.10}$$

is a standard Wiener process (see also [127, Chap. IX]). Using (21.0.7) and (21.0.8) it can be verified that $(\pi_t)_{t\ge 0}$ is a time-homogeneous (strong) Markov process under P_π with respect to the natural filtration. As the latter clearly coincides with $(\mathcal{F}_t^X)_{t\ge 0}$ it is also clear that the infimum in (21.0.5) can equivalently be taken over all stopping times of $(\pi_t)_{t\ge 0}$.

21.1. Infinite horizon

1. In order to solve the problem (21.0.5) above when the horizon is *infinite* let us consider the optimal stopping problem for the Markov process $(\pi_t)_{t\ge 0}$ given by

$$V(\pi) = \inf_\tau \mathsf{E}_\pi\big(M(\pi_\tau) + \tau\big) \tag{21.1.1}$$

where $\mathsf{P}_\pi(\pi_0 = \pi) = 1$, i.e. P_π is a probability measure under which the diffusion process $(\pi_t)_{t\ge 0}$ solving (21.0.9) starts at π, the infimum in (21.1.1) is taken over all stopping times τ of $(\pi_t)_{t\ge 0}$, and we set $M(\pi) = a\pi \wedge b(1-\pi)$ for $\pi \in [0,1]$. For further reference recall that the infinitesimal generator of $(\pi_t)_{t\ge 0}$ is given by

$$\mathbb{L} = \frac{\mu^2}{2\sigma^2}\, \pi^2(1-\pi)^2\, \frac{\partial^2}{\partial \pi^2}. \tag{21.1.2}$$

2. The optimal stopping problem (21.1.1) will be solved in two steps. In the first step we will make a guess for the solution. In the second step we will verify that the guessed solution is correct (Theorem 21.1).

From (21.1.1) and (21.0.9) above we see that the closer $(\pi_t)_{t\ge 0}$ gets to either 0 or 1 the less likely that the loss will decrease upon continuation. This suggests that there exist points $A \in [0,c]$ and $B \in [c,1]$ such that the stopping time

$$\tau_{A,B} = \inf\big\{t \ge 0 : \pi_t \notin (A,B)\big\} \tag{21.1.3}$$

is optimal in (21.1.1).

Standard arguments based on the strong Markov property (cf. Chap. III) lead to the following *free-boundary problem* for the unknown function V and the unknown points A and B:

$$\mathbb{L}V = -1 \quad \text{for } \pi \in (A,B), \tag{21.1.4}$$
$$V(A) = aA, \tag{21.1.5}$$
$$V(B) = b(1-B), \tag{21.1.6}$$
$$V'(A) = a \quad (\textit{smooth fit}), \tag{21.1.7}$$
$$V'(B) = -b \quad (\textit{smooth fit}), \tag{21.1.8}$$
$$V < M \quad \text{for } \pi \in (A,B), \tag{21.1.9}$$
$$V = M \quad \text{for } \pi \in [0,A) \cup (B,1]. \tag{21.1.10}$$

3. To solve the free-boundary problem (21.1.4)–(21.1.10) denote

$$\psi(\pi) = (1-2\pi)\log\Big(\frac{\pi}{1-\pi}\Big) \tag{21.1.11}$$

and with fixed $A \in (0,c)$ note by a direct verification that the function

$$V(\pi;A) = \frac{2\sigma^2}{\mu^2}\big(\psi(\pi)-\psi(A)\big) + \Big(a - \frac{2\sigma^2}{\mu^2}\,\psi'(A)\Big)(\pi - A) + aA \tag{21.1.12}$$

is the unique solution of the equation (21.1.4) for $\pi \ge A$ satisfying (21.1.5) and (21.1.7) at A.

When $A \in (0,c)$ is close to c, then $\pi \mapsto V(\pi;A)$ intersects $\pi \mapsto b(1-\pi)$ at some $B \in (c,1)$. The function $\pi \mapsto V(\pi;A)$ is concave on $(0,1)$, it satisfies $V(0+;A) = V(1-;A) = -\infty$, and $\pi \mapsto V(\pi;A')$ and $\pi \mapsto V(\pi;A'')$ do not intersect at any $\pi > A' \vee A''$ when $A' \neq A''$. For the latter note that $(\partial/\partial A)V(\pi;A) = (2\sigma^2/\mu^2)\psi''(A)(A-\pi) > 0$ for all $\pi > A$ since $\psi''(A) < 0$. Let π_A^0 denote the zero point of $\pi \mapsto V(\pi;A)$ on $(A,1)$. Then $\pi_A^0 \downarrow 0$ as $A \downarrow 0$ since clearly $\pi_A^0 \downarrow l$ while assuming $l > 0$ and passing to the limit for $A \downarrow 0$ in the equation $V(\pi_A^0;A) = 0$ leads to a contradiction. Finally, reducing A from c down to 0 and using the properties established above we get the existence of a unique point $A_* \in (0,c)$ for which there is $B_* \in (c,1)$ such that $V(B_*;A_*) = b(1-B_*)$ and $V'(B_*;A_*) = -b$ as required by (21.1.6) and (21.1.8) above. This establishes the existence of a unique solution $(V(\,\cdot\,;A_*),A_*,B_*)$ to the free-boundary problem (21.1.4)–(21.1.10). Note that $\pi \mapsto V(\pi;A_*)$ is C^2 on $(0,1)\setminus\{A,B\}$ but only C^1 at A_* and B_* when extended to be equal to M on $[0,A_*)$ and $(B_*,1]$. Note also that the extended function $\pi \mapsto V(\pi;A_*)$ is concave on $[0,1]$.

4. In this way we have arrived at the conclusions of the following theorem.

Theorem 21.1. *The value function* V *from* (21.1.1) *is explicitly given by*

$$V(\pi) = \begin{cases} \frac{2\sigma^2}{\mu^2}\big(\psi(\pi)-\psi(A)\big) + \Big(a - \frac{2\sigma^2}{\mu^2}\,\psi'(A_*)\Big) \\ \quad\times(\pi - A_*) + aA_* \quad \text{if } \pi \in (A_*, B_*), \\ a\pi \wedge b(1-\pi) \qquad\quad \text{if } \pi \in [0, A_*) \cup (B_*, 1] \end{cases} \tag{21.1.13}$$

where ψ *is given by* (21.1.11) *above while* $A_* \in (0, c)$ *and* $B_* \in (c, 1)$ *are the unique solution of the system of transcendental equations*

$$V(B_*; A_*) = b(1-B_*), \tag{21.1.14}$$

$$V'(B_*; A_*) = -b \tag{21.1.15}$$

where $\pi \mapsto V(\pi; A)$ *is given by* (21.1.12) *above. The stopping time* τ_{A_*,B_*} *given by* (21.1.3) *above is optimal in* (21.1.1).

Proof. Denote the function on the right-hand side of (21.1.13) by V_*. The properties of V_* stated in the end of paragraph 3 above show that Itô's formula (page 67) can be applied to $V(\pi_t)$ in its standard form (cf. Subsection 3.5). This gives

$$V_*(\pi_t) = V_*(\pi) + \int_0^t \mathbb{L}V_*(\pi_s) I(\pi_s \notin \{A, B\})\, ds \tag{21.1.16}$$

$$+ \frac{\mu}{\sigma}\int_0^t \pi_s(1-\pi_s)V_*'(\pi_s)\, d\bar{W}_s.$$

Recalling that $V_*(\pi) = M(\pi) = a\pi \wedge b(1-\pi)$ for $\pi \in [0, A_*) \cup (B_*, 1]$ and using that V_* satisfies (21.1.4) for $\pi \in (A_*, B_*)$, we see that

$$\mathbb{L}V_* \ge -1 \tag{21.1.17}$$

everywhere on $[0, 1]$ but A_* and B_*. By (21.1.9), (21.1.10), (21.1.16) and (21.1.17) it follows that

$$M(\pi_t) \ge V_*(\pi_t) \ge V_*(\pi) - t + M_t \tag{21.1.18}$$

where $M = (M_t)_{t\ge0}$ is a continuous local martingale given by

$$M_t = \frac{\mu}{\sigma}\int_0^t \pi_s(1-\pi_s)V_*'(\pi_s)\, d\bar{W}_s. \tag{21.1.19}$$

Using that $|V_*'(\pi)| \le a \vee b < \infty$ for all $\pi \in [0, 1]$ it is easily verified by standard means that M is a martingale. Moreover, by the optional sampling theorem (page 60) this bound also shows that $\mathsf{E}_\pi M_\tau = 0$ whenever $\mathsf{E}_\pi\sqrt{\tau} < \infty$ for a stopping time τ. In particular, the latter condition is satisfied if $\mathsf{E}_\pi\tau < \infty$. As clearly in

(21.1.1) it is enough to take the infimum only over stopping times τ satisfying $\mathsf{E}_\pi \tau < \infty$, we may insert τ in (21.1.18) instead of t, take E_π on both sides, and conclude that

$$\mathsf{E}_\pi \big(M(\pi_\tau) + \tau \big) \geq V_*(\pi) \tag{21.1.20}$$

for all $\pi \in [0,1]$. This shows that $V \geq V_*$. On the other hand, using (21.1.4) and the definition of τ_{A_*,B_*} in (21.1.3), we see from (21.1.16) that

$$M\big(\pi_{\tau_{A_*,B_*}}\big) = V_*\big(\pi_{\tau_{A_*,B_*}}\big) = V_*(\pi) - \tau_{A_*,B_*} + M_{\tau_{A_*,B_*}}. \tag{21.1.21}$$

Since $\mathsf{E}_\pi \tau_{A_*,B_*} < \infty$ (being true for any A and B) we see by taking E_π on both sides of (21.1.21) that equality in (21.1.20) is attained at $\tau = \tau_{A_*,B_*}$, and thus $V = V_*$. Combining this with the conclusions on the existence and uniqueness of A_* and B_* derived in paragraph 3 above, we see that the proof is complete. □

For more details on the Wiener sequential testing problem with infinite horizon (including the fixed probability error formulation) we refer to [196, Chap. 4, Sect. 1–2].

21.2. Finite horizon

1. In order to solve the problem (21.0.5) when the horizon T is *finite* let us consider the extended optimal stopping problem for the Markov process $(t, \pi_t)_{0 \leq t \leq T}$ given by

$$V(t,\pi) = \inf_{0 \leq \tau \leq T-t} \mathsf{E}_{t,\pi} G(t+\tau, \pi_{t+\tau}) \tag{21.2.1}$$

where $\mathsf{P}_{t,\pi}(\pi_t = \pi) = 1$, i.e. $\mathsf{P}_{t,\pi}$ is a probability measure under which the diffusion process $(\pi_{t+s})_{0 \leq s \leq T-t}$ solving (21.0.9) starts at π at time t, the infimum in (21.2.1) is taken over all stopping times τ of $(\pi_{t+s})_{0 \leq s \leq T-t}$, and we set $G(t,\pi) = t + a\pi \wedge b(1-\pi)$ for $(t,\pi) \in [0,T] \times [0,1]$. Since G is bounded and continuous on $[0,T] \times [0,1]$ it is possible to apply Corollary 2.9 (Finite horizon) with Remark 2.10 and conclude that an optimal stopping time exists in (21.2.1).

2. Let us now determine the structure of the optimal stopping time in the problem (21.2.1).

(i) It follows from (21.0.9) that the scale function of $(\pi_t)_{t \geq 0}$ is given by $S(x) = x$ for $x \in [0,1]$ and the speed measure of $(\pi_t)_{t \geq 0}$ is given by the equation $m(dx) = (2\sigma)/(\mu\, x\,(1-x))\, dx$ for $x \in (0,1)$. Hence the Green function of $(\pi_t)_{t \geq 0}$ on $[\pi_0, \pi_1] \subset (0,1)$ is given by $G_{\pi_0,\pi_1}(x,y) = (\pi_1 - x)(y - \pi_0)/(\pi_1 - \pi_0)$ for $\pi_0 \leq y \leq x$ and $G_{\pi_0,\pi_1}(x,y) = (\pi_1 - y)(x - \pi_0)/(\pi_1 - \pi_0)$ for $x \leq y \leq \pi_1$.

Set $H(\pi) = a\pi \wedge b(1-\pi)$ for $\pi \in [0,1]$ and let $d = H(c)$. Take $\varepsilon \in (0,d)$ and denote by $\pi_0 = \pi_0(\varepsilon)$ and $\pi_1 = \pi_1(\varepsilon)$ the unique points $0 < \pi_0 < c < \pi_1 < 1$ satisfying $H(\pi_0) = H(\pi_1) = d - \varepsilon$. Let $\sigma_\varepsilon = \inf\{\, t > 0 : \pi_t \notin (\pi_0, \pi_1) \,\}$ and set

$\sigma_\varepsilon^T = \sigma_\varepsilon \wedge T$. Then σ_ε and σ_ε^T are stopping times and it is easily verified that

$$\mathsf{E}_c\sigma_\varepsilon^T \le \mathsf{E}_c\sigma_\varepsilon = \int_{\pi_0}^{\pi_1} G_{\pi_0,\pi_1}(x,y)\, m(dy) \le \varepsilon^2 \tag{21.2.2}$$

for some $K > 0$ large enough (not depending on ε). Similarly, we find that

$$\begin{aligned}\mathsf{E}_c H(\pi_{\sigma_\varepsilon^T}) &= \mathsf{E}_c[H(\pi_{\sigma_\varepsilon})I(\sigma_\varepsilon < T)] + \mathsf{E}_c[H(\pi_T)I(\sigma_\varepsilon \ge T)] \\ &\le d - \varepsilon + d\,\mathsf{P}_c(\sigma_\varepsilon > T) \le d - \varepsilon + (d/T)\,\mathsf{E}_c\sigma_\varepsilon \\ &\le d - \varepsilon + L\,\varepsilon^2\end{aligned} \tag{21.2.3}$$

where $L = dK/T$.

Combining (21.2.2) and (21.2.3) we see that

$$\mathsf{E}_c G(\sigma_\varepsilon^T, \pi_{\sigma_\varepsilon^T}) = \mathsf{E}_c[\sigma_\varepsilon^T + H(\pi_{\sigma_\varepsilon^T})] \le d - \varepsilon + (K+L)\,\varepsilon^2 \tag{21.2.4}$$

for all $\varepsilon \in (0,d)$. If we choose $\varepsilon > 0$ in (21.2.4) small enough, we observe that $\mathsf{E}_c G(\sigma_\varepsilon^T, \pi_{\sigma_\varepsilon^T}) < d$. Using the fact that $G(t,\pi) = t + H(\pi)$ is linear in t, and $T > 0$ above is arbitrary, this shows that it is never optimal to stop in (21.2.1) when $\pi_{t+s} = c$ for $0 \le s < T-t$. In other words, this shows that all points (t,c) for $0 \le t < T$ belong to the continuation set

$$C = \{(t,\pi) \in [0,T)\times[0,1] : V(t,\pi) < G(t,\pi)\}. \tag{21.2.5}$$

(ii) Recalling the solution to the problem (21.0.5) in the case of infinite horizon, where the stopping time $\tau_* = \inf\{\, t > 0 : \pi_t \notin (A_*, B_*)\,\}$ is optimal and $0 < A_* < c < B_* < 1$ are uniquely determined from the system (21.1.14)–(21.1.15) (see also (4.85) in [196, p. 185]), we see that all points (t,π) for $0 \le t \le T$ with either $0 \le \pi \le A_*$ or $B_* \le \pi \le 1$ belong to the stopping set. Moreover, since $\pi \mapsto V(t,\pi)$ with $0 \le t \le T$ given and fixed is concave on $[0,1]$ (this is easily deduced using the same arguments as in [123, p. 105] or [196, p. 168]), it follows directly from the previous two conclusions about the continuation and stopping set that there exist functions g_0 and g_1 satisfying $0 < A_* \le g_0(t) < c < g_1(t) \le B_* < 1$ for all $0 \le t < T$ such that the continuation set is an open set of the form

$$C = \{(t,\pi) \in [0,T)\times[0,1] : \pi \in (g_0(t), g_1(t))\} \tag{21.2.6}$$

and the stopping set is the closure of the set

$$D = \{(t,\pi) \in [0,T)\times[0,1] : \pi \in [0, g_0(t)) \cup (g_1(t), 1]\}. \tag{21.2.7}$$

(Below we will show that V is continuous so that C is open indeed. We will also see that $g_0(T) = g_1(T) = c$.)

(iii) Since the problem (21.2.1) is time-homogeneous, in the sense that $G(t,\pi) = t + H(\pi)$ is linear in t and H depends on π only, it follows that the map $t \mapsto V(t,\pi) - t$ is increasing on $[0,T]$. Hence if (t,π) belongs to C for some $\pi \in (0,1)$ and we take any other $0 \le t' < t \le T$, then $V(t',\pi) - G(t',\pi) = V(t',\pi) - t' - H(\pi) \le V(t,\pi) - t - H(\pi) = V(t,\pi) - G(t,\pi) < 0$, showing that (t',π) belongs to C as well. From this we may conclude in (21.2.6)–(21.2.7) that the boundary $t \mapsto g_0(t)$ is *increasing* and the boundary $t \mapsto g_1(t)$ is *decreasing* on $[0,T]$.

(iv) Let us finally observe that the value function V from (21.2.1) and the boundaries g_0 and g_1 from (21.2.6)–(21.2.7) also depend on T and let them be denoted here by V^T, g_0^T and g_1^T, respectively. Using the fact that $T \mapsto V^T(t,\pi)$ is a decreasing function on $[t,\infty)$ and $V^T(t,\pi) = G(t,\pi)$ for all $\pi \in [0, g_0^T(t)] \cup [g_1^T(t), 1]$, we conclude that if $T < T'$, then $0 \le g_0^{T'}(t) \le g_0^T(t) < c < g_1^T(t) \le g_1^{T'}(t) \le 1$ for all $t \in [0,T)$. Letting T' in the previous expression go to ∞, we get that $0 < A_* \le g_0^T(t) < c < g_1^T(t) \le B_* < 1$ with $A_* \equiv \lim_{T\to\infty} g_0^T(t)$ and $B_* \equiv \lim_{T\to\infty} g_1^T(t)$ for all $t \ge 0$, where A_* and B_* are the optimal stopping points in the infinite horizon problem referred to above.

3. Let us now show that the value function $(t,\pi) \mapsto V(t,\pi)$ is *continuous* on $[0,T] \times [0,1]$. For this it is enough to prove that

$$\pi \mapsto V(t_0,\pi) \quad \text{is continuous at } \pi_0, \tag{21.2.8}$$

$$t \mapsto V(t,\pi) \quad \text{is continuous at } t_0 \text{ uniformly over } \pi \in [\pi_0 - \delta, \pi_0 + \delta] \tag{21.2.9}$$

for each $(t_0,\pi_0) \in [0,T] \times [0,1]$ with some $\delta > 0$ small enough (it may depend on π_0). Since (21.2.8) follows by the fact that $\pi \mapsto V(t,\pi)$ is concave on $[0,1]$, it remains to establish (21.2.9).

For this, let us fix arbitrary $0 \le t_1 < t_2 \le T$ and $0 < \pi < 1$, and let $\tau_1 = \tau_*(t_1,\pi)$ denote the optimal stopping time for $V(t_1,\pi)$. Set $\tau_2 = \tau_1 \wedge (T - t_2)$ and note since $t \mapsto V(t,\pi)$ is increasing on $[0,T]$ and $\tau_2 \le \tau_1$ that we have

$$\begin{aligned} 0 &\le V(t_2,\pi) - V(t_1,\pi) \\ &\le \mathsf{E}_\pi[(t_2+\tau_2) + H(\pi_{t_2+\tau_2})] - \mathsf{E}_\pi[(t_1+\tau_1) + H(\pi_{t_1+\tau_1})] \\ &\le (t_2 - t_1) + \mathsf{E}_\pi[H(\pi_{t_2+\tau_2}) - H(\pi_{t_1+\tau_1})] \end{aligned} \tag{21.2.10}$$

where we recall that $H(\pi) = a\pi \wedge b(1-\pi)$ for $\pi \in [0,1]$. Observe further that

$$\begin{aligned} &\mathsf{E}_\pi[H(\pi_{t_2+\tau_2}) - H(\pi_{t_1+\tau_1})] \\ &= \sum_{i=0}^{1} \frac{1 + (-1)^i(1-2\pi)}{2} \mathsf{E}_i\big[h(\varphi_{\tau_2}) - h(\varphi_{\tau_1})\big] \end{aligned} \tag{21.2.11}$$

where for each $\pi \in (0,1)$ given and fixed the function h is defined by

$$h(x) = \left| H\left(\frac{\frac{\pi}{1-\pi}x}{1 + \frac{\pi}{1-\pi}x} \right) \right| \tag{21.2.12}$$

for all $x > 0$. Then for any $0 < x_1 < x_2$ given and fixed it follows by the mean value theorem (note that h is C^1 on $(0,\infty)$ except one point) that there exists $\xi \in [x_1, x_2]$ such that

$$|h(x_2) - h(x_1)| \le |h'(\xi)|\,(x_2 - x_1) \tag{21.2.13}$$

where the derivative h' at ξ satisfies

$$|h'(\xi)| = \left|H'\left(\frac{\frac{\pi}{1-\pi}\xi}{1+\frac{\pi}{1-\pi}\xi}\right)\right| \frac{\pi(1-\pi)}{(1-\pi+\pi\xi)^2} \le K\,\frac{\pi(1-\pi)}{(1-\pi)^2} = K\,\frac{\pi}{1-\pi} \tag{21.2.14}$$

with some $K > 0$ large enough.

On the other hand, the explicit expression (21.0.7) yields

$$\begin{aligned} \varphi_{\tau_2} - \varphi_{\tau_1} &= \varphi_{\tau_2}\left(1 - \frac{\varphi_{\tau_1}}{\varphi_{\tau_2}}\right) \\ &= \varphi_{\tau_2}\left(1 - \exp\left(\frac{\mu}{\sigma^2}(X_{\tau_1} - X_{\tau_2}) - \frac{\mu^2}{2\sigma^2}(\tau_1 - \tau_2)\right)\right) \end{aligned} \tag{21.2.15}$$

and thus the strong Markov property (stationary independent increments) together with the representation (21.0.2) and the fact that $\tau_1 - \tau_2 \le t_2 - t_1$ imply

$$\begin{aligned} &\mathsf{E}_i|\varphi_{\tau_2} - \varphi_{\tau_1}| \\ &= \mathsf{E}_i\left|\varphi_{\tau_2}\left(1 - \exp\left(\frac{\mu}{\sigma}(W_{\tau_1} - W_{\tau_2}) - (-1)^i\frac{\mu^2}{2\sigma^2}(\tau_1 - \tau_2)\right)\right)\right| \\ &= \mathsf{E}_i\left[\varphi_{\tau_2}\,\mathsf{E}_i\left[\left|1 - \exp\left(\frac{\mu}{\sigma}(W_{\tau_1} - W_{\tau_2}) - (-1)^i\frac{\mu^2}{2\sigma^2}(\tau_1 - \tau_2)\right)\right| \,\Big|\, \mathcal{F}^X_{\tau_2}\right]\right] \\ &\le \mathsf{E}_i\varphi_{\tau_2}\,\mathsf{E}_i\left[\sup_{0\le t\le t_2-t_1}\exp\left(\frac{\mu}{\sigma}W_t + \frac{\mu^2}{2\sigma^2}t\right) - 1\right] \end{aligned} \tag{21.2.16}$$

for $i = 0, 1$. Since it easily follows that

$$\begin{aligned} \mathsf{E}_i\varphi_{\tau_2} &= \mathsf{E}_i\left[\exp\left(\frac{\mu}{\sigma}W_{\tau_2} - (-1)^i\frac{\mu^2}{2\sigma^2}\tau_2\right)\right] \\ &\le \exp\left(\frac{\mu^2}{\sigma^2}(T - t_2)\right) \le \exp\left(\frac{\mu^2}{\sigma^2}T\right) \end{aligned} \tag{21.2.17}$$

from (21.2.12)–(21.2.17) we get

$$\mathsf{E}_i|h(\varphi_{\tau_2}) - h(\varphi_{\tau_1})| \le K\,\frac{\pi}{1-\pi}\,\mathsf{E}_i|\varphi_{\tau_2} - \varphi_{\tau_1}| \le K\,\frac{\pi}{1-\pi}\,L(t_2 - t_1) \tag{21.2.18}$$

where the function L is defined by

$$L(t_2 - t_1) = \exp\left(\frac{\mu^2}{\sigma^2}T\right)\mathsf{E}_i\left[\sup_{0\le t\le t_2-t_1}\exp\left(\frac{\mu}{\sigma}W_t + \frac{\mu^2}{2\sigma^2}t\right) - 1\right]. \tag{21.2.19}$$

Therefore, combining (21.2.18) with (21.2.10)–(21.2.11) above, we obtain

$$V(t_2,\pi) - V(t_1,\pi) \le (t_2 - t_1) + K\,\frac{\pi}{1-\pi}\,L(t_2-t_1) \tag{21.2.20}$$

from where, by virtue of the fact that $L(t_2 - t_1) \to 0$ in (21.2.19) as $t_2 - t_1 \downarrow 0$, we easily conclude that (21.2.9) holds. In particular, this shows that the *instantaneous-stopping conditions* (21.2.40) below are satisfied.

4. In order to prove that the *smooth-fit* conditions (21.2.41) hold, i.e. that $\pi \mapsto V(t,\pi)$ is C^1 at $g_0(t)$ and $g_1(t)$, let us fix a point $(t,\pi) \in [0,T) \times (0,1)$ lying on the boundary g_0 so that $\pi = g_0(t)$. Then for all $\varepsilon > 0$ such that $\pi < \pi + \varepsilon < c$ we have

$$\frac{V(t,\pi+\varepsilon) - V(t,\pi)}{\varepsilon} \le \frac{G(t,\pi+\varepsilon) - G(t,\pi)}{\varepsilon} \tag{21.2.21}$$

and hence, taking the limit in (21.2.21) as $\varepsilon \downarrow 0$, we get

$$\frac{\partial^+ V}{\partial \pi}(t,\pi) \le \frac{\partial G}{\partial \pi}(t,\pi) \tag{21.2.22}$$

where the right-hand derivative in (21.2.22) exists (and is finite) by virtue of the concavity of $\pi \mapsto V(t,\pi)$ on $[0,1]$. Note that the latter will also be proved independently below.

Let us now fix some $\varepsilon > 0$ such that $\pi < \pi + \varepsilon < c$ and consider the stopping time $\tau_\varepsilon = \tau_*(t,\pi+\varepsilon)$ being optimal for $V(t,\pi+\varepsilon)$. Note that τ_ε is the first exit time of the process $(\pi_{t+s})_{0\le s\le T-t}$ from the set C in (21.2.6). Then by (21.0.1) and (21.0.8) it follows using the mean value theorem that there exists $\xi \in [\pi, \pi+\varepsilon]$ such that

$$\begin{aligned} V(t,\pi+\varepsilon) - V(t,\pi) &\ge \mathsf{E}_{\pi+\varepsilon} G(t+\tau_\varepsilon, \pi_{t+\tau_\varepsilon}) - \mathsf{E}_\pi G(t+\tau_\varepsilon, \pi_{t+\tau_\varepsilon}) \\ &= \sum_{i=0}^{1} \mathsf{E}_i [S_i(\pi+\varepsilon) - S_i(\pi)] = \varepsilon \sum_{i=0}^{1} \mathsf{E}_i S_i'(\xi_i) \end{aligned} \tag{21.2.23}$$

where the function S_i is defined by

$$S_i(\pi) = \frac{1+(-1)^i(1-2\pi)}{2}\, G\bigg(t+\tau_\varepsilon, \frac{(\xi_i/(1-\xi_i))\varphi_{\tau_\varepsilon}}{1+(\xi_i/(1-\xi_i))\varphi_{\tau_\varepsilon}}\bigg) \tag{21.2.24}$$

and its derivative S_i' at ξ_i is given by

$$\begin{aligned} S_i'(\xi_i) &= (-1)^{i+1}\, G\bigg(t+\tau_\varepsilon, \frac{(\xi_i/(1-\xi_i))\varphi_{\tau_\varepsilon}}{1+(\xi_i/(1-\xi_i))\varphi_{\tau_\varepsilon}}\bigg) \\ &+ \frac{1+(-1)^i(1-2\xi_i)}{2}\, \frac{\partial G}{\partial \pi}\bigg(t+\tau_\varepsilon, \frac{(\xi_i/(1-\xi_i))\varphi_{\tau_\varepsilon}}{1+(\xi_i/(1-\xi_i))\varphi_{\tau_\varepsilon}}\bigg) \frac{\varphi_{\tau_\varepsilon}}{(1-\xi_i+\xi_i\varphi_{\tau_\varepsilon})^2} \end{aligned} \tag{21.2.25}$$

for $i = 0, 1$. Since g_0 is increasing it is easily verified using (21.0.7)–(21.0.8) and the fact that $t \mapsto \pm \frac{\mu}{2\sigma} t$ is a lower function for the standard Wiener process W that $\tau_\varepsilon \to 0$ P_i-a.s. and thus $\varphi_{\tau_\varepsilon} \to 1$ P_i-a.s. as $\varepsilon \downarrow 0$ for $i = 0, 1$. Hence we easily find

$$S_i'(\xi_i) \to (-1)^{i+1} G(t, \pi) + \frac{1 + (-1)^i (1 - 2\pi)}{2} \frac{\partial G}{\partial \pi}(t, \pi) \quad \mathsf{P}_i\text{-a.s.} \tag{21.2.26}$$

as $\varepsilon \downarrow 0$, and clearly $|S_i'(\xi_i)| \le K_i$ with some $K_i > 0$ large enough for $i = 0, 1$.

It thus follows from (21.2.23) using (21.2.26) that

$$\frac{V(t, \pi + \varepsilon) - V(t, \pi)}{\varepsilon} \ge \sum_{i=0}^{1} \mathsf{E}_i S_i'(\xi_i) \to \frac{\partial G}{\partial \pi}(t, \pi) \tag{21.2.27}$$

as $\varepsilon \downarrow 0$ by the dominated convergence theorem. This combined with (21.2.21) above proves that $V_\pi^+(t, \pi)$ exists and equals $G_\pi(t, \pi)$. The smooth fit at the boundary g_1 is proved analogously.

5. We proceed by proving that the boundaries g_0 and g_1 are *continuous* on $[0, T]$ and that $g_0(T) = g_1(T) = c$.

(i) Let us first show that the boundaries g_0 and g_1 are right-continuous on $[0, T]$. For this, fix $t \in [0, T)$ and consider a sequence $t_n \downarrow t$ as $n \to \infty$. Since g_i is monotone, the right-hand limit $g_i(t+)$ exists for $i = 0, 1$. Because $(t_n, g_i(t_n)) \in \bar{D}$ for all $n \ge 1$, and $\bar{D}$ is closed, we see that $(t, g_i(t+)) \in \bar{D}$ for $i = 0, 1$. Hence by (21.2.7) we see that $g_0(t+) \le g_0(t)$ and $g_1(t+) \ge g_1(t)$. The reverse inequalities follow obviously from the fact that g_0 is increasing and g_1 is decreasing on $[0, T]$, thus proving the claim.

(ii) Suppose that at some point $t_* \in (0, T)$ the function g_1 makes a jump, i.e. let $g_1(t_*-) > g_1(t_*) \ge c$. Let us fix a point $t' < t_*$ close to t_* and consider the half-open set $R \subset C$ being a curved trapezoid formed by the vertices $(t', g_1(t'))$, $(t_*, g_1(t_*-))$, (t_*, π') and (t', π') with any π' fixed arbitrarily in the interval $(g_1(t_*), g_1(t_*-))$. Observe that the strong Markov property implies that the value function V from (21.2.1) is $C^{1,2}$ on C. Note also that the gain function G is $C^{1,2}$ in R so that by the Newton–Leibniz formula using (21.2.40) and (21.2.41) it follows that

$$V(t, \pi) - G(t, \pi) = \int_\pi^{g_1(t)} \int_u^{g_1(t)} \left(\frac{\partial^2 V}{\partial \pi^2} - \frac{\partial^2 G}{\partial \pi^2} \right)(t, v) \, dv \, du \tag{21.2.28}$$

for all $(t, \pi) \in R$.

Let us fix some $(t, \pi) \in C$ and take an arbitrary $\varepsilon > 0$ such that $(t+\varepsilon, \pi) \in C$. Then denoting by $\tau_\varepsilon = \tau_*(t+\varepsilon, \pi)$ the optimal stopping time for $V(t+\varepsilon, \pi)$, we have

$$
\begin{aligned}
&\frac{V(t+\varepsilon,\pi)-V(t,\pi)}{\varepsilon} \qquad (21.2.29)\\
&\ge \frac{\mathsf{E}_{t+\varepsilon,\pi}G(t+\varepsilon+\tau_\varepsilon,\pi_{t+\varepsilon+\tau_\varepsilon})-\mathsf{E}_{t,\pi}G(t+\tau_\varepsilon,\pi_{t+\tau_\varepsilon})}{\varepsilon}\\
&=\frac{\mathsf{E}_\pi[G(t+\varepsilon+\tau_\varepsilon,\pi_{\tau_\varepsilon})-G(t+\tau_\varepsilon,\pi_{\tau_\varepsilon})]}{\varepsilon}=1
\end{aligned}
$$

and thus, taking the limit in (21.2.29) as $\varepsilon\downarrow 0$, we get

$$\frac{\partial V}{\partial t}(t,\pi)\ge\frac{\partial G}{\partial t}(t,\pi)=1 \tag{21.2.30}$$

at each $(t,\pi)\in C$.

Since the strong Markov property implies that the value function V from (21.2.1) solves the equation (21.2.39), using (21.2.30) we obtain

$$\frac{\partial^2 V}{\partial\pi^2}(t,\pi)=-\frac{2\sigma^2}{\mu^2}\,\frac{1}{\pi^2(1-\pi)^2}\,\frac{\partial V}{\partial t}(t,\pi)\le-\varepsilon\,\frac{\sigma^2}{\mu^2} \tag{21.2.31}$$

for all $t'\le t<t_*$ and all $\pi'\le\pi<g_1(t')$ with $\varepsilon>0$ small enough.

Hence by (21.2.28) using that $G_{\pi\pi}=0$ we get

$$
\begin{aligned}
&V(t',\pi')-G(t',\pi') \qquad (21.2.32)\\
&\le-\varepsilon\,\frac{\sigma^2}{\mu^2}\,\frac{(g_1(t')-\pi')^2}{2}\to-\varepsilon\,\frac{\sigma^2}{\mu^2}\,\frac{(g_1(t_*-)-\pi')^2}{2}<0
\end{aligned}
$$

as $t'\uparrow t_*$. This implies that $V(t_*,\pi')<G(t_*,\pi')$ which contradicts the fact that (t_*,π') belongs to the stopping set $\bar{D}$. Thus $g_1(t_*-)=g_1(t_*)$ showing that g_1 is continuous at t_* and thus on $[0,T]$ as well. A similar argument shows that the function g_0 is continuous on $[0,T]$.

(iii) We finally note that the method of proof from the previous part (ii) also implies that $g_0(T)=g_1(T)=c$. To see this, we may let $t_*=T$ and likewise suppose that $g_1(T-)>c$. Then repeating the arguments presented above word by word we arrive to a contradiction with the fact that $V(T,\pi)=G(T,\pi)$ for all $\pi\in[c,g_1(T-)]$ thus proving the claim.

6. Summarizing the facts proved in paragraphs 5–8 above we may conclude that the following exit time is optimal in the extended problem (21.2.1):

$$\tau_*=\inf\{0\le s\le T-t:\pi_{t+s}\notin(g_0(t+s),g_1(t+s))\} \tag{21.2.33}$$

(the infimum of an empty set being equal to $T-t$) where the two boundaries

(g_0, g_1) satisfy the following properties (see *Figure* VI.1):

$$g_0 : [0,T] \to [0,1] \quad \text{is } \textit{continuous} \text{ and } \textit{increasing}, \tag{21.2.34}$$

$$g_1 : [0,T] \to [0,1] \quad \text{is } \textit{continuous} \text{ and } \textit{decreasing}, \tag{21.2.35}$$

$$A_* \le g_0(t) < c < g_1(t) \le B_* \quad \text{for all } 0 \le t < T, \tag{21.2.36}$$

$$g_i(T) = c \quad \text{for } i = 0, 1, \tag{21.2.37}$$

where A_* and B_* satisfying $0 < A_* < c < B_* < 1$ are the optimal stopping points for the infinite horizon problem uniquely determined from the system of transcendental equations (21.1.14)–(21.1.15) or [196, p. 185, (4.85)].

Standard arguments imply that the infinitesimal operator $\mathbb{L}$ of the process $(t, \pi_t)_{0 \le t \le T}$ acts on a function $f \in C^{1,2}([0,T) \times [0,1])$ according to the rule

$$(\mathbb{L}f)(t,\pi) = \left(\frac{\partial f}{\partial t} + \frac{\mu^2}{2\sigma^2} \pi^2 (1-\pi)^2 \frac{\partial^2 f}{\partial \pi^2} \right)(t,\pi) \tag{21.2.38}$$

for all $(t,\pi) \in [0,T) \times [0,1]$. In view of the facts proved above we are thus naturally led to formulate the following *free-boundary problem* for the *unknown value function* V from (21.2.1) and the *unknown boundaries* (g_0, g_1) from (21.2.6)–(21.2.7):

$$(\mathbb{L}V)(t,\pi) = 0 \quad \text{for } (t,\pi) \in C, \tag{21.2.39}$$

$$V(t,\pi)\big|_{\pi = g_0(t)+} = t + a g_0(t), \quad V(t,\pi)\big|_{\pi = g_1(t)-} = t + b(1 - g_1(t)), \tag{21.2.40}$$

$$\frac{\partial V}{\partial \pi}(t,\pi)\Big|_{\pi = g_0(t)+} = a, \quad \frac{\partial V}{\partial \pi}(t,\pi)\Big|_{\pi = g_1(t)-} = -b, \tag{21.2.41}$$

$$V(t,\pi) < G(t,\pi) \quad \text{for } (t,\pi) \in C, \tag{21.2.42}$$

$$V(t,\pi) = G(t,\pi) \quad \text{for } (t,\pi) \in D, \tag{21.2.43}$$

where C and D are given by (21.2.6) and (21.2.7), and the *instantaneous-stopping conditions* (21.2.40) are satisfied for all $0 \le t \le T$ and the *smooth-fit conditions* (21.2.41) are satisfied for all $0 \le t < T$.

Note that the superharmonic characterization of the value function (cf. Chapter I) implies that V from (21.2.1) is a *largest* function satisfying (21.2.39)–(21.2.40) and (21.2.42)–(21.2.43).

7. Making use of the facts proved above we are now ready to formulate the main result of the section. Below we set $\varphi(x) = (1/\sqrt{2\pi})e^{-x^2/2}$ and $\Phi(x) = \int_{-\infty}^{x} \varphi(y)\, dy$ for $x \in \mathbb{R}$.

Theorem 21.2. *In the Bayesian problem* (21.0.4)–(21.0.5) *of testing two simple hypotheses* (21.0.3) *the optimal decision rule* (τ_*, d_*) *is explicitly given by*

$$\tau_* = \inf\{0 \le t \le T : \pi_t \notin (g_0(t), g_1(t))\}, \tag{21.2.44}$$

$$d_* = \begin{cases} 1 \ (\textit{accept } H_1) & \textit{if } \pi_{\tau_*} = g_1(\tau_*), \\ 0 \ (\textit{accept } H_0) & \textit{if } \pi_{\tau_*} = g_0(\tau_*), \end{cases} \tag{21.2.45}$$

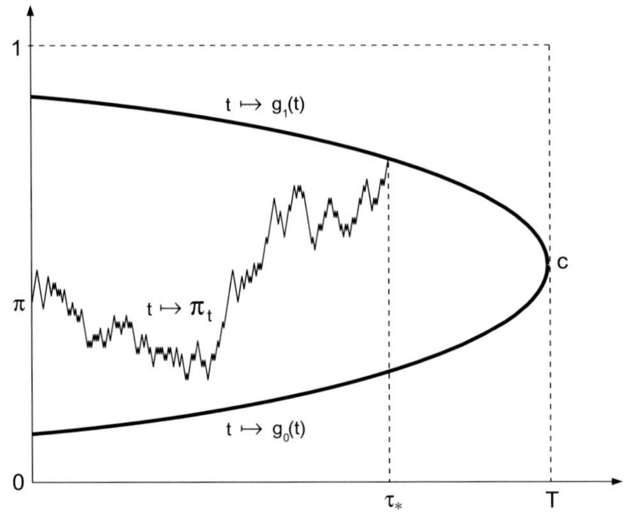

Figure VI.1: A computer drawing of the optimal stopping boundaries g_0 and g_1 from Theorem 21.2. In the case above it is optimal to accept the hypothesis H_1 .

where the two boundaries (g_0, g_1) *can be characterized as a unique solution of the coupled system of nonlinear integral equations*

$$\mathsf{E}_{t,g_i(t)}[a\pi_T \wedge b(1-\pi_T)] = ag_i(t) \wedge b(1-g_i(t)) + \sum_{j=0}^{1} \int_0^{T-t} (-1)^j \mathsf{P}_{t,g_i(t)}[\pi_{t+u} \le g_j(t+u)]\, du \quad (i=0,1) \tag{21.2.46}$$

for $0 \le t \le T$ *satisfying* (21.2.34)–(21.2.37) [*see Figure* VI.1].

More explicitly, the six terms in the system (21.2.46) *read as follows:*

$$\begin{aligned} &\mathsf{E}_{t,g_i(t)}[a\pi_T \wedge b(1-\pi_T)] \\ &= g_i(t) \int_{-\infty}^{\infty} \frac{1}{1-g_i(t)+g_i(t)\exp\left\{\frac{\mu z}{\sigma}\sqrt{T-t}+\frac{\mu^2}{2\sigma^2}(T-t)\right\}} \\ &\qquad \times \min\Big\{ag_i(t)\exp\left\{\frac{\mu z}{\sigma}\sqrt{T-t}+\frac{\mu^2}{2\sigma^2}(T-t)\right\}, b(1-g_i(t))\Big\}\,\varphi(z)\,dz \\ &\quad + (1-g_i(t)) \int_{-\infty}^{\infty} \frac{1}{1-g_i(t)+g_i(t)\exp\left\{\frac{\mu z}{\sigma}\sqrt{T-t}-\frac{\mu^2}{2\sigma^2}(T-t)\right\}} \\ &\qquad \times \min\Big\{ag_i(t)\exp\left\{\frac{\mu z}{\sigma}\sqrt{T-t}-\frac{\mu^2}{2\sigma^2}(T-t)\right\}, b(1-g_i(t))\Big\}\,\varphi(z)\,dz, \end{aligned} \tag{21.2.47}$$

$$\begin{aligned} &\mathsf{P}_{t,g_i(t)}\big(\pi_{t+u} \le g_j(t+u)\big) \qquad (21.2.48)\\ &= g_i(t)\,\Phi\bigg(\frac{\sigma}{\mu\sqrt{u}}\log\bigg(\frac{g_j(t+u)}{1-g_j(t+u)}\frac{1-g_i(t)}{g_i(t)}\bigg)-\frac{\mu\sqrt{u}}{2\sigma}\bigg)\\ &\quad + (1-g_i(t))\,\Phi\bigg(\frac{\sigma}{\mu\sqrt{u}}\log\bigg(\frac{g_j(t+u)}{1-g_j(t+u)}\frac{1-g_i(t)}{g_i(t)}\bigg)+\frac{\mu\sqrt{u}}{2\sigma}\bigg)\end{aligned}$$

for $0 \le u \le T-t$ *with* $0 \le t \le T$ *and* $i,j = 0,1$ *.*

[*Note that in the case when* $a=b$ *we have* $c=1/2$ *and the system* (21.2.46) *reduces to one equation only since* $g_1 = 1-g_0$ *by symmetry.*]

Proof. 1°. The existence of boundaries (g_0, g_1) satisfying (21.2.34)–21.2.37 such that τ_* from (21.2.44) is optimal in (21.0.4)–(21.0.5) was proved in paragraphs 2-6 above. By the local time-space formula (cf. Subsection 3.5) it follows that the boundaries (g_0, g_1) solve the system (21.2.46) (cf. (21.2.52)–(21.2.56) below). Thus it remains to show that the system (21.2.46) has no other solution in the class of functions (h_0, h_1) satisfying (21.2.34)–(21.2.37).

Let us thus assume that two functions (h_0, h_1) satisfying (21.2.34)–(21.2.37) solve the system (21.2.46), and let us show that these two functions (h_0, h_1) must then coincide with the optimal boundaries (g_0, g_1) . For this, let us introduce the function

$$V^h(t,\pi) = \begin{cases} U^h(t,\pi) & \text{if } (t,\pi) \in C_h, \\ G(t,\pi) & \text{if } (t,\pi) \in \bar{D}_h \end{cases} \qquad (21.2.49)$$

where the function U^h is defined by

$$U^h(t,\pi) = \mathsf{E}_{t,\pi} G(T,\pi_T) - \int_0^{T-t} \mathsf{P}_{t,\pi}\big((t+u, \pi_{t+u}) \in D_h\big)\, du \qquad (21.2.50)$$

for all $(t,\pi) \in [0,T) \times [0,1]$ and the sets C_h and D_h are defined as in (21.2.6) and (21.2.7) with h_i instead of g_i for $i=0,1$. Note that (21.2.50) with $G(t,\pi)$ instead of $U^h(t,\pi)$ on the left-hand side coincides with (21.2.46) when $\pi = g_i(t)$ and $h_j = g_j$ for $i,j=0,1$. Since (h_0,h_1) solve (21.2.46) this shows that V^h is continuous on $[0,T) \times [0,1]$. We need to verify that V^h coincides with the value function V from (21.2.1) and that h_i equals g_i for $i=0,1$.

2°. Using standard arguments based on the strong Markov property (or verifying directly) it follows that V^h i.e. U^h is $C^{1,2}$ on C_h and that

$$(\mathbb{L}V^h)(t,\pi) = 0 \quad \text{for } (t,\pi) \in C_h \, . \qquad (21.2.51)$$

Moreover, since $U^h_\pi := \partial U^h / \partial \pi$ is continuous on $[0,T) \times (0,1)$ (which is readily verified using the explicit expressions (21.2.47) and (21.2.48) above with π instead of $g_i(t)$ and h_j instead of g_j for $i,j=0,1$), we see that $V^h_\pi := \partial V^h / \partial \pi$ is continuous on $\bar{C}_h$. Finally, since $h_0(t) \in (0,c)$ and $h_1(t) \in (c,1)$ we see that V^h

i.e. G is $C^{1,2}$ on $\bar{D}_h$. Therefore, with $(t,\pi)\in[0,T)\times(0,1)$ given and fixed, the local time-space formula (cf. Subsection 3.5) can be applied, and in this way we get

$$\begin{aligned} V^h(t+s,\pi_{t+s}) &= V^h(t,\pi) \\ &+ \int_0^s (\mathbb{L}V^h)(t+u,\pi_{t+u})\, I\big(\pi_{t+u}\neq h_0(t+u), \pi_{t+u}\neq h_1(t+u)\big)\, du \\ &+ M_s^h + \frac{1}{2}\sum_{i=0}^{1}\int_0^s \Delta_\pi V_\pi^h(t+u,\pi_{t+u})\, I\big(\pi_{t+u}=h_i(t+u)\big)\, d\ell_u^{h_i} \end{aligned} \tag{21.2.52}$$

for $0\le s\le T-t$ where

$$\begin{aligned} &\Delta_\pi V_\pi^h(t+u,h_i(t+u)) \\ &= V_\pi^h(t+u,h_i(t+u)+) - V_\pi^h(t+u,h_i(t+u)-), \end{aligned} \tag{21.2.53}$$

the process $(\ell_s^{h_i})_{0\le s\le T-t}$ is the local time of $(\pi_{t+s})_{0\le s\le T-t}$ at the boundary h_i given by

$$\begin{aligned} \ell_s^{h_i} = \mathsf{P}\text{-}\lim_{\varepsilon\downarrow 0}\frac{1}{2\varepsilon}\int_0^s & I(h_i(t+u)-\varepsilon<\pi_{t+u}<h_i(t+u)+\varepsilon) \\ & \times \frac{\mu^2}{\sigma^2}\pi_{t+u}^2(1-\pi_{t+u})^2\, du \end{aligned} \tag{21.2.54}$$

for $i=0,1$, and $(M_s^h)_{0\le s\le T-t}$ defined by

$$\begin{aligned} M_s^h = \int_0^s V_\pi^h(t+u,\pi_{t+u})\, & I(\pi_{t+u}\neq h_0(t+u), \pi_{t+u}\neq h_1(t+u)) \\ & \times \frac{\mu}{\sigma}\pi_{t+u}\,(1-\pi_{t+u})\, d\bar{W}_u \end{aligned} \tag{21.2.55}$$

is a martingale under $\mathsf{P}_{t,\pi}$.

Setting $s=T-t$ in (21.2.52) and taking the $\mathsf{P}_{t,\pi}$ -expectation, using that V^h satisfies (21.2.51) in C_h and equals G in D_h , we get

$$\begin{aligned} \mathsf{E}_{t,\pi}G(T,\pi_T) &= V^h(t,\pi) \\ &+ \int_0^{T-t} \mathsf{P}_{t,\pi}\big((t+u,\pi_{t+u})\in D_h\big)\, du + \frac{1}{2}F(t,\pi) \end{aligned} \tag{21.2.56}$$

where (by the continuity of the integrand) the function F is given by

$$F(t,\pi) = \sum_{i=0}^{1}\int_0^{T-t} \Delta_\pi V_\pi^h(t+u,h_i(t+u))\, d_u \mathsf{E}_{t,\pi}\ell_u^{h_i} \tag{21.2.57}$$

for all $(t,\pi) \in [0,T) \times [0,1]$ and $i = 0,1$. Thus from (21.2.56) and (21.2.49) we see that

$$F(t,\pi) = \begin{cases} 0 & \text{if } (t,\pi) \in C_h, \\ 2\,(U^h(t,\pi) - G(t,\pi)) & \text{if } (t,\pi) \in \bar{D}_h \end{cases} \tag{21.2.58}$$

where the function U^h is given by (21.2.50).

3°. From (21.2.58) we see that if we are to prove that

$$\pi \mapsto V^h(t,\pi) \quad \text{is } C^1 \text{ at } h_i(t) \tag{21.2.59}$$

for each $0 \le t < T$ given and fixed and $i = 0,1$, then it will follow that

$$U^h(t,\pi) = G(t,\pi) \quad \text{for all } (t,\pi) \in \bar{D}_h. \tag{21.2.60}$$

On the other hand, if we know that (21.2.60) holds, then using the following general facts (obtained directly from the definition (21.2.49) above):

$$\frac{\partial}{\partial \pi}(U^h(t,\pi) - G(t,\pi))\Big|_{\pi = h_0(t)} \tag{21.2.61}$$
$$= V^h_\pi(t, h_0(t)+) - V^h_\pi(t, h_0(t)-) = \Delta_\pi V^h_\pi(t, h_0(t)),$$
$$\frac{\partial}{\partial \pi}(U^h(t,\pi) - G(t,\pi))\Big|_{\pi = h_1(t)} \tag{21.2.62}$$
$$= V^h_\pi(t, h_1(t)-) - V^h_\pi(t, h_1(t)+) = -\Delta_\pi V^h_\pi(t, h_1(t))$$

for all $0 \le t < T$, we see that (21.2.59) holds too. The equivalence of (21.2.59) and (21.2.60) suggests that instead of dealing with the equation (21.2.58) in order to derive (21.2.59) above we may rather concentrate on establishing (21.2.60) directly.

To derive (21.2.60) first note that using standard arguments based on the strong Markov property (or verifying directly) it follows that U^h is $C^{1,2}$ in D_h and that

$$(\mathbb{L}U^h)(t,\pi) = 1 \quad \text{for } (t,\pi) \in D_h. \tag{21.2.63}$$

It follows that (21.2.52) can be applied with U^h instead of V^h, and this yields

$$U^h(t+s, \pi_{t+s}) = U^h(t,\pi) + \int_0^s I((t+u, \pi_{t+u}) \in D_h)\, du + N^h_s \tag{21.2.64}$$

using (21.2.51) and (21.2.63) as well as that $\Delta_\pi U^h_\pi(t+u, h_i(t+u)) = 0$ for all $0 \le u \le s$ and $i = 0,1$ since U^h_π is continuous. In (21.2.64) we have

$$N^h_s = \int_0^s U^h_\pi(t+u, \pi_{t+u})\, I(\pi_{t+u} \ne h_0(t+u), \pi_{t+u} \ne h_1(t+u)) \tag{21.2.65}$$
$$\times \frac{\mu}{\sigma}\, \pi_{t+u}(1 - \pi_{t+u})\, d\bar{W}_u$$

from where we see that $(N_s^h)_{0\le s\le T-t}$ is a martingale under $\mathsf{P}_{t,\pi}$.

Next note that (21.2.52) applied to G instead of V^h yields

$$G(t+s,\pi_{t+s}) = G(t,\pi) + \int_0^s I(\pi_{t+u} \neq c)\, du - \frac{a+b}{2}\, \ell_s^c + M_s \tag{21.2.66}$$

using that $\mathbb{L}G = 1$ off $[0,T]\times\{c\}$ as well as that $\Delta_\pi G_\pi(t+u,c) = -b-a$ for $0 \le u \le s$. In (21.2.66) we have

$$\begin{aligned} M_s &= \int_0^s G_\pi(t+u,\pi_{t+u})\, I(\pi_{t+u} \neq c)\, \frac{\mu}{\sigma}\, \pi_{t+u}(1-\pi_{t+u})\, d\bar{W}_u \\ &= \int_0^s \big[a\, I(\pi_{t+u}<c) - b\, I(\pi_{t+u}>c)\big]\, \frac{\mu}{\sigma}\, \pi_{t+u}(1-\pi_{t+u})\, d\bar{W}_u \end{aligned} \tag{21.2.67}$$

from where we see that $(M_s)_{0\le s\le T-t}$ is a martingale under $\mathsf{P}_{t,\pi}$.

For $0 < \pi \le h_0(t)$ or $h_1(t) \le \pi < 1$ consider the stopping time

$$\sigma_h = \inf\{\, 0 \le s \le T-t : \pi_{t+s} \in [h_0(t+s), h_1(t+s)]\,\}. \tag{21.2.68}$$

Then using that $U^h(t,h_i(t)) = G(t,h_i(t))$ for all $0 \le t < T$ and $i=0,1$ since (h_0,h_1) solve (21.2.46)), and that $U^h(T,\pi) = G(T,\pi)$ for all $0 \le \pi \le 1$, we see that $U^h(t+\sigma_h,\pi_{t+\sigma_h}) = G(t+\sigma_h,\pi_{t+\sigma_h})$. Hence from (21.2.64) and (21.2.66) using the optional sampling theorem (page 60) we find

$$\begin{aligned} U^h(t,\pi) &= \mathsf{E}_{t,\pi} U^h(t+\sigma_h,\pi_{t+\sigma_h}) - \mathsf{E}_{t,\pi}\bigg[\int_0^{\sigma_h} I((t+u,\pi_{t+u}) \in D_h)\, du\bigg] \\ &= \mathsf{E}_{t,\pi} G(t+\sigma_h,\pi_{t+\sigma_h}) - \mathsf{E}_{t,\pi}\bigg[\int_0^{\sigma_h} I((t+u,\pi_{t+u}) \in D_h)\, du\bigg] \\ &= G(t,\pi) + \mathsf{E}_{t,\pi}\bigg[\int_0^{\sigma_h} I(\pi_{t+u} \neq c)\, du\bigg] \\ &\quad - \mathsf{E}_{t,\pi}\bigg[\int_0^{\sigma_h} I((t+u,\pi_{t+u}) \in D_h)\, du\bigg] = G(t,\pi) \end{aligned} \tag{21.2.69}$$

since $\pi_{t+u} \neq c$ and $(t+u,\pi_{t+u}) \in D_h$ for all $0 \le u < \sigma_h$. This establishes (21.2.60) and thus (21.2.59) holds as well.

It may be noted that a shorter but somewhat less revealing proof of (21.2.60) [and (21.2.59)] can be obtained by verifying directly (using the Markov property only) that the process

$$U^h(t+s,\pi_{t+s}) - \int_0^s I((t+u,\pi_{t+u}) \in D_h)\, du \tag{21.2.70}$$

is a martingale under $\mathsf{P}_{t,\pi}$ for $0 \le s \le T-t$. This verification moreover shows that the martingale property of (21.2.70) does not require that h_0 and h_1 are continuous and monotone (but only measurable). Taken together with the rest of the proof below this shows that the claim of *uniqueness* for the equations (21.2.46) holds in the *class of continuous functions* h_0 and h_1 from $[0,T]$ to $\mathbb{R}$ such that $0 < h_0(t) < c$ and $c < h_1(t) < 1$ for all $0 < t < T$.

4°. Let us consider the stopping time

$$\tau_h = \inf\{\, 0 \le s \le T-t : \pi_{t+s} \notin \big(h_0(t+s), h_1(t+s)\big) \,\}. \tag{21.2.71}$$

Observe that, by virtue of (21.2.59), the identity (21.2.52) can be written as

$$V^h(t+s, \pi_{t+s}) = V^h(t,\pi) + \int_0^s I((t+u, \pi_{t+u}) \in D_h)\, du + M_s^h \tag{21.2.72}$$

with $(M_s^h)_{0 \le s \le T-t}$ being a martingale under $\mathsf{P}_{t,\pi}$. Thus, inserting τ_h into (21.2.72) in place of s and taking the $\mathsf{P}_{t,\pi}$-expectation, by means of the optional sampling theorem (page 60) we get

$$V^h(t,\pi) = \mathsf{E}_{t,\pi} G(t+\tau_h, \pi_{t+\tau_h}) \tag{21.2.73}$$

for all $(t,\pi) \in [0,T) \times [0,1]$. Then comparing (21.2.73) with (21.2.1) we see that

$$V(t,\pi) \le V^h(t,\pi) \tag{21.2.74}$$

for all $(t,\pi) \in [0,T) \times [0,1]$.

5°. Let us now show that $g_0 \le h_0$ and $h_1 \le g_1$ on $[0,T]$. For this, recall that by the same arguments as for V^h we also have

$$V(t+s, \pi_{t+s}) = V(t,\pi) + \int_0^s I((t+u, \pi_{t+u}) \in D)\, du + M_s^g \tag{21.2.75}$$

where $(M_s^g)_{0 \le s \le T-t}$ is a martingale under $\mathsf{P}_{t,\pi}$. Fix some (t,π) belonging to both D and D_h (first below g_0 and h_0 and then above g_1 and h_1) and consider the stopping time

$$\sigma_g = \inf\{\, 0 \le s \le T-t : \pi_{t+s} \in [g_0(t+s), g_1(t+s)] \,\}. \tag{21.2.76}$$

Inserting σ_g into (21.2.72) and (21.2.75) in place of s and taking the $\mathsf{P}_{t,\pi}$-expectation, by means of the optional sampling theorem (page 60) we get

$$\begin{aligned} &\mathsf{E}_{t,\pi} V^h(t+\sigma_g, \pi_{t+\sigma_g}) \\ &= G(t,\pi) + \mathsf{E}_{t,\pi}\left[\int_0^{\sigma_g} I((t+u, \pi_{t+u}) \in D_h)\, du\right], \end{aligned} \tag{21.2.77}$$

$$\mathsf{E}_{t,\pi} V(t+\sigma_g, \pi_{t+\sigma_g}) = G(t,\pi) + \mathsf{E}_{t,\pi}\sigma_g. \tag{21.2.78}$$

Hence by means of (21.2.74) we see that

$$\mathsf{E}_{t,\pi}\left[\int_0^{\sigma_g} I((t+u, \pi_{t+u}) \in D_h)\, du\right] \ge \mathsf{E}_{t,\pi}\sigma_g \tag{21.2.79}$$

from where, by virtue of the continuity of h_i and g_i on $(0,T)$ for $i=0,1$, it readily follows that $D \subseteq D_h$, i.e. $g_0(t) \le h_0(t)$ and $h_1(t) \le g_1(t)$ for all $0 \le t \le T$.

6°. Finally, we show that h_i coincides with g_i for $i=0,1$. For this, let us assume that there exists some $t \in (0,T)$ such that $g_0(t) < h_0(t)$ or $h_1(t) < g_1(t)$ and take an arbitrary π from $(g_0(t), h_0(t))$ or $(h_1(t), g_1(t))$, respectively. Then inserting $\tau_* = \tau_*(t,\pi)$ from (21.2.33) into (21.2.72) and (21.2.75) in place of s and taking the $\mathsf{P}_{t,\pi}$-expectation, by means of the optional sampling theorem (page 60) we get

$$\mathsf{E}_{t,\pi} G(t+\tau_*, \pi_{t+\tau_*}) = V^h(t,\pi) + \mathsf{E}_{t,\pi}\left[\int_0^{\tau_*} I((t+u, \pi_{t+u}) \in D_h)\, du\right], \tag{21.2.80}$$

$$\mathsf{E}_{t,\pi} G(t+\tau_*, \pi_{t+\tau_*}) = V(t,\pi). \tag{21.2.81}$$

Hence by means of (21.2.74) we see that

$$\mathsf{E}_{t,\pi}\left[\int_0^{\tau_*} I((t+u, \pi_{t+u}) \in D_h)\, du\right] \le 0 \tag{21.2.82}$$

which is clearly impossible by the continuity of h_i and g_i for $i=0,1$. We may therefore conclude that V^h defined in (21.2.49) coincides with V from (21.2.1) and h_i is equal to g_i for $i=0,1$. This completes the proof of the theorem. □

Remark 21.3. Note that without loss of generality it can be assumed that $\mu > 0$ in (21.0.2). In this case the optimal decision rule (21.2.44)–(21.2.45) can be equivalently written as follows:

$$\tau_* = \inf\{\, 0 \le t \le T : X_t \notin (b_0^\pi(t), b_1^\pi(t))\,\}, \tag{21.2.83}$$

$$d_* = \begin{cases} 1 \text{ (accept } H_1) & \text{if } X_{\tau_*} = b_1^\pi(\tau_*), \\ 0 \text{ (accept } H_0) & \text{if } X_{\tau_*} = b_0^\pi(\tau_*), \end{cases} \tag{21.2.84}$$

where we set

$$b_i^\pi(t) = \frac{\sigma^2}{\mu} \log\left(\frac{1-\pi}{\pi}\, \frac{g_i(t)}{1-g_i(t)}\right) + \frac{\mu}{2} t \tag{21.2.85}$$

for $t \in [0,T]$, $\pi \in [0,1]$ and $i=0,1$. The result proved above shows that the following sequential procedure is optimal: *Observe* X_t *for* $t \in [0,T]$ *and stop the observation as soon as* X_t *becomes either greater than* $b_1^\pi(t)$ *or smaller than* $b_0^\pi(t)$ *for some* $t \in [0,T]$. *In the first case conclude that the drift equals* μ, *and in the second case conclude that the drift equals* 0.

Remark 21.4. In the preceding procedure we need to know the boundaries (b_0^π, b_1^π) i.e. the boundaries (g_0, g_1). We proved above that (g_0, g_1) is a unique solution of the system (21.2.46). This system cannot be solved analytically but can be dealt with numerically. The following simple method can be used to illustrate the latter (better methods are needed to achieve higher precision around the singularity point $t = T$ and to increase the speed of calculation).

Set $t_k = kh$ for $k = 0, 1, \ldots, n$ where $h = T/n$ and denote

$$J(t, g_i(t)) = \mathsf{E}_{t,g_i(t)}[a\pi_T \wedge b(1-\pi_T)] - ag_i(t) \wedge b(1-g_i(t)), \tag{21.2.86}$$

$$\begin{aligned} &K(t, g_i(t); t+u, g_0(t+u), g_1(t+u)) \\ &\qquad = \sum_{j=0}^{1} (-1)^j \mathsf{P}_{t,g_i(t)}(\pi_{t+u} \le g_j(t+u)) \end{aligned} \tag{21.2.87}$$

for $i = 0, 1$ upon recalling the explicit expressions (21.2.47) and (21.2.48) above. Note that K always depends on both g_0 and g_1. Then the following discrete approximation of the integral equations (21.2.46) is valid:

$$J(t_k, g_i(t_k)) = \sum_{l=k}^{n-1} K(t_k, g_i(t_k); t_{l+1}, g_0(t_{l+1}), g_1(t_{l+1}))\, h \tag{21.2.88}$$

for $k = 0, 1, \ldots, n-1$ and $i = 0, 1$. Setting $k = n-1$ and $g_0(t_n) = g_1(t_n) = c$ we can solve the system of two equations (21.2.88) numerically and get numbers $g_0(t_{n-1})$ and $g_1(t_{n-1})$. Setting $k = n-2$ and using the values $g_0(t_{n-1})$, $g_0(t_n)$, $g_1(t_{n-1})$, $g_1(t_n)$ we can solve (21.2.88) numerically and get numbers $g_0(t_{n-2})$ and $g_1(t_{n-2})$. Continuing the recursion we obtain $g_i(t_n), g_i(t_{n-1}), \ldots, g_i(t_1), g_i(t_0)$ as an approximation of the optimal boundary g_i at the points $T, T-h, \ldots, h, 0$ for $i = 0, 1$ (cf. *Figure* VI.1).

Notes. The problem of sequential testing of two simple hypotheses about the mean value of an observed Wiener process seeks to determine (as soon as possible and with minimal probability error) which of the given two values is a true mean. The problem admits two different formulations (cf. Wald [216]). In the Bayesian formulation it is assumed that the unknown mean has a given distribution, and in the variational formulation no probabilistic assumption about the unknown mean is made a priori. In this section we only study the Bayesian formulation.

The history of the problem is long and we only mention a few points starting with Wald and Wolfowitz [218]–[219] who used the Bayesian approach to prove the optimality of the sequential probability ratio test (SPRT) in the variational problem for i.i.d. sequences of observations. Dvoretzky, Kiefer and Wolfowitz [51] stated without proof that if the continuous-time log-likelihood ratio process has stationary independent increments, then the SPRT remains optimal in the variational problem. Mikhalevich [136] and Shiryaev [193] (see also [196, Chap. IV]) derived an explicit solution of the Bayesian and variational problem for a Wiener process with infinite horizon by reducing the initial optimal stopping problem to a free-boundary problem for a differential operator. A complete proof of the statement from [51] (under some mild assumptions) was given by Irle and Schmitz [96].

An explicit solution of the Bayesian and variational problem for a Poisson process with infinite horizon was derived in [168] by reducing the initial optimal stopping problem to a free-boundary problem for a differential-difference operator (see Section 23 below). The main aim of Subsection 21.2 above (following [71]) is to derive a solution of the Bayesian problem for a Wiener process with finite horizon.

22. Quickest detection of a Wiener process

In the Bayesian formulation of problem (proposed in [188] and [190]) it is assumed that we observe a trajectory of the Wiener process (Brownian motion) $X = (X_t)_{t \ge 0}$ with a drift changing from 0 to $\mu \neq 0$ at some random time θ taking the value 0 with probability π and being exponentially distributed with parameter $\lambda > 0$ given that $\theta > 0$.

1. For a precise probabilistic formulation of the Bayesian problem it is convenient to assume that all our considerations take place on a probability-statistical space $(\Omega; \mathcal{F}; \mathsf{P}_\pi, \pi \in [0,1])$ where the probability measure P_π has the following structure:

$$\mathsf{P}_\pi = \pi \mathsf{P}^0 + (1-\pi) \int_0^\infty \lambda e^{-\lambda s} \mathsf{P}^s \, ds \tag{22.0.1}$$

for $\pi \in [0,1]$ and P^s is a probability measure specified below for $s \ge 0$. Let θ be a non-negative random variable satisfying $\mathsf{P}_\pi(\theta = 0) = \pi$ and $\mathsf{P}_\pi(\theta > t \mid \theta > 0) = e^{-\lambda t}$ for all $t \ge 0$ and some $\lambda > 0$, and let $W = (W_t)_{t \ge 0}$ be a standard Wiener process started at zero under P_π for $\pi \in [0,1]$. It is assumed that θ and W are independent.

It is further assumed that we observe a process $X = (X_t)_{t \ge 0}$ satisfying the stochastic differential equation

$$dX_t = \mu I(t \ge \theta) \, dt + \sigma \, dW_t \qquad (X_0 = 0) \tag{22.0.2}$$

and thus being of the form

$$X_t = \begin{cases} \sigma W_t & \text{if } t < \theta, \\ \mu(t-\theta) + \sigma W_t & \text{if } t \ge \theta \end{cases} \tag{22.0.3}$$

where $\mu \neq 0$ and $\sigma^2 > 0$ are given and fixed. Thus $\mathsf{P}_\pi(X \in \cdot \mid \theta = s) = \mathsf{P}^s(X \in \cdot\,)$ is the distribution law of a Wiener process with the diffusion coefficient $\sigma > 0$ and a drift changing from 0 to μ at time $s \ge 0$. It is assumed that the time θ of *"disorder"* is unknown (i.e. it cannot be observed directly).

Being based upon the continuous observation of X, our task is to find a stopping time τ_* of X (i.e. a stopping time with respect to the natural filtration $\mathcal{F}_t^X = \sigma\{X_s : 0 \le s \le t\}$ generated by X for $t \ge 0$) that is *"as close as possible"* to the unknown time θ. More precisely, the *Wiener disorder problem* (or the *quickest detection problem for the Wiener process*) consists of computing the risk function

$$V(\pi) = \inf_{\tau} \Big(\mathsf{P}_\pi(\tau < \theta) + c\,\mathsf{E}_\pi[\tau - \theta]^+ \Big) \tag{22.0.4}$$

and finding the optimal stopping time τ_* at which the infimum in (22.0.4) is attained. Here $\mathsf{P}_\pi(\tau < \theta)$ is the probability of a *"false alarm"*, $\mathsf{E}_\pi[\tau - \theta]^+$ is the *"average delay"* in detecting the *"disorder"* correctly, and $c > 0$ is a given constant. Note that $\tau_* = T$ corresponds to the conclusion that $\theta \geq T$.

2. By means of standard arguments (see [196, pp. 195–197]) one can reduce the Bayesian problem (22.0.4) to the optimal stopping problem

$$V(\pi) = \inf_{\tau} \mathsf{E}_\pi \left[1 - \pi_\tau + c \int_0^\tau \pi_t \, dt \right] \tag{22.0.5}$$

for the *a posteriori probability process* $\pi_t = \mathsf{P}_\pi(\theta \leq t \mid \mathcal{F}_t^X)$ for $t \geq 0$ with $\mathsf{P}_\pi(\pi_0 = \pi) = 1$.

3. By the Bayes formula,

$$\pi_t = \pi \frac{d\mathsf{P}^0}{d\mathsf{P}_\pi}(t, X) + (1 - \pi) \int_0^t \frac{d\mathsf{P}^s}{d\mathsf{P}_\pi}(t, X) \lambda e^{-\lambda s} \, ds \tag{22.0.6}$$

where $(d\mathsf{P}^s / d\mathsf{P}_\pi)(t, X)$ is a Radon–Nikodým density of the measure $\mathsf{P}^s | \mathcal{F}_t^X$ with respect to the measure $\mathsf{P}_\pi | \mathcal{F}_t^X$.

Similarly

$$1 - \pi_t = (1 - \pi) e^{-\lambda t} \frac{d\mathsf{P}^t}{d\mathsf{P}_\pi}(t, X) = (1 - \pi) e^{-\lambda t} \frac{d\mathsf{P}^\infty}{d\mathsf{P}_\pi}(t, X) \tag{22.0.7}$$

where P^∞ is the probability law (measure) of the process $(\sigma W_t)_{t \geq 0}$. Hence for the *likelihood ratio process*

$$\varphi_t = \frac{\pi_t}{1 - \pi_t} \tag{22.0.8}$$

we get

$$\varphi_t = e^{\lambda t} Z_t \left(\varphi_0 + \lambda \int_0^t \frac{e^{-\lambda s}}{Z_s} \, ds \right) = e^{Y_t} \left(\frac{\pi}{1 - \pi} + \lambda \int_0^t e^{-Y_s} \, ds \right) \tag{22.0.9}$$

where (see Subsection 5.3)

$$Z_t = \frac{d\mathsf{P}^0}{d\mathsf{P}^\infty}(t, X) \equiv \frac{d(\mathsf{P}^0 | \mathcal{F}_t^X)}{d(\mathsf{P}^\infty | \mathcal{F}_t^X)} = \exp\!\left(\frac{\mu}{\sigma^2} \left(X_t - \frac{\mu}{2} t \right) \right) \tag{22.0.10}$$

and (for further reference) we set

$$Y_t = \lambda t + \frac{\mu}{\sigma^2} \left(X_t - \frac{\mu}{2} t \right). \tag{22.0.11}$$

By Itô's formula (page 67) one gets

$$dZ_t = \frac{\mu}{\sigma^2} Z_t \, dX_t \tag{22.0.12}$$

so that from (22.0.9) we find that

$$d\varphi_t = \lambda(1+\varphi_t)\,dt + \frac{\mu}{\sigma^2}\varphi_t\,dX_t. \tag{22.0.13}$$

Due to the identity

$$\pi_t = \frac{\varphi_t}{1+\varphi_t} \tag{22.0.14}$$

it follows that

$$d\pi_t = \Big(\lambda - \frac{\mu^2}{\sigma^2}\pi_t^2\Big)(1-\pi_t)\,dt + \frac{\mu}{\sigma^2}\pi_t(1-\pi_t)\,dX_t \tag{22.0.15}$$

or equivalently

$$d\pi_t = \lambda(1-\pi_t)\,dt + \frac{\mu}{\sigma}\pi_t(1-\pi_t)\,d\bar{W}_t \tag{22.0.16}$$

where the innovation process $(\bar{W}_t)_{t\ge 0}$ given by

$$\bar{W}_t = \frac{1}{\sigma}\Big(X_t - \mu\int_0^t \pi_s\,ds\Big) \tag{22.0.17}$$

is a standard Wiener process (see [127, Chap. IX]).

Using (22.0.9)+(22.0.11)+(22.0.14) it can be verified that $(\pi_t)_{t\ge 0}$ is a time-homogeneous (strong) Markov process under P_π for $\pi \in [0,1]$ with respect to the natural filtration. As the latter clearly coincides with $(\mathcal{F}_t^X)_{t\ge 0}$ it is also clear that the infimum in (22.0.5) can equivalently be taken over all stopping times of $(\pi_t)_{t\ge 0}$.

22.1. Infinite horizon

1. In order to solve the problem (22.0.5) when the horizon is *infinite* let us consider the optimal stopping problem for the Markov process $(\pi_t)_{t\ge 0}$ given by

$$V(\pi) = \inf_\tau \mathsf{E}_\pi\Big(M(\pi_\tau) + \int_0^\tau L(\pi_t)\,dt\Big) \tag{22.1.1}$$

where $\mathsf{P}_\pi(\pi_0 = \pi) = 1$, i.e. P_π is a probability measure under which the diffusion process $(\pi_t)_{t\ge 0}$ solving (22.0.16) above starts at π, the infimum in (22.1.1) is taken over all stopping times τ of $(\pi_t)_{t\ge 0}$, and we set $M(\pi) = 1-\pi$ and $L(\pi) = c\pi$ for $\pi \in [0,1]$. From (22.0.16) above we see that the infinitesimal generator of $(\pi_t)_{t\ge 0}$ is given by

$$\mathbb{L} = \lambda(1-\pi)\frac{\partial}{\partial \pi} + \frac{\mu^2}{2\sigma^2}\,\pi^2(1-\pi)^2\,\frac{\partial^2}{\partial \pi^2}. \tag{22.1.2}$$

2. The optimal stopping problem (22.1.1) will be solved in two steps. In the first step we will make a guess for the solution. In the second step we will verify that the guessed solution is correct (Theorem 22.1).

From (22.1.1) and (22.0.16) above we see that the closer $(\pi_t)_{t\ge 0}$ gets to 1 the less likely that the loss will decrease by continuation. This suggests that there exists a point $A \in (0,1)$ such that the stopping time

$$\tau_A = \inf\big\{t \ge 0 : \pi_t \ge A\big\} \tag{22.1.3}$$

is optimal in (22.1.1).

Standard arguments based on the strong Markov property (cf. Chapter III) lead to the following *free-boundary problem* for the *unknown function* V and the *unknown point* A:

$$\mathbb{L}V = -c\pi \quad \text{for } \pi \in (0,1), \tag{22.1.4}$$

$$V(A) = 1 - A, \tag{22.1.5}$$

$$V'(A) = -1 \quad (\textit{smooth fit}), \tag{22.1.6}$$

$$V < M \quad \text{for } \pi \in [0, A), \tag{22.1.7}$$

$$V = M \quad \text{for } \pi \in (A, 1]. \tag{22.1.8}$$

3. To solve the free-boundary problem (22.1.4)–(22.1.8) note that the equation (22.1.4) using (22.1.2) can be written as

$$V'' + \frac{\lambda}{\gamma}\,\frac{1}{\pi^2(1-\pi)}\,V' = -\frac{c}{\gamma}\,\frac{1}{\pi(1-\pi)^2} \tag{22.1.9}$$

where we set $\gamma = \mu^2/(2\sigma^2)$. This is a first order linear differential equation in V' and noting that

$$\int \frac{d\pi}{\pi^2(1-\pi)} = \log\Big(\frac{\pi}{1-\pi}\Big) - \frac{1}{\pi} =: \alpha(\pi) \tag{22.1.10}$$

the general solution of this equation is given by

$$V'(\pi) = e^{-\frac{\lambda}{\gamma}\,\alpha(\pi)}\bigg(C - \frac{c}{\gamma}\int_0^\pi \frac{e^{\alpha(\rho)}}{\rho(1-\rho)^2}\,d\rho\bigg) \tag{22.1.11}$$

where C is an undetermined constant. Since $e^{-(\lambda/\gamma)\alpha(\pi)} \to +\infty$ as $\pi \downarrow 0$, and $e^{\alpha(\rho)} \to 0$ exponentially fast as $\rho \downarrow 0$, we see from (22.1.11) that $V'(\pi) \to \pm\infty$ as $\pi \downarrow 0$ depending on if $C > 0$ or $C < 0$ respectively. We thus choose $C = 0$ in (22.1.11). Note that this is equivalent to the fact that $V'(0+) = 0$.

With this choice of C denote the right-hand side of (22.1.11) by $\psi(\pi)$, i.e. let

$$\psi(\pi) = -\frac{c}{\gamma}\,e^{-\frac{\lambda}{\gamma}\,\alpha(\pi)}\int_0^\pi \frac{e^{\alpha(\rho)}}{\rho(1-\rho)^2}\,d\rho \tag{22.1.12}$$

for $\pi \in (0,1)$. It is then easy verified that there exists a unique root A_* of the equation

$$\psi(A_*) = -1 \tag{22.1.13}$$

corresponding to (22.1.6) above. To meet (22.1.5) and (22.1.8) as well let us set

$$V_*(\pi) = \begin{cases} (1-A_*) + \int_{A_*}^{\pi} \psi(\rho)\, d\rho & \text{if } \pi \in [0, A_*), \\ 1-\pi & \text{if } \pi \in [A_*, 1] \end{cases} \tag{22.1.14}$$

for $\pi \in [0,1]$.

The preceding analysis shows that the function V_* defined by (22.1.14) is the unique solution of the free-boundary problem (22.1.4)–(22.1.8) satisfying $|V_*'(0+)| < \infty$ (or equivalently being bounded at zero). Note that V_* is C^2 on $[0, A_*) \cup (A_*, 1]$ but only C^1 at A_*. Note also that V_* is concave on $[0,1]$.

4. In this way we have arrived at the conclusions of the following theorem.

Theorem 22.1. *The value function* V *from* (22.1.1) *is given explicitly by* (22.1.14) *above. The stopping time* τ_{A_*} *given by* (22.1.3) *above is optimal in* (22.1.1).

Proof. The properties of V_* stated in the end of paragraph 3 above show that Itô's formula (page 67) can be applied to $V_*(\pi_t)$ in its standard form (cf. Subsection 3.5). This gives

$$\begin{aligned} V_*(\pi_t) = V_*(\pi) &+ \int_0^t \mathbb{L}V_*(\pi_s)\, I(\pi_s \neq A_*)\, ds \\ &+ \frac{\mu}{\sigma} \int_0^t \pi_s(1-\pi_s) V_*'(\pi_s)\, d\bar{W}_s. \end{aligned} \tag{22.1.15}$$

Recalling that $V(\pi) = 1-\pi$ for $\pi \in (A_*, 1]$ and using that V_* satisfies (22.1.4) for $\pi \in (0, A_*)$, we see that

$$\mathbb{L}V_*(\pi) \ge -c\,\pi \tag{22.1.16}$$

for all $\pi \in [\lambda/(\lambda+c), 1]$ and thus for all $\pi \in (0,1]$ since $A_* \ge \lambda/(\lambda+c)$ as is easily seen. By (22.1.7), (22.1.8), (22.1.15) and (22.1.16) it follows that

$$M(\pi_t) \ge V_*(\pi_t) \ge V_*(\pi) - \int_0^t L(\pi_s)\, ds + M_t \tag{22.1.17}$$

where $M = (M_t)_{t \ge 0}$ is a continuous local martingale given by

$$M_t = \frac{\mu}{\sigma} \int_0^t \pi_s(1-\pi_s) V'(\pi_s)\, d\bar{W}_s. \tag{22.1.18}$$

Using that $|V_*'(\pi)| \le 1 < \infty$ for all $\pi \in [0,1]$ it is easily verified by standard means that M is a martingale. Moreover, by the optional sampling theorem (page 60) this bound also shows that $\mathsf{E}_\pi M_\tau = 0$ whenever $\mathsf{E}_\pi \sqrt{\tau} < \infty$ for a stopping time τ. In particular, the latter condition is satisfied if $\mathsf{E}_\pi \tau < \infty$. As clearly in (22.1.1) it is enough to take the infimum only over stopping times τ satisfying $\mathsf{E}_\pi \tau < \infty$, we may insert τ in (22.1.17) instead of t, take E_π on both sides, and conclude that

$$\mathsf{E}_\pi \left(M(\pi_\tau) + \int_0^\tau L(X_t)\, dt \right) \ge V_*(\pi) \tag{22.1.19}$$

for all $\pi \in [0,1]$. This shows that $V \ge V_*$. On the other hand, using (22.1.4) and the definition of τ_{A_*} in (22.1.3), we see from (22.1.15) that

$$M\big(\pi_{\tau_{A_*}}\big) = V_*\big(\pi_{\tau_{A_*}}\big) = V_*(\pi) + \int_0^{\tau_{A_*}} L(X_t)\, dt + M_{\tau_{A_*}}. \tag{22.1.20}$$

Since $\mathsf{E}_\pi \tau_{A_*} < \infty$ (being true for any A) we see by taking E_π on both sides of (22.1.20) that equality in (22.1.19) is attained at $\tau = \tau_{A_*}$, and thus $V = V_*$. This completes the proof. □

For more details on the Wiener disorder problem with infinite horizon (including a fixed probability error formulation) we refer to [196, Chap. 4, Sect. 3–4].

22.2. Finite horizon

1. Solution of the Bayesian problem. In order to solve the problem (22.0.5) when the horizon T is *finite*, let us consider the extended optimal stopping problem for the Markov process $(t, \pi_t)_{0 \le t \le T}$ given by

$$V(t,\pi) = \inf_{0 \le \tau \le T-t} \mathsf{E}_{t,\pi} \left[G(\pi_{t+\tau}) + \int_0^\tau H(\pi_{t+s})\, ds \right] \tag{22.2.1}$$

where $\mathsf{P}_{t,\pi}(\pi_t = \pi) = 1$, i.e. $\mathsf{P}_{t,\pi}$ is a probability measure under which the diffusion process $(\pi_{t+s})_{0 \le s \le T-t}$ solving (22.0.16) starts at π at time t, the infimum in (22.2.1) is taken over all stopping times τ of $(\pi_{t+s})_{0 \le s \le T-t}$, and we set $G(\pi) = 1 - \pi$ and $H(\pi) = c\pi$ for all $\pi \in [0,1]$. Note that $(\pi_{t+s})_{0 \le s \le T-t}$ under $P_{t,\pi}$ is equally distributed as $(\pi_s)_{0 \le s \le T-t}$ under P_π. This fact will be frequently used in the sequel without further mention. Since G and H are bounded and continuous on $[0,1]$ it is possible to apply Corollary 2.9 (Finite horizon) with Remark 2.10 and conclude that an optimal stopping time exists in (22.2.1).

2. Let us now determine the structure of the optimal stopping time in the problem (22.2.1).

(i) Note that by (22.0.16) we get

$$G(\pi_{t+s}) = G(\pi) - \lambda \int_0^s (1 - \pi_{t+u})\, du + M_s \tag{22.2.2}$$

where the process $(M_s)_{0 \le s \le T-t}$ defined by $M_s = -\int_0^s (\mu/\sigma)\pi_{t+u}(1-\pi_{t+u})d\bar{W}_u$ is a continuous martingale under $\mathsf{P}_{t,\pi}$. It follows from (22.2.2) using the optional sampling theorem (page 60) that

$$\begin{aligned} &\mathsf{E}_{t,\pi}\left[G(\pi_{t+\sigma}) + \int_0^\sigma H(\pi_{t+u})\,du\right] \\ &\quad = G(\pi) + \mathsf{E}_{t,\pi}\left[\int_0^\sigma ((\lambda+c)\pi_{t+u} - \lambda)\,du\right] \end{aligned} \tag{22.2.3}$$

for each stopping time σ of $(\pi_{t+s})_{0 \le s \le T-t}$. Choosing σ to be the exit time from a small ball, we see from (22.2.3) that it is never optimal to stop when $\pi_{t+s} < \lambda/(\lambda+c)$ for $0 \le s < T-t$. In other words, this shows that all points (t,π) for $0 \le t < T$ with $0 \le \pi < \lambda/(\lambda+c)$ belong to the continuation set

$$C = \{(t,\pi) \in [0,T) \times [0,1] : V(t,\pi) < G(\pi)\}. \tag{22.2.4}$$

(ii) Recalling the solution to the problem (2.5) in the case of infinite horizon, where the stopping time $\tau_* = \inf\{\, t > 0 : \pi_t \ge A_* \}$ is optimal and $0 < A_* < 1$ is uniquely determined from the equation (22.1.13) (see also (4.147) in [196, p. 201]), we see that all points (t,π) for $0 \le t \le T$ with $A_* \le \pi \le 1$ belong to the stopping set. Moreover, since $\pi \mapsto V(t,\pi)$ with $0 \le t \le T$ given and fixed is concave on $[0,1]$ (this is easily deduced using the same arguments as in [196, pp. 197–198]), it follows directly from the previous two conclusions about the continuation and stopping set that there exists a function g satisfying $0 < \lambda/(\lambda+c) \le g(t) \le A_* < 1$ for all $0 \le t \le T$ such that the continuation set is an open set of the form

$$C = \{(t,\pi) \in [0,T) \times [0,1] : \pi < g(t)\} \tag{22.2.5}$$

and the stopping set is the closure of the set

$$D = \{(t,\pi) \in [0,T) \times [0,1] : \pi > g(t)\}. \tag{22.2.6}$$

(Below we will show that V is continuous so that C is open indeed. We will also see that $g(T) = \lambda/(\lambda+c)$.)

(iii) Since the problem (22.2.1) is time-homogeneous, in the sense that G and H are functions of space only (i.e. do not depend on time), it follows that the map $t \mapsto V(t,\pi)$ is increasing on $[0,T]$. Hence if (t,π) belongs to C for some $\pi \in [0,1]$ and we take any other $0 \le t' < t \le T$, then $V(t',\pi) \le V(t,\pi) < G(\pi)$, showing that (t',π) belongs to C as well. From this we may conclude in (22.2.5)–(22.2.6) that the boundary $t \mapsto g(t)$ is decreasing on $[0,T]$.

(iv) Let us finally observe that the value function V from (22.2.1) and the boundary g from (22.2.5)–(22.2.6) also depend on T and let them be denoted here by V^T and g^T, respectively. Using the fact that $T \mapsto V^T(t,\pi)$ is a decreasing function on $[t,\infty)$ and $V^T(t,\pi) = G(\pi)$ for all $\pi \in [g^T(t), 1]$, we conclude that if $T < T'$, then $0 \le g^T(t) \le g^{T'}(t) \le 1$ for all $t \in [0,T]$. Letting T' in the previous expression go to ∞, we get that $0 < \lambda/(\lambda+c) \le g^T(t) \le A_* < 1$ and $A_* \equiv \lim_{T\to\infty} g^T(t)$ for all $t \ge 0$, where A_* is the optimal stopping point in the infinite horizon problem referred to above (cf. Subsection 22.1).

3. Let us now show that the value function $(t,\pi) \mapsto V(t,\pi)$ is *continuous* on $[0,T]\times[0,1]$. For this it is enough to prove that

$$\pi \mapsto V(t_0,\pi) \quad \text{is continuous at } \pi_0, \tag{22.2.7}$$

$$t \mapsto V(t,\pi) \quad \text{is continuous at } t_0 \text{ uniformly over } \pi \in [\pi_0-\delta, \pi_0+\delta] \tag{22.2.8}$$

for each $(t_0,\pi_0) \in [0,T]\times[0,1]$ with some $\delta > 0$ small enough (it may depend on π_0). Since (22.2.7) follows by the fact that $\pi \mapsto V(t,\pi)$ is concave on $[0,1]$, it remains to establish (22.2.8).

For this, let us fix arbitrary $0 \le t_1 < t_2 \le T$ and $0 \le \pi \le 1$, and let $\tau_1 = \tau_*(t_1,\pi)$ denote the optimal stopping time for $V(t_1,\pi)$. Set $\tau_2 = \tau_1 \wedge (T-t_2)$ and note since $t \mapsto V(t,\pi)$ is increasing on $[0,T]$ and $\tau_2 \le \tau_1$ that we have

$$\begin{aligned} 0 &\le V(t_2,\pi) - V(t_1,\pi) \\ &\le \mathsf{E}_\pi\Big[1-\pi_{\tau_2} + c\int_0^{\tau_2} \pi_u\,du\Big] - \mathsf{E}_\pi\Big[1-\pi_{\tau_1} + c\int_0^{\tau_1} \pi_u\,du\Big] \\ &\le \mathsf{E}_\pi[\pi_{\tau_1} - \pi_{\tau_2}]. \end{aligned} \tag{22.2.9}$$

From (22.0.16) using the optional sampling theorem (page 60) we find that

$$\mathsf{E}_\pi \pi_\sigma = \pi + \lambda\,\mathsf{E}_\pi\left[\int_0^\sigma (1-\pi_t)\,dt\right] \tag{22.2.10}$$

for each stopping time σ of $(\pi_t)_{0\le t\le T}$. Hence by the fact that $\tau_1 - \tau_2 \le t_2 - t_1$ we get

$$\begin{aligned} \mathsf{E}_\pi[\pi_{\tau_1} - \pi_{\tau_2}] &= \lambda\,\mathsf{E}_\pi\left[\int_0^{\tau_1}(1-\pi_t)\,dt - \int_0^{\tau_2}(1-\pi_t)\,dt\right] \\ &= \lambda\,\mathsf{E}_\pi\left[\int_{\tau_2}^{\tau_1}(1-\pi_t)\,dt\right] \le \lambda\,\mathsf{E}_\pi[\tau_1-\tau_2] \le \lambda\,(t_2-t_1) \end{aligned} \tag{22.2.11}$$

for all $0 \le \pi \le 1$. Combining (22.2.9) with (22.2.11) we see that (22.2.8) follows. In particular, this shows that the *instantaneous-stopping condition* (22.2.33) below is satisfied.

4. In order to prove that the *smooth-fit condition* (22.2.34) below holds, i.e. that $\pi \mapsto V(t,\pi)$ is C^1 at $g(t)$, let us fix a point $(t,\pi) \in [0,T)\times(0,1)$ lying on the boundary g so that $\pi = g(t)$. Then for all $\varepsilon > 0$ such that $0 < \pi - \varepsilon < \pi$ we have

$$\frac{V(t,\pi) - V(t,\pi-\varepsilon)}{\varepsilon} \ge \frac{G(\pi) - G(\pi-\varepsilon)}{\varepsilon} = -1 \tag{22.2.12}$$

and hence, taking the limit in (22.2.12) as $\varepsilon \downarrow 0$, we get

$$\frac{\partial^- V}{\partial \pi}(t,\pi) \geq G'(\pi) = -1 \tag{22.2.13}$$

where the left-hand derivative in (22.2.13) exists (and is finite) by virtue of the concavity of $\pi \mapsto V(t,\pi)$ on $[0,1]$. Note that the latter will also be proved independently below.

Let us now fix some $\varepsilon > 0$ such that $0 < \pi - \varepsilon < \pi$ and consider the stopping time $\tau_\varepsilon = \tau_*(t, \pi - \varepsilon)$ being optimal for $V(t, \pi - \varepsilon)$. Note that τ_ε is the first exit time of the process $(\pi_{t+s})_{0 \leq s \leq T-t}$ from the set C in (22.2.5). Then from (22.2.1) using the equation (22.0.16) and the optional sampling theorem (page 60) we obtain

$$\begin{aligned}
& V(t,\pi) - V(t,\pi-\varepsilon) \qquad (22.2.14)\\
& \leq \mathsf{E}_\pi \left[1 - \pi_{\tau_\varepsilon} + c \int_0^{\tau_\varepsilon} \pi_u \, du \right] - \mathsf{E}_{\pi-\varepsilon} \left[1 - \pi_{\tau_\varepsilon} + c \int_0^{\tau_\varepsilon} \pi_u \, du \right] \\
& = \mathsf{E}_\pi \left[1 - \pi_{\tau_\varepsilon} + c \left(\tau_\varepsilon + \frac{\pi - \pi_{\tau_\varepsilon}}{\lambda} \right) \right] \\
& \quad - \mathsf{E}_{\pi-\varepsilon} \left[1 - \pi_{\tau_\varepsilon} + c \left(\tau_\varepsilon + \frac{\pi - \varepsilon - \pi_{\tau_\varepsilon}}{\lambda} \right) \right] \\
& = \left(\frac{c}{\lambda} + 1 \right) \left(\mathsf{E}_{\pi-\varepsilon} \pi_{\tau_\varepsilon} - \mathsf{E}_\pi \pi_{\tau_\varepsilon} \right) + c \Big(\mathsf{E}_\pi \tau_\varepsilon - \mathsf{E}_{\pi-\varepsilon} \tau_\varepsilon \Big) + \varepsilon \frac{c}{\lambda} .
\end{aligned}$$

By (22.0.1) and (22.0.9)+(22.0.11)+(22.0.14) it follows that

$$\begin{aligned}
& \mathsf{E}_{\pi-\varepsilon} \pi_{\tau_\varepsilon} - \mathsf{E}_\pi \pi_{\tau_\varepsilon} \qquad (22.2.15)\\
& = (\pi - \varepsilon) \mathsf{E}^0 S(\pi - \varepsilon) + (1 - \pi + \varepsilon) \int_0^\infty \lambda e^{-\lambda s} \mathsf{E}^s S(\pi - \varepsilon) \, ds \\
& \quad - \pi \mathsf{E}^0 S(\pi) - (1 - \pi) \int_0^\infty \lambda e^{-\lambda s} \mathsf{E}^s S(\pi) \, ds \\
& = \pi \mathsf{E}^0 [S(\pi - \varepsilon) - S(\pi)] + (1 - \pi) \int_0^\infty \lambda e^{-\lambda s} \mathsf{E}^s [S(\pi - \varepsilon) - S(\pi)] \, ds \\
& \quad - \varepsilon \mathsf{E}^0 S(\pi - \varepsilon) + \varepsilon \int_0^\infty \lambda e^{-\lambda s} \mathsf{E}^s S(\pi - \varepsilon) \, ds
\end{aligned}$$

where the function S is defined by

$$\begin{aligned}
S(\pi) = {} & e^{Y_{\tau_\varepsilon}} \left(\frac{\pi}{1-\pi} + \lambda \int_0^{\tau_\varepsilon} e^{-Y_u} \, du \right) \\
& \times \left(1 + e^{Y_{\tau_\varepsilon}} \left(\frac{\pi}{1-\pi} + \lambda \int_0^{\tau_\varepsilon} e^{-Y_u} \, du \right) \right)^{-1} .
\end{aligned} \tag{22.2.16}$$

By virtue of the mean value theorem there exists $\xi \in [\pi-\varepsilon, \pi]$ such that

$$\begin{aligned}&\pi \mathsf{E}^0[S(\pi-\varepsilon)-S(\pi)] + (1-\pi)\int_0^\infty \lambda e^{-\lambda s}\mathsf{E}^s[S(\pi-\varepsilon)-S(\pi)]\,ds \\ &= -\varepsilon\left(\pi\mathsf{E}^0 S'(\xi) + (1-\pi)\int_0^\infty \lambda e^{-\lambda s}\mathsf{E}^s S'(\xi)\,ds\right)\end{aligned} \tag{22.2.17}$$

where S' is given by

$$S'(\xi) = e^{Y_{\tau_\varepsilon}} \Big/ \left((1-\xi)^2\left[1+e^{Y_{\tau_\varepsilon}}\left(\frac{\xi}{1-\xi}+\lambda\int_0^{\tau_\varepsilon} e^{-Y_u}\,du\right)\right]^2\right). \tag{22.2.18}$$

Considering the second term on the right-hand side of (22.2.14) we find using (22.0.1) that

$$\begin{aligned}c\Big(\mathsf{E}_\pi\tau_\varepsilon - \mathsf{E}_{\pi-\varepsilon}\tau_\varepsilon\Big) &= c\varepsilon\left(\mathsf{E}^0\tau_\varepsilon + \int_0^\infty \lambda e^{-\lambda s}\mathsf{E}^s\tau_\varepsilon\,ds\right) \\ &= \frac{c\varepsilon}{1-\pi}\Big((1-2\pi)\mathsf{E}^0\tau_\varepsilon + \mathsf{E}_\pi\tau_\varepsilon\Big).\end{aligned} \tag{22.2.19}$$

Recalling that τ_ε is equally distributed as $\tilde{\tau}_\varepsilon = \inf\{\, 0 \le s \le T-t : \pi_s^{\pi-\varepsilon} \ge g(t+s)\,\}$, where we write $\pi_s^{\pi-\varepsilon}$ to indicate dependance on the initial point $\pi-\varepsilon$ through (22.0.9) in (22.0.14) above, and considering the hitting time σ_ε to the constant level $\pi = g(t)$ given by $\sigma_\varepsilon = \inf\{\, s \ge 0 : \pi_s^{\pi-\varepsilon} \ge \pi\,\}$, it follows that $\tilde{\tau}_\varepsilon \le \sigma_\varepsilon$ for every $\varepsilon > 0$ since g is decreasing, and $\sigma_\varepsilon \downarrow \sigma_0$ as $\varepsilon \downarrow 0$ where $\sigma_0 = \inf\{\, s > 0 : \pi_s^\pi \ge \pi\,\}$. On the other hand, since the diffusion process $(\pi_s^\pi)_{s\ge 0}$ solving (22.0.16) is regular (see e.g. [174, Chap. 7, Sect. 3]), it follows that $\sigma_0 = 0$ P_π-a.s. This in particular shows that $\tau_\varepsilon \to 0$ P_π-a.s. Hence we easily find that

$$S(\pi-\varepsilon) \to \pi, \quad S(\xi) \to \pi \quad \text{and} \quad S'(\xi) \to 1 \quad \mathsf{P}_\pi\text{-a.s.} \tag{22.2.20}$$

as $\varepsilon \downarrow 0$ for $s \ge 0$, and clearly $|S'(\xi)| \le K$ with some $K > 0$ large enough.

From (22.2.14) using (22.2.15)–(22.2.20) it follows that:

$$\begin{aligned}\frac{V(t,\pi)-V(t,\pi-\varepsilon)}{\varepsilon} &\le \Big(\frac{c}{\lambda}+1\Big)\Big(-1+o(1)\Big) + o(1) + \frac{c}{\lambda} \\ &= -1+o(1)\end{aligned} \tag{22.2.21}$$

as $\varepsilon \downarrow 0$ by the dominated convergence theorem and the fact that $\mathsf{P}^0 \ll \mathsf{P}_\pi$. This combined with (22.2.12) above proves that $V_\pi^-(t,\pi)$ exists and equals $G'(\pi) = -1$.

5. We proceed by proving that the boundary g is *continuous* on $[0,T]$ and that $g(T) = \lambda/(\lambda+c)$.

(i) Let us first show that the boundary g is right-continuous on $[0,T]$. For this, fix $t \in [0,T)$ and consider a sequence $t_n \downarrow t$ as $n \to \infty$. Since g is decreasing, the right-hand limit $g(t+)$ exists. Because $(t_n, g(t_n)) \in \bar{D}$ for all $n \geq 1$, and $\bar{D}$ is closed, we see that $(t, g(t+)) \in \bar{D}$. Hence by (22.2.6) we see that $g(t+) \geq g(t)$. The reverse inequality follows obviously from the fact that g is decreasing on $[0,T]$, thus proving the claim.

(ii) Suppose that at some point $t_* \in (0,T)$ the function g makes a jump, i.e. let $g(t_*-) > g(t_*) \geq \lambda/(\lambda+c)$. Let us fix a point $t' < t_*$ close to t_* and consider the half-open set $R \subset C$ being a curved trapezoid formed by the vertices $(t', g(t'))$, $(t_*, g(t_*-))$, (t_*, π') and (t', π') with any π' fixed arbitrarily in the interval $(g(t_*), g(t_*-))$. Observe that the strong Markov property implies that the value function V from (22.2.1) is $C^{1,2}$ on C. Note also that the gain function G is C^2 in R so that by the Newton–Leibniz formula using (22.2.33) and (22.2.34) it follows that

$$V(t,\pi) - G(\pi) = \int_\pi^{g(t)} \int_u^{g(t)} \left(\frac{\partial^2 V}{\partial \pi^2}(t,v) - \frac{\partial^2 G}{\partial \pi^2}(v) \right) dv\, du \tag{22.2.22}$$

for all $(t,\pi) \in R$.

Since $t \mapsto V(t,\pi)$ is increasing, we have

$$\frac{\partial V}{\partial t}(t,\pi) \geq 0 \tag{22.2.23}$$

for each $(t,\pi) \in C$. Moreover, since $\pi \mapsto V(t,\pi)$ is concave and (22.2.34) holds, we see that

$$\frac{\partial V}{\partial \pi}(t,\pi) \geq -1 \tag{22.2.24}$$

for each $(t,\pi) \in C$. Finally, since the strong Markov property implies that the value function V from (22.2.1) solves the equation (22.2.32), using (22.2.23) and (22.2.24) we obtain

$$\begin{aligned} \frac{\partial^2 V}{\partial \pi^2}(t,\pi) &= \frac{2\sigma^2}{\mu^2} \frac{1}{\pi^2(1-\pi)^2} \left(-c\pi - \lambda(1-\pi) \frac{\partial V}{\partial \pi}(t,\pi) - \frac{\partial V}{\partial t}(t,\pi) \right) \\ &\leq \frac{2\sigma^2}{\mu^2} \frac{1}{\pi^2(1-\pi)^2} \left(-c\pi + \lambda(1-\pi) \right) \leq -\varepsilon \frac{\sigma^2}{\mu^2} \end{aligned} \tag{22.2.25}$$

for all $t' \leq t < t_*$ and all $\pi' \leq \pi < g(t')$ with $\varepsilon > 0$ small enough. Note in (22.2.25) that $-c\pi + \lambda(1-\pi) < 0$ since all points (t,π) for $0 \leq t < T$ with $0 \leq \pi < \lambda/(\lambda+c)$ belong to C and consequently $g(t_*) \geq \lambda/(\lambda+c)$.

Hence by (22.2.22) using that $G_{\pi\pi} = 0$ we get

$$\begin{aligned} V(t',\pi') - G(\pi') &\leq -\varepsilon \frac{\sigma^2}{\mu^2} \frac{(g(t') - \pi')^2}{2} \\ &\to -\varepsilon \frac{\sigma^2}{\mu^2} \frac{(g(t_*-) - \pi')^2}{2} < 0 \end{aligned} \tag{22.2.26}$$

as $t' \uparrow t_*$. This implies that $V(t_*, \pi') < G(\pi')$ which contradicts the fact that (t_*, π') belongs to the stopping set $\bar{D}$. Thus $g(t_*-) = g(t_*)$ showing that g is continuous at t_* and thus on $[0, T]$ as well.

(iii) We finally note that the method of proof from the previous part (ii) also implies that $g(T) = \lambda/(\lambda + c)$. To see this, we may let $t_* = T$ and likewise suppose that $g(T-) > \lambda/(\lambda + c)$. Then repeating the arguments presented above word by word we arrive at a contradiction with the fact that $V(T, \pi) = G(\pi)$ for all $\pi \in [\lambda/(\lambda + c), g(T-)]$ thus proving the claim.

6. Summarizing the facts proved in paragraphs 2–5 above we may conclude that the following exit time is optimal in the extended problem (22.2.1):

$$\tau_* = \inf \{ 0 \le s \le T - t : \pi_{t+s} \ge g(t + s) \} \tag{22.2.27}$$

(the infimum of an empty set being equal $T - t$) where the boundary g satisfies the following properties (see *Figure* VI.2):

$$g : [0, T] \to [0, 1] \quad \text{is continuous and decreasing,} \tag{22.2.28}$$

$$\lambda/(\lambda + c) \le g(t) \le A_* \quad \text{for all } 0 \le t \le T, \tag{22.2.29}$$

$$g(T) = \lambda/(\lambda + c) \tag{22.2.30}$$

where A_* satisfying $0 < \lambda/(\lambda + c) < A_* < 1$ is the optimal stopping point for the infinite horizon problem uniquely determined from the transcendental equation (22.1.13) (or (4.147) in [196, p. 201]).

Standard arguments imply that the infinitesimal operator $\mathbb{L}$ of the process $(t, \pi_t)_{0 \le t \le T}$ acts on a function $f \in C^{1,2}([0, T) \times [0, 1])$ according to the rule

$$(\mathbb{L}f)(t, \pi) = \left(\frac{\partial f}{\partial t} + \lambda(1 - \pi) \frac{\partial f}{\partial \pi} + \frac{\mu^2}{2\sigma^2} \pi^2 (1 - \pi)^2 \frac{\partial^2 f}{\partial \pi^2} \right) (t, \pi) \tag{22.2.31}$$

for all $(t, \pi) \in [0, T) \times [0, 1]$. In view of the facts proved above we are thus naturally led to formulate the following *free-boundary problem* for the unknown value function V from (22.2.1) and the unknown boundary g from (22.2.5)–(22.2.6):

$$(\mathbb{L}V)(t, \pi) = -c\pi \quad \text{for } (t, \pi) \in C, \tag{22.2.32}$$

$$V(t, \pi)\big|_{\pi = g(t)-} = 1 - g(t) \quad (\textit{instantaneous stopping}), \tag{22.2.33}$$

$$\frac{\partial V}{\partial \pi}(t, \pi)\bigg|_{\pi = g(t)-} = -1 \quad (\textit{smooth fit}), \tag{22.2.34}$$

$$V(t, \pi) < G(\pi) \quad \text{for } (t, \pi) \in C, \tag{22.2.35}$$

$$V(t, \pi) = G(\pi) \quad \text{for } (t, \pi) \in D, \tag{22.2.36}$$

where C and D are given by (22.2.5) and (22.2.6), and the condition (22.2.33) is satisfied for all $0 \le t \le T$ and the condition (22.2.34) is satisfied for all $0 \le t < T$.

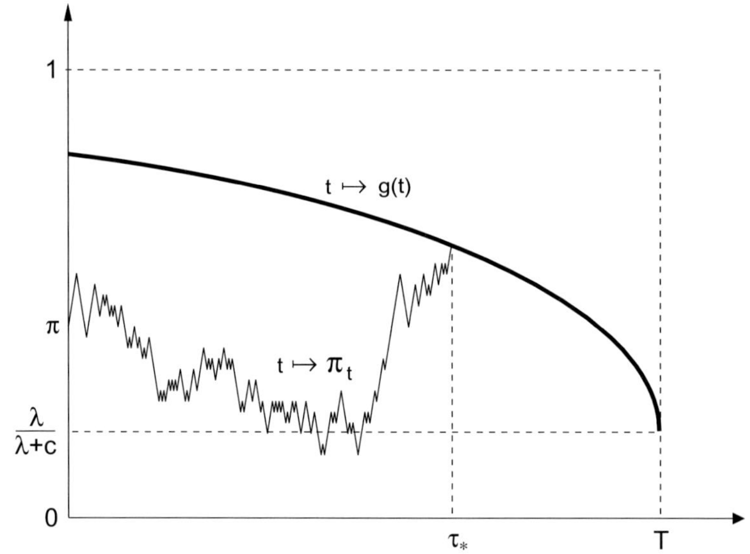

Figure VI.2: A computer drawing of the optimal stopping boundary g from Theorem 22.2. At time τ_* it is optimal to stop and conclude that the drift has been changed (from 0 to μ).

Note that the superharmonic characterization of the value function (cf. Chapter I) implies that V from (22.2.1) is a largest function satisfying (22.2.32)–(22.2.33) and (22.2.35)–(22.2.36).

7. Making use of the facts proved above we are now ready to formulate the main result of this subsection.

Theorem 22.2. *In the Bayesian formulation of the Wiener disorder problem* (22.0.4)–(22.0.5) *the optimal stopping time* τ_* *is explicitly given by*

$$\tau_* = \inf\{\, 0 \le t \le T : \pi_t \ge g(t) \,\} \tag{22.2.37}$$

where g *can be characterized as a unique solution of the nonlinear integral equation*

$$\begin{aligned} \mathsf{E}_{t,g(t)}\pi_T = g(t) &+ c \int_0^{T-t} \mathsf{E}_{t,g(t)} \big[\pi_{t+u}\, I(\pi_{t+u} < g(t+u)) \big]\, du \\ &+ \lambda \int_0^{T-t} \mathsf{E}_{t,g(t)} \big[(1-\pi_{t+u})\, I(\pi_{t+u} > g(t+u)) \big]\, du \end{aligned} \tag{22.2.38}$$

for $0 \le t \le T$ *satisfying* (22.2.28)–(22.2.30) [*see Figure* VI.2].

More explicitly, the three terms in the equation (22.2.38) *are given as follows:*

$$\mathsf{E}_{t,g(t)}\pi_T = g(t) + (1-g(t))\big(1-e^{-\lambda(T-t)}\big), \tag{22.2.39}$$

$$\mathsf{E}_{t,g(t)}\big[\pi_{t+u}\, I(\pi_{t+u} < g(t+u))\big] = \int_0^{g(t+u)} x\, p(g(t);u,x)\, dx, \tag{22.2.40}$$

$$\mathsf{E}_{t,g(t)}\big[(1-\pi_{t+u})\, I(\pi_{t+u} > g(t+u))\big] = \int_{g(t+u)}^{1} (1-x)\, p(g(t);u,x)\, dx \tag{22.2.41}$$

for $0 \le u \le T-t$ *with* $0 \le t \le T$, *where* p *is the transition density function of the process* $(\pi_t)_{0\le t\le T}$ *given in* (22.2.103) *below.*

Proof. 1°. The existence of a boundary g satisfying (22.2.28)–(22.2.30) such that τ_* from (22.2.37) is optimal in (22.0.4)–(22.0.5) was proved in paragraphs 2–6 above. By the local time-space formula (cf. Subsection 3.5) it follows that the boundary g solves the equation (22.2.38) (cf. (22.2.45)–(22.2.48) below). Thus it remains to show that the equation (22.2.38) has no other solution in the class of functions h satisfying (22.2.28)–(22.2.30).

Let us thus assume that a function h satisfying (22.2.28)–(22.2.30) solves the equation (22.2.38), and let us show that this function h must then coincide with the optimal boundary g. For this, let us introduce the function

$$V^h(t,\pi) = \begin{cases} U^h(t,\pi) & \text{if } \pi < h(t), \\ G(\pi) & \text{if } \pi \ge h(t), \end{cases} \tag{22.2.42}$$

where the function U^h is defined by

$$\begin{aligned} U^h(t,\pi) = \mathsf{E}_{t,\pi} G(\pi_T) &+ c\int_0^{T-t} \mathsf{E}_{t,\pi}\big[\pi_{t+u}\, I(\pi_{t+u} < h(t+u))\big]\, du \\ &+ \lambda \int_0^{T-t} \mathsf{E}_{t,\pi}\big[(1-\pi_{t+u})\, I(\pi_{t+u} > h(t+u))\big]\, du \end{aligned} \tag{22.2.43}$$

for all $(t,\pi) \in [0,T) \times [0,1]$. Note that (22.2.43) with $G(\pi)$ instead of $U^h(t,\pi)$ on the left-hand side coincides with (22.2.38) when $\pi = g(t)$ and $h = g$. Since h solves (22.2.38) this shows that V^h is continuous on $[0,T) \times [0,1]$. We need to verify that V^h coincides with the value function V from (22.2.1) and that h equals g.

2°. Using standard arguments based on the strong Markov property (or verifying directly) it follows that V^h i.e. U^h is $C^{1,2}$ on C_h and that

$$(\mathbb{L}V^h)(t,\pi) = -c\pi \quad \text{for } (t,\pi) \in C_h \tag{22.2.44}$$

where C_h is defined as in (22.2.5) with h instead of g. Moreover, since $U^h_\pi := \partial U^h/\partial\pi$ is continuous on $[0,T)\times(0,1)$ (which is readily verified using the explicit expressions (22.2.39)–(22.2.41) above with π instead of $g(t)$ and h instead of g), we see that $V^h_\pi := \partial V^h/\partial\pi$ is continuous on $\bar{C}_h$. Finally, it is clear that V^h i.e. G, is $C^{1,2}$ on $\bar{D}_h$, where D_h is defined as in (22.2.6) with h instead of g. Therefore, with $(t,\pi)\in[0,T)\times(0,1)$ given and fixed, the local time-space formula (cf. Subsection 3.5) can be applied, and in this way we get

$$\begin{aligned} V^h(t+s,\pi_{t+s}) = V^h(t,\pi) & \\ &+ \int_0^s (\mathbb{L}V^h)(t+u,\pi_{t+u})\, I(\pi_{t+u}\neq h(t+u))\, du \\ &+ M^h_s + \frac{1}{2}\int_0^s \Delta_\pi V^h_\pi(t+u,\pi_{t+u})\, I(\pi_{t+u}=h(t+u))\, d\ell^h_u \end{aligned} \tag{22.2.45}$$

for $0\le s\le T-t$ where $\Delta_\pi V^h_\pi(t+u,h(t+u)) = V^h_\pi(t+u,h(t+u)+) - V^h_\pi(t+u,h(t+u)-)$, the process $(\ell^h_s)_{0\le s\le T-t}$ is the local time of $(\pi_{t+s})_{0\le s\le T-t}$ at the boundary h given by

$$\begin{aligned} \ell^h_s = \mathsf{P}_{t,\pi}\text{-}\lim_{\varepsilon\downarrow 0}\frac{1}{2\varepsilon}\int_0^s & I(h(t+u)-\varepsilon<\pi_{t+u}<h(t+u)+\varepsilon) \\ &\times \frac{\mu^2}{\sigma^2}\pi^2_{t+u}(1-\pi_{t+u})^2\, du \end{aligned} \tag{22.2.46}$$

and $(M^h_s)_{0\le s\le T-t}$ defined by

$$M^h_s = \int_0^s V^h_\pi(t+u,\pi_{t+u})\, I(\pi_{t+u}\neq h(t+u))\,\frac{\mu}{\sigma}\,\pi_{t+u}(1-\pi_{t+u})\, d\bar{W}_u \tag{22.2.47}$$

is a martingale under $\mathsf{P}_{t,\pi}$.

Setting $s=T-t$ in (22.2.45) and taking the $\mathsf{P}_{t,\pi}$-expectation, using that V^h satisfies (22.2.44) in C_h and equals G in D_h, we get

$$\begin{aligned} \mathsf{E}_{t,\pi}G(\pi_T) = V^h(t,\pi) - c\int_0^{T-t} & \mathsf{E}_{t,\pi}\big[\pi_{t+u}\, I(\pi_{t+u}<h(t+u))\big]\, du \\ - \lambda\int_0^{T-t} & \mathsf{E}_{t,\pi}\big[(1-\pi_{t+u})\, I(\pi_{t+u}>h(t+u))\big]\, du + \frac{1}{2}F(t,\pi) \end{aligned} \tag{22.2.48}$$

where (by the continuity of the integrand) the function F is given by

$$F(t,\pi) = \int_0^{T-t} \Delta_\pi V^h_\pi(t+u,h(t+u))\, d_u \mathsf{E}_{t,\pi}\ell^h_u \tag{22.2.49}$$

for all $(t, \pi) \in [0, T) \times [0, 1]$. Thus from (22.2.48) and (22.2.42) we see that

$$F(t, \pi) = \begin{cases} 0 & \text{if } \pi < h(t), \\ 2\,(U^h(t, \pi) - G(\pi)) & \text{if } \pi \geq h(t) \end{cases} \tag{22.2.50}$$

where the function U^h is given by (22.2.43).

3°. From (22.2.50) we see that if we are to prove that

$$\pi \mapsto V^h(t, \pi) \quad \text{is} \quad C^1 \quad \text{at} \quad h(t) \tag{22.2.51}$$

for each $0 \leq t < T$ given and fixed, then it will follow that

$$U^h(t, \pi) = G(\pi) \quad \text{for all } \ h(t) \leq \pi \leq 1. \tag{22.2.52}$$

On the other hand, if we know that (22.2.52) holds, then using the general fact obtained directly from the definition (22.2.42) above,

$$\begin{aligned} \frac{\partial}{\partial \pi}(U^h(t, \pi) - G(\pi))\Big|_{\pi = h(t)} &= V^h_\pi(t, h(t)-) - V^h_\pi(t, h(t)+) \\ &= -\Delta_\pi V^h_\pi(t, h(t)) \end{aligned} \tag{22.2.53}$$

for all $0 \leq t < T$, we see that (22.2.51) holds too. The equivalence of (22.2.51) and (22.2.52) suggests that instead of dealing with the equation (22.2.50) in order to derive (22.2.51) above we may rather concentrate on establishing (22.2.52) directly.

To derive (22.2.52) first note that using standard arguments based on the strong Markov property (or verifying directly) it follows that U^h is $C^{1,2}$ in D_h and that

$$(\mathbb{L}U^h)(t, \pi) = -\lambda(1 - \pi) \quad \text{for } \ (t, \pi) \in D_h. \tag{22.2.54}$$

It follows that (22.2.45) can be applied with U^h instead of V^h, and this yields

$$\begin{aligned} U^h(t + s, \pi_{t+s}) = U^h(t, \pi) &- c \int_0^s \pi_{t+u}\, I(\pi_{t+u} < h(t + u))\, du \\ &- \lambda \int_0^s (1 - \pi_{t+u})\, I(\pi_{t+u} > h(t + u))\, du + N^h_s \end{aligned} \tag{22.2.55}$$

using (22.2.44) and (22.2.54) as well as that $\Delta_\pi U^h_\pi(t + u, h(t + u)) = 0$ for all $0 \leq u \leq s$ since U^h_π is continuous. In (22.2.55) we have $N^h_s = \int_0^s U^h_\pi(t + u, \pi_{t+u})\, I(\pi_{t+u} \neq h(t+u))\, (\mu/\sigma)\, \pi_{t+u}(1 - \pi_{t+u})\, d\bar{W}_u$ and $(N^h_s)_{0 \leq s \leq T-t}$ is a martingale under $\mathsf{P}_{t,\pi}$.

For $h(t) \leq \pi < 1$ consider the stopping time

$$\sigma_h = \inf\{\, 0 \leq s \leq T - t : \pi_{t+s} \leq h(t + s)\,\}. \tag{22.2.56}$$

Then using that $U^h(t,h(t)) = G(h(t))$ for all $0 \le t < T$ since h solves (22.2.38), and that $U^h(T,\pi) = G(\pi)$ for all $0 \le \pi \le 1$, we see that $U^h(t+\sigma_h, \pi_{t+\sigma_h}) = G(\pi_{t+\sigma_h})$. Hence from (22.2.55) and (22.2.2) using the optional sampling theorem (page 60) we find

$$
\begin{aligned}
U^h(t,\pi) &= \mathsf{E}_{t,\pi} U^h(t+\sigma_h, \pi_{t+\sigma_h}) \\
&\quad + c\mathsf{E}_{t,\pi}\Big[\int_0^{\sigma_h} \pi_{t+u}\, I(\pi_{t+u} < h(t+u))\, du\Big] \\
&\quad + \lambda\, \mathsf{E}_{t,\pi}\Big[\int_0^{\sigma_h} (1-\pi_{t+u})\, I(\pi_{t+u} > h(t+u))\, du\Big] \\
&= \mathsf{E}_{t,\pi} G(\pi_{t+\sigma_h}) + c\,\mathsf{E}_{t,\pi}\Big[\int_0^{\sigma_h} \pi_{t+u}\, I(\pi_{t+u} < h(t+u))\, du\Big] \\
&\quad + \lambda\, \mathsf{E}_{t,\pi}\Big[\int_0^{\sigma_h} (1-\pi_{t+u})\, I(\pi_{t+u} > h(t+u))\, du\Big] \\
&= G(\pi) - \lambda\, \mathsf{E}_{t,\pi}\Big[\int_0^{\sigma_h} (1-\pi_{t+u})\, du\Big] \\
&\quad + c\mathsf{E}_{t,\pi}\Big[\int_0^{\sigma_h} \pi_{t+u}\, I(\pi_{t+u} < h(t+u))\, du\Big] \\
&\quad + \lambda\, \mathsf{E}_{t,\pi}\Big[\int_0^{\sigma_h} (1-\pi_{t+u})\, I(\pi_{t+u} > h(t+u))\, du\Big] = G(\pi)
\end{aligned}
\tag{22.2.57}
$$

since $\pi_{t+u} > h(t+u)$ for all $0 \le u < \sigma_h$. This establishes (22.2.52) and thus (22.2.51) holds as well.

It may be noted that a shorter but somewhat less revealing proof of (22.2.52) [and (22.2.51)] can be obtained by verifying directly (using the Markov property only) that the process

$$
\begin{aligned}
U^h(t+s, \pi_{t+s}) &+ c\int_0^s \pi_{t+u}\, I(\pi_{t+u} < h(t+u))\, du \\
&+ \lambda \int_0^s (1-\pi_{t+u})\, I(\pi_{t+u} > h(t+u))\, du
\end{aligned}
\tag{22.2.58}
$$

is a martingale under $\mathsf{P}_{t,\pi}$ for $0 \le s \le T-t$. This verification moreover shows that the martingale property of (22.2.58) does not require that h is continuous and increasing (but only measurable). Taken together with the rest of the proof below this shows that the claim of *uniqueness* for the equation (22.2.38) holds in the *class of continuous functions* $h : [0,T] \to \mathbb{R}$ such that $0 \le h(t) \le 1$ for all $0 \le t \le T$.

4°. Let us consider the stopping time

$$
\tau_h = \inf\{\, 0 \le s \le T-t : \pi_{t+s} \ge h(t+s) \,\}. \tag{22.2.59}
$$

Observe that, by virtue of (22.2.51), the identity (22.2.45) can be written as

$$V^h(t+s, \pi_{t+s}) = V^h(t,\pi) - c\int_0^s \pi_{t+u}\, I(\pi_{t+u} < h(t+u))\, du \qquad (22.2.60)$$
$$- \lambda \int_0^s (1-\pi_{t+u})\, I(\pi_{t+u} > h(t+u))\, du + M_s^h$$

with $(M_s^h)_{0\le s\le T-t}$ being a martingale under $\mathsf{P}_{t,\pi}$. Thus, inserting τ_h into (22.2.60) in place of s and taking the $\mathsf{P}_{t,\pi}$-expectation, by means of the optional sampling theorem (page 60) we get

$$V^h(t,\pi) = \mathsf{E}_{t,\pi}\left[G(\pi_{t+\tau_h}) + c\int_0^{\tau_h} \pi_{t+u}\, du\right] \qquad (22.2.61)$$

for all $(t,\pi) \in [0,T)\times[0,1]$. Then comparing (22.2.61) with (22.2.1) we see that

$$V(t,\pi) \le V^h(t,\pi) \qquad (22.2.62)$$

for all $(t,\pi) \in [0,T)\times[0,1]$.

5°. Let us now show that $h \le g$ on $[0,T]$. For this, recall that by the same arguments as for V^h we also have

$$V(t+s, \pi_{t+s}) = V(t,\pi) - c\int_0^s \pi_{t+u}\, I(\pi_{t+u} < g(t+u))\, du \qquad (22.2.63)$$
$$- \lambda \int_0^s (1-\pi_{t+u})\, I(\pi_{t+u} > g(t+u))\, du + M_s^g$$

where $(M_s^g)_{0\le s\le T-t}$ is a martingale under $\mathsf{P}_{t,\pi}$. Fix some (t,π) such that $\pi > g(t)\vee h(t)$ and consider the stopping time

$$\sigma_g = \inf\{\, 0 \le s \le T-t : \pi_{t+s} \le g(t+s)\,\}. \qquad (22.2.64)$$

Inserting σ_g into (22.2.60) and (22.2.63) in place of s and taking the $\mathsf{P}_{t,\pi}$-expectation, by means of the optional sampling theorem (page 60) we get

$$\mathsf{E}_{t,\pi}\left[V^h(t+\sigma_g, \pi_{t+\sigma_g}) + c\int_0^{\sigma_g} \pi_{t+u}\, du\right] \qquad (22.2.65)$$
$$= G(\pi) + \mathsf{E}_{t,\pi}\left[\int_0^{\sigma_g} (c\pi_{t+u} - \lambda(1-\pi_{t+u}))\, I(\pi_{t+u} > h(t+u))\, du\right],$$
$$\mathsf{E}_{t,\pi}\left[V(t+\sigma_g, \pi_{t+\sigma_g}) + c\int_0^{\sigma_g} \pi_{t+u}\, du\right] \qquad (22.2.66)$$
$$= G(\pi) + \mathsf{E}_{t,\pi}\left[\int_0^{\sigma_g} (c\pi_{t+u} - \lambda(1-\pi_{t+u}))\, du\right].$$

Hence by means of (22.2.62) we see that

$$\begin{aligned}\mathsf{E}_{t,\pi}&\left[\int_0^{\sigma_g}(c\pi_{t+u}-\lambda(1-\pi_{t+u}))\,I(\pi_{t+u}>h(t+u))\,du\right]\\&\ge\mathsf{E}_{t,\pi}\left[\int_0^{\sigma_g}(c\pi_{t+u}-\lambda(1-\pi_{t+u}))\,du\right]\end{aligned}\tag{22.2.67}$$

from where, by virtue of the continuity of h and g on $(0,T)$ and the first inequality in (22.2.29), it readily follows that $h(t)\le g(t)$ for all $0\le t\le T$.

6°. Finally, we show that h coincides with g. For this, let us assume that there exists some $t\in(0,T)$ such that $h(t)<g(t)$ and take an arbitrary π from $(h(t),g(t))$. Then inserting $\tau_*=\tau_*(t,\pi)$ from (22.2.27) into (22.2.60) and (22.2.63) in place of s and taking the $\mathsf{P}_{t,\pi}$-expectation, by means of the optional sampling theorem (page 60) we get

$$\begin{aligned}\mathsf{E}_{t,\pi}&\left[G(\pi_{t+\tau_*})+c\int_0^{\tau_*}\pi_{t+u}\,du\right]=V^h(t,\pi)\\&+\mathsf{E}_{t,\pi}\left[\int_0^{\tau_*}(c\pi_{t+u}-\lambda(1-\pi_{t+u}))\,I(\pi_{t+u}>h(t+u))\,du\right],\end{aligned}\tag{22.2.68}$$

$$\mathsf{E}_{t,\pi}\left[G(\pi_{t+\tau_*})+c\int_0^{\tau_*}\pi_{t+u}\,du\right]=V(t,\pi).\tag{22.2.69}$$

Hence by means of (22.2.62) we see that

$$\mathsf{E}_{t,\pi}\left[\int_0^{\tau_*}(c\pi_{t+u}-\lambda(1-\pi_{t+u}))\,I(\pi_{t+u}>h(t+u))\,du\right]\le 0\tag{22.2.70}$$

which is clearly impossible by the continuity of h and g and the fact that $h\ge\lambda/(\lambda+c)$ on $[0,T]$. We may therefore conclude that V^h defined in (22.2.42) coincides with V from (22.2.1) and h is equal to g. This completes the proof of the theorem. □

Remark 22.3. Note that without loss of generality it can be assumed that $\mu>0$ in (22.0.2)–(22.0.3). In this case the optimal stopping time (22.2.37) can be equivalently written as follows:

$$\tau_*=\inf\{\,0\le t\le T: X_t\ge b^\pi(t,X_0^t)\,\}\tag{22.2.71}$$

where we set

$$\begin{aligned}b^\pi(t,X_0^t)=\frac{\sigma^2}{\mu}\log&\frac{g(t)/(1-g(t))}{\pi/(1-\pi)+\lambda\int_0^t e^{-\lambda s}e^{-\frac{\mu}{\sigma^2}(X_s-\frac{\mu s}{2})}\,ds}\\&+\left(\frac{\mu}{2}-\frac{\lambda\sigma^2}{\mu}\right)t\end{aligned}\tag{22.2.72}$$

for $(t,\pi) \in [0,T] \times [0,1]$ and X_0^t denotes the sample path $s \mapsto X_s$ for $s \in [0,t]$. The result proved above shows that the following sequential procedure is optimal: *Observe* X_t *for* $t \in [0,T]$ *and stop the observation as soon as* X_t *becomes greater than* $b^\pi(t, X_0^t)$ *for some* $t \in [0,T]$. *Then conclude that the drift has been changed from* 0 *to* μ.

Remark 22.4. In the preceding procedure we need to know the boundary b^π i.e. the boundary g. We proved above that g is a unique solution of the equation (22.2.38). This equation cannot be solved analytically but can be dealt with numerically. The following simple method can be used to illustrate the latter (better methods are needed to achieve higher precision around the singularity point $t = T$ and to increase the speed of calculation). See also paragraph 3 of Section 27 below for further remarks on numerics.

Set $t_k = kh$ for $k = 0, 1, \ldots, n$ where $h = T/n$ and denote

$$J(t, g(t)) = (1 - g(t))\big(1 - e^{-\lambda(T-t)}\big), \tag{22.2.73}$$

$$\begin{aligned} &K(t, g(t); t+u, g(t+u)) \\ &= \mathsf{E}_{t,g(t)}\big[c\pi_{t+u} I(\pi_{t+u} < g(t+u)) + \lambda(1-\pi_{t+u}) I(\pi_{t+u} > g(t+u))\big] \end{aligned} \tag{22.2.74}$$

upon recalling the explicit expressions (22.2.40) and (22.2.41) above. Then the following discrete approximation of the integral equation (22.2.38) is valid:

$$J(t_k, g(t_k)) = \sum_{l=k}^{n-1} K(t_k, g(t_k); t_{l+1}, g(t_{l+1}))\, h \tag{22.2.75}$$

for $k = 0, 1, \ldots, n-1$. Setting $k = n-1$ and $g(t_n) = \lambda/(\lambda+c)$ we can solve the equation (22.2.75) numerically and get a number $g(t_{n-1})$. Setting $k = n-2$ and using the values $g(t_{n-1})$, $g(t_n)$ we can solve (22.2.75) numerically and get a number $g(t_{n-2})$. Continuing the recursion we obtain $g(t_n), g(t_{n-1}), \ldots, g(t_1), g(t_0)$ as an approximation of the optimal boundary g at the points $T, T-h, \ldots, h, 0$ (cf. *Figure* VI.2).

8. Solution of the variational problem. In the variational problem with finite horizon (see [196, Chap. IV, Sect. 3–4] for the infinite horizon case) it is assumed that we observe a trajectory of the Wiener process (Brownian motion) $X = (X_t)_{0 \le t \le T}$ with a drift changing from 0 to $\mu \neq 0$ at some random time θ taking the value 0 with probability π and being exponentially distributed with parameter $\lambda > 0$ given that $\theta > 0$. (A more natural hypothesis may be that θ is uniformly distributed on $[0,T]$.)

1°. Adopting the setting and notation of paragraph 1 above, let $\mathfrak{M}(\alpha, \pi, T)$ denote the class of stopping times τ of X satisfying $0 \le \tau \le T$ and

$$\mathsf{P}_\pi(\tau < \theta) \le \alpha \tag{22.2.76}$$

where $0 \le \alpha \le 1$ and $0 \le \pi \le 1$ are given and fixed. The variational problem seeks to determine a stopping time $\widehat{\tau}$ in the class $\mathfrak{M}(\alpha,\pi,T)$ such that

$$\mathsf{E}_\pi[\widehat{\tau}-\theta]^+ \le \mathsf{E}_\pi[\tau-\theta]^+ \tag{22.2.77}$$

for any other stopping time τ from $\mathfrak{M}(\alpha,\pi,T)$. The stopping time $\widehat{\tau}$ is then said to be optimal in the variational problem (22.2.76)–(22.2.77).

2°. To solve the variational problem (22.2.76)–(22.2.77) we will follow the train of thought from [196, Chap. IV, Sect. 3] which is based on exploiting the solution of the Bayesian problem found in Theorem 22.2 above. For this, let us first note that if $\alpha \ge 1-\pi$, then letting $\widehat{\tau} \equiv 0$ we see that $\mathsf{P}_\pi(\widehat{\tau}<\theta) = \mathsf{P}_\pi(0<\theta) = 1-\pi \le \alpha$ and clearly $\mathsf{E}_\pi[\widehat{\tau}-\theta]^+ = \mathsf{E}_\pi[-\theta]^+ = 0 \le \mathsf{E}[\tau-\theta]^+$ for every $\tau \in \mathfrak{M}(\alpha,\pi,T)$ showing that $\widehat{\tau}\equiv 0$ is optimal in (22.2.76)–(22.2.77). Similarly, if $\alpha = e^{-\lambda T}(1-\pi)$, then letting $\widehat{\tau}\equiv T$ we see that $\mathsf{P}_\pi(\widehat{\tau}<\theta) = \mathsf{P}_\pi(T<\theta) = e^{-\lambda T}(1-\pi) = \alpha$ and clearly $\mathsf{E}_\pi[\widehat{\tau}-\theta]^+ = \mathsf{E}_\pi[T-\theta]^+ \le \mathsf{E}[\tau-\theta]^+$ for every $\tau \in \mathfrak{M}(\alpha,\pi,T)$ showing that $\widehat{\tau}\equiv T$ is optimal in (22.2.76)–(22.2.77). The same argument also shows that $\mathfrak{M}(\alpha,\pi)$ is empty if $\alpha < e^{-\lambda T}(1-\pi)$. We may thus conclude that the set of admissible α which lead to a nontrivial optimal stopping time $\widehat{\tau}$ in (22.2.76)–(22.2.77) equals $(e^{-\lambda T}(1-\pi), 1-\pi)$ where $\pi \in [0,1)$.

3°. To describe the key technical points in the argument below leading to the solution of (22.2.76)–(22.2.77), let us consider the optimal stopping problem (22.2.1) with $c>0$ given and fixed. In this context set $V(t,\pi)=V(t,\pi;c)$ and $g(t)=g(t;c)$ to indicate the dependence on c and recall that $\tau_* = \tau_*(c)$ given in (22.2.37) is an optimal stopping time in (22.2.1). We then have:

$$g(t;c) \le g(t;c') \quad \text{for all } t\in[0,T] \text{ if } c>c', \tag{22.2.78}$$

$$g(t;c)\uparrow 1 \quad \text{if } c\downarrow 0 \text{ for each } t\in[0,T], \tag{22.2.79}$$

$$g(t;c)\downarrow 0 \quad \text{if } c\uparrow\infty \text{ for each } t\in[0,T]. \tag{22.2.80}$$

To verify (22.2.78) let us assume that $g(t;c) > g(t;c')$ for some $t\in[0,T)$ and $c>c'$. Then for any $\pi \in (g(t;c'), g(t;c))$ given and fixed we have $V(t,\pi;c) < 1-\pi = V(t,\pi;c')$ contradicting the obvious fact that $V(t,\pi;c) \ge V(t,\pi;c')$ as it is clearly seen from (22.2.1). The relations (22.2.79) and (22.2.80) are verified in a similar manner.

4°. Finally, to exhibit the optimal stopping time $\widehat{\tau}$ in (22.2.76)–(22.2.77) when $\alpha \in (e^{-\lambda T}(1-\pi), 1-\pi)$ and $\pi\in[0,1)$ are given and fixed, let us introduce the function

$$u(c;\pi) = \mathsf{P}_\pi(\tau_* < \theta) \tag{22.2.81}$$

for $c>0$ where $\tau_* = \tau_*(c)$ from (22.2.37) is an optimal stopping time in (22.0.5). Using that $\mathsf{P}_\pi(\tau_*<\theta) = \mathsf{E}_\pi[1-\pi_{\tau_*}]$ and (22.2.78) above it is readily verified that $c\mapsto u(c;\pi)$ is continuous and strictly increasing on $(0,\infty)$. [Note that a strict increase follows from the fact that $g(T;c) = \lambda/(\lambda+c)$.] From (22.2.79) and

(22.2.80) we moreover see that $u(0+;\pi) = e^{-\lambda T}(1-\pi)$ due to $\tau_*(0+) \equiv T$ and $u(+\infty;\pi) = 1-\pi$ due to $\tau_*(+\infty) \equiv 0$. This implies that the equation

$$u(c;\pi) = \alpha \tag{22.2.82}$$

has a unique root $c = c(\alpha)$ in $(0,\infty)$.

5°. The preceding conclusions can now be used to formulate the main result of this paragraph.

Theorem 22.5. *In the variational formulation of the Wiener disorder problem* (22.2.76)–(22.2.77) *there exists a nontrivial optimal stopping time* $\widehat{\tau}$ *if and only if*

$$\alpha \in (e^{-\lambda T}(1-\pi), 1-\pi) \tag{22.2.83}$$

where $\pi \in [0,1)$. *In this case* $\widehat{\tau}$ *may be explicitly identified with* $\tau_* = \tau_*(c)$ *in* (22.2.37) *where* $g(t) = g(t;c)$ *is the unique solution of the integral equation* (22.2.38) *and* $c = c(\alpha)$ *is a unique root of the equation* (22.2.82) *on* $(0,\infty)$.

Proof. It remains to show that $\widehat{\tau} = \tau_*(c)$ with $c = c(\alpha)$ and $\alpha \in (e^{-\lambda T}(1-\pi), 1-\pi)$ for $\pi \in [0,1)$ satisfies (22.2.77). For this note that since $\mathsf{P}_\pi(\widehat{\tau} < \theta) = \alpha$ by construction, it follows by the optimality of $\tau_*(c)$ in (22.0.4) that

$$\alpha + c\mathsf{E}_\pi[\widehat{\tau} - \theta]^+ \le \mathsf{P}_\pi(\tau < \theta) + c\mathsf{E}_\pi[\tau - \theta]^+ \tag{22.2.84}$$

for any other stopping time τ with values in $[0,T]$. Moreover, if τ belongs to $\mathfrak{M}(\alpha,\pi)$, then $\mathsf{P}_\pi(\tau < \theta) \le \alpha$ and from (22.2.84) we see that $\mathsf{E}_\pi[\widehat{\tau} - \theta]^+ \le \mathsf{E}_\pi[\tau - \theta]^+$ establishing (22.2.77). The proof is complete. □

Remark 22.6. Recall from part (iv) of paragraph 2 above that $g(t;c) \le A_*(c)$ for all $0 \le t \le T$ where $0 < A_*(c) < 1$ is uniquely determined from the equation (22.1.13) (or (4.147) in [196, p. 201]). Since $A_*(c(\alpha)) = 1-\alpha$ by Theorem 10 in [196, p. 205] it follows that the optimal stopping boundary $t \mapsto g(t;c(\alpha))$ in (22.2.76)–(22.2.77) satisfies $g(t;c(\alpha)) \le 1-\alpha$ for all $0 \le t \le T$.

9. Appendix. In this appendix we exhibit an explicit expression for the transition density function of the a posteriori probability process $(\pi_t)_{0\le t\le T}$ given in (22.0.14)–(22.0.16) above.

1°. Let $B = (B_t)_{t\ge 0}$ be a standard Wiener process defined on a probability space $(\Omega, \mathcal{F}, \mathsf{P})$. With $t > 0$ and $\nu \in \mathbb{R}$ given and fixed recall from [224, p. 527] that the random variable $A_t^{(\nu)} = \int_0^t e^{2(B_s+\nu s)}\,ds$ has the conditional distribution:

$$\mathsf{P}\big(A_t^{(\nu)} \in dz \mid B_t + \nu t = y\big) = a(t,y,z)\,dz \tag{22.2.85}$$

where the density function a for $z > 0$ is given by

$$\begin{aligned} a(t,y,z) = {} & \frac{1}{\pi z^2} \exp\left(\frac{y^2+\pi^2}{2t} + y - \frac{1}{2z}\Big(1+e^{2y}\Big)\right) \\ & \times \int_0^\infty \exp\left(-\frac{w^2}{2t} - \frac{e^y}{z}\cosh(w)\right)\sinh(w)\,\sin\Big(\frac{\pi w}{t}\Big)\,dw. \end{aligned} \tag{22.2.86}$$

This implies that the random vector $\big(2(B_t+\nu t), A_t^{(\nu)}\big)$ has the distribution

$$\mathsf{P}\Big(2(B_t+\nu t)\in dy, A_t^{(\nu)}\in dz\Big) = b(t,y,z)\,dy\,dz \tag{22.2.87}$$

where the density function b for $z>0$ is given by

$$\begin{aligned} b(t,y,z) &= a\left(t,\frac{y}{2},z\right)\frac{1}{2\sqrt{t}}\varphi\left(\frac{y-2\nu t}{2\sqrt{t}}\right) \\ &= \frac{1}{(2\pi)^{3/2}z^2\sqrt{t}}\exp\left(\frac{\pi^2}{2t}+\Big(\frac{\nu+1}{2}\Big)y-\frac{\nu^2}{2}t-\frac{1}{2z}\Big(1+e^y\Big)\right) \\ &\quad\times\int_0^\infty \exp\left(-\frac{w^2}{2t}-\frac{e^{y/2}}{z}\cosh(w)\right)\sinh(w)\,\sin\left(\frac{\pi w}{t}\right)dw \end{aligned} \tag{22.2.88}$$

and we set $\varphi(x)=(1/\sqrt{2\pi})e^{-x^2/2}$ for $x\in\mathbb{R}$ (for related expressions in terms of Hermite functions see [46] and [181]).

Denoting $I_t=\alpha B_t+\beta t$ and $J_t=\int_0^t e^{\alpha B_s+\beta s}\,ds$ with $\alpha\neq 0$ and $\beta\in\mathbb{R}$ given and fixed, and using that the scaling property of B implies

$$\begin{aligned} &\mathsf{P}\left(\alpha B_t+\beta t\le y, \int_0^t e^{\alpha B_s+\beta s}\,ds\le z\right) \\ &\qquad = \mathsf{P}\left(2(B_{t'}+\nu t')\le y, \int_0^{t'} e^{2(B_s+\nu s)}\,ds\le\frac{\alpha^2}{4}z\right) \end{aligned} \tag{22.2.89}$$

with $t'=\alpha^2 t/4$ and $\nu=2\beta/\alpha^2$, it follows by applying (22.2.87) and (22.2.88) that the random vector (I_t,J_t) has the distribution

$$\mathsf{P}\big(I_t\in dy, J_t\in dz\big) = f(t,y,z)\,dy\,dz \tag{22.2.90}$$

where the density function f for $z>0$ is given by

$$\begin{aligned} f(t,y,z) &= \frac{\alpha^2}{4}\,b\left(\frac{\alpha^2}{4}t,\,y,\,\frac{\alpha^2}{4}z\right) \\ &= \frac{2\sqrt{2}}{\pi^{3/2}\alpha^3}\,\frac{1}{z^2\sqrt{t}}\exp\left\{\frac{2\pi^2}{\alpha^2 t}+\Big(\frac{\beta}{\alpha^2}+\frac{1}{2}\Big)y-\frac{\beta^2}{2\alpha^2}t-\frac{2}{\alpha^2 z}(1+e^y)\right\} \\ &\quad\times\int_0^\infty \exp\left(-\frac{2w^2}{\alpha^2 t}-\frac{4e^{y/2}}{\alpha^2 z}\cosh(w)\right)\sinh(w)\,\sin\left(\frac{4\pi w}{\alpha^2 t}\right)dw. \end{aligned} \tag{22.2.91}$$

2°. Letting $\alpha=-\mu/\sigma$ and $\beta=-\lambda-\mu^2/(2\sigma^2)$ it follows from the explicit expressions (22.0.9)+(22.0.11) and (22.0.3) that

$$\mathsf{P}^0(\varphi_t\in dx) = \mathsf{P}\Big(e^{-I_t}\Big(\frac{\pi}{1-\pi}+\lambda J_t\Big)\in dx\Big) = g(\pi;t,x)\,dx \tag{22.2.92}$$

where the density function g for $x > 0$ is given by

$$g(\pi;t,x) = \frac{d}{dx}\int_{-\infty}^{\infty}\int_0^{\infty} I\Big(e^{-y}\Big(\frac{\pi}{1-\pi}+\lambda z\Big) \le x\Big) f(t,y,z)\,dy\,dz \tag{22.2.93}$$
$$= \int_{-\infty}^{\infty} f\Big(t,y,\frac{1}{\lambda}\Big(xe^{y}-\frac{\pi}{1-\pi}\Big)\Big)\frac{e^{y}}{\lambda}\,dy.$$

Moreover, setting

$$\widetilde{I}_{t-s} = \alpha(B_t - B_s) + \beta(t-s) \quad \text{and} \quad \widetilde{J}_{t-s} = \int_s^t e^{\alpha(B_u-B_s)+\beta(u-s)}\,du \tag{22.2.94}$$

as well as $\widehat{I}_s = \alpha B_s + \widehat{\beta} s$ and $\widehat{J}_s = \int_0^s e^{\alpha B_u + \widehat{\beta} u} du$ with $\widehat{\beta} = -\lambda + \mu^2/(2\sigma^2)$, it follows from the explicit expressions (22.0.9)+(22.0.11) and (22.0.3) that

$$\begin{aligned}
&\mathsf{P}^s(\varphi_t \in dx) \\
&= \mathsf{P}\Big(e^{-\gamma s}e^{-\widetilde{I}_{t-s}}\big(e^{(\widehat{\beta}-\beta)s}e^{-\widehat{I}_s}\big(\tfrac{\pi}{1-\pi}+\lambda\widehat{J}_s\big)+\lambda e^{\gamma s}\widetilde{J}_{t-s}\big) \in dx\Big) \\
&= h(s;\pi;t,x)\,dx
\end{aligned} \tag{22.2.95}$$

for $0 < s < t$ where $\gamma = \mu^2/\sigma^2$. Since stationary independent increments of B imply that the random vector $(\widetilde{I}_{t-s}, \widetilde{J}_{t-s})$ is independent of $(\widehat{I}_s, \widehat{J}_s)$ and equally distributed as (I_{t-s}, J_{t-s}), we see upon recalling (22.2.92)–(22.2.93) that the density function h for $x > 0$ is given by

$$h(s;\pi;t,x) = \frac{d}{dx}\int_{-\infty}^{\infty}\int_0^{\infty}\int_0^{\infty} I\Big(e^{-\gamma s}e^{-y}\big(e^{(\widehat{\beta}-\beta)s}w+\lambda e^{\gamma s}z\big) \le x\Big) \tag{22.2.96}$$
$$\times f(t-s,y,z)\,\widehat{g}(\pi;s,w)\,dy\,dz\,dw$$
$$= \int_{-\infty}^{\infty}\int_0^{\infty} f\Big(t-s,y,\frac{xe^{y}-e^{(\widehat{\beta}-\beta-\gamma)s}w}{\lambda}\Big)\,\widehat{g}(\pi;s,w)\,\frac{e^{y}}{\lambda}\,dy\,dw$$

where the density function $\widehat{g}$ for $w > 0$ equals

$$\widehat{g}(\pi;s,w) = \frac{d}{dx}\int_{-\infty}^{\infty}\int_0^{\infty} I\Big(e^{-y}\Big(\frac{\pi}{1-\pi}+\lambda z\Big) \le w\Big)\widehat{f}(s,y,z)\,dy\,dz \tag{22.2.97}$$
$$= \int_{-\infty}^{\infty} \widehat{f}\Big(s,y,\frac{1}{\lambda}\Big(we^{y}-\frac{\pi}{1-\pi}\Big)\Big)\frac{e^{y}}{\lambda}\,dy$$

and the density function $\widehat{f}$ for $z > 0$ is defined as in (22.2.90)–(22.2.91) with $\widehat{\beta}$ instead of β.

Finally, by means of the same arguments as in (22.2.92)–(22.2.93) it follows from the explicit expressions (22.0.9)+(22.0.11) and (22.0.3) that

$$\mathsf{P}^t(\varphi_t \in dx) = \mathsf{P}\Big(e^{-\widehat{I}_t}\Big(\frac{\pi}{1-\pi}+\lambda\widehat{J}_t\Big) \in dx\Big) = \widehat{g}(\pi;t,x)\,dx \tag{22.2.98}$$

where the density function $\widehat{g}$ for $x > 0$ is given by (22.2.97).

3°. Noting by (22.0.1) that

$$\begin{aligned} \mathsf{P}_\pi(\varphi_t \in dx) = \pi \mathsf{P}^0(\varphi_t \in dx) + (1-\pi) \int_0^t \lambda e^{-\lambda s} \mathsf{P}^s(\varphi_t \in dx)\, ds \\ + (1-\pi)\, e^{-\lambda t} \mathsf{P}^t(\varphi_t \in dx) \end{aligned} \tag{22.2.99}$$

we see by (22.2.92)+(22.2.95)+(22.2.98) that the process $(\varphi_t)_{0 \le t \le T}$ has the marginal distribution

$$\mathsf{P}_\pi(\varphi_t \in dx) = q(\pi; t, x)\, dx \tag{22.2.100}$$

where the transition density function q for $x > 0$ is given by

$$\begin{aligned} q(\pi; t, x) = \pi\, g(\pi; t, x) + (1-\pi) \int_0^t \lambda e^{-\lambda s}\, h(s; \pi; t, x)\, ds \\ + (1-\pi)\, e^{-\lambda t}\, \widehat{g}(\pi; t, x) \end{aligned} \tag{22.2.101}$$

with g, h, $\widehat{g}$ from (22.2.93), (22.2.96), (22.2.97) respectively.

Hence by (22.0.14) we easily find that the process $(\pi_t)_{0 \le t \le T}$ has the marginal distribution

$$\mathsf{P}_\pi(\pi_t \in dx) = p(\pi; t, x)\, dx \tag{22.2.102}$$

where the transition density function p for $0 < x < 1$ is given by

$$p(\pi; t, x) = \frac{1}{(1-x)^2}\, q\Big(\pi; t, \frac{x}{1-x}\Big). \tag{22.2.103}$$

This completes the Appendix.

Notes. The quickest detection problems considered in this chapter belong to the class of the "disorder/change point" problems that can be described as follows.

We have two "statistically different" processes $X^1 = (X_t^1)_{t \ge 0}$ and $X^2 = (X_t^2)_{t \ge 0}$ that form the observable process $X = (X_t)_{t \ge 0}$ as

$$X_t = \begin{cases} X_t^1 & \text{if } t < \theta, \\ X_{t-\theta}^2 & \text{if } t \ge \theta \end{cases} \tag{22.2.104}$$

where θ is either a random variable or an unknown parameter that we want to estimate on basis of the observations of X. There are two formulations of the problem:

(a) If we have *all observations* of X_t on an admissible time interval $[0, T]$ and we try to construct an $\mathcal{F}_T^X$-measurable estimate $\hat{\theta} = \hat{\theta}(X_s, s \le T)$ of θ, then we speak of a *"change-point"* problem. It is clear that this "a posteriori" problem is essentially a classical estimation problem of mathematical statistics.

(b) We speak of a "disorder" problem if observations are arriving sequentially in time and we want to construct an alarm time τ (i.e. stopping time) that in some sense is "as close as possible" to the disorder time θ (when a "good" process X^1 turns into a "bad" process X^2).

Mathematical formulations of the "disorder" problem first of all depend on assumptions about the disorder time θ. *Parametric* formulations assume simply that θ is an unknown (unobservable) parameter taking values in a subset of $\mathbb{R}_+$. *Bayesian* formulations (which we address in the present monograph) assume that θ is a random variable with distribution F_θ. In our text we suppose that F_θ is an exponential distribution on $[0,\infty)$ and we consider two formulations of the "quickest detection" problem: "infinite horizon" and "finite horizon". In the first case we admit for τ all values from $\mathbb{R}_+ = [0,\infty)$. In the second case we admit for τ only values from the time interval $[0,T]$ (it explains the terminology "finite horizon").

We refer to [113] for many theoretical investigations of the "disorder/change point" problems as well as for a number of important applications (detection of "breaks" in geological data; quickest detection of the beginning of earthquakes, tsunamis, and general "spontaneously appearing effects", see also [26]). Applications in financial data analysis (detection of arbitrage) are recently discussed in [198]. For quickest detection problems with exponential penalty for delay see [173] and [12]. See also [200] for the criterion $\inf_\tau \mathsf{E}|\tau - \theta|$.

From the standpoint of applications it is also interesting to consider problems where the disorder appears on a finite time interval or a decision should be made before a certain finite time. Similar to the problems of testing statistical hypotheses on finite time intervals, the corresponding quickest detection problems in the *finite horizon* formulation are more difficult than in the case of infinite horizon (because additional "sufficient" statistics time t should be taken into account for finite horizon problems). Clearly, among all processes that can be considered in the problem, the Wiener process and the Poisson process take a central place. Once these problems are understood sufficiently well, the study of problems including other processes may follow a similar line of arguments.

Shiryaev in [188, 1961] (see also [187], [189]–[193], [196, Chap. IV]) derived an explicit solution of the Bayesian and variational problem for a Wiener process with infinite horizon by reducing the initial optimal stopping problem to a free-boundary problem for a differential operator (see also [208]). Some particular cases of the Bayesian problem for a Poisson process with infinite horizon were solved by Gal'chuk and Rozovskii [73] and Davis [35]. A complete solution of the latter problem was given in [169] by reducing the initial optimal stopping problem to a free-boundary problem for a differential-difference operator (see Subsection 24.1 below). The main aim of Subsection 22.2 above (following [72]) is to derive a solution of the Bayesian and variational problem for a Wiener process with finite horizon.

23. Sequential testing of a Poisson process

In this section we continue our study of sequential testing problems considered in Section 21 above. Instead of the Wiener process we now deal with the Poisson process.

23.1. Infinite horizon

1. Description of the problem. Suppose that at time $t=0$ we begin to observe a Poisson process $X = (X_t)_{t\geq 0}$ with intensity $\lambda > 0$ which is either λ_0 or λ_1 where $\lambda_0 < \lambda_1$. Assuming that the true value of λ is not known to us, our problem is then to decide as soon as possible and with a minimal error probability (both specified later) if the true value of λ is either λ_0 or λ_1.

Depending on the hypotheses about the unknown intensity λ, this problem admits two formulations. The *Bayesian* formulation relies upon the hypothesis that an a priori probability distribution of λ is given to us, and that λ takes either of the values λ_0 and λ_1 at time $t=0$ according to this distribution. The *variational* formulation (sometimes also called a *fixed error probability* formulation) involves no probabilistic assumptions on the unknown intensity λ.

2. Solution of the Bayesian problem. In the Bayesian formulation of the problem (see [196, Chap. 4]) it is assumed that at time $t=0$ we begin observing a trajectory of the *point* process $X=(X_t)_{t\geq 0}$ with the compensator $A=(A_t)_{t\geq 0}$ (see [128, Chap. 18]) where $A_t = \lambda t$ and a random intensity $\lambda = \lambda(\omega)$ takes two values λ_1 and λ_0 with probabilities π and $1-\pi$. (We assume that $\lambda_1 > \lambda_0 > 0$ and $\pi \in [0,1]$.)

2.1. For a precise probability-statistical description of the Bayesian sequential testing problem it is convenient to assume that all our considerations take place on a probability-statistical space $(\Omega, \mathcal{F}; \mathsf{P}_\pi, \pi \in [0,1])$ where P_π has the special structure

$$\mathsf{P}_\pi = \pi \mathsf{P}_1 + (1-\pi)\mathsf{P}_0 \tag{23.1.1}$$

for $\pi \in [0,1]$. We further assume that the $\mathcal{F}_0$-measurable random variable $\lambda = \lambda(\omega)$ takes two values λ_1 and λ_0 with probabilities $\mathsf{P}_\pi(\lambda=\lambda_1) = \pi$ and $\mathsf{P}_\pi(\lambda=\lambda_0) = 1-\pi$. Concerning the observable point process $X = (X_t)_{t\geq 0}$, we assume that $\mathsf{P}_\pi(X \in \cdot \mid \lambda = \lambda_i) = \mathsf{P}_i(X \in \cdot\,)$, where $\mathsf{P}_i(X \in \cdot\,)$ coincides with the distribution of a Poisson process with intensity λ_i for $i=0,1$.

Probabilities π and $1-\pi$ play a role of *a priori* probabilities of the statistical hypotheses

$$H_1\colon\ \lambda = \lambda_1, \tag{23.1.2}$$

$$H_0\colon\ \lambda = \lambda_0. \tag{23.1.3}$$

2.2. Based upon information which is continuously updated through observation of the point process X, our problem is to test sequentially the hypotheses

H_1 and H_0. For this it is assumed that we have at our disposal a class of sequential *decision rules* (τ, d) consisting of *stopping times* $\tau = \tau(\omega)$ with respect to $(\mathcal{F}_t^X)_{t\ge 0}$ where $\mathcal{F}_t^X = \sigma\{X_s : s \le t\}$, and $\mathcal{F}_\tau^X$-measurable functions $d = d(\omega)$ which take values 0 and 1. Stopping the observation of X at time τ, the *terminal decision function* d indicates that either the hypothesis H_1 or the hypothesis H_0 should be accepted; if $d = 1$ we accept H_1, and if $d = 0$ we accept that H_0 is true.

2.3. Each decision rule (τ, d) implies losses of two kinds: the loss due to a cost of the observation, and the loss due to a wrong terminal decision. The average loss of the first kind may be naturally identified with $c\,\mathsf{E}_\pi(\tau)$, and the average loss of the second kind can be expressed as $a\,\mathsf{P}_\pi(d=0, \lambda=\lambda_1) + b\,\mathsf{P}_\pi(d=1, \lambda=\lambda_0)$, where $c, a, b > 0$ are some constants. It will be clear from (23.1.8) below that there is no restriction to assume that $c = 1$, as the case of general $c > 0$ follows by replacing a and b with a/c and b/c respectively. Thus, the *total average loss* of the decision rule (τ, d) is given by

$$L_\pi(\tau, d) = \mathsf{E}_\pi\Big(\tau + a\,1_{(d=0,\lambda=\lambda_1)} + b\,1_{(d=1,\lambda=\lambda_0)}\Big). \tag{23.1.4}$$

Our problem is then to compute

$$V(\pi) = \inf_{(\tau,d)} L_\pi(\tau, d) \tag{23.1.5}$$

and to find the optimal decision rule (τ_*, d_*), called *the π-Bayes decision rule,* at which the infimum in (23.1.5) is attained.

Observe that for any decision rule (τ, d) we have

$$a\,\mathsf{P}_\pi(d=0, \lambda=\lambda_1) + b\,\mathsf{P}_\pi(d=1, \lambda=\lambda_0) = a\pi\,\alpha(d) + b(1-\pi)\,\beta(d) \tag{23.1.6}$$

where $\alpha(d) = \mathsf{P}_1(d=0)$ is called *the probability of an error of the first kind,* and $\beta(d) = \mathsf{P}_0(d=1)$ is called *the probability of an error of the second kind.*

2.4. The problem (23.1.5) can be reduced to an optimal stopping problem for the *a posteriori probability process* defined by

$$\pi_t = \mathsf{P}_\pi\big(\lambda = \lambda_1 \mid \mathcal{F}_t^X\big) \tag{23.1.7}$$

with $\pi_0 = \pi$ under P_π. Standard arguments (see [196, pp. 166–167]) show that

$$V(\pi) = \inf_\tau \mathsf{E}_\pi\big(\tau + g_{a,b}(\pi_\tau)\big) \tag{23.1.8}$$

where $g_{a,b}(\pi) = a\pi \wedge b(1-\pi)$ (recall that $x \wedge y = \min\{x, y\}$), the optimal stopping time τ_* in (23.1.8) is also optimal in (23.1.5), and the optimal decision function d_* is obtained by setting

$$d_* = \begin{cases} 1 & \text{if } \pi_{\tau_*} \ge b/(a+b), \\ 0 & \text{if } \pi_{\tau_*} < b/(a+b). \end{cases} \tag{23.1.9}$$

Our main task in the sequel is therefore reduced to solving the optimal stopping problem (23.1.8).

2.5. Another natural process, which is in a one-to-one correspondence with the process $(\pi_t)_{t\ge 0}$, is the *likelihood ratio process*; it is defined as the Radon–Nikodým density

$$\varphi_t = \frac{d(\mathsf{P}_1 \,|\, \mathcal{F}_t^X)}{d(\mathsf{P}_0 \,|\, \mathcal{F}_t^X)} \tag{23.1.10}$$

where $\mathsf{P}_i \,|\, \mathcal{F}_t^X$ denotes the restriction of P_i to $\mathcal{F}_t^X$ for $i = 0, 1$. Since

$$\pi_t = \pi \, \frac{d(\mathsf{P}_1 \,|\, \mathcal{F}_t^X)}{d(\mathsf{P}_\pi \,|\, \mathcal{F}_t^X)} \tag{23.1.11}$$

where $\mathsf{P}_\pi \,|\, \mathcal{F}_t^X = \pi \, \mathsf{P}_1 \,|\, \mathcal{F}_t^X + (1-\pi)\, \mathsf{P}_0 \,|\, \mathcal{F}_t^X$, it follows that

$$\pi_t = \left(\frac{\pi}{1-\pi} \, \varphi_t \right) \Big/ \left(1 + \frac{\pi}{1-\pi} \, \varphi_t \right) \tag{23.1.12}$$

as well as that

$$\varphi_t = \frac{1-\pi}{\pi} \, \frac{\pi_t}{1-\pi_t}. \tag{23.1.13}$$

Moreover, the following explicit expression is known to be valid (see e.g. [51] or [128, Theorem 19.7]):

$$\varphi_t = \exp\left(X_t \log \frac{\lambda_1}{\lambda_0} - (\lambda_1 - \lambda_0)\, t \right). \tag{23.1.14}$$

This representation may now be used to reveal the *Markovian* structure in the problem. Since the process $X = (X_t)_{t\ge 0}$ is a time-homogeneous Markov process having stationary independent increments (Lévy process) under both P_0 and P_1, from the representation (23.1.14), and due to the one-to-one correspondence (23.1.12), we see that $(\varphi_t)_{t\ge 0}$ and $(\pi_t)_{t\ge 0}$ are time-homogeneous Markov processes under both P_0 and P_1 with respect to natural filtrations which clearly coincide with $(\mathcal{F}_t^X)_{t\ge 0}$. Using then further that $\mathsf{E}_\pi(H \mid \mathcal{F}_t^X) = \mathsf{E}_1(H \mid \mathcal{F}_t^X)\, \pi_t + \mathsf{E}_0(H \mid \mathcal{F}_t^X)\,(1-\pi_t)$ for any (bounded) measurable H, it follows that $(\pi_t)_{t\ge 0}$, and thus $(\varphi_t)_{t\ge 0}$ as well, is a time-homogeneous Markov process under each P_π for $\pi \in [0,1]$. (Observe, however, that although the same argument shows that X is a Markov process under each P_π for $\pi \in (0,1)$, it is not a time-homogeneous Markov process unless π equals 0 or 1.) Note also directly from (23.1.7) that $(\pi_t)_{t\ge 0}$ is a martingale under each P_π for $\pi \in [0,1]$. Thus, the optimal stopping problem (23.1.8) falls into the class of optimal stopping problems for Markov processes (cf. Chapter I), and we therefore proceed by finding the infinitesimal operator of $(\pi_t)_{t\ge 0}$. A slight modification of the arguments above shows that all these processes possess a strong Markov property actually.

2.6. By Itô's formula (page 67) one can verify (cf. [106, Ch. I, § 4]) that processes $(\varphi_t)_{t\ge0}$ and $(\pi_t)_{t\ge0}$ solve the following stochastic equations respectively:

$$d\varphi_t = \Big(\frac{\lambda_1}{\lambda_0} - 1\Big)\varphi_{t-}\, d\big(X_t - \lambda_0 t\big), \tag{23.1.15}$$

$$d\pi_t = \frac{(\lambda_1 - \lambda_0)\,\pi_{t-}(1-\pi_{t-})}{\lambda_1\pi_{t-} + \lambda_0(1-\pi_{t-})}\Big(dX_t - \big(\lambda_1\pi_{t-} + \lambda_0(1-\pi_{t-})\big)\,dt\Big) \tag{23.1.16}$$

(cf. formula (19.86) in [128]). The equation (23.1.16) may now be used to determine the infinitesimal operator of the Markov process $(\pi_t, \mathcal{F}_t^X, \mathsf{P}_\pi)_{t\ge0}$ for $\pi\in[0,1]$. For this, let $f = f(\pi)$ from $C^1[0,1]$ be given. Then by Itô's formula (page 67) we find

$$\begin{aligned} f(\pi_t) &= f(\pi_0) \\ &\quad + \int_0^t f'(\pi_{s-})\,d\pi_s + \sum_{0<s\le t}\big(f(\pi_s) - f(\pi_{s-}) - f'(\pi_{s-})\,\Delta\pi_s\big) \\ &= f(\pi_0) + \int_0^t f'(\pi_{s-})\big(-(\lambda_1-\lambda_0)\,\pi_{s-}(1-\pi_{s-})\big)\,ds \\ &\qquad + \sum_{0<s\le t}\big(f(\pi_s) - f(\pi_{s-})\big) \\ &= f(\pi_0) + \int_0^t f'(\pi_{s-})\big(-(\lambda_1-\lambda_0)\,\pi_{s-}(1-\pi_{s-})\big)\,ds \\ &\qquad + \int_0^t\!\!\int_0^1 \big(f(\pi_{s-}+y) - f(\pi_{s-})\big)\,\mu^\pi(ds,dy) \\ &= f(\pi_0) + \int_0^t f'(\pi_{s-})\big(-(\lambda_1-\lambda_0)\,\pi_{s-}(1-\pi_{s-})\big)\,ds \\ &\qquad + \int_0^t\!\!\int_0^1 \big(f(\pi_{s-}+y) - f(\pi_{s-})\big)\,\nu^\pi(ds,dy) \\ &\qquad + \int_0^t\!\!\int_0^1 \big(f(\pi_{s-}+y) - f(\pi_{s-})\big)\,\big(\mu^\pi(ds,dy) - \nu^\pi(ds,dy)\big) \\ &= f(\pi_0) + \int_0^t (\mathbb{L}f)(\pi_{s-})\,ds + M_t \end{aligned} \tag{23.1.17}$$

where μ^π is the random measure of jumps of the process $(\pi_t)_{t\ge0}$ and ν^π is a compensator of μ^π (see e.g. [129, Chap. 3] or [106, Chap. II]), the operator $\mathbb{L}$ is given as in (23.1.19) below, and $M = (M_t)_{t\ge0}$ defined as

$$M_t = \int_0^t\!\!\int_0^1 \big(f(\pi_{s-}+y) - f(\pi_{s-})\big)\,\big(\mu^\pi(ds,dy) - \nu^\pi(ds,dy)\big) \tag{23.1.18}$$

is a local martingale with respect to $(\mathcal{F}_t^X)_{t\ge0}$ and P_π for every $\pi \in [0,1]$. It follows from (23.1.17) that the infinitesimal operator of $(\pi_t)_{t\ge0}$ acts on $f \in C^1[0,1]$ like

$$(\mathbb{L}f)(\pi) = -(\lambda_1-\lambda_0)\pi(1-\pi)f'(\pi) + \big(\lambda_1\pi+\lambda_0(1-\pi)\big)\left(f\left(\frac{\lambda_1\pi}{\lambda_1\pi+\lambda_0\,(1-\pi)}\right)-f(\pi)\right). \tag{23.1.19}$$

2.7. Looking back at (23.1.5) and using explicit expressions (23.1.4) and (23.1.6) with (23.1.1), it is easily verified (cf. [123, p. 105]) that the value function $\pi \mapsto V(\pi)$ is concave on $[0,1]$, and thus it is continuous on $(0,1)$. Evidently, this function is pointwise dominated by $\pi \mapsto g_{a,b}(\pi)$. From these facts and from the general theory of optimal stopping for Markov processes (cf. Chapter I) we may guess that the value function $\pi \mapsto V(\pi)$ from (23.1.8) should solve the following *free-boundary problem* (for a differential-difference equation defined by the infinitesimal operator):

$$(\mathbb{L}V)(\pi) = -1, \quad A_* < \pi < B_*, \tag{23.1.20}$$

$$V(\pi) = a\pi \wedge b(1-\pi), \quad \pi \notin (A_*,B_*), \tag{23.1.21}$$

$$V(A_*+) = V(A_*), \;\; V(B_*-) = V(B_*) \quad (\textit{continuous fit}), \tag{23.1.22}$$

$$V'(A_*) = a \quad (\textit{smooth fit}) \tag{23.1.23}$$

for some $0 < A_* < b/(a+b) < B_* < 1$ to be found. Observe that (23.1.21) contains two conditions relevant for the system: (i) $V(A_*) = aA_*$ and (ii) $V(\pi) = b(1-\pi)$ for $\pi \in [B_*, S(B_*)]$ with $S = S(\pi)$ from (23.1.24) below. These conditions are in accordance with the fact that if the process $(\pi_t)_{t\ge0}$ starts or ends up at some π outside (A_*,B_*), we must stop it instantly.

Note from (23.1.16) that the process $(\pi_t)_{t\ge0}$ moves continuously towards 0 and only jumps towards 1 at times of jumps of the point process X. This provides some intuitive support for the principle of smooth fit to hold at A_* (cf. Subsection 9.1). However, without a concavity argument it is not a priori clear why the condition $V(B_*-) = V(B_*)$ should hold at B_*. As *Figure* VI.3 shows, this is a rare property shared only by exceptional pairs (A,B) (cf. Subsection 9.2), and one could think that once A_* is fixed through the "smooth fit", the unknown B_* will be determined uniquely through the "continuous fit". While this train of thoughts sounds perfectly logical, we shall see quite opposite below that the equation (23.1.19) dictates our travel to solution from B_* to A_*.

Our next aim is to show that the three conditions in (23.1.22) and (23.1.23) are sufficient to determine a unique solution of the free-boundary problem which in turn leads to the solution of the optimal stopping problem (23.1.8).

2.8. *Solution of the free-boundary problem* (23.1.20)–(23.1.23). Consider the equation (23.1.20) on $(0,B]$ with some $B > b/(a+b)$ given and fixed. Introduce

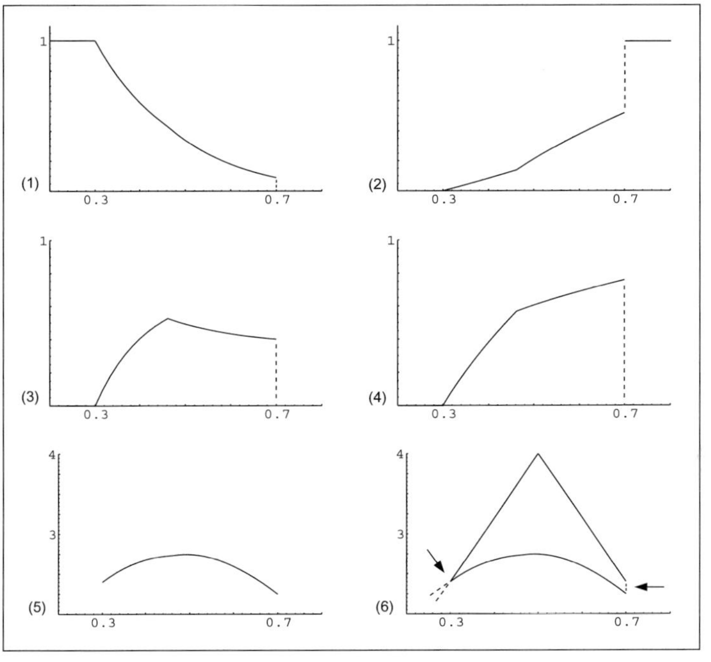

Figure VI.3: In view of the problem (23.1.8) and its decomposition via (23.1.4) and (23.1.6) with (23.1.1), we consider $\tau = \inf\{\, t \geq 0 : \pi_t \notin (A,B) \,\}$ for $(\pi_t)_{t\geq 0}$ from (23.1.7)+(23.1.12)+(23.1.14) with $\pi \in (A,B)$ given and fixed, so that $\pi_0 = \pi$ under P_0 and P_1; the computer drawings above show the following functions respectively: (1) $\pi \mapsto \mathsf{P}_1(\pi_\tau = A)$; (2) $\pi \mapsto \mathsf{P}_0(\pi_\tau \geq B)$; (3) $\pi \mapsto \mathsf{E}_1\tau$; (4) $\pi \mapsto \mathsf{E}_0\tau$; (5) $\pi \mapsto \pi\mathsf{E}_1\tau + (1-\pi)\mathsf{E}_0\tau + a\pi\mathsf{P}_1(\pi_\tau = A) + b(1-\pi)\,\mathsf{P}_0(\pi_\tau \geq B) = \mathsf{E}_\pi(\tau + g_{a,b}(\pi_\tau))$; (6) $\pi \mapsto \mathsf{E}_\pi(\tau + g_{a,b}(\pi_\tau))$ and $\pi \mapsto g_{a,b}(\pi)$, where $A = 0.3$, $B = 0.7$, $\lambda_0 = 1$, $\lambda_1 = e$ and $a = b = 8$. Functions (1)–(4) are found by solving systems analogous to the system (23.1.64)–(23.1.66); their discontinuity at B should be noted, as well as the discontinuity of their first derivative at $B_1 = 0.46\ldots$ from (23.1.25); observe that the function (5) is a superposition of functions (1)–(4), and thus the same discontinuities carry over to the function (5), unless something special occurs. The crucial fact to be observed is that if the function (5) is to be the value function (23.1.8), and thus extended by the gain function $\pi \mapsto g_{a,b}(\pi)$ outside (A,B), then such an extension would generally be discontinuous at B and have a discontinuous first derivative at A; this is depicted in the final picture (6). It is a matter of fact that the optimal A_* and B_* are to be chosen in such a way that both of these discontinuities disappear; these are the *principles of continuous and smooth fit* respectively. Observe that in this case the discontinuity of the first derivative of (5) also disappears at B_1, and the extension obtained is C^1 everywhere but at B_* where it is only C^0 generally (see *Figure* VI.5 below).

the "step" function

$$S(\pi) = \frac{\lambda_1 \pi}{\lambda_1 \pi + \lambda_0 (1-\pi)} \tag{23.1.24}$$

for $\pi \le B$. Observe that $\pi \mapsto S(\pi)$ is increasing, and find points $\cdots < B_2 < B_1 < B_0 := B$ such that $S(B_n) = B_{n-1}$ for $n \ge 1$. It is easily verified that

$$B_n = \frac{(\lambda_0)^n B}{(\lambda_0)^n B + (\lambda_1)^n (1-B)} \qquad (n = 0, 1, 2, \ldots). \tag{23.1.25}$$

Denote $I_n = (B_n, B_{n-1}]$ for $n \ge 1$, and introduce the "distance" function

$$d(\pi, B) = 1 + \left[\log\left(\frac{B}{1-B} \frac{1-\pi}{\pi} \right) \Big/ \log\left(\frac{\lambda_1}{\lambda_0} \right) \right] \tag{23.1.26}$$

for $\pi \le B$, where $[x]$ denotes the integer part of x. Observe that d is defined to satisfy

$$\pi \in I_n \iff d(\pi, B) = n \tag{23.1.27}$$

for all $0 < \pi \le B$.

Consider the equation (23.1.20) on I_1 upon setting $V(\pi) = b(1-\pi)$ for $\pi \in (B, S(B)]$; this is then a first-order linear differential equation which can be solved explicitly, and imposing a continuity condition at B which is in agreement with (23.1.22), we obtain a *unique* solution $\pi \mapsto V(\pi; B)$ on I_1; move then further and consider the equation (23.1.20) on I_2 upon using the solution found on I_1; this is then a first-order linear differential equation which can be solved explicitly, and imposing a continuity condition over $I_2 \cup I_1$ at B_1, we obtain a *unique* solution $\pi \mapsto V(\pi; B)$ on I_2; continuing this process by induction, we find the following formula:

$$\begin{aligned} V(\pi; B) = {} & \frac{(1-\pi)^{\gamma_1}}{\pi^{\gamma_0}} \sum_{k=0}^{n-1} \left(C_{n-k} \, \frac{\beta^k}{k!} \, \log^k \left(\left(\frac{\lambda_1}{\lambda_0} \right)^{k-1} \frac{\pi}{1-\pi} \right) \right) \\ & - \left(n \, \frac{\lambda_1 - \lambda_0}{\lambda_0 \lambda_1} + b \right) \pi + \left(\frac{n}{\lambda_0} + b \right) \end{aligned} \tag{23.1.28}$$

for $\pi \in I_n$, where $C_1, \ldots, C_n$ are constants satisfying the following recurrent relation:

$$C_{p+1} = \sum_{k=0}^{p-1} \left(C_{p-k} \left(f_k^{(p)} - f_{k+1}^{(p)} \right) \right) + \frac{(B_p)^{\gamma_0}}{(1-B_p)^{\gamma_1}} \left(\frac{\lambda_1 - \lambda_0}{\lambda_0 \lambda_1} \, B_p - \frac{1}{\lambda_0} \right) \tag{23.1.29}$$

for $p = 0, 1, \ldots, n-1$, with

$$f_k^{(p)} = \frac{\beta^k}{k!} \, \log^k \left(\left(\frac{\lambda_1}{\lambda_0} \right)^{k-p-1} \frac{B}{1-B} \right) \tag{23.1.30}$$

and where we set

$$\gamma_0 = \frac{\lambda_0}{\lambda_1 - \lambda_0}, \qquad \gamma_1 = \frac{\lambda_1}{\lambda_1 - \lambda_0}, \qquad \beta = \frac{1}{(\lambda_1 - \lambda_0)} \frac{(\lambda_0)^{\gamma_1}}{(\lambda_1)^{\gamma_0}}. \tag{23.1.31}$$

Making use of the distance function (23.1.26), we may now write down the unique solution of (23.1.20) on $(0, B]$ satisfying (23.1.21) on $[B, S(B)]$ and the second part of (23.1.22) at B:

$$\begin{aligned} V(\pi; B) = {} & \frac{(1-\pi)^{\gamma_1}}{\pi^{\gamma_0}} \sum_{k=0}^{d(\pi,B)-1} \left(C_{d(\pi,B)-k} \, \frac{\beta^k}{k!} \log^k \left(\left(\frac{\lambda_1}{\lambda_0} \right)^{k-1} \frac{\pi}{1-\pi} \right) \right) \\ & - \left(d(\pi, B) \, \frac{\lambda_1 - \lambda_0}{\lambda_0 \lambda_1} + b \right) \pi + \left(\frac{d(\pi, B)}{\lambda_0} + b \right) \end{aligned} \tag{23.1.32}$$

for $0 < \pi \le B$. It is clear from our construction above that $\pi \mapsto V(\pi; B)$ is C^1 on $(0, B)$ and C^0 at B.

Observe that when computing the first derivative of $\pi \mapsto V(\pi; B)$, we can treat $d(\pi, B)$ in (23.1.32) as not depending on π. This then gives the following explicit expression:

$$\begin{aligned} V'(\pi; B) = {} & \frac{(1-\pi)^{\gamma_1 - 1}}{\pi^{\gamma_0 + 1}} \\ & \times \sum_{k=0}^{d(\pi,B)-1} \left(C_{d(\pi,B)-k} \, \frac{\beta^k}{k!} \log^k \left(\left(\frac{\lambda_1}{\lambda_0} \right)^{k-1} \frac{\pi}{1-\pi} \right) \right. \\ & \times \left. \left(k \Big/ \log \left(\left(\frac{\lambda_1}{\lambda_0} \right)^{k-1} \frac{\pi}{1-\pi} \right) - (\pi + \gamma_0) \right) \right) - \left(d(\pi, B) \, \frac{\lambda_1 - \lambda_0}{\lambda_0 \lambda_1} + b \right) \end{aligned} \tag{23.1.33}$$

for $0 < \pi \le B$.

Setting $C = b/(a+b)$ elementary calculations show that $\pi \mapsto V(\pi; B)$ is concave on $(0, B)$, as well as that $V(\pi; B) \to -\infty$ as $\pi \downarrow 0$, for all $B \in [C, 1]$. Moreover, it is easily seen from (23.1.28) (with $n = 1$) that $V(\pi; 1) < 0$ for all $0 < \pi < 1$. Thus, if for some $\widehat{B} > C$, close to C, it happens that $\pi \mapsto V(\pi; \widehat{B})$ crosses $\pi \mapsto a\pi$ when π moves to the left from $\widehat{B}$, then a uniqueness argument presented in Remark 23.2 below (for different B's the curves $\pi \mapsto V(\pi; B)$ do not intersect) shows that there exists $B_* \in (C, 1)$, obtained by moving B from $\widehat{B}$ to 1 or vice versa, such that for some $A_* \in (0, C)$ we have $V(A_*; B_*) = a A_*$ and $V'(A_*; B_*) = a$ (see *Figure* VI.4). Observe that the first identity captures part (i) of (23.1.22), while the second settles (23.1.23).

These considerations show that the system (23.1.20)–(23.1.23) has a unique (nontrivial) solution consisting of A_*, B_* and $\pi \mapsto V(\pi; B_*)$, if and only if

$$\lim_{B \downarrow C} V'(B-; B) < a. \tag{23.1.34}$$

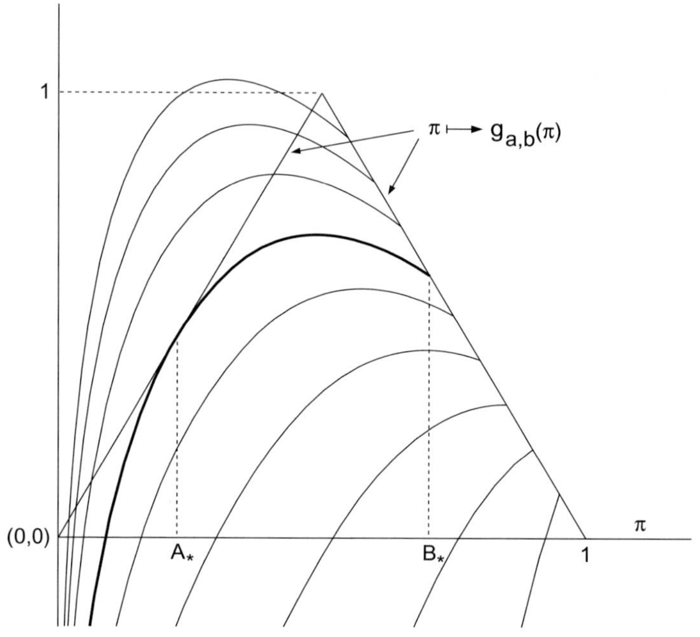

Figure VI.4: A computer drawing of "continuous fit" solutions $\pi \mapsto V(\pi; B)$ of (23.1.20), satisfying (23.1.21) on $[B, S(B)]$ and the second part of (23.1.22) at B, for different B in $(b/(a+b), 1)$; in this particular case we took $B = 0.95, 0.80, 0.75, \ldots, 0.55$, with $\lambda_0 = 1$, $\lambda_1 = 5$ and $a = b = 2$. The unique B_* is obtained through the requirement that the map $\pi \mapsto V(\pi; B_*)$ hits "smoothly" the gain function $\pi \mapsto g_{a,b}(\pi)$ at A_*; as shown above, this happens for $A_* = 0.22\ldots$ and $B_* = 0.70\ldots$; such obtained A_* and B_* are a unique solution of the system (23.1.38)–(23.1.39). The solution $\pi \mapsto V(\pi; B_*)$ leads to the explicit form of the value function (23.1.8) as shown in *Figure* VI.5 below.

Geometrically this is the case when for $B > C$, close to C, the solution $\pi \mapsto V(\pi; B)$ intersects $\pi \mapsto a\pi$ at some $\pi < B$. It is now easily verified by using (23.1.28) (with $n = 1$) that (23.1.34) holds if and only if the following condition is satisfied:

$$\lambda_1 - \lambda_0 > \frac{1}{a} + \frac{1}{b}. \tag{23.1.35}$$

In this process one should observe that B_1 from (23.1.25) tends to a number strictly less than C when $B \downarrow C$, so that all calculations are actually performed on I_1.

Thus, the condition (23.1.35) is necessary and sufficient for the existence of a unique nontrivial solution of the system (23.1.20)–(23.1.23); in this case the optimal A_* and B_* are uniquely determined as the solution of the system of transcendental equations $V(A_*; B_*) = a\,A_*$ and $V'(A_*; B_*) = a$, where $\pi \mapsto V(\pi; B)$ and $\pi \mapsto V'(\pi; B)$ are given by (23.1.32) and (23.1.33) respectively; once A_* and B_* are fixed, the solution $\pi \mapsto V(\pi; B_*)$ is given for $\pi \in [A_*, B_*]$ by means of (23.1.32).

2.9. *Solution of the optimal stopping problem* (23.1.8). We shall now show that the solution of the free-boundary problem (23.1.20)–(23.1.23) found above coincides with the solution of the optimal stopping problem (23.1.8). This in turn leads to the solution of the Bayesian problem (23.1.5).

Theorem 23.1. (I): *Suppose that the condition* (23.1.35) *holds. Then the* π*-Bayes decision rule* (τ_*, d_*) *in the problem* (23.1.5) *of testing two simple hypotheses* H_1 *and* H_0 *is explicitly given by* (*see Remark* 23.3 *below*):

$$\tau_* = \inf\big\{t \ge 0 : \pi_t \notin (A_*, B_*)\big\}, \tag{23.1.36}$$

$$d_* = \begin{cases} 1 \ (\textit{accept } H_1) & \textit{if } \ \pi_{\tau_*} \ge B_*, \\ 0 \ (\textit{accept } H_0) & \textit{if } \ \pi_{\tau_*} = A_* \end{cases} \tag{23.1.37}$$

where the constants A_* *and* B_* *satisfying* $0 < A_* < b/(a+b) < B_* < 1$ *are uniquely determined as solutions of the system of transcendental equations*:

$$V(A_*; B_*) = a\,A_*, \tag{23.1.38}$$

$$V'(A_*; B_*) = a \tag{23.1.39}$$

with $\pi \mapsto V(\pi; B)$ *and* $\pi \mapsto V'(\pi; B)$ *in* (23.1.32) *and* (23.1.33) *respectively.*

(II): *In the case when the condition* (23.1.35) *fails to hold, the* π*-Bayes decision rule is trivial*: *Accept* H_1 *if* $\pi > b/(a+b)$, *and accept* H_0 *if* $\pi < b/(a+b)$; *either decision is equally good if* $\pi = b/(a+b)$.

Proof. (I): 1°. We showed above that the free-boundary problem (23.1.20)–(23.1.23) is solvable if and only if (23.1.35) holds, and in this case the solution $\pi \mapsto V_*(\pi)$ is given explicitly by $\pi \mapsto V(\pi; B_*)$ in (23.1.32) for $A_* \le \pi \le B_*$, where A_* and B_* are a unique solution of (23.1.38)–(23.1.39).

In accordance with the interpretation of the free-boundary problem, we extend $\pi \mapsto V_*(\pi)$ to the whole of $[0,1]$ by setting $V_*(\pi) = a\pi$ for $0 \le \pi < A_*$ and $V_*(\pi) = b(1-\pi)$ for $B_* < \pi \le 1$ (see *Figure* VI.5). Note that $\pi \mapsto V_*(\pi)$ is C^1 on $[0,1]$ everywhere but at B_* where it is C^0. To complete the proof it is enough to show that such defined map $\pi \mapsto V_*(\pi)$ equals the value function defined in (23.1.8), and that τ_* defined in (23.1.36) is an optimal stopping time.

2°. Since $\pi \mapsto V_*(\pi)$ is not C^1 only at one point at which it is C^0, recalling also that $\pi \mapsto V_*(\pi)$ is concave, we can apply Itô's formula (page 67) to $V_*(\pi_t)$. In exactly the same way as in (23.1.17) this gives

$$V_*(\pi_t) \;=\; V_*(\pi) + \int_0^t (\mathbb{L}V_*)(\pi_{s-})\,ds + M_t \tag{23.1.40}$$

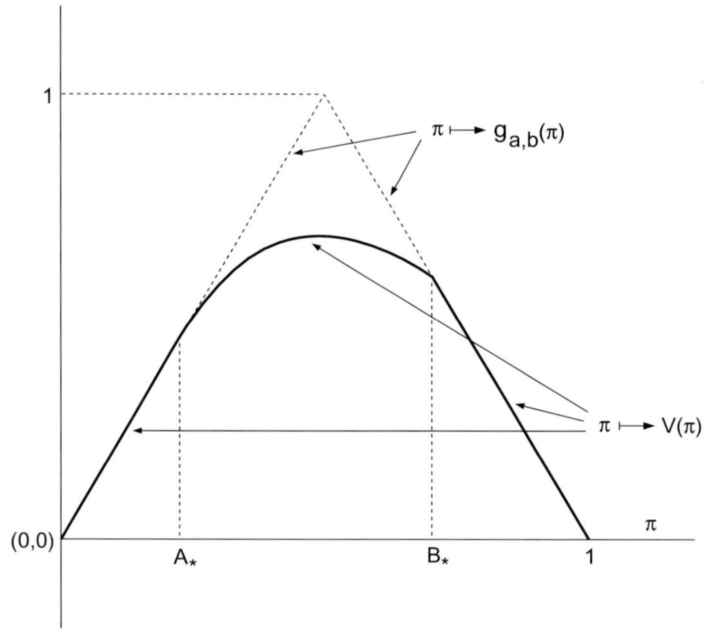

Figure VI.5: A computer drawing of the value function (23.1.8) in the case $\lambda_0 = 1$, $\lambda_1 = 5$ and $a = b = 2$ as indicated in *Figure* VI.4 above. The interval (A_*, B_*) is the set of continued observation of the process $(\pi_t)_{t\geq 0}$, while its complement in $[0,1]$ is the stopping set. Thus, as indicated in (23.1.36), the observation should be stopped as soon as the process $(\pi_t)_{t\geq 0}$ enters $[0,1]\setminus(A_*, B_*)$, and this stopping time is optimal in the problem (23.1.8). The optimal decision function is then given by (23.1.37).

where $M = (M_t)_{t\geq 0}$ is a martingale given by

$$M_t = \int_0^t \Big(V_*(\pi_{s-} + \Delta\pi_s) - V_*(\pi_{s-}) \Big)\, d\widehat{X}_s \tag{23.1.41}$$

and $\widehat{X}_t = X_t - \int_0^t \mathsf{E}_\pi(\lambda \,|\, \mathcal{F}^X_{s-})\, ds = X_t - \int_0^t (\lambda_1 \pi_{s-} + \lambda_0(1-\pi_{s-}))\, ds$ is the so-called *innovation process* (see e.g. [128, Theorem 18.3]) which is a martingale with respect to $(\mathcal{F}^X_t)_{t\geq 0}$ and P_π whenever $\pi \in [0,1]$. Note in (23.1.40) that we may extend V'_* arbitrarily to B_* as the time spent by the process $(\pi_t)_{t\geq 0}$ at B_* is of Lebesgue measure zero.

3°. Recall that $(\mathbb{L}V_*)(\pi) = -1$ for $\pi \in (A_*, B_*)$, and note that due to the smooth fit (23.1.23) we also have $(\mathbb{L}V_*)(\pi) \geq -1$ for all $\pi \in [0,1]\setminus(A_*, B_*]$.

To verify this claim first note that $(\mathbb{L}V_*)(\pi) = 0$ for $\pi \in (0, S^{-1}(A_*)) \cup (B_*, 1)$, since $\mathbb{L}f \equiv 0$ if $f(\pi) = a\pi$ or $f(\pi) = b(1-\pi)$. Observe also that $(\mathbb{L}V_*)(S^{-1}(A_*)) = 0$ and $(\mathbb{L}V_*)(A_*) = -1$ both due to the smooth fit (23.1.23). Thus, it is enough to verify that $(\mathbb{L}V_*)(\pi) \geq -1$ for $\pi \in (S^{-1}(A_*), A_*)$.

For this, consider the equation $\mathbb{L}V = -1$ on $(S^{-1}(A_*), A_*]$ upon imposing $V(\pi) = V(\pi; B_*)$ for $\pi \in (A_*, S(A_*)]$, and solve it under the initial condition $V(A_*) = V(A_*; B_*) + c$ where $c \geq 0$. This generates a unique solution $\pi \mapsto V_c(\pi)$ on $(S^{-1}(A_*), A_*]$, and from (23.1.28) we read that $V_c(\pi) = V(\pi; B_*) + K_c(1-\pi)^{\gamma_1}/\pi^{\gamma_0}$ for $\pi \in (S^{-1}(A_*), A_*]$ where $K_c = c(A_*)^{\gamma_0}/(1-A_*)^{\gamma_1}$. (Observe that the curves $\pi \mapsto V_c(\pi)$ do not intersect on $(S^{-1}(A_*), A_*]$ for different c's.) Hence we see that there exists $c_0 > 0$ large enough such that for each $c > c_0$ the curve $\pi \mapsto V_c(\pi)$ lies strictly above the curve $\pi \mapsto a\pi$ on $(S^{-1}(A_*), A_*]$, and for each $c < c_0$ the two curves intersect. For $c \in [0, c_0)$ let π_c denote the (closest) point (to A_*) at which $\pi \mapsto V_c(\pi)$ intersects $\pi \mapsto a\pi$ on $(S^{-1}(A_*), A_*]$. Then $\pi_0 = A_*$ and π_c decreases (continuously) in the direction of $S^{-1}(A_*)$ when c increases from 0 to c_0. Observe that the points π_c are "good" points since by $V_c(\pi_c) = a\pi_c = V_*(\pi_c)$ with $V_c'(\pi_c) > a = V_*'(\pi_c)$ and $V_c(S(\pi_c)) = V(S(\pi_c); B_*) = V_*(S(\pi_c))$ we see from (23.1.19) that $(\mathbb{L}V_*)(\pi_c) \geq (\mathbb{L}V_c)(\pi_c) = -1$. Thus, if we show that π_c reaches $S^{-1}(A_*)$ when $c \uparrow c_0$, then the proof of the claim will be complete. Therefore assume on the contrary that this is not the case. Then $V_{c_1}(S^{-1}(A_*)-) = aS^{-1}(A_*)$ for some $c_1 < c_0$, and $V_c(S^{-1}(A_*)-) > aS^{-1}(A_*)$ for all $c > c_1$. Thus by choosing $c > c_1$ close enough to c_1, we see that a point $\widetilde{\pi}_c > S^{-1}(A_*)$ arbitrarily close to $S^{-1}(A_*)$ is obtained at which $V_c(\widetilde{\pi}_c) = a\widetilde{\pi}_c = V_*(\widetilde{\pi}_c)$ with $V_c'(\widetilde{\pi}_c) < a = V_*'(\widetilde{\pi}_c)$ and $V_c(S(\widetilde{\pi}_c)) = V(S(\widetilde{\pi}_c); B_*) = V_*(S(\widetilde{\pi}_c))$, from where we again see by (23.1.19) that $(\mathbb{L}V_*)(\widetilde{\pi}_c) \leq (\mathbb{L}V_c)(\widetilde{\pi}_c) = -1$. This however leads to a contradiction because $\pi \mapsto (\mathbb{L}V_*)(\pi)$ is continuous at $S^{-1}(A_*)$ (due to the smooth fit) and $(\mathbb{L}V_*)(S^{-1}(A_*)) = 0$ as already stated earlier. Thus, we have $(\mathbb{L}V_*)(\pi) \geq -1$ for all $\pi \in [0,1]$ (upon setting $V_*'(B_*){:=}0$ for instance).

4°. Recall further that $V_*(\pi) \leq g_{a,b}(\pi)$ for all $\pi \in [0,1]$. Moreover, since $\pi \mapsto V_*(\pi)$ is bounded, and $(X_t - \lambda_i t)_{t \geq 0}$ is a martingale under P_i for $i = 0, 1$, it is easily seen from (23.1.41) with (23.1.17) upon using the optional sampling theorem (page 60), that $\mathsf{E}_\pi M_\tau = 0$ whenever τ is a stopping time of X such that $\mathsf{E}_\pi \tau < \infty$. Thus, taking the expectation on both sides in (23.1.40), we obtain

$$V_*(\pi) \leq \mathsf{E}_\pi \big(\tau + g_{a,b}(\pi_\tau) \big) \tag{23.1.42}$$

for all such stopping times, and hence $V_*(\pi) \leq V(\pi)$ for all $\pi \in [0,1]$.

5°. On the other hand, the stopping time τ_* from (23.1.36) clearly satisfies $V_*(\pi_{\tau_*}) = g_{a,b}(\pi_{\tau_*})$. Moreover, a direct analysis of τ_* based on (23.1.12)–(23.1.14) (see Remark 23.3 below), together with the fact that for any Poisson process $N = (N_t)_{t \geq 0}$ the exit time of the process $(N_t - \mu t)_{t \geq 0}$ from $[\widetilde{A}, \widetilde{B}]$ has a finite expectation for any real μ, shows that $\mathsf{E}_\pi \tau_* < \infty$ for all $\pi \in [0,1]$. Taking then the expectation on both sides in (23.1.40), we get

$$V_*(\pi) = \mathsf{E}_\pi\big(\tau_* + g_{a,b}(\pi_{\tau_*})\big) \tag{23.1.43}$$

for all $\pi \in [0,1]$. This fact and the consequence of (23.1.42) stated above show that $V_* = V$, and that τ_* is an optimal stopping time. The proof of the first part is complete.

(II): Although, in principle, it is clear from our construction above that the second part of the theorem holds as well, we shall present a formal argument for completeness.

Suppose that the π-Bayes decision rule is not trivial. In other words, this means that $V(\pi) < g_{a,b}(\pi)$ for some $\pi \in (0,1)$. Since $\pi \mapsto V(\pi)$ is concave, this implies that there are $0 < A_* < b/(a+b) < B_* < 1$ such that $\tau_* = \inf\{\, t > 0 : \pi_t \notin (A_*, B_*)\,\}$ is optimal for the problems (23.1.8) and (23.1.5) respectively, with d_* from (23.1.9) in the latter case. Thus $V(\pi) = \mathsf{E}_\pi(\tau_* + g_{a,b}(\pi_{\tau_*}))$ for $\pi \in [0,1]$, and therefore by the general Markov processes theory, and due to the strong Markov property of $(\pi_t)_{t \ge 0}$, we know that $\pi \mapsto V(\pi)$ solves (23.1.20) and satisfies (23.1.21) and (23.1.22); a priori we do not know if the smooth fit condition (23.1.23) is satisfied. Nevertheless, these arguments show the existence of a solution to (23.1.20) on $(0, B_*]$ which is $b(1-B_*)$ at B_* and which crosses $\pi \mapsto a\pi$ at (some) $A_* < b/(a+b)$. But then the same uniqueness argument used in paragraph 2.8 above (see Remark 23.2 below) shows that there must exist points $\widehat{A}_* \le A_*$ and $\widehat{B}_* \ge B_*$ such that the solution $\pi \mapsto \widehat{V}(\pi; \widehat{B}_*)$ of (23.1.20) satisfying $\widehat{V}(\widehat{B}_*; \widehat{B}_*) = b(1-\widehat{B}_*)$ hits $\pi \mapsto a\pi$ smoothly at $\widehat{A}_*$. The first part of the proof above then shows that the stopping time $\widehat{\tau}_* = \inf\{\, t > 0 : \pi_t \notin (\widehat{A}_*, \widehat{B}_*)\,\}$ is optimal. As this stopping time is known to be P_π-a.s. pointwise the smallest possible optimal stopping time (cf. Chapter I or see the proof of Theorem 23.4 below), this shows that τ_* cannot be optimal unless the smooth fit condition holds at A_*, that is, unless $\widehat{A}_* = A_*$ and $\widehat{B}_* = B_*$. In any case, however, this argument implies the existence of a nontrivial solution to the system (23.1.20)–(23.1.23), and since this fact is equivalent to (23.1.35) as shown above, we see that condition (23.1.35) cannot be violated.

Observe that we have actually proved that if the optimal stopping problem (23.1.8) has a nontrivial solution, then the principle of smooth fit holds at A_*. An alternative proof of the statement could be done by using Lemma 3 in [196, p. 118]. The proof of the theorem is complete. □

Remark 23.2. The following probabilistic argument can be given to show that the two curves $\pi \mapsto V(\pi, B')$ and $\pi \mapsto V(\pi, B'')$ from (23.1.32) do not intersect on $(0, B']$ whenever $0 < B' < B'' \le 1$.

Assume that the two curves do intersect at some $Z < B'$. Let $\pi \mapsto \alpha\pi + \beta$ denote the tangent of the map $V(\,\cdot\,; B')$ at Z. Define a map $\pi \mapsto g(\pi)$ by setting $g(\pi) = (\alpha\pi + \beta) \wedge b(1-\pi)$ for $\pi \in [0,1]$, and consider the optimal stopping problem (23.1.8) with g instead of $g_{a,b}$. Let $V = V(\pi)$ denote the value function. Consider also the map $\pi \mapsto V_*(\pi)$ defined by $V_*(\pi) = V(\pi; B')$ for $\pi \in [Z, B']$ and $V_*(\pi) = g(\pi)$ for $\pi \in [0,1] \setminus [Z, B']$. As $\pi \mapsto V_*(\pi)$ is C^0 at B' and C^1 at Z, then in exactly the same way as in paragraphs 3°–5° of the proof above we find that $V_*(\pi) = V(\pi)$ for all $\pi \in [0,1]$. However, if we consider the stopping time $\sigma_* = \inf\{\, t>0 : \pi_t \notin (Z, B'') \,\}$, then it follows in the same way as in paragraph 5° of the proof above that $V(\pi; B'') = \mathsf{E}_\pi(\sigma_* + g(\pi_{\sigma_*}))$ for all $\pi \in [Z, B'']$. As $V(\pi; B'') < V_*(\pi)$ for $\pi \in (Z, B']$, this is a contradiction. Thus, the curves do not intersect.

Remark 23.3. 1. Observe that the optimal decision rule (23.1.36)–(23.1.37) can be equivalently rewritten as follows:

$$\tau_* = \inf\big\{t \geq 0 : Z_t \notin (\widetilde{A}_*, \widetilde{B}_*)\big\}, \tag{23.1.44}$$

$$d_* = \begin{cases} 1 \text{ (accept } H_1) & \text{if } Z_{\tau_*} \geq \widetilde{B}_*, \\ 0 \text{ (accept } H_0) & \text{if } Z_{\tau_*} = \widetilde{A}_* \end{cases} \tag{23.1.45}$$

where we use the following notation:

$$Z_t = X_t - \mu t, \tag{23.1.46}$$

$$\widetilde{A}_* = \log\left(\frac{A_*}{1-A_*}\,\frac{1-\pi}{\pi}\right) \Big/ \log\left(\frac{\lambda_1}{\lambda_0}\right), \tag{23.1.47}$$

$$\widetilde{B}_* = \log\left(\frac{B_*}{1-B_*}\,\frac{1-\pi}{\pi}\right) \Big/ \log\left(\frac{\lambda_1}{\lambda_0}\right), \tag{23.1.48}$$

$$\mu = (\lambda_1 - \lambda_0) \Big/ \log\left(\frac{\lambda_1}{\lambda_0}\right). \tag{23.1.49}$$

2. The representation (23.1.44)–(23.1.45) reveals the structure and applicability of the optimal decision rule in a clearer manner. The result proved above shows that the following sequential procedure is optimal: *While observing X_t, monitor Z_t, and stop the observation as soon as Z_t enters either $(-\infty, \widetilde{A}_*]$ or $[\widetilde{B}_*, \infty)$; in the first case conclude $\lambda = \lambda_0$, in the second conclude $\lambda = \lambda_1$.* In this process the condition (23.1.35) must be satisfied, and the constants A_* and B_* should be determined as a unique solution of the system (23.1.38)–(23.1.39). This system can be successfully treated by means of standard numerical methods if one mimics our travel from B_* to A_* in the construction of our solution in paragraph 2.8 above. A pleasant fact is that only a few steps by (23.1.24) will be often needed to recapture A_* if one starts from B_*.

3. Note that the same problem of testing two statistical hypotheses for a Poisson process was treated by different methods in [179]. One may note that the necessary and sufficient condition (23.1.35) of Theorem 23.1 is different from the condition $a\lambda_1 + b(\lambda_0 + \lambda_1) < b/a$ found in [179].

3. Solution of the variational problem. In the variational formulation of the problem it is assumed that the sequentially observed process $X = (X_t)_{t \ge 0}$ is a Poisson process with intensity λ_0 or λ_1, and no probabilistic assumption is made about the outcome of λ_0 and λ_1 at time 0. To formulate the problem we shall adopt the setting and notation from the previous part. Thus P_i is a probability measure on $(\Omega, \mathcal{F})$ under which $X = (X_t)_{t \ge 0}$ is a Poisson process with intensity λ_i for $i = 0, 1$.

3.1. Given the numbers α, $\beta > 0$ such that $\alpha + \beta < 1$, let $\Delta(\alpha, \beta)$ denote the class of all decision rules (τ, d) satisfying

$$\alpha(d) \le \alpha \quad \text{and} \quad \beta(d) \le \beta \tag{23.1.50}$$

where $\alpha(d) = \mathsf{P}_1(d = 0)$ and $\beta(d) = \mathsf{P}_0(d = 1)$. The variational problem is then to find a decision rule $(\widehat{\tau}, \widehat{d}\,)$ in the class $\Delta(\alpha, \beta)$ such that

$$\mathsf{E}_0 \widehat{\tau} \le \mathsf{E}_0 \tau \quad \text{and} \quad \mathsf{E}_1 \widehat{\tau} \le \mathsf{E}_1 \tau \tag{23.1.51}$$

for any other decision rule (τ, d) from the class $\Delta(\alpha, \beta)$. Note that the main virtue of the requirement (23.1.51) is its simultaneous validity for both P_0 and P_1.

Our main aim below is to show how the solution of the variational problem together with a precise description of all admissible pairs (α, β) can be obtained from the Bayesian solution (Theorem 23.1). The sequential procedure which leads to the optimal decision rule $(\widehat{\tau}, \widehat{d})$ in this process is a SPRT (sequential probability ratio test). We now describe a procedure of passing from the Bayesian solution to the variational solution.

3.2. It is useful to note that the explicit procedure of passing from the Bayesian solution to the variational solution presented in the next three steps is not confined to a Poissonian case but is also valid in greater generality including the Wiener case (for details in the case of discrete time see [123]).

Step 1 (*Construction*): Given $\alpha, \beta > 0$ with $\alpha + \beta < 1$, find constants A and B satisfying $A < 0 < B$ such that the stopping time

$$\widehat{\tau} = \inf\big\{ t \ge 0 : Z_t \notin (A, B) \big\} \tag{23.1.52}$$

satisfies the following identities:

$$\mathsf{P}_1\big(Z_{\widehat{\tau}} = A\big) = \alpha, \tag{23.1.53}$$

$$\mathsf{P}_0\big(Z_{\widehat{\tau}} \ge B\big) = \beta \tag{23.1.54}$$

where $(Z_t)_{t \ge 0}$ is as in (23.1.46). Associate with $\widehat{\tau}$ the following decision function:

$$\widehat{d} = \begin{cases} 1 \text{ (accept } H_1) & \text{if } Z_{\widehat{\tau}} \ge B, \\ 0 \text{ (accept } H_0) & \text{if } Z_{\widehat{\tau}} = A. \end{cases} \tag{23.1.55}$$

We will actually see below that not for all values α and β do such A and B exist; a function $G:(0,1)\to(0,1)$ is displayed in (23.1.73) such that the solution (A,B) to (23.1.53)–(23.1.54) exists only for $\beta\in(0,G(\alpha))$ if $\alpha\in(0,1)$. Such values α and β will be called *admissible*.

Step 2 (*Embedding*): Once A and B are found for admissible α and β, we may respectively identify them with $\widetilde{A}_*$ and $\widetilde{B}_*$ from (23.1.47) and (23.1.48). Then, for any $\widehat{\pi}\in(0,1)$ given and fixed, we can uniquely determine A_* and B_* satisfying $0<A_*<B_*<1$ such that (23.1.47) and (23.1.48) hold with $\pi=\widehat{\pi}$. Once A_* and B_* are given, we can choose $a>0$ and $b>0$ in the Bayesian problem (23.1.4)–(23.1.5) such that the optimal stopping time in (23.1.8) is exactly the exit time τ_* of $(\pi_t)_{t\ge0}$ from (A_*,B_*) as given in (23.1.36). Observe that this is possible to achieve since the optimal A_* and B_* range through all $(0,1)$ when a and b satisfying (23.1.35) range through $(0,\infty)$. (For this, let any $B_*\in(0,1)$ be given and fixed, and choose $\widetilde{a}>0$ and $\widetilde{b}>0$ such that $B_*=\widetilde{b}/(\widetilde{a}+\widetilde{b})$ with $\lambda_1-\lambda_0=1/\widetilde{a}+1/\widetilde{b}$. Then consider the solution $V(\,\cdot\,;B_*):=V_b(\,\cdot\,;B_*)$ of (23.1.20) on $(0,B_*)$ upon imposing $V_b(\pi;B_*)=b(1-\pi)$ for $\pi\in[B_*,S(B_*)]$ where $b\ge\widetilde{b}$. To each such a solution there corresponds $a>0$ such that $\pi\mapsto a\pi$ hits $\pi\mapsto V_b(\pi;B_*)$ smoothly at some $A=A(b)$. When b increases from $\widetilde{b}$ to ∞, then $A(b)$ decreases from B_* to zero. This is easily verified by a simple comparison argument upon noting that $\pi\mapsto V_b(\pi;B_*)$ stays strictly above $\pi\mapsto V(\pi;B_*)+V_b(B_*;B_*)$ on $(0,B_*)$ (recall the idea used in Remark 23.3 above). As each $A(b)$ obtained (in the pair with B_*) is optimal (recall the arguments used in paragraphs 3°–5° of the proof of Theorem 23.1), the proof of the claim is complete.)

Step 3 (*Verification*): Consider the process $(\widehat{\pi}_t)_{t\ge0}$ defined by (23.1.12)+(23.1.14) with $\pi=\widehat{\pi}$, and denote by $(\widehat{\tau}_*,\widehat{d}_*)$ the optimal decision rule (23.1.36)–(23.1.37) associated with it. From our construction above note that $\widehat{\tau}$ from (23.1.52) actually coincides with $\widehat{\tau}_*$, as well as that $\{\widehat{\pi}_{\widehat{\tau}_*}=A_*\}=\{Z_{\widehat{\tau}}=A\}$ and $\{\widehat{\pi}_{\widehat{\tau}_*}\ge B_*\}=\{Z_{\widehat{\tau}}\ge B\}$. Thus (23.1.53) and (23.1.54) show that

$$\mathsf{P}_1\big(\widehat{d}_*=0\big)=\alpha, \tag{23.1.56}$$

$$\mathsf{P}_0\big(\widehat{d}_*=1\big)=\beta \tag{23.1.57}$$

for the admissible α and β. If now any decision rule (τ,d) from $\Delta(\alpha,\beta)$ is given, then either $\mathsf{P}_1(d=0)=\alpha$ and $\mathsf{P}_0(d=1)=\beta$, or at least one strict inequality holds. In both cases, however, from (23.1.4)–(23.1.6) and (23.1.56)–(23.1.57) we easily see that $\mathsf{E}_{\widehat{\pi}}\widehat{\tau}_*\le\mathsf{E}_{\widehat{\pi}}\tau$, since otherwise $\widehat{\tau}_*$ would not be optimal. Since $\widehat{\tau}_*=\widehat{\tau}$, it follows that $\mathsf{E}_{\widehat{\pi}}\widehat{\tau}\le\mathsf{E}_{\widehat{\pi}}\tau$, and letting $\widehat{\pi}$ first go to 0 and then to 1, we obtain (23.1.51) in the case when $\mathsf{E}_0\tau<\infty$ and $\mathsf{E}_1\tau<\infty$. If either $\mathsf{E}_0\tau$ or $\mathsf{E}_1\tau$ equals ∞, then (23.1.51) follows by the same argument after a simple truncation (e.g. if $\mathsf{E}_0\tau<\infty$ but $\mathsf{E}_1\tau=\infty$, choose $n\ge1$ such that $\mathsf{P}_0(\tau>n)\le\varepsilon$, apply the same argument to $\tau_n:=\tau\wedge n$ and $d_n:=d1_{\{\tau\le n\}}+1_{\{\tau>n\}}$, and let ε go to zero in the end.) This solves the variational problem posed above for all admissible α and β.

3.3. The preceding arguments also show:

$$\text{If either } \mathsf{P}_1(d=0) < \alpha \text{ or } \mathsf{P}_0(d=1) < \beta \text{ for some } (\tau, d) \in \Delta(\alpha,\beta) \text{ with admissible } \alpha \text{ and } \beta \text{, then at least one strict inequality in (23.1.51) holds.} \tag{23.1.58}$$

Moreover, since $\widehat{\tau}_*$ is known to be $\mathsf{P}_{\widehat{\pi}}$-a.s. the smallest possible optimal stopping time (cf. Chapter I or see the proof of Theorem 23.4 below), from the arguments above we also get

$$\text{If } \mathsf{P}_1(d=0) = \alpha \text{ and } \mathsf{P}_0(d=1) = \beta \text{ for some } (\tau,d) \in \Delta(\alpha,\beta) \text{ with admissible } \alpha \text{ and } \beta \text{, and both equalities in (23.1.51) hold, then } \tau = \widehat{\tau} \;\; \mathsf{P}_0\text{-a.s. and } \mathsf{P}_1\text{-a.s.} \tag{23.1.59}$$

The property (23.1.59) characterizes $\hat{\tau}$ as *a unique stopping time of the decision rule with maximal admissible error probabilities having both* P_0 *and* P_1 *expectation at minimum.*

3.4. It remains to determine admissible α and β in (23.1.53) and (23.1.54) above. For this, consider $\widehat{\tau}$ defined in (23.1.52) for some $A < 0 < B$, and note from (23.1.14) that $\varphi_t = \exp\big(Z_t \log(\lambda_1/\lambda_0)\big)$. By means of (23.1.10) we find

$$\begin{aligned} \mathsf{P}_1\big(Z_{\hat{\tau}} = A\big) &= \mathsf{P}_1\bigg(\varphi_{\hat{\tau}} = \exp\bigg\{A\log\Big(\frac{\lambda_1}{\lambda_0}\Big)\bigg\}\bigg) \\ &= \exp\bigg\{A\log\Big(\frac{\lambda_1}{\lambda_0}\Big)\bigg\}\mathsf{P}_0\big(Z_{\hat{\tau}} = A\big) \\ &= \exp\bigg\{A\log\Big(\frac{\lambda_1}{\lambda_0}\Big)\bigg\}\Big(1 - \mathsf{P}_0\big(Z_{\hat{\tau}} \ge B\big)\Big). \end{aligned} \tag{23.1.60}$$

Using (23.1.53)–(23.1.54), from (23.1.60) we see that

$$A = \log\Big(\frac{\alpha}{1-\beta}\Big)\Big/\log\Big(\frac{\lambda_1}{\lambda_0}\Big). \tag{23.1.61}$$

To determine B, let P_0^z be a probability measure under which $X = (X_t)_{t\ge 0}$ is a Poisson process with intensity λ_0 and $Z = (Z_t)_{t\ge 0}$ starts at z. It is easily seen that the infinitesimal operator of Z under $(\mathsf{P}_0^z)_{z\in\mathbb{R}}$ acts like

$$(\mathbb{L}_0 f)(z) = -\mu f'(z) + \lambda_0\big(f(z+1) - f(z)\big). \tag{23.1.62}$$

In view of (23.1.54), introduce the function

$$u(z) = \mathsf{P}_0^z\big(Z_{\hat{\tau}} \ge B\big). \tag{23.1.63}$$

Strong Markov arguments then show that $z \mapsto u(z)$ solves the following system:

$$(\mathbb{L}_0 u)(z) = 0 \quad \text{if } z \in (A, B) \setminus \{B-1\}, \tag{23.1.64}$$

$$u(A) = 0, \tag{23.1.65}$$

$$u(z) = 1 \quad \text{if } z \geq B. \tag{23.1.66}$$

The solution of this system is given in (4.15) of [51]. To display it, introduce the function

$$F(x; B) = \sum_{k=0}^{\delta(x,B)} \frac{(-1)^k}{k!} \Big((B - x - k)\, \rho\, e^{-\rho} \Big)^k \tag{23.1.67}$$

for $x \leq B$, where we denote

$$\delta(x, B) = -[x - B + 1], \tag{23.1.68}$$

$$\rho = \log\left(\frac{\lambda_1}{\lambda_0}\right) \Big/ \left(\frac{\lambda_1}{\lambda_0} - 1\right). \tag{23.1.69}$$

Setting $J_n = [B - n - 1, B - n)$ for $n \geq 0$, observe that $\delta(x, B) = n$ if and only if $x \in J_n$.

It is then easily verified that the solution of the system (23.1.64)–(23.1.66) is given by

$$u(z) = 1 - e^{-\rho(z-A)} \frac{F(z; B)}{F(A; B)} \tag{23.1.70}$$

for $A \leq z < B$. Note that $z \mapsto u(z)$ is C^1 everywhere in (A, B) but at $B - 1$ where it is only C^0; note also that $u(A+) = u(A) = 0$, but $u(B-) < u(B) = 1$ (see *Figure* VI.6).

Going back to (23.1.54), and using (23.1.70), we see that

$$\mathsf{P}_0\big(Z_{\hat{\tau}} \geq B\big) = 1 - e^{\rho A} \frac{F(0; B)}{F(A; B)}. \tag{23.1.71}$$

Letting $B \downarrow 0$ in (23.1.71), and using the fact that the expression (23.1.71) is continuous in B and decreases to 0 as $B \uparrow \infty$, we clearly obtain a necessary and sufficient condition on β to satisfy (23.1.54), once $A = A(\alpha, \beta)$ is fixed through (23.1.61); as $F(0; 0) = 1$, this condition reads

$$\beta < 1 - \frac{e^{\rho A(\alpha,\beta)}}{F\big(A(\alpha, \beta); 0\big)}. \tag{23.1.72}$$

Note, however, if β increases, then the function on the right-hand side in (23.1.72) decreases, and thus there exists a unique $\beta_* = \beta_*(\alpha) > 0$ at which equality in

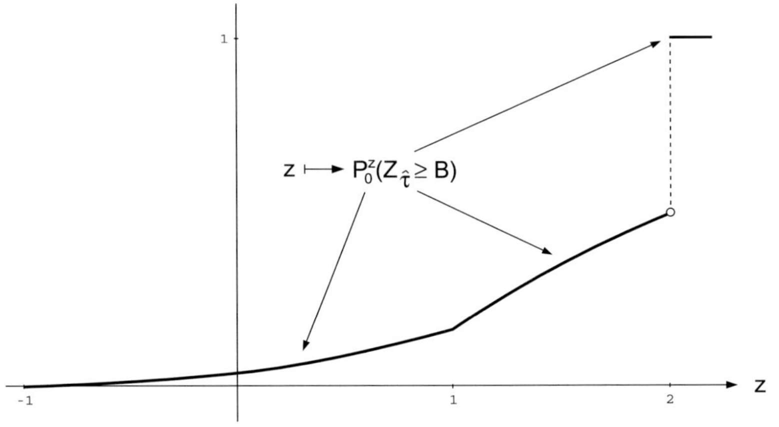

Figure VI.6: A computer drawing of the map $u(z) = \mathsf{P}_0^z(Z_{\hat\tau} \ge B)$ from (23.1.63) in the case $A = -1$, $B = 2$ and $\lambda_0 = 0.5$. This map is a unique solution of the system (23.1.64)–(23.1.66). Its discontinuity at B should be noted, as well as the discontinuity of its first derivative at $B-1$. Observe also that $u(A+) = u(A) = 0$. The case of general A, B and λ_0 looks very much the same.

(23.1.72) is attained. (This value can easily be computed by means of standard numerical methods.) Setting

$$G(\alpha) = 1 - \frac{e^{\rho A(\alpha,\beta_*(\alpha))}}{F\big(A(\alpha,\beta_*(\alpha));0\big)} \tag{23.1.73}$$

we see that admissible α and β are characterized by $0 < \beta < G(\alpha)$ (see *Figure* VI.7). In this case A is given by (23.1.61), and B is uniquely determined from the equation

$$F(0;B) - (1-\beta)\, F(A;B)\, e^{-\rho A} = 0. \tag{23.1.74}$$

The set of all admissible α and β will be denoted by $\mathcal{A}$. Thus, we have

$$\mathcal{A} = \big\{(\alpha,\beta) : 0 < \alpha < 1,\ 0 < \beta < G(\alpha)\big\}. \tag{23.1.75}$$

3.5. The preceding considerations may be summarised as follows (see also Remark 23.5 below).

Theorem 23.4. *In the problem* (23.1.50)–(23.1.51) *of testing two simple hypotheses* (23.1.2)–(23.1.3) *based upon sequential observations of the Poisson process* $X = (X_t)_{t\ge0}$ *under* P_0 *or* P_1*, there exists a unique decision rule* $(\widehat\tau, \widehat d\,) \in \Delta(\alpha,\beta)$ *satisfying* (23.1.51) *for any other decision rule* $(\tau,d) \in \Delta(\alpha,\beta)$ *whenever* $(\alpha,\beta) \in \mathcal{A}$*. The decision rule* $(\widehat\tau, \widehat d\,)$ *is explicitly given by* (23.1.52)+(23.1.55) *with* A

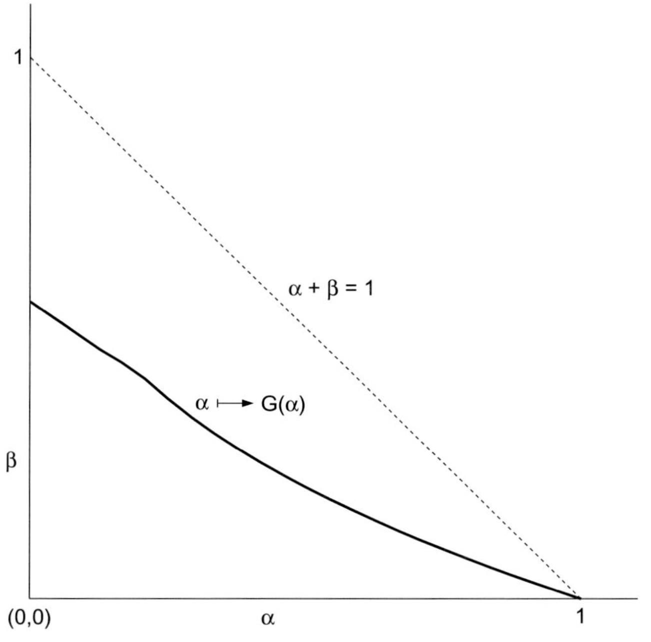

Figure VI.7: A computer drawing of the map $\alpha \mapsto G(\alpha)$ from (23.1.73) in the case $\lambda_0 = 1$ and $\lambda_1 = 3$. The area $\mathcal{A}$ which lies below the graph of G determines the set of all admissible α and β. The case of general λ_0 and λ_1 looks very much the same; it can also be shown that $G(0+)$ decreases if the difference $\lambda_1 - \lambda_0$ increases, as well as that $G(0+)$ increases if both λ_0 and λ_1 increase so that the difference $\lambda_1 - \lambda_0$ remains constant; in all cases $G(1-) = 0$. It may seem somewhat surprising that $G(0+) < 1$; observe, however, this is in agreement with the fact that $(Z_t)_{t \geq 0}$ from (23.1.46) is a supermartingale under P_0. (A little peak on the graph, at $\hat{\alpha} = 0.19\ldots$ and $\hat{\beta} = 0.42\ldots$ in this particular case, corresponds to the disturbance when A from (23.1.61) passes through -1 while $B = 0+$; it is caused by a discontinuity of the first derivative of the map from (23.1.71) at $B - 1$ (see *Figure* VI.6).)

in (23.1.61) *and* B *from* (23.1.74)*, it satisfies* (23.1.58)*, and is characterized by* (23.1.59)*.*

Proof. It only remains to prove (23.1.59). For this, in the notation used above, assume that τ is a stopping time of X satisfying the hypotheses of (23.1.59). Then clearly τ is an optimal stopping time in (23.1.8) for $\pi = \widehat{\pi}$ with a and b as in Step 2 above.

Recall that $V_*(\pi) \le g_{a,b}(\pi)$ for all π, and observe that $\widehat{\tau}$ can be written as

$$\widehat{\tau} = \inf\big\{t \ge 0 : V_*(\widehat{\pi}_t) \ge g_{a,b}(\widehat{\pi}_t)\big\} \tag{23.1.76}$$

where $\pi \mapsto V_*(\pi)$ is the value function (23.1.8) appearing in the proof of Theorem 23.1. Supposing now that $\mathsf{P}_{\widehat{\pi}}(\tau < \widehat{\tau}) > 0$, we easily find by (23.1.76) that

$$\mathsf{E}_{\widehat{\pi}}\big(\tau + g_{a,b}(\widehat{\pi}_\tau)\big) > \mathsf{E}_{\widehat{\pi}}\big(\tau + V_*(\widehat{\pi}_\tau)\big). \tag{23.1.77}$$

On the other hand, it is clear from (23.1.40) with $\mathbb{L}V_* \ge -1$ that $(t + V_*(\widehat{\pi}_t))_{t \ge 0}$ is a submartingale. Thus by the optional sampling theorem (page 60) it follows that

$$\mathsf{E}_{\widehat{\pi}}\big(\tau + V_*(\widehat{\pi}_\tau)\big) \ge V_*(\widehat{\pi}). \tag{23.1.78}$$

However, from (23.1.77) and (23.1.78) we see that τ cannot be optimal, and thus we must have $\mathsf{P}_{\widehat{\pi}}(\tau \ge \widehat{\tau}) = 1$. Moreover, since it follows from our assumption that $\mathsf{E}_{\widehat{\pi}}\tau = \mathsf{E}_{\widehat{\pi}}\widehat{\tau}$, this implies that $\tau = \widehat{\tau}$ $\mathsf{P}_{\widehat{\pi}}$-a.s. Finally, as $\mathsf{P}_i \ll \mathsf{P}_{\widehat{\pi}}$ for $i = 0, 1$, we see that $\tau = \widehat{\tau}$ both P_0-a.s. and P_1-a.s. The proof of the theorem is complete. □

Observe that the sequential procedure of the optimal decision rule $(\widehat{\tau}, \widehat{d}\,)$ from Theorem 23.4 is precisely the SPRT. The explicit formulae for $\mathsf{E}_0\widehat{\tau}$ and $\mathsf{E}_1\widehat{\tau}$ are given in (4.22) of [51].

Remark 23.5. If $(\alpha, \beta) \notin \mathcal{A}$, that is, if $\beta \ge G(\alpha)$ for some $\alpha, \beta > 0$ such that $\alpha + \beta < 1$, then no decision rule given by the SPRT-form (23.1.52)+(23.1.55) can solve the variational problem (23.1.50)–(23.1.51).

To see this, let such $(\alpha, \beta^*) \notin \mathcal{A}$ be given, and let (τ, d) be a decision rule satisfying (23.1.52)+(23.1.55) for some $A < 0 < B$. Denote $\beta = \mathsf{P}_0(Z_\tau \ge B)$ and choose α to satisfy (23.1.61). Then $\beta < G(\alpha) \le \beta^*$ by definition of the map G. Given $\beta' \in (\beta, G(\alpha))$, let B' be taken to satisfy (23.1.54) with β', and let α' be determined from (23.1.61) with β' so that A remains unchanged. Clearly $0 < B' < B$ and $0 < \alpha' < \alpha$, and (23.1.53) holds with A and α' respectively. But then (τ', d') satisfying (23.1.52)+(23.1.55) with $A < 0 < B'$ still belongs to $\Delta(\alpha, \beta^*)$, while clearly $\tau' < \tau$ both under P_0 and P_1. This shows that (τ, d) does not solve the variational problem.

The preceding argument shows that the admissible class $\mathcal{A}$ from (23.1.75) is exactly the class of all error probabilities (α, β) for which the SPRT is optimal. A pleasant fact is that $\mathcal{A}$ always contains a neighborhood around $(0, 0)$ in $[0, 1] \times [0, 1]$, which is the most interesting case from the standpoint of statistical applications.

Notes. The main aim of this section (following [168]) was to present an *explicit solution* of the problem of testing two statistical hypotheses about the intensity of an observed Poisson process in the context of a *Bayesian formulation*, and then apply this result to deduce the optimality of the method (SPRT) in the context of a *variational formulation*, providing a precise description of the set of all admissible probabilities of a wrong decision ("errors of the first and second kind").

Despite the fact that the Bayesian approach to sequential analysis of problems on testing two statistical hypotheses has gained a considerable interest in the last fifty or so years (see e.g. [216], [217], [18], [123], [31], [196], [203]), it turns out that a very few problems of that type have been solved explicitly (by obtaining a solution in closed form). In this respect the case of testing two simple hypotheses about the mean value of a Wiener process with drift is exceptional, as the explicit solution to the problem has been obtained in both Bayesian and variational formulation (cf. Section 21). These solutions (including the proof of the optimality of the SPRT) were found by reducing the initial problem to a free-boundary problem (for a second-order differential operator) which could be solved explicitly. It is clear from the material above that the Poisson free-boundary problem is more delicate, since in this case one needs to deal with a differential-difference operator, the appearance of which is a consequence of the discontinuous character of the observed (Poisson) process.

The variational problem formulation (23.1.51) is due to Wald [216]. In the papers [218] and [219] Wald and Wolfowitz proved the optimality of the SPRT in the case of i.i.d. observations and under special assumptions on the admissibility of (α, β) (see [218], [219], [5], [123] for more details and compare it with the admissability notion given above). In the paper [51] Dvoretzky, Kiefer and Wolfowitz considered the problem of optimality of the SPRT in the case of continuous time and satisfied themselves with the remark that "a careful examination of the results of [218] and [219] shows that their conclusions in no way require that the processes be discrete in time" omitting any further detail and concentrating their attention on the problem of finding the error probabilities $\alpha(d)$ and $\beta(d)$ with expectations $\mathsf{E}_0\tau$ and $\mathsf{E}_1\tau$ for the given SPRT (τ, d) defined by "stopping boundaries" A and B in the case of a Wiener or Poisson process.

The SPRT is known to be optimal in the variational formulation for a large class of observable processes (see [51], [96], [17]). For the general problem of the minimax optimality of the SPRT (in the sense (23.1.51)) in the case of continuous time see [96].

24. Quickest detection of a Poisson process

In this section we continue our study of quickest detection problems considered in Section 22 above. Instead of the Wiener process we now deal with the Poisson process.

24.1. Infinite horizon

1. Description of the problem. The Poisson disorder problem is less formally stated as follows. Suppose that at time $t = 0$ we begin observing a trajectory of the

Poisson process $X = (X_t)_{t \ge 0}$ whose intensity changes from λ_0 to λ_1 at some random (unknown) time θ which is assumed to take value 0 with probability π, and is exponentially distributed with parameter λ given that $\theta > 0$. Based upon the information which is continuously updated through our observation of the trajectory of X, our problem is to terminate the observation (and declare the alarm) at a time τ_* which is as close as possible to θ (measured by a cost function with parameter $c > 0$ specified below).

1.1. The problem can be formally stated as follows. Let $N^{\lambda_0} = (N_t^{\lambda_0})_{t \ge 0}$, $N^{\lambda_1} = (N_t^{\lambda_1})_{t \ge 0}$ and $L = (L_t)_{t \ge 0}$ be three independent stochastic processes defined on a probability-statistical space $(\Omega, \mathcal{F}; \mathsf{P}_\pi, \pi \in [0,1])$ such that:

N^{λ_0} is a Poisson process with intensity $\lambda_0 > 0$; (24.1.1)

N^{λ_1} is a Poisson process with intensity $\lambda_1 > 0$; (24.1.2)

L is a continuous Markov chain with two states λ_0 and λ_1, initial distribution $[1-\pi; \pi]$, and transition-probability matrix $[e^{-\lambda t}, 1-e^{-\lambda t}; 0, 1]$ for $t > 0$ where $\lambda > 0$. (24.1.3)

Thus $\mathsf{P}_\pi(L_0 = \lambda_1) = 1 - \mathsf{P}_\pi(L_0 = \lambda_0) = \pi$, and given that $L_0 = \lambda_0$, there is a single passage of L from λ_0 to λ_1 at a random time $\theta > 0$ satisfying $\mathsf{P}_\pi(\theta > t) = e^{-\lambda t}$ for all $t > 0$.

The process $X = (X_t)_{t \ge 0}$ observed is given by

$$X_t = \int_0^t I(L_{s-} = \lambda_0)\, dN_s^{\lambda_0} + \int_0^t I(L_{s-} = \lambda_1)\, dN_s^{\lambda_1} \tag{24.1.4}$$

and we set $\mathcal{F}_t^X = \sigma\{X_s : 0 \le s \le t\}$ for $t \ge 0$. Denoting $\theta = \inf\{\, t \ge 0 : L_t = \lambda_1 \,\}$ we see that $\mathsf{P}_\pi(\theta = 0) = \pi$ and $\mathsf{P}_\pi(\theta > t \,|\, \theta > 0) = e^{-\lambda t}$ for all $t > 0$. It is assumed that *the time* θ *of "disorder"* is unknown (i.e. it cannot be observed directly).

The *Poisson disorder problem* (or the *quickest detection problem for the Poisson process*) seeks to find a stopping time τ_* of X that is "as close as possible" to θ as a solution of the following optimal stopping problem:

$$V(\pi) = \inf_\tau \Big(\mathsf{P}_\pi(\tau < \theta) + c\, \mathsf{E}_\pi(\tau - \theta)^+ \Big) \tag{24.1.5}$$

where $\mathsf{P}_\pi(\tau < \theta)$ is interpreted as the probability of a *"false alarm"*, $\mathsf{E}_\pi(\tau - \theta)^+$ is interpreted as the *"average delay"* in detecting the occurrence of "disorder" correctly, $c > 0$ is a given constant, and the infimum in (24.1.5) is taken over all stopping times τ of X (compare this with the "Wiener disorder problem" in Section 22 above). A stopping time of X means a stopping time with respect to the natural filtration $(\mathcal{F}_t^X)_{t \ge 0}$ generated by X. The same terminology will be used for other processes in the sequel as well.

1.2. Introducing the *a posteriori probability process*

$$\pi_t = \mathsf{P}_\pi(\theta \le t \,|\, \mathcal{F}_t^X) \tag{24.1.6}$$

for $t \ge 0$, it is easily seen that $\mathsf{P}_\pi(\tau < \theta) = \mathsf{E}_\pi(1-\pi_\tau)$ and $\mathsf{E}_\pi(\tau-\theta)^+ = \mathsf{E}_\pi\big(\int_0^\tau \pi_t\,dt\big)$ for all stopping times τ of X, so that (24.1.5) can be rewritten as follows:

$$V(\pi) = \inf_\tau \mathsf{E}_\pi\bigg((1-\pi_\tau) + c\int_0^\tau \pi_t\,dt\bigg) \tag{24.1.7}$$

where the infimum is taken over all stopping times τ of $(\pi_t)_{t\ge0}$ (as shown following (24.1.12) below).

Define the *likelihood ratio process*

$$\varphi_t = \frac{\pi_t}{1-\pi_t}. \tag{24.1.8}$$

Similarly to the case of a Wiener process (see (22.0.9)) we find that

$$\varphi_t = e^{\lambda t} Z_t\bigg(\varphi_0 + \lambda\int_0^t \frac{e^{-\lambda s}}{Z_s}\,ds\bigg) \tag{24.1.9}$$

where the likelihood process

$$Z_t = \frac{d\mathsf{P}^0}{d\mathsf{P}^\infty}(t,X) = \frac{d(\mathsf{P}^0\,|\,\mathcal{F}_t^X)}{d(\mathsf{P}^\infty\,|\,\mathcal{F}_t^X)} = \exp\bigg(\log\Big(\frac{\lambda_1}{\lambda_0}\Big)X_t - (\lambda_1-\lambda_0)t\bigg) \tag{24.1.10}$$

and the measures P^0 and P^∞ (as well as P^s) are defined analogously to the Wiener process case (thus P^s is the probability law (measure) of the process X given that $\theta = s$ for $s \in [0,\infty]$). From (24.1.9)–(24.1.10) by Itô's formula (page 67) one finds that the processes $(\varphi_t)_{t\ge0}$ and $(\pi_t)_{t\ge0}$ solve the following stochastic equations respectively:

$$d\varphi_t = \lambda(1+\varphi_t)\,dt + \Big(\frac{\lambda_1}{\lambda_0}-1\Big)\varphi_{t-}\;d(X_t-\lambda_0 t), \tag{24.1.11}$$

$$\begin{aligned} d\pi_t = \lambda(1-\pi_t)\,dt &+ \frac{(\lambda_1-\lambda_0)\,\pi_{t-}(1-\pi_{t-})}{\lambda_1\pi_{t-}+\lambda_0(1-\pi_{t-})} \\ &\times\Big(dX_t - \big(\lambda_1\pi_{t-}+\lambda_0(1-\pi_{t-})\big)\,dt\Big). \end{aligned} \tag{24.1.12}$$

It follows that $(\varphi_t)_{t\ge0}$ and $(\pi_t)_{t\ge0}$ are time-homogeneous (strong) Markov processes under P_π with respect to the natural filtrations which clearly coincide with $(\mathcal{F}_t^X)_{t\ge0}$ respectively. Thus, the infimum in (24.1.7) may indeed be viewed as taken over all stopping times τ of $(\pi_t)_{t\ge0}$, and the optimal stopping problem (24.1.7) falls into the class of optimal stopping problems for Markov processes (cf. Chapter I). We thus proceed by finding the infinitesimal operator of the Markov process $(\pi_t)_{t\ge0}$.

1.3. Noting that

$$\mathsf{P}_\pi = \pi\,\mathsf{P}^0 + (1-\pi)\int_0^\infty \lambda e^{-\lambda s}\mathsf{P}^s\,ds \tag{24.1.13}$$

it follows that the so-called *innovation process* $(\bar{X}_t)_{t\ge 0}$ defined by

$$\bar{X}_t = X_t - \int_0^t \mathsf{E}_\pi(L_s \,|\, \mathcal{F}_{s-}^X)\,ds = X_t - \int_0^t \big(\lambda_1\pi_{s-} + \lambda_0(1-\pi_{s-})\big)\,ds \tag{24.1.14}$$

is a martingale under P_π with respect to $(\mathcal{F}_t^X)_{t\ge 0}$ for $\pi \in [0,1]$. Moreover, from (24.1.12) and (24.1.14) we get

$$d\pi_t = \lambda(1-\pi_t)\,dt + \frac{(\lambda_1-\lambda_0)\,\pi_{t-}(1-\pi_{t-})}{\lambda_1\,\pi_{t-} + \lambda_0\,(1-\pi_{t-})}\,d\bar{X}_t. \tag{24.1.15}$$

This implies that the infinitesimal operator of $(\pi_t)_{t\ge 0}$ acts on $f \in C^1[0,1]$ according to the rule

$$\begin{aligned}(\mathbb{L}f)(\pi) &= \big(\lambda - (\lambda_1-\lambda_0)\pi\big)\,(1-\pi)\,f'(\pi) \\ &\quad + \big(\lambda_1\,\pi + \lambda_0\,(1-\pi)\big)\Big(f\Big(\frac{\lambda_1\,\pi}{\lambda_1\,\pi+\lambda_0\,(1-\pi)}\Big) - f(\pi)\Big).\end{aligned} \tag{24.1.16}$$

Note that for $\lambda = 0$ the equations (24.1.11)–(24.1.12) and (24.1.16) reduce to (23.1.15)–(23.1.16) and (23.1.19) respectively.

1.4. Using (24.1.13) it is easily verified that the following facts are valid:

The map $\pi \mapsto V(\pi)$ is concave (continuous) and decreasing on $[0,1]$; (24.1.17)

The stopping time $\tau_* = \inf\{\, t\ge 0 : \pi_t \ge B_* \,\}$ is optimal in the problem (24.1.5)+(24.1.7), where B_* is the smallest π from $[0,1]$ satisfying $V(\pi) = 1-\pi$. (24.1.18)

Thus $V(\pi) < 1-\pi$ for all $\pi \in [0, B_*)$ and $V(\pi) = 1-\pi$ for all $\pi \in [B_*, 1]$. It should be noted in (24.1.18) that $\pi_t = \varphi_t/(1+\varphi_t)$, and hence by (24.1.9)–(24.1.10) we see that π_t is a (path-dependent) functional of the process X observed up to time t. Thus, by observing a trajectory of X it is possible to decide when to stop in accordance with the rule τ_* given in (24.1.18).

The question arises, however, to determine the optimal threshold B_* in terms of the four parameters $\lambda_0, \lambda_1, \lambda, c$ as well as to compute the value $V(\pi)$ for $\pi \in [0, B_*)$ (especially for $\pi = 0$). We tackle these questions by forming a free-boundary problem.

2. The free-boundary problem. Being aided by the general optimal stopping theory of Markov processes (cf. Chapter I), and making use of the preceding facts, we are naturally led to formulate the following *free-boundary problem* for $\pi \mapsto V(\pi)$ and B_* defined above:

$$(\mathbb{L}V)(\pi) = -c\pi \quad (0 < \pi < B_*), \tag{24.1.19}$$
$$V(\pi) = 1 - \pi \quad (B_* \le \pi \le 1), \tag{24.1.20}$$
$$V(B_*-) = 1 - B_* \quad (\textit{continuous fit}). \tag{24.1.21}$$

In some cases (specified below) the following condition will be satisfied as well:

$$V'(B_*) = -1 \quad (\textit{smooth fit}). \tag{24.1.22}$$

However, we will also see below that this condition may fail.

Finally, it is easily verified by passing to the limit for $\pi \downarrow 0$ that each continuous solution of the system (24.1.19)–(24.1.20) must necessarily satisfy

$$V'(0+) = 0 \quad (\textit{normal entrance}) \tag{24.1.23}$$

whenever $V(0+)$ is finite. This condition proves useful in the case when $\lambda_1 < \lambda_0$.

2.1. *Solving the free-boundary problem.* It turns out that the case $\lambda_1 < \lambda_0$ is much different from the case $\lambda_1 > \lambda_0$. Thus assume first that $\lambda_1 > \lambda_0$ and consider the equation (24.1.19) on $(0, B]$ for some $0 < B < 1$ given and fixed. Introduce the "step" function

$$S(\pi) = \frac{\lambda_1 \pi}{\lambda_1 \pi + \lambda_0 (1 - \pi)} \tag{24.1.24}$$

for $\pi \le B$ (cf. (23.1.24)). Observe that $S(\pi) > \pi$ for all $0 < \pi < 1$ and find points $\cdots < B_2 < B_1 < B_0 := B$ such that $S(B_n) = B_{n-1}$ for $n \ge 1$. It is easily verified that

$$B_n = \frac{(\lambda_0)^n B}{(\lambda_0)^n B + (\lambda_1)^n (1 - B)} \quad (n = 0, 1, 2, \ldots). \tag{24.1.25}$$

Denote $I_n = (B_n, B_{n-1}]$ for $n \ge 1$, and introduce the "distance" function

$$d(\pi, B) = 1 + \left[\log\left(\frac{B}{1-B} \, \frac{1-\pi}{\pi} \right) \Big/ \log\left(\frac{\lambda_1}{\lambda_0} \right) \right] \tag{24.1.26}$$

for $\pi \le B$ (cf. (23.1.26)), where $[x]$ denotes the integer part of x. Observe that d is defined to satisfy

$$\pi \in I_n \iff d(\pi, B) = n \tag{24.1.27}$$

for all $0 < \pi \le B$.

Now consider the equation (24.1.19) first on I_1 upon setting $V(\pi) = 1 - \pi$ for $\pi \in (B, S(B)]$. This is then a first-order linear differential equation which can be solved explicitly. Imposing a continuity condition at B (which is in agreement with (24.1.21) above) we obtain a *unique* solution $\pi \mapsto V(\pi; B)$ on I_1. It is possible to verify that the following formula holds:

$$V(\pi; B) = c_1(B)\, V_g(\pi) + V_{p,1}(\pi; B) \qquad (\pi \in I_1) \tag{24.1.28}$$

where $\pi \mapsto V_{p,1}(\pi; B)$ is a (bounded) *particular* solution of the *nonhomogeneous* equation in (24.1.19):

$$V_{p,1}(\pi; B) = -\frac{\lambda_0(\lambda_1 - c)}{\lambda_1(\lambda_0+\lambda)}\,\pi + \frac{\lambda_0\lambda_1+\lambda c}{\lambda_1(\lambda_0+\lambda)} \tag{24.1.29}$$

and $\pi \mapsto V_g(\pi)$ is a *general* solution of the *homogeneous* equation in (24.1.19):

$$V_g(\pi) = \begin{cases} \dfrac{(1-\pi)^{\gamma_1}}{|\lambda - (\lambda_1 - \lambda_0)\pi|^{\gamma_0}}, & \text{if } \lambda \neq \lambda_1 - \lambda_0, \\ (1-\pi)\exp\Big(\dfrac{\lambda_1}{(\lambda_1-\lambda_0)(1-\pi)}\Big), & \text{if } \lambda = \lambda_1 - \lambda_0, \end{cases} \tag{24.1.30}$$

where $\gamma_1 = \lambda_1/(\lambda_1 - \lambda_0 - \lambda)$ and $\gamma_0 = (\lambda_0+\lambda)/(\lambda_1 - \lambda_0 - \lambda)$, and the constant $c_1(B)$ is determined by the continuity condition $V(B-; B) = 1 - B$ leading to

$$c_1(B) = -\frac{1}{V_g(B)}\left(\frac{\lambda_1\lambda+\lambda_0 c}{\lambda_1(\lambda_0+\lambda)}\,B - \frac{\lambda(\lambda_1 - c)}{\lambda_1(\lambda_0+\lambda)}\right) \tag{24.1.31}$$

where $V_g(B)$ is obtained by replacing π in (24.1.30) by B. [We observe from (24.1.29)–(24.1.31) however that the continuity condition at B cannot be met when B equals $\widehat{B}$ from (24.1.34) below unless $\widehat{B}$ equals $\lambda(\lambda_1 - c)/(\lambda\lambda_1 + c\lambda_0)$ from (24.1.41) below (the latter is equivalent to $c = \lambda_1 - \lambda_0 - \lambda$). Thus, if $B = \widehat{B} \neq \lambda(\lambda_1 - c)/(\lambda\lambda_1 + c\lambda_0)$ then there is no solution $\pi \mapsto V(\pi; B)$ on I_1 that satisfies $V(\pi; B) = 1 - \pi$ for $\pi \in (B, S(B)]$ and is continuous at B. It turns out, however, that this analytic fact has no significant implication for the solution of (24.1.5)+(24.1.7).]

Next consider the equation (24.1.19) on I_2 upon using the solution found on I_1 and setting $V(\pi) = c_1(B)\, V_g(\pi) + V_{p,1}(\pi; B)$ for $\pi \in (B_1, S(B_1)]$. This is then again a first-order linear differential equation which can be solved explicitly. Imposing a continuity condition over $I_2 \cup I_1$ at B_1 (which is in agreement with (24.1.17) above) we obtain a unique solution $\pi \mapsto V(\pi; B)$ on I_2. It turns out, however, that the general solution of this equation cannot be expressed in terms of elementary functions (unless $\lambda = 0$ as shown in Subsection 23.1 above) but one needs, for instance, the Gauss hypergeometric function. As these expressions are increasingly complex to record, we omit the explicit formulae in the sequel.

Continuing the preceding procedure by induction as long as possible (considering the equation (24.1.19) on I_n upon using the solution found on I_{n-1} and imposing a continuity condition over $I_n \cup I_{n-1}$ at B_{n-1}) we obtain a unique solution $\pi \mapsto V(\pi; B)$ on I_n given as

$$V(\pi; B) = c_n(B)\, V_g(\pi) + V_{p,n}(\pi; B) \qquad (\pi \in I_n) \tag{24.1.32}$$

where $\pi \mapsto V_{p,n}(\pi; B)$ is a (bounded) particular solution, $\pi \mapsto V_g(\pi)$ is a general solution given by (24.1.30), and $B \mapsto c_n(B)$ is a function of B (and the four parameters). [We will see however in Theorem 24.1 below that in the case $B > \widehat{B} > 0$ with $\widehat{B}$ from (24.1.34) below the solution (24.1.32) exists for $\pi \in (\widehat{B}, B]$ but explodes at $\widehat{B}$ unless $B = B_*$.]

The key difference in the case $\lambda_1 < \lambda_0$ is that $S(\pi) < \pi$ for all $0 < \pi < 1$ so that we need to deal with points $B := B_0 < B_1 < B_2 < \cdots$ such that $S(B_n) = B_{n-1}$ for $n \geq 1$. Then the facts (24.1.25)–(24.1.27) remain preserved provided that we set $I_n = [B_{n-1}, B_n)$ for $n \geq 1$. In order to prescribe the initial condition when considering the equation (24.1.19) on I_1 , we can take $B = \varepsilon > 0$ small and make use of (24.1.23) upon setting $V(\pi) = v$ for all $\pi \in [S(B), B)$ where $v \in (0,1)$ is a given number satisfying $V(B) = v$. Proceeding by induction as earlier (considering the equation (24.1.19) on I_n upon using the solution found on I_{n-1} and imposing a continuity condition over $I_{n-1} \cup I_n$ at B_{n-1}) we obtain a unique solution $\pi \mapsto V(\pi; \varepsilon, v)$ on I_n given as

$$V(\pi; \varepsilon, v) = c_n(\varepsilon)\, V_g(\pi) + V_{p,n}(\pi; \varepsilon, v) \qquad (\pi \in I_n) \tag{24.1.33}$$

where $\pi \mapsto V_{p,n}(\pi; \varepsilon, v)$ is a particular solution, $\pi \mapsto V_g(\pi)$ is a general solution given by (24.1.30), and $\varepsilon \mapsto c_n(\varepsilon)$ is a function of ε (and the four parameters). We shall see in Theorem 24.1 below how these solutions can be used to determine the optimal $\pi \mapsto V(\pi)$ and B_* .

2.2. *Two key facts about the solution.* Both of these facts hold only in the case when $\lambda_1 > \lambda_0$ and they will be used in the proof of Theorem 24.1 stated below. The first fact to be observed is that

$$\widehat{B} = \frac{\lambda}{\lambda_1 - \lambda_0} \tag{24.1.34}$$

is a *singularity point* of the equation (24.1.19) whenever $\lambda < \lambda_1 - \lambda_0$. This is clearly seen from (24.1.30) where $V_g(\pi) \to \infty$ for $\pi \to \widehat{B}$. The second fact of interest is that

$$\widetilde{B} = \frac{\lambda}{\lambda + c} \tag{24.1.35}$$

is a *smooth-fit point* of the system (24.1.19)–(24.1.21) whenever $\lambda_1 > \lambda_0$ and $c \neq \lambda_1 - \lambda_0 - \lambda$, i.e. $V'(\widetilde{B}-; \widetilde{B}) = -1$ in the notation of (24.1.32) above. This can be verified by (24.1.28) using (24.1.29)–(24.1.31). It means that $\widetilde{B}$ is the unique point which in addition to (24.1.19)–(24.1.21) has the power of satisfying the smooth-fit condition (24.1.22).

It may also be noted in the verification above that the equation $V'(B-;B) = -1$ has no solution when $c = \lambda_1 - \lambda_0 - \lambda$ as the only candidate $\bar{B} := \widetilde{B} = \widehat{B}$ satisfies

$$V'(\bar{B}-;\bar{B}) = -\frac{\lambda_0}{\lambda_1}. \tag{24.1.36}$$

This identity follows readily from (24.1.28)–(24.1.31) upon noticing that $c_1(\bar{B}) = 0$. Thus, when c runs from $+\infty$ to $\lambda_1 - \lambda_0 - \lambda$, the smooth-fit point $\widetilde{B}$ runs from 0 to the singularity point $\widehat{B}$, and once $\widetilde{B}$ has reached $\widehat{B}$ for $c = \lambda_1 - \lambda_0 - \lambda$, the smooth-fit condition (24.1.22) breaks down and gets replaced by the condition (24.1.36) above. We will soon attest below that in all these cases the smooth-fit point $\widetilde{B}$ is actually equal to the optimal-stopping point B_* from (24.1.18) above.

Observe that the equation (24.1.19) has no singularity points when $\lambda_1 < \lambda_0$. This analytic fact reveals a key difference between the two cases.

3. Conclusions. In parallel to the two analytic properties displayed above we begin this part by stating the relevant *probabilistic properties* of the a posteriori probability process.

3.1. *Sample-path properties of* $(\pi_t)_{t \ge 0}$. First consider the case $\lambda_1 > \lambda_0$. Then from (24.1.12) we see that $(\pi_t)_{t \ge 0}$ can only jump towards 1 (at times of the jumps of the process X). Moreover, the sign of the drift term $\lambda(1-\pi) - (\lambda_1 - \lambda_0)\pi(1-\pi) = (\lambda_1 - \lambda_0)(\widehat{B} - \pi)(1-\pi)$ is determined by the sign of $\widehat{B} - \pi$. Hence we see that $(\pi_t)_{t \ge 0}$ has a positive drift in $[0, \widehat{B})$, a negative drift in $(\widehat{B}, 1]$, and a zero drift at $\widehat{B}$. Thus, if $(\pi_t)_{t \ge 0}$ starts or ends up at $\widehat{B}$, it is trapped there until the first jump of the process X occurs. At that time $(\pi_t)_{t \ge 0}$ finally leaves $\widehat{B}$ by jumping towards 1. This also shows that once $(\pi_t)_{t \ge 0}$ leaves $[0, \widehat{B})$ it never comes back. The sample-path behaviour of $(\pi_t)_{t \ge 0}$ when $\lambda_1 > \lambda_0$ is depicted in *Figure* VI.8 (Part i).

Next consider the case $\lambda_1 < \lambda_0$. Then from (24.1.12) we see that $(\pi_t)_{t \ge 0}$ can only jump towards 0 (at times of the jumps of the process X). Moreover, the sign of the drift term $\lambda(1-\pi) - (\lambda_1 - \lambda_0)\pi(1-\pi) = (\lambda + (\lambda_0 - \lambda_1)\pi)(1-\pi)$ is always positive. Thus $(\pi_t)_{t \ge 0}$ always moves continuously towards 1 and can only jump towards 0. The sample-path behaviour of $(\pi_t)_{t \ge 0}$ when $\lambda_1 < \lambda_0$ is depicted in *Figure* VI.8 (Part ii).

3.2. *Sample-path behaviour and the principles of smooth and continuous fit.* With a view to (24.1.18), and taking $0 < B < 1$ given and fixed, we shall now examine the manner in which the process $(\pi_t)_{t \ge 0}$ enters $[B, 1]$ if starting at $B - d\pi$ where $d\pi$ is infinitesimally small (or equivalently enters $(B, 1]$ if starting at B). Our previous analysis then shows the following (see *Figure* VI.8).

If $\lambda_1 > \lambda_0$ and $B < \widehat{B}$, or $\lambda_1 < \lambda_0$, then $(\pi_t)_{t \ge 0}$ enters $[B, 1]$ by passing through B continuously. If, however, $\lambda_1 > \lambda_0$ and $B > \widehat{B}$, then the only way for

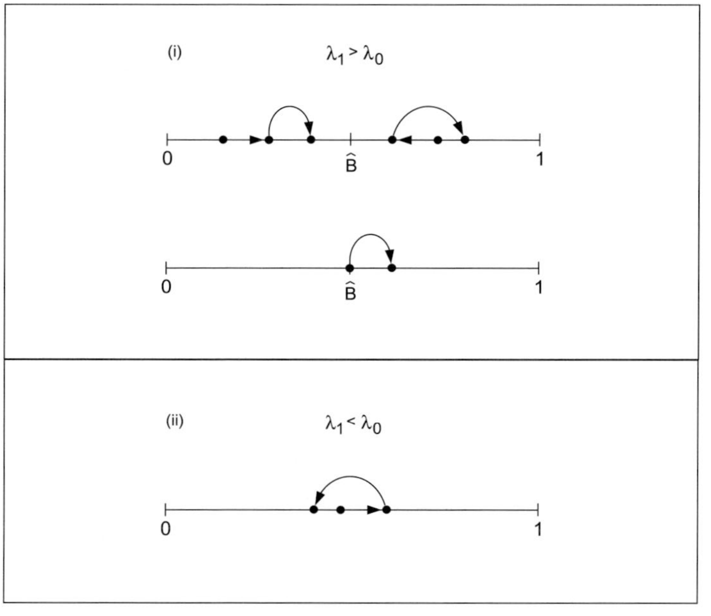

Figure VI.8: Sample-path properties of the a posteriori probability process $(\pi_t)_{t\ge 0}$ from (24.1.6)+(24.1.12). The point $\widehat{B}$ is a singularity point (24.1.34) of the free-boundary equation (24.1.19).

$(\pi_t)_{t\ge 0}$ to enter $[B,1]$ is by jumping over B. (Jumping exactly at B happens with probability zero.)

The case $\lambda_1 > \lambda_0$ and $B = \widehat{B}$ is special. If starting outside $[B,1]$ then $(\pi_t)_{t\ge 0}$ travels towards $\widehat{B}$ by either moving continuously or by jumping. However, the closer $(\pi_t)_{t\ge 0}$ gets to $\widehat{B}$ the smaller the drift to the right becomes, and if there were no jump over $\widehat{B}$ eventually, the process $(\pi_t)_{t\ge 0}$ would never reach $\widehat{B}$ as the drift to the right tends to zero together with the distance of $(\pi_t)_{t\ge 0}$ to $\widehat{B}$. This fact can be formally verified by analysing the explicit representation of $(\varphi_t)_{t\ge 0}$ in (24.1.9)–(24.1.10) and using that $\pi_t = \varphi_t/(1+\varphi_t)$ for $t \ge 0$. Thus, in this case as well, the only way for $(\pi_t)_{t\ge 0}$ to enter $[\widehat{B},1]$ after starting at $B - d\pi$ is by jumping over to $(\widehat{B},1]$.

We will demonstrate below that the sample-path behaviour of the process $(\pi_t)_{t\ge 0}$ during the entrance of $[B_*,1]$ has a precise analytic counterpart in terms of the free-boundary problem (24.1.19). If the process $(\pi_t)_{t\ge 0}$ may enter $[B_*,1]$ by passing through B_* *continuously*, then the *smooth-fit condition* (24.1.22) *holds* at B_* ; if, however, the process $(\pi_t)_{t\geq 0}$ enters $[B_*, 1]$ exclusively by *jumping* over B_* , then the *smooth-fit condition* (24.1.22) *breaks down.* In this case the continuous-fit condition (24.1.21) still holds at B_* , and the existence of a singularity point $\widehat{B}$ can be used to determine the optimal B_* as shown below.

Due to the fact that the times of jumps of the process $(\pi_t)_{t\geq 0}$ are 'sufficiently apart' it is evident that the preceding two sample-path behaviors can be rephrased in terms of *regularity* of the boundary point B_* as discussed in Section 7 above.

3.3. The preceding considerations may now be summarized as follows.

Theorem 24.1. *Consider the Poisson disorder problem* (24.1.5) *and the equivalent optimal-stopping problem* (24.1.7) *where the process* $(\pi_t)_{t\geq 0}$ *from* (24.1.6) *solves* (24.1.12) *and* $\lambda_0, \lambda_1, \lambda, c > 0$ *are given and fixed.*

Then there exists $B_* \in (0,1)$ *such that the stopping time*

$$\tau_* = \inf\{\, t \geq 0 : \pi_t \geq B_* \,\} \tag{24.1.37}$$

is optimal in (24.1.5) *and* (24.1.7). *Moreover, the optimal cost function* $\pi \mapsto V(\pi)$ *from* (24.1.5)+(24.1.7) *solves the free-boundary problem* (24.1.19)–(24.1.21), *and the optimal threshold* B_* *is determined as follows.*

(i): *If* $\lambda_1 > \lambda_0$ *and* $c > \lambda_1 - \lambda_0 - \lambda$, *then the smooth-fit condition* (24.1.22) *holds at* B_* , *and the following explicit formula is valid*:

$$B_* = \frac{\lambda}{\lambda + c}. \tag{24.1.38}$$

In this case $B_* < \widehat{B}$ *where* $\widehat{B}$ *is a singularity point of the free-boundary equation* (24.1.19) *given in* (24.1.34) *above* (*see* Figure VI.9).

(ii): *If* $\lambda_1 > \lambda_0$ *and* $c = \lambda_1 - \lambda_0 - \lambda$, *then the smooth-fit condition breaks down at* B_* *and gets replaced by the condition* (24.1.36) *above* ($V'(B_*-) = -\lambda_0/\lambda_1$) . *The optimal threshold* B_* *is still given by* (24.1.38), *and in this case* $B_* = \widehat{B}$ (*see* Figure VI.10).

(iii): *If* $\lambda_1 > \lambda_0$ *and* $c < \lambda_1 - \lambda_0 - \lambda$, *then the smooth-fit condition does not hold at* B_* , *and the optimal threshold* B_* *is determined as a unique solution in* $(\widehat{B}, 1)$ *of the following equation*:

$$c_{d(\widehat{B}, B_*)}(B_*) = 0 \tag{24.1.39}$$

where the map $B \mapsto d(\widehat{B}, B)$ *is defined in* (24.1.26), *and the map* $B \mapsto c_n(B)$ *is defined by* (24.1.31) *and* (24.1.32) *above* (*see Figure* VI.11). *In particular, when* c *satisfies*

$$\frac{\lambda_1 \lambda_0\, (\lambda_1 - \lambda_0 - \lambda)}{\lambda_1 \lambda_0 + (\lambda_1 - \lambda_0)(\lambda - \lambda_0)} \leq c < \lambda_1 - \lambda_0 - \lambda, \tag{24.1.40}$$

then the following explicit formula is valid:

$$B_* = \frac{\lambda\,(\lambda_1 - c)}{\lambda\lambda_1 + c\lambda_0} \tag{24.1.41}$$

which in the case $c = \lambda_1 - \lambda_0 - \lambda$ *reduces again to* (24.1.38) *above.*

In the cases (i)–(iii) *the optimal cost function* $\pi \mapsto V(\pi)$ *from* (24.1.5)+(24.1.7) *is given by* (24.1.32) *with* B_* *in place of* B *for all* $0 < \pi \le B_*$ *(with* $V(0) = V(0+)$ *) and* $V(\pi) = 1 - \pi$ *for* $B_* \le \pi \le 1$.

(iv): *If* $\lambda_1 < \lambda_0$ *then the smooth-fit condition holds at* B_* *, and the optimal threshold* B_* *can be determined using the normal entrance condition* (24.1.23) *as follows (see Figure* VI.12*). For* $\varepsilon > 0$ *small let* v_ε *denote a unique number in* $(0,1)$ *for which the map* $\pi \mapsto V(\pi;\varepsilon,v_\varepsilon)$ *from* (24.1.33) *hits the map* $\pi \mapsto 1-\pi$ *smoothly at some* B_*^ε *from* $(0,1)$ *. Then we have*

$$B_* = \lim_{\varepsilon\downarrow 0} B_*^\varepsilon, \tag{24.1.42}$$

$$V(\pi) = \lim_{\varepsilon\downarrow 0} V(\pi;\varepsilon,v_\varepsilon) \tag{24.1.43}$$

for all $0 < \pi \le B_*$ *(with* $V(0) = V(0+)$ *) and* $V(\pi) = 1 - \pi$ *for* $B_* \le \pi \le 1$.

Proof. We have already established in (24.1.18) above that τ_* from (24.1.37) is optimal in (24.1.5) and (24.1.7) for some $B_* \in [0,1]$ to be found. It thus follows by the strong Markov property of the process $(\pi_t)_{t\ge 0}$ together with (24.1.17) above that the optimal cost function $\pi \mapsto V(\pi)$ from (24.1.5)+(24.1.7) solves the free-boundary problem (24.1.19)–(24.1.21). Some of these facts will also be reproved below.

First consider the case $\lambda_1 > \lambda_0$. In paragraph 2.1 above it was shown that for each given and fixed $B \in (0,\widehat{B})$ the problem (24.1.19)–(24.1.21) with B in place of B_* has a unique continuous solution given by the formula (24.1.32). Moreover, this solution is (at least) C^1 everywhere but possibly at B where it is (at least) C^0 . As explained following (24.1.31) above, these facts also hold for $B = \widehat{B}$ when $\widehat{B}$ equals $\lambda(\lambda_1 - c)/(\lambda\lambda_1 + c\lambda_0)$ from (24.1.41) above. We will now show how the optimal threshold B_* is determined among all these candidates B when $c \ge \lambda_1 - \lambda_0 - \lambda$.

(i)+(ii): Since the innovation process (24.1.14) is a martingale under P_π with respect to $(\mathcal{F}_t^X)_{t\ge 0}$, it follows by (24.1.15) that

$$\pi_t = \pi + \lambda \int_0^t (1 - \pi_{s-})\, ds + M_t \tag{24.1.44}$$

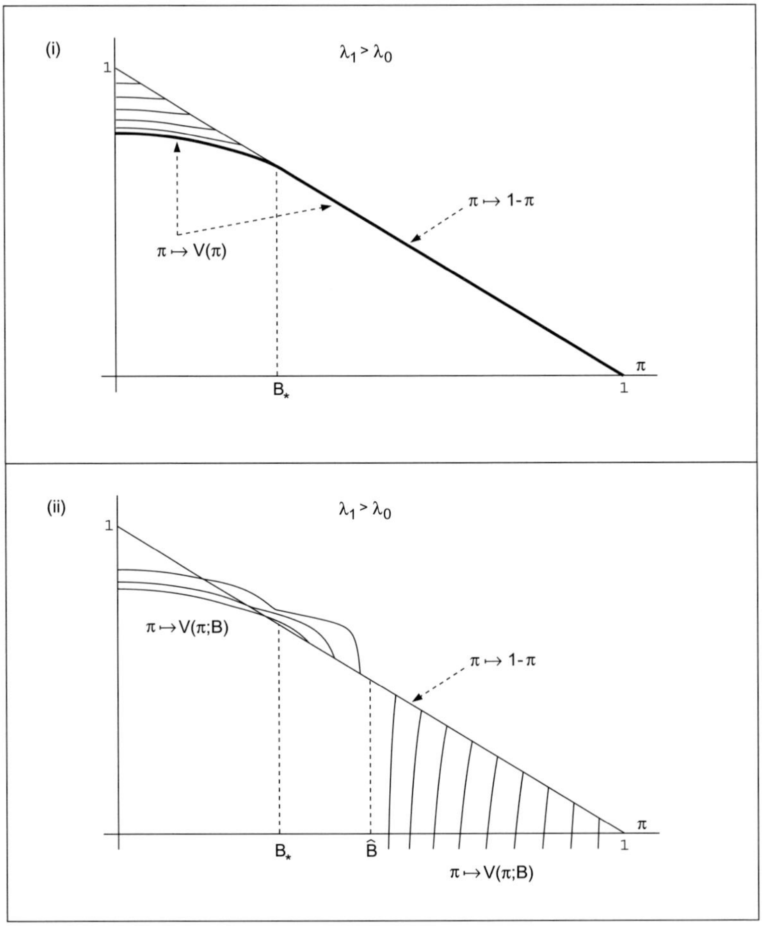

Figure VI.9: A computer drawing of the maps $\pi \mapsto V(\pi; B)$ from (24.1.32) for different B from $(0,1)$ in the case $\lambda_1 = 4$, $\lambda_0 = 2$, $\lambda = 1$, $c = 2$. The singularity point $\widehat{B}$ from (24.1.34) equals $1/2$, and the smooth-fit point $\widetilde{B}$ from (24.1.35) equals $1/3$. The optimal threshold B_* coincides with the smooth-fit point $\widetilde{B}$. The value function $\pi \mapsto V(\pi)$ from (24.1.5)+(24.1.7) equals $\pi \mapsto V(\pi; B_*)$ for $0 \le \pi \le B_*$ and $1 - \pi$ for $B_* \le \pi \le 1$. (This is presented in Part (i) above.) The solutions $\pi \mapsto V(\pi; B)$ for $B > B_*$ are ruled out since they fail to satisfy $0 \le V(\pi) \le 1 - \pi$ for all $\pi \in [0,1]$. (This is shown in Part (ii) above.) The general case $\lambda_1 > \lambda_0$ with $c > \lambda_1 - \lambda_0 - \lambda$ looks very much the same.

where $M = (M_t)_{t \geq 0}$ is a martingale under P_π with respect to $(\mathcal{F}_t^X)_{t \geq 0}$. Hence by the optional sampling theorem (page 60) we easily find

$$\begin{aligned} &\mathsf{E}_\pi\bigg((1-\pi_\tau) + c \int_0^\tau \pi_t \, dt \bigg) \qquad (24.1.45) \\ &\quad = (1-\pi) + (\lambda + c)\, \mathsf{E}_\pi \bigg(\int_0^\tau \Big(\pi_t - \frac{\lambda}{\lambda + c} \Big)\, dt \bigg) \end{aligned}$$

for all stopping times τ of $(\pi_t)_{t \geq 0}$. Recalling the sample-path behaviour of $(\pi_t)_{t \geq 0}$ in the case $\lambda_1 > \lambda_0$ as displayed in paragraph 3.1 above (cf. *Figure* VI.8 (Part i)), and the definition of $V(\pi)$ in (24.1.7) together with the fact that $\widetilde{B} = \lambda/(\lambda+c) \leq \widehat{B}$ when $c \geq \lambda_1 - \lambda_0 - \lambda$, we clearly see from (24.1.45) that it is never optimal to stop $(\pi_t)_{t \geq 0}$ in $[0, \widetilde{B})$, as well as that $(\pi_t)_{t \geq 0}$ must be stopped immediately after entering $[\widetilde{B}, 1]$ as it will never return to the "favourable" set $[0, \widetilde{B})$ again. This proves that $\widetilde{B}$ equals the optimal threshold B_*, i.e. that τ_* from (24.1.37) with B_* from (24.1.38) is optimal in (24.1.5) and (24.1.7). The claim about the breakdown of the smooth-fit condition (24.1.22) when $c = \lambda_1 - \lambda_0 - \lambda$ has been already established in paragraph 2.2 above (cf. *Figure* VI.10).

(iii): It was shown in paragraph 2.1 above that for each given and fixed $B \in (\widehat{B}, 1)$ the problem (24.1.19)–(24.1.21) with B in place of B_* has a unique continuous solution on $(\widehat{B}, 1]$ given by the formula (24.1.32). We will now show that there exists a unique point $B_* \in (\widehat{B}, 1)$ such that $\lim_{\pi \downarrow \widehat{B}} V(\pi; B) = \pm\infty$ if $B \in (\widehat{B}, B_*) \cup (B_*, 1)$ and $\lim_{\pi \downarrow \widehat{B}} V(\pi; B_*)$ is finite. This point is the optimal threshold, i.e. the stopping time τ_* from (24.1.37) is optimal in (24.1.5) and (24.1.7). Moreover, the point B_* can be characterized as a unique solution of the equation (24.1.39) in $(\widehat{B}, 1)$.

In order to verify the preceding claims we will first state the following observation which proves useful. Setting $g(\pi) = 1 - \pi$ for $0 < \pi < 1$ we have

$$(\mathbb{L}g)(\pi) \geq -c\pi \iff \pi \geq \widetilde{B} \qquad (24.1.46)$$

where $\widetilde{B}$ is given in (24.1.35). This is verified straightforwardly using (24.1.16).

Now since $\widehat{B}$ is a singularity point of the equation (24.1.19) (recall our discussion in paragraph 2.2 above), and moreover $\pi \mapsto V(\pi)$ from (24.1.5)+(24.1.7) solves (24.1.19)–(24.1.21), we see that the optimal threshold B_* from (24.1.18) must satisfy (24.1.39). This is due to the fact that a particular solution $\pi \mapsto V_{p,n}(\pi; B_*)$ for $n = d(\widehat{B}, B_*)$ in (24.1.32) above is taken bounded. The key remaining fact to be established is that there cannot be two (or more) points in $(\widehat{B}, 1)$ satisfying (24.1.39).

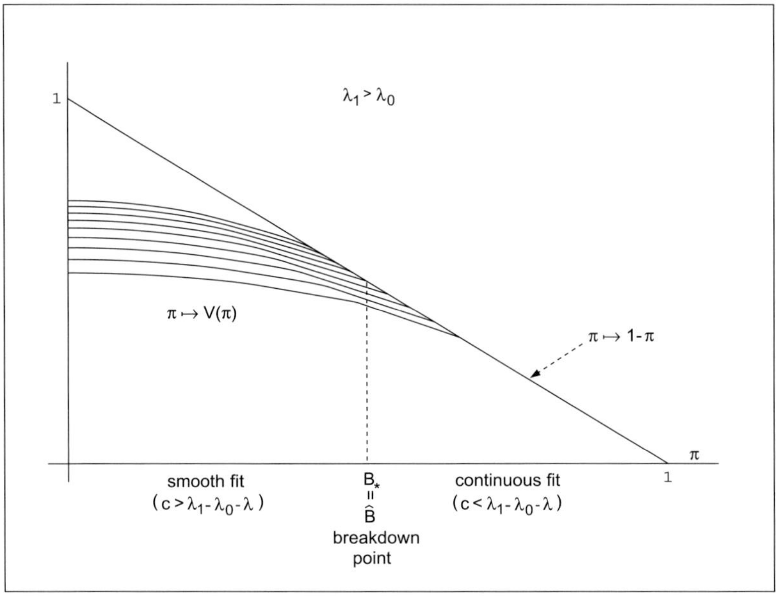

Figure VI.10: A computer drawing of the value functions $\pi \mapsto V(\pi)$ from (24.1.5)+(24.1.7) in the case $\lambda_1 = 4$, $\lambda_0 = 2$, $\lambda = 1$ and $c = 1.4,\ 1.3,\ 1.2,\ 1.1,\ 1,\ 0.9,\ 0.8,\ 0.7,\ 0.6$. The given $V(\pi)$ equals $V(\pi; B_*)$ from (24.1.32) for all $0 < \pi \le B_*$ where B_* as a function of c is given by (24.1.38) and (24.1.41). The smooth-fit condition (24.1.22) holds in the cases $c = 1.4,\ 1.3,\ 1.2,\ 1.1$. The point $c = 1$ is a breakdown point when the optimal threshold B_* equals the singularity point $\widehat{B}$ from (24.1.34), and the smooth-fit condition gets replaced by the condition (24.1.36) with $\bar{B} = B_* = \widehat{B} = 0.5$ in this case. For $c = 0.9,\ 0.8,\ 0.7,\ 0.6$ the smooth-fit condition (24.1.22) does not hold. In these cases the continuous-fit condition (24.1.21) is satisfied. Moreover, numerical computations suggest that the mapping $B_* \mapsto V'(B_*-; B_*)$ which equals -1 for $0 < B_* < \widehat{B}$ and jumps to $-\lambda_0/\lambda_1 = -0.5$ for $B_* = \widehat{B}$ is decreasing on $[\widehat{B}, 1)$ and tends to a value slightly larger than -0.6 when $B_* \uparrow 1$ that is $c \downarrow 0$. The general case $\lambda_1 > \lambda_0$ looks very much the same.

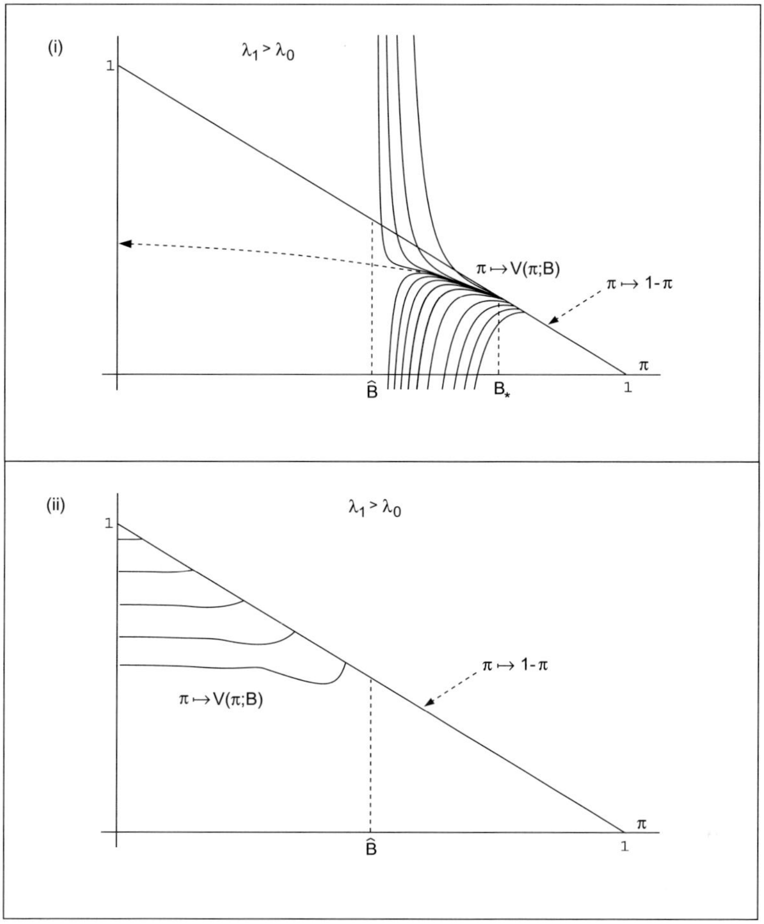

Figure VI.11: A computer drawing of the maps $\pi \mapsto V(\pi;B)$ from (24.1.32) for different B from $(0,1)$ in the case $\lambda_1 = 4$, $\lambda_0 = 2$, $\lambda = 1$, $c = 2/5$. The singularity point $\widehat{B}$ from (24.1.34) equals $1/2$. The optimal threshold B_* can be determined from the fact that all solutions $\pi \mapsto V(\pi;B)$ for $B > B_*$ hit zero for some $\pi > \widehat{B}$, and all solutions $\pi \mapsto V(\pi;B)$ for $B < B_*$ hit $1-\pi$ for some $\pi > \widehat{B}$. (This is shown in Part (i) above.) A simple numerical method based on the preceding fact suggests the estimates $0.750 < B_* < 0.752$. The value function $\pi \mapsto V(\pi)$ from (24.1.5)+(24.1.7) equals $\pi \mapsto V(\pi;B_*)$ for $0 \le \pi \le B_*$ and $1-\pi$ for $B_* \le \pi \le 1$. The solutions $\pi \mapsto V(\pi;B)$ for $B \le \widehat{B}$ are ruled out since they fail to be concave. (This is shown in Part (ii) above.) The general case $\lambda_1 > \lambda_0$ with $c < \lambda_1 - \lambda_0 - \lambda$ looks very much the same.

Assume on the contrary that there are two such points B_1 and B_2. We may however assume that both B_1 and B_2 are larger than $\widetilde{B}$ since for $B \in (\widehat{B}, \widetilde{B})$ the solution $\pi \mapsto V(\pi; B)$ is ruled out by the fact that $V(\pi; B) > 1 - \pi$ for $\pi \in (B - \varepsilon, B)$ with $\varepsilon > 0$ small. This fact is verified directly using (24.1.28)–(24.1.31). Thus, each map $\pi \mapsto V(\pi; B_i)$ solves (24.1.19)–(24.1.21) on $(0, B_i]$ and is continuous (bounded) at $\widehat{B}$ for $i = 1, 2$. Since $S(\pi) > \pi$ for all $0 < \pi < 1$ when $\lambda_1 > \lambda_0$, it follows easily from (24.1.16) that each solution $\pi \mapsto V(\pi; B_i)$ of (24.1.19)–(24.1.21) must also satisfy $-\infty < V(0+; B_i) < +\infty$ for $i = 1, 2$.

In order to make use of the preceding fact we shall set $h_\beta(\pi) = (1 + (\beta - 1)\widehat{B}) - \beta\pi$ for $0 \le \pi \le \widehat{B}$ and $h_\beta(\pi) = 1 - \pi$ for $\widehat{B} \le \pi \le 1$. Since both maps $\pi \mapsto V(\pi; B_i)$ are bounded on $(0, \widehat{B})$ we can fix $\beta > 0$ large enough so that $V(\pi; B_i) \le h_\beta(\pi)$ for all $0 < \pi \le \widehat{B}$ and $i = 1, 2$. Consider then the auxiliary optimal stopping problem

$$W(\pi) := \inf_{\tau} \mathsf{E}_\pi \left(h_\beta(\pi_\tau) + c \int_0^\tau \pi_t \, dt \right) \tag{24.1.47}$$

where the supremum is taken over all stopping times τ of $(\pi_t)_{t \ge 0}$. Extend the map $\pi \mapsto V(\pi; B_i)$ on $[B_i, 1]$ by setting $V(\pi; B_i) = 1 - \pi$ for $B_i \le \pi \le 1$ and denote the resulting (continuous) map on $[0, 1]$ by $\pi \mapsto V_i(\pi)$ for $i = 1, 2$. Then $\pi \mapsto V_i(\pi)$ satisfies (24.1.19)–(24.1.21), and since $B_i \ge \widetilde{B}$, we see by means of (24.1.46) that the following condition is also satisfied:

$$(\mathbb{L}V_i)(\pi) \ge -c\pi \tag{24.1.48}$$

for $\pi \in [B_i, 1]$ and $i = 1, 2$. We will now show that the preceding two facts have the power of implying that $V_i(\pi) = W(\pi)$ for all $\pi \in [0, 1]$ with either $i \in \{1, 2\}$ given and fixed.

It follows by Itô's formula (page 67) that

$$V_i(\pi_t) = V_i(\pi) + \int_0^t (\mathbb{L}V_i)(\pi_{s-}) \, ds + M_t \tag{24.1.49}$$

where $M = (M_t)_{t \ge 0}$ is a martingale (under P_π) given by

$$M_t = \int_0^t \Big(V_i(\pi_{s-} + \Delta\pi_s) - V_i(\pi_{s-}) \Big) \, d\widehat{X}_s \tag{24.1.50}$$

and $\widehat{X}_t = X_t - \int_0^t \big(\lambda_1 \pi_{s-} + \lambda_0(1 - \pi_{s-})\big) \, ds$ is the innovation process. By the optional sampling theorem (page 60) it follows from (24.1.49) using (24.1.48) and the fact that $V_i(\pi) \le h_\beta(\pi)$ for all $\pi \in [0, 1]$ that $V_i(\pi) \le W(\pi)$ for all $\pi \in [0, 1]$. Moreover, defining $\tau_i = \inf\{\, t \ge 0 : \pi_t \ge B_i \,\}$ it is easily seen e.g. by (24.1.44) that $\mathsf{E}_\pi \tau_i < \infty$. Using then that $\pi \mapsto V_i(\pi)$ is bounded on $[0, 1]$, it follows easily by the optional sampling theorem (page 60) that $\mathsf{E}_\pi(M_{\tau_i}) = 0$. Since

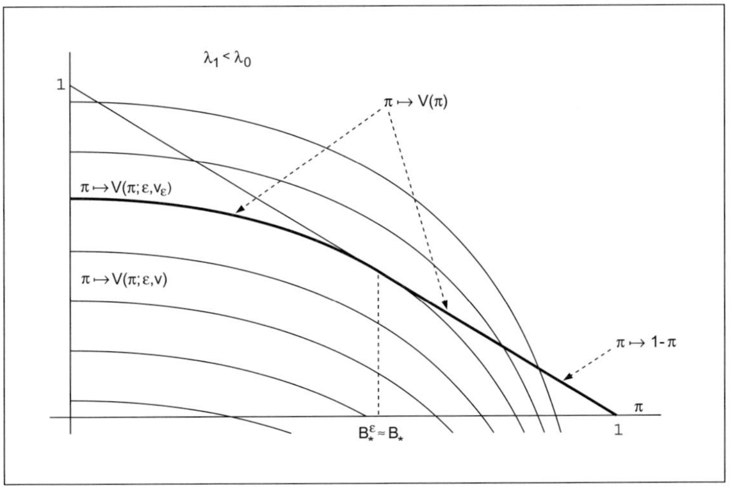

Figure VI.12: A computer drawing of the maps $\pi \mapsto V(\pi;\varepsilon,v)$ from (24.1.33) for different v from $(0,1)$ with $\varepsilon = 0.001$ in the case $\lambda_1 = 2$, $\lambda_0 = 4$, $\lambda = 1$, $c = 1$. For each $\varepsilon > 0$ there is a unique number $v_\varepsilon \in (0,1)$ such that the map $\pi \mapsto V(\pi;\varepsilon,v_\varepsilon)$ hits the map $\pi \mapsto 1-\pi$ smoothly at some $B_*^\varepsilon \in (0,1)$. Letting $\varepsilon \downarrow 0$ we obtain $B_*^\varepsilon \to B_*$ and $V(\pi;\varepsilon,v_\varepsilon) \to V(\pi)$ for all $\pi \in [0,1]$ where B_* is the optimal threshold from (24.1.18) and $\pi \mapsto V(\pi)$ is the value function from (24.1.5)+(24.1.7).

moreover $V_i(\pi_{\tau_i}) = h_\beta(\pi_{\tau_i})$ and $(\mathbb{L}V_i)(\pi_{s-}) = -c\pi_{s-}$ for all $s \le \tau_i$, we see from (24.1.49) that the inequality $V_i(\pi) \le W(\pi)$ derived above is actually equality for all $\pi \in [0,1]$. This proves that $V(\pi;B_1) = V(\pi;B_2)$ for all $\pi \in [0,1]$, or in other words, that there cannot be more than one point B_* in $(\widehat{B},1)$ satisfying (24.1.39). Thus, there is only one solution $\pi \mapsto V(\pi)$ of (24.1.19)–(24.1.21) which is finite at $\widehat{B}$ (see *Figure* VI.11), and the proof of the claim is complete.

(iv): It was shown in paragraph 2.1 above that the map $\pi \mapsto V(\pi;\varepsilon,v)$ from (24.1.33) is a unique continuous solution of the equation $(\mathbb{L}V)(\pi) = -c\pi$ for $\varepsilon < \pi < 1$ satisfying $V(\pi) = v$ for all $\pi \in [S(\varepsilon),\varepsilon]$. It can be checked using (24.1.30) that

$$V_{p,1}(\pi;\varepsilon,v) = \frac{c\lambda_0}{\lambda_1(\lambda_0+\lambda)}\,\pi + \frac{c\lambda}{\lambda_1(\lambda_0+\lambda)} + v, \tag{24.1.51}$$

$$c_1(\varepsilon) = -\,\frac{1}{V_g(\varepsilon)}\left(\frac{c\lambda_0}{\lambda_1(\lambda_0+\lambda)}\,\varepsilon + \frac{c\lambda}{\lambda_1(\lambda_0+\lambda)}\right) \tag{24.1.52}$$

for $\pi \in I_1 = [\varepsilon, \varepsilon_1)$ where $S(\varepsilon_1) = \varepsilon$. Moreover, it may be noted directly from (24.1.16) above that $\mathbb{L}(f+c) = \mathbb{L}(f)$ for every constant c, and thus $V(\pi;\varepsilon,v) = V(\pi;\varepsilon,0) + v$ for all $\pi \in [S(\varepsilon),1)$. Consequently, the two maps $\pi \mapsto V(\pi;\varepsilon,v')$ and $\pi \mapsto V(\pi;\varepsilon,v'')$ do not intersect in $[S(\varepsilon),1)$ when v' and v'' are different.

Each map $\pi \mapsto V(\pi;\varepsilon,v)$ is concave on $[S(\varepsilon),1)$. This fact can be proved by a probabilistic argument using (24.1.13) upon considering the auxiliary optimal stopping problem (24.1.47) where the map $\pi \mapsto h_\beta(\pi)$ is replaced by the concave map $h_v(\pi) = v \wedge (1-\pi)$. [It is a matter of fact that $\pi \mapsto W(\pi)$ from (24.1.47) is concave on $[0,1]$ whenever $\pi \mapsto h_\beta(\pi)$ is so.] Moreover, using (24.1.30) and (24.1.51)–(24.1.52) in (24.1.33) with $n=1$ it is possible to see that for v close to 0 we have $V(\pi;\varepsilon,v) < 0$ for some $\pi > \varepsilon$, and for v close to 1 we have $V(\pi;\varepsilon,v) > 1-\pi$ for some $\pi > \varepsilon$ (see *Figure* VI.12). Thus a simple concavity argument implies the existence of a unique point $B_*^\varepsilon \in (0,1)$ at which $\pi \mapsto V(\pi;\varepsilon,v_\varepsilon)$ for some $v_\varepsilon \in (0,1)$ hits $\pi \mapsto 1-\pi$ smoothly. The key nontrivial point in the verification that $V(\pi;\varepsilon,v_\varepsilon)$ equals the value function $W(\pi)$ of the optimal stopping problem (24.1.47) with $\pi \mapsto h_{v_\varepsilon}(\pi)$ in place of $\pi \mapsto h_\beta(\pi)$ is to establish that $(\mathbb{L}(V(\,\cdot\,;\varepsilon,v_\varepsilon)))(\pi) \ge -c\pi$ for all $\pi \in (B_*^\varepsilon, S^{-1}(B_*^\varepsilon))$. Since B_*^ε is a smooth-fit point, however, this can be done using the same method which we applied in paragraph 3° of the proof of Theorem 23.1. Moreover, when $\varepsilon \downarrow 0$ then clearly (24.1.42) and (24.1.43) are valid (recall (24.1.17) and (24.1.23) above), and the proof of the theorem is complete. □

Notes. The Poisson disorder problem seeks to determine a stopping time which is as close as possible to the (unknown) time of "disorder" when the intensity of an observed Poisson process changes from one value to another. The problem was first studied in [73] where a solution has been found in the case when $\lambda + c \ge \lambda_1 > \lambda_0$. This result has been extended in [35] to the case when $\lambda + c \ge \lambda_1 - \lambda_0 > 0$. Many other authors have also studied the problem from a different standpoint (see e.g. [131]). The main purpose of the present section (following [169]) is to describe the structure of the solution in the general case. The method of proof consists of reducing the initial (optimal stopping) problem to a *free-boundary* differential-difference problem. The key point in the solution is reached by specifying when the principle of smooth fit breaks down and gets superseded by the principle of continuous fit. This can be done in probabilistic terms (by describing the sample path behaviour of the a posteriori probability process) and in analytic terms (via the existence of a singularity point of the free-boundary equation).

The Wiener process version of the disorder problem (where the drift changes) appeared earlier (see [188]) and is now well understood (cf. Section 22 above). The method of proof consists of reducing the initial optimal stopping problem to a free-boundary differential problem which can be solved explicitly. The principle of smooth fit plays a key role in this context.

In this section we adopt the same methodology as in the Wiener process case. A discontinuous character of the observed (Poisson) process in the present case, however, forces us to deal with a differential-difference equation forming a free-boundary problem which is more delicate. This in turn leads to a new effect of the breakdown of the smooth fit principle (and its replacement by the principle of continuous fit), and the key issue in the solution is to understand and specify when exactly this happens. This can be done, on one hand, in terms of the a posteriori probability process (i.e. its jump structure and sample path behaviour), and on the other hand, in terms of a singularity point of the equation from the free-boundary problem. Moreover, it turns out that the existence of such a singularity point makes explicit computations feasible.

The facts on the principles of continuous and smooth fit presented above complement and further extend our findings in Section 23 above.

For more general problems of Poisson disorder see [9] and [38] and the references therein.

Chapter VII.

Optimal stopping in mathematical finance

25. The American option

25.1. Infinite horizon

1. According to theory of modern finance (see e.g. [197]) the arbitrage-free price of the *American put option* with *infinite horizon* (perpetual option) is given by

$$V(x) = \sup_{\tau} \mathsf{E}_x \big(e^{-r\tau} (K - X_\tau)^+ \big) \tag{25.1.1}$$

where the supremum is taken over all stopping times τ of the geometric Brownian motion $X = (X_t)_{t \ge 0}$ solving

$$dX_t = rX_t\,dt + \sigma X_t\,dB_t \tag{25.1.2}$$

with $X_0 = x > 0$ under P_x. We recall that $B = (B_t)_{t \ge 0}$ is a standard Brownian motion process started at zero, $r > 0$ is the interest rate, $K > 0$ is the strike (exercise) price, and $\sigma > 0$ is the volatility coefficient.

The equation (25.1.2) under P_x has a unique (strong) solution given by

$$X_t = x \exp\big(\sigma B_t + (r - \sigma^2/2)\,t\big) \tag{25.1.3}$$

for $t \ge 0$ and $x > 0$. The process X is strong Markov (diffusion) with the infinitesimal generator given by

$$\mathbb{L}_X = rx\,\frac{\partial}{\partial x} + \frac{\sigma^2}{2}\,x^2\,\frac{\partial^2}{\partial x^2}. \tag{25.1.4}$$

The aim of this subsection is to compute the arbitrage-free price V from (25.1.1) and to determine the optimal exercise time τ_* (at which the supremum in (25.1.1) is attained).

2. The optimal stopping problem (25.1.1) will be solved in two steps. In the first step we will make a guess for the solution. In the second step we will verify that the guessed solution is correct (Theorem 25.1).

From (25.1.1) and (25.1.3) we see that the closer X gets to 0 the less likely that the gain will increase upon continuation. This suggests that there exists a point $b \in (0, K)$ such that the stopping time

$$\tau_b = \inf\{t \ge 0 : X_t \le b\} \tag{25.1.5}$$

is optimal in the problem (25.1.1). [In (25.1.5) we use the standard convention that $\inf(\varnothing) = \infty$ (see Remark 25.2 below).]

Standard arguments based on the strong Markov property (cf. Chapter III) lead to the following *free-boundary problem* for the *unknown value function* V and the *unknown point* b:

$$\mathbb{L}_X V = rV \quad \text{for } x > b, \tag{25.1.6}$$

$$V(x) = (K - x)^+ \quad \text{for } x = b, \tag{25.1.7}$$

$$V'(x) = -1 \quad \text{for } x = b \quad \text{(smooth fit)}, \tag{25.1.8}$$

$$V(x) > (K - x)^+ \quad \text{for } x > b, \tag{25.1.9}$$

$$V(x) = (K - x)^+ \quad \text{for } 0 < x < b. \tag{25.1.10}$$

3. To solve the free-boundary problem note that the equation (25.1.6) using (25.1.4) reads

$$Dx^2 V'' + rxV' - rV = 0 \tag{25.1.11}$$

where we set $D = \sigma^2/2$. One may now recognize (25.1.11) as the Cauchy–Euler equation. Let us thus seek a solution in the form

$$V(x) = x^p. \tag{25.1.12}$$

Inserting (25.1.12) into (25.1.11) we get

$$p^2 - \Big(1 - \frac{r}{D}\Big)p - \frac{r}{D} = 0. \tag{25.1.13}$$

The quadratic equation (25.1.13) has two roots, $p_1 = 1$ and $p_2 = -r/D$. Thus the general solution of (25.1.11) can be written as

$$V(x) = C_1\, x + C_2\, x^{-r/D} \tag{25.1.14}$$

where C_1 and C_2 are undetermined constants. From the fact that $V(x) \le K$ for all $x > 0$, we see that C_1 must be zero. Thus (25.1.7) and (25.1.8) become two algebraic equations in two unknowns C_2 and b (free-boundary). Solving this system one gets

$$C_2 = \frac{D}{r}\Big(\frac{K}{1+D/r}\Big)^{1+r/D}, \tag{25.1.15}$$

$$b = \frac{K}{1+D/r}. \tag{25.1.16}$$

Inserting (25.1.15) into (25.1.14) upon using that $C_1 = 0$ we conclude that

$$V(x) = \begin{cases} \frac{D}{r}\Big(\frac{K}{1+D/r}\Big)^{1+r/D} x^{-r/D} & \text{if } x \in [b,\infty), \\ K - x & \text{if } x \in (0,b]. \end{cases} \tag{25.1.17}$$

Note that V is C^2 on $(0,b) \cup (b,\infty)$ but only C^1 at b. Note also that V is convex on $(0,\infty)$.

4. In this way we have arrived at the two conclusions of the following theorem.

Theorem 25.1. *The arbitrage-free price* V *from* (25.1.1) *is given explicitly by* (25.1.17) *above. The stopping time* τ_b *from* (25.1.5) *with* b *given by* (25.1.16) *above is optimal in the problem* (25.1.1).

Proof. To distinguish the two functions let us denote the value function from (25.1.1) by $V_*(x)$ for $x > 0$. We need to prove that $V_*(x) = V(x)$ for all $x > 0$ where $V(x)$ is given by (25.1.17) above.

1°. The properties of V stated following (25.1.17) above show that Itô's formula (page 67) can be applied to $e^{-rt}V(X_t)$ in its standard form (cf. Subsection 3.5). This gives

$$e^{-rt}V(X_t) = V(x) + \int_0^t e^{-rs}(\mathbb{L}_X V - rV)(X_s) I(X_s \neq b)\, ds \tag{25.1.18}$$
$$+ \int_0^t e^{-rs}\sigma X_s V'(X_s)\, dB_s.$$

Setting $G(x) = (K-x)^+$ we see that $(\mathbb{L}_X G - rG)(x) = -rK < 0$ so that together with (25.1.6) we have

$$(\mathbb{L}_X V - rV) \le 0 \tag{25.1.19}$$

everywhere on $(0,\infty)$ but b. Since $\mathsf{P}_x(X_s = b) = 0$ for all s and all x, we see that (25.1.7), (25.1.9)–(25.1.10) and (25.1.18)–(25.1.19) imply that

$$e^{-rt}(K - X_t)^+ \le e^{-rt}V(X_t) \le V(x) + M_t \tag{25.1.20}$$

where $M = (M_t)_{t\ge 0}$ is a continuous local martingale given by

$$M_t = \int_0^t e^{-rs}\sigma X_s V'(X_s)\, dB_s. \tag{25.1.21}$$

(Using that $|V'(x)| \le 1$ for all $x > 0$ it is easily verified by standard means that M is a martingale.)

Let $(\tau_n)_{n \ge 1}$ be a localization sequence of (bounded) stopping times for M (for example $\tau_n \equiv n$ will do). Then for every stopping time τ of X we have by (25.1.20) above

$$e^{-r(\tau \wedge \tau_n)}(K - X_{\tau \wedge \tau_n})^+ \le V(x) + M_{\tau \wedge \tau_n} \tag{25.1.22}$$

for all $n \ge 1$. Taking the P_x-expectation, using the optional sampling theorem (page 60) to conclude that $\mathsf{E}_x M_{\tau \wedge \tau_n} = 0$ for all n, and letting $n \to \infty$, we find by Fatou's lemma that

$$\mathsf{E}_x\big(e^{-r\tau}(K - X_\tau)^+\big) \le V(x). \tag{25.1.23}$$

Taking the supremum over all stopping times τ of X we find that $V_*(x) \le V(x)$ for all $x > 0$.

2°. To prove the reverse inequality (equality) we observe from (25.1.18) upon using (25.1.6) (and the optional sampling theorem as above) that

$$\mathsf{E}_x\Big(e^{-r(\tau_b \wedge \tau_n)} V(X_{\tau_b \wedge \tau_n})\Big) = V(x) \tag{25.1.24}$$

for all $n \ge 1$. Letting $n \to \infty$ and using that $e^{-r\tau_b} V(X_{\tau_b}) = e^{-r\tau_b}(K - X_{\tau_b})^+$ (both expressions being 0 when $\tau_b = \infty$), we find by the dominated convergence theorem that

$$\mathsf{E}_x\big(e^{-r\tau_b}(K - X_{\tau_b})^+\big) = V(x). \tag{25.1.25}$$

This shows that τ_b is optimal in (25.1.1). Thus $V_*(x) = V(x)$ for all $x > 0$ and the proof is complete. □

Remark 25.2. It is evident from the definition of τ_b in (25.1.5) and the explicit representation of X in (25.1.3) that τ_b is not always finite. Using the well-known Doob formula (see e.g. [197, Chap. VIII, § 2a, (51)])

$$\mathsf{P}\Big(\sup_{t \ge 0}(B_t - \alpha t) \ge \beta\Big) = e^{-2\alpha\beta} \tag{25.1.26}$$

for $\alpha > 0$ and $\beta > 0$, it is straightforwardly verified that

$$\mathsf{P}_x(\tau_b < \infty) = \begin{cases} 1 & \text{if } r \le D \text{ or } x \in (0, b], \\ \left(\frac{b}{x}\right)^{(r/D)-1} & \text{if } r > D \text{ and } x \in (b, \infty) \end{cases} \tag{25.1.27}$$

for $x > 0$.

25.2. Finite horizon

1. The arbitrage-free price of the *American put option* with *finite horizon* (cf. (25.1.1) above) is given by

$$V(t,x) = \sup_{0 \le \tau \le T-t} \mathsf{E}_{t,x}\big(e^{-r\tau}(K - X_{t+\tau})^+\big) \tag{25.2.1}$$

where τ is a stopping time of the geometric Brownian motion $X = (X_{t+s})_{s \ge 0}$ solving

$$dX_{t+s} = rX_{t+s}\,ds + \sigma X_{t+s}\,dB_s \tag{25.2.2}$$

with $X_t = x > 0$ under $\mathsf{P}_{t,x}$. We recall that $B = (B_s)_{s \ge 0}$ denotes a standard Brownian motion process started at zero, $T > 0$ is the expiration date (maturity), $r > 0$ is the interest rate, $K > 0$ is the strike (exercise) price, and $\sigma > 0$ is the volatility coefficient. Similarly to (25.1.2) the strong solution of (25.2.2) under $\mathsf{P}_{t,x}$ is given by

$$X_{t+s} = x \exp\big(\sigma B_s + (r - \sigma^2/2)\,s\big) \tag{25.2.3}$$

whenever $t \ge 0$ and $x > 0$ are given and fixed. The process X is strong Markov (diffusion) with the infinitesimal generator given by

$$\mathbb{L}_X = rx\,\frac{\partial}{\partial x} + \frac{\sigma^2}{2}\,x^2\,\frac{\partial^2}{\partial x^2}. \tag{25.2.4}$$

We refer to [197] for more information on the derivation and economic meaning of (25.2.1).

2. Let us determine the structure of the optimal stopping time in the problem (25.2.1).

(i) First note that since the gain function $G(x) = (K-x)^+$ is continuous, it is possible to apply Corollary 2.9 (Finite horizon) with Remark 2.10 and conclude that there exists an optimal stopping time in the problem (25.2.1). From our earlier considerations we may therefore conclude that the *continuation set* equals

$$C = \{\, (t,x) \in [0,T) \times (0,\infty) : V(t,x) > G(x) \,\} \tag{25.2.5}$$

and the *stopping set* equals

$$\bar{D} = \{\, (t,x) \in [0,T] \times (0,\infty) : V(t,x) = G(x) \,\}. \tag{25.2.6}$$

It means that the stopping time $\tau_{\bar{D}}$ defined by

$$\tau_{\bar{D}} = \inf\{\, 0 \le s \le T-t : X_{t+s} \in \bar{D} \,\} \tag{25.2.7}$$

is optimal in (25.2.1).

(ii) We claim that all points (t,x) with $x \geq K$ for $0 \leq t < T$ belong to the continuation set C. Indeed, this is easily verified by considering $\tau_\varepsilon = \inf\{\, 0 \leq s \leq T-t : X_{t+s} \leq K-\varepsilon \,\}$ for $0 < \varepsilon < K$ and noting that $\mathsf{P}_{t,x}(0 < \tau_\varepsilon < T-t) > 0$ if $x \geq K$ with $0 \leq t < T$. The strict inequality implies that $\mathsf{E}_{t,x}(e^{-r\tau_\varepsilon}(K - X_{t+\tau_\varepsilon})^+) > 0$ so that (t,x) with $x \geq K$ for $0 \leq t < T$ cannot belong to the stopping set $\bar{D}$ as claimed.

(iii) Recalling the solution to the problem (25.2.1) in the case of infinite horizon, where the stopping time $\tau_* = \inf\{\, s > 0 : X_s \leq A_* \,\}$ is optimal and $0 < A_* < K$ is explicitly given by Theorem 25.1 above, we see that all points (t,x) with $0 < x \leq A_*$ for $0 \leq t \leq T$ belong to the stopping set $\bar{D}$. Moreover, since $x \mapsto V(t,x)$ is convex on $(0,\infty)$ for each $0 \leq t \leq T$ given and fixed (the latter is easily verified using (25.2.1) and (25.2.3) above), it follows directly from the previous two conclusions about C and $\bar{D}$ that there exists a function $b : [0,T] \to \mathbb{R}$ satisfying $0 < A_* \leq b(t) < K$ for all $0 \leq t < T$ (later we will see that $b(T) = K$ as well) such that the continuation set C equals

$$C = \{(t,x) \in [0,T) \times (0,\infty) : x > b(t)\} \tag{25.2.8}$$

and the stopping set $\bar{D}$ is the closure of the set

$$D = \{(t,x) \in [0,T] \times (0,\infty) : x < b(t)\} \tag{25.2.9}$$

joined with remaining points (T,x) for $x \geq b(T)$. (Below we will show that V is continuous so that C is open.)

(iv) Since the problem (25.2.1) is time-homogeneous, in the sense that the gain function $G(x) = (K-x)^+$ is a function of space only (i.e. does not depend on time), it follows that the map $t \mapsto V(t,x)$ is decreasing on $[0,T]$ for each $x \in (0,\infty)$. Hence if (t,x) belongs to C for some $x \in (0,\infty)$ and we take any other $0 \leq t' < t \leq T$, then $V(t',x) - G(x) \geq V(t,x) - G(x) > 0$, showing that (t',x) belongs to C as well. From this we may conclude that the boundary $t \mapsto b(t)$ in (25.2.8) and (25.2.9) is *increasing* on $[0,T]$.

3. Let us show that the value function $(t,x) \mapsto V(t,x)$ is *continuous* on $[0,T] \times (0,\infty)$.

For this, it is enough to prove that

$$x \mapsto V(t,x) \quad \text{is continuous at } x_0, \tag{25.2.10}$$

$$t \mapsto V(t,x) \quad \text{is continuous at } t_0 \text{ uniformly over } x \in [x_0-\delta, x_0+\delta] \tag{25.2.11}$$

for each $(t_0,x_0) \in [0,T] \times (0,\infty)$ with some $\delta > 0$ small enough (it may depend on x_0).

Since (25.2.10) follows from the fact that $x \mapsto V(t,x)$ is convex on $(0,\infty)$, it remains to establish (25.2.11).

For this, let us fix arbitrary $0 \le t_1 < t_2 \le T$ and $x \in (0,\infty)$, and let $\tau_1 = \tau_*(t_1, x)$ denote the optimal stopping time for $V(t_1, x)$. Set $\tau_2 = \tau_1 \wedge (T - t_2)$ and note, since $t \mapsto V(t,x)$ is decreasing on $[0,T]$, that upon denoting $S_t = \exp(\sigma B_t + \gamma t)$ with $\gamma = r - \sigma^2/2$ we have

$$\begin{aligned} 0 &\le V(t_1,x) - V(t_2,x) \\ &\le \mathsf{E}\big(e^{-r\tau_1}(K - xS_{\tau_1})^+\big) - \mathsf{E}\big(e^{-r\tau_2}(K - xS_{\tau_2})^+\big) \\ &\le \mathsf{E}\big(e^{-r\tau_2}\big[(K - xS_{\tau_1})^+ - (K - xS_{\tau_2})^+\big]\big) \\ &\le x\,\mathsf{E}\,(S_{\tau_2} - S_{\tau_1})^+ \end{aligned} \tag{25.2.12}$$

where we use that $\tau_2 \le \tau_1$ and that $(K-y)^+ - (K-z)^+ \le (z-y)^+$ for $y, z \in \mathbb{R}$.

Set $Z_t = \sigma B_t + \gamma t$ and recall that stationary independent increments of $Z = (Z_t)_{t \ge 0}$ imply that $(Z_{\tau_2 + t} - Z_{\tau_2})_{t \ge 0}$ is a version of Z, i.e. the two processes have the same law. Using that $\tau_1 - \tau_2 \le t_2 - t_1$ hence we get

$$\begin{aligned} \mathsf{E}\,(S_{\tau_2} - S_{\tau_1})^+ &= \mathsf{E}\Big(\mathsf{E}\big((S_{\tau_2} - S_{\tau_1})^+ \mid \mathcal{F}_{\tau_2}\big)\Big) \\ &= \mathsf{E}\Big(S_{\tau_2}\,\mathsf{E}\big((1 - S_{\tau_1}/S_{\tau_2})^+ \mid \mathcal{F}_{\tau_2}\big)\Big) \\ &= \mathsf{E}\Big(S_{\tau_2}\,\mathsf{E}\big((1 - e^{Z_{\tau_1} - Z_{\tau_2}})^+ \mid \mathcal{F}_{\tau_2}\big)\Big) \\ &= \mathsf{E}\,(S_{\tau_2})\,\mathsf{E}\big(1 - e^{Z_{\tau_1} - Z_{\tau_2}}\big)^+ \\ &= \mathsf{E}\,(S_{\tau_2})\,\mathsf{E}\Big(1 - \inf_{0 \le t \le t_2 - t_1} e^{Z_{\tau_2 + t} - Z_{\tau_2}}\Big) \\ &= \mathsf{E}\,(S_{\tau_2})\,\mathsf{E}\Big(1 - \inf_{0 \le t \le t_2 - t_1} e^{Z_t}\Big) =: \mathsf{E}\,(S_{\tau_2})\,L(t_2 - t_1) \end{aligned} \tag{25.2.13}$$

where we also used that $Z_{\tau_1} - Z_{\tau_2}$ is independent from $\mathcal{F}_{\tau_2}$. By basic properties of Brownian motion it is easily seen that $L(t_2 - t_1) \to 0$ as $t_2 - t_1 \to 0$.

Combining (25.2.12) and (25.2.13) we find by the martingale property of $\big(\exp(\sigma B_t - (\sigma^2/2)t)\big)_{t \ge 0}$ that

$$0 \le V(t_1,x) - V(t_2,x) \le x\,\mathsf{E}\,(S_{\tau_2})\,L(t_2 - t_1) \le x\,e^{rT} L(t_2 - t_1) \tag{25.2.14}$$

from where (25.2.11) becomes evident. This completes the proof.

4. In order to prove that the *smooth-fit* condition (25.2.28) holds, i.e. that $x \mapsto V(t,x)$ is C^1 at $b(t)$, let us fix a point $(t,x) \in (0,T) \times (0,\infty)$ lying on the boundary b so that $x = b(t)$. Then $x < K$ and for all $\varepsilon > 0$ such that $x + \varepsilon < K$ we have

$$\frac{V(t,x+\varepsilon) - V(t,x)}{\varepsilon} \ge \frac{G(x+\varepsilon) - G(x)}{\varepsilon} = -1 \tag{25.2.15}$$

and hence, taking the limit in (25.2.15) as $\varepsilon \downarrow 0$, we get

$$\frac{\partial^+ V}{\partial x}(t,x) \ge G'(x) = -1 \tag{25.2.16}$$

where the right-hand derivative exists (and is finite) by virtue of the convexity of the mapping $x \mapsto V(t,x)$ on $(0,\infty)$. (Note that the latter will also be proved independently below.)

To prove the converse inequality, let us fix $\varepsilon > 0$ such that $x+\varepsilon < K$, and consider the stopping time $\tau_\varepsilon = \tau_*(t, x+\varepsilon)$ being optimal for $V(t,x+\varepsilon)$. Then we have

$$\begin{aligned} &V(t,x+\varepsilon) - V(t,x) \qquad (25.2.17)\\ &\quad \le \mathsf{E}\left(e^{-r\tau_\varepsilon}(K-(x+\varepsilon)S_{\tau_\varepsilon})^+\right) - \mathsf{E}\left(e^{-r\tau_\varepsilon}(K - xS_{\tau_\varepsilon})^+\right)\\ &\quad \le \mathsf{E}\Big(e^{-r\tau_\varepsilon}\big[(K-(x+\varepsilon)S_{\tau_\varepsilon})^+ - (K-xS_{\tau_\varepsilon})^+\big] I\big((x+\varepsilon)S_{\tau_\varepsilon} < K\big)\Big)\\ &\quad = -\varepsilon\, \mathsf{E}\Big(e^{-r\tau_\varepsilon} S_{\tau_\varepsilon} I\big((x+\varepsilon)S_{\tau_\varepsilon} < K\big)\Big). \end{aligned}$$

Using that $s \mapsto -\frac{\gamma}{\sigma}s$ is a lower function of B at zero and the fact that the optimal boundary $s \mapsto b(s)$ is increasing on $[t,T]$, it is not difficult to verify that $\tau_\varepsilon \to 0$ P-a.s. as $\varepsilon \downarrow 0$. In particular, this implies that

$$\mathsf{E}\left(e^{-r\tau_\varepsilon} S_{\tau_\varepsilon} I((x+\varepsilon)S_{\tau_\varepsilon} < K)\right) \to 1 \tag{25.2.18}$$

as $\varepsilon \downarrow 0$ by the dominated convergence theorem.

Combining (25.2.17) and (25.2.18) we see that

$$\frac{\partial^+ V}{\partial x}(t,x) \le G'(x) = -1 \tag{25.2.19}$$

which together with (25.2.16) completes the proof.

5. We proceed to prove that the boundary b is *continuous* on $[0,T]$ and that $b(T) = K$.

(i) Let us first show that the boundary b is right-continuous on $[0,T]$. For this, fix $t \in (0,T]$ and consider a sequence $t_n \downarrow t$ as $n \to \infty$. Since b is increasing, the right-hand limit $b(t+)$ exists. Because $(t_n, b(t_n)) \in \bar{D}$ for all $n \ge 1$, and $\bar{D}$ is closed, we get that $(t, b(t+)) \in \bar{D}$. Hence by (25.2.9) we see that $b(t+) \le b(t)$. The reverse inequality follows obviously from the fact that b is increasing on $[0,T]$, thus proving the claim.

(ii) Suppose that at some point $t_* \in (0,T)$ the function b makes a jump, i.e. let $b(t_*) > b(t_*-)$. Let us fix a point $t' < t_*$ close to t_* and consider the half-open set $R \subseteq C$ being a curved trapezoid formed by the vertices $(t', b(t'))$, $(t_*, b(t_*-))$, (t_*, x') and (t', x') with any x' fixed arbitrarily in the interval $(b(t_*-), b(t_*))$.

Recall that the strong Markov property (cf. Chapter III) implies that the value function V is $C^{1,2}$ in C. Note also that the gain function G is C^2 in

R so that by the Newton–Leibniz formula using (25.2.27) and (25.2.28) it follows that

$$V(t,x) - G(x) = \int_{b(t)}^{x} \int_{b(t)}^{u} (V_{xx}(t,v) - G_{xx}(v))\, dv\, du \tag{25.2.20}$$

for all $(t,x) \in R$. Moreover, the strong Markov property (cf. Chapter III) implies that the value function V solves the equation (25.2.26) from where using that $t \mapsto V(t,x)$ and $x \mapsto V(t,x)$ are decreasing so that $V_t \le 0$ and $V_x \le 0$ in C, we obtain

$$\begin{aligned} V_{xx}(t,x) &= \frac{2}{\sigma^2 x^2}\big(rV(t,x) - V_t(t,x) - rxV_x(t,x)\big) \\ &= \frac{2}{\sigma^2 x^2}\, r(K-x)^+ \ge c > 0 \end{aligned} \tag{25.2.21}$$

for each $(t,x) \in R$ where $c > 0$ is small enough.

Hence by (25.2.20) using that $G_{xx} = 0$ in R we get

$$V(t',x') - G(x') \ge c\,\frac{(x' - b(t'))^2}{2} \longrightarrow c\,\frac{(x' - b(t_*))^2}{2} > 0 \tag{25.2.22}$$

as $t' \uparrow t_*$. This implies that $V(t_*,x') > G(x')$ which contradicts the fact that (t_*,x') belong to the stopping set $\bar{D}$. Thus $b(t_*+) = b(t_*)$ showing that b is continuous at t_* and thus on $[0,T]$ as well.

(iii) We finally note that the method of proof from the previous part (ii) also implies that $b(T) = K$. To see this, we may let $t_* = T$ and likewise suppose that $b(T) < K$. Then repeating the arguments presented above word by word we arrive at a contradiction with the fact that $V(t,x) = G(x)$ for all $x \in [b(T), K]$.

6. Summarizing the facts proved in paragraphs 1–5 above we may conclude that the following hitting time is optimal in the problem (25.2.1):

$$\tau_b = \inf\{\, 0 \le s \le T-t : X_{t+s} \le b(t+s) \,\} \tag{25.2.23}$$

(the infimum of an empty set being equal to $T-t$) where the boundary b satisfies the properties

$$b : [0,T] \to (0,K] \quad \text{is continuous and increasing,} \tag{25.2.24}$$

$$b(T) = K. \tag{25.2.25}$$

(see Figure VII.1).

Standard arguments based on the strong Markov property (cf. Chapter III) lead to the following *free-boundary problem* for the *unknown value function* V

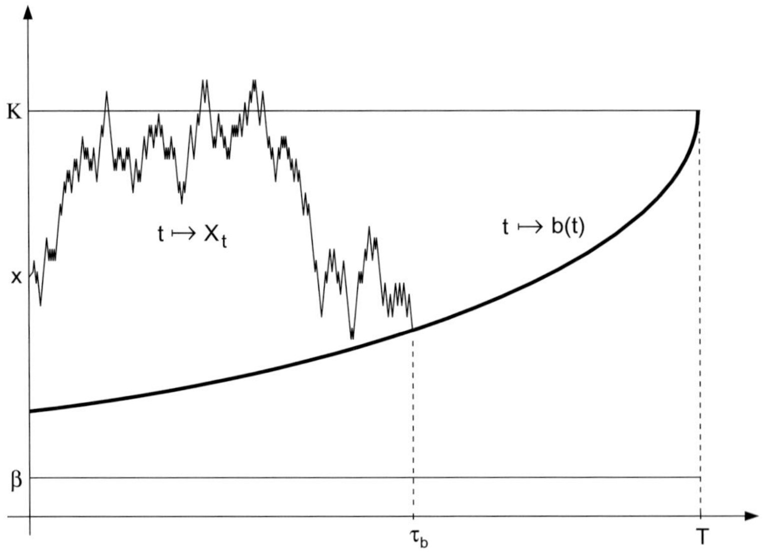

Figure VII.1: A computer drawing of the optimal stopping boundary b from Theorem 25.3. The number β is the optimal stopping point in the case of infinite horizon (Theorem 25.1).

and the *unknown boundary* b:

$$V_t + \mathbb{L}_X V = rV \quad \text{in } C, \tag{25.2.26}$$

$$V(t,x) = (K-x)^+ \quad \text{for } x = b(t), \tag{25.2.27}$$

$$V_x(t,x) = -1 \quad \text{for } x = b(t) \quad (\textit{smooth fit}), \tag{25.2.28}$$

$$V(t,x) > (K-x)^+ \quad \text{in } C, \tag{25.2.29}$$

$$V(t,x) = (K-x)^+ \quad \text{in } D \tag{25.2.30}$$

where the continuation set C is defined in (25.2.8) above and the stopping set $\bar{D}$ is the closure of the set D in (25.2.9) above.

7. The following properties of V and b were verified above:

$$V \text{ is continuous on } [0,T]\times\mathbb{R}_+, \tag{25.2.31}$$

$$V \text{ is } C^{1,2} \text{ on } C \text{ (and } C^{1,2} \text{ on } \bar{D}), \tag{25.2.32}$$

$$x \mapsto V(t,x) \text{ is decreasing and convex with } V_x(t,x) \in [-1,0], \tag{25.2.33}$$

$$t \mapsto V(t,x) \text{ is decreasing with } V(T,x) = (K-x)^+, \tag{25.2.34}$$

$$t \mapsto b(t) \text{ is increasing and continuous with } 0 < b(0+) < K \tag{25.2.35}$$

and $b(T-) = K$.

Note also that (25.2.28) means that $x \mapsto V(t,x)$ is C^1 at $b(t)$.

Once we know that V satisfying (25.2.28) is "sufficiently regular" (cf. footnote 14 in [27] when $t \mapsto V(t,x)$ is known to be C^1 for all x), we can apply Itô's formula (page 67) to $e^{-rs}V(t+s, X_{t+s})$ in its standard form and take the $\mathsf{P}_{t,x}$-expectation on both sides in the resulting identity. The martingale term then vanishes by the optional sampling theorem (page 60) using the final part of (25.2.33) above, so that by (25.2.26) and (25.2.27)+(25.2.30) upon setting $s = T-t$ (being the key advantage of the finite horizon) one obtains the *early exercise premium representation* of the value function

$$V(t,x) = e^{-r(T-t)}\mathsf{E}_{t,x}G(X_T) - \int_0^{T-t} e^{-ru}\mathsf{E}_{t,x}\Big(H(t-u, X_{t+u})\, I\big(X_{t+u} \le b(t+u)\big)\Big)\, du$$
$$= e^{-r(T-t)}\mathsf{E}_{t,x}G(X_T) + rK\int_0^{T-t} e^{-ru}\mathsf{P}_{t,x}\big(X_{t+u} \le b(t+u)\big)\, du \tag{25.2.36}$$

for all $(t,x) \in [0,T]\times\mathbb{R}_+$ where we set $G(x) = (K-x)^+$ and $H = G_t + \mathbb{L}_X G - rG$ so that $H = -rK$ for $x < b(t)$.

A detail worth mentioning in this derivation (see (25.2.47) below) is that (25.2.36) follows from (3.5.9) with $F(t+s, X_{t+s}) = e^{-rs}V(t+s, X_{t+s})$ without knowing a priori that $t \mapsto V(t,x)$ is C^1 at $b(t)$ as required under the condition of "sufficiently regular" recalled prior to (25.2.36) above. This approach is more direct since the sufficient conditions (3.5.10)–(3.5.13) for (3.5.9) are easier verified than sufficient conditions [such as b is C^1 or (locally) Lipschitz] for $t \mapsto V(t,x)$ to be C^1 at $b(t)$. This is also more in the spirit of the free-boundary equation (25.2.39) to be derived below where neither differentiability nor a Lipschitz property of b plays a role in the formulation.

Since $V(t,x) = G(x) = (K-x)^+$ in $\bar{D}$ by (25.2.27)+(25.2.30), we see that (25.2.36) reads

$$K - x = e^{-r(T-t)}\,\mathsf{E}_{t,x}(K - X_T)^+ + rK\int_0^{T-t} e^{-ru}\mathsf{P}_{t,x}\big(X_{t+u} \le b(t+u)\big)\, du \tag{25.2.37}$$

for all $x \in (0, b(t)]$ and all $t \in [0,T]$.

8. A natural candidate equation is obtained by inserting $x = b(t)$ in (25.2.37). This leads to the *free-boundary equation* (cf. Subsection 14.1)

$$K - b(t) = e^{-r(T-t)}\,\mathsf{E}_{t,b(t)}(K - X_T)^+ + rK\int_0^{T-t} e^{-ru}\,\mathsf{P}_{t,b(t)}\big(X_{t+u} \le b(t+u)\big)\, du \tag{25.2.38}$$

which upon using (25.2.3) more explicitly reads as follows:

$$
\begin{aligned}
&K - b(t) \qquad\qquad (25.2.39)\\
&= e^{-r(T-t)} \int_0^K \Phi\Big(\frac{1}{\sigma\sqrt{T-t}} \Big(\log \frac{K-z}{b(t)} - \Big(r - \frac{\sigma^2}{2}\Big)(T-t)\Big)\Big)\, dz \\
&\quad + rK \int_0^{T-t} e^{-ru}\, \Phi\Big(\frac{1}{\sigma\sqrt{u}} \Big(\log \frac{b(t+u)}{b(t)} - \Big(r - \frac{\sigma^2}{2}\Big) u\Big)\Big)\, du
\end{aligned}
$$

for all $t \in [0,T]$ where $\Phi(x) = (1/\sqrt{2\pi}) \int_{-\infty}^{x} e^{-z^2/2} dz$ for $x \in \mathbb{R}$. It is a *nonlinear Volterra integral equation of the second kind* (see [212]).

9. The main result of the present subsection may now be stated as follows (see also Remark 25.5 below).

Theorem 25.3. *The optimal stopping boundary in the American put problem* (25.2.1) *can be characterized as the unique solution of the free-boundary equation* (25.2.39) *in the class of continuous increasing functions* $c : [0,T] \to \mathbb{R}$ *satisfying* $0 < c(t) < K$ *for all* $0 < t < T$.

Proof. The fact that the optimal stopping boundary b solves (25.2.38) i.e. (25.2.39) was derived above. The main emphasis of the theorem is thus on the claim of *uniqueness.* Let us therefore assume that a continuous increasing $c : [0,T] \to \mathbb{R}$ solving (25.2.39) is given such that $0 < c(t) < K$ for all $0 < t < T$, and let us show that this c must then coincide with the optimal stopping boundary b. The proof of this implication will be presented in the nine steps as follows.

1°. In view of (25.2.36) and with the aid of calculations similar to those leading from (25.2.38) to (25.2.39), let us introduce the function

$$
\begin{aligned}
&U^c(t,x) \qquad\qquad (25.2.40)\\
&= e^{-r(T-t)}\, \mathsf{E}_{t,x} G(X_T) + rK \int_0^{T-t} e^{-ru}\, \mathsf{P}_{t,x}\big(X_{t+u} \le c(t+u)\big)\, du \\
&= e^{-r(T-t)}\, U_1^c(t,x) + rK\, U_2^c(t,x)
\end{aligned}
$$

where U_1^c and U_2^c are defined as follows:

$$
U_1^c(t,x) = \int_0^K \Phi\Big(\frac{1}{\sigma\sqrt{T-t}} \Big(\log \frac{K-z}{x} - \gamma\,(T-t)\Big)\Big)\, dz, \tag{25.2.41}
$$

$$
U_2^c(t,x) = \int_t^T e^{-r(v-t)}\, \Phi\Big(\frac{1}{\sigma\sqrt{v-t}} \Big(\log \frac{c(v)}{x} - \gamma\,(v-t)\Big)\Big)\, dv \tag{25.2.42}
$$

for all $(t,x) \in [0,T) \times (0,\infty)$ upon setting $\gamma = r - \sigma^2/2$ and substituting $v = t+u$.

Denoting $\varphi = \Phi'$ we then have

$$\frac{\partial U_1^c}{\partial x}(t,x) = -\frac{1}{\sigma x\sqrt{T-t}}\int_0^K \varphi\Big(\frac{1}{\sigma\sqrt{T-t}}\Big(\log\frac{K-z}{x} - \gamma\,(T-t)\Big)\Big)\,dz, \tag{25.2.43}$$

$$\frac{\partial U_2^c}{\partial x}(t,x) = -\frac{1}{\sigma x}\int_t^T \frac{e^{-r(v-t)}}{\sqrt{v-t}}\,\varphi\Big(\frac{1}{\sigma\sqrt{v-t}}\Big(\log\frac{c(v)}{x} - \gamma\,(v-t)\Big)\Big)\,dv \tag{25.2.44}$$

for all $(t,x) \in [0,T) \times (0,\infty)$ where the interchange of differentiation and integration is justified by standard means. From (25.2.43) and (25.2.44) we see that $\partial U_1^c/\partial x$ and $\partial U_2^c/\partial x$ are continuous on $[0,T)\times(0,\infty)$, which in view of (25.2.40) implies that U_x^c is continuous on $[0,T)\times(0,\infty)$.

2°. In accordance with (25.2.36) define a function $V^c : [0,T)\times(0,\infty) \to \mathbb{R}$ by setting $V^c(t,x) = U^c(t,x)$ for $x > c(t)$ and $V^c(t,x) = G(x)$ for $x \le c(t)$ when $0 \le t < T$. Note that since c solves (25.2.39) we have that V^c is continuous on $[0,T)\times(0,\infty)$, i.e. $V^c(t,x) = U^c(t,x) = G(x)$ for $x = c(t)$ when $0 \le t < T$. Let C_1 and C_2 be defined by means of c as in (3.5.3) and (3.5.4) with $[0,T)$ instead of $\mathbb{R}_+$, respectively.

Standard arguments based on the Markov property (or a direct verification) show that V^c i.e. U^c is $C^{1,2}$ on C_1 and that

$$V_t^c + \mathbb{L}_X V^c = rV^c \quad \text{in} \quad C_1. \tag{25.2.45}$$

Moreover, since U_x^c is continuous on $[0,T)\times(0,\infty)$ we see that V_x^c is continuous on $\bar{C}_1$. Finally, since $0 < c(t) < K$ for $0 < t < T$ we see that V^c i.e. G is $C^{1,2}$ on $\bar{C}_2$.

3°. Summarizing the preceding conclusions one can easily verify that with $(t,x) \in [0,T)\times(0,\infty)$ given and fixed, the function $F : [0,T-t)\times(0,\infty) \to \mathbb{R}$ defined by

$$F(s,y) = e^{-rs}V^c(t+s,xy) \tag{25.2.46}$$

satisfies (3.5.10)–(3.5.13) (in the relaxed form) so that (3.5.9) can be applied. In this way we get

$$\begin{aligned} e^{-rs}V^c(t+s,X_{t+s}) &= V^c(t,x) \\ &+ \int_0^s e^{-ru}\Big(V_t^c + \mathbb{L}_X V^c - rV^c\Big)(t+u,X_{t+u})\,I(X_{t+u} \ne c(t+u))\,du \\ &+ M_s^c + \frac{1}{2}\int_0^s e^{-ru}\Delta_x V_x^c(t+u,c(t+u))\,d\ell_u^c(X) \end{aligned} \tag{25.2.47}$$

where $M_s^c = \int_0^s e^{-ru}V_x^c(t+u,X_{t+u})\,\sigma X_{t+u}\,I(X_{t+u} \ne c(t+u))\,dB_u$ and we set $\Delta_x V_x^c(v,c(v)) = V_x^c(v,c(v)+) - V_x^c(v,c(v)-)$ for $t \le v \le T$. Moreover, it is easily seen from (25.2.43) and (25.2.44) that $(M_s^c)_{0\le s\le T-t}$ is a martingale under $\mathsf{P}_{t,x}$ so that $\mathsf{E}_{t,x}M_s^c = 0$ for each $0 \le s \le T-t$.

4°. Setting $s = T-t$ in (25.2.47) and then taking the $\mathsf{P}_{t,x}$-expectation, using that $V^c(T,x) = G(x)$ for all $x > 0$ and that V^c satisfies (25.2.45) in C_1, we get

$$e^{-r(T-t)}\mathsf{E}_{t,x}G(X_T) = V^c(t,x) \tag{25.2.48}$$
$$+ \int_0^{T-t} e^{-ru}\mathsf{E}_{t,x}\big(H(t+u, X_{t+u})\, I(X_{t+u} \le c(t+u))\big)\, du$$
$$+ \frac{1}{2}\int_0^{T-t} e^{-ru}\Delta_x V_x^c(t+u, c(t+u))\, d_u \mathsf{E}_{t,x}(\ell_u^c(X))$$

for all $(t,x) \in [0,T) \times (0,\infty)$ where $H = G_t + \mathbb{L}_X G - rG = -rK$ for $x \le c(t)$. From (25.2.48) we thus see that

$$V^c(t,x) = e^{-r(T-t)}\mathsf{E}_{t,x}G(X_T) \tag{25.2.49}$$
$$+\, rK \int_0^{T-t} e^{-ru}\mathsf{P}_{t,x}(X_{t+u} \le c(t+u))\, du$$
$$- \frac{1}{2}\int_0^{T-t} e^{-ru}\Delta_x V_x^c(t+u, c(t+u))\, d_u \mathsf{E}_{t,x}(\ell_u^c(X))$$

for all $(t,x) \in [0,T) \times (0,\infty)$. Comparing (25.2.49) with (25.2.40), and recalling the definition of V^c in terms of U^c and G, we get

$$\int_0^{T-t} e^{-ru}\Delta_x V_x^c(t+u, c(t+u))\, d_u \mathsf{E}_{t,x}(\ell_u^c(X)) \tag{25.2.50}$$
$$= 2\,\big(U^c(t,x) - G(x)\big)\, I(x \le c(t))$$

for all $0 \le t < T$ and $x > 0$, where $I(x \le c(t))$ equals 1 if $x \le c(t)$ and 0 if $x > c(t)$.

5°. From (25.2.50) we see that if we are to prove that

$$x \mapsto V^c(t,x) \quad \text{is} \quad C^1 \quad \text{at} \quad c(t) \tag{25.2.51}$$

for each $0 \le t < T$ given and fixed, then it will follow that

$$U^c(t,x) = G(x) \quad \text{for all} \quad 0 < x \le c(t). \tag{25.2.52}$$

On the other hand, if we know that (25.2.52) holds, then using the general fact

$$\frac{\partial}{\partial x}\Big(U^c(t,x) - G(x)\Big)\Big|_{x=c(t)} = V_x^c(t, c(t)+) - V_x^c(t, c(t)-) \tag{25.2.53}$$
$$= \Delta_x V_x^c(t, c(t))$$

for all $0 \le t < T$, we see that (25.2.51) holds too (since U_x^c is continuous). The equivalence of (25.2.51) and (25.2.52) just explained then suggests that instead of dealing with the equation (25.2.50) in order to derive (25.2.51) above (which was the content of an earlier proof) we may rather concentrate on establishing (25.2.52) directly. [To appreciate the simplicity and power of the probabilistic argument to be given shortly below one may differentiate (25.2.50) with respect to x, compute the left-hand side explicitly (taking care of a jump relation), and then try to prove the uniqueness of the zero solution to the resulting (weakly singular) Volterra integral equation using any of the known analytic methods (see e.g. [212]).]

6°. To derive (25.2.52) first note that standard arguments based on the Markov property (or a direct verification) show that U^c is $C^{1,2}$ on C_2 and that

$$U_t^c + \mathbb{L}_X U^c - rU^c = -rK \quad \text{in} \quad C_2 . \tag{25.2.54}$$

Since F in (25.2.46) with U^c instead of V^c is continuous and satisfies (3.5.10)–(3.5.13) (in the relaxed form), we see that (3.5.9) can be applied just as in (25.2.47), and this yields

$$\begin{aligned} & e^{-rs} U^c(t+s, X_{t+s}) \\ & = U^c(t,x) - rK \int_0^s e^{-ru} I(X_{t+u} \le c(t+u))\, du + \widetilde{M}_s^c \end{aligned} \tag{25.2.55}$$

upon using (25.2.45) and (25.2.54) as well as that $\Delta_x U_x^c(t+u, c(t+u)) = 0$ for all $0 \le u \le s$ since U_x^c is continuous. In (25.2.55) we have $\widetilde{M}_s^c = \int_0^s e^{-ru} U_x^c(t+u, X_{t+u})\, \sigma X_{t+u}\, I(X_{t+u} \ne c(t+u))\, dB_u$ and $(\widetilde{M}_s^c)_{0 \le s \le T-t}$ is a martingale under $\mathsf{P}_{t,x}$.

Next note that (3.5.9) applied to F in (25.2.46) with G instead of V^c yields

$$\begin{aligned} e^{-rs} G(X_{t+s}) = G(x) - rK & \int_0^s e^{-ru} I(X_{t+u} < K)\, du \\ & + M_s^K + \frac{1}{2} \int_0^s e^{-ru}\, d\ell_u^K(X) \end{aligned} \tag{25.2.56}$$

upon using that $G_t + \mathbb{L}_X G - rG$ equals $-rK$ on $(0,K)$ and 0 on (K,∞) as well as that $\Delta_x G_x(t+u, K) = 1$ for $0 \le u \le s$. In (25.2.56) we have $M_s^K = \int_0^s e^{-ru}\, G'(X_{t+u})\, \sigma X_{t+u}\, I(X_{t+u} \ne K)\, dB_u = -\int_0^s e^{-ru}\, \sigma X_{t+u}\, I(X_{t+u} < K)\, dB_u$ and $(M_s^K)_{0 \le s \le T-t}$ is a martingale under $\mathsf{P}_{t,x}$.

For $0 < x \le c(t)$ consider the stopping time

$$\sigma_c = \inf\{\, 0 \le s \le T-t : X_{t+s} \ge c(t+s) \,\}. \tag{25.2.57}$$

Then using that $U^c(t, c(t)) = G(c(t))$ for all $0 \le t < T$ since c solves (25.2.9), and that $U^c(T,x) = G(x)$ for all $x > 0$ by (25.2.40), we see that $U^c(t+\sigma_c, X_{t+\sigma_c}) = G(X_{t+\sigma_c})$. Hence from (25.2.55) and (25.2.56) using the optional sampling theorem (page 60) we find

$$\begin{aligned}U^c(t,x) &= \mathsf{E}_{t,x}\Big(e^{-r\sigma_c}U^c(t+\sigma_c, X_{t+\sigma_c})\Big) \qquad (25.2.58)\\ &\quad + rK\,\mathsf{E}_{t,x}\Big(\int_0^{\sigma_c} e^{-ru} I(X_{t+u} \le c(t+u))\,du\Big)\\ &= \mathsf{E}_{t,x}\big(e^{-r\sigma_c}G(X_{t+\sigma_c})\big) + rK\,\mathsf{E}_{t,x}\Big(\int_0^{\sigma_c} e^{-ru} I(X_{t+u} \le c(t+u))\,du\Big)\\ &= G(x) - rK\,\mathsf{E}_{t,x}\Big(\int_0^{\sigma_c} e^{-ru} I(X_{t+u} < K)\,du\Big)\\ &\quad + rK\,\mathsf{E}_{t,x}\Big(\int_0^{\sigma_c} e^{-ru} I(X_{t+u} \le c(t+u))\,du\Big) = G(x)\end{aligned}$$

since $X_{t+u} < K$ and $X_{t+u} \le c(t+u)$ for all $0 \le u < \sigma_c$. This establishes (25.2.52) and thus (25.2.51) holds as well as explained above.

7°. Consider the stopping time

$$\tau_c = \inf\{\, 0 \le s \le T-t : X_{t+s} \le c(t+s)\,\}. \qquad (25.2.59)$$

Note that (25.2.47) using (25.2.45) and (25.2.51) reads

$$\begin{aligned}e^{-rs}V^c(t+s, X_{t+s}) &= V^c(t,x) \qquad (25.2.60)\\ &\quad + \int_0^s e^{-ru} H(t+u, X_{t+u})\, I(X_{t+u} \le c(t+u))\,du + M_s^c\end{aligned}$$

where $H = G_t + \mathbb{L}_X G - rG = -rK$ for $x \le c(t)$ and $(M_s^c)_{0\le s\le T-t}$ is a martingale under $\mathsf{P}_{t,x}$. Thus $\mathsf{E}_{t,x} M_{\tau_c}^c = 0$, so that after inserting τ_c in place of s in (25.2.60), it follows upon taking the $\mathsf{P}_{t,x}$-expectation that

$$V^c(t,x) = \mathsf{E}_{t,x}\Big(e^{-r\tau_c}(K - X_{t+\tau_c})^+\Big) \qquad (25.2.61)$$

for all $(t,x) \in [0,T) \times (0,\infty)$ where we use that $V^c(t,x) = G(x) = (K-x)^+$ for $x \le c(t)$ or $t = T$. Comparing (25.2.61) with (25.2.1) we see that

$$V^c(t,x) \le V(t,x) \qquad (25.2.62)$$

for all $(t,x) \in [0,T) \times (0,\infty)$.

8°. Let us now show that $c \ge b$ on $[0,T]$. For this, recall that by the same arguments as for V^c we also have

$$\begin{aligned}e^{-rs}V(t+s, X_{t+s}) &= V(t,x) \qquad (25.2.63)\\ &\quad + \int_0^s e^{-ru} H(t+u, X_{t+u})\, I(X_{t+u} \le b(t+u))\,du + M_s^b\end{aligned}$$

where $H = G_t + \mathbb{L}_X G - rG = -rK$ for $x \le b(t)$ and $(M_s^b)_{0 \le s \le T-t}$ is a martingale under $\mathsf{P}_{t,x}$. Fix $(t,x) \in (0,T) \times (0,\infty)$ such that $x < b(t) \wedge c(t)$ and consider the stopping time

$$\sigma_b = \inf \{ \, 0 \le s \le T-t : X_{t+s} \ge b(t+s) \, \}. \tag{25.2.64}$$

Inserting σ_b in place of s in (25.2.60) and (25.2.63) and taking the $\mathsf{P}_{t,x}$-expectation, we get

$$\begin{aligned} \mathsf{E}_{t,x}\big(e^{-r\sigma_b} V^c(t+\sigma_b, X_{t+\sigma_b})\big) = G(x) \\ - rK\, \mathsf{E}_{t,x}\Big(\int_0^{\sigma_b} e^{-ru} I(X_{t+u} \le c(t+u))\, du \Big), \end{aligned} \tag{25.2.65}$$

$$\mathsf{E}_{t,x}\big(e^{-r\sigma_b} V(t+\sigma_b, X_{t+\sigma_b})\big) = G(x) - rK\, \mathsf{E}_{t,x}\Big(\int_0^{\sigma_b} e^{-ru}\, du \Big). \tag{25.2.66}$$

Hence by (25.2.62) we see that

$$\mathsf{E}_{t,x}\Big(\int_0^{\sigma_b} e^{-ru} I(X_{t+u} \le c(t+u))\, du \Big) \ge \mathsf{E}_{t,x}\Big(\int_0^{\sigma_b} e^{-ru}\, du \Big) \tag{25.2.67}$$

from where it follows by the continuity of c and b that $c(t) \ge b(t)$ for all $0 \le t \le T$.

9°. Finally, let us show that c must be equal to b. For this, assume that there is $t \in (0,T)$ such that $c(t) > b(t)$, and pick $x \in (b(t), c(t))$. Under $\mathsf{P}_{t,x}$ consider the stopping time τ_b from (25.2.23). Inserting τ_b in place of s in (25.2.60) and (25.2.63) and taking the $\mathsf{P}_{t,x}$-expectation, we get

$$\begin{aligned} \mathsf{E}_{t,x}\big(e^{-r\tau_b} G(X_{t+\tau_b})\big) = V^c(t,x) \\ - rK\, \mathsf{E}_{t,x}\Big(\int_0^{\tau_b} e^{-ru} I(X_{t+u} \le c(t+u))\, du \Big), \end{aligned} \tag{25.2.68}$$

$$\mathsf{E}_{t,x}\big(e^{-r\tau_b} G(X_{t+\tau_b})\big) = V(t,x). \tag{25.2.69}$$

Hence by (25.2.62) we see that

$$\mathsf{E}_{t,x}\Big(\int_0^{\tau_b} e^{-ru} I(X_{t+u} \le c(t+u))\, du \Big) \le 0 \tag{25.2.70}$$

from where it follows by the continuity of c and b that such a point x cannot exist. Thus c must be equal to b, and the proof is complete. □

Remark 25.4. The fact that U^c defined in (25.2.40) must be equal to G below c when c solves (25.2.39) is truly remarkable. The proof of this fact given above (paragraphs 2°– 6°) follows the way which led to its discovery. A shorter but somewhat less revealing proof can also be obtained by introducing U^c as in (25.2.40) and then verifying directly (using the Markov property only) that

$$e^{-rs}U^c(t+s, X_{t+s}) + rK \int_0^s e^{-ru} I(X_{t+u} \le c(t+u))\, du \tag{25.2.71}$$

is a martingale under $\mathsf{P}_{t,x}$ for $0 \le s \le T-t$. In this way it is possible to circumvent the material from paragraphs 2°–4° and carry out the rest of the proof starting with (25.2.56) onward. Moreover, it may be noted that the martingale property of (25.2.71) does not require that c is increasing (but only measurable). This shows that the claim of uniqueness in Theorem 25.3 holds in the class of continuous (or left-continuous) functions $c : [0,T] \to \mathbb{R}$ such that $0 < c(t) < K$ for all $0 < t < T$. It also identifies some limitations of the approach based on the local time-space formula (cf. Subsection 3.5) as initially undertaken (where c needs to be of bounded variation).

Remark 25.5. Note that in Theorem 25.3 above we do not assume that the solution starts (ends) at a particular point. The equation (25.2.39) is highly nonlinear and seems to be out of the scope of any existing theory on nonlinear integral equations (the kernel having four arguments). Similar equations arise in the first-passage problem for Brownian motion (cf. Subsection 14.2).

Notes. According to theory of modern finance (see e.g. [197]) the arbitrage-free price of the American put option with a strike price K coincides with the value function V of the optimal stopping problem with the gain function $G = (K-x)^+$. The optimal stopping time in this problem is the first time when the price process (geometric Brownian motion) falls below the value of a time-dependent boundary b. When the option's expiration date T is finite, the mathematical problem of finding V and b is inherently two-dimensional and therefore analytically more difficult (for infinite T the problem is one-dimensional and b is constant).

The first mathematical analysis of the problem is due to McKean [133] who considered a "discounted" American call with the gain function $G = e^{-\beta t}(x-K)^+$ and derived a free-boundary problem for V and b. He further expressed V in terms of b so that b itself solves a countable system of nonlinear integral equations (p. 39 in [133]). The approach of expressing V in terms of b was in line with the ideas coming from earlier work of Kolodner [114] on free-boundary problems in mathematical physics (such as Stefan's ice melting problem). The existence and uniqueness of a solution to the system for b derived by McKean was left open in [133].

McKean's work was taken further by van Moerbeke [215] who derived a single nonlinear integral equation for b (pp. 145–146 in [215]). The connection to the physical problem is obtained by introducing the auxiliary function $\widetilde{V} = \partial(V-G)/\partial t$ so that the "smooth-fit condition" from the optimal stopping problem translates into the "condition of heat balance" (i.e. the law of conservation of energy) in the physical problem. A motivation for the latter may be seen from the fact that in the mathematical physics literature at the time it was realized that the existence and local uniqueness of a solution to such nonlinear integral equations can be proved by applying the contraction principle (fixed point theorem), first for a small time interval and then extending it to any interval of time by induction (see [137] and [70]). Applying this method, van Moerbeke has proved the existence and local uniqueness of a solution to the integral equations of a general optimal stopping problem (see Sections 3.1 and 3.2 in [215]) while the proof of the same claim in the context of the discounted American call [133] is merely indicated (see Section 4.4 in [215]). One of the technical difficulties in this context is that the derivative b' of the optimal boundary b is not bounded at the initial point T as used in the general proof (cf. Sections 3.1 and 3.2 in [215]).

The fixed point method usually results in a long and technical proof with an indecisive end where the details are often sketched or omitted. Another consequence of the approach is the fact that the integral equations in [133] and [215] involve both b and its derivative b', so that either the fixed point method results in proving that b is differentiable, or this needs to be proved a priori if the existence is claimed simply by identifying b with the boundary of the set where $V = G$. The latter proof, however, appears difficult to give directly, so that if one is only interested in the actual values of b which indicate when to stop, it seems that the differentiability of b plays a minor role. Finally, since it is not obvious to see (and it was never explicitly addressed) how the "condition of heat balance" relates to the economic mechanism of "no-arbitrage" behind the American option, one is led to the conclusion that the integral equations derived by McKean and van Moerbeke, being motivated purely by the mathematical tractability arising from the work in mathematical physics, are perhaps more complicated then needed from the standpoint of optimal stopping.

This was to be confirmed in the beginning of the 1990's when Kim [110], Jacka [102] and Carr, Jarrow, Myneni [27] independently arrived at a nonlinear integral equation for b that is closely linked to the *early exercise premium representation* of V having a clear economic meaning (see Section 1 in [27] and Corollary 3.1 in [142]). In fact, the equation is obtained by inserting $x = b(t)$ in this representation, and for this reason it is called the *free-boundary equation* (see (25.2.39) above). The early exercise premium representation for V follows transparently from the free-boundary formulation (given that the smooth-fit condition holds) and moreover corresponds to the decomposition of the superharmonic function V into its harmonic and its potential part (the latter being the basic principle of optimal stopping established in the works of Snell [206] and Dynkin [52]).

The superharmonic characterization of the value function V (cf. Chapter I) implies that $e^{-rs}V(t-s, X_{t+s})$ is the smallest supermartingale dominating $e^{-rs}(K - X_{t+s})^+$ on $[0, T-t]$, i.e. that $V(t,x)$ is the smallest superharmonic function (relative to $\partial/\partial t + \mathbb{L}_X - rI$) dominating $(K-x)^+$ on $[0,T] \times \mathbb{R}_+$. The two requirements (i.e. smallest and superharmonic) manifest themselves in the single analytic condition of smooth fit (25.2.28).

The derivation of the smooth-fit condition given in Myneni [142] upon integrating the second formula on p. 15 and obtaining the third one seems to violate the Newton–Leibniz formula unless $x \mapsto V(t,x)$ is smooth at $b(t)$ so that there is nothing to prove. Myneni writes that this proof is essentially from McKean [133]. A closer inspection of his argument on p. 38 in [133] reveals the same difficulty. Alternative derivations of the smooth-fit principle for Brownian motion and diffusions are given in Grigelionis & Shiryaev [88] and Chernoff [30] by a Taylor expansion of V at $(t, b(t))$ and in Bather [11] and van Moerbeke [215] by a Taylor expansion of G at $(t, b(t))$. The latter approach seems more satisfactory generally since V is not known a priori. Jacka [104] (Corollary 7) develops a different approach which he applies in [102] (Proposition 2.8) to verify (25.2.28).

It follows from the preceding that the optimal stopping boundary b satisfies the free-boundary equation, however, as pointed out by Myneni [142] (p. 17) "the uniqueness and regularity of the stopping boundary from this integral equation remain open". This attempt is in line with McKean [133] (p. 33) who wrote that "another inviting unsolved problem is to discuss the integral equation for the free-boundary of section 6", concluding the paper (p. 39) with the words "even the existence and uniqueness of solutions is still unproved". McKean's integral equations [133] (p. 39) are more complicated (involving b' as well) than the equation (25.2.37). Thus the simplification of his equations to the equations (25.2.37) and finally the equation (25.2.39) may be viewed as a step to the solution of the problem. Theorem 4.3 of Jacka [102] states that if $c : [0,T] \to \mathbb{R}$ is a "left-continuous" solution of (25.2.37) for all $x \in (0, c(t)]$ satisfying $0 < c(t) < K$ for all $t \in (0,T)$, then c is the optimal stopping boundary b. Since (25.2.37) is a different equation for each new $x \in (0, c(t)]$, we see that this assumption in effect corresponds to c solving a countable system of nonlinear integral equations (obtained by letting x in $(0, c(t)]$ run through rationals for instance). From the standpoint of numerical calculation it is therefore of interest to reduce the number of these equations.

The main purpose of the present section (following [164]) is to show that the question of Myneni can be answered affirmatively and that the *free-boundary equation alone does indeed characterize the optimal stopping boundary* b. The key argument in the proof is based on the local time-space formula [163] (see Subsection 3.5). The same method of proof can be applied to more general continuous Markov processes (diffusions) in problems of optimal stopping with finite horizon. For example, in this way it is also possible to settle the somewhat more complicated problem of the Russian option with finite horizon [165] (see Section 26 below).

With reference to [133] and [215] it is claimed in [142] (and used in some other papers too) that b is C^1 but we could not find a complete/transparent proof in either of these papers (nor anywhere else). If it is known that b is C^1, then the proof above shows that C in (25.2.32) can be replaced by $\bar{C}$, implying also that $s \mapsto V(s, b(t))$ is C^1 at t. For both, in fact, it is sufficient to know that b is (locally) Lipschitz, but it seems that this fact is no easier to establish directly, and we do not know of any transparent proof.

For more information on the American option problem we refer to the survey paper [142], the book [197] and Sections 2.5–2.8 in the book [107] where further references can also be found. For a numerical discussion of the free-boundary equation and possible improvements along these lines see e.g. [93]. For asymptotics of the optimal stopping boundary see [121], and for a proof that it is convex see [58]. For random walks and Lévy processes see [33], [140] and [2].

26. The Russian option

26.1. Infinite horizon

1. The arbitrage-free price of the *Russian option* with *infinite horizon* (perpetual option) is given by

$$V = \sup_{\tau} \mathsf{E}\left(e^{-(r+\lambda)\tau} M_\tau\right) \tag{26.1.1}$$

where the supremum is taken over all stopping times τ of the geometric Brownian motion $S = (S_t)_{t\geq 0}$ solving

$$dS_t = rS_t\,dt + \sigma S_t\,dB_t \quad (S_0 = s) \tag{26.1.2}$$

and $M = (M_t)_{t\geq 0}$ is the *maximum* process given by

$$M_t = \Big(\max_{0\leq u\leq t} S_u\Big) \vee m \tag{26.1.3}$$

where $m \geq s > 0$ are given and fixed. We recall that $B = (B_t)_{t\geq 0}$ is a standard Brownian motion process started at zero, $r > 0$ is the interest rate, $\lambda > 0$ is the discounting rate, and $\sigma > 0$ is the volatility coefficient.

2. The problem (26.1.1) is two-dimensional since the underlying Markov process may be identified with (S, M). Using the same method as in Section 13 it is possible to solve the problem (26.1.1) explicitly. Instead we will follow a different route to the explicit solution using a *change of measure* (cf. Subsection 15.3) which reduces the initial *two-dimensional* problem to an equivalent *one-dimensional* problem (cf. Subsection 6.2). This reduction becomes especially handy in the case when the horizon in (26.1.1) is finite (cf. Subsection 26.2 below).

Recalling that the strong solution of (26.1.2) is given by (26.1.9) below and writing M_τ in (26.1.1) as $S_\tau(M_\tau/S_\tau)$, we see by change of measure that

$$V = s \sup_{\tau} \widetilde{\mathsf{E}}(e^{-\lambda\tau} X_\tau) \tag{26.1.4}$$

where we set

$$X_t = \frac{M_t}{S_t} \tag{26.1.5}$$

and $\widetilde{\mathsf{P}}$ is a probability measure satisfying $d\widetilde{\mathsf{P}} = \exp\big(\sigma B_t - (\sigma^2/2)\,t\big)\,d\mathsf{P}$ when restricted to $\mathcal{F}_t^B = \sigma(B_s : 0 \le s \le t)$ for $t \ge 0$. By Girsanov's theorem (see [106] or [197]) we see that the process $\widetilde{B} = (\widetilde{B}_t)_{t\ge 0}$ given by $\widetilde{B}_t = B_t - \sigma t$ is a standard Brownian motion under $\widetilde{\mathsf{P}}$ for $t \ge 0$. By Itô's formula (page 67) one finds that the process $X = (X_t)_{t\ge 0}$ solves

$$dX_t = -rX_t\,dt + \sigma X_t\,d\widehat{B}_t + dR_t \quad (X_0 = x) \tag{26.1.6}$$

under $\widetilde{\mathsf{P}}$ where $\widehat{B} = -\widetilde{B}$ is a standard Brownian motion, and we set

$$R_t = \int_0^t I(X_s = 1)\,\frac{dM_s}{S_s} \tag{26.1.7}$$

for $t \ge 0$. It follows that X is a diffusion process in $[1,\infty)$ having 1 as a boundary point of *instantaneous reflection*. The infinitesimal generator of X is therefore given by

$$\begin{aligned} \mathbb{L}_X &= -rx\,\frac{\partial}{\partial x} + \frac{\sigma^2}{2}x^2\,\frac{\partial^2}{\partial x^2} \quad \text{in } (1,\infty), \\ \frac{\partial}{\partial x} &= 0 \quad \text{at } 1+. \end{aligned} \tag{26.1.8}$$

The latter means that the infinitesimal generator of X is acting on a space of C^2 functions f defined on $[1,\infty)$ such that $f'(1+) = 0$.

3. For further reference recall that the strong solution of (26.1.2) is given by

$$S_t = s\exp\Big(\sigma B_t + \Big(r - \frac{\sigma^2}{2}\Big)t\Big) = s\exp\Big(\sigma \widetilde{B}_t + \Big(r + \frac{\sigma^2}{2}\Big)t\Big) \tag{26.1.9}$$

for $t \ge 0$ where B and $\widetilde{B}$ are standard Brownian motions with respect to P and $\widetilde{\mathsf{P}}$ respectively. When dealing with the process X on its own, however, note that there is no restriction to assume that $s = 1$ and $m = x$ with $x \ge 1$.

4. Summarizing the preceding facts we see that the *Russian option problem* with *infinite horizon* reduces to solving the following optimal stopping problem:

$$V(x) = \sup_\tau \mathsf{E}_x\big(e^{-\lambda\tau}X_\tau\big) \tag{26.1.10}$$

where τ is a stopping time of the diffusion process X satisfying (26.1.5)–(26.1.8) above and $X_0 = x$ under P_x with $x \ge 1$ given and fixed.

5. The optimal stopping problem (26.1.10) will be solved in two steps. In the first step we will make a guess for the solution. In the second step we will verify that the guessed solution is correct (Theorem 26.1).

From (26.1.6) and (26.1.10) we see that the further away X gets from 1 the less likely that the gain will increase upon continuation. This suggests that there exists a point $b \in (1, \infty]$ such that the stopping time

$$\tau_b = \inf\{\, t \geq 0 : X_t \geq b \,\} \tag{26.1.11}$$

is optimal in the problem (26.1.10).

Standard arguments based on the strong Markov property (cf. Chapter III) lead to the following *free-boundary problem* for the *unknown value function* V and the *unknown point* b:

$$\mathbb{L}_X V = \lambda V \quad \text{for } x \in [1, \infty), \tag{26.1.12}$$
$$V(x) = x \quad \text{for } x = b, \tag{26.1.13}$$
$$V'(x) = 1 \quad \text{for } x = b \quad (\textit{smooth fit}), \tag{26.1.14}$$
$$V'(x) = 0 \quad \text{for } x = 1 \quad (\textit{normal reflection}), \tag{26.1.15}$$
$$V(x) > x \quad \text{for } x \in [1, b), \tag{26.1.16}$$
$$V(x) = x \quad \text{for } x \in (b, \infty). \tag{26.1.17}$$

6. To solve the free-boundary problem (26.1.12)–(26.1.17) note that the equation (26.1.12) using (26.1.8) reads as

$$Dx^2 V'' - rxV' - \lambda V = 0 \tag{26.1.18}$$

where we set $D = \sigma^2/2$. One may now recognize (26.1.18) as the Cauchy–Euler equation. Let us thus seek a solution in the form

$$V(x) = x^p. \tag{26.1.19}$$

Inserting (26.1.19) into (26.1.18) we get

$$p^2 - \Big(1 + \frac{r}{D}\Big)p - \frac{\lambda}{D} = 0. \tag{26.1.20}$$

The quadratic equation (26.1.20) has two roots:

$$p_{1,2} = \frac{\big(1 + \frac{r}{D}\big) \pm \sqrt{\big(1 + \frac{r}{D}\big)^2 + \frac{4\lambda}{D}}}{2}. \tag{26.1.21}$$

Thus the general solution of (26.1.18) can be written as

$$V(x) = C_1 x^{p_1} + C_2 x^{p_2} \tag{26.1.22}$$

where C_1 and C_2 are undetermined constants. The three conditions (26.1.13)–(26.1.15) can be used to determine C_1, C_2 and b (free boundary) uniquely. This gives

$$C_1 = -\frac{p_2}{p_1 b^{p_2-1} - p_2 b^{p_1-1}}, \tag{26.1.23}$$

$$C_2 = \frac{p_1}{p_1 b^{p_2-1} - p_2 b^{p_1-1}}, \tag{26.1.24}$$

$$b = \left(\frac{p_1(p_2-1)}{p_2(p_1-1)}\right)^{1/(p_1-p_2)}. \tag{26.1.25}$$

Note that $C_1 > 0$, $C_2 > 0$ and $b > 1$. Inserting (26.1.23) and (26.1.24) into (26.1.22) we obtain

$$V(x) = \begin{cases} \dfrac{1}{p_1 b^{p_2-1} - p_2 b^{p_1-1}} \left(p_1 x^{p_2} - p_2 x^{p_1}\right) & \text{if } x \in [1,b], \\ x & \text{if } x \in [b,\infty) \end{cases} \tag{26.1.26}$$

where b is given by (26.1.25). Note that V is C^2 on $[1,b) \cup (b,\infty)$ but only C^1 at b. Note also that V is convex and increasing on $[1,\infty)$ and that (26.1.16) is satisfied.

7. In this way we have arrived at the conclusions in the following theorem.

Theorem 26.1. *The arbitrage-free price* V *from* (26.1.10) *is given explicitly by* (26.1.26) *above. The stopping time* τ_b *from* (26.1.11) *with* b *given by* (26.1.25) *above is optimal in the problem* (26.1.10).

Proof. To distinguish the two functions let us denote the value function from (26.1.10) by $V_*(x)$ for $x \geq 1$. We need to prove that $V_*(x) = V(x)$ for all $x \geq 1$ where $V(x)$ is given by (26.1.26) above.

1°. The properties of V stated following (26.1.26) above show that Itô's formula (page 67) can be applied to $e^{-\lambda t} V(X_t)$ in its standard form (cf. Subsection 3.5). This gives

$$\begin{aligned} e^{-\lambda t} V(X_t) &= V(x) + \int_0^t e^{-\lambda s} \left(\mathbb{L}_X V - \lambda V\right)(X_s)\, I(X_s \neq b)\, ds \\ &\quad + \int_0^t e^{-\lambda s} V'(X_s)\, dR_s + \int_0^t e^{-\lambda s} \sigma X_s V'(X_s)\, d\widehat{B}_s \\ &= V(x) + \int_0^t e^{-\lambda s} (\mathbb{L}_X V - \lambda V)(X_s) I(X_s \neq b)\, ds \\ &\quad + \int_0^t e^{-\lambda s} \sigma X_s V'(X_s)\, d\widehat{B}_s \end{aligned} \tag{26.1.27}$$

upon using (26.1.7) and (26.1.15) to conclude that the integral with respect to dR_s equals zero. Setting $G(x) = x$ we see that $(\mathbb{L}_X G - \lambda G)(x) = -(r+\lambda)x < 0$ so that together with (26.1.12) we have

$$(\mathbb{L}_X V - \lambda V) \leq 0 \tag{26.1.28}$$

everywhere on $[1,\infty)$ but b. Since $\mathsf{P}_x(X_s=b)=0$ for all s and all x, we see that (26.1.13), (26.1.16)–(26.1.17) and (26.1.27)–(26.1.28) imply that

$$e^{-\lambda t}X_t \le e^{-\lambda t}V(X_t) \le V(x) + M_t \tag{26.1.29}$$

where $M=(M_t)_{t\ge 0}$ is a continuous local martingale given by

$$M_t = \int_0^t e^{-\lambda s}\sigma X_s V'(X_s)\, d\widehat{B}_s. \tag{26.1.30}$$

(Using that $0\le V'(x)\le 1$ for all $x\ge 1$ it is easily verified by standard means that M is a martingale.)

Let $(\tau_n)_{n\ge1}$ be a localizations sequence of (bounded) stopping times for M (for example $\tau_n \equiv n$ will do). Then for every stopping time τ of X we have by (26.1.29) above:

$$e^{-\lambda(\tau\wedge\tau_n)}X_{\tau\wedge\tau_n} \le V(x) + M_{\tau\wedge\tau_n} \tag{26.1.31}$$

for all $n\ge 1$. Taking the P_x-expectation, using the optional sampling theorem (page 60) to conclude that $\mathsf{E}_x M_{\tau\wedge\tau_n}=0$ for all n, and letting $n\to\infty$, we find by Fatou's lemma that

$$\mathsf{E}_x\big(e^{-\lambda\tau}X_\tau\big) \le V(x). \tag{26.1.32}$$

Taking the supremum over all stopping times τ of X we conclude that $V_*(x)\le V(x)$ for all $x\in[1,\infty)$.

2°. To prove the reverse inequality (equality) we may observe from (26.1.27) upon using (26.1.12) (and the optional sampling theorem as above) that

$$\mathsf{E}_x\Big(e^{-\lambda(\tau_b\wedge\tau_n)}V(X_{\tau_b\wedge\tau_n})\Big) = V(x) \tag{26.1.33}$$

for all $n\ge 1$. Letting $n\to\infty$ and using that $e^{-\lambda\tau_b}V(X_{\tau_b}) = e^{-\lambda\tau_b}X_{\tau_b}$, we find by the dominated convergence theorem that

$$\mathsf{E}_x\big(e^{-\lambda\tau_b}X_{\tau_b}\big) = V(x). \tag{26.1.34}$$

This shows that τ_b is optimal in (26.1.10). Thus $V_*(x)=V(x)$ for all $x\in[1,\infty)$ and the proof is complete. □

Remark 26.2. In the notation of Theorem 26.1 above set

$$u(x) = \mathsf{E}_x\tau_b \tag{26.1.35}$$

for $x\in[1,b]$. Standard arguments based on the strong Markov property (cf. Section 7) imply that u solves

$$\mathbb{L}_X u = -1 \quad \text{on } (1,b), \tag{26.1.36}$$

$$u(b) = 0, \tag{26.1.37}$$

$$u'(1) = 0. \tag{26.1.38}$$

The general solution of (26.1.1) is given by

$$u(x) = C_1 + C_2\, x^{1+r/D} + \frac{1}{r+D} \log x. \tag{26.1.39}$$

Using (26.1.2) and (26.1.3) we find

$$C_1 = \frac{b^{1+r/D}}{(1+r/D)(r+D)} - \frac{1}{r+D} \log\Big(1 + \frac{r}{D}\Big), \tag{26.1.40}$$

$$C_2 = -\frac{1}{(1+r/D)(r+D)}. \tag{26.1.41}$$

It can easily be verified using standard means (Itô's formula and the optional sampling theorem) that (26.1.39) with (26.1.40) and (26.1.41) give the correct expression for (26.1.35). In particular, this also shows that $\tau_b < \infty$ P_x-a.s. for every $x \in [1, \infty)$, where $b > 1$ is given and fixed (arbitrary). Thus τ_b in (26.1.11) is indeed a (finite) stopping time under every P_x with $x \in [1, \infty)$.

26.2. Finite horizon

1. The arbitrage-free price of the *Russian option* with *finite horizon* (cf. (26.1.1) above) is given by

$$V = \sup_{0 \le \tau \le T} \mathsf{E}\big(e^{-r\tau} M_\tau\big) \tag{26.2.1}$$

where the supremum is taken over all stopping times τ of the geometric Brownian motion $S = (S_t)_{0 \le t \le T}$ solving

$$dS_t = rS_t\, dt + \sigma S_t\, dB_t \quad (S_0 = s) \tag{26.2.2}$$

and $M = (M_t)_{0 \le t \le T}$ is the maximum process given by

$$M_t = \Big(\max_{0 \le u \le t} S_u\Big) \vee m \tag{26.2.3}$$

where $m \ge s > 0$ are given and fixed. We recall that $B = (B_t)_{t \ge 0}$ is a standard Brownian motion process started at zero, $T > 0$ is the expiration date (maturity), $r > 0$ is the interest rate, and $\sigma > 0$ is the volatility coefficient.

The first part of this subsection is analogous to the first part of the previous subsection (cf. paragraphs 1–3) and we will briefly repeat all the details merely for completeness and ease of reference.

2. For the purpose of comparison with the infinite-horizon results from the previous subsection we will also introduce a discounting rate $\lambda \ge 0$ so that M_τ in (26.2.1) is to be replaced by $e^{-\lambda\tau} M_\tau$. By change of measure as in (26.1.4) above it then follows that

$$V = s \sup_{0 \le \tau \le T} \widetilde{\mathsf{E}}\big(e^{-\lambda\tau} X_\tau\big) \tag{26.2.4}$$

where we set

$$X_t = \frac{M_t}{S_t} \tag{26.2.5}$$

and $\widetilde{\mathsf{P}}$ is defined following (26.1.5) above so that $\widetilde{B}_t = B_t - \sigma t$ is a standard Brownian motion under $\widetilde{\mathsf{P}}$ for $0 \le t \le T$. As in (26.1.6) above one finds that X solves

$$dX_t = -rX_t\,dt + \sigma X_t\,d\widehat{B}_t + dR_t \quad (X_0 = x) \tag{26.2.6}$$

under $\widetilde{\mathsf{P}}$ where $\widehat{B} = -\widetilde{B}$ is a standard Brownian motion, and we set

$$R_t = \int_0^t I(X_s = 1)\,\frac{dM_s}{S_s} \tag{26.2.7}$$

for $0 \le t \le T$. Recall that X is a diffusion process in $[1, \infty)$ with 1 being instantaneously reflecting, and the infinitesimal generator of X is given by

$$\begin{aligned} &\mathbb{L}_X = -rx\,\frac{\partial}{\partial x} + \frac{\sigma^2}{2}x^2\frac{\partial^2}{\partial x^2} \quad \text{in } (1,\infty), \\ &\frac{\partial}{\partial x} = 0 \quad \text{at } 1+ . \end{aligned} \tag{26.2.8}$$

For more details on the derivation of (26.2.4)–(26.2.8) see the text of (26.1.4)-(26.1.8) above.

3. For further reference recall that the strong solution of (26.2.2) is given by

$$S_t = s\exp\Big(\sigma B_t + \Big(r - \frac{\sigma^2}{2}\Big)t\Big) = s\exp\Big(\sigma \widetilde{B}_t + \Big(r + \frac{\sigma^2}{2}\Big)t\Big) \tag{26.2.9}$$

for $0 \le t \le T$ where B and $\widetilde{B}$ are standard Brownian motions under P and $\widetilde{\mathsf{P}}$ respectively. Recall also when dealing with the process X on its own that there is no restriction to assume that $s = 1$ and $m = x$ with $x \ge 1$.

4. Summarizing the preceding facts we see that the *Russian option problem* with *finite horizon* reduces to solving the following optimal stopping problem:

$$V(t,x) = \sup_{0 \le \tau \le T-t} \widetilde{\mathsf{E}}_{t,x}\big(e^{-\lambda\tau}X_{t+\tau}\big) \tag{26.2.10}$$

where τ is a stopping time of the diffusion process X satisfying (26.2.5)–(26.2.8) above and $X_t = x$ under $\widetilde{\mathsf{P}}_{t,x}$ with $(t,x) \in [0,T] \times [1,\infty)$ given and fixed.

5. Standard Markovian arguments (cf. Chapter III) indicate that V from (26.2.10) solves the following *free-boundary problem*:

$$
\begin{aligned}
&V_t + \mathbb{L}_X V = \lambda V \quad \text{in } C, && (26.2.11)\\
&V(t,x) = x \quad \text{for } x = b(t) \text{ or } t = T, && (26.2.12)\\
&V_x(t,x) = 1 \quad \text{for } x = b(t) \quad (\textit{smooth fit}), && (26.2.13)\\
&V_x(t,1+) = 0 \quad (\textit{normal reflection}), && (26.2.14)\\
&V(t,x) > x \quad \text{in } C, && (26.2.15)\\
&V(t,x) = x \quad \text{in } D && (26.2.16)
\end{aligned}
$$

where the continuation set C and the stopping set $\bar{D}$ (as the closure of the set D below) are defined by

$$
C = \{\, (t,x) \in [0,T) \times [1,\infty) : x < b(t) \,\}, \tag{26.2.17}
$$

$$
D = \{\, (t,x) \in [0,T) \times [1,\infty) : x > b(t) \,\} \tag{26.2.18}
$$

and $b : [0,T] \to \mathbb{R}$ is the (unknown) optimal stopping boundary, i.e. the stopping time

$$
\tau_b = \inf \{\, 0 \le s \le T-t : X_{t+s} \ge b(t+s) \,\} \tag{26.2.19}
$$

is optimal in the problem (26.2.10).

6. It will follow from the result of Theorem 26.3 below that the free-boundary problem (26.2.11)–(26.2.16) characterizes the value function V and the optimal stopping boundary b in a unique manner. Our main aim, however, is to follow the train of thought where V is first expressed in terms of b, and b itself is shown to satisfy a nonlinear integral equation. A particularly simple approach for achieving this goal in the case of the American put option has been exposed in Subsection 25.2 above and we will take it up in the present subsection as well. We will moreover see (as in the case of the American put option above) that the nonlinear equation derived for b cannot have other solutions.

7. Below we will make use of the following functions:

$$
F(t,x) = \widetilde{\mathsf{E}}_{0,x}(X_t) = \int_1^\infty \int_0^m \left(\tfrac{m \vee x}{s}\right) f(t,s,m)\, ds\, dm, \tag{26.2.20}
$$

$$
\begin{aligned}
G(t,x,y) &= \widetilde{\mathsf{E}}_{0,x}\big(X_t\, I(X_t \ge y)\big) && (26.2.21)\\
&= \int_1^\infty \int_0^m \left(\tfrac{m \vee x}{s}\right) I\Big(\left(\tfrac{m \vee x}{s}\right) \ge y\Big)\, f(t,s,m)\, ds\, dm
\end{aligned}
$$

for $t > 0$ and $x, y \ge 1$, where $(s,m) \mapsto f(t,s,m)$ is the probability density function of (S_t, M_t) under $\widetilde{\mathsf{P}}$ with $S_0 = M_0 = 1$ given by (see e.g. [107, p. 368]):

$$
f(t,s,m) = \frac{2}{\sigma^3 \sqrt{2\pi t^3}}\, \frac{\log(m^2/s)}{sm} \exp\left(-\frac{\log^2(m^2/s)}{2\sigma^2 t} + \frac{\beta}{\sigma} \log s - \frac{\beta^2}{2}\, t \right) \tag{26.2.22}
$$

for $0 < s \le m$ and $m \ge 1$ with $\beta = r/\sigma + \sigma/2$, and is equal to 0 otherwise.

8. The main result of the present subsection may now be stated as follows.

Theorem 26.3. *The optimal stopping boundary in the Russian option problem* (26.2.10) *can be characterized as the unique continuous decreasing solution* $b : [0,T] \to \mathbb{R}$ *of the nonlinear integral equation*

$$b(t) = e^{-\lambda(T-t)} F(T-t, b(t)) + (r+\lambda) \int_0^{T-t} e^{-\lambda u}\, G(u, b(t), b(t+u))\, du \quad (26.2.23)$$

satisfying $b(t) > 1$ *for all* $0 < t < T$. [*The solution* b *satisfies* $b(T-) = 1$ *and the stopping time* τ_b *from* (26.2.19) *is optimal in* (26.2.10) (*see* Figure VII.2).]

The arbitrage-free price of the Russian option (26.2.10) *admits the following "early exercise premium" representation*:

$$V(t,x) = e^{-\lambda(T-t)} F(T-t, x) + (r+\lambda) \int_0^{T-t} e^{-\lambda u}\, G(u, x, b(t+u))\, du \quad (26.2.24)$$

for all $(t,x) \in [0,T] \times [1,\infty)$. [*Further properties of* V *and* b *are exhibited in the proof below.*]

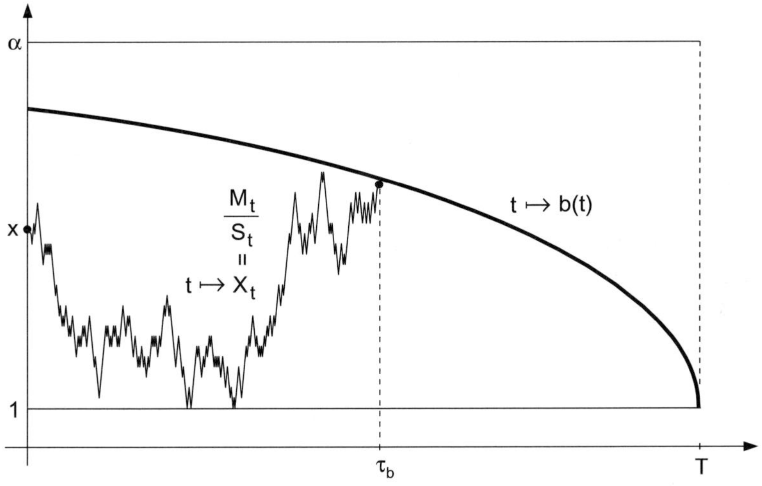

Figure VII.2: A computer drawing of the optimal stopping boundary b from Theorem 26.3. The number α is the optimal stopping point in the case of infinite horizon (Theorem 26.1). If the discounting rate λ is zero, then α is infinite (i.e. it is never optimal to stop), while b is still finite and looks as above.

Proof. The proof will be carried out in several steps. We begin by stating some general remarks which will be freely used below without further mention.

It is easily seen that $\mathsf{E}(\max_{0 \le t \le T} X_t) < \infty$ so that $V(t,x) < \infty$ for all $(t,x) \in [0,T] \times [1,\infty)$. Recall that it is no restriction to assume that $s = 1$ and

$m=x$ so that $X_t=(M_t \vee x)/S_t$ with $S_0=M_0=1$. We will write X_t^x instead of X_t to indicate the dependence on x when needed. Since $M_t \vee x=(x-M_t)^+ + M_t$ we see that V admits the following representation:

$$V(t,x)=\sup_{0\le\tau\le T-t} \widetilde{\mathsf{E}}\left(e^{-\lambda\tau}\,\frac{(x-M_\tau)^+ + M_\tau}{S_\tau}\right) \tag{26.2.25}$$

for $(t,x)\in[0,T]\times[1,\infty)$. It follows immediately from (26.2.25) that

$$x\mapsto V(t,x) \quad \text{is increasing and convex on } [1,\infty) \tag{26.2.26}$$

for each $t\ge 0$ fixed. It is also obvious from (26.2.25) that $t\mapsto V(t,x)$ is decreasing on $[0,T]$ with $V(T,x)=x$ for each $x\ge 1$ fixed.

1°. We show that $V:[0,T]\times[1,\infty)\to\mathbb{R}$ is *continuous*. For this, using $\sup(f)-\sup(g)\le\sup(f-g)$ and $(y-z)^+ - (x-z)^+ \le (y-x)^+$ for $x,y,z\in\mathbb{R}$, it follows that

$$V(t,y)-V(t,x)\le(y-x)\sup_{0\le\tau\le T-t}\widetilde{\mathsf{E}}\left(\frac{e^{-\lambda\tau}}{S_\tau}\right)\le y-x \tag{26.2.27}$$

for $1\le x<y$ and all $t\ge 0$, where in the second inequality we used (26.2.9) to deduce that $1/S_t=\exp(\sigma\widehat{B}_t-(r+\sigma^2/2)t)\le\exp(\sigma\widehat{B}_t-(\sigma^2/2)t)$ and the latter is a martingale under $\widetilde{\mathsf{P}}$. From (26.2.27) with (26.2.26) we see that $x\mapsto V(t,x)$ is continuous uniformly over $t\in[0,T]$. Thus to prove that V is continuous on $[0,T]\times[1,\infty)$ it is enough to show that $t\mapsto V(t,x)$ is continuous on $[0,T]$ for each $x\ge 1$ given and fixed. For this, take any $t_1<t_2$ in $[0,T]$ and $\varepsilon>0$, and let τ_1^ε be a stopping time such that $\widetilde{\mathsf{E}}(e^{-\lambda\tau_1^\varepsilon}X^x_{t_1+\tau_1^\varepsilon})\ge V(t_1,x)-\varepsilon$. Setting $\tau_2^\varepsilon=\tau_1^\varepsilon\wedge(T-t_2)$ we see that $V(t_2,x)\ge\widetilde{\mathsf{E}}(e^{-\lambda\tau_2^\varepsilon}X^x_{t_2+\tau_2^\varepsilon})$. Hence we get

$$0\le V(t_1,x)-V(t_2,x)\le\widetilde{\mathsf{E}}\left(e^{-\lambda\tau_1^\varepsilon}X^x_{t_1+\tau_1^\varepsilon}-e^{-\lambda\tau_2^\varepsilon}X^x_{t_2+\tau_2^\varepsilon}\right)+\varepsilon. \tag{26.2.28}$$

Letting first $t_2-t_1\to 0$ using $\tau_1^\varepsilon-\tau_2^\varepsilon\to 0$ and then $\varepsilon\downarrow 0$ we see that $V(t_1,x)-V(t_2,x)\to 0$ by dominated convergence. This shows that $t\mapsto V(t,x)$ is continuous on $[0,T]$, and the proof of the initial claim is complete.

Denote $G(x)=x$ for $x\ge 1$ and introduce the continuation set $C=\{\,(t,x)\in[0,T)\times[1,\infty):V(t,x)>G(x)\,\}$ and the stopping set $\bar{D}=\{(t,x)\in[0,T)\times[1,\infty):V(t,x)=G(x)\}$. Since V and G are continuous, we see that C is open (and $\bar{D}$ is closed indeed) in $[0,T)\times[1,\infty)$. Standard arguments based on the strong Markov property [see Corollary 2.9 (Finite horizon) with Remark 2.10] show that the first hitting time $\tau_{\bar{D}}=\inf\{\,0\le s\le T-t:(t+s,X_{t+s})\in\bar{D}\,\}$ is optimal in (26.2.10).

2°. We show that the continuation set C just defined is given by (26.2.17) for some *decreasing* function $b:[0,T)\to(1,\infty)$. It follows in particular that the stopping set coincides with the closure $\bar{D}$ in $[0,T) \times [1,\infty)$ of the set D in (26.2.18) as claimed. To verify the initial claim, note that by Itô's formula (page 67) and (26.2.6) we have

$$e^{-\lambda s} X_{t+s} = X_t - (r+\lambda) \int_0^s e^{-\lambda u} X_{t+u}\, du + \int_0^s e^{-\lambda u} \frac{dM_{t+u}}{S_{t+u}} + N_s \tag{26.2.29}$$

where $N_s = \sigma \int_0^s e^{-\lambda u} X_{t+u}\, d\widehat{B}_{t+u}$ is a martingale for $0 \le s \le T-t$. Let $t \in [0,T]$ and $x > y \ge 1$ be given and fixed. We will first show that $(t,x) \in C$ implies that $(t,y) \in C$. For this, let $\tau_* = \tau_*(t,x)$ denote the optimal stopping time for $V(t,x)$. Taking the expectation in (26.2.29) stopped at τ_*, first under $\widetilde{\mathsf{P}}_{t,y}$ and then under $\widetilde{\mathsf{P}}_{t,x}$, and using the optional sampling theorem (page 60) to get rid of the martingale part, we find

$$\begin{aligned} V(t,y) - y &\ge \widetilde{\mathsf{E}}_{t,y}\Big(e^{-\lambda\tau_*} X_{t+\tau_*}\Big) - y \qquad\qquad (26.2.30)\\ &= -(r+\lambda)\, \widetilde{\mathsf{E}}_{t,y}\Big(\int_0^{\tau_*} e^{-\lambda u} X_{t+u}\, du\Big) + \widetilde{\mathsf{E}}_{t,y}\Big(\int_0^{\tau_*} e^{-\lambda u} \frac{dM_{t+u}}{S_{t+u}}\Big)\\ &\ge -(r+\lambda)\, \widetilde{\mathsf{E}}_{t,x}\Big(\int_0^{\tau_*} e^{-\lambda u} X_{t+u}\, du\Big) + \widetilde{\mathsf{E}}_{t,x}\Big(\int_0^{\tau_*} e^{-\lambda u} \frac{dM_{t+u}}{S_{t+u}}\Big)\\ &= \widetilde{\mathsf{E}}_{t,x}\Big(e^{-\lambda\tau_*} X_{t+\tau_*}\Big) - x = V(t,x) - x > 0 \end{aligned}$$

proving the claim. To explain the second inequality in (26.2.30) note that the process X under $\widetilde{\mathsf{P}}_{t,z}$ can be realized as the process $X^{t,z}$ under P where we set $X^{t,z}_{t+u} = (S^*_u \vee z)/S_u$ with $S^*_u = \max_{0 \le v \le u} S_v$. Then note that $X^{t,y}_{t+u} \le X^{t,x}_{t+u}$ and $d(S^*_u \vee y) \ge d(S^*_u \vee x)$ whenever $y \le x$, and thus each of the two terms on the left-hand side of the inequality is larger than the corresponding term on the right-hand side, implying the inequality. The fact just proved establishes the existence of a function $b : [0,T] \to [1,\infty]$ such that the continuation set C is given by (26.2.17) above.

Let us show that b is decreasing. For this, with $x \ge 1$ and $t_1 < t_2$ in $[0,T]$ given and fixed, it is enough to show that $(t_2,x) \in C$ implies that $(t_1,x) \in C$. To verify this implication, recall that $t \mapsto V(t,x)$ is decreasing on $[0,T]$, so that we have

$$V(t_1,x) \ge V(t_2,x) > x \tag{26.2.31}$$

proving the claim.

Let us show that b does not take the value ∞. For this, assume that there exists $t_0 \in (0,T]$ such that $b(t) = \infty$ for all $0 \le t \le t_0$. It implies that $(0,x) \in C$ for any $x \ge 1$ given and fixed, so that if $\tau_* = \tau_*(0,x)$ denote the optimal stopping time for $V(0,x)$, we have $V(0,x) > x$ which by (26.2.29) is equivalent to

$$\widetilde{\mathsf{E}}_{0,x}\Big(\int_0^{\tau_*} e^{-\lambda u} \frac{dM_u}{S_u}\Big) > (r+\lambda)\, \widetilde{\mathsf{E}}_{0,x}\Big(\int_0^{\tau_*} e^{-\lambda u} X_u\, du\Big). \tag{26.2.32}$$

Recalling that $M_u = S^*_u \vee x$ we see that

$$\widetilde{\mathsf{E}}_{0,x}\Big(\int_0^{\tau_*} e^{-\lambda u}\,\frac{dM_u}{S_u}\Big) \le \widetilde{\mathsf{E}}\Big(\Big(\max_{0\le u\le T}(1/S_u)\Big)\Big((S^*_T\vee x)-x\Big)\Big) \tag{26.2.33}$$
$$\le \widetilde{\mathsf{E}}\Big(\Big(\max_{0\le u\le T}(1/S_u)\Big)S^*_T\, I(S^*_T > x)\Big) \to 0$$

as $x\to\infty$. Recalling that $X_u = (S^*_u\vee x)/S_u$ and noting that $\tau_* > t_0$ we see that

$$\widetilde{\mathsf{E}}_{0,x}\Big(\int_0^{\tau_*} e^{-\lambda u} X_u\,du\Big) \ge e^{-\lambda t_0}\, x\,\widetilde{\mathsf{E}}\Big(\int_0^{t_0}\frac{du}{S_u}\Big) \to \infty \tag{26.2.34}$$

as $x\to\infty$. From (26.2.33) and (26.2.34) we see that the strict inequality in (26.2.32) is violated if x is taken large enough, thus proving that b does not take the value ∞ on $(0,T]$. To disprove the case $b(0+)=\infty$, i.e. $t_0=0$ above, we may note that the gain function $G(x)=x$ in (26.2.10) is independent of time, so that $b(0+)=\infty$ would also imply that $b(t)=\infty$ for all $0\le t\le\delta$ in the problem (26.2.10) with the horizon $T+\delta$ instead of T where $\delta>0$. Applying the same argument as above to the $T+\delta$ problem (26.2.10) we again arrive at a contradiction. We thus may conclude that $b(0+)<\infty$ as claimed. Yet another quick argument for b to be finite in the case $\lambda>0$ can be given by noting that $b(t)<\alpha$ for all $t\in[0,T]$ where $\alpha\in(1,\infty)$ is the optimal stopping point in the infinite horizon problem given explicitly by the right-hand side of (26.1.25) above. Clearly $b(t)\uparrow\alpha$ as $T\to\infty$ for each $t\ge 0$, where we set $\alpha=\infty$ in the case $\lambda=0$.

Let us show that b cannot take the value 1 on $[0,T)$. This fact is equivalent to the fact that the process (S_t, M_t) in (26.2.1) (with $r+\lambda$ instead of r) cannot be optimally stopped at the diagonal $s=m$ in $(0,\infty)\times(0,\infty)$. The latter fact is well known for diffusions in the maximum process problems of optimal stopping with linear cost (see Proposition 13.1) and only minor modifications are needed to extend the argument to the present case. For this, set $Z_t=\sigma B_t+(r-\sigma^2/2)t$ and note that the exponential case of (26.2.1) (with $r+\lambda$ instead of r) reduces to the linear case of Proposition 13.1 for the diffusion Z and $c=r+\lambda$ by means of Jensen's inequality as follows:

$$\mathsf{E}\Big(e^{-(r+\lambda)\tau}M_\tau\Big) = \mathsf{E}\Big(\exp\Big(\max_{0\le t\le\tau} Z_t - c\tau\Big)\Big) \tag{26.2.35}$$
$$\ge \exp\Big(\mathsf{E}\Big(\max_{0\le t\le\tau} Z_t - c\tau\Big)\Big).$$

Denoting $\tau_n=\inf\{t>0 : Z_t\notin(-1/n,1/n)\}$ it is easily verified (see the proof of Proposition 13.1) that

$$\mathsf{E}\Big(\max_{0\le t\le\tau_n} Z_t\Big) \ge \frac{\delta}{n} \quad\text{and}\quad \mathsf{E}(\tau_n) \le \frac{\kappa}{n^2} \tag{26.2.36}$$

for all $n \geq 1$ with some constants $\delta > 0$ and $\kappa > 0$. Choosing n large enough, upon recalling (26.2.35), we see that (26.2.36) shows that it is never optimal to stop at the diagonal in the case of infinite horizon. To derive the same conclusion in the finite horizon case, replace τ_n by $\sigma_n = \tau_n \wedge T$ and note by the Markov inequality and (26.2.36) that

$$\mathsf{E}\Big(\max_{0\le t\le\tau_n} Z_t - \max_{0\le t\le\sigma_n} Z_t\Big) \le \frac{1}{n}\,\mathsf{P}(\tau_n > T) \tag{26.2.37}$$
$$\le \frac{\mathsf{E}(\tau_n)}{n\,T} \le \frac{\kappa}{n^3\,T} = O(n^{-3})$$

which together with (26.2.35) and (26.2.36) shows that

$$\mathsf{E}\Big(e^{-(r+\lambda)\sigma_n} M_{\sigma_n}\Big) \geq \exp\Big(\mathsf{E}\Big(\max_{0\le t\le\sigma_n} Z_t - c\sigma_n\Big)\Big) > 1 \tag{26.2.38}$$

for n large enough. From (26.2.38) we see that it is never optimal to stop at the diagonal in the case of finite horizon either, and thus b does not take the value 1 on $[0,T)$ as claimed.

Since the stopping set equals $\bar{D} = \{(t,x) \in [0,T) \times [1,\infty) : x \geq b(t)\}$ and b is decreasing, it is easily seen that b is right-continuous on $[0,T)$. Before we pass to the proof of its continuity we first turn to the key principle of optimal stopping in problem (26.2.10).

3°. We show that the *smooth-fit* condition (26.2.13) holds. For this, let $t \in [0,T)$ be given and fixed and set $x = b(t)$. We know that $x > 1$ so that there exists $\varepsilon > 0$ such that $x - \varepsilon > 1$ too. Since $V(t,x) = G(x)$ and $V(t,x-\varepsilon) > G(x-\varepsilon)$, we have:

$$\frac{V(t,x) - V(t,x-\varepsilon)}{\varepsilon} \le \frac{G(x) - G(x-\varepsilon)}{\varepsilon} = 1 \tag{26.2.39}$$

so that by letting $\varepsilon \downarrow 0$ in (26.2.39) and using that the left-hand derivative $V_x^-(t,x)$ exists since $y \mapsto V(t,y)$ is convex, we get $V_x^-(t,x) \le 1$. To prove the reverse inequality, let $\tau_\varepsilon = \tau_\varepsilon^*(t,x-\varepsilon)$ denote the optimal stopping time for $V(t,x-\varepsilon)$. We then have:

$$\frac{V(t,x) - V(t,x-\varepsilon)}{\varepsilon} \tag{26.2.40}$$
$$\ge \frac{1}{\varepsilon}\,\widetilde{\mathsf{E}}\bigg(e^{-\lambda\tau_\varepsilon}\bigg(\frac{(x-M_{\tau_\varepsilon})^+ + M_{\tau_\varepsilon}}{S_{\tau_\varepsilon}} - \frac{(x-\varepsilon-M_{\tau_\varepsilon})^+ + M_{\tau_\varepsilon}}{S_{\tau_\varepsilon}}\bigg)\bigg)$$
$$= \frac{1}{\varepsilon}\,\widetilde{\mathsf{E}}\bigg(\frac{e^{-\lambda\tau_\varepsilon}}{S_{\tau_\varepsilon}}\Big((x-M_{\tau_\varepsilon})^+ - (x-\varepsilon-M_{\tau_\varepsilon})^+\Big)\bigg)$$
$$\ge \frac{1}{\varepsilon}\,\widetilde{\mathsf{E}}\bigg(\frac{e^{-\lambda\tau_\varepsilon}}{S_{\tau_\varepsilon}}\Big((x-M_{\tau_\varepsilon})^+ - (x-\varepsilon-M_{\tau_\varepsilon})^+\Big) I(M_{\tau_\varepsilon} \le x-\varepsilon)\bigg)$$
$$= \widetilde{\mathsf{E}}\bigg(\frac{e^{-\lambda\tau_\varepsilon}}{S_{\tau_\varepsilon}}\, I(M_{\tau_\varepsilon} \le x-\varepsilon)\bigg) \longrightarrow 1$$

as $\varepsilon \downarrow 0$ by bounded convergence, since $\tau_\varepsilon \to 0$ so that $M_{\tau_\varepsilon} \to 1$ with $1 < x - \varepsilon$ and likewise $S_{\tau_\varepsilon} \to 1$. It thus follows from (26.2.40) that $V_x^-(t,x) \geq 1$ and therefore $V_x^-(t,x) = 1$. Since $V(t,y) = G(y)$ for $y > x$, it is clear that $V_x^+(t,x) = 1$. We may thus conclude that $y \mapsto V(t,y)$ is C^1 at $b(t)$ and $V_x(t,b(t)) = 1$ as stated in (26.2.13).

4°. We show that b is *continuous* on $[0,T]$ and that $b(T-) = 1$. For this, note first that since the supremum in (26.2.10) is attained at the first exit time τ_b from the open set C, standard arguments based on the strong Markov property (cf. Chapter III) imply that V is $C^{1,2}$ on C and satisfies (26.2.11). Suppose that there exists $t \in (0,T]$ such that $b(t-) > b(t)$ and fix any $x \in [b(t), b(t-))$. Note that by (26.2.13) we have

$$V(s,x) - x = \int_x^{b(s)} \int_y^{b(s)} V_{xx}(s,z)\, dz\, dy \tag{26.2.41}$$

for each $s \in (t-\delta, t)$ where $\delta > 0$ with $t - \delta > 0$. Since $V_t - rx\, V_x + (\sigma^2/2)\, x^2\, V_{xx} - \lambda V = 0$ in C we see that $(\sigma^2/2)\, x^2\, V_{xx} = -V_t + rx\, V_x + \lambda V \geq rV_x$ in C since $V_t \leq 0$ and $V_x \geq 0$ upon recalling also that $x \geq 1$ and $\lambda V \geq 0$. Hence we see that there exists $c > 0$ such that $V_{xx} \geq c\, V_x$ in $C \cap \{(t,x) \in [0,T) \times [1,\infty) : x \leq b(0)\}$, so that this inequality applies in particular to the integrand in (26.2.41). In this way we get

$$V(s,x) - x \geq c \int_x^{b(s)} \int_y^{b(s)} V_x(s,z)\, dz\, dy = c \int_x^{b(s)} \big(b(s) - V(s,y)\big)\, dy \tag{26.2.42}$$

for all $s \in (t-\delta, t)$. Letting $s \uparrow t$ we find that

$$V(t,x) - x \geq c \int_x^{b(t-)} \big(b(t-) - y\big)\, dy = \frac{c}{2} \big(b(t-) - x\big)^2 > 0 \tag{26.2.43}$$

which is a contradiction since (t,x) belongs to the stopping set $\bar{D}$. This shows that b is continuous on $[0,T]$. Note also that the same argument with $t = T$ shows that $b(T-) = 1$.

5°. We show that the *normal reflection* condition (26.2.14) holds. For this, note first that since $x \mapsto V(t,x)$ is increasing (and convex) on $[1,\infty)$ it follows that $V_x(t,1+) \geq 0$ for all $t \in [0,T)$. Suppose that there exists $t \in [0,T)$ such that $V_x(t,1+) > 0$. Recalling that V is $C^{1,2}$ on C so that $t \mapsto V_x(t,1+)$ is continuous on $[0,T)$, we see that there exists $\delta > 0$ such that $V_x(s,1+) \geq \varepsilon > 0$ for all $s \in [t, t+\delta]$ with $t + \delta < T$. Setting $\tau_\delta = \tau_b \wedge (t+\delta)$ it follows by Itô's formula (page 67) that

$$\begin{aligned} \widetilde{\mathsf{E}}_{t,1}\Big(e^{-\lambda\tau_\delta}\, V(t+\tau_\delta, X_{t+\tau_\delta})\Big) &= V(t,1) \\ &\quad + \widetilde{\mathsf{E}}_{t,1}\bigg(\int_0^{\tau_\delta} e^{-\lambda u}\, V_x(t+u, X_{t+u})\, dR_{t+u}\bigg) \end{aligned} \tag{26.2.44}$$

using (26.2.11) and the optional sampling theorem (page 60) since V_x is bounded.

Since $(e^{-\lambda(s\wedge\tau_b)}V(t+(s\wedge\tau_b), X_{t+(s\wedge\tau_b)}))_{0\le s\le T-t}$ is a martingale under $\widetilde{\mathsf{P}}_{t,1}$, we find that the expression on the left-hand side in (26.2.44) equals the first term on the right-hand side, and thus

$$\widetilde{\mathsf{E}}_{t,1}\Big(\int_0^{\tau_\delta} e^{-\lambda u}\, V_x(t+u, X_{t+u})\, dR_{t+u}\Big) = 0. \tag{26.2.45}$$

On the other hand, since $V_x(t+u, X_{t+u})\, dR_{t+u} = V_x(t+u, 1+)\, dR_{t+u}$ by (26.2.7), and $V_x(t+u, 1+) \ge \varepsilon > 0$ for all $u \in [0, \tau_\delta]$, we see that (26.2.45) implies that

$$\widetilde{\mathsf{E}}_{t,1}\Big(\int_0^{\tau_\delta} dR_{t+u}\Big) = 0. \tag{26.2.46}$$

By (26.2.6) and the optional sampling theorem (page 60) we see that (26.2.46) is equivalent to

$$\widetilde{\mathsf{E}}_{t,1}(X_{t+\tau_\delta}) - 1 + r\,\widetilde{\mathsf{E}}_{t,1}\Big(\int_0^{\tau_\delta} X_{t+u}\, du\Big) = 0. \tag{26.2.47}$$

Since $X_s \ge 1$ for all $s \in [0, T]$ we see that (26.2.47) implies that $\tau_\delta = 0$ $\widetilde{\mathsf{P}}_{t,1}$-a.s. As clearly this is impossible, we see that $V_x(t, 1+) = 0$ for all $t \in [0, T)$ as claimed in (26.2.14).

6°. We show that b solves the equation (26.2.23) on $[0, T]$. For this, set $F(t, x) = e^{-\lambda t} V(t, x)$ and note that $F : [0, T) \times [1, \infty) \to \mathbb{R}$ is a continuous function satisfying the following conditions:

$$F \text{ is } C^{1,2} \text{ on } C \cup D, \tag{26.2.48}$$
$$F_t + \mathbb{L}_X F \text{ is locally bounded,} \tag{26.2.49}$$
$$x \mapsto F(t, x) \text{ is convex,} \tag{26.2.50}$$
$$t \mapsto F_x(t, b(t)\pm) \text{ is continuous.} \tag{26.2.51}$$

To verify these claims, note first that $F(t, x) = e^{-\lambda t} G(x) = e^{-\lambda t} x$ for $(t, x) \in D$ so that the second part of (26.2.48) is obvious. Similarly, since $F(t, x) = e^{-\lambda t}\ V(t, x)$ and V is $C^{1,2}$ on C, we see that the same is true for F, implying the first part of (26.2.48). For (26.2.49), note that $(F_t + \mathbb{L}_X F)(t, x) = e^{-\lambda t}(V_t + \mathbb{L}_X V - \lambda V)(t, x) = 0$ for $(t, x) \in C$ by means of (26.2.11), and $(F_t + \mathbb{L}_X F)(t, x) = e^{-\lambda t}(G_t + \mathbb{L}_X G - \lambda G)(t, x) = -(r + \lambda)\, x\, e^{-\lambda t}$ for $(t, x) \in D$, implying the claim. [When we say in (26.2.49) that $F_t + \mathbb{L}_X F$ is locally bounded, we mean that $F_t + \mathbb{L}_X F$ is bounded on $K \cap (C \cup D)$ for each compact set K in $[0, T) \times [1, \infty)$.] The condition (26.2.50) follows by (26.2.26) above. Finally, recall by (26.2.13) that $x \mapsto V(t, x)$ is C^1 at $b(t)$ with $V_x(t, b(t)) = 1$ so that $F_x(t, b(t)\pm) = e^{-\lambda t}$ implying (26.2.51). Let us also note that the condition (26.2.50) can further be relaxed to the form where $F_{xx} = F_1 + F_2$ on $C \cup D$ where F_1 is non-negative and F_2 is continuous on $[0,T) \times [1,\infty)$. This will be referred to below as the relaxed form of (26.2.48)–(26.2.51).

Having a continuous function $F : [0,T) \times [1,\infty) \to \mathbb{R}$ satisfying (26.2.48)–(26.2.51) one finds (cf. Subsection 3.5) that for $t \in [0,T)$ the following change-of-variable formula holds:

$$
\begin{aligned}
F(t,X_t) &= F(0,X_0) + \int_0^t (F_t + \mathbb{L}_X F)(s,X_s)\, I(X_s \neq b(s))\, ds \qquad (26.2.52)\\
&+ \int_0^t F_x(s,X_s)\, \sigma X_s I(X_s \neq b(s))\, d\widehat{B}_s + \int_0^t F_x(s,X_s)\, I(X_s \neq b(s))\, dR_s \\
&+ \frac{1}{2}\int_0^t \Big(F_x(s,X_s+) - F_x(s,X_s-)\Big) I(X_s = b(s))\, d\ell_s^b(X)
\end{aligned}
$$

where $\ell_s^b(X)$ is the local time of X at the curve b given by

$$
\ell_s^b(X) = \mathsf{P}\text{-}\lim_{\varepsilon \downarrow 0} \frac{1}{2\varepsilon} \int_0^s I(b(r) - \varepsilon < X_r < b(r) + \varepsilon)\, \sigma^2 X_r^2\, dr \qquad (26.2.53)
$$

and $d\ell_s^b(X)$ refers to the integration with respect to the continuous increasing function $s \mapsto \ell_s^b(X)$. Note also that formula (26.2.52) remains valid if b is replaced by any other continuous function of bounded variation $c : [0,T] \to \mathbb{R}$ for which (26.2.48)–(26.2.51) hold with C and D defined in the same way.

Applying (26.2.52) to $e^{-\lambda s} V(t+s, X_{t+s})$ under $\widetilde{\mathsf{P}}_{t,x}$ with $(t,x) \in [0,T) \times [1,\infty)$ yields

$$
\begin{aligned}
e^{-\lambda s} V(t+s, X_{t+s}) &= V(t,x) \qquad (26.2.54)\\
&\quad + \int_0^s e^{-\lambda u} \big(V_t + \mathbb{L}_X V - \lambda V\big)(t+u, X_{t+u})\, du + M_s \\
&= V(t,x) + \int_0^s e^{-\lambda u} \big(G_t + \mathbb{L}_X G - \lambda G\big)(t+u, X_{t+u}) \\
&\qquad \times I(X_{t+u} \geq b(t+u))\, du + M_s \\
&= V(t,x) - (r+\lambda) \int_0^s e^{-\lambda u} X_{t+u}\, I(X_{t+u} \geq b(t+u))\, du + M_s
\end{aligned}
$$

upon using (26.2.11), (26.2.12)+(26.2.16), (26.2.14), (26.2.13) and $G_t + \mathbb{L}_X G - \lambda G = -(r+\lambda) G$, where we set $M_s = \int_0^s e^{-\lambda u} V_x(t+u, X_{t+u})\, \sigma X_{t+u}\, d\widehat{B}_{t+u}$ for $0 \leq s \leq T-t$. Since $0 \leq V_x \leq 1$ on $[0,T] \times [1,\infty)$, it is easily verified that $(M_s)_{0 \leq s \leq T-t}$ is a martingale, so that $\widetilde{\mathsf{E}}_{t,x} M_s = 0$ for all $0 \leq s \leq T-t$. Inserting $s = T-t$ in (26.2.54), using that $V(T,x) = G(x) = x$ for all $x \in [1,\infty)$, and taking the $\widetilde{\mathsf{P}}_{t,x}$-expectation in the resulting identity, we get

$$
\begin{aligned}
e^{-\lambda(T-t)} \widetilde{\mathsf{E}}_{t,x} X_T &= V(t,x) \qquad (26.2.55)\\
&\quad - (r+\lambda) \int_0^{T-t} e^{-\lambda u} \widetilde{\mathsf{E}}_{t,x}\Big(X_{t+u}\, I(X_{t+u} \geq b(t+u))\Big) du
\end{aligned}
$$

for all $(t,x) \in [0,T) \times [1,\infty)$. By (26.2.20) and (26.2.21) we see that (26.2.55) is the early exercise premium representation (26.2.24). Recalling that $V(t,x) = G(x) = x$ for $x \geq b(t)$, and setting $x = b(t)$ in (26.2.55), we see that b satisfies the equation (26.2.23) as claimed.

7°. We show that b is the *unique* solution of the equation (26.2.23) in the class of continuous decreasing functions $c : [0,T] \to \mathbb{R}$ satisfying $c(t) > 1$ for all $0 \leq t < T$. The proof of this fact will be carried out in several remaining paragraphs to the end of the main proof. Let us thus assume that a function c belonging to the class described above solves (26.2.23), and let us show that this c must then coincide with the optimal stopping boundary b.

For this, in view of (26.2.55), let us introduce the function

$$\begin{aligned} U^c(t,x) &= e^{-\lambda(T-t)} \widetilde{\mathsf{E}}_{t,x} X_T \\ &\quad + (r+\lambda) \int_0^{T-t} e^{-\lambda u} \widetilde{\mathsf{E}}_{t,x} \Big(X_{t+u}\, I(X_{t+u} \geq c(t+u)) \Big)\, du \end{aligned} \tag{26.2.56}$$

for $(t,x) \in [0,T) \times [1,\infty)$. Using (26.2.20) and (26.2.21) as in (26.2.24) we see that (26.2.56) reads

$$U^c(t,x) = e^{-\lambda(T-t)} F(T-t,x) + (r+\lambda) \int_0^{T-t} e^{-\lambda u}\, G(u,x,c(t+u))\, du \tag{26.2.57}$$

for $(t,x) \in [0,T) \times [1,\infty)$. A direct inspection of the expressions in (26.2.57) using (26.2.20)–(26.2.22) shows that U_x^c is continuous on $[0,T) \times [1,\infty)$.

8°. In accordance with (26.2.24) define a function $V^c : [0,T) \times [1,\infty) \to \mathbb{R}$ by setting $V^c(t,x) = U^c(t,x)$ for $x < c(t)$ and $V^c(t,x) = G(x)$ for $x \geq c(t)$ when $0 \leq t < T$. Note that since c solves (26.2.23) we have that V^c is continuous on $[0,T) \times [1,\infty)$, i.e. $V^c(t,x) = U^c(t,x) = G(x)$ for $x = c(t)$ when $0 \leq t < T$. Let C and D be defined by means of c as in (26.2.17) and (26.2.18) respectively.

Standard arguments based on the Markov property (or a direct verification) show that V^c i.e. U^c is $C^{1,2}$ on C and that

$$V_t^c + \mathbb{L}_X V^c = \lambda V^c \quad \text{in } C, \tag{26.2.58}$$

$$V_x^c(t,1+) = 0 \tag{26.2.59}$$

for all $t \in [0,T)$. Moreover, since U_x^c is continuous on $[0,T) \times [1,\infty)$ we see that V_x^c is continuous on $\bar{C}$. Finally, it is obvious that V^c i.e. G is $C^{1,2}$ on $\bar{D}$.

9°. Summarizing the preceding conclusions one can easily verify that the function $F : [0,T) \times [1,\infty) \to \mathbb{R}$ defined by $F(t,x) = e^{-\lambda t} V^c(t,x)$ satisfies (26.2.48)–(26.2.51) (in the relaxed form) so that (26.2.52) can be applied. In this way, under $\mathsf{P}_{t,x}$ with $(t,x) \in [0,T) \times [1,\infty)$ given and fixed, using (26.2.59) we get

$$e^{-\lambda s}V^c(t{+}s, X_{t+s}) = V^c(t,x) \tag{26.2.60}$$
$$+ \int_0^s e^{-\lambda u}\Big(V_t^c{+}\mathbb{L}_X V^c - \lambda V^c\Big)(t{+}u, X_{t+u}) I(X_{t+u} \neq c(t{+}u))\, du$$
$$+ M_s^c + \frac{1}{2}\int_0^s e^{-\lambda u}\Delta_x V_x^c(t{+}u, c(t{+}u))\, d\ell_u^c(X)$$

where $M_s^c = \int_0^s e^{-\lambda u} V_x^c(t+u, X_{t+u})\, \sigma X_{t+u}\, I(X_{t+u} \neq c(t+u))\, d\widehat{B}_{t+u}$ and we set $\Delta_x V_x^c(v, c(v)) = V_x^c(v, c(v)+) - V_x^c(v, c(v)-)$ for $t \le v \le T$. Moreover, it is readily seen from the explicit expression for V_x^c obtained using (26.2.57) above that $(M_s^c)_{0\le s\le T-t}$ is a martingale under $\widetilde{\mathsf{P}}_{t,x}$ so that $\widetilde{\mathsf{E}}_{t,x}(M_s^c) = 0$ for each $0{\le}s{\le}T{-}t$.

10°. Setting $s = T - t$ in (26.2.60) and then taking the $\widetilde{\mathsf{P}}_{t,x}$-expectation, using that $V^c(T,x) = G(x)$ for all $x \ge 1$ and that V^c satisfies (26.2.58) in C, we get

$$e^{-\lambda(T-t)}\widetilde{\mathsf{E}}_{t,x}X_T = V^c(t,x) \tag{26.2.61}$$
$$-\ (r{+}\lambda)\int_0^{T-t} e^{-\lambda u}\widetilde{\mathsf{E}}_{t,x}\Big(X_{t+u}\, I(X_{t+u}{\ge}c(t{+}u))\Big)du$$
$$+\ \frac{1}{2}\int_0^{T-t} e^{-\lambda u}\Delta_x V_x^c(t{+}u, c(t{+}u))\, d_u\widetilde{\mathsf{E}}_{t,x}(\ell_u^c(X))$$

for all $(t,x) \in [0,T) \times [1,\infty)$. Comparing (26.2.61) with (26.2.56), and recalling the definition of V^c in terms of U^c and G, we get

$$\int_0^{T-t} e^{-\lambda u}\Delta_x V_x^c(t{+}u, c(t{+}u))\, d_u\widetilde{\mathsf{E}}_{t,x}(\ell_u^c(X)) \tag{26.2.62}$$
$$= 2\,\big(U^c(t,x) - G(x)\big)\, I(x{\ge}c(t))$$

for all $0 \le t < T$ and $x \ge 1$, where $I(x{\ge}c(t))$ equals 1 if $x \ge c(t)$ and 0 if $x < c(t)$.

11°. From (26.2.62) we see that if we are to prove that

$$x \mapsto V^c(t,x) \text{ is } C^1 \text{ at } c(t) \text{ for each } 0 \le t < T \tag{26.2.63}$$

given and fixed, then it will follow that

$$U^c(t,x) = G(x) \text{ for all } x \ge c(t)\,. \tag{26.2.64}$$

On the other hand, if we know that (26.2.64) holds, then using the general fact

$$\frac{\partial}{\partial x}\Big(U^c(t,x) - G(x)\Big)\Big|_{x=c(t)} = V_x^c(t, c(t)-) - V_x^c(t, c(t)+) \tag{26.2.65}$$
$$= -\Delta_x V_x^c(t, c(t))$$

for all $0 \le t < T$, we see that (26.2.63) holds too (since U^c_x is continuous). The equivalence of (26.2.63) and (26.2.64) suggests that instead of dealing with the equation (26.2.62) in order to derive (26.2.62) above we may rather concentrate on establishing (26.2.63) directly.

12°. To derive (26.2.64) first note that standard arguments based on the Markov property (or a direct verification) show that U^c is $C^{1,2}$ on D and that

$$U^c_t + \mathbb{L}_X U^c - \lambda U^c = -(r+\lambda)G \quad \text{in } D. \tag{26.2.66}$$

Since the function $F : [0,T) \times [1,\infty) \to \mathbb{R}$ defined by $F(t,x) = e^{-\lambda t}U^c(t,x)$ is continuous and satisfies (26.2.48)–(26.2.51) (in the relaxed form), we see that (26.2.52) can be applied just like in (26.2.60) with U^c instead of V^c, and this yields

$$\begin{aligned} e^{-\lambda s}U^c(t+s, X_{t+s}) &= U^c(t,x) \\ &\quad - (r+\lambda)\int_0^s e^{-\lambda u} X_{t+u}\, I(X_{t+u} \ge c(t+u))\, du + \widetilde{M}^c_s \end{aligned} \tag{26.2.67}$$

upon using (26.2.58)–(26.2.59) and (26.2.66) as well as that $\Delta_x U^c_x(t+u, c(t+u)) = 0$ for $0 \le u \le s$ since U^c_x is continuous. In (26.2.67) we have $\widetilde{M}^c_s = \int_0^s e^{-\lambda u}\, U^c_x(t+u, X_{t+u})\, \sigma X_{t+u}\, I(X_{t+u} \neq c(t+u))\, d\widehat{B}_{t+u}$ and $(\widetilde{M}^c_s)_{0 \le s \le T-t}$ is a martingale under $\mathsf{P}_{t,x}$.

Next note that Itô's formula (page 67) implies

$$\begin{aligned} e^{-\lambda s}G(X_{t+s}) &= G(x) - (r+\lambda)\int_0^s e^{-\lambda u} X_{t+u}\, du + M_s \\ &\quad + \int_0^s e^{-\lambda u}\, dR_{t+u} \end{aligned} \tag{26.2.68}$$

upon using that $G_t + \mathbb{L}_X G - rG = -(r+\lambda)G$ as well as that $G_x(t+u, X_{t+u}) = 1$ for $0 \le u \le s$. In (26.2.68) we have $M_s = \int_0^s e^{-\lambda u} \sigma X_{t+u} d\widehat{B}_{t+u}$ and $(M_s)_{0 \le s \le T-t}$ is a martingale under $\mathsf{P}_{t,x}$.

For $x \ge c(t)$ consider the stopping time

$$\sigma_c = \inf\{\, 0 \le s \le T-t : X_{t+s} \le c(t+s) \,\}. \tag{26.2.69}$$

Then using that $U^c(t, c(t)) = G(c(t))$ for all $0 \le t < T$ since c solves (26.2.23), and that $U^c(T,x) = G(x)$ for all $x \ge 1$ by (26.2.56), we see that $U^c(t+\sigma_c, X_{t+\sigma_c}) = G(X_{t+\sigma_c})$. Hence from (26.2.67) and (26.2.68) using the optional sampling theorem (page 60) we find

$$\begin{aligned}U^c(t,x) &= \mathsf{E}_{t,x}\Big(e^{-\lambda\sigma_c}U^c(t+\sigma_c, X_{t+\sigma_c})\Big) \\ &\quad + (r+\lambda)\,\mathsf{E}_{t,x}\bigg(\int_0^{\sigma_c} e^{-\lambda u}X_{t+u}\, I(X_{t+u}\geq c(t+u))\, du\bigg) \\ &= \mathsf{E}_{t,x}\Big(e^{-r\sigma_c}G(X_{t+\sigma_c})\Big) \\ &\quad + (r+\lambda)\,\mathsf{E}_{t,x}\bigg(\int_0^{\sigma_c} e^{-\lambda u}X_{t+u}\, I(X_{t+u}\geq c(t+u))\, du\bigg) \\ &= G(x) - (r+\lambda)\,\mathsf{E}_{t,x}\bigg(\int_0^{\sigma_c} e^{-\lambda u}X_{t+u}\, du\bigg) \\ &\quad + (r+\lambda)\mathsf{E}_{t,x}\bigg(\int_0^{\sigma_c} e^{-\lambda u}X_{t+u}\, I(X_{t+u}\geq c(t+u))\, du\bigg) = G(x)\end{aligned} \tag{26.2.70}$$

since $X_{t+u} \geq c(t+u) > 1$ for all $0 \leq u \leq \sigma_c$. This establishes (26.2.64) and thus (26.2.63) holds too.

It may be noted that a shorter but somewhat less revealing proof of (26.2.64) [and (26.2.63)] can be obtained by verifying directly (using the Markov property only) that the process

$$e^{-\lambda s}U^c(t+s, X_{t+s}) + (r+\lambda)\int_0^s e^{-\lambda u}X_{t+u}\, I(X_{t+u}\geq c(t+u))\, du \tag{26.2.71}$$

is a martingale under $\mathsf{P}_{t,x}$ for $0 \leq s \leq T-t$. This verification moreover shows that the martingale property of (26.2.71) does not require that c is increasing but only measurable. Taken together with the rest of the proof below this shows that the claim of uniqueness for the equation (26.2.23) holds in the class of continuous functions $c : [0,T] \rightarrow \mathbb{R}$ such that $c(t) > 1$ for all $0 < t < T$.

13°. Consider the stopping time

$$\tau_c = \inf\{\, 0 \leq s \leq T-t : X_{t+s} \geq c(t+s)\,\}. \tag{26.2.72}$$

Note that (26.2.60) using (26.2.58) and (26.2.63) reads

$$\begin{aligned}e^{-\lambda s}V^c(t+s, X_{t+s}) &= V^c(t,x) \\ &\quad - (r+\lambda)\int_0^s e^{-\lambda u}X_{t+u}\, I(X_{t+u}\geq c(t+u))\, du + M_s^c\end{aligned} \tag{26.2.73}$$

where $(M_s^c)_{0\leq s\leq T-t}$ is a martingale under $\widetilde{\mathsf{P}}_{t,x}$. Thus $\widetilde{\mathsf{E}}_{t,x}M_{\tau_c}^c = 0$, so that after inserting τ_c in place of s in (26.2.73), it follows upon taking the $\widetilde{\mathsf{P}}_{t,x}$-expectation that

$$V^c(t,x) = \widetilde{\mathsf{E}}_{t,x}\Big(e^{-\lambda\tau_c}X_{t+\tau_c}\Big) \tag{26.2.74}$$

for all $(t,x) \in [0,T) \times [1,\infty)$ where we use that $V^c(t,x) = G(x) = x$ for $x \geq c(t)$ or $t = T$. Comparing (26.2.74) with (26.2.10) we see that

$$V^c(t,x) \leq V(t,x) \tag{26.2.75}$$

for all $(t,x) \in [0,T) \times [1,\infty)$.

14°. Let us now show that $b \geq c$ on $[0,T]$. For this, recall that by the same arguments as for V^c we also have

$$\begin{aligned} e^{-\lambda s} V(t+s, X_{t+s}) = V(t,x) \\ - (r+\lambda) \int_0^s e^{-\lambda u} X_{t+u}\, I(X_{t+u} \geq b(t+u))\, du + M_s^b \end{aligned} \tag{26.2.76}$$

where $(M_s^b)_{0 \leq s \leq T-t}$ is a martingale under $\widetilde{\mathsf{P}}_{t,x}$. Fix $(t,x) \in [0,T) \times [1,\infty)$ such that $x > b(t) \vee c(t)$ and consider the stopping time

$$\sigma_b = \inf\{\, 0 \leq s \leq T-t : X_{t+s} \leq b(t+s) \}. \tag{26.2.77}$$

Inserting σ_b in place of s in (26.2.73) and (26.2.76) and taking the $\widetilde{\mathsf{P}}_{t,x}$-expectation, we get

$$\begin{aligned} &\widetilde{\mathsf{E}}_{t,x}\big(e^{-\lambda \sigma_b} V^c(t+\sigma_b, X_{t+\sigma_b})\big) \\ &\quad = x - (r+\lambda)\, \widetilde{\mathsf{E}}_{t,x}\bigg(\int_0^{\sigma_b} e^{-\lambda u} X_{t+u}\, I(X_{t+u} \geq c(t+u))\, du \bigg), \end{aligned} \tag{26.2.78}$$

$$\widetilde{\mathsf{E}}_{t,x}\big(e^{-\lambda \sigma_b} V(t+\sigma_b, X_{t+\sigma_b})\big) = x - (r+\lambda)\, \widetilde{\mathsf{E}}_{t,x}\bigg(\int_0^{\sigma_b} e^{-\lambda u} X_{t+u}\, du \bigg). \tag{26.2.79}$$

Hence by (26.2.75) we see that

$$\begin{aligned} \widetilde{\mathsf{E}}_{t,x}\bigg(\int_0^{\sigma_b} e^{-\lambda u} X_{t+u}\, I(X_{t+u} \geq c(t+u))\, du \bigg) \\ \geq \widetilde{\mathsf{E}}_{t,x}\bigg(\int_0^{\sigma_b} e^{-\lambda u} X_{t+u}\, du \bigg) \end{aligned} \tag{26.2.80}$$

from where it follows by the continuity of c and b, using $X_{t+u} > 0$, that $b(t) \geq c(t)$ for all $t \in [0,T]$.

15°. Finally, let us show that c must be equal to b. For this, assume that there is $t \in (0,T)$ such that $b(t) > c(t)$, and pick $x \in (c(t), b(t))$. Under $\widetilde{\mathsf{P}}_{t,x}$ consider the stopping time τ_b from (26.2.19). Inserting τ_b in place of s in (26.2.73) and (26.2.76) and taking the $\widetilde{\mathsf{P}}_{t,x}$-expectation, we get

$$\begin{aligned} \widetilde{\mathsf{E}}_{t,x}\big(e^{-\lambda \tau_b} X_{t+\tau_b}\big) = V^c(t,x) \\ - (r+\lambda)\, \widetilde{\mathsf{E}}_{t,x}\bigg(\int_0^{\tau_b} e^{-\lambda u} X_{t+u}\, I(X_{t+u} \geq c(t+u))\, du \bigg), \end{aligned} \tag{26.2.81}$$

$$\widetilde{\mathsf{E}}_{t,x}\big(e^{-\lambda \tau_b} X_{t+\tau_b}\big) = V(t,x). \tag{26.2.82}$$

Hence by (26.2.75) we see that

$$\widetilde{\mathsf{E}}_{t,x}\bigg(\int_0^{\tau_b} e^{-\lambda u} X_{t+u}\, I(X_{t+u} \ge c(t+u))\, du\bigg) \le 0 \tag{26.2.83}$$

from where it follows by the continuity of c and b using $X_{t+u} > 0$ that such a point x cannot exist. Thus c must be equal to b, and the proof is complete. □

Notes. According to theory of modern finance (see e.g. [197]) the arbitrage-free price of the Russian option (first introduced and studied in [185] and [186]) is given by (26.2.1) above where M denotes the maximum of the stock price S. This option is characterized by "reduced regret" because its owner is paid the maximum stock price up to the time of exercise and hence feels less remorse for not having exercised the option earlier.

In the case of *infinite* horizon T, and when M_τ in (26.2.1) is replaced by $e^{-\lambda\tau}M_\tau$, the problem was solved in [185] and [186]. The original derivation [185] was two-dimensional (see Section 13 for a general principle in this context) and the subsequent derivation [186] reduced the problem to one dimension using a change of measure. The latter methodology was also adopted in the present section.

Note that the infinite horizon formulation requires the discounting rate $\lambda > 0$ to be present (i.e. non-zero), since otherwise the option price would be infinite. Clearly, such a discounting rate is not needed (i.e. can be taken zero) when the horizon T is finite, so that the most attractive feature of the option — no regret — remains fully preserved.

The fact that the Russian option problem becomes one-dimensional (after a change of measure is applied) sets the mathematical problem on an equal footing with the American option problem (put or call) with finite horizon. The latter problem, on the other hand, has been studied since the 1960's, and for more details and references we refer to Section 25 above. The main aim of the present section is to extend these results to the Russian option with finite horizon.

We showed above (following [165]) that the optimal stopping boundary for the Russian option with finite horizon can be characterized as the unique solution of a nonlinear integral equation arising from the early exercise premium representation (an explicit formula for the arbitrage-free price in terms of the optimal stopping boundary having a clear economic interpretation). The results obtained stand in a complete parallel with the best known results on the American put option with finite horizon (cf. Subsection 25.2 above). The key argument in the proof relies upon a local time-space formula (cf. Subsection 3.5). Papers [57] and [47] provide useful additions to the main results of the present section.

27. The Asian option

Unlike in the case of the American option (Section 25) and the Russian option (Section 26) it turns out that the *infinite horizon* formulation of the Asian option problem considered below leads to a trivial solution: the value function is constant and it is never optimal to stop (see the text following (27.1.31) below). This is hardly a rule for Asian options as their infinite horizon formulations, contrary to what one could expect generally, are more difficult than finite horizon ones. The reason for this unexpected twist is twofold. Firstly, the integral functional is more complicated than the maximum functional after the state variable is added to make it Markovian (recall our discussions in Chapter III). Secondly, the existence of a finite horizon (i.e. the end of time) enables one to use backward induction upon taking the horizon as an initial point. Nonlinear integral equations (derived in the present chapter) may be viewed as a continuous-time analogue of the method of backward induction considered in Chapter I above. The fact that these equations have unique solutions constitutes the key element which makes finite horizons more amenable.

27.1. Finite horizon

1. According to financial theory (see e.g. [197]) the arbitrage-free price of the *early* exercise *Asian call option* with *floating strike* is given by

$$V = \sup_{0<\tau\le T} \mathsf{E}\Big(e^{-r\tau}\big(S_\tau - \tfrac{1}{\tau}I_\tau\big)^+\Big) \tag{27.1.1}$$

where τ is a stopping time of the geometric Brownian motion $S = (S_t)_{0\le t\le T}$ solving

$$dS_t = rS_t\,dt + \sigma S_t\,dB_t \quad (S_0 = s) \tag{27.1.2}$$

and $I = (I_t)_{0\le t\le T}$ is the integral process given by

$$I_t = a + \int_0^t S_s\,ds \tag{27.1.3}$$

where $s > 0$ and $a \ge 0$ are given and fixed. We recall that $B = (B_t)_{t\ge 0}$ is a standard Brownian motion started at zero, $T > 0$ is the expiration date (maturity), $r > 0$ is the interest rate, and $\sigma > 0$ is the volatility coefficient.

By change of measure (cf. Subsection 5.3 above) we may write

$$V = \sup_{0<\tau\le T} \mathsf{E}\bigg(e^{-r\tau}S_\tau\Big(1 - \frac{1}{\tau}\,X_\tau\Big)^+\bigg) = s \sup_{0<\tau\le T} \widetilde{\mathsf{E}}\bigg(\Big(1 - \frac{1}{\tau}\,X_\tau\Big)^+\bigg) \tag{27.1.4}$$

where we set

$$X_t = \frac{I_t}{S_t} \tag{27.1.5}$$

and $\widetilde{\mathsf{P}}$ is a probability measure defined by $d\widetilde{\mathsf{P}} = \exp(\sigma B_T - (\sigma^2/2)T)\,d\mathsf{P}$ so that $\widetilde{B}_t = B_t - \sigma t$ is a standard Brownian motion under $\widetilde{\mathsf{P}}$ for $0 \le t \le T$. By Itô's formula (page 67) one finds that

$$dX_t = (1 - rX_t)\,dt + \sigma X_t\,d\widehat{B}_t \qquad (X_0 = x) \tag{27.1.6}$$

under $\widetilde{\mathsf{P}}$ where $\widehat{B} = -\widetilde{B}$ is a standard Brownian motion and $x = a/s$. The infinitesimal generator of X is therefore given by

$$\mathbb{L}_X = (1 - rx)\,\frac{\partial}{\partial x} + \frac{\sigma^2}{2}\,x^2\,\frac{\partial^2}{\partial x^2}\,. \tag{27.1.7}$$

For further reference recall that the strong solution of (27.1.2) is given by

$$S_t = s\exp\Big(\sigma B_t + \Big(r - \frac{\sigma^2}{2}\Big)t\Big) = s\exp\Big(\sigma \widetilde{B}_t + \Big(r + \frac{\sigma^2}{2}\Big)t\Big) \tag{27.1.8}$$

for $0 \le t \le T$ where B and $\widetilde{B}$ are standard Brownian motions under P and $\widetilde{\mathsf{P}}$ respectively. When dealing with the process X on its own, however, note that there is no restriction to assume that $s = 1$ and $a = x$ with $x \ge 0$.

Summarizing the preceding facts we see that the early exercise Asian call problem reduces to solving the following optimal stopping problem:

$$V(t,x) = \sup_{0<\tau\le T-t} \widetilde{\mathsf{E}}_{t,x}\Big(\Big(1 - \frac{1}{t+\tau}\,X_{t+\tau}\Big)^+\Big) \tag{27.1.9}$$

where τ is a stopping time of the diffusion process X solving (27.1.6) above and $X_t = x$ under $\widetilde{\mathsf{P}}_{t,x}$ with $(t,x) \in [0,T] \times [0,\infty)$ given and fixed.

Standard Markovian arguments (cf. Chapter III) indicate that V from (27.1.9) solves the following *free-boundary problem*:

$$V_t + \mathbb{L}_X V = 0 \;\; \text{in } C, \tag{27.1.10}$$

$$V(t,x) = \Big(1 - \frac{x}{t}\Big)^+ \;\; \text{for } x = b(t) \text{ or } t = T, \tag{27.1.11}$$

$$V_x(t,x) = -\frac{1}{t} \;\; \text{for } x = b(t) \; (\textit{smooth fit}), \tag{27.1.12}$$

$$V(t,x) > \Big(1 - \frac{x}{t}\Big)^+ \;\; \text{in } C, \tag{27.1.13}$$

$$V(t,x) = \Big(1 - \frac{x}{t}\Big)^+ \;\; \text{in } D \tag{27.1.14}$$

where the continuation set C and the stopping set $\bar{D}$ (as the closure of the set D below) are defined by

$$C = \{\,(t,x) \in [0,T)\times[0,\infty) : x > b(t)\,\}, \tag{27.1.15}$$

$$D = \{\,(t,x) \in [0,T)\times[0,\infty) : x < b(t)\,\}, \tag{27.1.16}$$

and $b : [0,T] \to \mathbb{R}$ is the (unknown) optimal stopping boundary, i.e. the stopping time

$$\tau_b = \inf\,\{\,0 \le s \le T - t : X_{t+s} \le b(t+s)\,\} \tag{27.1.17}$$

is optimal in (27.1.9) (i.e. the supremum is attained at this stopping time). It follows from the result of Theorem 27.1 below that the free-boundary problem (27.1.10)–(27.1.14) characterizes the value function V and the optimal stopping boundary b in a unique manner (proving also the existence of the latter).

2. In the sequel we make use of the following functions:

$$F(t,x) = \widetilde{\mathsf{E}}_{0,x}\Big(1-\frac{X_t}{T}\Big)^+ = \int_0^\infty\!\!\int_0^\infty \Big(1-\frac{x+a}{Ts}\Big)^+ f(t,s,a)\,ds\,da, \tag{27.1.18}$$

$$\begin{aligned} G(t,x,y) &= \widetilde{\mathsf{E}}_{0,x}\big(X_t\, I(X_t\le y)\big) \qquad\qquad (27.1.19)\\ &= \int_0^\infty\!\!\int_0^\infty \Big(\frac{x+a}{s}\Big) I\Big(\frac{x+a}{s}\le y\Big) f(t,s,a)\,ds\,da, \end{aligned}$$

$$H(t,x,y) = \widetilde{\mathsf{P}}_{0,x}(X_t\le y) = \int_0^\infty\!\!\int_0^\infty I\Big(\frac{x+a}{s}\le y\Big) f(t,s,a)\,ds\,da, \tag{27.1.20}$$

for $t>0$ and $x,y\ge 0$, where $(s,a)\mapsto f(t,s,a)$ is the probability density function of (S_t,I_t) under $\widetilde{\mathsf{P}}$ with $S_0=1$ and $I_0=0$ given by

$$\begin{aligned} f(t,s,a) &= \frac{2\sqrt{2}}{\pi^{3/2}\sigma^3}\,\frac{s^{r/\sigma^2}}{a^2\sqrt{t}}\,\exp\Big(\frac{2\pi^2}{\sigma^2 t}-\frac{(r+\sigma^2/2)^2}{2\sigma^2}\,t-\frac{2}{\sigma^2 a}(1+s)\Big) \qquad (27.1.21)\\ &\quad\times\int_0^\infty \exp\Big(-\frac{2z^2}{\sigma^2 t}-\frac{4\sqrt{s}}{\sigma^2 a}\cosh(z)\Big)\sinh(z)\,\sin\Big(\frac{4\pi z}{\sigma^2 t}\Big)\,dz \end{aligned}$$

for $s>0$ and $a>0$. (For a derivation of the right-hand side in (27.1.21) see the Appendix below.)

The main result of the present section may now be stated as follows.

Theorem 27.1. *The optimal stopping boundary in the Asian call problem* (27.1.9) *can be characterized as the unique continuous increasing solution* $b:[0,T]\to\mathbb{R}$ *of the nonlinear integral equation*

$$\begin{aligned} 1-\frac{b(t)}{t} &= F(T-t,b(t)) \qquad\qquad (27.1.22)\\ &\quad-\int_0^{T-t}\frac{1}{t+u}\bigg(\Big(\frac{1}{t+u}+r\Big)cG(u,b(t),b(t+u))\\ &\qquad\qquad - H(u,b(t),b(t+u))\bigg)\,du \end{aligned}$$

satisfying $0<b(t)<t/(1+rt)$ *for all* $0<t<T$. *The solution* b *satisfies* $b(0+)=0$ *and* $b(T-)=T/(1+rT)$, *and the stopping time* τ_b *from* (27.1.17) *is optimal in* (27.1.9).

The arbitrage-free price of the Asian call option (27.1.9) *admits the following "early exercise premium" representation*:

$$V(t,x) = F(T-t,x) - \int_0^{T-t} \frac{1}{t+u}\left(\left(\frac{1}{t+u}+r\right)G(u,x,b(t+u)) - H(u,x,b(t+u))\right) du \tag{27.1.23}$$

for all $(t,x) \in [0,T] \times [0,\infty)$. [*Further properties of* V *and* b *are exhibited in the proof below.*]

Proof. The proof will be carried out in several steps. We begin by stating some general remarks which will be freely used below without further mention.

1°. The reason that we take the supremum in (27.1.1) and (27.1.9) over $\tau > 0$ is that the ratio $1/(t+\tau)$ is not well defined for $\tau = 0$ when $t = 0$. Note however in (27.1.1) that $I_\tau/\tau \to \infty$ as $\tau \downarrow 0$ when $I_0 = a > 0$ and that $I_\tau/\tau \to s$ as $\tau \downarrow 0$ when $I_0 = a = 0$. Similarly, note in (27.1.9) that $X_\tau/\tau \to \infty$ as $\tau \downarrow 0$ when $X_0 = x > 0$ and $X_\tau/\tau \to 1$ as $\tau \downarrow 0$ when $X_0 = x = 0$. Thus in both cases the gain process (the integrand in (27.1.1) and (27.1.9)) tends to 0 as $\tau \downarrow 0$. This shows that in either (27.1.1) or (27.1.9) it is never optimal to stop at $t = 0$. To avoid similar (purely technical) complications in the proof to follow we will equivalently consider $V(t,x)$ only for $t > 0$ with the supremum taken over $\tau \geq 0$. The case of $t = 0$ will become evident (by continuity) at the end of the proof.

2°. Recall that it is no restriction to assume that $s = 1$ and $a = x$ so that $X_t = (x + I_t)/S_t$ with $I_0 = 0$ and $S_0 = 1$. We will write X_t^x instead of X_t to indicate the dependence on x when needed. It follows that V admits the following representation:

$$V(t,x) = \sup_{0 \leq \tau \leq T-t} \widetilde{\mathsf{E}}\left(1 - \frac{x + I_\tau}{(t+\tau)\, S_\tau}\right)^+ \tag{27.1.24}$$

for $(t,x) \in (0,T] \times [0,\infty)$. From (27.1.24) we immediately see that

$$x \mapsto V(t,x) \text{ is decreasing and convex on } [0,\infty) \tag{27.1.25}$$

for each $t > 0$ fixed.

3°. We show that $V : (0,T] \times [0,\infty) \to \mathbb{R}$ is continuous. For this, using $\sup f - \sup g \leq \sup(f-g)$ and $(z-x)^+ - (z-y)^+ \leq (y-x)^+$ for $x, y, z \in \mathbb{R}$, we get

$$
\begin{aligned}
&V(t,x)-V(t,y) \qquad (27.1.26)\\
&\le \sup_{0\le\tau\le T-t}\left(\widetilde{\mathsf{E}}\left(1-\frac{x+I_\tau}{(t+\tau)S_\tau}\right)^+-\widetilde{\mathsf{E}}\left(1-\frac{y+I_\tau}{(t+\tau)S_\tau}\right)^+\right)\\
&\le (y-x)\sup_{0\le\tau\le T-t}\widetilde{\mathsf{E}}\left(\frac{1}{(t+\tau)\,S_\tau}\right)\le\frac{1}{t}\,(y-x)
\end{aligned}
$$

for $0\le x\le y$ and $t>0$, where in the last inequality we used (27.1.8) to deduce that $1/S_t=\exp(\sigma\widehat{B}_t-(r+\sigma^2/2)t)\le\exp(\sigma\widehat{B}_t-(\sigma^2/2)t)$ and the latter is a martingale under $\widetilde{\mathsf{P}}$. From (27.1.26) with (27.1.25) we see that $x\mapsto V(t,x)$ is continuous at x_0 uniformly over $t\in[t_0-\delta,t_0+\delta]$ for some $\delta>0$ (small enough) whenever $(t_0,x_0)\in(0,T]\times[0,\infty)$ is given and fixed. Thus to prove that V is continuous on $(0,T]\times[0,\infty)$ it is enough to show that $t\mapsto V(t,x)$ is continuous on $(0,T]$ for each $x\ge0$ given and fixed. For this, take any $t_1<t_2$ in $(0,T]$ and $\varepsilon>0$, and let τ_1^ε be a stopping time such that $\widetilde{\mathsf{E}}((1-(X^x_{t_1+\tau_1^\varepsilon})/(t_1+\tau_1^\varepsilon))^+)\ge V(t_1,x)-\varepsilon$. Setting $\tau_2^\varepsilon=\tau_1^\varepsilon\wedge(T-t_2)$ we see that $V(t_2,x)\ge\widetilde{\mathsf{E}}((1-(X_{t_2+\tau_2^\varepsilon})/(t_2+\tau_2^\varepsilon))^+)$. Hence we get

$$
\begin{aligned}
&V(t_1,x)-V(t_2,x) \qquad (27.1.27)\\
&\le\widetilde{\mathsf{E}}\left(\left(1-\frac{X^x_{t_1+\tau_1^\varepsilon}}{t_1+\tau_1^\varepsilon}\right)^+\right)-\widetilde{\mathsf{E}}\left(\left(1-\frac{X^x_{t_2+\tau_2^\varepsilon}}{t_2+\tau_2^\varepsilon}\right)^+\right)+\varepsilon\\
&\le\widetilde{\mathsf{E}}\left(\left(\frac{X^x_{t_2+\tau_2^\varepsilon}}{t_2+\tau_2^\varepsilon}-\frac{X^x_{t_1+\tau_1^\varepsilon}}{t_1+\tau_1^\varepsilon}\right)^+\right)+\varepsilon.
\end{aligned}
$$

Letting first $t_2-t_1\to0$ using $\tau_1^\varepsilon-\tau_2^\varepsilon\to0$ and then $\varepsilon\downarrow0$ we see that $\limsup_{t_2-t_1\to0}(V(t_1,x)-V(t_2,x))\le0$ by dominated convergence. On the other hand, let τ_2^ε be a stopping time such that $\widetilde{\mathsf{E}}((1-(X^x_{t_2+\tau_2^\varepsilon})/(t_2+\tau_2^\varepsilon))^+)\ge V(t_2,x)-\varepsilon$. Then we have

$$
\begin{aligned}
&V(t_1,x)-V(t_2,x) \qquad (27.1.28)\\
&\ge\widetilde{\mathsf{E}}\left(\left(1-\frac{X^x_{t_1+\tau_2^\varepsilon}}{t_1+\tau_2^\varepsilon}\right)^+\right)-\widetilde{\mathsf{E}}\left(\left(1-\frac{X^x_{t_2+\tau_2^\varepsilon}}{t_2+\tau_2^\varepsilon}\right)^+\right)-\varepsilon.
\end{aligned}
$$

Letting first $t_2-t_1\to0$ and then $\varepsilon\downarrow0$ we see that $\liminf_{t_2-t_1\to0}(V(t_1,x)-V(t_2,x))\ge0$. Combining the two inequalities we find that $t\mapsto V(t,x)$ is continuous on $(0,T]$. This completes the proof of the initial claim.

4°. Denote the gain function by $G(t,x)=(1-x/t)^+$ for $(t,x)\in(0,T]\times[0,\infty)$ and introduce the continuation set $C=\{\,(t,x)\in(0,T)\times[0,\infty):V(t,x)>G(t,x)\,\}$ and the stopping set $\bar{D}=\{\,(t,x)\in(0,T)\times[0,\infty):V(t,x)=G(t,x)\,\}$. Since V and G are continuous, we see that C is open (and $\bar{D}$ is closed indeed) in $(0,T)\times[0,\infty)$. Standard arguments based on the strong Markov property [see Corollary 2.9 (Finite horizon) with Remark 2.10] show that the first hitting time $\tau_{\bar{D}} = \inf\{\, 0 \le s \le T-t : (t+s, X_{t+s}) \in \bar{D}\,\}$ is optimal in (27.1.9) as well as that V is $C^{1,2}$ on C and satisfies (27.1.10). In order to determine the structure of the optimal stopping time $\tau_{\bar{D}}$ (i.e. the shape of the sets C and D) we will first examine basic properties of the diffusion process X solving (27.1.6) under $\widetilde{\mathsf{P}}$.

5°. The state space of X equals $[0,\infty)$ and it is clear from the representation (27.1.5) with (27.1.8) that 0 is an entrance boundary point. The drift of X is given by $b(x) = 1 - rx$ and the diffusion coefficient of X is given by $\sigma(x) = \sigma x$ for $x \ge 0$. Hence we see that $b(x)$ is greater/less than 0 if and only if x is less/greater than $1/r$. This shows that there is a permanent push (drift) of X towards the constant level $1/r$ (when X is above $1/r$ the push of X is downwards and when X is below $1/r$ the push of X is upwards). The scale function of X is given by

$$S(x) = \int_1^x y^{2r/\sigma^2} e^{2/\sigma^2 y}\, dy \tag{27.1.29}$$

for $x > 0$, and the speed measure of X is given by

$$m(dx) = (2/\sigma^2)\, x^{-2(1+r/\sigma^2)}\, e^{-2/\sigma^2 x}\, dx \tag{27.1.30}$$

on the Borel σ-algebra of $(0,\infty)$. Since $S(0) = -\infty$ and $S(\infty) = +\infty$ we see that X is recurrent. Moreover, since $\int_0^\infty m(dx)(2/\sigma^2)^{-2r/\sigma^2}\,\Gamma(1{+}2r/\sigma^2)$ is finite we find that X has an invariant probability density function given by

$$f(x) = \frac{(2/\sigma^2)^{1+2r/\sigma^2}}{\Gamma(1{+}2r/\sigma^2)}\, \frac{1}{x^{2(1+r/\sigma^2)}}\, e^{-2/\sigma^2 x} \tag{27.1.31}$$

for $x > 0$. In particular, it follows that $X_t/t \to 0$ $\widetilde{\mathsf{P}}$-a.s. as $t \to \infty$. This fact has an important consequence for the optimal stopping problem (27.1.9): *If the horizon T is infinite, then it is never optimal to stop.* Indeed, in this case letting $\tau \equiv t$ and passing to the limit for $t \to \infty$ we see that $V \equiv 1$ on $(0,\infty)\times[0,\infty)$. This shows that the infinite horizon formulation of the problem (27.1.9) provides no useful information to the finite horizon formulation (unlike in the cases of American and Russian options above). To examine the latter beyond the trivial fact that all points (t,x) with $x \ge t$ belong to C (which is easily seen by considering the hitting times $\tau_\varepsilon = \inf\{\, 0 \le s \le T-t : X_{t+s} \le (t+s) - \varepsilon\,\}$ and noting that $\widetilde{\mathsf{P}}_{t,x}(0 < \tau_\varepsilon < T-t) > 0$ if $x \ge t$ with $0 < t < T$) we will examine the gain process in the problem (27.1.9) using stochastic calculus as follows.

6°. Setting $\alpha(t) = t$ for $0 \le t \le T$ to denote the diagonal in the state space and applying the local time-space formula (cf. Subsection 3.5) under $\widetilde{\mathsf{P}}_{t,x}$ when

$(t,x) \in (0,T) \times [0,\infty)$ is given and fixed, we get

$$\begin{aligned} G(t+s, X_{t+s}) &= G(t,x) + \int_0^s G_t(t+u, X_{t+u})\, du \qquad (27.1.32)\\ &+ \int_0^s G_x(t+u, X_{t+u})\, dX_{t+u} + \frac{1}{2}\int_0^s G_{xx}(t+u, X_{t+u})\, d\langle X,X\rangle_{t+u}\\ &+ \frac{1}{2}\int_0^s \Big(G_x(t+u, \alpha(t+u)+) - G_x(t+u, \alpha(t+u)-)\Big)\, d\ell^{\alpha}_{t+u}(X)\\ &= G(t,x) + \int_0^s \left(\frac{X_{t+u}}{(t+u)^2} - \frac{1-rX_{t+u}}{(t+u)}\right) I\big(X_{t+u} < \alpha(t+u)\big)\, du\\ &- \sigma \int_0^s \frac{X_{t+u}}{t+u}\, I\big(X_{t+u} < \alpha(t+u)\big)\, d\widehat{B}_u + \frac{1}{2}\int_0^s \frac{d\ell^{\alpha}_{t+u}(X)}{t+u} \end{aligned}$$

where $\ell^{\alpha}_{t+u}(X)$ is the local time of X on the curve α given by

$$\begin{aligned} &\ell^{\alpha}_{t+u}(X) \qquad (27.1.33)\\ &= \widetilde{\mathsf{P}}\text{-}\lim_{\varepsilon\downarrow 0} \frac{1}{2\varepsilon}\int_0^u I\big(\alpha(t+v)-\varepsilon < X_{t+v} < \alpha(t+v)+\varepsilon\big)\, d\langle X,X\rangle_{t+v}\\ &= \widetilde{\mathsf{P}}\text{-}\lim_{\varepsilon\downarrow 0} \frac{1}{2\varepsilon}\int_0^u I\big(\alpha(t+v)-\varepsilon < X_{t+v} < \alpha(t+v)+\varepsilon\big)\, \frac{\sigma^2}{2}\, X^2_{t+v}\, dv \end{aligned}$$

and $d\ell^{\alpha}_{t+u}(X)$ refers to the integration with respect to the continuous increasing function $u \mapsto \ell^{\alpha}_{t+u}(X)$. From (27.1.32) we respectively read

$$G(t+s, X_{t+s}) = G(t,x) + A_s + M_s + L_s \qquad (27.1.34)$$

where A and L are processes of bounded variation (L is increasing) and M is a continuous (local) martingale. We note moreover that $s \mapsto L_s$ is strictly increasing only when $X_s = \alpha(s)$ for $0 \le s \le T-t$ i.e. when X visits α. On the other hand, when X is below α then the integrand $a(t+u, X_{t+u})$ of A_s may be either positive or negative. To determine both sets exactly we need to examine the sign of the expression $a(t,x) = x/t^2 - (1-rx)/t$. It follows that $a(t,x)$ is larger/less than 0 if and only if x is larger/less than $\gamma(t)$ where $\gamma(t) = t/(1+rt)$ for $0 \le t \le T$. By considering the exit times from small balls in $(0,T) \times [0,\infty)$ with centre at (t,x) and making use of (27.1.32) with the optional sampling theorem (page 60) to get rid of the martingale part, upon observing that $\gamma(t) < \alpha(t)$ for all $0 < t \le T$ so that the local time part is zero, we see that all points (t,x) lying above the curve γ (i.e. $x > \gamma(t)$ for $0 < t < T$) belong to the continuation set C. Exactly the same arguments (based on the fact that the favourable sets above γ and on α are far away from X) show that for each $x < \gamma(T) = T/(1+rT)$ given and fixed, all points (t,x) belong to the stopping set $\bar{D}$ when t is close to T. Moreover, recalling (27.1.25) and the fact that $V(t,x) \ge G(t,x)$ for all $x \ge 0$ with $t \in (0,T)$ fixed, we see that for each $t \in (0,T)$ there is a point

$b(t) \in [0, \gamma(t)]$ such that $V(t,x) > G(t,x)$ for $x > b(t)$ and $V(t,x) = G(t,x)$ for $x \in [0, b(t)]$. Combining it with the previous conclusion on $\bar{D}$ we find that $b(T-) = \gamma(T) = T/(1+rT)$. (Yet another argument for this identity will be given below. Note that this identity is different from the identity $b(T-) = T$ used in [89, p. 1126].) This establishes the existence of the nontrivial (nonzero) optimal stopping boundary b on a left-neighbourhood of T. We will now show that b extends (continuously and decreasingly) from the initial neighbourhood of T backward in time as long as it visits 0 at some time $t_0 \in [0, T)$, and later in the second part of the proof below we will deduce that this t_0 is equal to 0. The key argument in the proof is provided by the following inequality. Notice that this inequality is not obvious a priori (unlike in the cases of American and Russian options above) since $t \mapsto G(t,x)$ is increasing and the supremum in (27.1.9) is taken over a smaller class of stopping times $\tau \in [0, T-t]$ when t is larger.

7°. We show that the *inequality*

$$V_t(t,x) \le G_t(t,x) \tag{27.1.35}$$

is satisfied for all $(t,x) \in C$. (It may be noted from (27.1.10) that $V_t = -(1-rx)V_x - (\sigma^2/2)x^2\,V_{xx} \le (1-rx)/t$ since $V_x \ge -1/t$ and $V_{xx} \ge 0$ by (27.1.25), so that $V_t \le G_t$ holds above γ because $(1-rx)/t \le x/t^2$ if and only if $x \ge t/(1+rt)$. Hence the main issue is to show that (27.1.35) holds below γ and above b. Any analytic proof of this fact seems difficult and we resort to probabilistic arguments.)

To prove (27.1.35) fix $0 < t < t+h < T$ and $x \ge 0$ so that $x \le \gamma(t)$. Let $\tau = \tau_S(t+h, x)$ be the optimal stopping time for $V(t+h, x)$. Since $\tau \in [0, T-t-h] \subseteq [0, T-t]$ we see that $V(t,x) \ge \widetilde{\mathsf{E}}_{t,x}((1 - X_{t+\tau}/(t+\tau))^+)$ so that using the inequality stated prior to (27.1.26) above (and the convenient refinement by an indicator function), we get

$$\begin{aligned}
&V(t+h,x) - V(t,x) - \Big(G(t+h,x) - G(t,x)\Big) \qquad\qquad (27.1.36)\\
&\le \widetilde{\mathsf{E}}\left(\left(1 - \frac{x+I_\tau}{(t+h+\tau)S_\tau}\right)^+\right) - \widetilde{\mathsf{E}}\left(\left(1 - \frac{x+I_\tau}{(t+\tau)S_\tau}\right)^+\right) - \left(\frac{x}{t} - \frac{x}{t+h}\right)\\
&\le \widetilde{\mathsf{E}}\left(\left(\frac{x+I_\tau}{(t+\tau)S_\tau} - \frac{x+I_\tau}{(t+h+\tau)\,S_\tau}\right) I\left(\frac{x+I_\tau}{(t+h+\tau)S_\tau} \le 1\right)\right) - \frac{xh}{t\,(t+h)}\\
&= \widetilde{\mathsf{E}}\left(\frac{x+I_\tau}{S_\tau}\left(\frac{1}{t+\tau} - \frac{1}{t+h+\tau}\right) I\left(\frac{x+I_\tau}{(t+h+\tau)\,S_\tau} \le 1\right)\right) - \frac{xh}{t\,(t+h)}\\
&= \widetilde{\mathsf{E}}\left(\frac{x+I_\tau}{(t+h+\tau)\,S_\tau}\,\frac{h}{t+\tau}\, I\left(\frac{x+I_\tau}{(t+h+\tau)\,S_\tau} \le 1\right)\right) - \frac{xh}{t\,(t+h)}\\
&\le \frac{h}{t}\,\widetilde{\mathsf{E}}\left(\frac{x+I_\tau}{(t+h+\tau)\,S_\tau}\, I\left(\frac{x+I_\tau}{(t+h+\tau)\,S_\tau} \le 1\right)\right) - \frac{xh}{t\,(t+h)} \le 0
\end{aligned}$$

where the final inequality follows from the fact that with $Z := (x + I_\tau)/((t + h + \tau)S_\tau)$ we have $V(t+h,x) = \widetilde{\mathsf{E}}((1-Z)^+) = \widetilde{\mathsf{E}}((1-Z)\, I(Z \le 1)) = \widetilde{\mathsf{P}}(Z \le 1) - \widetilde{\mathsf{E}}(Z\, I(Z \le 1)) \ge G(t+h,x) = 1 - x/(t+h)$ so that $\widetilde{\mathsf{E}}(Z\, I(Z \le 1)) \le \widetilde{\mathsf{P}}(Z \le 1) - 1 + x/(t+h) \le x/(t+h)$ as claimed. Dividing the initial expression in (27.1.36) by h and letting $h \downarrow 0$ we obtain (27.1.35) for all $(t,x) \in C$ such that $x \le \gamma(t)$. Since $V_t \le G_t$ above γ (as stated following (27.1.35) above) this completes the proof of (27.1.35).

8°. We show that $t \mapsto b(t)$ is *increasing* on $(0,T)$. This is an immediate consequence of (27.1.36). Indeed, if (t_1,x) belongs to C and t_0 from $(0,T)$ satisfies $t_0 < t_1$, then by (27.1.36) we have that $V(t_0,x) - G(t_0,x) \ge V(t_1,x) - G(t_1,x) > 0$ so that (t_0,x) must belong to C. It follows that b cannot be strictly decreasing thus proving the claim.

9°. We show that the *smooth-fit* condition (27.1.12) holds, i.e. that $x \mapsto V(t,x)$ is C^1 at $b(t)$. For this, fix a point $(t,x) \in (0,T) \times (0,\infty)$ lying at the boundary so that $x = b(t)$. Then $x \le \gamma(t) < \alpha(t)$ and for all $\varepsilon > 0$ such that $x + \varepsilon < \alpha(t)$ we have

$$\frac{V(t,x+\varepsilon) - V(t,x)}{\varepsilon} \ge \frac{G(t,x+\varepsilon) - G(t,x)}{\varepsilon} = -\frac{1}{t}. \tag{27.1.37}$$

Letting $\varepsilon \downarrow 0$ and using that the limit on the left-hand side exists (since $x \mapsto V(t,x)$ is convex), we get the inequality

$$\frac{\partial^+ V}{\partial x}(t,x) \ge \frac{\partial G}{\partial x}(t,x) = -\frac{1}{t}. \tag{27.1.38}$$

To prove the converse inequality, fix $\varepsilon > 0$ such that $x + \varepsilon < \alpha(t)$, and consider the stopping times $\tau_\varepsilon = \tau_S(t, x+\varepsilon)$ being optimal for $V(t,x+\varepsilon)$. Then we have

$$\begin{aligned} &\frac{V(t,x{+}\varepsilon) - V(t,x)}{\varepsilon} \\ &\le \frac{1}{\varepsilon}\left(\widetilde{\mathsf{E}}\left[\left(1 - \frac{x{+}\varepsilon{+}I_{\tau_\varepsilon}}{(t{+}\tau_\varepsilon)S_{\tau_\varepsilon}}\right)^+ - \left(1 - \frac{x{+}I_{\tau_\varepsilon}}{(t{+}\tau_\varepsilon)S_{\tau_\varepsilon}}\right)^+\right]\right) \\ &\le \frac{1}{\varepsilon}\,\widetilde{\mathsf{E}}\left(\frac{x + I_{\tau_\varepsilon}}{(t+\tau_\varepsilon)\, S_{\tau_\varepsilon}} - \frac{x+\varepsilon+I_{\tau_\varepsilon}}{(t+\tau_\varepsilon)\, S_{\tau_\varepsilon}}\right) = -\,\widetilde{\mathsf{E}}\left(\frac{1}{(t+\tau_\varepsilon)\, S_{\tau_\varepsilon}}\right). \end{aligned} \tag{27.1.39}$$

Since each point x in $(0,\infty)$ is regular for X, and the boundary b is increasing, it follows that $\tau_\varepsilon \downarrow 0$ $\widetilde{\mathsf{P}}$-a.s. as $\varepsilon \downarrow 0$. Letting $\varepsilon \downarrow 0$ in (27.1.39) we get

$$\frac{\partial^+ V}{\partial x}(t,x) \le -\frac{1}{t} \tag{27.1.40}$$

by dominated convergence. It follows from (27.1.38) and (27.1.40) that $(\partial^+ V/\partial x)(t,x) = -1/t$ implying the claim.

10°. We show that b is *continuous.* Note that the same proof also shows that $b(T-) = T/(1+rT)$ as already established above by a different method.

Let us first show that b is right-continuous. For this, fix $t \in (0,T)$ and consider a sequence $t_n \downarrow t$ as $n \to \infty$. Since b is increasing, the right-hand limit $b(t+)$ exists. Because $(t_n, b(t_n)) \in \bar{D}$ for all $n \geq 1$, and $\bar{D}$ is closed, it follows that $(t, b(t+)) \in \bar{D}$. Hence by (27.1.16) we see $b(t+) \leq b(t)$. Since the reverse inequality follows obviously from the fact that b is increasing, this completes the proof of the first claim.

Let us next show that b is left-continuous. Suppose that there exists $t \in (0,T)$ such that $b(t-) < b(t)$. Fix a point x in $(b(t-), b(t)]$ and note by (27.1.12) that for $s < t$ we have

$$V(s,x) - G(s,x) = \int_{b(s)}^{x} \int_{b(s)}^{y} \big(V_{xx}(s,z) - G_{xx}(s,z)\big)\, dz\, dy \tag{27.1.41}$$

upon recalling that V is $C^{1,2}$ on C. Note that $G_{xx} = 0$ below α so that if $V_{xx} \geq c$ on $R = \{\, (u,y) \in C : s \leq u < t \text{ and } b(u) < y \leq x \,\}$ for some $c > 0$ (for all $s < t$ close enough to t and some $x > b(t-)$ close enough to $b(t-)$) then by letting $s \uparrow t$ in (27.1.41) we get

$$V(t,x) - G(t,x) \geq c\, \frac{(x - b(t))^2}{2} > 0 \tag{27.1.42}$$

contradicting the fact that (t,x) belongs to $\bar{D}$ and thus is an optimal stopping point. Hence the proof reduces to showing that $V_{xx} \geq c$ on small enough R for some $c > 0$.

To derive the latter fact we may first note from (27.1.10) upon using (27.1.35) that $V_{xx} = (2/(\sigma^2 x^2))(-V_t - (1-rx)V_x) \geq (2/(\sigma^2 x^2))(-x/t^2 - (1-rx)V_x)$. Suppose now that for each $\delta > 0$ there is $s < t$ close enough to t and there is $x > b(t-)$ close enough to $b(t-)$ such that $V_x(u,y) \leq -1/u + \delta$ for all $(u,y) \in R$ (where we recall that $-1/u = G_x(u,y)$ for all $(u,y) \in R$). Then from the previous inequality we find that $V_{xx}(u,y) \geq (2/(\sigma^2 y^2))(-y/u^2 + (1-ry)(1/u - \delta)) = (2/(\sigma^2 y^2))((u - y(1+ru))/u^2 - \delta(1-ru)) \geq c > 0$ for $\delta > 0$ small enough since $y < u/(1+ru) = \gamma(u)$ and $y < 1/r$ for all $(u,y) \in R$. Hence the proof reduces to showing that $V_x(u,y) \leq -1/u + \delta$ for all $(u,y) \in R$ with R small enough when $\delta > 0$ is given and fixed.

To derive the latter inequality we can make use of the estimate (27.1.39) to conclude that

$$\frac{V(u, y+\varepsilon) - V(u,y)}{\varepsilon} \leq -\widetilde{\mathsf{E}}\left(\frac{1}{(u + \sigma_\varepsilon)\, M_{\sigma_\varepsilon}} \right) \tag{27.1.43}$$

where $\sigma_\varepsilon = \inf\{\, 0 \leq v \leq T - u : X_{u+v}^{y+\varepsilon} = b(u) \,\}$ and $M_t = \sup_{0 \leq s \leq t} S_s$. A simple comparison argument (based on the fact that b is increasing) shows that the supremum over all $(u,y) \in R$ on the right-hand side of (27.1.43) is attained at $(s, x+\varepsilon)$. Letting $\varepsilon \downarrow 0$ in (27.1.43) we thus get

$$V_x(u,y) \le -\widetilde{\mathsf{E}}\left(\frac{1}{(u+\sigma)\,M_\sigma}\right) \tag{27.1.44}$$

for all $(u,y) \in R$ where $\sigma = \inf\{\, 0 \le v \le T-s : X^x_{s+v} = b(s)\,\}$. Since by regularity of X we find that $\sigma \downarrow 0$ $\widetilde{\mathsf{P}}$-a.s. as $s \uparrow t$ and $x \downarrow b(t-)$, it follows from (27.1.44) that

$$V_x(u,y) \le -\frac{1}{u} + \widetilde{\mathsf{E}}\left(\frac{(u+\sigma)\,M_\sigma - u}{u\,(u+\sigma)\,M_\sigma}\right) \le -\frac{1}{u} + \delta \tag{27.1.45}$$

for all $s<t$ close enough to t and some $x > b(t-)$ close enough to $b(t-)$. This completes the proof of the second claim, and thus the initial claim is proved as well.

11°. We show that V is given by the formula (27.1.23) and that b solves equation (27.1.22). For this, note that V satisfies the following conditions:

$$V \text{ is } C^{1,2} \text{ on } C \cup D, \tag{27.1.46}$$

$$V_t + \mathbb{L}_X V \text{ is locally bounded}, \tag{27.1.47}$$

$$x \mapsto V(t,x) \text{ is convex}, \tag{27.1.48}$$

$$t \mapsto V_x(t, b(t)\pm) \text{ is continuous.} \tag{27.1.49}$$

Indeed, the conditions (27.1.46) and (27.1.47) follow from the facts that V is $C^{1,2}$ on C and $V=G$ on D upon recalling that D lies below γ so that $G(t,x) = 1 - x/t$ for all $(t,x) \in D$ and thus G is $C^{1,2}$ on D. [When we say in (27.1.47) that $V_t + \mathbb{L}_X V$ is locally bounded, we mean that $V_t + \mathbb{L}_X V$ is bounded on $K \cap (C \cup D)$ for each compact set K in $[0,T] \times \mathbb{R}_+$.] The condition (27.1.48) was established in (27.1.25) above. The condition (27.1.49) follows from (27.1.12) since according to the latter we have $V_x(t, b(t)\pm) = -1/t$ for $t > 0$.

Since (27.1.46)–(27.1.49) are satisfied we know that the local time-space formula (cf. Subsection 3.5) can be applied. This gives

$$\begin{aligned} V(t{+}s, X_{t+s}) &= V(t,x) \\ &\quad + \int_0^s \big(V_t + \mathbb{L}_X V\big)(t{+}u, X_{t+u})\, I\big(X_{t+u} \ne b(t{+}u)\big)\, du \\ &\quad + \int_0^s \sigma\, X_{t+u}\, V_x(t+u, X_{t+u})\, I\big(X_{t+u} \ne b(t{+}u)\big)\, dB_u \\ &\quad + \frac{1}{2} \int_0^s \big(V_x(t{+}u, X_{t+u}{+}) - V_x(t{+}u, X_{t+u}{-})\big) \\ &\qquad\qquad \times I\big(X_{t+u} = b(t{+}u)\big)\, d\ell^b_{t+u}(X) \\ &= \int_0^s \big(G_t + \mathbb{L}_X G\big)(t+u, X_{t+u})\, I\big(X_{t+u} < b(t{+}u)\big)\, du + M_s \end{aligned} \tag{27.1.50}$$

where the final equality follows by the smooth-fit condition (27.1.12) and $M_s = \int_0^s \sigma X_{t+u} V_x(t+u, X_{t+u})\, I\big(X_{t+u} \neq b(t+u)\big)\, dB_u$ is a continuous martingale for $0 \le s \le T-t$ with $t > 0$. Noting that $(G_t + \mathbb{L}_X G)(t,x) = x/t^2 - (1-rx)/t$ for $x < t$ we see that (27.1.50) yields

$$\begin{aligned} V(t+s, X_{t+s}) &= V(t,x) \qquad (27.1.51)\\ &+ \int_0^s \left(\frac{X_{t+u}}{(t+u)^2} - \frac{1-rX_{t+u}}{(t+u)} \right) I\big(X_{t+u} < b(t+u)\big)\, du + M_s. \end{aligned}$$

Setting $s = T-t$, using that $V(T,x) = G(T,x)$ for all $x \ge 0$, and taking the $\widetilde{\mathsf{P}}_{t,x}$-expectation in (27.1.51), we find by the optional sampling theorem (page 60) that

$$\begin{aligned} \widetilde{\mathsf{E}}_{t,x}\Big(1 - \frac{X_T}{T}\Big)^+ &= V(t,x) \qquad (27.1.52)\\ &+ \int_0^{T-t} \widetilde{\mathsf{E}}_{t,x}\left(\left(\frac{X_{t+u}}{(t+u)^2} - \frac{1-rX_{t+u}}{(t+u)} \right) I\big(X_{t+u} < b(t+u)\big) \right) du. \end{aligned}$$

Making use of (27.1.18)–(27.1.20) we see that (27.1.52) is the formula (27.1.23). Moreover, inserting $x = b(t)$ in (27.1.52) and using that $V(t,b(t)) = G(t,b(t)) = 1 - b(t)/t$, we see that b satisfies the equation (27.1.22) as claimed.

12°. We show that $b(t) > 0$ for all $0 < t \le T$ and that $b(0+) = 0$. For this, suppose that $b(t_0) = 0$ for some $t_0 \in (0,T)$ and fix $t \in (0,t_0)$. Then $(t,x) \in C$ for all $x > 0$ as small as desired. Taking any such $(t,x) \in C$ and denoting by $\tau_{\bar{D}} = \tau_{\bar{D}}(t,x)$ the first hitting time to $\bar{D}$ under $\widetilde{\mathsf{P}}_{t,x}$, we find by (27.1.51) that

$$\begin{aligned} V(t+\tau_{\bar{D}}, X_{t+\tau_{\bar{D}}}) = G(t+\tau_{\bar{D}}, X_{t+\tau_{\bar{D}}}) &= \left(1 - \frac{X_{t+\tau_{\bar{D}}}}{t+\tau_{\bar{D}}}\right)^+ \qquad (27.1.53)\\ = V(t,x) + M_{t+\tau_{\bar{D}}} &= 1 - \frac{x}{t} + M_{t+\tau_{\bar{D}}}. \end{aligned}$$

Taking the $\widetilde{\mathsf{P}}_{t,x}$-expectation and letting $x \downarrow 0$ we get

$$\widetilde{\mathsf{E}}_{t,0}\left(1 - \frac{X_{t+\tau_{\bar{D}}}}{t+\tau_{\bar{D}}}\right)^+ = 1 \qquad (27.1.54)$$

where $\tau_{\bar{D}} = \tau_{\bar{D}}(t,0)$. As clearly $\widetilde{\mathsf{P}}_{t,0}(X_{t+\tau_{\bar{D}}} \ge T) > 0$ we see that the left-hand side of (27.1.54) is strictly smaller than 1 thus contradicting the identity. This shows that $b(t)$ must be strictly positive for all $0 < t \le T$. Combining this conclusion with the known inequality $b(t) \le \gamma(t)$ which is valid for all $0 < t \le T$ we see that $b(0+) = 0$ as claimed.

13°. We show that b is the *unique* solution of the nonlinear integral equation (27.1.22) in the class of continuous functions $c : (0,T) \to \mathbb{R}$ satisfying $0 < c(t) < t/(1+rt)$ for all $0 < t < T$. (Note that this class is larger than the class of functions having the established properties of b which is moreover known to be increasing.) The proof of the uniqueness will be presented in the final three steps of the main proof as follows.

14°. Let $c : (0,T] \to \mathbb{R}$ be a continuous solution of the equation (27.1.22) satisfying $0 < c(t) < t$ for all $0 < t < T$. We want to show that this c must then be equal to the optimal stopping boundary b.

Motivated by the derivation (27.1.50)–(27.1.52) which leads to the formula (27.1.55), let us consider the function $U^c : (0,T] \times [0,\infty) \to \mathbb{R}$ defined as follows:

$$U^c(t,x) = \widetilde{\mathsf{E}}_{t,x}\Big(\Big(1-\frac{X_T}{T}\Big)^+\Big) \tag{27.1.55}$$
$$-\int_0^{T-t} \widetilde{\mathsf{E}}_{t,x}\Big(\Big(\frac{X_{t+u}}{(t+u)^2}-\frac{1-rX_{t+u}}{(t+u)}\Big)I\big(X_{t+u}<c(t+u)\big)\Big)\,du$$

for $(t,x) \in (0,T] \times [0,\infty)$. In terms of (27.1.18)–(27.1.20) note that U^c is explicitly given by

$$U^c(t,x) = F(T-t,x) \tag{27.1.56}$$
$$-\int_0^{T-t} \tfrac{1}{t+u}\Big(\Big(\tfrac{1}{t+u}+r\Big)G\big(u,x,c(t+u)\big)-H\big(u,x,c(t+u)\big)\Big)\,du$$

for $(t,x) \in (0,T] \times [0,\infty)$. Observe that the fact that c solves (27.1.22) on $(0,T)$ means exactly that $U^c(t,c(t)) = G(t,c(t))$ for all $0 < t < T$. We will now moreover show that $U^c(t,x) = G(t,x)$ for all $x \in [0,c(t)]$ with $t \in (0,T)$. This is the key point in the proof (cf. Subsections 25.2 and 26.2 above) that can be derived using a martingale argument as follows.

If $X = (X_t)_{t\ge 0}$ is a Markov process (with values in a general state space) and we set $F(t,x) = \mathsf{E}_x G(X_{T-t})$ for a (bounded) measurable function G with $\mathsf{P}_x(X_0 = x) = 1$, then the Markov property of X implies that $F(t,X_t)$ is a martingale under P_x for $0 \le t \le T$. Similarly, if we set $F(t,x) = \mathsf{E}_x(\int_0^{T-t} H(X_u)\,du)$ for a (bounded) measurable function H with $\mathsf{P}_x(X_0 = x) = 1$, then the Markov property of X implies that $F(t,X_t) + \int_0^t H(X_u)\,du$ is a martingale under P_x for $0 \le t \le T$. Combining these two martingale facts applied to the time-space Markov process $(t+s, X_{t+s})$ instead of X_s, we find that

$$U^c(t+s,X_{t+s}) - \int_0^s \Big(\frac{X_{t+u}}{(t+u)^2}-\frac{1-rX_{t+u}}{(t+u)}\Big)I\big(X_{t+u}<c(t+u)\big)\,du \tag{27.1.57}$$

is a martingale under $\widetilde{\mathsf{P}}_{t,x}$ for $0 \le s \le T-t$. We may thus write

$$U^c(t+s,X_{t+s}) \tag{27.1.58}$$
$$-\int_0^s \Big(\frac{X_{t+u}}{(t+u)^2}-\frac{1-rX_{t+u}}{(t+u)}\Big)I\big(X_{t+u}<c(t+u)\big)\,du = U^c(t,x)+N_s$$

where $(N_s)_{0\le s\le T-t}$ is a martingale with $N_0=0$ under $\widetilde{\mathsf{P}}_{t,x}$.

On the other hand, we know from (27.1.32) that

$$
\begin{aligned}
G(t+s,X_{t+s}) &= G(t,x) \\
&+\int_0^s \left(\frac{X_{t+u}}{(t+u)^2}-\frac{1-rX_{t+u}}{(t+u)}\right) I\big(X_{t+u}<\alpha(t+u)\big)\,du + M_s + L_s
\end{aligned}
\tag{27.1.59}
$$

where $M_s=-\sigma\int_0^s (X_{t+u}/(t+u))\,I(X_{t+u}<\alpha(t+u))\,d\widehat{B}_u$ is a continuous martingale under $\widetilde{\mathsf{P}}_{t,x}$ and $L_s=(1/2)\int_0^s d\ell^{\alpha}_{t+u}(X)/(t+u)$ is an increasing process for $0\le s\le T-t$.

For $0\le x\le c(t)$ with $t\in(0,T)$ given and fixed, consider the stopping time

$$
\sigma_c=\inf\{\,0\le s\le T-t : X_{t+s}\ge c(t+s)\,\}. \tag{27.1.60}
$$

Using that $U^c(t,c(t))=G(t,c(t))$ for all $0<t<T$ (since c solves (27.1.22) as pointed out above) and that $U^c(T,x)=G(T,x)$ for all $x\ge 0$, we see that $U^c(t+\sigma_c,X_{t+\sigma_c})=G(t+\sigma_c,X_{t+\sigma_c})$. Hence from (27.1.58) and (27.1.59) using the optional sampling theorem (page 60) we find

$$
\begin{aligned}
U^c(t,x) &= \widetilde{\mathsf{E}}_{t,x}\Big(U^c(t+\sigma_c,X_{t+\sigma_c})\Big) \\
&\qquad - \widetilde{\mathsf{E}}_{t,x}\bigg(\int_0^{\sigma_c}\left(\frac{X_{t+u}}{(t+u)^2}-\frac{1-rX_{t+u}}{(t+u)}\right) I\big(X_{t+u}<c(t+u)\big)\,du\bigg) \\
&= \widetilde{\mathsf{E}}_{t,x}\Big(G(t+\sigma_c,X_{t+\sigma_c})\Big) \\
&\quad - \widetilde{\mathsf{E}}_{t,x}\bigg(\int_0^{\sigma_c}\left(\frac{X_{t+u}}{(t+u)^2}-\frac{1-rX_{t+u}}{(t+u)}\right) I\big(X_{t+u}<c(t+u)\big)\,du\bigg) \\
&= G(t,x)+\widetilde{\mathsf{E}}_{t,x}\bigg(\int_0^{\sigma_c}\left(\frac{X_{t+u}}{(t+u)^2}-\frac{1-rX_{t+u}}{(t+u)}\right) I\big(X_{t+u}<\alpha(t+u)\big)\,du\bigg) \\
&\quad - \widetilde{\mathsf{E}}_{t,x}\bigg(\int_0^{\sigma_c}\left(\frac{X_{t+u}}{(t+u)^2}-\frac{1-rX_{t+u}}{(t+u)}\right) I\big(X_{t+u}<c(t+u)\big)\,du\bigg) \\
&= G(t,x)
\end{aligned}
\tag{27.1.61}
$$

since $X_{t+u}<\alpha(t+u)$ and $X_{t+u}<c(t+u)$ for all $0\le u<\sigma_c$. This proves that $U^c(t,x)=G(t,x)$ for all $x\in[0,c(t)]$ with $t\in(0,T)$ as claimed.

15°. We show that $U^c(t,x)\le V(t,x)$ for all $(t,x)\in(0,T]\times[0,\infty)$. For this, consider the stopping time

$$
\tau_c=\inf\{\,0\le s\le T-t : X_{t+s}\le c(t+s)\,\} \tag{27.1.62}
$$

under $\widetilde{\mathsf{P}}_{t,x}$ with $(t,x) \in (0,T] \times [0,\infty)$ given and fixed. The same arguments as those given following (27.1.60) above show that $U^c(t+\tau_c, X_{t+\tau_c}) = G(t+\tau_c, X_{t+\tau_c})$. Inserting τ_c instead of s in (27.1.58) and using the optional sampling theorem (page 60) we get

$$U^c(t,x) = \widetilde{\mathsf{E}}_{t,x}\, U^c(t+\tau_c, X_{t+\tau_c}) = \widetilde{\mathsf{E}}_{t,x}\, G(t+\tau_c, X_{t+\tau_c}) \le V(t,x) \tag{27.1.63}$$

where the final inequality follows from the definition of V proving the claim.

16°. We show that $c \ge b$ on $[0,T]$. For this, consider the stopping time

$$\sigma_b = \inf\,\{\, 0 \le s \le T-t : X_{t+s} \ge b(t+s)\,\} \tag{27.1.64}$$

under $\widetilde{\mathsf{P}}_{t,x}$ where $(t,x) \in (0,T) \times [0,\infty)$ such that $x < b(t) \wedge c(t)$. Inserting σ_b in place of s in (27.1.51) and (27.1.58) and using the optional sampling theorem (page 60) we get

$$\begin{aligned} &\widetilde{\mathsf{E}}_{t,x}\, V(t+\sigma_b, X_{t+\sigma_b}) = G(t,x) && (27.1.65)\\ &\quad + \widetilde{\mathsf{E}}_{t,x}\Big(\int_0^{\sigma_b} \Big(\frac{X_{t+u}}{(t+u)^2} - \frac{1-rX_{t+u}}{(t+u)} \Big)\, du \Big), \\ &\widetilde{\mathsf{E}}_{t,x}\, U^c(t+\sigma_b, X_{t+\sigma_b}) = G(t,x) && (27.1.66)\\ &\quad + \widetilde{\mathsf{E}}_{t,x}\Big(\int_0^{\sigma_b} \Big(\frac{X_{t+u}}{(t+u)^2} - \frac{1-rX_{t+u}}{(t+u)} \Big) I\big(X_{t+u} < c(t+u)\big)\, du \Big) \end{aligned}$$

where we also use that $V(t,x) = U^c(t,x) = G(t,x)$ for $x < b(t) \wedge c(t)$. Since $U^c \le V$ it follows from (27.1.65) and (27.1.66) that

$$\widetilde{\mathsf{E}}_{t,x}\Big(\int_0^{\sigma_b} \Big(\frac{X_{t+u}}{(t+u)^2} - \frac{1-rX_{t+u}}{(t+u)} \Big) I\big(X_{t+u} \ge c(t+u)\big)\, du \Big) \ge 0. \tag{27.1.67}$$

Due to the fact that $b(t) < t/(1+rt)$ for all $0 < t < T$, we see that $X_{t+u}/(t+u)^2 - (1-rX_{t+u})/(t+u) < 0$ in (27.1.67) so that by the continuity of b and c it follows that $c \ge b$ on $[0,T]$ as claimed.

17°. We show that c must be equal to b. For this, let us assume that there is $t \in (0,T)$ such that $c(t) > b(t)$. Pick $x \in (b(t), c(t))$ and consider the stopping time τ_b from (27.1.17). Inserting τ_b instead of s in (27.1.51) and (27.1.58) and using the optional sampling theorem (page 60) we get

$$\begin{aligned} &\widetilde{\mathsf{E}}_{t,x}\, G(t+\tau_b, X_{t+\tau_b}) = V(t,x), && (27.1.68)\\ &\widetilde{\mathsf{E}}_{t,x}\, (G(t+\tau_b, X_{t+\tau_b}) = U^c(t,x) && (27.1.69)\\ &\quad + \widetilde{\mathsf{E}}_{t,x}\Big(\int_0^{\tau_b} \Big(\frac{X_{t+u}}{(t+u)^2} - \frac{1-rX_{t+u}}{(t+u)} \Big) I\big(X_{t+u} < c(t+u)\big)\, du \Big) \end{aligned}$$

where we also use that $V(t+\tau_b, X_{t+\tau_b}) = U^c(t+\tau_b, X_{t+\tau_b}) G(t+\tau_b, X_{t+\tau_b})$ upon recalling that $c \geq b$ and $U^c = G$ either below c or at T. Since $U^c \leq V$ we see from (27.1.68) and (27.1.69) that

$$\widetilde{\mathsf{E}}_{t,x}\bigg(\int_0^{\tau_b}\bigg(\frac{X_{t+u}}{(t+u)^2} - \frac{1 - rX_{t+u}}{(t+u)}\bigg) I\big(X_{t+u} < c(t+u)\big)\, du\bigg) \geq 0. \tag{27.1.70}$$

Due to the fact that $c(t) < t/(1+rt)$ for all $0 < t < T$ by assumption, we see that $X_{t+u}/(t+u)^2 - (1 - rX_{t+u})/(t+u) < 0$ in (27.1.70) so that by the continuity of b and c it follows that such a point (t, x) cannot exist. Thus c must be equal to b, and the proof is complete. □

3. Remarks on numerics. 1°. The following method can be used to calculate the optimal stopping boundary b numerically by means of the integral equation (27.1.22). Note that the formula (27.1.23) can be used to calculate the arbitrage-free price V when b is known.

Set $t_i = ih$ for $i = 0, 1, \ldots, n$ where $h = T/n$ and denote

$$J(t, b(t)) = 1 - \tfrac{b(t)}{t} - F(T-t, b(t)), \tag{27.1.71}$$

$$\begin{aligned} &K(t, b(t); t+u, b(t+u)) \qquad\qquad (27.1.72)\\ &\quad = \tfrac{1}{t+u}\Big(\Big(\tfrac{1}{t+u} + r\Big) G(u, b(t), b(t+u)) - H(u, b(t), b(t+u))\Big). \end{aligned}$$

Then the following discrete approximation of the integral equation (27.1.22) is valid:

$$J(t_i, b(t_i)) = \sum_{j=i+1}^{n} K(t_i, b(t_i); t_j, b(t_j))\, h \tag{27.1.73}$$

for $i = 0, 1, \ldots, n-1$. Letting $i = n-1$ and $b(t_n) = T/(1+rT)$ we can solve equation (27.1.73) numerically and get a number $b(t_{n-1})$. Letting $i = n-2$ and using the values of $b(t_{n-1})$ and $b(t_n)$ we can solve equation (27.1.73) numerically and get a number $b(t_{n-2})$. Continuing the recursion we obtain $b(t_n)$, $b(t_{n-1}), \ldots, b(t_1), b(t_0)$ as an approximation of the optimal stopping boundary b at points $0, h, \ldots, T-h, T$.

It is an interesting numerical problem to show that the approximation converges to the true function b on $[0, T]$ as $h \downarrow 0$. Another interesting problem is to derive the rate of convergence.

2°. To perform the previous recursion we need to compute the functions F, G, H from (27.1.18)–(27.1.20) as efficiently as possible. Simply by observing the expressions (27.1.18)–(27.1.21) it is apparent that finding these functions numerically is not trivial. Moreover, the nature of the probability density function f in (27.1.21) presents a further numerical challenge. Part of this probability density function is the Hartman–Watson density discussed in [8]. As t tends to zero, the numerical estimate of the Hartman–Watson density oscillates, with the oscillations increasing rapidly in both amplitude and frequency as t gets closer to zero. The authors of [8] mention that this may be a consequence of the fact that $t \mapsto \exp(2\pi^2/\sigma^2 t)$ rapidly increases to infinity while $z \mapsto \sin(4\pi z/\sigma^2 t)$ oscillates more and more frequently. This rapid oscillation makes accurate estimation of $f(t,s,a)$ with t close to zero very difficult.

The problems when dealing with t close to zero are relevant to pricing the early exercise Asian call option. To find the optimal stopping boundary b as the solution to the implicit equation (27.1.73) it is necessary to work backward from T to 0. Thus to get an accurate estimate for b when $b(T)$ is given, the next estimate of $b(u)$ must be found for some value of u close to T so that $t = T - u$ will be close to zero.

Even if we get an accurate estimate for f, to solve (27.1.18)–(27.1.20) we need to evaluate two nested integrals. This is slow computationally. A crude attempt has been made at storing values for f and using these to estimate F, G, H in (27.1.18)–(27.1.20) but this method has not produced reliable results.

3°. Another approach to finding the functions F, G, H from (27.1.18)–(27.1.20) can be based on numerical solutions of partial differential equations. Two distinct methods are available.

Consider the transition probability density of the process X given by

$$p(s,x;t,y) = \frac{d}{dy}\,\widetilde{\mathsf{P}}(X_t \le y \mid X_s = x) \tag{27.1.74}$$

where $0 \le s < t$ and $x, y \ge 0$. Since $p(s,x;t,y) = p(0,x;t-s,y)$ we see that there is no restriction to assume that $s = 0$ in the sequel.

The *forward equation* approach leads to the initial-value problem

$$p_t = -((1-ry)p)_y + (Dyp)_{yy} \quad (t>0\,,\ y>0\,), \tag{27.1.75}$$

$$p(0,x;0+,y) = \delta(y-x) \quad (\,y \ge 0\,) \tag{27.1.76}$$

where $D = \sigma^2/2$ and $x \ge 0$ is given and fixed (recall that δ denotes the Dirac delta function). Standard results (cf. [64]) imply that there is a unique non-negative solution $(t,y) \mapsto p(0,x;t,y)$ of (27.1.75)–(27.1.76). The solution p satisfies the following boundary conditions:

$$p(0,x;t,0+) = 0 \quad (0 \text{ is entrance}\,), \tag{27.1.77}$$

$$p(0,x;t,\infty-) = 0 \quad (\,\infty \text{ is normal}\,). \tag{27.1.78}$$

The solution p satisfies the following integrability condition:

$$\int_0^\infty p(0,x;t,y)\,dy = 1 \tag{27.1.79}$$

for all $x \geq 0$ and all $t \geq 0$. Once the solution $(t, y) \mapsto p(0, x; t, y)$ of (27.1.75)–(27.1.76) has been found, the functions F, G, H from (27.1.18)–(27.1.20) can be computed using the general formula

$$\widetilde{\mathsf{E}}_{0,x}\big(g(X_t)\big) = \int_0^\infty g(y)\, p(0, x; t, y)\, dy \tag{27.1.80}$$

upon choosing the appropriate function $g : \mathbb{R}_+ \to \mathbb{R}_+$.

The *backward equation* approach leads to the terminal-value problem

$$q_t = (1 - rx)\, q_x + D\, x^2\, q_{xx} \quad (\, t > 0,\, x > 0\,), \tag{27.1.81}$$

$$q(T, x) = h(x) \quad (\, x \geq 0\,) \tag{27.1.82}$$

where $h : \mathbb{R}_+ \to \mathbb{R}_+$ is a given function. Standard results (cf. [64]) imply that there is a unique non-negative solution $(t, x) \mapsto q(t, x)$ of (27.1.81)–(27.1.82). Taking $x \mapsto h(x)$ to be $x \mapsto (1 - x/T)^+$ (with T fixed), $x \mapsto x\, I(x \leq y)$ (with y fixed), $x \mapsto I(x \leq y)$ (with y fixed) it follows that the unique non-negative solution q of (27.1.81)–(27.1.82) coincides with F, G, H from (27.1.18)–(27.1.20) respectively. (For numerical results of a similar approach see [177].)

4°. It is an interesting numerical problem to carry out either of the two methods described above and produce approximations to the optimal stopping boundary b using (27.1.73). Another interesting problem is to derive the rate of convergence.

4. Appendix. In this appendix we exhibit an explicit expression for the probability density function f of (S_t, I_t) under $\widetilde{\mathsf{P}}$ with $S_0 = 1$ and $I_0 = 0$ given in (27.1.21) above.

Let $B = (B_t)_{t \geq 0}$ be a standard Brownian motion defined on a probability space $(\Omega, \mathcal{F}, \mathsf{P})$. With $t > 0$ and $\nu \in \mathbb{R}$ given and fixed recall from [224, p. 527] that the random variable $A_t^{(\nu)} = \int_0^t e^{2(B_s + \nu s)}\, ds$ has the conditional distribution

$$\mathsf{P}\big(A_t^{(\nu)} \in dy \,\big|\, B_t + \nu t = x\big) = a(t, x, y)\, dy \tag{27.1.83}$$

where the density function a for $y > 0$ is given by

$$\begin{aligned} a(t, x, y) = {} & \frac{1}{\pi y^2} \exp\left(\frac{x^2 + \pi^2}{2t} + x - \frac{1}{2y}\big(1 + e^{2x}\big)\right) \\ & \times \int_0^\infty \exp\left(-\frac{z^2}{2t} - \frac{e^x}{y}\cosh(z)\right) \sinh(z)\, \sin\left(\frac{\pi z}{t}\right) dz. \end{aligned} \tag{27.1.84}$$

This implies that the random vector $\big(2(B_t + \nu t), A_t^{(\nu)}\big)$ has the distribution

$$\mathsf{P}\Big(2(B_t + \nu t) \in dx,\, A_t^{(\nu)} \in dy\Big) = b(t, x, y)\, dx\, dy \tag{27.1.85}$$

where the density function b for $y > 0$ is given by

$$
\begin{aligned}
b(t,x,y) &= a\Big(t,\frac{x}{2},y\Big)\,\frac{1}{2\sqrt{t}}\,\varphi\Big(\frac{x-2\nu t}{2\sqrt{t}}\Big) \\
&= \frac{1}{(2\pi)^{3/2}y^2\sqrt{t}}\,\exp\Big(\frac{\pi^2}{2t}+\Big(\frac{\nu+1}{2}\Big)x-\frac{\nu^2}{2}\,t-\frac{1}{2y}(1+e^x)\Big) \\
&\quad\times\int_0^\infty \exp\Big(-\frac{z^2}{2t}-\frac{e^{x/2}}{y}\cosh(z)\Big)\sinh(z)\,\sin\Big(\frac{\pi z}{t}\Big)\,dz
\end{aligned}
\tag{27.1.86}
$$

and we set $\varphi(z) = (1/\sqrt{2\pi})e^{-z^2/2}$ for $z \in \mathbb{R}$ (for related expressions in terms of Hermite functions see [46] and [181]).

Denoting $K_t = \alpha B_t + \beta t$ and $L_t = \int_0^t e^{\alpha B_s + \beta s} ds$ with $\alpha \neq 0$ and $\beta \in \mathbb{R}$ given and fixed, and using that the scaling property of B implies

$$
\begin{aligned}
&\mathsf{P}\Big(\alpha B_t+\beta t\le x,\ \int_0^t e^{\alpha B_s+\beta s}\,ds\le y\Big) \\
&\qquad\qquad = \mathsf{P}\Big(2(B_{t'}+\nu t')\le x,\ \int_0^{t'} e^{2(B_s+\nu s)}\,ds\le\frac{\alpha^2}{4}\,y\Big)
\end{aligned}
\tag{27.1.87}
$$

with $t' = \alpha^2 t/4$ and $\nu = 2\beta/\alpha^2$, it follows by applying (27.1.85) and (27.1.86) that the random vector (K_t, L_t) has the distribution

$$
\mathsf{P}\big(K_t \in dx, L_t \in dy\big) = c(t,x,y)\,dx\,dy \tag{27.1.88}
$$

where the density function c for $y > 0$ is given by

$$
\begin{aligned}
c(t,x,y) &= \frac{\alpha^2}{4}\,b\Big(\frac{\alpha^2}{4}\,t,\ x,\ \frac{\alpha^2}{4}\,y\Big) \\
&= \frac{2\sqrt{2}}{\pi^{3/2}\alpha^3}\,\frac{1}{y^2\sqrt{t}}\,\exp\Big(\frac{2\pi^2}{\alpha^2 t}+\Big(\frac{\beta}{\alpha^2}+\frac{1}{2}\Big)x-\frac{\beta^2}{2\alpha^2}\,t-\frac{2}{\alpha^2 y}(1+e^x)\Big) \\
&\quad\times\int_0^\infty \exp\Big(-\frac{2z^2}{\alpha^2 t}-\frac{4e^{x/2}}{\alpha^2 y}\cosh(z)\Big)\sinh(z)\,\sin\Big(\frac{4\pi z}{\alpha^2 t}\Big)\,dz.
\end{aligned}
\tag{27.1.89}
$$

From (27.1.8) and (27.1.3) we see that

$$
f(t,s,a) = \frac{1}{s}\,c(t,\log s,a) = \frac{1}{s}\,\frac{\alpha^2}{4}\,b\Big(\frac{\alpha^2}{4}\,t,\ \log s,\ \frac{\alpha^2}{4}\,a\Big) \tag{27.1.90}
$$

with $\alpha = \sigma$ and $\beta = r + \sigma^2/2$. Hence (27.1.21) follows by the final expression in (27.1.86).

Notes. According to financial theory (see e.g. [197]) the arbitrage-free price of the *early* exercise *Asian call option* with *floating strike* is given as V in (27.1.1) above where I_τ/τ denotes the *arithmetic* average of the stock price S up to time τ. The problem was first studied in [89] where approximations to the value function V and the optimal boundary b were derived. The main aim of the present section (following [170]) is to derive exact expressions for V and b.

The optimal stopping problem (27.1.1) is three-dimensional. When a change of measure is applied (as in [186] and [115]) the problem reduces to (27.1.9) and becomes two-dimensional. The problem (27.1.9) is more complicated than the well-known problems of American and Russian options (cf. Sections 25 and 26 above) since the gain function depends on time in a nonlinear way. From the result of Theorem 27.1 above it follows that the free-boundary problem (27.1.10)–(27.1.14) characterizes the value function V and the optimal stopping boundary b in a unique manner. Our main aim, however, is to follow the train of thought initiated by Kolodner [114] where V is initially expressed in terms of b, and b itself is then shown to satisfy a nonlinear integral equation. A particularly simple approach for achieving this goal in the case of the American put option has been suggested in [110], [102], [27] and we take it up in the present section. We moreover see (as in [164] and [165]) that the nonlinear equation derived for b cannot have other solutions. The key argument in the proof relies upon a local time-space formula (see Subsection 3.5).

The latter fact of *uniqueness* may be seen as the principal result of the section. The same method of proof can also be used to show the uniqueness of the optimal stopping boundary solving nonlinear integral equations derived in [89] and [223] where this question was not explicitly addressed. These equations arise from the early exercise Asian options (call or put) with floating strike based on *geometric* averaging. The early exercise Asian *put* option with floating strike can be dealt with analogously to the Asian call option treated here. For financial interpretations of the early exercise Asian options and other references on the topic see [89] and [223].

Chapter VIII.

Optimal stopping in financial engineering

28. Ultimate position

The problem to be discussed in this section is motivated by the optimal stopping problem studied in Section 30 below and our wish to cover the *Mayer formulation* of the same problem (cf. Section 6). Since the gain process in the optimal stopping problem depends on the future, we refer to it as an *optimal prediction* problem. These problems appear to be of particular interest in financial engineering.

1. Let $B = (B_t)_{0 \le t \le 1}$ be a standard Brownian motion defined on a probability space $(\Omega, \mathcal{F}, \mathsf{P})$, and let $M : \mathbb{R} \to \mathbb{R}$ be a measurable (continuous) function such that $\mathsf{E}\, M(B_1)^2 < \infty$.

Consider the optimal prediction problem

$$V = \inf_{0 \le \tau \le 1} \mathsf{E} \big(M(B_1) - B_\tau \big)^2 \tag{28.0.1}$$

where the infimum is taken over all stopping times τ of B (satisfying $0 \le \tau \le 1$). Note that $M(B_1)$ is not adapted to the natural filtration $\mathcal{F}_t^B = \sigma(B_s : 0 \le s \le t)$ of B for $t \in [0,1\rangle$ so that the problem (28.0.1) falls outside the scope of general theory of optimal stopping from Chapter I.

The following simple arguments reduce the optimal prediction problem (28.0.1) to an optimal stopping problem (in terms of the general optimal stopping theory). For this, note that

$$\begin{aligned} \mathsf{E}\Big(\big(M(B_1) - B_t\big)^2 \,\big|\, \mathcal{F}_t^B \Big) &= \mathsf{E}\Big(\big(M(B_1 - B_t + B_t) - B_t\big)^2 \,\big|\, \mathcal{F}_t^B \Big) \\ &= \mathsf{E}\big(M(B_{1-t} + x) - x \big)^2 \big|_{x = B_t} \end{aligned} \tag{28.0.2}$$

upon using that $B_1 - B_t$ is independent from $\mathcal{F}_t^B$ and equally distributed as B_{1-t}.

Let

$$G(t,x) = \mathsf{E}\big(M(B_{1-t}+x)-x\big)^2 = \mathsf{E}\big(M(\sqrt{1-t}\,B_1+x)-x\big)^2 \qquad (28.0.3)$$
$$= \int_{\mathbb{R}} \big(M(\sqrt{1-t}\,y+x)-x\big)^2 \varphi(y)\,dy$$

where we use that $B_{1-t} =^{\text{law}} \sqrt{1-t}\,B_1$ and set

$$\varphi(y) = \frac{1}{\sqrt{2\pi}}\, e^{-y^2/2} \qquad (28.0.4)$$

for $y \in \mathbb{R}$ to denote the standard normal density function. We get from (28.0.2) and (28.0.3) that

$$\mathsf{E}\Big(\big(M(B_1)-B_t\big)^2 \,\big|\, \mathcal{F}_t^B\Big) = G(t,B_t) \qquad (28.0.5)$$

for $0 \le t \le 1$.

2. Standard arguments based on the fact that each stopping time is the limit of a decreasing sequence of discrete stopping times imply that (28.0.5) extends as follows:

$$\mathsf{E}\Big(\big(M(B_1)-B_\tau\big)^2 \,\big|\, \mathcal{F}_\tau^B\Big) = G(\tau,B_\tau) \qquad (28.0.6)$$

for all stopping times τ of B. Taking E in (28.0.6) we find that the optimal prediction problem (28.0.1) is equivalent to the optimal stopping problem

$$V = \inf_{0\le\tau\le 1} \mathsf{E}\, G(\tau,B_\tau) \qquad (28.0.7)$$

where the infimum is taken over all stopping times τ of B (satisfying $0 \le \tau \le 1$).

This problem can be treated by the methods of Chapters VI and VII. We will omit further details. (Note that when $M(x) = x$ for all $x \in \mathbb{R}$, then the optimal stopping time τ_* is trivial as it equals 1 identically.)

29. Ultimate integral

The problem to be discussed in this section (similarly to the previous section) is motivated by the optimal prediction problem studied in Section 30 below and our wish to cover the *Lagrange formulation* of the same problem (cf. Section 6).

1. Let $B = (B_t)_{0\le t\le 1}$ be a standard Brownian motion defined on a probability space $(\Omega,\mathcal{F},\mathsf{P})$, and let $L : \mathbb{R} \to \mathbb{R}$ be a measurable (continuous) function such that $\mathsf{E}\big(\int_0^1 L(B_t)\,dt\big)^2 < \infty$.

Consider the optimal prediction problem

$$V = \inf_{0\le\tau\le 1} \mathsf{E}\Big(\int_0^1 L(B_t)\,dt - B_\tau\Big)^2 \qquad (29.0.8)$$

where the infimum is taken over all stopping times τ of B (satisfying $0 \le \tau \le 1$). Note that $\int_0^1 L(B_t)\,dt$ is not adapted to the natural filtration $\mathcal{F}_t^B = \sigma(B_s : 0 \le s \le t)$ of B for $t \in [0,1)$ so that the problem (29.0.8) falls outside the scope of general theory of optimal stopping from Chapter I.

The following simple arguments reduce the optimal prediction problem (29.0.8) to an optimal stopping problem (in terms of the general optimal stopping theory). In the sequel we will assume that L is continuous. Set $M(x) = \int_0^x \int_0^y L(z)\,dz\,dy$ for $x \in \mathbb{R}$. Then M is C^2 and Itô's formula (page 67) yields

$$M(B_t) = M(0) + \int_0^t M'(B_s)\,dB_s + \frac{1}{2}\int_0^t M''(B_s)\,ds. \tag{29.0.9}$$

Hence we find

$$\begin{aligned}\int_0^1 L(B_s)\,ds &= \int_0^1 M''(B_s)\,ds \\ &= 2\Big(M(B_1) - M(0) - \int_0^1 M'(B_s)\,dB_s\Big).\end{aligned} \tag{29.0.10}$$

Inserting (29.0.10) into (29.0.8) we get

$$\begin{aligned}&\mathsf{E}\Big[\int_0^1 L(B_t)\,dt - B_\tau\Big]^2 \\ &\quad= \mathsf{E}\Big[2\Big(M(B_1) - M(0) - \int_0^1 M'(B_t)\,dB_t\Big) - B_\tau\Big]^2 \\ &\quad= 4\,\mathsf{E}\Big[M(B_1) - M(0) - \int_0^1 M'(B_t)\,dB_t\Big]^2 \\ &\qquad - 4\,\mathsf{E}(M(B_1)B_\tau) - 4\,\mathsf{E}\Big[\Big(\int_0^1 M'(B_t)\,dB_t\Big)B_\tau\Big] + \mathsf{E}\,B_\tau^2.\end{aligned} \tag{29.0.11}$$

By (29.0.10) we have

$$\begin{aligned}&\mathsf{E}\Big(M(B_1) - M(0) - \int_0^1 M'(B_t)\,dB_t\Big)^2 \\ &\quad= \frac{1}{4}\,\mathsf{E}\Big(\int_0^1 L(B_t)\,dt\Big)^2 =: C_L.\end{aligned} \tag{29.0.12}$$

By stationary and independent increments of B (just as in (28.0.2)–(28.0.7) in Section 28 above) we get

$$\begin{aligned}\mathsf{E}(M(B_1)B_\tau) &= \mathsf{E}\big(M(B_1 - B_\tau + B_\tau)B_\tau\big) \\ &= \mathsf{E}\Big(M\big(\sqrt{1-t}\,B_1 + x\big)\Big)\Big|_{t=\tau,x=B_\tau} \\ &= G(\tau, B_\tau)\end{aligned} \tag{29.0.13}$$

for all stopping times τ of B (satisfying $0 \le \tau \le 1$). Finally, by the martingale property of $\int_0^t M'(B_s)\,dB_s$ for $t \in [0,1]$ we obtain

$$\begin{aligned}\mathsf{E}\Big[\Big(\int_0^1 M'(B_t)\,dB_t\Big)B_\tau\Big] &= \mathsf{E}\Big[\Big(\int_0^\tau M'(B_t)\,dB_t\Big)B_\tau\Big] \\ &= \mathsf{E}\int_0^\tau M'(B_t)\,dt\end{aligned} \tag{29.0.14}$$

as long as $\mathsf{E}\int_0^1 (M'(B_t))^2\,dt < \infty$ for instance. Inserting (29.0.12)–(29.0.14) into (29.0.11) and using that $\mathsf{E}B_\tau^2 = \mathsf{E}\tau$, we get

$$\begin{aligned}&\mathsf{E}\Big[\int_0^1 L(B_t)\,dt - B_\tau\Big]^2 \\ &= 4\Big(C_L - \mathsf{E}G(\tau,B_\tau) - \mathsf{E}\int_0^\tau M'(B_t)\,dt\Big) + \mathsf{E}\tau.\end{aligned} \tag{29.0.15}$$

Setting $H = M' - \frac{1}{4}$ this shows that the optimal prediction problem (29.0.8) is equivalent to the optimal stopping problem

$$\widetilde{V} = \sup_{0\le\tau\le 1} \mathsf{E}\Big(G(\tau,B_\tau) + \int_0^\tau H(B_t)\,dt\Big) \tag{29.0.16}$$

where the supremum is taken over all stopping times τ of B (satisfying $0 \le \tau \le 1$).

2. Consider the case when $L(x) = x$ for all $x \in \mathbb{R}$ in the problem (29.0.8). Setting $I_1 = \int_0^1 B_t\,dt$ we find by the integration by parts formula (or Itô's formula applied to tB_t and letting $t = 1$ in the result) that the following analogue of the formula (30.1.7) below is valid:

$$I_1 = \int_0^1 (1-t)\,dB_t. \tag{29.0.17}$$

Denoting $M_t = \int_0^t (1-s)\,dB_s$, it follows by the martingale property of the latter for $t \in [0,1]$ that

$$\begin{aligned}\mathsf{E}(I_1 - B_\tau)^2 &= \mathsf{E}|I_1|^2 - 2\,\mathsf{E}(I_1 B_\tau) + \mathsf{E}|B_\tau|^2 \\ &= \frac{1}{3} - 2\,\mathsf{E}\Big[\int_0^\tau (1-s)\,ds\Big] + \mathsf{E}\tau \\ &= \frac{1}{3} + \mathsf{E}(\tau^2 - 2\tau) + \mathsf{E}\tau = \frac{1}{3} + \mathsf{E}(\tau^2 - \tau)\end{aligned} \tag{29.0.18}$$

for all stopping time τ of B (satisfying $0 \le \tau \le 1$). Hence we see that

$$V = \inf_{0\le\tau\le 1} \mathsf{E}(I_1 - B_\tau)^2 = \frac{1}{12} = 0.08\ldots \tag{29.0.19}$$

and that the infimum is attained at $\tau_* \equiv 1/2$. This shows that the problem (29.0.8) with $L(x) = x$ for $x \in \mathbb{R}$ has a trivial solution.

30. Ultimate maximum

Imagine the real-line movement of a Brownian particle started at 0 during the time interval $[0,1]$. Let S_1 denote the *maximal* positive height that the particle ever reaches during this time interval. As S_1 is a random quantity whose values depend on the *entire* Brownian path over the time interval, its *ultimate* value is at any given time $t \in [0,1)$ unknown. Following the Brownian particle from the initial time 0 onward, the question arises naturally of how to determine a time when the movement should be terminated so that the position of the particle at that time is as "close" as possible to the *ultimate maximum* S_1. In the next two subsections we present the solution to this problem if "closeness" is measured by a mean-square distance.

30.1. Free Brownian motion

1. To formulate the problem above more precisely, let $B = (B_t)_{0\le t\le 1}$ be a standard Brownian motion defined on a probability space $(\Omega, \mathcal{F}, \mathsf{P})$, and let $(\mathcal{F}_t^B)_{0\le t\le 1}$ denote the natural filtration generated by B. Letting $\mathfrak{M}$ denote the family of all stopping (Markov) times τ with respect to $(\mathcal{F}_t^B)_{0\le t\le 1}$ satisfying $0 \le \tau \le 1$, the problem is to compute

$$V_* = \inf_{\tau\in\mathfrak{M}} \mathsf{E}\Big(B_\tau - \max_{0\le t\le 1} B_t\Big)^2 \tag{30.1.1}$$

and to find an optimal stopping time (the one at which the infimum in (30.1.1) is attained).

The solution of this problem is presented in Theorem 30.1 below. It turns out that the *maximum* process $S = (S_t)_{0\le t\le 1}$ given by

$$S_t = \sup_{0\le s\le t} B_s \tag{30.1.2}$$

and the *CUSUM-type* (reflected) process $S - B = (S_t - B_t)_{0\le t\le 1}$ play a key role in the solution.

The optimal stopping problem (30.1.1) is of interest, for example, in financial engineering where an optimal decision (i.e. optimal stopping time) should be based on a prediction of the time when the observed process take its maximal value (over a given time interval). The argument also carries over to many other applied problems where such predictions play a role.

2. The main result of this subsection is contained in the following theorem. Below we let

$$\varphi(x) = \frac{1}{\sqrt{2\pi}}\, e^{-x^2/2} \;\;\&\;\; \Phi(x) = \int_{-\infty}^{x} \varphi(y)\, dy \qquad (x \in \mathbb{R}) \tag{30.1.3}$$

denote the density and distribution function of a standard normal variable.

Theorem 30.1. *Consider the optimal stopping problem* (30.1.1) *where* $B = (B_t)_{0 \le t \le 1}$ *is a standard Brownian motion. Then the value* V_* *is given by the formula*

$$V_* = 2\Phi(z_*) - 1 = 0.73\ldots \tag{30.1.4}$$

where $z_* = 1.12\ldots$ *is the unique root of the equation*

$$4\Phi(z_*) - 2z_*\varphi(z_*) - 3 = 0 \tag{30.1.5}$$

and the following stopping time is optimal (*see Figures* VIII.2–VIII.5):

$$\tau_* = \inf\left\{ 0 \le t \le 1 : S_t - B_t \ge z_* \sqrt{1-t} \right\} \tag{30.1.6}$$

where S_t *is given by* (30.1.2) *above.*

Proof. Since $S_1 = \sup_{0 \le s \le 1} B_s$ is a square-integrable functional of the Brownian path on $[0,1]$, by the Itô–Clark representation theorem (see e.g. [174, p. 199]) there exists a unique $(\mathcal{F}_t^B)_{0 \le t \le 1}$-adapted process $H = (H_t)_{0 \le t \le 1}$ satisfying $\mathsf{E}(\int_0^1 H_t^2\, dt) < \infty$ such that

$$S_1 = a + \int_0^1 H_t \, dB_t \tag{30.1.7}$$

where $a = \mathsf{E} S_1$. Moreover, the following explicit formula is known to be valid:

$$H_t = 2\left(1 - \Phi\left(\frac{S_t - B_t}{\sqrt{1-t}}\right)\right) \tag{30.1.8}$$

for $0 \le t \le 1$ (see e.g. [178, p. 93] and [107, p. 365] or paragraph 3 below for a direct argument).

1°. Associate with H the square-integrable martingale $M = (M_t)_{0 \le t \le 1}$ given by

$$M_t = \int_0^t H_s \, dB_s. \tag{30.1.9}$$

By the martingale property of M and the optional sampling theorem (page 60), we obtain

$$\begin{aligned} \mathsf{E}(B_\tau - S_1)^2 &= \mathsf{E}|B_\tau|^2 - 2\mathsf{E}(B_\tau M_1) + \mathsf{E}|S_1|^2 \\ &= \mathsf{E}\tau - 2\mathsf{E}(B_\tau M_\tau) + 1 = \mathsf{E}\left(\int_0^\tau (1 - 2H_t)\, dt \right) + 1 \end{aligned} \tag{30.1.10}$$

for all $\tau \in \mathfrak{M}$ (recall that $S_1 \overset{\text{law}}{=} |B_1|$). Inserting (30.1.8) into (30.1.10) we see that (30.1.1) can be rewritten as

$$V_* = \inf_{\tau \in \mathfrak{M}} \mathsf{E}\left(\int_0^\tau F\left(\frac{S_t - B_t}{\sqrt{1-t}} \right) dt \right) + 1 \tag{30.1.11}$$

where we denote $F(x) = 4\Phi(x) - 3$.

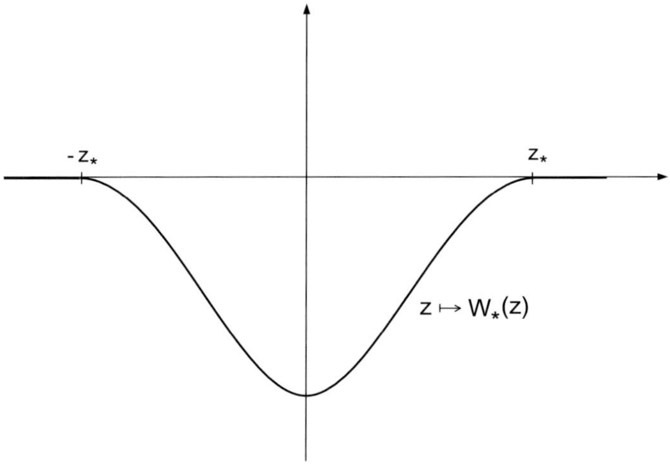

Figure VIII.1: A computer drawing of the map (30.1.19). The smooth fit (30.1.23) holds at $-z_*$ and z_* .

Since $S - B = (S_t - B_t)_{0 \le t \le 1}$ is a Markov process for which the natural filtration $(\mathcal{F}_t^{S-B})_{t \ge 0}$ coincides with the natural filtration $(\mathcal{F}^B)_{t \ge 0}$, it follows from general theory of optimal stopping (see Subsection 2.2) that in (30.1.11) we need only consider stopping times which are hitting times for $S - B$. Recalling moreover that $S - B \stackrel{\text{law}}{=} |B|$ by Lévy's distributional theorem (see (4.4.24)) and once more appealing to general theory, we see that (30.1.11) is equivalent to the optimal stopping problem

$$V_* = \inf_{\tau \in \mathfrak{M}} \mathsf{E}\bigg(\int_0^\tau F\bigg(\frac{|B_t|}{\sqrt{1-t}} \bigg)\, dt \bigg) + 1. \tag{30.1.12}$$

In our treatment of this problem, we first make use of a deterministic change of time (cf. Subsection 5.1 and Section 10).

2°. Motivated by the form of (30.1.12), consider the process $Z = (Z_t)_{t \ge 0}$ given by

$$Z_t = e^t B_{1-e^{-2t}}. \tag{30.1.13}$$

By Itô's formula (page 67) we find that Z is a (strong) solution of the linear stochastic differential equation

$$dZ_t = Z_t\, dt + \sqrt{2}\, d\beta_t \tag{30.1.14}$$

where the process $\beta = (\beta_t)_{0 \le t \le 1}$ is given by

$$\beta_t = \frac{1}{\sqrt{2}} \int_0^t e^s\, dB_{1-e^{-2s}} = \frac{1}{\sqrt{2}} \int_0^{1-e^{-2t}} \frac{1}{\sqrt{1-s}}\, dB_s. \tag{30.1.15}$$

As β is a continuous Gaussian martingale with mean zero and variance equal to t, it follows by Lévy's characterization theorem (see e.g. [174, p. 150]) that β is a standard Brownian motion. We thus may conclude that Z is a diffusion process with the infinitesimal generator given by

$$\mathbb{L}_Z = z\frac{d}{dz} + \frac{d^2}{dz^2}. \tag{30.1.16}$$

Substituting $t = 1 - e^{-2s}$ in (30.1.12) and using (30.1.13), we obtain

$$V_* = 2 \inf_{\tau \in \mathfrak{M}} \mathsf{E}\Big(\int_0^{\sigma_\tau} e^{-2s} F\big(|Z_s|\big)\, ds \Big) + 1 \tag{30.1.17}$$

upon setting $\sigma_\tau = \log(1/\sqrt{1-\tau})$. It is clear from (30.1.13) that τ is a stopping time with respect to $(\mathcal{F}_t^B)_{0 \le t \le 1}$ if and only if σ_τ is a stopping time with respect to $(\mathcal{F}_s^Z)_{s \ge 0}$. This shows that our initial problem (30.1.1) reduces to solving

$$W_* = \inf_{\sigma} \mathsf{E}\Big(\int_0^{\sigma} e^{-2s} F\big(|Z_s|\big)\, ds \Big) \tag{30.1.18}$$

where the infimum is taken over all $(\mathcal{F}_s^Z)_{s \ge 0}$-stopping times σ with values in $[0, \infty]$. This problem belongs to the general theory of optimal stopping for *time-homogeneous* Markov processes (see Subsection 2.2).

3°. To calculate (30.1.18) define

$$W_*(z) = \inf_{\sigma} \mathsf{E}_z\Big(\int_0^{\sigma} e^{-2s} F\big(|Z_s|\big)\, ds \Big) \tag{30.1.19}$$

for $z \in \mathbb{R}$, where $Z_0 = z$ under P_z, and the infimum is taken as above. General theory combined with basic properties of the map $z \mapsto F(|z|)$ prompts that the stopping time

$$\sigma_* = \inf\{\, t > 0 : |Z_t| \ge z_* \,\} \tag{30.1.20}$$

should be optimal in (30.1.19), where $z_* > 0$ is a constant to be found.

To determine z_* and compute the value function $z \mapsto W_*(z)$ in (30.1.19), it is a matter of routine to formulate the following free-boundary problem:

$$\big(\mathbb{L}_Z - 2\big) W(z) = -F(|z|) \quad \text{for } z \in (-z_*, z_*), \tag{30.1.21}$$

$$W(\pm z_*) = 0 \quad \text{(instantaneous stopping)}, \tag{30.1.22}$$

$$W'(\pm z_*) = 0 \quad \text{(smooth fit)} \tag{30.1.23}$$

where $\mathbb{L}_Z$ is given by (30.1.16) above. We shall extend the solution of (30.1.21)–(30.1.23) by setting its value equal to 0 for $z \notin (-z_*, z_*)$, and thus the map so obtained will be C^2 everywhere on $\mathbb{R}$ but at $-z_*$ and z_* where it is C^1.

Inserting $\mathbb{L}_Z$ from (30.1.16) into (30.1.21) leads to the equation

$$W''(z) + zW'(z) - 2W(z) = -F(|z|) \tag{30.1.24}$$

for $z \in (-z_*, z_*)$. The form of the equation (30.1.14) and the value (30.1.18) indicates that $z \mapsto W_*(z)$ should be even; thus we shall additionally impose

$$W'(0) = 0 \tag{30.1.25}$$

and consider (30.1.24) only for $z \in [0, z_*)$.

The general solution of the equation (30.1.24) for $z \geq 0$ is given by

$$W(z) = C_1(1+z^2) + C_2\big(z\varphi(z) + (1+z^2)\Phi(z)\big) + 2\Phi(z) - \tfrac{3}{2}. \tag{30.1.26}$$

The three conditions $W(z_*) = W'(z_*) = W'(0) = 0$ determine constants C_1, C_2 and z_* uniquely; it is easily verified that $C_1 = \Phi(z_*)$, $C_2 = -1$, and z_* is the unique root of the equation (30.1.5). Inserting this back into (30.1.24), we obtain the following candidate for the value (30.1.19):

$$W(z) = \Phi(z_*)(1+z^2) - z\varphi(z) + (1 - z^2)\Phi(z) - \tfrac{3}{2} \tag{30.1.27}$$

when $z \in [0, z_*]$, upon extending it to an even function on $\mathbb{R}$ as indicated above (see *Figure* VIII.1).

To verify that this solution $z \mapsto W(z)$ coincides with the value function (30.1.19), and that σ_* from (30.1.20) is an optimal stopping time, we shall note that $z \mapsto W(z)$ is C^2 everywhere but at $\pm z_*$ where it is C^1. Thus by the Itô–Tanaka–Meyer formula (page 68) we find

$$\begin{aligned} e^{-2t}W(Z_t) = W(Z_0) &+ \int_0^t e^{-2s}\Big(\mathbb{L}_Z W(Z_s) - 2W(Z_s)\Big)\, ds \\ &+ \sqrt{2}\int_0^t e^{-2s}W'(Z_s)\, d\beta_s. \end{aligned} \tag{30.1.28}$$

Hence by (30.1.24) and the fact that $\mathbb{L}_Z W(z) - 2W(z) = 0 > -F(|z|)$ for $z \notin [-z_*, z_*]$, upon extending W'' to $\pm z_*$ as we please and using that the Lebesgue measure of those $t > 0$ for which $Z_t = \pm z_*$ is zero, we get

$$e^{-2t}W(Z_t) \geq W(Z_0) - \int_0^t e^{-2s}F(|Z_s|)\, ds + M_t \tag{30.1.29}$$

where $M_t = \sqrt{2}\int_0^t e^{-2s}W'(Z_s)\, d\beta_s$ is a continuous local martingale for $t \geq 0$. Using further that $W(z) \leq 0$ for all z, a simple application of the optional sampling theorem (page 60) in the stopped version of (30.1.29) under P_z shows that $W_*(z) \geq W(z)$ for all z. To prove equality one may note that the passage from (30.1.28) to (30.1.29) also yields

$$0 = W(Z_0) - \int_0^{\sigma_*} e^{-2s}F(|Z_s|)\, ds + M_{\sigma_*} \tag{30.1.30}$$

upon using (30.1.21) and (30.1.22). Since clearly $\mathsf{E}_z\sigma_* < \infty$ and thus $\mathsf{E}_z\sqrt{\sigma_*} < \infty$ as well, and $z \mapsto W'(z)$ is bounded on $[-z_*, z_*]$, we can again apply the optional sampling theorem and conclude that $\mathsf{E}_z M_{\sigma_*} = 0$. Taking the expectation under P_z on both sides in (30.1.30) enables one therefore to conclude $W_*(z) = W(z)$ for all z, and the proof of the claim is complete.

From (30.1.17)–(30.1.19) and (30.1.27) we find that $V_* = 2W_*(0) + 1 = 2(\Phi(z_*) - 1) + 1 = 2\Phi(z_*) - 1$. This establishes (30.1.4). Transforming σ_* from (30.1.20) back to the initial problem via the equivalence of (30.1.11), (30.1.12) and (30.1.17), we see that τ_* from (30.1.6) is optimal. The proof is complete. □

Remark 30.2. Recalling that $S - B \overset{\text{law}}{=} |B|$ we see that τ_* is identically distributed as the stopping time $\widetilde{\tau} = \inf\{t > 0 : |B_t| = z_*\sqrt{1-t}\}$. This implies $\mathsf{E}\tau_* = \mathsf{E}\widetilde{\tau} = \mathsf{E}|B_{\widetilde{\tau}}|^2 = (z_*)^2\mathsf{E}(1-\widetilde{\tau}) = (z_*)^2(1-\mathsf{E}\tau_*)$, and hence we obtain

$$\mathsf{E}\tau_* = \frac{(z_*)^2}{1+(z_*)^2} = 0.55\ldots. \tag{30.1.31}$$

Moreover, using that $(B_t^4 - 6tB_t^2 + 3t^2)_{t\geq 0}$ is a martingale, similar arguments show that

$$\mathsf{E}(\tau_*)^2 = \frac{(z_*)^6 + 5(z_*)^4}{(1+(z_*)^2)(3+6(z_*)^2+(z_*)^4)} = 0.36\ldots. \tag{30.1.32}$$

From (30.1.31) and (30.1.32) we find

$$\mathrm{Var}(\tau_*) = \frac{2(z_*)^4}{(1+(z_*)^2)^2(3+6(z_*)^2+(z_*)^4)} = 0.05\ldots. \tag{30.1.33}$$

Remark 30.3. For the sake of comparison with (30.1.4) and (30.1.31) it is interesting to note that

$$V_0 = \inf_{0\leq t\leq 1} \mathsf{E}\Big(\Big(B_t - \max_{0\leq s\leq 1} B_s\Big)^2\Big) = \frac{1}{\pi} + \frac{1}{2} = 0.81\ldots \tag{30.1.34}$$

with the infimum being attained at $t = \frac{1}{2}$. For this, recall from (30.1.10) and (30.1.8) that

$$\mathsf{E}(B_t - S_1)^2 = \mathsf{E}\bigg(\int_0^t F\Big(\frac{S_s - B_s}{\sqrt{1-s}}\Big)\,ds\bigg) + 1 \tag{30.1.35}$$

where $F(x) = 4\Phi(x) - 3$. Using further that $S - B \overset{\text{law}}{=} |B|$, elementary calculations show

$$\begin{aligned}
\mathsf{E}(B_t - S_1)^2 &= 4\bigg(\int_0^t \mathsf{E}\Big(\Phi\Big(\frac{|B_s|}{\sqrt{1-s}}\Big)\Big)\,ds\bigg) - 3t + 1 \qquad (30.1.36)\\
&= 4\int_0^t \bigg(1 - \frac{1}{\pi}\arctan\sqrt{\frac{1-s}{s}}\bigg)\,ds - 3t + 1\\
&= -\frac{4}{\pi}\bigg(t\arctan\sqrt{\frac{1-t}{t}} + \frac{1}{2}\arctan\sqrt{\frac{t}{1-t}} - \frac{1}{2}\sqrt{t\,(1-t)}\bigg) + t + 1.
\end{aligned}$$

Hence (30.1.34) is easily verified by standard means.

Remark 30.4. In view of the fact that σ_* from (30.1.20) with $z_* = 1.12\dots$ from (30.1.5) is optimal in the problem (30.1.19), it is interesting to observe that the unique solution of the equation $F(\hat{z}) = 0$ is given by $\hat{z} = 0.67\dots$. Noting moreover that the map $z \mapsto F(z)$ is increasing on $[0,\infty)$ and satisfies $F(0) = -1$, we see that $F(z) < 0$ for all $z \in [0,\hat{z})$ and $F(z) > 0$ for all $z > \hat{z}$. The size of the gap between $\hat{z}$ and z_* quantifies the tendency of the process $|Z|$ to return to the "favourable" set $[0,\hat{z})$ where clearly it is never optimal to stop.

Remark 30.5. The case of a general time interval $[0,T]$ easily reduces to the case of a unit time interval treated above by using the scaling property of Brownian motion implying

$$\inf_{0\le\tau\le T} \mathsf{E}\Big(B_\tau - \max_{0\le t\le T} B_t\Big)^2 = T \inf_{0\le\tau\le 1} \mathsf{E}\Big(B_\tau - \max_{0\le t\le 1} B_t\Big)^2 \tag{30.1.37}$$

which further equals to $T(2\Phi(z_*)-1)$ by (30.1.4). Moreover, the same argument shows that the optimal stopping time in (30.1.37) is given by

$$\tau_* = \inf\big\{\, 0 \le t \le T : B_t \ge z_*\sqrt{T-t}\,\big\} \tag{30.1.38}$$

where z_* is the same as in Theorem 30.1.

Remark 30.6. From the point of view of mathematical statistics, the "estimator" B_τ of S_1 is *biased*, since $\mathsf{E}B_\tau = 0$ for all $0 \le \tau \le 1$ but at the same time $\mathsf{E}S_1 \ne 0$. Instead of V_* and V_0 it is thus desirable to consider the values

$$\widetilde{V}_* = \inf_{a\in\mathbb{R},\, \tau\in\mathfrak{M}} \mathsf{E}\big(a + B_\tau - S_1\big)^2 \quad \& \quad \widetilde{V}_0 = \inf_{a\in\mathbb{R},\, 0\le t\le 1} \mathsf{E}\big(a + B_t - S_1\big)^2 \tag{30.1.39}$$

and compare them with the values from (30.1.1) and (30.1.34). However, by using that $\mathsf{E}B_\tau = 0$ we also find at once that $a_* = \mathsf{E}S_1$ is optimal in (30.1.39) with $\widetilde{V}_* = V_* - \frac{2}{\pi} = 0.09\dots$ and $\widetilde{V}_0 = V_0 - \frac{2}{\pi} = 0.18\dots$.

3. Stochastic integral representation of the maximum process. In this paragraph we present a *direct* derivation of the stochastic integral representation (30.1.7) and (30.1.8) (cf. [178, pp. 89–93] and [107, pp. 363–369]). For the sake of comparison we shall deal with a standard Brownian motion with *drift*

$$B_t^\mu = B_t + \mu t \tag{30.1.40}$$

for $0 \le t \le 1$ where μ is a real number. The maximum process $S^\mu = (S_t^\mu)_{0\le t\le 1}$ associated with $B^\mu = (B_t^\mu)_{0\le t\le 1}$ is given by

$$S_t^\mu = \sup_{0\le s\le t} B_s^\mu. \tag{30.1.41}$$

1°. To derive the analogue of (30.1.7) and (30.1.8) in this case, we shall first note that stationary independent increments of B^μ imply

$$\begin{aligned}\mathsf{E}\big(S_1^\mu \mid \mathcal{F}_t^B\big) &= S_t^\mu + \mathsf{E}\Big(\Big(\sup_{t\le s\le 1} B_s^\mu - S_t^\mu\Big)^+ \Big| \, \mathcal{F}_t^B\Big) \\ &= S_t^\mu + \mathsf{E}\Big(\Big(\sup_{t\le s\le 1} (B_s^\mu - B_t^\mu) - (S_t^\mu - B_t^\mu)\Big)^+ \Big| \, \mathcal{F}_t^B\Big) \\ &= S_t^\mu + \mathsf{E}\Big(S_{1-t}^\mu - (z-x)\Big)^+ \Big|_{z=S_t^\mu,\, x=B_t^\mu}.\end{aligned} \tag{30.1.42}$$

Using further the formula $\mathsf{E}(X-c)^+ = \int_c^\infty \mathsf{P}(X>z)\,dz$, we see that (30.1.42) reads

$$\mathsf{E}\big(S_1^\mu \mid \mathcal{F}_t^B\big) = S_t^\mu + \int_{S_t^\mu - B_t^\mu}^\infty \big(1 - F_{1-t}^\mu(z)\big)\,dz := f(t, B_t^\mu, S_t^\mu) \tag{30.1.43}$$

where we use the notation

$$F_{1-t}^\mu(z) = \mathsf{P}\big(S_{1-t}^\mu \le z\big), \tag{30.1.44}$$

and the map $f = f(t,x,s)$ is defined accordingly.

2°. Applying Itô's formula (page 67) to the right-hand side of (30.1.43), and using that the left-hand side defines a continuous martingale, we find upon setting $a_\mu = \mathsf{E}S_1^\mu$ that

$$\begin{aligned}\mathsf{E}\big(S_1^\mu \mid \mathcal{F}_t^B\big) &= a_\mu + \int_0^t \frac{\partial f}{\partial x}(s, B_s^\mu, S_s^\mu)\,dB_s \\ &= a_\mu + \int_0^t \big(1 - F_{1-s}^\mu(S_s^\mu - B_s^\mu)\big)\,dB_s,\end{aligned} \tag{30.1.45}$$

as a nontrivial continuous martingale cannot have paths of bounded variation. This reduces the initial problem to the problem of calculating (30.1.44).

3°. The following explicit formula is well known (see (4.4.21)):

$$F_{1-t}^\mu(z) = \Phi\Big(\frac{z - \mu(1-t)}{\sqrt{1-t}}\Big) - e^{2\mu z}\,\Phi\Big(\frac{-z-\mu(1-t)}{\sqrt{1-t}}\Big). \tag{30.1.46}$$

Inserting this into (30.1.45) we obtain the representation

$$S_1^\mu = a_\mu + \int_0^1 H_t^\mu \, dB_t \tag{30.1.47}$$

where the process H^μ is explicitly given by

$$\begin{aligned}H_t^\mu = 1 &- \Phi\Big(\frac{(S_t^\mu - B_t^\mu) - \mu(1-t)}{\sqrt{1-t}}\Big) \\ &+ e^{2\mu(S_t^\mu - B_t^\mu)}\,\Phi\Big(\frac{-(S_t^\mu - B_t^\mu) - \mu(1-t)}{\sqrt{1-t}}\Big).\end{aligned} \tag{30.1.48}$$

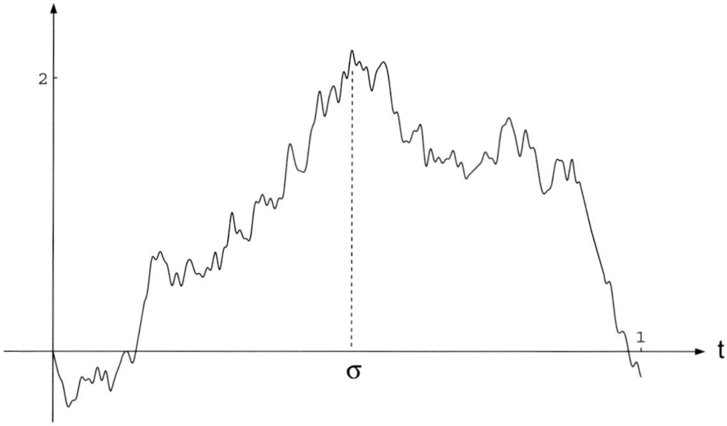

Figure VIII.2: A computer simulation of a Brownian path $(B_t(\omega))_{0\le t\le 1}$ with the maximum being attained at $\sigma = 0.51$.

Setting $\mu = 0$ in this expression, we recover (30.1.7) and (30.1.8).

4°. Note that the argument above extends to a large class of processes with stationary independent increments (Lévy processes) by reducing the initial problem to calculating the analogue of (30.1.44). In particular, the following "prediction" result deserves a special note. It is derived in exactly the same way as (30.1.43) above.

Let $X = (X_t)_{0\le t\le T}$ be a process with stationary independent increments started at zero, and let us denote $S_t = \sup_{0\le s\le t} X_s$ for $0 \le t \le T$. If $\mathsf{E}S_T < \infty$ then the predictor $\mathsf{E}(S_T \mid \mathcal{F}_t^X)$ of S_T based on the observations $\{X_s : 0 \le s \le t\}$ is given by the formula

$$\mathsf{E}\big(S_T \mid \mathcal{F}_t^X\big) = S_t + \int_{S_t - X_t}^{\infty} \big(1 - F_{T-t}(z)\big)\, dz \tag{30.1.49}$$

where $F_{T-t}(z) = \mathsf{P}(S_{T-t} \le z)$.

4. In the setting of the optimal prediction problem (30.1.1) above the following remarkable identity holds:

$$\mathsf{E}\,|\tau - \theta| = \mathsf{E}\,(B_\tau - B_\theta)^2 - \tfrac{1}{2} \tag{30.1.50}$$

for all stopping times τ of B (satisfying $0 \le \tau \le 1$) where θ is the (P -a.s. unique) time at which the maximum of B on $[0,1]$ is attained (i.e. $B_\theta = S_1$).

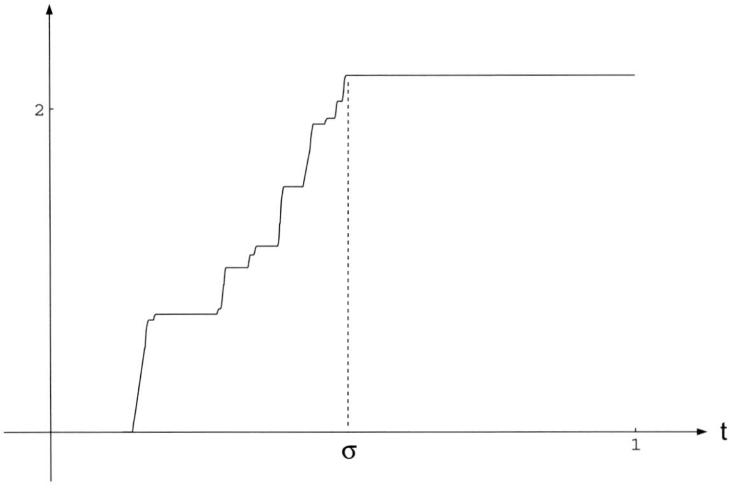

Figure VIII.3: A computer drawing of the maximum process $(S_t(\omega))_{0 \le t \le 1}$ associated with the Brownian path from *Figure* VIII.2.

To verify (30.1.50) note that

$$
\begin{aligned}
|\tau - \theta| &= (\tau - \theta)^+ + (\tau - \theta)^- = (\tau - \theta)^+ + \theta - \tau \wedge \theta \\
&= \int_0^\tau I(\theta \le t)\, dt + \theta - \int_0^\tau I(\theta > t)\, dt \\
&= \theta + \int_0^\tau (2I(\theta \le t) - 1)\, dt.
\end{aligned}
\tag{30.1.51}
$$

Taking E on both sides we get

$$
\begin{aligned}
\mathsf{E}\,|\tau - \theta| &= \mathsf{E}\theta + \mathsf{E} \int_0^\tau (2I(\theta \le t) - 1)\, dt \\
&= \frac{1}{2} + \mathsf{E} \int_0^\infty \big(2\,\mathsf{E}(\theta \le t \,|\, \mathcal{F}_t^B) - 1\big)\, I(t < \tau)\, dt \\
&= \frac{1}{2} + \mathsf{E} \int_0^\tau (2\pi_t - 1)\, dt
\end{aligned}
\tag{30.1.52}
$$

where we set

$$
\pi_t = \mathsf{P}(\theta \le t \,|\, \mathcal{F}_t^B). \tag{30.1.53}
$$

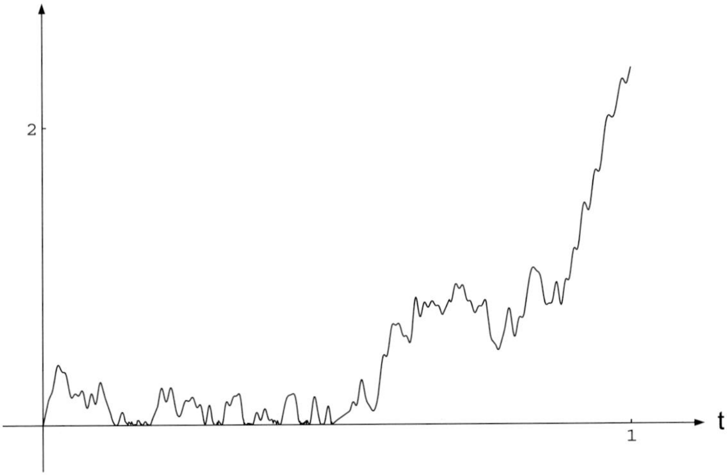

Figure VIII.4: A computer drawing of the process $(S_t(\omega) - B_t(\omega))_{0\le t\le 1}$ from *Figures* VIII.2 and VIII.3.

By stationary and independent increments of B, upon using (30.1.46) above with $\mu = 0$, we get

$$\begin{aligned}\pi_t &= \mathsf{P}\Big(S_t \le \max_{t\le s\le 1} B_s \,\Big|\, \mathcal{F}_t^B\Big) = \mathsf{P}\Big(S_t - B_t \le \max_{t\le s\le 1} B_s - B_t \,\Big|\, \mathcal{F}_t^B\Big) \\ &= \mathsf{P}(z - x \le S_{1-t})\big|_{z=S_t,\, x=B_t} = 1 - F_{1-t}(S_t - B_t) \\ &= 2\Phi\Big(\frac{S_t - B_t}{\sqrt{1-t}}\Big) - 1.\end{aligned} \tag{30.1.54}$$

Inserting (30.1.54) into (30.1.52) and using (30.1.8)+(30.1.10) above we see that (30.1.50) holds as claimed.

Finally, taking the infimum on both sides of (30.1.50) over all stopping times τ of B (satisfying $0 \le \tau \le 1$), we see that the stopping time τ_* given in (30.1.6) above is optimal for both (30.1.1) as well as the optimal prediction problem

$$W = \inf_{0\le\tau\le 1} \mathsf{E}\,|\tau - \theta| \tag{30.1.55}$$

where the infimum is taken over all stopping times τ of B (satisfying $0 \le \tau \le 1$). Thus, *not only is τ_* the optimal time to stop as close as possible to the ultimate maximum, but also τ_* is the optimal time to stop as close as possible to the time at which the ultimate maximum is attained.* This is indeed the most extraordinary feature of this particular stopping time.

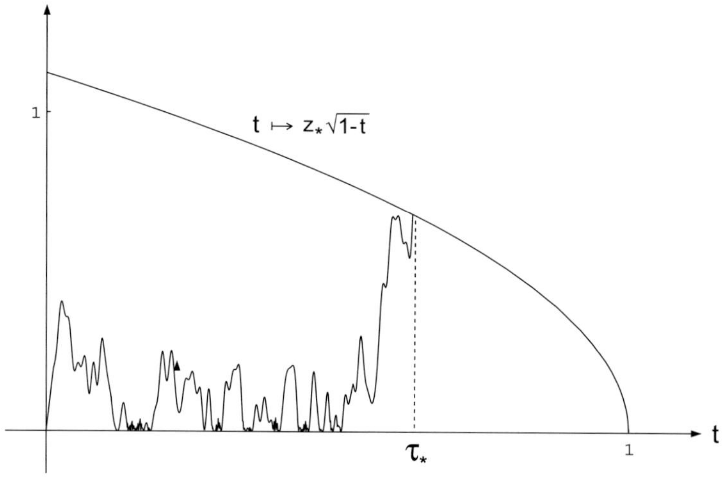

Figure VIII.5: A computer drawing of of the optimal stopping strategy (30.1.6) for the Brownian path from *Figures* VIII.2–VIII.4. It turns out that $\tau_* = 0.62$ in this case (cf. *Figure* VIII.2).

30.2. Brownian motion with drift

Let $B = (B_t)_{t \ge 0}$ be a standard Brownian motion defined on a probability space $(\Omega, \mathcal{F}, \mathsf{P})$ where $B_0 = 0$ under P. Set

$$B_t^\mu = B_t + \mu t \tag{30.2.1}$$

for $t \ge 0$ where $\mu \in \mathbb{R}$ is given and fixed. Then $B^\mu = (B_t^\mu)_{t \ge 0}$ is a standard Brownian motion with drift μ. Define

$$S_t^\mu = \max_{0 \le s \le t} B_s^\mu \tag{30.2.2}$$

for $t \ge 0$. Then $S^\mu = (S_t^\mu)_{t \ge 0}$ is the maximum process associated with B^μ.

1. The optimal prediction problem. Given $T > 0$ we consider the optimal prediction problem

$$V = \inf_{0 \le \tau \le T} \mathsf{E}\,(B_\tau^\mu - S_T^\mu)^2 \tag{30.2.3}$$

where the infimum is taken over all stopping times τ of B^μ (the latter means that τ is a stopping time with respect to the natural filtration of B^μ that in turn is the same as the natural filtration of B given by $\mathcal{F}_t^B = \sigma(B_s : 0 \le s \le t)$ for $t \in [0, T]$). The problem (30.2.3) consists of finding an optimal stopping time (at which the infimum is attained) and computing V as explicitly as possible.

1°. The identity (30.2.4) below reduces the optimal prediction problem (30.2.3) above (where the gain process $(B_t^\mu - S_T^\mu)_{0 \le t \le T}$ is not adapted to the natural filtration of B^μ) to the optimal stopping problem (30.2.10) below (where the gain process is adapted). Similar arguments in the case case $\mu = 0$ were used in Subsection 30.1 above.

Lemma 30.7. *The following identity holds* [*see* (30.1.43) *above*]:

$$\mathsf{E}\left((S_T^\mu - B_t^\mu)^2 \,|\, \mathcal{F}_t^B\right) = (S_t^\mu - B_t^\mu)^2 + 2\int_{S_t^\mu - B_t^\mu}^\infty z\big(1 - F_{T-t}^\mu(z)\big)\,dz \tag{30.2.4}$$

for all $0 \le t \le T$ *where*

$$F_{T-t}^\mu(z) = \mathsf{P}(S_{T-t}^\mu \le z) = \Phi\Big(\frac{z - \mu(T-t)}{\sqrt{T-t}}\Big) - e^{2\mu z}\Phi\Big(\frac{-z - \mu(T-t)}{\sqrt{T-t}}\Big) \tag{30.2.5}$$

for $z \ge 0$.

Proof. By stationary independent increments of B^μ we have (cf. (30.1.42))

$$\begin{aligned}\mathsf{E}\left((S_T^\mu - B_t^\mu)^2 \,|\, \mathcal{F}_t^B\right) &= \mathsf{E}\Big[\Big(S_t^\mu + \Big(\max_{t\le s\le T} B_s^\mu - S_t^\mu\Big)^+ - B_t^\mu\Big)^2 \,|\, \mathcal{F}_t^B\Big] \qquad (30.2.6)\\ &= \mathsf{E}\Big[\Big(S_t^\mu - B_t^\mu + \Big(\max_{t\le s\le T} B_s^\mu - B_t^\mu - (S_t^\mu - B_t^\mu)\Big)^+\Big)^2 \,|\, \mathcal{F}_t^B\Big]\\ &= \Big[\mathsf{E}\left(x + (S_{T-t}^\mu - x)^+\right)^2\Big]\Big|_{x = S_t^\mu - B_t^\mu}\end{aligned}$$

for $0 \le t \le T$ given and fixed. Integration by parts gives

$$\begin{aligned}\mathsf{E}\left(x + (S_{T-t}^\mu - x)^+\right)^2 &= \mathsf{E}\left(x^2 I(S_{T-t}^\mu \le x)\right) + \mathsf{E}\left((S_{T-t}^\mu)^2 I(S_{T-t}^\mu > x)\right) \qquad (30.2.7)\\ &= x^2 \mathsf{P}(S_{T-t}^\mu \le x) + \int_x^\infty z^2\, F_{T-t}^\mu(dz)\\ &= x^2 F_{T-t}^\mu(x) + \Big(z^2\big(F_{T-t}^\mu(z) - 1\big)\Big)\Big|_x^\infty + 2\int_x^\infty z\,\big(1 - F_{T-t}^\mu(z)\big)\,dz\\ &= x^2 + 2\int_x^\infty z\,\big(1 - F_{T-t}^\mu(z)\big)\,dz\end{aligned}$$

for all $x \ge 0$. Combining (30.2.6) and (30.2.7) we get (30.2.4). (The identity (30.2.5) is a well-known result of [39, p. 397] and [130, p. 526].) □

2°. Denoting $F_{T-t}^\mu(z) = F^\mu(T-t, z)$, standard arguments based on the fact that each stopping time is the limit of a decreasing sequence of discrete stopping times imply that (30.2.4) extends as follows:

$$\mathsf{E}\left((S_T^\mu - B_\tau^\mu)^2 \,|\, \mathcal{F}_\tau^B\right) = (S_\tau^\mu - B_\tau^\mu)^2 + 2\int_{S_\tau^\mu - B_\tau^\mu}^\infty z\big(1 - F^\mu(T-\tau, z)\big)\,dz \tag{30.2.8}$$

for all stopping times τ of B^μ with values in $[0, T]$. Setting

$$X_t = S_t^\mu - B_t^\mu \tag{30.2.9}$$

for $t \geq 0$ and taking expectations in (30.2.8) we find that the optimal prediction problem (30.2.3) is equivalent to the optimal stopping problem

$$V = \inf_{0 \leq \tau \leq T} \mathsf{E}\left(X_\tau^2 + 2 \int_{X_\tau}^{\infty} z \left(1 - F^\mu(T-\tau, z)\right) dz \right) \tag{30.2.10}$$

where the infimum is taken over all stopping times τ of X (upon recalling that the natural filtrations of B^μ and X coincide). The process $X = (X_t)_{t \geq 0}$ is strong Markov so that (30.2.10) falls into the class of optimal stopping problems for Markov processes (cf. Subsection 2.2). The structure of (30.2.10) is complicated since the gain process depends on time in a highly nonlinear way.

3°. A successful treatment of (30.2.10) requires that the problem be extended so that the process X can start at arbitrary points in the state space $[0, \infty)$. For this, recall that (cf. [84]) the following identity in law holds:

$$X \overset{\text{law}}{=} |Y| \tag{30.2.11}$$

where $|Y| = (|Y_t|)_{t \geq 0}$ and the process $Y = (Y_t)_{t \geq 0}$ is a unique strong solution to the ("bang-bang") stochastic differential equation

$$dY_t = -\mu \operatorname{sign}(Y_t)\, dt + dB_t \tag{30.2.12}$$

with $Y_0 = 0$. Moreover, it is known (cf. [84]) that under $Y_0 = x$ in (30.2.12) the process $|Y|$ has the same law as a Brownian motion with drift $-\mu$ started at $|x|$ and reflected at 0. The infinitesimal operator of $|Y|$ acts on functions $f \in C_b^2([0, \infty))$ satisfying $f'(0+) = 0$ as $-\mu f'(x) + \frac{1}{2} f''(x)$. Since an optimal stopping time in (30.2.10) is the first entry time of the process to a closed set (this follows by general optimal stopping results of Chapter I and will be made more precise below) it is possible to replace the process X in (30.2.10) by the process $|Y|$. On the other hand, since it is difficult to solve the equation (30.2.12) explicitly so that the dependence of X on x is clearly expressed, we will take a different route based on the following fact.

Lemma 30.8. *The process $X^x = (X_t^x)_{t \geq 0}$ defined by*

$$X_t^x = x \vee S_t^\mu - B_t^\mu \tag{30.2.13}$$

is Markov under P making $\mathsf{P}_x = \mathrm{Law}(X^x \mid \mathsf{P})$ for $x \geq 0$ a family of probability measures on the canonical space $(C_+, \mathcal{B}(C_+))$ under which the coordinate process $X = (X_t)_{t \geq 0}$ is Markov with $\mathsf{P}_x(X_0 = x) = 1$.

Proof. Let $x \geq 0$, $t \geq 0$ and $h > 0$ be given and fixed. We then have:

$$\begin{aligned} X_{t+h}^x &= x \vee S_{t+h}^\mu - B_{t+h}^\mu \\ &= (x \vee S_t^\mu) \vee \Big(\max_{t \leq s \leq t+h} B_s^\mu \Big) - (B_{t+h}^\mu - B_t^\mu) - B_t^\mu \\ &= \left(x \vee S_t^\mu - B_t^\mu \right) \vee \Big(\max_{t \leq s \leq t+h} B_s^\mu - B_t^\mu \Big) - (B_{t+h}^\mu - B_t^\mu). \end{aligned} \tag{30.2.14}$$

Hence by stationary independent increments of B^μ we get:

$$\mathsf{E}\big(f(X^x_{t+h})\,|\,\mathcal{F}^B_t\big) = \mathsf{E}\big(f(z \vee S^\mu_h - B^\mu_h)\big)\big|_{z=X^x_t} \tag{30.2.15}$$

for every bounded Borel function f. This shows that X^x is a Markov process under P. Moreover, the second claim follows from (30.2.15) by a basic transformation theorem for integrals upon using that the natural filtrations of B and X^x coincide. This completes the proof. □

4°. By means of Lemma 30.8 we can now extend the optimal stopping problem (30.2.10) where $X_0 = 0$ under P to the optimal stopping problem

$$V(t,x) = \inf_{0 \le \tau \le T-t} \mathsf{E}_{t,x}\bigg(X^2_{t+\tau} + 2\int_{X_{t+\tau}}^\infty z\,\big(1 - F^\mu(T-t-\tau, z)\big)\,dz\bigg) \tag{30.2.16}$$

where $X_t = x$ under $\mathsf{P}_{t,x}$ with $(t,x) \in [0,T]\times[0,\infty)$ given and fixed. The infimum in (30.2.16) is taken over all stopping times τ of X.

In view of the fact that B^μ has stationary independent increments, it is no restriction to assume that the process X under $\mathsf{P}_{t,x}$ is explicitly given as

$$X^x_{t+s} = x \vee S^\mu_s - B^\mu_s \tag{30.2.17}$$

under P for $s \in [0, T-t]$. Setting

$$R(t,z) = 1 - F^\mu_{T_t}(z) \tag{30.2.18}$$

and introducing the gain function

$$G(t,x) = x^2 + 2\int_x^\infty z\,R(t,z)\,dz \tag{30.2.19}$$

we see that (30.2.16) can be written as follows:

$$V(t,x) = \inf_{0 \le \tau \le T-t} \mathsf{E}_{t,x} G(t+\tau, X_{t+\tau}) \tag{30.2.20}$$

for $(t,x) \in [0,T]\times[0,\infty)$.

5°. The preceding analysis shows that the optimal prediction problem (30.2.3) reduces to solving the optimal stopping problem (30.2.20). Introducing the continuation set $C = \{\,(t,x) \in [0,T]\times[0,\infty) : V(t,x) < G(t,x)\,\}$ and the stopping set $D = \{\,(t,x) \in [0,T]\times[0,\infty) : V(t,x) = G(t,x)\,\}$, we may infer from general theory of optimal stopping for Markov processes (cf. Chapter I) that the optimal stopping time in (30.2.20) is given by

$$\tau_D = \inf\{\,0 \le s \le T-t : (t+s, X_{t+s}) \in D\,\}. \tag{30.2.21}$$

It then follows using (30.2.9) that the optimal stopping time in (30.2.3) is given by

$$\tau_* = \inf\{\,0 \le t \le T : (t, S^\mu_t - B^\mu_t) \in D\,\}. \tag{30.2.22}$$

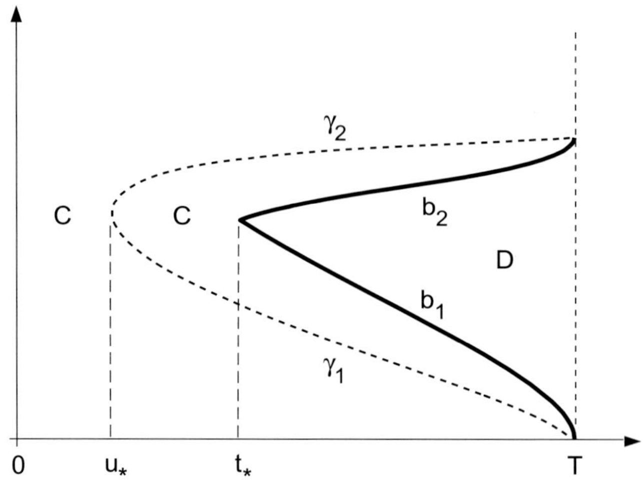

Figure VIII.6: (*The "black-hole" effect.*) A computer drawing of the optimal stopping boundaries b_1 and b_2 when $\mu > 0$ is away from 0.

The problems (30.2.20) and (30.2.3) are therefore reduced to determining D and V (outside D). We will see below that this task is complicated primarily because the gain function G depends on time in a highly nonlinear way. The main aim of the present subsection is to expose solutions to the problems formulated.

2. The free-boundary problem. Consider the optimal stopping problem (30.2.20). Recall that the problem reduces to determining the stopping set D and the value function V outside D. It turns out that the shape of D depends on the sign of μ.

1°. **The case** $\mu > 0$. It will be shown in the proof below that $D = \{ (t,x) \in [t_*,T) \times [0,\infty) : b_1(t) \le x \le b_2(t) \} \cup \{ (T,x) : x \in 0,\infty) \}$ where $t_* \in [0,T)$, the function $t \mapsto b_1(t)$ is continuous and decreasing on $[t_*,T]$ with $b_1(T) = 0$, and the function $t \mapsto b_2(t)$ is continuous and increasing on $[t_*,T]$ with $b_2(T) = 1/2\mu$. If $t_* \ne 0$ then $b_1(t_*) = b_2(t_*)$, and if $t_* = 0$ then $b_1(t_*) \le b_2(t_*)$. We also have $b_1(t) < b_2(t)$ for all $t_* < t \le T$. See *Figures* VIII.6+VIII.7.

It follows that the optimal stopping time (30.2.21) can be written as

$$\tau_D = \inf \{ t_* \le t \le T : b_1(t) \le X_t \le b_2(t) \}. \tag{30.2.23}$$

Inserting this expression into (30.2.20) and recalling that C equals D^c in $[0,T] \times [0,\infty)$, we can use Markovian arguments to formulate the following free-boundary

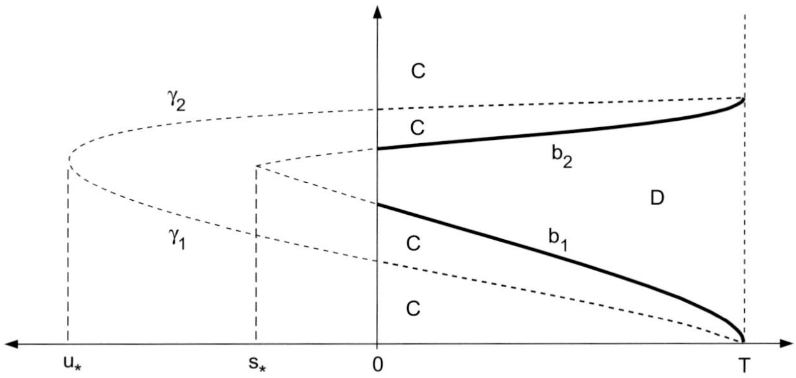

Figure VIII.7: A computer drawing of the optimal stopping boundaries b_1 and b_2 when $\mu \geq 0$ is close to 0.

problem:

$$V_t - \mu V_x + \tfrac{1}{2} V_{xx} = 0 \quad \text{in } C, \tag{30.2.24}$$
$$V(t, b_1(t)) = G(t, b_1(t)) \quad \text{for } t_* \leq t \leq T, \tag{30.2.25}$$
$$V(t, b_2(t)) = G(t, b_2(t)) \quad \text{for } t_* \leq t \leq T, \tag{30.2.26}$$
$$V_x(t, b_1(t)-) = G_x(t, b_1(t)) \quad \text{for } t_* \leq t < T \quad \text{(smooth fit)}, \tag{30.2.27}$$
$$V_x(t, b_2(t)+) = G_x(t, b_2(t)) \quad \text{for } t_* \leq t < T \quad \text{(smooth fit)}, \tag{30.2.28}$$
$$V_x(t, 0+) = 0 \quad \text{for } 0 \leq t < T \quad \text{(normal reflection)}, \tag{30.2.29}$$
$$V < G \quad \text{in } C, \tag{30.2.30}$$
$$V = G \quad \text{in } D. \tag{30.2.31}$$

Note that the conditions (30.2.27)–(30.2.29) will be derived in the proof below while the remaining conditions are obvious.

2°. **The case** $\mu \leq 0$. It will be seen in the proof below that $D = \{ (t, x) \in [0, T) \times [0, \infty) : x \geq b_1(t) \} \cup \{ (T, x) : x \in [0, \infty) \}$ where the continuous function $t \mapsto b_1(t)$ is decreasing on $[z_*, T]$ with $b_1(T) = 0$ and increasing on $[0, z_*)$ for some $z_* \in [0, T)$ (with $z_* = 0$ if $\mu = 0$). See *Figures* VIII.8+VIII.9.

It follows that the optimal stopping time (30.2.21) can be written as

$$\tau_D = \inf \{ 0 \leq t \leq T : X_t \geq b_1(t) \}. \tag{30.2.32}$$

Inserting this expression into (30.2.20) and recalling again that C equals D^c in $[0, T] \times [0, \infty)$, we can use Markovian arguments to formulate the following

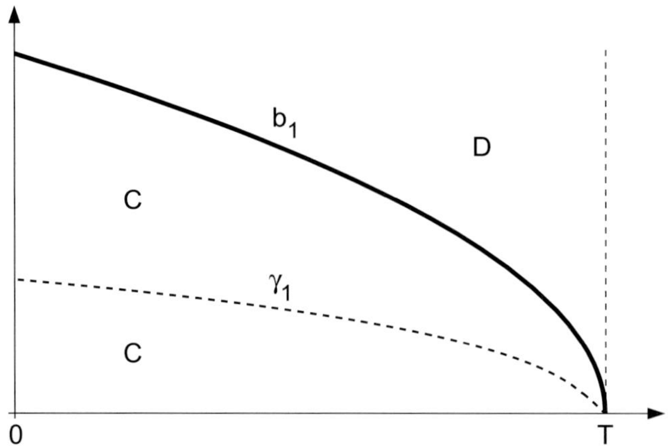

Figure VIII.8: A computer drawing of the optimal stopping boundary b_1 when $\mu \le 0$ is close to 0.

free-boundary problem:

$$V_t - \mu V_x + \tfrac{1}{2} V_{xx} = 0 \quad \text{in } C, \tag{30.2.33}$$

$$V(t, b_1(t)) = G(t, b_1(t)) \quad \text{for } 0 \le t \le T, \tag{30.2.34}$$

$$V_x(t, b_1(t)-) = G_x(t, b_1(t)) \quad \text{for } 0 \le t < T \quad \text{(smooth fit)}, \tag{30.2.35}$$

$$V_x(t, 0+) = 0 \quad \text{for } 0 \le t < T \quad \text{(normal reflection)}, \tag{30.2.36}$$

$$V < G \quad \text{in } C, \tag{30.2.37}$$

$$V = G \quad \text{in } D. \tag{30.2.38}$$

Note that the conditions (30.2.35) and (30.2.36) can be derived similarly to the conditions (30.2.27) and (30.2.29) above while the remaining conditions are obvious.

3°. It will be clear from the proof below that the case $\mu \le 0$ may be viewed as the case $\mu > 0$ with $b_2 \equiv \infty$ (and $t_* = 0$). This is in accordance with the facts that $b_2 \uparrow \infty$ as $\mu \downarrow 0$ and the point $s_* < T$ at which $b_1(s_*) = b_2(s_*)$ tends to $-\infty$ as $\mu \downarrow 0$. (Note that t_* equals $s_* \vee 0$ and that extending the time interval $[0, T]$ to negative values in effect corresponds to enlarging the terminal value T in the problem (30.2.20) above.) Since the case $\mu > 0$ is richer and more interesting we will only treat this case in complete detail. The case $\mu \le 0$ can be dealt with analogously and most of the details will be omitted.

4°. It will follow from the result of Theorem 30.9 below that the free-boundary problem (30.2.24)–(30.2.31) characterizes the value function V and the optimal stopping boundaries b_1 and b_2 in a unique manner. Motivated by

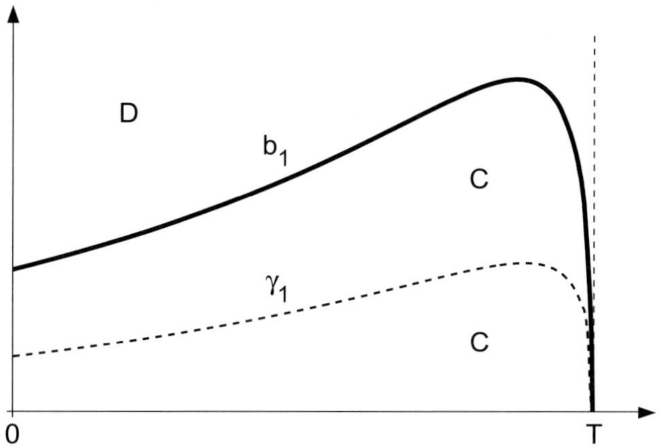

Figure VIII.9: (*The "hump" effect.*) A computer drawing of the optimal stopping boundary b_1 when $\mu<0$ is away from 0.

wider application, however, our main aim will be to express V in terms of b_1 and b_2 and show that b_1 and b_2 themselves satisfy a coupled system of nonlinear integral equations (which may then be solved numerically). Such an approach was applied in Subsections 25.2, 26.2 and 27.1 above. The present problem, however, is in many ways different and more complicated. We will nonetheless succeed in proving (as in the cases above with one boundary) that the coupled system of nonlinear equations derived for b_1 and b_2 cannot have other solutions. The key argument in the proof relies upon a local time-space formula (see Subsection 3.5). The analogous facts hold for the free-boundary problem (30.2.33)–(30.2.38) and the optimal stopping boundary b_1 (see Theorem 30.9 below).

3. Solution to the problem. To solve the problems (30.2.3) and (30.2.20) let us introduce the function

$$H = G_t - \mu\, G_x + \tfrac{1}{2}\, G_{xx} \tag{30.2.39}$$

on $[0,T]\times[0,\infty)$ where G is given in (30.2.19). A lengthy but straightforward calculation shows that

$$\begin{aligned} H(t,x) = {} & \big(2\mu^2(T-t) - 2\mu x + 3\big)\, \Phi\Big(\frac{x-\mu(T-t)}{\sqrt{T-t}}\Big) \\ & - 2\mu\sqrt{T-t}\,\varphi\Big(\frac{x-\mu(T-t)}{\sqrt{T-t}}\Big) \\ & - e^{2\mu x}\,\Phi\Big(\frac{-x-\mu(T-t)}{\sqrt{T-t}}\Big) - 2\big(1+\mu^2(T-t)\big) \end{aligned} \tag{30.2.40}$$

for $(t,x)\in[0,T]\times[0,\infty)$.

Let $P = \{(t,x) \in [0,T] \times [0,\infty) : H(t,x) \geq 0\}$ and $N = \{(t,x) \in [0,T] \times [0,\infty) : H(t,x) < 0\}$. A direct analysis based on (30.2.40) shows that in the case $\mu > 0$ we have $P = \{(t,x) \in [u_*,T] \times [0,\infty) : \gamma_1(t) \leq x \leq \gamma_2(t)\}$ where $u_* \in [0,T)$, the function $t \mapsto \gamma_1(t)$ is continuous and decreasing on $[u_*,T]$ with $\gamma_1(T) = 0$, and the function $t \mapsto \gamma_2(t)$ is continuous and increasing on $[u_*,T]$ with $\gamma_2(T) = 1/2\mu$. If $u_* \neq 0$ then $\gamma_1(u_*) = \gamma_2(u_*)$, and if $u_* = 0$ then $\gamma_1(u_*) \leq \gamma_2(u_*)$. We also have $\gamma_1(t) < \gamma_2(t)$ for all $u_* < t \leq T$. See *Figures* VIII.6+VIII.7. Similarly, a direct analysis based on (30.2.40) shows that in the case $\mu \leq 0$ we have $P = \{(t,x) \in [0,T] \times [0,\infty) : x \geq \gamma_1(t)\}$ where the continuous function $t \mapsto \gamma_1(t)$ is decreasing on $[w_*,T]$ with $\gamma_1(T) = 0$ and increasing on $[0,w_*)$ for some $w_* \in [0,T)$ (with $w_* = 0$ if $\mu = 0$). See *Figures* VIII.8+VIII.9.

Below we will make use of the following functions:

$$J(t,x) = \mathsf{E}_x G(T, X_{T-t}) = \int_0^\infty ds \int_{-\infty}^s db\, G(T, x \vee s - b)\, f(T-t, b, s), \tag{30.2.41}$$

$$K(t,x,t+u,y,z) = \mathsf{E}_x\big(H(t+u, X_u)\, I(y < X_u < z)\big) = \int_0^\infty ds \int_{-\infty}^s db\, H(t+u, x \vee s - b)\, I(y < x \vee s - b < z)\, f(u,b,s), \tag{30.2.42}$$

$$L(t,x,t+u,y) = \mathsf{E}_x\big(H(t+u, X_u)\, I(X_u > y)\big) = \int_0^\infty ds \int_{-\infty}^s db\, H(t+u, x \vee s - b)\, I(x \vee s - b > y)\, f(u,b,s) \tag{30.2.43}$$

for $(t,x) \in [0,T] \times [0,\infty)$, $u \geq 0$ and $0 < y < z$, where $(b,s) \mapsto f(t,b,s)$ is the probability density function of (B_t^μ, S_t^μ) under P given by

$$f(t,b,s) = \sqrt{\frac{2}{\pi}}\, \frac{1}{t^{3/2}}\, (2s - b)\, \exp\left\{ -\frac{(2s-b)^2}{2t} + \mu\Big(b - \frac{\mu t}{2}\Big) \right\} \tag{30.2.44}$$

for $t > 0$, $s \geq 0$ and $b \leq s$ (see e.g. [107, p. 368]).

The main results of the present subsection may now be stated as follows.

Theorem 30.9. *Consider the problems* (30.2.3) *and* (30.2.20). *We can then distinguish the following two cases.*

1. **The case $\boldsymbol{\mu > 0}$.** *The optimal stopping boundaries in* (30.2.20) *can be characterized as the unique solution to the coupled system of nonlinear Volterra integral equations*

$$J(t, b_1(t)) = G(t, b_1(t)) + \int_0^{T-t} K(t, b_1(t), t+u, b_1(t+u), b_2(t+u))\, du, \tag{30.2.45}$$

$$J(t, b_2(t)) = G(t, b_2(t)) + \int_0^{T-t} K(t, b_2(t), t+u, b_1(t+u), b_2(t+u))\, du \tag{30.2.46}$$

in the class of functions $t \mapsto b_1(t)$ *and* $t \mapsto b_2(t)$ *on* $[t_*, T]$ *for* $t_* \in [0, T)$ *such that the function* $t \mapsto b_1(t)$ *is continuous and decreasing on* $[t_*, T]$ *, the function* $t \mapsto b_2(t)$ *is continuous and increasing on* $[t_*, T]$ *, and* $\gamma_1(t) \le b_1(t) < b_2(t) \le \gamma_2(t)$ *for all* $t \in (t_*, T]$ *. The solutions* b_1 *and* b_2 *satisfy* $b_1(T) = 0$ *and* $b_2(T) = 1/2\mu$ *, and the stopping time* τ_D *from* (30.2.23) *is optimal in* (30.2.20)*. The stopping time* (30.2.22) *given by*

$$\tau_* = \inf \{ 0 \le t \le T : b_1(t) \le S_t^\mu - B_t^\mu \le b_2(t) \} \tag{30.2.47}$$

is optimal in (30.2.3)*. The value function* V *from* (30.2.20) *admits the following representation*:

$$V(t,x) = J(t,x) - \int_0^{T-t} K(t, x, t+u, b_1(t+u), b_2(t+u))\, du \tag{30.2.48}$$

for $(t,x) \in [0,T] \times [0,\infty)$ *. The value* V *from* (30.2.3) *equals* $V(0,0)$ *in* (30.2.48)*.*

2. **The case** $\boldsymbol{\mu \le 0}$. *The optimal stopping boundary in* (30.2.20) *can be characterized as the unique solution to the nonlinear Volterra integral equation*

$$J(t, b_1(t)) = G(t, b_1(t)) + \int_0^{T-t} L(t, b_1(t), t+u, b_1(t+u))\, du \tag{30.2.49}$$

in the class of continuous functions $t \mapsto b_1(t)$ *on* $[0,T]$ *that are decreasing on* $[z_*, T]$ *and increasing on* $[0, z_*)$ *for some* $z_* \in [0, T)$ *and satisfy* $b_1(t) \ge \gamma_1(t)$ *for all* $t \in [0,T]$ *. The solution* b_1 *satisfies* $b_1(T) = 0$ *and the stopping time* τ_D *from* (30.2.32) *is optimal in* (30.2.20)*. The stopping time* (30.2.22) *given by*

$$\tau_* = \inf \{ 0 \le t \le T : S_t^\mu - B_t^\mu \ge b_1(t) \} \tag{30.2.50}$$

is optimal in (30.2.3)*. The value function* V *from* (30.2.20) *admits the following representation*:

$$V(t,x) = J(t,x) - \int_0^{T-t} L(t, x, t+u, b_1(t+u))\, du \tag{30.2.51}$$

for $(t,x) \in [0,T] \times [0,\infty)$ *. The value* V *from* (30.2.3) *equals* $V(0,0)$ *in* (30.2.51)*.*

Proof. The proof will be carried out in several steps. We will only treat the case $\mu > 0$ in complete detail. The case $\mu \le 0$ can be dealt with analogously and details in this direction will be omitted. Thus we will assume throughout that $\mu > 0$ is given and fixed. We begin by invoking a result from general theory of optimal stopping for Markov processes (cf. Chapter I).

1°. We show that the stopping time τ_D in (30.2.21) is optimal in the problem (30.2.20). For this, recall that it is no restriction to assume that the process X under $\mathsf{P}_{t,x}$ is given explicitly by (30.2.17) under P. Since clearly $(t,x) \mapsto \mathsf{E}\, G(t+\tau, X_\tau^x)$ is continuous (and thus usc) for each stopping time τ, it follows that $(t,x) \mapsto V(t,x)$ is usc (recall that the infimum of usc functions defines an usc function). Since $(t,x) \mapsto G(t,x)$ is continuous (and thus lsc) by general theory [see Corollary 2.9 (Finite horizon) with Remark 2.10] it follows that τ_D is optimal in (30.2.20) as claimed. Note also that C is open and D is closed in $[0,T]\times[0,\infty)$.

2°. The initial insight into the shape of D is provided by stochastic calculus as follows. By Itô's formula (page 67) we have

$$G(t+s, X_{t+s}) = G(t,x) + \int_0^s G_t(t+u, X_{t+u})\, du + \int_0^s G_x(t+u, X_{t+u})\, dX_{t+u} + \frac{1}{2}\int_0^s G_{xx}(t+u, X_{t+u})\, d\langle X, X\rangle_{t+u} \tag{30.2.52}$$

for $0 \le s \le T-t$ and $x \ge 0$ given and fixed. By the Itô–Tanaka formula (page 68), recalling (30.2.11) and (30.2.12) above, we have

$$\begin{aligned} X_t = |Y_t| &= x + \int_0^t \operatorname{sign}(Y_s)\, I(Y_s \ne 0)\, dY_s + \ell_t^0(Y) \\ &= x - \mu \int_0^t I(Y_s \ne 0)\, ds + \int_0^t \operatorname{sign}(Y_s)\, I(Y_s \ne 0)\, dB_s + \ell_t^0(Y) \end{aligned} \tag{30.2.53}$$

where $\operatorname{sign}(0) = 0$ and $\ell_t^0(Y)$ is the local time of Y at 0 given by

$$\ell_t^0(Y) = \mathsf{P}\text{-}\lim_{\varepsilon \downarrow 0} \frac{1}{2\varepsilon} \int_0^t I(-\varepsilon < Y_s < \varepsilon)\, ds \tag{30.2.54}$$

upon using that $d\langle Y, Y\rangle_s = ds$. It follows from (30.2.53) that

$$dX_t = -\mu\, I(Y_t \ne 0)\, dt + \operatorname{sign}(Y_t)\, I(Y_t \ne 0)\, dB_t + d\ell_t^0(Y). \tag{30.2.55}$$

Inserting (30.2.55) into (30.2.52), using that $d\langle X, X\rangle_t = I(Y_t \ne 0)\, dt$ and $\mathsf{P}(Y_t = 0) = 0$, we get

$$\begin{aligned} G(t+s, X_{t+s}) &= G(t,x) + \int_0^s \big(G_t - \mu G_x + \tfrac{1}{2} G_{xx}\big)(t+u, X_{t+u})\, du \\ &\quad + \int_0^s G_x(t+u, X_{t+u})\, \operatorname{sign}(Y_{t+u})\, dB_{t+u} + \int_0^s G_x(t+u, X_{t+u})\, d\ell_{t+u}^0(Y) \\ &= G(t,x) + \int_0^s H(t+u, X_{t+u})\, du + M_s \end{aligned} \tag{30.2.56}$$

where H is given by (30.2.39) above and $M_s = \int_0^s G_x(t+u, X_{t+u})\, \mathrm{sign}\,(Y_{t+u}) dB_{t+u}$ is a continuous (local) martingale for $s \geq 0$. In the last identity in (30.2.56) we use that (quite remarkably) $G_x(t,0) = 0$ while $d\ell^0_{t+u}(Y)$ is concentrated at 0 so that the final integral in (30.2.56) vanishes.

From the final expression in (30.2.56) we see that the initial insight into the shape of D is gained by determining the sets P and N as introduced following (30.2.40) above. By considering the exit times from small balls in $[0,T) \times [0,\infty)$ and making use of (30.2.56) with the optional sampling theorem (page 60), we see that it is never optimal to stop in N. We thus conclude that $D \subseteq P$.

A deeper insight into the shape of D is provided by the following arguments. Due to the fact that P is bounded by γ_1 and γ_2 as described following (30.2.40) above, it is readily verified using (30.2.56) above and simple comparison arguments that for each $x \in (0, 1/2\mu)$ there exists $t = t(x) \in (0,T)$ close enough to T such that every point (x,u) belongs to D for $u \in [t,T]$. Note that this fact is fully in agreement with intuition since after starting at (u,x) close to (T,x) there will not be enough time to reach either of the favourable sets below γ_1 or above γ_2 to compensate for the loss incurred by strictly positive H via (30.2.56). These arguments in particular show that $D \setminus \{ (T,x) : x \in \mathbb{R}_+ \}$ is nonempty.

To prove the existence claim above, let $x \in (0, 1/2\mu)$ be given and fixed. If the claim is not true, then one can find $a > 0$ small enough such that the rectangle $R = [t,T] \times [x-a, x+a]$ is contained in $C \cap P$. Indeed, if there are $x' < x$ and $x'' > x$ such that $(u', x') \in D$ and $(u'', x'') \in D$ for some $u', u'' \in [t,T]$, then taking the larger of u' and u'', say u', we know that all the points (x,u) for $u \geq u'$ must belong to D, violating the fact that the existence claim is not true. The former conclusion follows by combining the facts stated (and independently verified) in the first sentence of paragraphs following (30.2.62) and (30.2.63) below. On the other hand, if there is no such $x' < x$ as above, then we can replace x by $x - \varepsilon$ for $\varepsilon > 0$ small enough, find a rectangle R for $x - \varepsilon$ in place of x as claimed above, prove the initial existence claim for $x - \varepsilon$, and then apply the fact of the first sentence following (30.2.63) below to derive the initial existence claim for x. Likewise, if there is no such $x'' > x$, we can proceed in a symmetric way to derive the initial existence claim for x. Thus, we may assume that the rectangle R above is contained in $C \cap P$ for small enough $a > 0$.

Define the stopping time

$$\tau_a = \inf \{ s \geq 0 : X_{t+s} \notin (x-a, x+a) \} \tag{30.2.57}$$

under the measure $\mathsf{P}_{t,x}$. Using equation (30.2.56) together with the optional sampling theorem (page 60) we see that

$$
\begin{aligned}
V(t,x)-G(t,x) &= \mathsf{E}_{t,x}\big(G(t+\tau_D, X_{t+\tau_D})\big)-G(t,x) \qquad (30.2.58)\\
&= \mathsf{E}_{t,x}\bigg(\int_0^{\tau_D} H(t+u, X_{t+u})\,du\bigg)\\
&= \mathsf{E}_{t,x}\bigg(\int_0^{\tau_a\wedge(T-t)} H(t+u, X_{t+u})\,du\bigg)\\
&\quad + \mathsf{E}_{t,x}\bigg(\int_{\tau_a\wedge(T-t)}^{\tau_D} H(t+u, X_{t+u})\,du\bigg)\\
&\geq c\,\mathsf{E}_{t,x}\big(\tau_a\wedge(T-t)\big) - d\,\mathsf{E}_{t,x}\bigg(\int_{\tau_a\wedge(T-t)}^{\tau_D}(1+X_{t+u})\,du\bigg).
\end{aligned}
$$

since $H \geq c > 0$ on R by continuity of H, and $H(u,x) \geq -d\,(1+x)$ for $(u,x) \in [t,T]\times\mathbb{R}_+$ with $d>0$ large enough by (30.2.40) above. Considering the final term, we find from the Hölder inequality that

$$
\begin{aligned}
&\mathsf{E}_{t,x}\bigg(\int_{\tau_a\wedge(1-t)}^{\tau_D}(1+X_{t+u})\,du\bigg) \qquad (30.2.59)\\
&\leq \mathsf{E}_{t,x}\Big(\Big(1+\max_{t\leq s\leq T} X_s\Big)\Big((T-t)-\tau_a\wedge(T-t)\Big)\Big)\\
&\leq \sqrt{\mathsf{E}_{t,x}\Big(1+\max_{0\leq s\leq T} X_s\Big)^2}\sqrt{\mathsf{E}_{t,x}\Big((T-t)-\tau_a\wedge(T-t)\Big)^2}\\
&\leq C\sqrt{\mathsf{E}_{t,x}\Big(\big(T-t-\tau_a\big)^2 I(\tau_a<T-t)\Big)}
\end{aligned}
$$

where C is a positive constant independent of t. Since $(T-t-\tau_a)^2 \leq (T-t)^2 \leq T-t$ on the set $\{\tau_a < T-t\}$ for $t<T$ close to T, we may conclude that

$$
\begin{aligned}
V(t,x)-G(t,x) &\geq c\,\mathsf{E}_{t,x}\big((T-t)\,I(\tau_a>T-t)\big) \qquad (30.2.60)\\
&\quad - D\sqrt{(T-t)\,\mathsf{P}_{t,x}(\tau_a<T-t)}\\
&= (T-t)\bigg(c\,\mathsf{P}_{t,x}(\tau_a\geq T-t) - D\sqrt{\frac{\mathsf{P}_{t,x}(\tau_a<T-t)}{T-t}}\,\bigg)
\end{aligned}
$$

where $c>0$ and $D>0$ are constants independent of t. Since clearly $\mathsf{P}_{t,x}(\tau_a \geq T-t) \to 1$ and $\mathsf{P}_{t,x}(\tau_a<T-t)/(T-t) \to 0$ as $t\uparrow T$ (the distribution function of the first exit time of $(-\mu t+B_t)_{t\geq 0}$ from $(-a,a)$ has a zero derivative at zero), it thus follows that $V(t,x)-G(t,x)>0$ for $t<T$ close to T. As this is impossible we see that the initial existence claim must be true.

The final insight into the shape of D is obtained by the following fortunate fact:

$$
t \mapsto H(t,x) \text{ is increasing on } [0,T] \qquad (30.2.61)
$$

whenever $x \geq 0$. Indeed, this can be verified by a direct differentiation in (30.2.40) which yields

$$H_t(t,x) = 2\Big(\frac{x+\mu(T-t)}{(T-t)^{3/2}}\Big)\,\varphi\Big(\frac{x-\mu(T-t)}{\sqrt{T-t}}\Big) + 2\mu^2\Big(1-\Phi\Big(\frac{x-\mu(T-t)}{\sqrt{T-t}}\Big)\Big) \tag{30.2.62}$$

from where one sees that $H_t \geq 0$ on $[0,T)\times[0,\infty)$ upon recalling that $\mu > 0$ by assumption.

We next show that $(t_1,x) \in D$ implies that $(t_2,x) \in D$ whenever $0 \leq t_1 \leq t_2 \leq T$ and $x \geq 0$. For this, assume that $(t_2,x) \in C$ for some $t_2 \in (t_1,T)$. Let $\tau_* = \tau_D(t_2,x)$ denote the optimal stopping time for $V(t_2,x)$. Then by (30.2.56) and (30.2.61) using the optional sampling theorem (page 60) we have

$$\begin{aligned} V(t_1,x) - G(t_1,x) &\leq \mathsf{E}\,G(t_1{+}\tau_*, X^x_{\tau_*}) - G(t_1,x) \\ &= \mathsf{E}\Big(\int_0^{\tau_*} H(t_1{+}u, X^x_u)\,du\Big) \leq \mathsf{E}\Big(\int_0^{\tau_*} H(t_2{+}u, X^x_u)\,du\Big) \\ &= \mathsf{E}\big(G(t_2{+}\tau_*, X^x_{\tau_*}) - G(t_2,x)\big) = V(t_2,x) - G(t_2,x) < 0. \end{aligned} \tag{30.2.63}$$

Hence (t_1,x) belongs to C which is a contradiction. This proves the initial claim.

Finally we show that for $(t,x_1) \in D$ and $(t,x_2) \in D$ with $x_1 \leq x_2$ in $(0,\infty)$ we have $(t,z) \in D$ for every $z \in [x_1,x_2]$. For this, fix $z \in (x_1,x_2)$ and let $\tau_* = \tau_D(t,z)$ denote the optimal stopping time for $V(t,z)$. Since (u,x_1) and (u,x_2) belong to D for all $u \in [t,T]$ we see that τ_* must be smaller than or equal to the exit time from the rectangle R with corners at (t,x_1), (t,x_2), (T,x_1) and (T,x_2). However, since $H > 0$ on R we see from (30.2.56) upon using the optional sampling theorem (page 60) that $V(t,z) > G(t,z)$. This shows that (t,z) cannot belong to C, thus proving the initial claim.

Summarising the facts derived above we can conclude that D equals the set of all (t,x) in $[t_*,T]\times[0,\infty)$ with $t_* \in [0,T)$ such that $b_1(t) \leq x \leq b_2(t)$, where the function $t \mapsto b_1(t)$ is decreasing on $[t_*,T]$ with $b_1(T) = 0$, the function $t \mapsto b_2(t)$ is increasing on $[t_*,T]$ with $b_2(T) = 1/2\mu$, and $\gamma_1(t) \leq b_1(t) \leq b_2(t) \leq \gamma_2(t)$ for all $t \in [t_*,T]$. See *Figures* VIII.6+VIII.7. It follows in particular that the stopping time τ_D from (30.2.23) is optimal in (30.2.20) and the stopping time from (30.2.47) is optimal in (30.2.3).

3°. We show that V is continuous on $[0,T]\times[0,\infty)$. For this, we will first show that $x \mapsto V(t,x)$ is continuous on $[0,\infty)$ uniformly over $t \in [0,T]$. Indeed, if $x < y$ in $[0,\infty)$ are given and fixed, we then have

$$\begin{aligned} V(t,x) - V(t,y) &= \inf_{0\leq\tau\leq T-t} \mathsf{E}\,G(t{+}\tau, X^x_\tau) - \inf_{0\leq\tau\leq T-t} \mathsf{E}\,G(t{+}\tau, X^y_\tau) \\ &\geq \inf_{0\leq\tau\leq T-t} \mathsf{E}\big(G(t{+}\tau, X^x_\tau) - G(t{+}\tau, X^y_\tau)\big) \end{aligned} \tag{30.2.64}$$

for all $t \in [0,T]$. It is easily verified that $x \mapsto G(t,x)$ is increasing so that $x \mapsto V(t,x)$ is increasing on $[0,\infty)$ for every $t \in [0,T]$. Hence it follows from (30.2.64) that

$$0 \le V(t,y) - V(t,x) \le \sup_{0\le\tau\le T-t} \mathsf{E}\big(G(t+\tau, X_\tau^y) - G(t+\tau, X_\tau^x)\big) \tag{30.2.65}$$

for all $t \in [0,T]$. Using (30.2.19) we find

$$\begin{aligned} G(t+\tau, X_\tau^y) - G(t+\tau, X_\tau^x) &= (X_\tau^y)^2 - (X_\tau^x)^2 - 2\int_{X_\tau^x}^{X_\tau^y} z R(t+\tau, z)\, dt \\ &\le \big(X_\tau^y - X_\tau^x\big)\big(X_\tau^y + X_\tau^x + 2c\big) \\ &= \big(y \vee S_\tau^\mu - x \vee S_\tau^\mu\big)\big(y \vee S_\tau^\mu - B_\tau^\mu + x \vee S_\tau^\mu - B_\tau^\mu + 2c\big) \\ &\le (y-x)Z \end{aligned} \tag{30.2.66}$$

where $c = \sup_{z\ge 0} zR(t,z) \le \mathsf{E}\, S_{T-t}^\mu \le \mathsf{E}\, S_T^\mu < \infty$ by Markov's inequality and $Z = 2(y+1) + 4\max_{0\le t\le T} |B_t^\mu| + 2c$ belongs to $L^1(\mathsf{P})$. From (30.2.65) and (30.2.66) we find

$$0 \le V(t,y) - V(t,x) \le (y-x)\mathsf{E}\, Z \tag{30.2.67}$$

for all $t \in [0,T]$ implying that $x \mapsto V(t,x)$ is continuous on $[0,\infty)$ uniformly over $t \in [0,T]$.

To complete the proof of the initial claim it is sufficient to show that $t \mapsto V(t,x)$ is continuous on $[0,T]$ for each $x \in [0,\infty)$ given and fixed. For this, fix x in $[0,\infty)$ and $t_1 < t_2$ in $[0,T]$. Let $\tau_1 = \tau_D(t_1,x)$ and $\tau_2 = \tau_D(t_2,x)$ be optimal for $V(t_1,x)$ and $V(t_2,x)$ respectively. Setting $\tau_1^\varepsilon = \tau_1 \wedge (T-t_2)$ with $\varepsilon = t_2 - t_1$ we have

$$\begin{aligned} &\mathsf{E}\big(G(t_2+\tau_2, X_{\tau_2}^x) - G(t_1+\tau_2, X_{\tau_2}^x)\big) \\ &\le V(t_2,x) - V(t_1,x) \\ &\le \mathsf{E}\big(G(t_2+\tau_1^\varepsilon, X_{\tau_1^\varepsilon}^x) - G(t_1+\tau_1, X_{\tau_1}^x)\big). \end{aligned} \tag{30.2.68}$$

Note that

$$G_t(t,x) = -2\int_x^\infty z f_{T-t}^\mu(z)\, dz \tag{30.2.69}$$

where $f_{T-t}^\mu(z) = (dF_{T-t}^\mu/dz)(z)$ so that

$$|G_t(t,x)| \le 2\int_0^\infty z f_{T-t}^\mu(z)\, dz = 2\,\mathsf{E}\, S_{T-t}^\mu \le 2\,\mathsf{E}\, S_T^\mu \tag{30.2.70}$$

for all $t \in [0,T]$. Hence setting $\beta = 2\,\mathsf{E}\, S_T^\mu$ by the mean value theorem we get

$$|G(u_2,x) - G(u_1,x)| \le \beta(u_2 - u_1) \tag{30.2.71}$$

for all $u_1 < u_2$ in $[0,T]$. Using (30.2.71) in (30.2.68) upon subtracting and adding $G(t_1+\tau_1, X^x_{\tau_1^\varepsilon})$ we obtain

$$\begin{aligned} -\beta(t_2-t_1) &\le V(t_2,x) - V(t_1,x) \\ &\le 2\beta(t_2-t_1) + \mathsf{E}\big(G(t_1+\tau_1, X^x_{\tau_1^\varepsilon}) - G(t_1+\tau_1, X^x_{\tau_1})\big). \end{aligned} \tag{30.2.72}$$

Note that

$$G_x(t,x) = 2x F^\mu_{T-t}(x) \le 2x \tag{30.2.73}$$

so that the mean value theorem implies

$$\begin{aligned} |G(t_1+\tau_1, X^x_{\tau_1^\varepsilon}) - G(t_1+\tau_1, X^x_{\tau_1})| &= |G_x(t_1+\tau_1,\xi)|\,|X^x_{\tau_1^\varepsilon} - X^x_{\tau_1}| \\ &\le 2\big(X^x_{\tau_1^\varepsilon} \vee X^x_{\tau_1}\big)\,|X^x_{\tau_1^\varepsilon} - X^x_{\tau_1}| \end{aligned} \tag{30.2.74}$$

where ξ lies between $X^x_{\tau_1^\varepsilon}$ and $X^x_{\tau_1}$. Since X^x_τ is dominated by the random variable $x + 2\max_{0\le t\le T}|B^\mu_t|$ which belongs to $L^1(\mathsf{P})$ for every stopping time τ, letting $t_2 - t_1 \to 0$ and using that $\tau_1^\varepsilon - \tau_1 \to 0$ we see from (30.2.72) and (30.2.74) that $V(t_2,x) - V(t_1,x) \to 0$ by dominated convergence. This shows that $t \mapsto V(t,x)$ is continuous on $[0,T]$ for each $x \in [0,\infty)$, and thus V is continuous on $[0,T]\times[0,\infty)$ as claimed. Standard arguments based on the strong Markov property and classic results from PDEs (cf. Chapter III) show that V is $C^{1,2}$ on C and satisfies (30.2.24). These facts will be freely used below.

4°. We show that $x \mapsto V(t,x)$ is differentiable at $b_i(t)$ for $i = 1,2$ and that $V_x(t,b_i(t)) = G_x(t,b_i(t))$ for $t \in [t_*,T)$. For this, fix $t \in [t_*,T)$ and set $x = b_2(t)$ (the case $x = b_1(t)$ can be treated analogously). We then have

$$\frac{V(t,x+\varepsilon) - V(t,x)}{\varepsilon} \le \frac{G(t,x+\varepsilon) - G(t,x)}{\varepsilon} \tag{30.2.75}$$

for all $\varepsilon > 0$. Letting $\varepsilon \downarrow 0$ in (30.2.75) we find

$$\limsup_{\varepsilon\downarrow 0} \frac{V(t,x+\varepsilon) - V(t,x)}{\varepsilon} \le G_x(t,x). \tag{30.2.76}$$

Let $\tau_\varepsilon = \tau_D(t,x+\varepsilon)$ be optimal for $V(t,x+\varepsilon)$. Then by the mean value theorem we have

$$\begin{aligned} \frac{V(t,x+\varepsilon) - V(t,x)}{\varepsilon} &\ge \frac{1}{\varepsilon}\Big(\mathsf{E}\,G(t+\tau_\varepsilon, X^{x+\varepsilon}_{\tau_\varepsilon}) - \mathsf{E}\,G(t+\tau_\varepsilon, X^x_{\tau_\varepsilon})\Big) \\ &= \frac{1}{\varepsilon}\,\mathsf{E}\big(G_x(t+\tau_\varepsilon,\xi_\varepsilon)(X^{x+\varepsilon}_{\tau_\varepsilon} - X^x_{\tau_\varepsilon})\big) \end{aligned} \tag{30.2.77}$$

where ξ_ε lies between $X^x_{\tau_\varepsilon}$ and $X^{x+\varepsilon}_{\tau_\varepsilon}$. Using that $t \mapsto b_2(t)$ is increasing and that $t \mapsto \lambda t$ is a lower function for B at $0+$ for every $\lambda \in \mathbb{R}$, it is possible to verify that $\tau_\varepsilon \to 0$ as $\varepsilon \downarrow 0$. Hence it follows that $\xi_\varepsilon \to x$ as $\varepsilon \downarrow 0$ so that $G_x(t+\tau_\varepsilon, \xi_\varepsilon) \to G_x(t,x)$ as $\varepsilon \downarrow 0$. Moreover, using (30.2.73) we find

$$\begin{aligned} G_x(t+\tau_\varepsilon, \xi_\varepsilon) &\le 2\xi_\varepsilon \le 2X_{\tau_\varepsilon}^{x+\varepsilon} = 2\big((x+\varepsilon) \vee S_{\tau_\varepsilon}^\mu - B_{\tau_\varepsilon}^\mu\big) \\ &\le 2\Big(x+\varepsilon+2 \max_{0\le t\le T} |\, B_t^\mu \,|\Big) \end{aligned} \tag{30.2.78}$$

where the final expression belongs to $L^1(\mathsf{P})$ (recall also that $G_x \ge 0$). Finally, we have

$$\tfrac{1}{\varepsilon}\left(X_{\tau_\varepsilon}^{x+\varepsilon} - X_{\tau_\varepsilon}^{x}\right) = \tfrac{1}{\varepsilon}\left((x+\varepsilon) \vee S_{\tau_\varepsilon}^\mu - x \vee S_{\tau_\varepsilon}^\mu\right) \to 1 \tag{30.2.79}$$

when $\varepsilon \downarrow 0$ as well as

$$0 \le \tfrac{1}{\varepsilon}\left(X_{\tau_\varepsilon}^{x+\varepsilon} - X_{\tau_\varepsilon}^{x}\right) \le 1 \tag{30.2.80}$$

for all $\varepsilon > 0$. Letting $\varepsilon \downarrow 0$ in (30.2.77) and using (30.2.78)–(30.2.80), we may conclude that

$$\liminf_{\varepsilon \downarrow 0} \frac{V(t,x+\varepsilon) - V(t,x)}{\varepsilon} \ge G_x(t,x) \tag{30.2.81}$$

by dominated convergence. Combining (30.2.76) and (30.2.81) we see that $x \mapsto V(t,x)$ is differentiable at $b_2(t)$ with $V_x(t,b_2(t)) = G_x(t,b_2(t))$ as claimed. Analogously one finds that $x \mapsto V(t,x)$ is differentiable at $b_1(t)$ with $V_x(t,b_1(t)) = G_x(t,b_1(t))$ and further details of this derivation will be omitted.

A small modification of the proof above shows that $x \mapsto V(t,x)$ is C^1 at $b_2(t)$. Indeed, let $\tau_\delta = \tau_D(t,x+\delta)$ be optimal for $V(t,x+\delta)$ where $\delta > 0$ is given and fixed. Instead of (30.2.75) above we have by the mean value theorem that

$$\begin{aligned} \frac{V(t,x+\delta+\varepsilon) - V(t,x+\delta)}{\varepsilon} &\le \frac{1}{\varepsilon}\Big(\mathsf{E}\, G(t+\tau_\delta, X_{\tau_\delta}^{x+\delta+\varepsilon}) - \mathsf{E}\, G(t+\tau_\delta, X_{\tau_\delta}^{x+\delta})\Big) \\ &= \frac{1}{\varepsilon}\, \mathsf{E}\Big(G_x(t+\tau_\delta, \eta_\varepsilon)\left(X_{\tau_\delta}^{x+\delta+\varepsilon} - X_{\tau_\delta}^{x+\delta}\right)\Big) \end{aligned} \tag{30.2.82}$$

where η_ε lies between $X_{\tau_\delta}^{x+\delta}$ and $X_{\tau_\delta}^{x+\delta+\varepsilon}$ for $\varepsilon > 0$. Clearly $\eta_\varepsilon \to X_{\tau_\delta}^{x+\delta}$ as $\varepsilon \downarrow 0$. Letting $\varepsilon \downarrow 0$ in (30.2.82) and using the same arguments as in (30.2.78)–(30.2.80) we can conclude that

$$V_x(t,x+\delta) \le \mathsf{E}\, G_x(t+\tau_\delta, X_{\tau_\delta}^{x+\delta}). \tag{30.2.83}$$

Moreover, in exactly the same way as in (30.2.77)–(30.2.81) we find that the reverse inequality in (30.2.83) also holds, so that we have

$$V_x(t,x+\delta) = \mathsf{E}\, G_x(t+\tau_\delta, X_{\tau_\delta}^{x+\delta}). \tag{30.2.84}$$

Letting $\delta \downarrow 0$ in (30.2.84), recalling that $\tau_\delta \to 0$, and using the same arguments as in (30.2.78), we find by dominated convergence that

$$\lim_{\delta \downarrow 0} V_x(t,x+\delta) = G_x(t,x) = V_x(t,x). \tag{30.2.85}$$

Thus $x \mapsto V(t,x)$ is C^1 at $b_2(t)$ as claimed. Similarly one finds that $x \mapsto V(t,x)$ is C^1 at $b_1(t)$ with $V_x(t,b_1(t)+) = G_x(t,b_1(t))$ and further details of this derivation will be omitted. This establishes the smooth fit conditions (30.2.27)–(30.2.28) and (30.2.35) above.

5°. We show that $t \mapsto b_1(t)$ and $t \mapsto b_2(t)$ are continuous on $[t_*,T]$. Again we only consider the case of b_2 in detail, since the case of b_1 can be treated similarly. Note that the same proof also shows that $b_2(T-) = 1/2\mu$ and that $b_1(T-) = 0$.

Let us first show that b_2 is right-continuous. For this, fix $t \in [t_*,T)$ and consider a sequence $t_n \downarrow t$ as $n \to \infty$. Since b_2 is increasing, the right-hand limit $b_2(t+)$ exists. Because $(t_n, b_2(t_n))$ belongs to D for all $n \geq 1$, and D is closed, it follows that $(t, b_2(t+))$ belongs to D. Hence by (30.2.23) we may conclude that $b_2(t+) \leq b_2(t)$. Since the fact that b_2 is increasing gives the reverse inequality, it follows that b_2 is right-continuous as claimed.

Let us next show that b_2 is left-continuous. For this, suppose that there exists $t \in (t_*,T)$ such that $b_2(t-) < b_2(t)$. Fix a point $x \in (b_2(t-), b_2(t))$ and note by (30.2.28) that we have

$$V(s,x) - G(s,x) = \int_{b_2(s)}^{x} \int_{b_2(s)}^{y} \big(V_{xx}(s,z) - G_{xx}(s,z)\big)\, dz\, dy \tag{30.2.86}$$

for any $s \in (t_*,t)$. By (30.2.24) and (30.2.39) we find that

$$\tfrac{1}{2}(V_{xx} - G_{xx}) = G_t - V_t + \mu(V_x - G_x) - H. \tag{30.2.87}$$

From (30.2.63) we derive the key inequality

$$V_t(t,x) \geq G_t(t,x) \tag{30.2.88}$$

for all $(t,x) \in [0,T)\times[0,\infty)$. Inserting (30.2.87) into (30.2.86) and using (30.2.88) and (30.2.26) we find

$$\begin{aligned} V(s,x) - G(s,x) &\leq \int_{b_2(s)}^{x} \int_{b_2(s)}^{y} 2\big(\mu(V_x - G_x)(s,z) - H(s,z)\big)\, dz\, dy \\ &= \int_{b_2(s)}^{x} 2\mu\big(V(s,y) - G(s,y)\big)\, dy - \int_{b_2(s)}^{x} \int_{b_2(s)}^{y} 2H(s,z)\, dz\, dy \\ &\leq - \int_{b_2(s)}^{x} \int_{b_2(s)}^{y} 2H(s,z)\, dz\, dy \end{aligned} \tag{30.2.89}$$

for any $s \in (t_*,t)$. From the properties of the function γ_2 it follows that there exists $s_* < t$ close enough to t such that (s,z) belongs to P for all $s \in [s_*,t)$ and $z \in [b_2(s), x]$. Moreover, since H is continuous and thus attains its infimum on a compact set, it follows that $2H(s,z) \geq m > 0$ for all $s \in [s_*, t)$ and $z \in [b_2(s), x]$. Using this fact in (30.2.89) we get

$$V(s,x) - G(s,x) \leq -m \, \frac{(x - b_2(s))^2}{2} < 0 \tag{30.2.90}$$

for all $s \in [s_*, t)$. Letting $s \uparrow t$ in (30.2.90) we conclude that $V(t,x) < G(t,x)$ violating the fact that $(t,x) \in D$. This shows that b_2 is left-continuous and thus continuous. The continuity of b_1 is proved analogously.

6°. We show that the normal reflection condition (30.2.29) holds. For this, note first since $x \mapsto V(t,x)$ is increasing on $[0,\infty)$ that $V_x(t,0+) \geq 0$ for all $t \in [0,T)$ (note that the limit exists since V is $C^{1,2}$ on C). Suppose that there exists $t \in [0,T)$ such that $V_x(t,0+) > 0$. Recalling that V is $C^{1,2}$ on C so that $t \mapsto V_x(t,0+)$ is continuous on $[0,T)$, we see that there exists $\delta > 0$ such that $V_x(s,0+) \geq \varepsilon > 0$ for all $s \in [t, t+\delta]$ with $t+\delta < T$. Setting $\tau_\delta = \tau_D \wedge \delta$ it follows by the Itô–Tanaka formula (as in (30.2.56) above) upon using (30.2.24) and the optional sampling theorem (recall (30.2.83) and (30.2.73) for the latter) that we have

$$\begin{aligned} \mathsf{E}_{t,0}\big(V(t{+}\tau_\delta, X_{t+\tau_\delta})\big) &= V(t,0) + \mathsf{E}_{t,0}\bigg(\int_0^{\tau_\delta} V_x(t{+}u, X_{t+u})\, d\ell^0_{t+u}(Y)\bigg) \\ &\geq V(t,0) + \varepsilon\, \mathsf{E}_{t,0}\big(\ell^0_{t+\tau_\delta}(Y)\big). \end{aligned} \tag{30.2.91}$$

Since $(V(t{+}s{\wedge}\tau_D, X_{t+s\wedge\tau_D}))_{0\leq s\leq T-t}$ is a martingale under $\mathsf{P}_{t,0}$ by general theory of optimal stopping for Markov processes (cf. Chapter I) we see from (30.2.91) that $\mathsf{E}_{t,0}\big(\ell^0_{t+\tau_\delta}(Y)\big)$ must be equal to 0. Since however properties of the local time clearly exclude this, we must have $V(t,0+)$ equal to 0 as claimed in (30.2.29) above.

7°. We show that V is given by the formula (30.2.48) and that b_1 and b_2 solve the system (30.2.45)–(30.2.46). For this, note that by (30.2.24) and (30.2.88) we have $\frac{1}{2}V_{xx} = -V_t + \mu V_x \leq -G_t + \mu V_x$ in C. It is easily verified using (30.2.73) and (30.2.83) that $V_x(t,x) \leq M/2\mu$ for all $t \in [0,T)$ and all $x \in [0,(1/2\mu){+}1]$ with some $M > 0$ large enough. Using this inequality in the previous inequality we get $V_{xx} \leq -G_t + M$ in $A = C \cap ([0,T) \times [0,(1/2\mu){+}1])$. Setting $h(t,x) = \int_0^x \int_0^y (-G_t(t,z) + M)\, dz\, dy$ we easily see that h is $C^{1,2}$ on $[0,T) \times [0,\infty)$ and that $h_{xx} = -G_t + M$. Thus the previous inequality reads $V_{xx} \leq h_{xx}$ in A, and setting $F = V - h$ we see that $x \mapsto F(t,x)$ is concave on $[0, b_1(t)]$ and $[b_2(t), (1/2\mu){+}1]$ for $t \in [t_*, T)$. We also see that F is $C^{1,2}$ on C and $D^\circ = \{\, (t,x) \in [t_*, T) \times [0,\infty) : b_1(t) < x < b_2(t) \,\}$ since both V and G are so. Moreover, it is also clear that $F_t - \mu F_x + \frac{1}{2}F_{xx}$ is locally bounded on $C \cup D^\circ$ in the sense that the function is bounded on $K \cap (C \cup D^\circ)$ for each compact set K in $[0,T) \times [0,\infty)$. Finally, we also see using (30.2.27) and (30.2.28) that $t \mapsto F_x(t, b_i(t)\mp) = V_x(t, b_i(t)\mp) - h_x(t, b_i(t)\mp) = G_x(t, b_i(t)) - h_x(t, b_i(t))$ is continuous on $[t_*, T)$ since b_i is continuous for $i = 1, 2$.

Since the previous conditions are satisfied we know that the local time-space formula (cf. Subsection 3.5) can be applied to $F(t+s, X_{t+s})$. Since h is $C^{1,2}$ on $[0,T)\times[0,\infty)$ we know that the Itô–Tanaka–Meyer formula (page 68) can be applied to $h(t+s, X_{t+s})$ as in (30.2.56) above (upon noting that $h_x(t,0+)=0$). Adding the two formulae, using in the former that $F_x(t,0+) = -h_x(t,0+) = 0$ since $V_x(t,0+)=0$ by (30.2.29) above, we get

$$\begin{aligned} V(t+s, X_{t+s}) &= V(t,x) \\ &+ \int_0^s \big(V_t - \mu V_x + \tfrac{1}{2} V_{xx}\big)(t+u, X_{t+u}) \\ &\qquad\qquad \times I\big(X_{t+u} \notin \{b_1(t+u), b_2(t+u)\}\big)\, du \\ &+ \int_0^s V_x(t+u, X_{t+u})\, \mathrm{sign}\,(Y_{t+u})\, I\big(X_{t+u} \notin \{b_1(t+u), b_2(t+u)\}\big)\, dB_{t+u} \\ &+ \sum_{i=1}^{2} \int_0^s \big(V_x(t+u, X_{t+u}+) - V_x(t+u, X_{t+u}-)\big) \\ &\qquad\qquad \times I\big(X_{t+u} = b_i(t+u)\big)\, d\ell_{t+u}^{b_i}(X) \end{aligned} \tag{30.2.92}$$

for $t \in [0,T)$ and $x \in [0,\infty)$. Making use of (30.2.24)+(30.2.31) in the first integral and (30.2.27)–(30.2.28) in the final integral (which consequently vanishes), we obtain

$$\begin{aligned} V(t+s, X_{t+s}) &= V(t,x) \\ &+ \int_0^s H(t+u, X_{t+u}) I\big(b_1(t+u) < X_{t+u} < b_2(t+u)\big)\, du + M_s \end{aligned} \tag{30.2.93}$$

for $t \in [0,T)$ and $x \in [0,\infty)$ where $M_s = \int_0^s V_x(t+u, X_{t+u})\, dB_{t+u}$ is a continuous (local) martingale for $s \geq 0$.

Setting $s = T-t$, using that $V(T,x) = G(T,x)$ for all $x \geq 0$, and taking the $\mathsf{P}_{t,x}$-expectation in (30.2.93), we find by the optional sampling theorem (page 60) that

$$\begin{aligned} V(t,x) &= \mathsf{E}_{t,x} G(T, X_T) \\ &- \int_0^{T-t} \mathsf{E}_{t,x}\big[H(t+u, X_{t+u}) I\big(b_1(t+u) < X_{t+u} < b_2(t+u)\big)\big]\, du \end{aligned} \tag{30.2.94}$$

for $t \in [0,T)$ and $x \in [0,\infty)$. Making use of (30.2.41) and (30.2.42) we see that (30.2.94) is the formula (30.2.48). Moreover, inserting $x = b_i(t)$ in (30.2.94) and using that $V(t, b_i(t)) = G(t, b_i(t))$ for $i = 1,2$ we see that b_1 and b_2 satisfy the system (30.2.45)–(30.2.46) as claimed.

8°. We show that b_1 and b_2 are the unique solution to the system (30.2.45)–(30.2.46) in the class of continuous functions $t \mapsto b_1(t)$ and $t \mapsto b_2(t)$ on $[t_*, T]$ for $t_* \in [0,T)$ such that $\gamma_1(t) \leq b_1(t) < b_2(t) \leq \gamma_2(t)$ for all $t \in (t_*, T]$. Note that there is no need to assume that b_1 is decreasing and b_2 is increasing as established above. The proof of uniqueness will be presented in the final three steps of the main proof below.

9°. Let $c_1 : [t_*, T] \to \mathbb{R}$ and $c_2 : [t_*, T] \to \mathbb{R}$ be a solution to the system (30.2.45)–(30.2.46) for $t_* \in [0, T)$ such that c_1 and c_2 are continuous and satisfy $\gamma_1(t) \le c_1(t) < c_2(t) \le \gamma_2(t)$ for all $t \in (t_*, T]$. We need to show that these c_1 and c_2 must then be equal to the optimal stopping boundaries b_1 and b_2 respectively.

Motivated by the derivation (30.2.92)–(30.2.94) which leads to the formula (30.2.48), let us consider the function $U^c : [0, T) \times [0, \infty) \to \mathbb{R}$ defined as follows:

$$U^c(t,x) = \mathsf{E}_{t,x} G(T, X_T) - \int_0^{T-t} \mathsf{E}_{t,x}\Big(H(t+u, X_{t+u})\, I\big(c_1(t+u) < X_{t+u} < c_2(t+u)\big)\Big)\, du \tag{30.2.95}$$

for $(t,x) \in [0,T) \times [0,\infty)$. In terms of (30.2.41) and (30.2.42) note that U^c is explicitly given by

$$U^c(t,x) = J(t,x) - \int_0^{T-t} K\big(t, x, t+u, c_1(t+u), c_2(t+u)\big)\, du \tag{30.2.96}$$

for $(t,x) \in [0,T) \times [0,\infty)$. Observe that the fact that c_1 and c_2 solve the system (30.2.45)–(30.2.46) means exactly that $U^c(t, c_i(t)) = G(t, c_i(t))$ for all $t \in [t_*, T]$ and $i = 1, 2$. We will moreover show that $U^c(t,x) = G(t,x)$ for all $x \in [c_1(t), c_2(t)]$ with $t \in [t_*, T]$. This is the key point in the proof (cf. Subsections 25.2, 26.2, 27.1) that can be derived using martingale arguments as follows.

If $X = (X_t)_{t \ge 0}$ is a Markov process (with values in a general state space) and we set $F(t,x) = \mathsf{E}_x G(X_{T-t})$ for a (bounded) measurable function G with $\mathsf{P}(X_0 = x) = 1$, then the Markov property of X implies that $F(t, X_t)$ is a martingale under P_x for $0 \le t \le T$. Similarly, if we set $F(t,x) = \mathsf{E}_x\big(\int_0^{T-t} H(X_s)\, ds\big)$ for a (bounded) measurable function H with $\mathsf{P}(X_0 = x) = 1$, then the Markov property of X implies that $F(t, X_t) + \int_0^t H(X_s)\, ds$ is a martingale under P_x for $0 \le t \le T$.

Combining the two martingale facts applied to the time-space Markov process $(t+s, X_{t+s})$ instead of X_s, we find that

$$U^c(t+s, X_{t+s}) - \int_0^s H(t+u, X_{t+u})\, I\big(c_1(t+u) < X_{t+u} < c_2(t+u)\big)\, du \tag{30.2.97}$$

is a martingale under $\mathsf{P}_{t,x}$ for $0 \le s \le T-t$. We may thus write

$$U^c(t+s, X_{t+s}) - \int_0^s H(t+u, X_{t+u})\, I\big(c_1(t+u) < X_{t+u} < c_2(t+u)\big)\, du = U^c(t,x) + N_s \tag{30.2.98}$$

where $(N_s)_{0 \le s \le T-t}$ is a martingale under $\mathsf{P}_{t,x}$. On the other hand, we know from (30.2.56) that

$$G(t+s, X_{t+s}) = G(t,x) + \int_0^s H(t+u, X_{t+u})\, du + M_s \tag{30.2.99}$$

where $M_s = \int_0^s G_x(t+u, X_{t+u}) \operatorname{sign}(Y_{t+u})\, dB_{t+u}$ is a continuous (local) martingale under $\mathsf{P}_{t,x}$ for $0 \le s \le T-t$.

For $x \in [c_1(t), c_2(t)]$ with $t \in [t_*, T]$ given and fixed, consider the stopping time

$$\sigma_c = \inf\{\, 0 \le s \le T-t : X_{t+s} \le c_1(t+s) \text{ or } X_{t+s} \ge c_2(t+s)\,\} \tag{30.2.100}$$

under $\mathsf{P}_{t,x}$. Using that $U^c(t, c_i(t)) = G(t, c_i(t))$ for all $t \in [t_*, T]$ (since c_1 and c_2 solve the system (30.2.45)–(30.2.46) as pointed out above) and that $U^c(T,x) = G(T,x)$ for all $x \ge 0$, we see that $U^c(t+\sigma_c, X_{t+\sigma_c}) = G(t+\sigma_c, X_{t+\sigma_c})$. Hence from (30.2.98) and (30.2.99) using the optional sampling theorem (page 60) we find

$$\begin{aligned} U^c(t,x) &= \mathsf{E}_{t,x} U^c(t+\sigma_c, X_{t+\sigma_c}) \\ &\quad - \mathsf{E}_{t,x}\Big[\int_0^{\sigma_c} H(t+u, X_{t+u})\, I\big(c_1(t+u) < X_{t+u} < c_2(t+u)\big)\, du\Big] \\ &= \mathsf{E}_{t,x} G(t+\sigma_c, X_{t+\sigma_c}) - \mathsf{E}_{t,x}\Big[\int_0^{\sigma_c} H(t+u, X_{t+u})\, du\Big] = G(t,x) \end{aligned} \tag{30.2.101}$$

since $X_{t+u} \in (c_1(t+u), c_2(t+u))$ for all $u \in [0, \sigma_c)$. This proves that $U^c(t,x) = G(t,x)$ for all $x \in [c_1(t), c_2(t)]$ with $t \in [t_*, T]$ as claimed.

10°. We show that $U^c(t,x) \ge V(t,x)$ for all $(t,x) \in [0,T] \times [0,\infty)$. For this, consider the stopping time

$$\tau_c = \inf\{\, 0 \le s \le T-t : c_1(t+s) \le X_{t+s} \le c_2(t+s)\,\} \tag{30.2.102}$$

under $\mathsf{P}_{t,x}$ with $(t,x) \in [0,T] \times [0,\infty)$ given and fixed. The same arguments as those following (30.2.100) above show that $U^c(t+\tau_c, X_{t+\tau_c}) = G(t+\tau_c, X_{t+\tau_c})$. Inserting τ_c instead of s in (30.2.98) and using the optional sampling theorem (page 60), we get

$$U^c(t,x) = \mathsf{E}_{t,x}\big(U^c(t+\tau_c, X_{t+\tau_c})\big) = \mathsf{E}_{t,x}\big(G(t+\tau_c, X_{t+\tau_c})\big) \ge V(t,x) \tag{30.2.103}$$

proving the claim.

11°. We show that $c_1 \le b_1$ and $c_2 \ge b_2$ on $[t_*, T]$. For this, suppose that there exists $t \in [t_*, T)$ such that $c_2(t) < b_2(t)$ and examine first the case when $c_2(t) > b_1(t)$. Choose a point $x \in (b_1(t) \vee c_1(t), c_2(t)]$ and consider the stopping time

$$\sigma_b = \inf\{\, 0 \le s \le T-t : X_{t+s} \le b_1(t+s) \text{ or } X_{t+s} \ge b_2(t+s)\,\} \tag{30.2.104}$$

under $\mathsf{P}_{t,x}$. Inserting σ_b in the place of s in (30.2.93) and (30.2.98) and using the optional sampling theorem (page 60), we get

$$\mathsf{E}_{t,x}\big(V(t+\sigma_b, X_{t+\sigma_b})\big) = V(t,x) + \mathsf{E}_{t,x}\bigg(\int_0^{\sigma_b} H(t+u, X_{t+u})\, du\bigg), \tag{30.2.105}$$

$$\begin{aligned}\mathsf{E}_{t,x}\big(U^c(t+\sigma_b, X_{t+\sigma_b})\big) &= U^c(t,x) \\ &+ \mathsf{E}_{t,x}\bigg(\int_0^{\sigma_b} H(t+u, X_{t+u})\, I\big(c_1(t+u) < X_{t+u} < c_2(t+u)\big)\, du\bigg).\end{aligned} \tag{30.2.106}$$

Since $U^c \geq V$ and $V(t,x) = U^c(t,x) = G(t,x)$ for $x \in [b_1(t) \vee c_1(t), b_2(t) \wedge c_2(t)]$ with $t \in [t_*, T]$, it follows from (30.2.105) and (30.2.106) that

$$\mathsf{E}_{t,x}\bigg[\int_0^{\sigma_b} H(t+u, X_{t+u}) I\big(X_{t+u} \leq c_1(t+u) \text{ or } X_{t+u} \geq c_2(t+u)\big)\, du\bigg] \leq 0. \tag{30.2.107}$$

Due to the fact that $H(t+u, X_{t+u}) > 0$ for $u \in [0, \sigma_b)$ we see by the continuity of b_i and c_i for $i = 1,2$ that (30.2.107) is not possible. Thus under $c_2(t) < b_2(t)$ we cannot have $c_2(t) > b_1(t)$. If however $c_2(t) \leq b_1(t)$, then due to the facts that b_1 is decreasing with $b_1(T) = 0$ and $c_2(T) > 0$, there must exist $u \in (t,T)$ such that $c_2(u) \in (b_1(u), b_2(u))$. Applying then the preceding arguments at time u instead of time t, we again arrive at a contradiction. Hence we can conclude that $c_2(t) \geq b_2(t)$ for all $t \in [t_*, T]$. In exactly the same way (or by symmetry) one can derive that $c_1(t) \leq b_1(t)$ for $t \in [t_*, T]$ completing the proof of the initial claim.

12°. We show that c_1 must be equal to b_1 and c_2 must be equal to b_2. For this, let us assume that there exists $t \in [t_*, T)$ such that $c_1(t) < b_1(t)$ or $c_2(t) > b_2(t)$. Pick an arbitrary point x from $(c_1(t), b_1(t))$ or $(b_2(t), c_2(t))$ and consider the stopping time τ_D from (30.2.23) under $\mathsf{P}_{t,x}$. Inserting τ_D instead of s in (30.2.93) and (30.2.98), and using the optional sampling theorem (page 60), we get

$$\mathsf{E}_{t,x}\big(G(t+\tau_D, X_{t+\tau_D})\big) = V(t,x), \tag{30.2.108}$$

$$\begin{aligned}\mathsf{E}_{t,x}\big(G(t+\tau_D, X_{t+\tau_D})\big) &= U^c(t,x) \\ &+ \mathsf{E}_{t,x}\bigg(\int_0^{\tau_D} H(t+u, X_{t+u})\, I\big(c_1(t+u) < X_{t+u} < c_2(t+u)\big)\, du\bigg)\end{aligned} \tag{30.2.109}$$

where we also use that $V(t+\tau_D, X_{t+\tau_D}) = U^c(t+\tau_D, X_{t+\tau_D}) = G(t+\tau_D, X_{t+\tau_D})$ upon recalling that $c_1 \leq b_1$ and $c_2 \geq b_2$, and $U^c = G$ either between c_1 and c_2 or at T. Since $U^c \geq V$ we see from (30.2.108) and (30.2.109) that

$$\mathsf{E}_{t,x}\bigg(\int_0^{\tau_D} H(t+u, X_{t+u})\, I\big(c_1(t+u) < X_{t+u} < c_2(t+u)\big)\, du\bigg) \leq 0. \tag{30.2.110}$$

Due to the fact that $H(t+u, X_{t+u}) > 0$ for $X_{t+u} \in (c_1(t+u), c_2(t+u))$ we see from (30.2.110) by the continuity of b_i and c_i for $i = 1,2$ that such a point

(t, x) cannot exist. Thus c_i must be equal to b_i for $i = 1, 2$ and the proof is complete. □

Remark 30.10. The following simple method can be used to solve the system (30.2.45)–(30.2.46) numerically. Better methods are needed to achieve higher precision around the singularity point $t = T$ and to increase the speed of calculation. These issues are worthy of further consideration.

Set $t_k = kh$ for $k = 0, 1, \dots, n$ where $h = T/n$ and denote (recalling (30.2.41) and (30.2.42) above for more explicit expressions)

$$I(t, b_i(t)) = J(t, b_i(t)) - G(t, b_i(t)) \tag{30.2.111}$$

$$= \mathsf{E}_{b_i(t)}\big(G(T, X_{T-t})\big) - G(t, b_i(t)), \tag{30.2.112}$$

$$K(t, b_i(t), t+u, b_1(t+u), b_2(t+u)) \tag{30.2.113}$$

$$= \mathsf{E}_{b_i(t)}\big(H(t+u, X_{t+u})\, I\big(b_1(t+u) < X_{t+u} < b_2(t+u)\big)\big)$$

for $i = 1, 2$. Note that K always depends on both b_1 and b_2.

The following discrete approximation of the integral equations (30.2.45) and (30.2.46) is then valid:

$$I(t_k, b_i(t_k)) = \sum_{j=k}^{n-1} K\big(t_k, b_i(t_k), t_{j+1}, b_1(t_{j+1}), b_2(t_{j+1})\big)\, h \tag{30.2.114}$$

for $k = 0, 1, \dots, n-1$ where $i = 1, 2$. Setting $k = n-1$ with $b_1(t_n) = 0$ and $b_2(t_n) = 1/2\mu$ we can solve the system (30.2.114) for $i = 1, 2$ numerically and get numbers $b_1(t_{n-1})$ and $b_2(t_{n-1})$. Setting $k = n-2$ and using values $b_1(t_{n-1})$, $b_1(t_n)$, $b_2(t_{n-1})$, $b_2(t_n)$ we can solve (30.2.114) numerically and get numbers $b_1(t_{n-2})$ and $b_2(t_{n-2})$. Continuing the recursion we obtain $g_i(t_n)$, $g_i(t_{n-1}), \dots, g_i(t_1)$, $g_i(t_0)$ as an approximation of the optimal boundary b_i at the points $T, T-h, \dots, h, 0$ for $i = 1, 2$ (see *Figures* VIII.6+VIII.7). The equation (30.2.49) can be treated analogously (see *Figures* VIII.8+VIII.9).

Notes. Stopping a stochastic process as close as possible to its ultimate maximum is an undertaking of great practical and theoretical interest (e.g. in "financial engineering"). Mathematical problems of this type may be referred to as *optimal prediction* problems. Variants of these problems have appeared in the past under different names (the optimal selection problem, the best choice problem, the secretary problem, the house selling problem) concerning which the older papers [108], [76], [21], [86] are interesting to consult. Most of this work has been done in the case of discrete time.

The case of continuous time (Subsection 30.1) has been studied in the paper [85] when the process is a standard Brownian motion (see also [152] for a related problem). This hypothesis leads to an explicit solution of the problem using the method of time change. Motivated by wider applications, our aim in Subsection 30.2 (following [42]) is to continue this study when the process is a standard Brownian motion with drift. It turns out that this extension is not only far from being routine, but also requires a different line of argument to be developed, which in turn is applicable to a broader class of diffusions and Markov processes. The identity (30.1.50) was observed by Urusov [213].

The continuation set of the problem turns out to be "humped" when the drift is negative. This is rather unexpected and indicates that the problem is strongly time dependent. The most surprising discovery revealed by the solution, however, is the existence of a nontrivial stopping set (a "black hole" as we call it) when the drift is positive. This fact is not only counter-intuitive but also has important practical implications. For example, in a growing economy where the appreciation rate of a stock price is strictly positive, any financial strategy based on optimal prediction of the ultimate maximum should be thoroughly re-examined in the light of this new phenomenon.

Bibliography

[1] Abramowitz, M. *and* Stegun, I. A. (eds.) (1964). *Handbook of Mathematical Functions with Formulas, Graphs, and Mathematical Tables.* U.S. Department of Commerce, Washington.

[2] Alili, L. *and* Kyprianou, A. E. (2005). Some remarks on first passage of Lévy processes, the American put and pasting principles. *Ann. Appl. Probab.* **15** (2062–2080).

[3] Alvarez, L. H. R. (2001). Reward functionals, salvage values, and optimal stopping. *Math. Methods Oper. Res.* **54** (315–337).

[4] André, D. (1887). Solution directe du problème résolu par M. Bertrand. *C. R. Acad. Sci. Paris* **105** (436–437).

[5] Arrow, K. J., Blackwell, D. *and* Girshick, M. A. (1949). Bayes and minimax solutions of sequential decision problems. *Econometrica* **17** (213–244).

[6] Azéma, J. *and* Yor, M. (1979). Une solution simple au problème de Skorokhod. *Sém. Probab.* XIII, *Lecture Notes in Math.* **721**, Springer, Berlin (90–115).

[7] Bachelier, L. (1900). Théorie de la spéculation. *Ann. Sci. École Norm. Sup.* (3) **17** (21–86). English translation "*Theory of Speculation*" in "*The Random Character of Stock Market Prices*", MIT Press, Cambridge, Mass. 1964 (ed. P. H. Cootner) (17–78).

[8] Barrieu, P., Rouault, A. *and* Yor, M. (2004). A study of the Hartman–Watson distribution motivated by numerical problems related to the pricing of asian options. *J. Appl. Probab.* **41** (1049–1058).

[9] Bayraktar, E., Dayanik, S. *and* Karatzas, I. (2006). Adaptive Poisson disorder problem. To appear in *Ann. Appl. Probab.*

[10] Bather, J. A. (1962). Bayes procedures for deciding the sign of a normal mean. *Proc. Cambridge Philos. Soc.* **58** (599–620).

[11] Bather, J. (1970). Optimal stopping problems for Brownian motion. *Adv. in Appl. Probab.* **2** (259–286).

[12] Beibel, M. (2000). A note on sequential detection with exponential penalty for the delay. *Ann. Statist.* **28** (1696–1701).

[13] Beibel, M. *and* Lerche, H. R. (1997). A new look at optimal stopping problems related to mathematical finance. *Empirical Bayes, sequential analysis and related topics in statistics and probability* (New Brunswick, NJ, 1995). *Statist. Sinica* **7** (93–108).

[14] Beibel, M. *and* Lerche, H. R. (2002). A note on optimal stopping of regular diffusions under random discounting. *Theory Probab. Appl.* **45** (547–557).

[15] Bellman, R. (1952). On the theory of dynamic programming. *Proc. Natl. Acad. Sci. USA* **38** (716–719).

[16] Bellman, R. (1957). *Dynamic Programming.* Princeton Univ. Press, Princeton.

[17] Bhat, B. R. (1988). Optimal properties of SPRT for some stochastic processes. *Contemp. Math.* **80** (285–299).

[18] Blackwell, D. *and* Girshick M. A. (1954). *Theory of Games and Statistical Decisions.* Wiley, New York; Chapman & Hall, London.

[19] Bliss, G. A. (1946). *Lectures on the Calculus of Variations.* Univ. Chicago Press, Chicago.

[20] Bolza, O. (1913). Über den "Anormalen Fall" beim Lagrangeschen und Mayerschen Problem mit gemischten Bedingungen und variablen Endpunkten. *Math. Ann.* **74** (430–446).

[21] Boyce, W. M. (1970). Stopping rules for selling bonds. *Bell J. Econom. Management Sci.* **1** (27–53).

[22] Breakwell, J. *and* Chernoff H. (1964). Sequential tests for the mean of a normal distribution. II (Large t). *Ann. Math. Statist.* **35** (162–173).

[23] Brekke, K. A. *and* Øksendal, B. (1991). The high contact principle as a sufficiency condition for optimal stopping. *Stochastic Models and Option Values* (Loen, 1989), *Contrib. Econom. Anal.* 200, North-Holland (187–208).

[24] Buonocore, A., Nobile, A. G. *and* Ricciardi, L. M. (1987). A new integral equation for the evaluation of first-passage-time probability densities. *Adv. in Appl. Probab.* **19** (784–800).

[25] BURKHOLDER, D. L. (1991). *Explorations in martingale theory and its applications.* École d'Été de Probabilités de Saint-Flour XIX—1989, *Lecture Notes in Math.* **1464**, Springer-Verlag, Berlin (1–66).

[26] CARLSTEIN, E., MÜLLER, H.-G. *and* SIEGMUND, D. (eds.) (1994). *Change-Point Problems.* IMS Lecture Notes Monogr. Ser. **23**. Institute of Mathematical Statistics, Hayward.

[27] CARR, P., JARROW, R. *and* MYNENI, R. Alternative characterizations of American put options. *Math. Finance* **2** (78–106).

[28] CHAPMAN, S. (1928). On the Brownian displacements and thermal diffusion of grains suspended in a non-uniform fluid. *Proc. R. Soc. Lond. Ser. A* **119** (34–54).

[29] CHERNOFF, H. (1961). Sequential tests for the mean of a normal distribution. *Proceedings of the 4th Berkeley Symposium on Mathematical Statistics and Probability.* Vol. I. Univ. California Press, Berkeley, Calif. (79–91).

[30] CHERNOFF, H. (1968). Optimal stochastic control. *Sankhyā Ser. A* **30** (221–252).

[31] CHOW, Y. S., ROBBINS, H. *and* SIEGMUND, D. (1971). *Great Expectations: The Theory of Optimal Stopping.* Houghton Mifflin, Boston, Mass.

[32] COX, D. C. (1984). Some sharp martingale inequalities related to Doob's inequality. *IMS Lecture Notes Monograph Ser.* **5** (78–83).

[33] DARLING, D. A., LIGGETT, T. *and* TAYLOR, H. M. (1972). Optimal stopping for partial sums. *Ann. Math. Statist.* **43** (1363–1368).

[34] DAVIS, B. (1976). On the L^p norms of stochastic integrals and other martingales. *Duke Math. J.* **43** (697–704).

[35] DAVIS, M. H. A. (1976). A note on the Poisson disorder problem. *Math. Control Theory,* Proc. Conf. Zakopane 1974, *Banach Center Publ.* **1** (65–72).

[36] DAVIS, M. H. A. *and* KARATZAS, I. (1994). A deterministic approach to optimal stopping. *Probability, statistics and optimisation.* Wiley Ser. Probab. Math. Statist., Wiley, Chichester (455–466).

[37] DAYANIK, S. *and* KARATZAS, I. (2003). On the optimal stopping problem for one-dimensional diffusions. *Stochastic Process. Appl.* **107** (173–212).

[38] DAYANIK, S. *and* SEZER, S. O. (2006). Compound Poisson disorder problem. To appear in *Math. Oper. Res.*

[39] DOOB, J. L. (1949). Heuristic approach to the Kolmogorov–Smirnov theorems. *Ann. Math. Statist.* **20** (393–403).

[40] Doob, J. L. (1953). *Stochastic Processes.* Wiley, New York.

[41] Dupuis, P. *and* Wang, H. (2005). On the convergence from discrete to continuous time in an optimal stopping problem. *Ann. Appl. Probab.* **15** (1339–1366).

[42] Du Toit, J. *and* Peskir, G. (2006). The trap of complacency in predicting the maximum. To appear in *Ann. Probab.*

[43] Dubins, L. E. *and* Gilat, D. (1978). On the distribution of maxima of martingales. *Proc. Amer. Math. Soc.* **68** (337–338).

[44] Dubins, L. E. *and* Schwarz, G. (1988). A sharp inequality for submartingales and stopping times. *Astérisque* **157–158** (129–145).

[45] Dubins, L. E., Shepp, L. A. *and* Shiryaev, A. N. (1993). Optimal stopping rules and maximal inequalities for Bessel processes. *Theory Probab. Appl.* **38** (226–261).

[46] Dufresne, D. (2001). The integral of geometric Brownian motion. *Adv. in Appl. Probab.* **33** (223–241).

[47] Duistermaat, J. J., Kyprianou, A. E. *and* van Schaik, K. (2005). Finite expiry Russian options. *Stochastic Process. Appl.* **115** (609–638).

[48] Durbin, J. (1985). The first-passage density of a continuous Gaussian process to a general boundary. *J. Appl. Probab.* **22** (99–122).

[49] Durbin, J. (1992). The first-passage density of the Brownian motion process to a curved boundary (with an appendix by D. Williams). *J. Appl. Probab.* **29** (291–304).

[50] Durrett, R. (1984). *Brownian Motion and Martingales in Analysis.* Wadsworth, Belmont.

[51] Dvoretzky, A., Kiefer, J. *and* Wolfowitz, J. (1953). Sequential decision problems for processes with continuous time parameter. Testing hypotheses. *Ann. Math. Statist.* **24** (254–264).

[52] Dynkin, E. B. (1963). The optimum choice of the instant for stopping a Markov process. *Soviet Math. Dokl.* **4** (627–629).

[53] Dynkin, E. B. (1965). *Markov Processes.* Vols. I, II. Academic Press, New York; Springer-Verlag, Berlin-Gottingen-Heidelberg. (Russian edition published in 1963 by "Fizmatgiz".)

[54] Dynkin, E. B. (2002). *Diffusions, Superdiffusions and Partial Differential Equations.* Amer. Math. Soc., Providence.

[55] Dynkin, E. B. *and* Yushkevich A. A. (1969). *Markov Processes: Theorems and Problems.* Plenum Press, New York. (Russian edition published in 1967 by "Nauka".)

[56] Einstein, A. (1905). Über die von der molekularkinetischen Theorie der Wärme geforderte Bewegung von in ruhenden Flüssigkeiten suspendierten Teilchen. *Ann. Phys. (4)* **17** (549–560). English translation *"On the motion of small particles suspended in liquids at rest required by the molecular-kinetic theory of heat"* in the book *'Einstein's Miraculous Year'* by Princeton Univ. Press 1998 (85–98).

[57] Ekström, E. (2004). Russian options with a finite time horizon. *J. Appl. Probab.* **41** (313–326).

[58] Ekström, E. (2004). Convexity of the optimal stopping boundary for the American put option. *J. Math. Anal. Appl.* **299** (147–156).

[59] Engelbert H. J. (1973). On the theory of optimal stopping rules for Markov processes. *Theory Probab. Appl.* **18** (304–311).

[60] Engelbert H. J. (1974). On optimal stopping rules for Markov processes with continuous time. *Theory Probab. Appl.* **19** (278–296).

[61] Engelbert H. J. (1975). On the construction of the payoff $s(x)$ in the problem of optimal stopping of a Markov sequence. (Russian) *Math. Oper. Forschung und Statistik.* **6** (493–498).

[62] Fakeev, A. G. (1970). Optimal stopping rules for stochastic processes with continuous parameter. *Theory Probab. Appl.* **15** (324–331).

[63] Fakeev, A. G. (1971). Optimal stopping of a Markov process. *Theory Probab. Appl.* **16** (694–696).

[64] Feller, W. (1952). The parabolic differential equations and the associated semi-groups of transformations. *Ann. of Math. (2)* **55** (468–519).

[65] Ferebee, B. (1982). The tangent approximation to one-sided Brownian exit densities. *Z. Wahrscheinlichkeitstheor. Verw. Geb.* **61** (309–326).

[66] Fick, A. (1885). Ueber diffusion. (*Poggendorff's*) *Annalen der Physik und Chemie* **94** (59–86).

[67] Fleming, W. H. *and* Rishel, R. W. (1975). *Deterministic and Stochastic Optimal Control.* Springer-Verlag, Berlin–New York.

[68] Fokker, A. D. (1914). Die mittlere Energie rotierender elektrischer Dipole im Strahlungsfeld. *Ann. Phys.* **43** (810–820).

[69] FORTET, R. (1943). Les fonctions aléatoires du type de Markoff associées à certaines équations linéaires aux dérivées partielles du type parabolique. *J. Math. Pures Appl. (9)* **22** (177–243).

[70] FRIEDMAN, A. (1959). Free boundary problems for parabolic equations. I. Melting of solids. *J. Math. Mech.* **8** (499–517).

[71] GAPEEV, P. V. *and* PESKIR, G. (2004). The Wiener sequential testing problem with finite horizon. *Stoch. Stoch. Rep.* **76** (59–75).

[72] GAPEEV, P. V. *and* PESKIR, G. (2006). The Wiener disorder problem with finite horizon. To appear in *Stochastic Process. Appl.*

[73] GAL'CHUK, L. I. *and* ROZOVSKII, B. L. (1972). The 'disorder' problem for a Poisson process. *Theory Probab. Appl.* **16** (712–716).

[74] GILAT, D. (1986). The best bound in the $L \log L$ inequality of Hardy and Littlewood and its martingale counterpart. *Proc. Amer. Math. Soc.* **97** (429–436).

[75] GILAT, D. (1988). On the ratio of the expected maximum of a martingale and the L_p-norm of its last term. *Israel J. Math.* **63** (270–280).

[76] GILBERT, J. P. *and* MOSTELLER, F. (1966). Recognizing the maximum of a sequence. *J. Amer. Statist. Assoc.* **61** (35–73).

[77] GIRSANOV, I. V. (1960). On transforming a certain class of stochastic processes by absolutely continuous substitution of measures. *Theory Probab. Appl.* **5** (285–301).

[78] GRAVERSEN, S. E. *and* PESKIR, G. (1997). On Wald-type optimal stopping for Brownian motion. *J. Appl. Probab.* **34** (66–73).

[79] GRAVERSEN, S. E. *and* PESKIR, G. (1997). On the Russian option: The expected waiting time. *Theory Probab. Appl.* **42** (564–575).

[80] GRAVERSEN, S. E. *and* PESKIR, G. (1997). On Doob's maximal inequality for Brownian motion. *Stochastic Process. Appl.* **69** (111–125).

[81] GRAVERSEN, S. E. *and* PESKIR, G. (1998). Optimal stopping and maximal inequalities for geometric Brownian motion. *J. Appl. Probab.* **35** (856–872).

[82] GRAVERSEN, S. E. *and* PESKIR, G. (1998). Optimal stopping and maximal inequalities for linear diffusions. *J. Theoret. Probab.* **11** (259–277).

[83] GRAVERSEN, S. E. *and* PESKIR, G. (1998). Optimal stopping in the $L \log L$-inequality of Hardy and Littlewood. *Bull. London Math. Soc.* **30** (171–181).

[84] Graversen, S. E. *and* Shiryaev, A. N. (2000). An extension of P. Lévy's distributional properties to the case of a Brownian motion with drift. *Bernoulli* **6** (615–620).

[85] Graversen, S. E. Peskir, G. *and* Shiryaev, A. N. (2001). Stopping Brownian motion without anticipation as close as possible to its ultimate maximum. *Theory Probab. Appl.* **45** (125–136).

[86] Griffeath, D. *and* Snell, J. L. (1974). Optimal stopping in the stock market. *Ann. Probab.* **2** (1–13).

[87] Grigelionis, B. I. (1967). The optimal stopping of Markov processes. (Russian) *Litovsk. Mat. Sb.* **7** (265–279).

[88] Grigelionis, B. I. *and* Shiryaev, A. N. (1966). On Stefan's problem and optimal stopping rules for Markov processes. *Theory Probab. Appl.* **11** (541–558).

[89] Hansen, A. T. *and* Jørgensen, P. L. (2000). Analytical valuation of American-style Asian options. *Management Sci.* **46** (1116–1136).

[90] Hardy, G. H. *and* Littlewood, J. E. (1930). A maximal theorem with function-theoretic applications. *Acta Math.* **54** (81–116).

[91] Haugh, M. B. *and* Kogan, L. (2004). Pricing American options: a duality approach. *Oper. Res.* **52** (258–270).

[92] Hochstadt, H. (1973). *Integral Equations.* Wiley, New York–London–Sydney.

[93] Hou, C., Little, T. *and* Pant, V. (2000). A new integral representation of the early exercise boundary for American put options. *J. Comput. Finance* **3** (73–96).

[94] Ikeda, N. *and* Watanabe, S. (1981). *Stochastic Differential Equations and Diffusion Processes.* North-Holland, Amsterdam-New York; Kodansha, Tokyo.

[95] Irle, A. *and* Paulsen, V. (2004). Solving problems of optimal stopping with linear costs of observations. *Sequential Anal.* **23** (297–316).

[96] Irle, A. *and* Schmitz, N. (1984). On the optimality of the SPRT for processes with continuous time parameter. *Math. Operationsforsch. Statist. Ser. Statist.* **15** (91–104).

[97] Itô, K. (1944). Stochastic integral. *Imperial Academy. Tokyo. Proceedings.* **20** (519–524).

[98] Itô, K. (1946). On a stochastic integral equation. *Japan Academy. Proceedings* **22** (32–35).

[99] Itô, K. (1951). On stochastic differential equations. *Mem. Amer. Math. Soc.* **4** (1–51).

[100] Itô, K. *and* McKean, H. P., Jr. (1965). *Diffusion Processes and Their Sample Paths.* Academic Press, New York; Springer-Verlag, Berlin–New York. Reprinted in 1996 by Springer-Verlag.

[101] Jacka, S. D. (1988). Doob's inequalities revisited: A maximal H^1-embedding. *Stochastic Process. Appl.* **29** (281–290).

[102] Jacka, S. D. (1991). Optimal stopping and the American put. *Math. Finance* **1** (1–14).

[103] Jacka, S. D. (1991). Optimal stopping and best constants for Doob-like inequalities I: The case $p = 1$. *Ann. Probab.* **19** (1798–1821).

[104] Jacka, S. D. (1993). Local times, optimal stopping and semimartingales. *Ann. Probab.* **21** (329–339).

[105] Jacka, S. D. *and* Lynn, J. R. (1992). Finite-horizon optimal stopping obstacle problems and the shape of the continuation region. *Stochastics Stochastics Rep.* **39** (25–42).

[106] Jacod, J. *and* Shiryaev, A. N. (1987). *Limit Theorems for Stochastic Processes.* Springer-Verlag, Berlin. Second ed.: 2003.

[107] Karatzas, I. *and* Shreve, S. E. (1998). *Methods of Mathematical Finance.* Springer-Verlag, New York.

[108] Karlin, S. (1962). Stochastic models and optimal policy for selling an asset. *Studies in Applied Probability and Management Science*, Stanford Univ. Press, Standord (148–158).

[109] Karlin, S. *and* Taylor, H. M. (1981). *A Second Course in Stochastic Processes.* Academic Press, New York–London.

[110] Kim, I. J. (1990). The analytic valuation of American options. *Rev. Financial Stud.* **3** (547–572).

[111] Kolmogoroff, A. (1931). Über die analytischen Methoden in der Wahrscheinlichkeitsrechnung. *Math. Ann.* **104** (415–458). English translation "*On analytical methods in probability theory*" in "*Selected works of A. N. Kolmogorov*" Vol. II (ed. A. N. Shiryayev), Kluwer Acad. Publ., Dordrecht, 1992 (62–108).

[112] Kolmogoroff, A. (1933). Zur Theorie der stetigen zufälligen Prozesse. *Math. Ann.* **108** (149–160). English translation "*On the theory of continuous random processes*" in "*Selected works of A. N. Kolmogorov*" Vol. II (ed. A. N. Shiryayev) Kluwer Acad. Publ., Dordrecht, 1992 (156–168).

[113] Kolmogorov, A. N., Prokhorov, Yu. V. *and* Shiryaev, A. N. (1990). Probabilistic-statistical methods of detecting spontaneously occuring effects. *Proc. Steklov Inst. Math.* **182** (1–21).

[114] Kolodner, I. I. (1956). Free boundary problem for the heat equation with applications to problems of change of phase. I. General method of solution. *Comm. Pure Appl. Math.* **9** (1–31).

[115] Kramkov, D. O. *and* Mordecki, E. (1994). Integral option. *Theory Probab. Appl.* **39** (162–172).

[116] Kramkov, D. O. *and* Mordecki, E. (1999). Optimal stopping and maximal inequalities for Poisson processes. *Publ. Mat. Urug.* **8** (153–178).

[117] Krylov, N. V. (1970). On a problem with two free boundaries for an elliptic equation and optimal stopping of a Markov process. *Soviet Math. Dokl.* **11** (1370–1372).

[118] Kuznetsov, S. E. (1980). Any Markov process in a Borel space has a transition function. *Theory Probab. Appl.* **25** (384–388).

[119] Kyprianou, A. E. *and* Surya, B. A. (2005). On the Novikov-Shiryaev optimal stopping problems in continuous time. *Electron. Comm. Probab.* **10** (146–154).

[120] Lamberton, D. *and* Rogers, L. C. G. (2000). Optimal stopping and embedding. *J. Appl. Probab.* **37** (1143–1148).

[121] Lamberton, D. *and* Villeneuve, S. (2003). Critical price near maturity for an American option on a dividend-paying stock. *Ann. Appl. Probab.* **13** (800–815).

[122] Lawler G. F. (1991). *Intersections of Random Walks.* Birkhäuser, Boston.

[123] Lehmann, E. L. (1959). *Testing Statistical Hypotheses.* Wiley, New York; Chapman & Hall, London.

[124] Lerche, H. R. (1986). *Boundary Crossing of Brownian Motion.* Lecture Notes in Statistics **40**. Springer-Verlag, Berlin–Heidelberg.

[125] Lévy, P. (1939). Sur certains processus stochastiques homogènes. *Compositio Math.* **7** (283–339).

[126] Lindley, D. V. (1961). Dynamic programming and decision theory. *Appl. Statist.* **10** (39–51).

[127] Liptser, R. S. *and* Shiryayev, A. N. (1977). *Statistics of Random Processes I.* Springer-Verlag, New York–Heidelberg. (Russian edition published in 1974 by "Nauka".) Second, revised and expanded English edition: 2001.

[128] Liptser, R. S. *and* Shiryayev, A. N. (1978). *Statistics of Random Processes II.* Springer-Verlag, New York–Heidelberg. (Russian edition published in 1974 by "Nauka".) Second, revised and expanded English edition: 2001.

[129] Liptser, R. S. *and* Shiryayev, A. N. (1989). *Theory of Martingales.* Kluwer Acad. Publ., Dordrecht. (Russian edition published in 1986 by "Nauka".)

[130] Malmquist, S. (1954). On certain confidence contours for distribution functions. *Ann. Math. Statist.* **25** (523–533).

[131] Marcellus, R. L. (1990). A Markov renewal approach to the Poisson disorder problem. *Comm. Statist. Stochastic Models* **6** (213–228).

[132] McKean, H. P., Jr. (1960/1961). The Bessel motion and a singular integral equation. *Mem. Coll. Sci. Univ. Kyoto Ser. A. Math.* **6** (317–322).

[133] McKean, H. P., Jr. (1965). Appendix: A free boundary problem for the heat equation arising from a problem of mathematical economics. *Ind. Management Rev.* **6** (32–39).

[134] Meyer, P.-A. (1966). *Probability and Potentials.* Blaisdell Publishing Co., Ginn and Co., Waltham–Toronto–London.

[135] Mikhalevich, V. S. (1956). Sequential Bayes solutions and optimal methods of statistical acceptance control. *Theory Probab. Appl.* **1** (395–421).

[136] Mikhalevich, V. S. (1958). A Bayes test of two hypotheses concerning the mean of a normal process. (Ukrainian) *Vīsn. Kiïv. Unīv.* No. **1** (254–264).

[137] Miranker, W. L. (1958). A free boundary value problem for the heat equation. *Quart. Appl. Math.* **16** (121–130).

[138] Miroshnichenko, T. P. (1975). Optimal stopping of the integral of a Wiener process. *Theory Probab. Apll.* **20** (397–401).

[139] Mordecki, E. (1999). Optimal stopping for a diffusion with jumps. *Finance Stoch.* **3** (227–236).

[140] Mordecki, E. (2002). Optimal stopping and perpetual options for Lévy processes. *Finance Stoch.* **6** (473–493).

[141] MUCCI, A. G. (1978). Existence and explicit determination of optimal stopping times. *Stochastic Process. Appl.* **8** (33–58).

[142] MYNENI, R. (1992). The pricing of the American option. *Ann. Appl. Probab.* **2** (1–23).

[143] NOVIKOV, A. A. (1971). On stopping times for a Wiener process. *Theory Probab. Appl.* **16** (449–456).

[144] NOVIKOV, A. A. *and* SHIRYAEV, A. N. (2004). On an effective solution of the optimal stopping problem for random walks. *Theory Probab. Appl.* **49** (344–354).

[145] ØKSENDAL, B. (1990). The high contact principle in optimal stopping and stochastic waves. *Seminar on Stochastic Processes*, 1989 (San Diego, CA, 1989). *Progr. Probab.* 18, Birkhäuser, Boston, MA (177–192).

[146] ØKSENDAL, B. *and* REIKVAM, K. (1998). Viscosity solutions of optimal stopping problems. *Stochastics Stochastics Rep.* **62** (285–301).

[147] ØKSENDAL, B. (2005). Optimal stopping with delayed information. *Stoch. Dyn.* **5** (271–280).

[148] PARK, C. *and* PARANJAPE, S. R. (1974). Probabilities of Wiener paths crossing differentiable curves. *Pacific J. Math.* **53** (579–583).

[149] PARK, C. *and* SCHUURMANN, F. J. (1976). Evaluations of barrier-crossing probabilities of Wiener paths. *J. Appl. Probab.* **13** (267–275).

[150] PEDERSEN, J. L. (1997). Best bounds in Doob's maximal inequality for Bessel processes. *J. Multivariate Anal.* **75**, 2000 (36–46).

[151] PEDERSEN, J. L. (2000). Discounted optimal stopping problems for the maximum process. *J. Appl. Probab.* **37** (972–983).

[152] PEDERSEN, J. L. (2003). Optimal prediction of the ultimate maximum of Brownian motion. *Stochastics Stochastics Rep.* **75** (205–219).

[153] PEDERSEN, J. L. (2005). Optimal stopping problems for time-homogeneous diffusions: a review. *Recent advances in applied probability*, Springer, New York (427–454).

[154] PEDERSEN, J. L. *and* PESKIR, G. (1998). Computing the expectation of the Azéma–Yor stopping times. *Ann. Inst. H. Poincaré Probab. Statist.* **34** (265–276).

[155] PEDERSEN, J. L. *and* PESKIR, G. (2000). Solving non-linear optimal stopping problems by the method of time-change. *Stochastic Anal. Appl.* **18** (811–835).

[156] PESKIR, G. (1998). Optimal stopping inequalities for the integral of Brownian paths. *J. Math. Anal. Appl.* **222** (244–254).

[157] PESKIR, G. (1998). The integral analogue of the Hardy–Littlewood $L \log L$-inequality for Brownian motion. *Math. Inequal. Appl.* **1** (137–148).

[158] PESKIR, G. (1998). The best Doob-type bounds for the maximum of Brownian paths. *High Dimensional Probability* (Oberwolfach 1996), *Progr. Probab.* **43**, Birkhäuser, Basel (287–296).

[159] PESKIR, G. (1998). Optimal stopping of the maximum process: The maximality principle. *Ann. Probab.* **26** (1614–1640).

[160] PESKIR, G. (1999). Designing options given the risk: The optimal Skorokhod-embedding problem. *Stochastic Process. Appl.* **81** (25–38).

[161] PESKIR, G. (2002). On integral equations arising in the first-passage problem for Brownian motion. *J. Integral Equations Appl.* **14** (397–423).

[162] PESKIR, G. (2002). Limit at zero of the Brownian first-passage density. *Probab. Theory Related Fields* **124** (100–111).

[163] PESKIR, G. (2005). A change-of-variable formula with local time on curves. *J. Theoret. Probab.* **18** (499–535).

[164] PESKIR, G. (2005). On the American option problem. *Math. Finance* **15** (169–181).

[165] PESKIR, G. (2005). The Russian option: Finite horizon. *Finance Stoch.* **9** (251–267).

[166] PESKIR, G. (2006). A change-of-variable formula with local time on surfaces. To appear in *Sém. de Probab.* (Lecture Notes in Math.) Springer.

[167] PESKIR, G. (2006). Principle of smooth fit and diffusions with angles. *Research Report* No. **7**, *Probab. Statist. Group Manchester* (11 pp).

[168] PESKIR, G. *and* SHIRYAEV, A. N. (2000). Sequential testing problems for Poisson processes. *Ann. Statist.* **28** (837–859).

[169] PESKIR, G. *and* SHIRYAEV, A. N. (2002). Solving the Poisson disorder problem. *Advances in Finance and Stochastics.* Essays in Honour of Dieter Sondermann. Springer, Berlin (295–312).

[170] PESKIR, G. *and* UYS, N. (2005). On Asian options of American type. *Exotic Option Pricing and Advanced Lévy Models* (Eindhoven, 2004), Wiley (217–235).

[171] PHAM, H. (1997). Optimal stopping, free boundary, and American option in a jump-diffusion model. *Appl. Math. Optim.* 35 (145–164).

[172] PLANCK, M. (1917). Über einen Satz der statistischen Dynamik und seine Erweiterung in der Quantentheorie. *Sitzungsber. Preuß. Akad. Wiss.* **24** (324–341).

[173] POOR, H. V. (1998). Quickest detection with exponential penalty for delay. *Ann. Statist.* **26** (2179–2205).

[174] REVUZ, D. *and* YOR, M. (1999). *Continuous Martingales and Brownian Motion.* Springer, Berlin.

[175] RICCIARDI, L. M., SACERDOTE, L. *and* SATO, S. (1984). On an integral equation for first-passage-time probability densities. *J. Appl. Probab.* **21** (302–314).

[176] ROGERS, L. C. G. (2002). Monte Carlo valuation of American options. *Math. Finance* **12** (271–286).

[177] ROGERS, L. C. G. *and* SHI, Z. (1995). The value of an Asian option. *J. Appl. Probab.* **32** (1077–1088).

[178] ROGERS, L. C. G. *and* WILLIAMS, D. (1987). *Diffusions, Markov Processes, and Martingales; Vol. 2: Itô Calculus.* Wiley, New York.

[179] ROMBERG, H. F. (1972). Continuous sequential testing of a Poisson process to minimize the Bayes risk. *J. Amer. Statist. Assoc.* **67** (921–926).

[180] SALMINEN, P. (1985). Optimal stopping of one-dimensional diffusions. *Math. Nachr.* **124** (85–101).

[181] SCHRÖDER, M. (2003). On the integral of geometric Brownian motion. *Adv. in Appl. Probab.* **35** (159–183).

[182] SCHRÖDINGER, E. (1915). Zur Theorie der Fall- und Steigversuche an Teilchen mit Brownscher Bewegung. *Physik. Zeitschr.* **16** (289–295).

[183] SHEPP, L. A. (1967). A first passage time for the Wiener process. *Ann. Math. Statist.* **38** (1912–1914).

[184] SHEPP, L. A. (1969). Explicit solutions of some problems of optimal stopping. *Ann. Math. Statist.* **40** (993–1010).

[185] SHEPP, L. A. *and* SHIRYAEV, A. N. (1993). The Russian option: Reduced regret. *Ann. Appl. Probab.* **3** (631–640).

[186] SHEPP, L. A. *and* SHIRYAEV, A. N. (1994). A new look at pricing of the "Russian option". *Theory Probab. Appl.* **39** (103–119).

[187] Shiryaev, A. N. (1961). The detection of spontaneous effects. *Soviet Math. Dokl.* **2** (740–743).

[188] Shiryaev, A. N. (1961). The problem of the most rapid detection of a disturbance of a stationary regime. *Soviet Math. Dokl.* **2** (795–799).

[189] Shiryaev, A. N. (1961). A problem of quickest detection of a disturbance of a stationary regime. (Russian) PhD Thesis. Steklov Institute of Mathematics, Moscow. 130 pp.

[190] Shiryaev, A. N. (1963). On optimal methods in quickest detection problems. *Theory Probab. Appl.* **8** (22–46).

[191] Shiryaev, A. N. (1966). On the theory of decision functions and control of a process of observation based on incomplete information. *Select. Transl. Math. Statist. Probab.* **6** (162–188).

[192] Shiryaev, A. N. (1965). Some exact formulas in a "disorder" problem. *Theory Probab. Appl.* **10** (349–354).

[193] Shiryaev, A. N. (1967). Two problems of sequential analysis. *Cybernetics* **3** (63–69).

[194] Shiryaev, A. N. (1969). Optimal stopping rules for Markov processes with continuous time. (With discussion.) *Bull. Inst. Internat. Statist.* **43** (1969), book 1 (395–408).

[195] Sirjaev, A. N. (1973). *Statistical Sequential Analysis: Optimal Stopping Rules.* American Mathematical Society, Providence. (First Russian edition published by "Nauka" in 1969.)

[196] Shiryayev, A. N. (1978). *Optimal Stopping Rules.* Springer, New York–Heidelberg. (Russian editions published by "Nauka": 1969 (first ed.), 1976 (second ed.).)

[197] Shiryaev, A. N. (1999). *Essentials of Stochastic Finance. Facts, Models, Theory.* World Scientific, River Edge. (Russian edition published by FASIS in 1998.)

[198] Shiryaev, A. N. (2002). Quickest detection problems in the technical analysis of the financial data. *Mathematical Finance—Bachelier Congress* (Paris, 2000), Springer, Berlin (487–521).

[199] Shiryaev, A. N. (2004). *Veroyatnost'.* Vol. 1, 2. MCCME, Moscow (Russian). English translation: *Probability.* To appear in Springer.

[200] Shiryaev, A. N. (2004). A remark on the quickest detection problems. *Statist. Decisions* **22** (79–82).

[201] SIEGERT, A. J. F. (1951). On the first passage time probability problem. *Phys. Rev. II* **81** (617–623).

[202] SIEGMUND, D. O. (1967). Some problems in the theory of optimal stopping rules. *Ann. Math. Statist.* **38** (1627–1640).

[203] SIEGMUND, D. O. (1985). *Sequential Analysis. Tests and Confidence Intervals.* Springer, New York.

[204] SMOLUCHOWSKI, M. v. (1913). Einige Beispiele Brown'scher Molekularbewegung unter Einfluß äußerer Kräfte. *Bull. Intern. Acad. Sc. Cracovie A* (418–434).

[205] SMOLUCHOWSKI, M. v. (1915). Notiz über die Berechnung der Brownschen Molekularbewegung bei der Ehrenhaft-Millikanschen Versuchsanordnung. *Physik. Zeitschr.* **16** (318–321).

[206] SNELL, J. L. (1952). Applications of martingale system theorems. *Trans. Amer. Math. Soc.* **73** (293–312).

[207] STRASSEN, V. (1967). Almost sure behavior of sums of independent random variables and martingales. *Proceedings of the Fifth Berkeley Symposium on Mathematical Statistics and Probability* (Berkeley, 1965/66) **II**, Part 1, Univ. California Press, Berkeley (315–343).

[208] STRATONOVICH, R. L. (1962). Some extremal problems in mathematical statistics and conditional Markov processes. *Theory Probab. Appl.* **7** (216–219).

[209] STROOCK D. W., VARADHAN S. R. S. (1979) *Multidimensional Diffusion Processes.* Springer, Berlin–New York.

[210] TAYLOR, H. M. (1968). Optimal stopping in a Markov process. *Ann. Math. Statist.* **39** (1333–1344).

[211] THOMPSON, M. E. (1971). Continuous parameter optimal stopping problems. *Z. Wahrscheinlichkeitstheor. verw. Geb.* **19** (302–318).

[212] TRICOMI, F. G. (1957). *Integral Equations.* Interscience Publishers, New York–London.

[213] URUSOV, M. On a property of the moment at which Brownian motion attains its maximum and some optimal stopping problems. *Theory Probab. Appl.* **49** (2005) (169–176).

[214] VAN MOERBEKE, P. (1974). Optimal stopping and free boundary problems. *Rocky Mountain J. Math.* **4** (539–578).

[215] van Moerbeke, P. (1976). On optimal stopping and free boundary problems. *Arch. Ration. Mech. Anal.* **60** (101–148).

[216] Wald, A. (1947). *Sequential Analysis.* Wiley, New York; Chapman & Hall, London.

[217] Wald, A. (1950). *Statistical Decision Functions.* Wiley, New York; Chapman & Hall, London.

[218] Wald, A. *and* Wolfowitz, J. (1948). Optimum character of the sequential probability ratio test. *Ann. Math. Statist.* **19** (326–339).

[219] Wald, A. *and* Wolfowitz, J. (1949). Bayes solutions of sequential decision problems. *Proc. Natl. Acad. Sci. USA* **35** (99–102). *Ann. Math. Statist.* **21**, 1950 (82–99).

[220] Walker, L. H. (1974). Optimal stopping variables for Brownian motion. *Ann. Probab.* **2** (317–320).

[221] Wang, G. (1991). Sharp maximal inequalities for conditionally symmetric martingales and Brownian motion. *Proc. Amer. Math. Soc.* **112** (579–586).

[222] Whittle, P. (1964). Some general results in sequential analysis. *Biometrika* **51** (123–141).

[223] Wu, L., Kwok, Y. K. *and* Yu, H. (1999). Asian options with the American early exercise feature. *Int. J. Theor. Appl. Finance* **2** (101–111).

[224] Yor, M. (1992). On some exponential functionals of Brownian motion. *Adv. in Appl. Probab.* **24** (509–531).

Subject Index

List of Symbols